ENTREPRENEURSHIP IN RENEWABLE ENERGY TECHNOLOGIES

The book focusses on various options of taking up ventures for starting entrepreneurship in small/ large scale in the field of renewable energy technologies. The book covers the fundamentals of entrepreneurship, renewable energy resources, their technologies involved and applications along with financial evaluations. The book will cater to the needs of students, researchers, various stakeholders, entrepreneurs etc. by providing valuable information on renewable energy technologies and their applications in developing entrepreneurship and establishing enterprise at individual level, specifically focusing on low carbon technology for sustenance of environment which is becoming increasingly important.

Dr. M. K. Ghosal, presently working as Professor in Odisha University of Agriculture and Technology (OUAT), Bhubaneswar, Odisha, India. His areas of expertise are Renewable Energy Engineering, Solar Thermal Engineering, Solar Photovoltaic Technology and Greenhouse Technology. Author of six books (four in Renewable Energy Technology and two in Sources of Farm Power). He did his B.Tech (Agricultural Engineering) from OUAT, Bhubaneswar, Odisha, M. Tech. (Farm Power and Machinery) from Indian Institute of Technology (IIT), Kharagpur, West Bengal and Ph.D. (Energy Engineering) from Indian Institute of Technology (IIT), Delhi. To his credit, he has Published 140 research articles in various national and international scientific journals. He is presently the Associate Editor of two agricultural based Journals i.e. International Journal of Tropical Agriculture and International Journal of Agricultural Engineering. He has been actively involved in teaching Renewable Energy Engineering subjects both in UG and PG level of Agricultural University since last 25 years.

ENTREPRENEURSHIP IN RENEWABLE ENERGY TECHNOLOGIES

MANOJ KUMAR GHOSAL

Professor

Department of Farm Machinery and Power Engineering,
Odisha University of Agriculture and Technology,
Bhubaneswar-751003, Odisha, India

CRC Press
Taylor & Francis Group
Boca Raton London New York

CRC Press is an imprint of the
Taylor & Francis Group, an **informa** business

NARENDRA PUBLISHING HOUSE
DELHI (INDIA)

First published 2023
by CRC Press
4 Park Square, Milton Park, Abingdon, Oxon, OX14 4RN

and by CRC Press
6000 Broken Sound Parkway NW, Suite 300, Boca Raton, FL 33487-2742

© 2023 M.K. Ghosal and Narendra Publishing House

CRC Press is an imprint of Informa UK Limited

British Library Cataloguing-in-Publication Data
A catalogue record for this book is available from the British Library

ISBN: 9781032388915 (hbk)
ISBN: 9781032388922 (pbk)
ISBN: 9781003347316 (ebk)

DOI: 10.4324/9781003347316

Typeset in Times New Roman, Arial, Symbol, Calibri and Dingbat by Amrit Graphics, Delhi 110032

CONTENTS

भा.कृ.अनु.प. – केन्द्रीय कृषि अभियांत्रिकी संस्थान

ICAR-CENTRAL INSTITUTE OF AGRICULTURAL ENGINEERING

नबी बाग, बैरसिया रोड, भोपाल – 462 038 (म.प्र.), भारत

Nabi Bagh, Berasia Road, Bhopal - 462 038 (M.P.), India

Phone: (0755) 2737191, Fax: (0755) 2734016, http://www.ciae.nic.in

e-mail: director.ciae@icar.gov.in, directorciae@gmail.com

डॉ. सी. आर. मेहता

निदेशक

Dr. C.R. Mehta

Director

FOREWORD

In the present scenario, the use of renewable energy is gaining more importance due to over-exploitation and increasing price of conventional or fossil fuels besides growing concerns of environment by their extensive usages. In the coming years, effective and efficient harnessing of energy from the renewable energy sources will play a crucial role in providing environment-friendly options in the energy sector to fulfil the rising demands of power in domestic, agricultural, industrial, transport and other needs of the mankind. Renewable energy is even more relevant for developing countries whose energy requirements are increasing rapidly as a result of large scale industrialization and growing population. There have been significant developments in the field of renewable energy technology during the last decade. For the developing countries, providing sustainable sources of energy to its citizens in an efficient and cost effective manner is a challenging task. Hence, the role of entrepreneurs is very important for dissemination of available technologies at the micro-level not only to meet the needs of the society but also to switch over slowly on the uses of renewable sources of energy and thus, creating employment opportunities for the upcoming young generation.

India is richly endowed with renewable energy sources. The government is aiming to achieve 227 GW of renewable energy capacity (including 114 GW of solar capacity addition and 67 GW of wind power capacity) by 2022. There have also been the continued efforts at the Government level to popularize and support renewable energy technologies to enhance power generation for supplementing conventional sources of energy. The renewable energy technologies covered in this book entitled "Entrepreneurship in Renewable Energy Technologies" are power generation using small scale solar photovoltaic and wind energy systems, solar drying, solar water pumping, and bio-energy applications. The basic principles and typical practical applications of technologies have been explained with numerical problems in each chapter for easy understanding by the practising energy professionals. Further, the economic considerations to harness energy from various renewable sources are discussed in details and efforts have been made to highlight the importance of popularizing renewable energy sources as cost-effective power generation alternatives. This book would be helpful to the individuals interested in acquiring knowledge and adopting renewable energy technologies.

This book "Entrepreneurship in Renewable Energy Technologies" authored by Dr. Manoj Kumar Ghosal, Professor, Dept. of Farm Machinery and Power Engineering, Odisha University of Agriculture and Technology and published by Jaya Publishing House, New Delhi would be useful for those who do not have prior technical knowledge and skills in the field of renewable energy technologies but would still like to use/promote the related technologies on entrepreneurship basis. I hope that it will help students, researchers, energy professionals and even the general masses to understand the various aspects of renewable energy technologies and their applications to address the current challenges of the society. I sincerely convey my best wishes to the author.

13/5/21

CR Mehta

PREFACE

Entrepreneurship in any domain depends on the availability of opportunities that entrepreneurs discover and transform into realized products and services. Entrepreneurs are the prime agents to solve the barriers such as developing skilled man-power, simplification of technologies, creating awareness among the users, providing services at the door steps, establishing liaison with the Government and financial institutions, creating a market base, providing relevant information to policy makers and planners for successful implementation of technologies etc. looking into the availability of resources locally. It is an established fact that renewable energy technologies have a large potential of meeting energy demands of the society in an environment friendly manner. However, these technologies are yet to gain market acceptance and facing a variety of barriers (technical, financial and socio-cultural) against their large-scale dissemination. This may be due to the lack of awareness of the technologies among the users, non-availability of suitable skilled man-power to install the devices and deficiency in the existence of entrepreneurial spirit among the existing young generation to take up the venture either in manufacturing the devices or providing necessary services in the renewable energy sector. Hence, to make the renewable energy systems easily and quickly accessible in the society, there is the necessity of conceive of new ideas, innovative thinking and risk taking ability of starting a new venture that can be obviously expected from the entrepreneurs. Evidence across the globe indicates that renewable energy (RE) is gradually substituting the mature energy systems that are often based on fossil fuels. RE currently accounts for 18.2% of global energy consumption and 26.5% of global electricity production (IEA, 2019).

This book aims at focusing the various possible options for taking up ventures for starting an enterprise based on renewable energy technologies looking into the needs of the society and availability of resources locally. The formulation of a detailed project report in relation to the establishment of an enterprise based on its technical feasibility, economic viability and environmental sustainability of technologies has been discussed.

The book contains nine chapters. Chapter 1 deals with basic concept, meaning, need and function of entrepreneurship. Chapter 2 deals with the various sources of energy with their current status and utilization from both renewable and non-renewable resources. Chapters 3 and 4 discuss the sizing, installation and economic viability of generating power both from roof top based on/off grid solar photovoltaic system. Chapter 5 discusses the importance and sizing of solar-wind hybrid system to improve the system efficiency by using the strength of one source to overcome the weakness of the other. Chapter 6 presents the basic concepts of drying and importance of hybrid solar photovoltaic/thermal drying system and its application in effective drying of agricultural and horticultural produces.

Chapter 7 deals with the novel technology of undertaking thermo-chemical conversion of agricultural residues, following microwave assisted pyrolysis for producing value added products. Chapter 8 discusses the feasibility of adopting solar photovoltaic water pumping system for achieving assured irrigation for about 1 hectare of land to enhance productivity of crops and improve water use efficiency. Finally, chapter 9 deals with the details of financial evaluation for establishing an enterprise for assessing its economic viability to take up entrepreneurship in a profitable manner.

The book can also be used as a reference material on an interdisciplinary approach for the entrepreneurs, students, professionals, stake holders etc. who want to establish enterprise in any field other than the renewable energy sector. Important definitions and glossary of related terms have been given towards the end of the book for their clear and appropriate meaning. Above all, the book is mainly intended to be used by entrepreneur, technicians, manufacturers, students and professionals who are engaged in the profession of renewable energy engineering.

The author feels highly indebted to all those who have directly and indirectly helped in bringing the book to this shape and would also like to express appreciations to my family members for continued patience, understanding and support throughout the preparation of the manuscript.

Special thanks are to M/S Jaya Publishing House, New Delhi for taking interest in publishing the book.

Despite careful scrutiny of the proofs, there may still be some errors. The author would feel obliged if the readers point out the errors for rectification in the successive editions.

Suggestions for further improvement of the book by the readers will be appreciated.

May 2021   **Manoj Kumar Ghosal**

CHAPTER - 1

ENTREPRENEURSHIP
CONCEPT, MEANING, NEED AND FUNCTION

INTRODUCTION

The quality of human resource plays a vital role in the progress of a nation. Innovative, creative, hardworking and honest people lead the nation in appropriate direction for the overall development of the society resulting into accelerating its economic growth. Entrepreneurship is the yardstick to measure the level of development of a country. Entrepreneurship development clearly indicates the availability of quality human resource, a nation possesses. In addition to land, labour and capital, entrepreneurship is considered to be a major part and fourth factor of economic advancement.Its role is very important in the developing country rather than the developed countries with respect to creation of self-employment and reduction of unemployment, bringing about an improvement in the quality of life in the community by the way of emphasizing mostly the need of the society and contributing eventually to the growth in GDP. It has been existing since time immemorial and has been viewed in different manner as per the conditions prevailing in the economy of the globe. The present day's initiatives are more idea-centric and need-based approaching rather than product based ventures to create more wealth and constant innovation in improving the prevailing practice to the next best one. Accordingly, various sizes and types of entrepreneurial endeavour have been emerged all over the country. It covers touching into all sectors such as education, agriculture, medicine, law, research, social work, engineering etc. with a view to fulfil the needs of the people and achieving economic sustainability. With the growing education particularly in the field of technical education, a large mass of youths is now being attracted to start with various new ventures to earn more profit through self –employment and creating employment opportunities for the others. Based on the importance and role of entrepreneurship in improving the economy of the nation, many persons are interested now-a-days to build up their career in this direction. However, it was believed earlier that entrepreneurship is a born-quality. That notion has gone and it has become the key necessity and major point of attraction for the Government to strengthen the economy by formulating so many planning and policies, launching various schemes and providing required facilities for entrepreneurship development among the people. In a nutshell, entrepreneurship provides opportunity for the individual to create wealth by making products through the optimal ways of using available resources to reduce costs and increase in their profit resulting into the growth of the economy. It basically involves the systematic innovation consisting of purposeful and organised search for changes, and a systematic analysis of the opportunities with a view to achieve economic and social transformations. The education and training for this also explores the path of success among the new entrants by understanding and developing their traits and attitudes. It has been established that infusion of this entrepreneurial culture in a society can tackle the fast growing problem of unemployment due to rapid growth in population. This is mostly needed for underdeveloped and developing countries for alleviating poverty, raising per capita income and ultimately supplementing the economic growth.

Entrepreneurship does not emerge spontaneously. Rather, it is the outcome of a dynamic process of interaction between the person and the environment of economic development. Ultimately the choice of entrepreneurship as a career lies with the individual, yet he/she must see it as a desirable as well as a feasible option. In this regard, it becomes imperative to look at both the factors in the environment as well as the factors in the individual for the entrepreneurship development. Entrepreneurship may be considered to be a catalytic process needed for building of nation and creation of wealth. It can be viewed at the two levels, one at the level of individual i.e. developing entrepreneurial qualities in an individual and another at the level of community for creating an environment in which entrepreneurship activities can grow. Various programmes of entrepreneurship development have therefore been initiated now both at Government and non- Government level for its massive promotion in the society with a view to encourage self-employment, providing employment for the others and enhancing economic development.

1.1 ENTREPRENEURSHIP: CONCEPT AND MEANING

The word, "entrepreneur" is derived from the French word, 'entreprendre' and also the German word 'unternehmen' both meaning to 'undertake' or 'to do something' i.e. a person who undertakes the risk of a new enterprise. The word "entrepreneur", therefore, first appeared in the French language in the beginning of the sixteenth century. The word was applicable not to the economic activities but to the undertaking of military expeditions. The term 'entrepreneur' was first introduced in economics by the early 18th century through French economist Richard Cantillon. In his writings, he formally defined the entrepreneur, a person who purchases the means of production for combining them into marketable products at a profit motive in future or is one who buys service-providing factors at certain prices in order to sell their products at uncertain prices in the future. Since then, a perusal of the usage of the term in economics shows that entrepreneurship is a dynamic process with risk/uncertainty bearing, coordination of productive resources, introduction of innovations and creating wealth. Entrepreneurship is the set of activities performed by an entrepreneur which therefore, precedes entrepreneurship.

In short, entrepreneurship is an operational process initiated by an entrepreneur, by which he/she reaches to ultimate objective of establishing an enterprise and manages it profitably facing all risks and uncertainties. This entire process can be termed as 'Entrepreneurship' which has been shown in the following way.

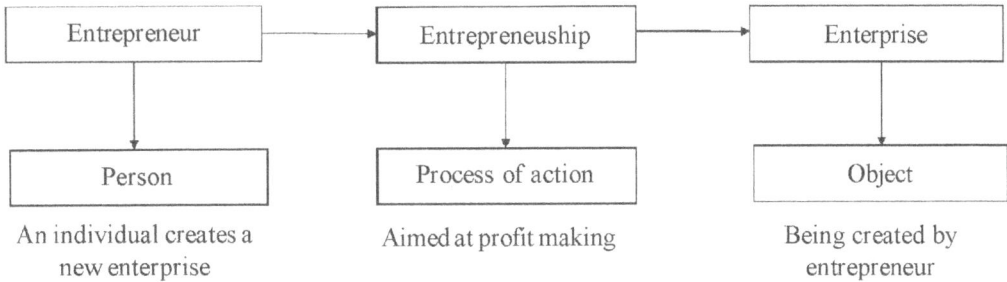

```
┌─────────────────┐      ┌─────────────────┐      ┌─────────────────┐
│   Entrepreneur  │ ───> │ Entrepreneuship │ ──┐  │   Enterprise    │
└─────────────────┘      └─────────────────┘   │  └─────────────────┘
         │                        │             │           │
         ▼                        ▼             │           ▼
┌─────────────────┐      ┌─────────────────┐    │  ┌─────────────────┐
│     Person      │      │ Process of action│   │  │     Object      │
└─────────────────┘      └─────────────────┘      └─────────────────┘

An individual creates a    Aimed at profit making    Being created by
   new enterprise                                       entrepreneur
```

Figure 1.1: Relationship among entrepreneur, entrepreneurship and enterprise.

The terms 'entrepreneur,' 'entrepreneurship' and 'enterprise' can be understood by drawing an analogy with the structure of a sentence in English language. Entrepreneur is the person (the subject), entrepreneurship is the process (the verb) and enterprise is the creation of the person and the output of the process (the object). Entrepreneurs play important roles both in relation to economic development and for their enterprise. In relation to economic development, entrepreneurs contribute to growth in GDP, capital formation and employment generation besides creating business opportunities for others and bringing about an improvement in the quality of life in the community in which they operate. In relation to the enterprise, they perform a number of roles, right from the conception of a business idea, examining its feasibility and mobilisation of resources for its eventual realisation as a business firm. They bear the uncertainties and risks associated with the business activity, introduce product, market, technological and a host of other innovations.

Enterprise

An enterprise is a project or undertaking that fulfils a need of the society which no one has ever addressed.

Entrepreneur

An entrepreneur is a person responsible for setting up a business or an enterprise. He/she has the interest, skill for innovation and looks for high achievements. He/she is a catalytic agent of change and works for the good of people. He/she tries with new projects that create wealth, opens up many employment opportunities and leads to growth of other sectors. International Labour Organization (ILO) defines entrepreneurs as those people who have the ability to see and evaluate business opportunities, together with the necessary resources to take advantage of them and to initiate appropriate action to ensure success.

1.2 DISTINCTION BETWEEN ENTREPRENEURSHIP, ENTREPRENEUR, ENTERPRISE AND WAGE-BASED EMPLOYEE

The terms such as entrepreneur is used synonymously with entrepreneurship, but they are completely different in their meaning and connotation and also in case of entrepreneur and wage based employee (Manager). Their differences are given in the following tables (1.1, 1.2, 1.3 and 1.4).

Table 1.1: Distinctions between Entrepreneur and Entrepreneurship

Entrepreneur	Entrepreneurship
1. Entrepreneur is a person	1. Entrepreneurship is a process.
2. Entrepreneur is an organizer	2. Entrepreneurship is the organized from of initiative
3. Entrepreneur is a risk-taker.	3. Entrepreneurship is a risk-taking activity
4. Entrepreneur is an innovator	4. Entrepreneurship is the process of innovation
5. Entrepreneur is a good planner	5. Entrepreneurship is the planning for successful performance
6. Entrepreneur is a leader	6. Entrepreneurship is the crux of leadership
7. Entrepreneur is a decision-maker	7. Entrepreneurship is nothing but a decision-making activity
8. Entrepreneur is a visualizer	8. Entrepreneurships is the vision
9. Entrepreneur is an administrator	9. Entrepreneurship is the administration.
10. Entrepreneur is an initiator	10. Entrepreneurship is taking an initiative

Table 1.2: Distinction between Entrepreneur and Enterprise

Entrepreneur	Enterprise
1. Entrepreneur is a person	1. Enterprise is a business unit.
2. Entrepreneur is risk-taker.	2. Enterprise is the unit involving risk and uncertainty.
3. Entrepreneur is decision-maker	3. Enterprise serves as the framework within which decision concerning what to produce, how much to produce, where to produce are taken by the entrepreneur.
4. Entrepreneur engages himself in producing and selling the product	4. Enterprise implies the harmonious interrelation or service of functions and staff primarily for the purpose of making/selling product or service.
5. Entrepreneur procures raw materials and other inputs for production.	5. Enterprise utilises the raw materials and other inputs in the process of production.

Table 1.3 Distinctions between Entrepreneur and Wage based Employee (Manager)

Entrepreneur	Wage based Employee (Manager)
1. Entrepreneur is an innovator	1. Manager keeps in managing the business on established rules, policies and procedures.
2. Entrepreneur is a moderate risk-taker.	2. Manager does not undertake risk.
3. Entrepreneur is not guided by the motive of profit. He continuously puts his efforts for achieving the goal-reward of the entire process in connection.	3. Manager is interested in salary. His salary is fixed and certain.
4. Entrepreneur is self-employed and he/she is his/her own boss.	4. Manager is a salaried person and dependent upon his/her employer

Intrapreneur

An Intrapreneur is someone who has entrepreneurial skills but chooses to align his or her talents with a large organisation in place of creating his or her own. He is considered to be an entrepreneur inside the enterprise, or an entrepreneur within a large firm, who uses entrepreneurial skills without incurring the risks associated with those activities. Intrapreneurs are usually employees within a company who are assigned to create special idea for developing the project like an entrepreneur does. Intrapreneur is also an innovator like entrepreneur and introduces new products, ideas and services within the frame work of a company for its growth and success in a changing environment. The intrapreneur's main job is to turn his/her special idea into a profitable venture for the company. Intrapreneurship is necessary as it is the best way to retain talented staff. Otherwise, most of them will just quit and develop these ideas on their own. It will be a win-win situation for both the organisation and the talented employee.

Table 1.4: Distinction between Entrepreneur and Intrapreneur

S.N.	Entrepreneur	Intrapreneur
1	Entrepreneur is the owner of the enterprise	Intrapreneur is not the owner of the enterprise but works for the business
2	Entrepreneur mostly controls the outside affairs of the organization	Intrapreneur looks mostly the internal affairs of the organization
3	Entrepreneur bears the full risk of his/her business	Intrapreneur bears a small part of the entire business
4	Entrepreneur converts the ideas of the intrapreneur into viable opportunities	Intrapreneur takes the responsibility of creating innovation of any kind from within the organization
5	Entrepreneur is independent	Intrapreneur is semi-independent
6	Entrepreneur takes the profit of the business	Intrapreneur does not take profit of the organization but he can be provided with a variety of prerequisites for their innovation.

1.3 DISTINCTION BETWEEN SELF-EMPLOYMENT AND WAGE BASED EMPLOYMENT

Usually after completion of formal education, two options lie before a person to earn his/her livelihood. One may be the employment in Government service, public and private sectors with pre-decided wage/salary. The another one is the route for self-employment in which one is to innovate an idea as per the demands of the society, organizes production/services by marshalling resources and finally markets the products and services. Such type of person is categorized as entrepreneur and he/she works with a challenging attitude to become successful in the ventures. Some of the comparative features between self-employment and wage based employment are therefore presented as follows (Table 1.5)

Table 1.5: Distinction between Wage-employment and Self-employment

S.N.	Features	Wage-employment	Self-employment
1	Nature	Self-saturating	Self-generating
2	Scope	Limited	Unlimited
3	Orientation	Routine and dependent type work	Creative and independent type of work
4	Satisfaction	Sometimes not satisfied with work	Satisfied with pursuing own idea and creative urge
5	Contribution	Consumes national wealth	Generates national wealth
6	Earning	Fixed	Growing and generating surplus
7	Status	Employee	Employer

1.4 DISTINCTION BETWEEN CREATIVITY AND INNOVATION

The terms creativity and innovation are often used to mean the same thing in the usage of language. But they are not the same, though they are closely related, both have a unique combination. Creativity is the ability to bring something new. Ideas have little value until they are converted into innovations, new products, service or processes. Therefore, innovation is the transmission of creative ideas into useful application, but creativity is a prerequisite to innovation.

- Creativity is the ability to bring something new into existence. It is the ability to see (or to be aware) and to respond
- Creativity involves the application of a person's metal ability and curiosity to some ideas, with the creation or discovery of something new as a result.

1.5 DISTINCTION BETWEEN INVENTION AND INNOVATION

Invention is the creation of new products, processes or technologies not previously known to exist. On the other hand, innovation is the transformation of creative/new ideas or into useful applications by combining resources in new or unusual ways to provide value to society for improved products, technology, or services.Invention means discovering new methods and new materials and innovation means utilizing or applying invention and discovering into better quality goods that give greater satisfaction to the consumer and higher profit to entrepreneurs. In a nutshell, an inventor produces ideas where as an innovator implements them for economic gain. Innovative cycle for an enterprise is presented in Figure 1.2.

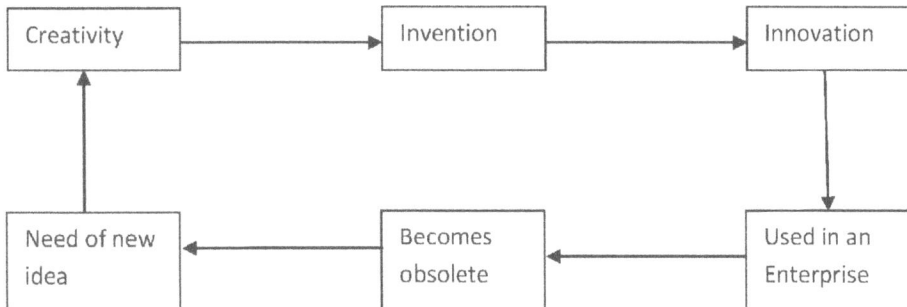

Fig. 1.2. Innovative cycle in an enterprise

1.6 NEED FOR ENTREPRENEURSHIP

Every country, whether developed or developing, needs entrepreneurs. Whereas, a developing country needs entrepreneurs to initiate the process of development, the developed one needs entrepreneurship to sustain it. With the advancement of science and technology, entrepreneurship has emerged as a critical input for socio-economic development. It relates to the purposeful activity of an individual or group of associated individuals, undertaken to initiate, maintain, or earn profit by production and distribution of economic goods and services. Entrepreneurship is neither a science nor an art. It is a practice based on knowledge. In developing country like India, entrepreneurship movement is gathering momentum now-a-days and a large number of support agencies are coming forward to drive such movement a long way for success. A high sense of responsibility is an essential ingredient for development of entrepreneurship. The need for entrepreneurship can thus be highlighted as follows;

Life-line of a nation

No country can progress without the development of entrepreneurship. Every country is trying to promote its trade so that it is able to share the benefits of development. Therefore, entrepreneurship is the yardstick to measure the level of development of a country.

Providing Innovation

Entrepreneurship provides new ideas, imagination and vision to the enterprise. An entrepreneur is an innovator as he/she tries to find new technology, products and markets.

He/she increases the productivity of various resources. The entrepreneur stands at the centre of the whole process of economic development. He/she conceives business ideas and puts them into effect, to enhance the process of economic development.

Change of Growth / Inclusive Growth

An enterprise operates in a changing environment. The entrepreneur moulds the enterprise in such a changing environment that the latter moulds not only the enterprise, but also alters the environment itself, to ensure the success of the enterprise. In order to meet the challenge of automation and the complexities of advanced technology, there is a need for the development of entrepreneurship.

Increased Profits

Increase of profits is the main aim in any enterprise, either by increasing the sales revenue or reducing cost. It is beyond the control of an enterprise to increase the sales revenue. Entrepreneurship, by reducing costs, increases its profits and provides opportunities for future growth and development.

Employment Opportunities

Entrepreneurship and its activities provide the maximum employment potential. Large numbers of persons are employed in entrepreneurial activities in the country. The growths in these activities bring more and more employment opportunities.

Social Benefits

It is not only beneficial to the business enterprise, but to the society at large. It raises the standard of living by providing good quality products and services at the lowest possible cost. It also makes the optimum use of scarce resources and promotes peace and prosperity in the society

1.7 PROCESS OF ENTREPRENEURSHIP

Although enterprises are different and unique, the entrepreneurs who create them, most of them appear to work through a process. The step by step processes through which an entrepreneur is to proceed are as follows;

Self-Discovery: Learning what they enjoy doing; examining their strengths and weaknesses. Examining work experience and relating it to potential opportunities.

Identifying opportunities: Looking for needs, wants, problems, and challenges that are not yet being met, or dealt effectively.

Generating and evaluating ideas: Using creativity and past experience to devise new and innovative ways to solve a problem, or meet a need, and then narrowing the field to one best idea.

Planning: Researching and identifying resources needed to turn the idea into a viable venture. Doing the research in the form of a written business plan prepare marketing strategies.

Raising Start-up capital: Using the business plan to attract investors, venture capitalists and partners. This stage can involve producing prototypes or test-marketing services.

 Start-Up: Launching the venture, developing a customer base and adjusting marketing and operational plans as required.

Growth: Growing the business, developing and following strategic plans, adapting to new circumstances.

Harvest: Selling the business and harvesting the rewards. For many entrepreneurs, this also means moving on to new venture and new challenges

1.8 ENTREPRENEURIAL CHARACTERISTICS

Entrepreneur is a key component in economic progress. He/she does the job of combining elements such as innovativeness, readiness to take risk, sensing opportunities, identifying and mobilizing potential resources, concerns for excellence and is persistent in achieving the goal. The entrepreneurs are known for their special characteristics which make them entrepreneur. An entrepreneur may not possess all the characteristics as mentioned below, but may possess own set of characteristics. It may be noted that there is no sex discrimination in becoming an entrepreneur; males and females are equally competent to become entrepreneur in any field. However, to be successful, an entrepreneur should have the following characteristic features.

Risk taking: Risk means the condition of not knowing the outcome of an activity or decision. A successful entrepreneur prefers a situation where there is moderate challenge, which can be overcome by his/her efforts. He/she prefers to take calculated risk in dealing with a challenging situation. He/she enjoys the excitement of challenge, but does not gamble. This characteristic is one of the most important features. Success or failure of business depends on how decisions are made. A risk situation occurs when one is required to make a choice between two or more alternatives whose potential outcomes are not known. A risk situation involves potential gain or loss. The greater the possible loss, the greater is the risk involved.

Independence: Most of the entrepreneurs start on their own because they dislike to work under others. They prefer to be their own boss and want to be responsible for their own decisions.

Achievement motivation: Entrepreneurs have high achievement motivation, i.e., an urge for excellence and desire for success in competing with standards set by others as well as one sets for oneself.

Self-confidence: The entrepreneurs believe that they are capable of doing things and have the ability to control the actions and activities of other people. They have confidence in their own capabilities to achieve the goals set by them. They also believe that events in their lives are mainly self-determined and need to be tackled.

Taking initiative: The entrepreneurs are independent and highly self-reliant persons. Entrepreneurs actively seek and take initiatives. When they undertake a task, they make it sure to complete it. They prefer situations where the result depends on their ability or efforts rather than a chance or other factors beyond their control.

Creativity and Innovation: Competency in creativity and innovation are sometimes basic traits of certain individuals. He/she might not have any new ideas. He/she may use the creative ideas and innovative products and services to meet the challenges of a situation, take advantage of the utility of an idea or a product to create wealth. Example, changes in the packaging of potato chips.

Problem Solving: Once an entrepreneur is aware that he has ventured on a new area and has taken certain calculated risks, he/she should also be aware that many problems are bound to come in the path of progress. He/she should understand that there is more than one way of solving problems, look for alternative strategies or resources that would help to solve the problem, generate new ideas, products, services etc. For example, when an entrepreneur faces cash crunch, he should look for alternative sources for receiving funds. For example, Ratan Tata shifted the manufacturing plant of Nano cars from Singur to Sanand due to unforeseen complexities.

Leadership: An entrepreneur should also be an effective leader who should be able to guide and motivate his/her entire team. Whenever a company faces problem, it is the will power and effective business acumen and communication skills which oversees the success of the corporation.

Persistence: In most cases, the entrepreneurial pursuits are new and need very close attention. Creating a need in the market for the enterprise, is one of the main requirements of the entrepreneur. This calls for intense perseverance on the part of the entrepreneur. Roadblocks to success should not deter the entrepreneur.

Stress takers: Entrepreneurs are capable of working for long hours and solving different complexities at the same time. As the captain of an enterprise an entrepreneur faces a number of problems and in right moment, he takes right decisions which may involve physical as well as mental stress. He can face these challenges if he has the capability to work for long hours and keep himself cool under monotony.

Goal setter: Entrepreneurs are very goal oriented. They have an ability to set goals and commitment to work towards them. Entrepreneurs are goal and action oriented. They have high need for achievement. They set clear, measureable goals and accordingly they set priorities, measure and guide their time allocations.

Team Worker: True entrepreneurs bring people together as a team and lead them in right direction for achieving success.

Decision making: Decision-making skill is a fundamental characteristic of an entrepreneur. This implies the function of choosing a particular course of action out of several alternative courses for the purpose of achieving specified goals. It reflects the quality and competence of the entrepreneur and is very much necessary at every stage of creation of an enterprise. The stages are such as (i) setting of goals (ii) formulation of policies (iii) designing organizational structure (iv) motivation (v) communication (vi) control, etc. Hence, decision making is necessary at all-times mostly during the conditions of uncertainty and risk.

Apart from these, while planning for setting up a small enterprise, entrepreneurs have to follow the procedures and regulations such as: (a) product identification and project selection (b) preparation of techno-economic feasibility report (c) accommodation and other infrastructural facilities (d) power/water connection (e) procurement of machinery and raw materials (f) financial support (g) marketing of the products or services (h) availing incentives and subsidies declared by the Government for development of SSI sector (i) procurement of pollution control certificates, etc.

Thus, the planning can act as bridge between the present positions and expected future shape of an enterprise. It provides a sense of vision to the entrepreneurs to cope with risky and uncertain situation. Table 1.6 below depicts and summarises the characteristics and traits of an entrepreneur

Table 1.6: Characteristics and traits entrepreneurs

Characteristics	Traits
Self-confidence	Confidence, independence, individuality and optimism
Task-oriented	Need for achievement, profit-oriented persistence, perseverance, determination, hard work and initiative
Risk bearer	Risk-taking ability, likes and challenges
Leadership	Leadership behaviour and responsive to suggestions and criticisms
Originality	Innovative, creative, flexible (openness of mind), resourceful, versatile and knowledgeable
Future-oriented	Foresight perceptive

1.9 FACTORS AFFECTING ENTREPRENEURSHIP

It is an established fact that the progress of a nation entirely depends on the growth of the entrepreneurship, which essentially looks into the economy of the society. The developing country is therefore now-a-days emphasizing more on the growth of the entrepreneurship and encouraging among the people. Hence factors affecting entrepreneurship need to be studied and discussed thoroughly in view of enhancing the economy of a country. These factors are presented as below.

(i) **Availability of Capital:** Availability of capital is one of the major factors for an entrepreneur to start an enterprise. The development of entrepreneurship depends on the degree of strength of the capital market of the nation. In the initial stage, the entrepreneur requires capital to start a risky activity and need instant capital to scale up the ongoing activity if the business moves in right direction. Therefore, countries which have a well-developed system of providing capital at every stage i.e. seed capital, venture capital, private equity and well developed stock and bond markets experience a higher degree of economic growth led by entrepreneurship.

(ii) **Political Factors:** Political factors also play a huge role in the development of entrepreneurship in a given geographical area. This is because politicians decide the type of market that should be in their locality. They are the key persons to frame the policy at the Government level in order to encourage more and more number of entrepreneurs. Due to globalization, the restrictions on industries have been minimized. Thus political environment, having less interference of state and central govt and less restriction on industries, encourage entrepreneurship development.

(iii) **Raw Materials:** Easy and cheap availability of raw materials from the natural resources in a particular locality is an essential component for the viability of an enterprise to be decided by the entrepreneur. In some countries the raw material is available through the market by paying at a fair price. However, in some countries seller lobbies to gain complete control over these natural resources. They sell the raw materials at inflated prices and therefore grab most of the profit that the entrepreneur can obtain. Therefore, countries where the supply of raw material faces such issues, the entrepreneurial ventures would not last long and the entrepreneur would face losses to their activities. Government initiatives need to be taken to solve such problems looking into the growth and progress of the entrepreneurship.

(iv) **Labour Markets:** Labour is an important factor for the sustainability of any enterprise in order to manufacture the products and offering need-based services to the people. The success of the entrepreneurs is therefore dependent on the availability of skilled labour at reasonable prices. However, in many countries, labour market is not well organized. Labourers demand higher wages from the entrepreneurs and prohibit other workers from working at a lower price. This creates an upward surge in the costs required to produce and as such has a negative effect on entrepreneurship. Interference of local authority along with implementation of strict guidelines at Government level need to be prioritised to solve such issues.

(v) Infrastructure: There are some services which are required by almost every enterprise to flourish. These services would include transport, electricity, communication, water supply etc. Since these services are so basic and essential for the survival of the enterprise, they can be referred to as the infrastructure which is required to develop any business. Therefore, if any country focuses on increasing the efficiency of these services, they are likely to impact the businesses of almost all entrepreneurs in the region. Hence, countries which have a well-developed infrastructure system, witness high growth of entrepreneurship.

(vi) Technological Environment: Technology is an art of converting the natural resource into goods and services, more beneficial to the society. Higher the technological development, more is the entrepreneurship development, universally accepted. Due to technological development, new product and new production process are evolved, new raw materials are explored for utilization and new researches are encouraged for modernization. So it can be said that the country, in which technological environment is more suitable, favours remarkably the entrepreneurship development. Due to this reason, presently rapid entrepreneurship development has been reported in the countries like Japan, America, and China.

(vii) Entrepreneurial Quality of the Individual: Entrepreneurship is a human skill which can be developed or upgraded to produce skilled man power whose existence is essentially required for the success of a particular enterprise. Entrepreneur is an individual having specific knowledge, skills and efficiency. Any new enterprise is created by an individual or group of individuals. The creativity of an individual encourages the entrepreneur to establish a new enterprise. Creativity consists of innovation, search and research. Such skills are not shown in all individuals. Supporting society, higher education, training etc. play an important role in developing such skills. Thus characteristics needed for an individual with respect to his/her skills, motives, attitudes, social-culture conditions etc. are important for the growth of the enterprise.

Of course, the above list of factors is not exhaustive. Entrepreneurship is far too complex phenomenon to discuss and enlist in a few bullet points. However, the above list does provide an indication towards the type of factors that can play an important role for the development of the enterprise.

1.10 TYPES OF ENTREPRENEUR

In present days, different types of entrepreneurs are found engaged in various types of activities. Therefore, entrepreneurs can be of different types. Some may prefer to take up the venture alone or share the risk in groups with others. They are found in every economic system and every form of economic activity as well as in other social and cultural activities. They are seen from amongst farmers, labourers, fishermen, tribes, artisans, artists, importers, exporters, bankers, professionals, politicians, bureaucrats and so many others. The classifications of entrepreneurs are discussed as below.

(i) **Innovative Entrepreneur:** This category of entrepreneurs is characterized by the smell of innovativeness. These entrepreneurs sense the opportunities for the introduction of new products, new methods of production techniques, or discovers a new market or a new service or reorganises the enterprise. They have initiative to start new ventures and find innovative ways to start an enterprise. They are very much helpful for the society because of bringing the transformation in the lifestyle of the people based on their needs.

(ii) **Imitative or adoptive Entrepreneur:** This type of entrepreneur usually imitates or adopt the existing practices already made by innovative entrepreneurs. They are adaptive and more flexible. They are organisers of factors of production rather than creators. The imitative entrepreneurs are also revolutionary and important. Adopting the tested technology, they generate ample employment avenues for the country. They contribute to the development of underdeveloped economies. The example is that the local mobile companies using the same technology as big companies do, manufacture their products for wide adoption and popularization.

(iii) **Fabian Entrepreneur:** These entrepreneurs are very much sceptical in their approach in adopting or innovating new technology in their enterprise. They do not venture or take risks and not adaptable to the changing environment. They are rigid and fundamental in their approach. They follow the footsteps of their predecessors. They love to remain in the existing business with the age-old techniques of production. They imitate only when they are sure that failure to do so would result in a loss or collapse of the enterprise.

(iv) **Drone Entrepreneur:** This type of entrepreneur refuses to copy or use opportunities that come in their way. They are conventional in their approach. They never like to get rid of their traditional business and traditional machinery or systems of the business. These entrepreneurs are conservative or orthodox in outlook. They are not ready to make changes in their existing production methods even if they suffer losses. They always feel comfortable with their old-fashioned technology of production even though the environment as well as society have undergone considerable changes. They may be termed as laggards as they continue to operate in their traditional way and resist changes.

Other Types of Entrepreneurs

Apart from the above types of entrepreneur, there are also other classifications of entrepreneur based on the nature of activity, type of business, use of technology etc. These categories are discussed as below.

- **Business Entrepreneur:** Business entrepreneurs are those who don't concern with the manufacturing activity but only undertake trading as their enterprise or business. They identify viable market opportunities and convert them into profitable business to create the demand for their product in the market. The examples of such types of enterprises are cloth stores, medicine shop, stationary shops etc. Most of the entrepreneurs belong to this category because majority of entrepreneurs are found in the field of small trading as a means of earning profit.

- **Service Entrepreneur:** This category of entrepreneur is mainly involved with the service activities such as engineering workshop, repair and maintenance, laundry, hair cutting saloon, beauty parlour etc. These entrepreneurs acquire relevant skills to become successful in their activity by providing satisfactory services to the people.

- **Industrial Entrepreneur:** This category of entrepreneur is mainly involved with manufacturing products and offering services for the people. They have the ability to convert the available resources and technology into a profitable venture. These entrepreneurs operate large, medium, small and tiny enterprises for manufacturing activities and selling the products.

- **Agricultural Entrepreneur:** This category of entrepreneur is mainly involved with the agricultural and allied activities for crop cultivation, animal husbandry, pisciculture, dairy, poultry farming etc. They mostly follow the improved methods of agricultural operations/practices to increase production and productivity. They engage in raising crops and marketing of agricultural produces, fertilisers and other inputs of agriculture through employment of modern techniques, machines and irrigation.

- **Corporate Entrepreneur:** This category of entrepreneurs establishes corporation based on their intelligence and skills. The corporation is formed and registered under statute and operates under legal entity. Usually, they are the promoters of the undertakings/corporations, engaged in business, trade or enterprise.

- **Technical Entrepreneur:** Entrepreneurs of this category are technically educated and experienced, acquire necessary skills and craftsmanship to develop new and improved quality of goods and services. They are the greatest resource persons of a country to strengthen its entreprencurial development.

- **Non-technical Entrepreneur:** Entrepreneurs of this category are involved with the non-technical aspects of the products of the enterprise. They are mainly concerned with developing alternative marketing and distribution strategies to promote their business. They work hard for increasing the demand of their goods/services.

1.11 MARKET SURVEY AND MARKET RESEARCH AND MARKETING RESEARCH FOR AN ENTERPRISE

Marketing of the product is the biggest problem faced by the entrepreneur and the problem is more prevalent in case of small enterprises. Reduction in the cost of production and maintaining high quality are the key to the successful marketing. While launching an enterprise, the first attention needs to be given to the marketing. Markets are becoming increasingly dynamic, competitive and risky. Therefore, entrepreneur must know and understand thoroughly the market where his/her transaction would be carried out, otherwise, there may be every chance of failure of loss for the venture. The knowledge of the market gives him/her the relevant data which is useful in forecasting the probable sales and in formulating the desired market strategy. Hence, market survey, market research and marketing research are the important components for an entrepreneur to reduce the risk factor in setting up a new business venture and also to manage it successfully if risk relating to marketing arises in future.

The word 'market' is derived from the Latin word 'marcatus' meaning merchandise, trade or a place where business is conducted. It is responsible for making the economy strong and stable and is a very dynamic concept. It has been now-a-days steadily acquiring greater dynamism and robustness.

Market survey is the systematic and organized collection of primary data directly from the respondents for a particular product/service which includes the investigation into customers' inclinations. It is a study of various customers' capabilities such as investment attributes and buying potential. Market surveys are tools to directly collect feedback from the target audience to understand their characteristics, expectations, and requirements. However, *market research* is referred as the systematic study and evaluation of all factors bearing on any business operation which involves the transfer of goods from a producer to a consumer. Surveys collect only the primary data, i.e., directly from respondents whereas research, on the other hand, comprises both primary and secondary data. The researchers can refer the data from multiple sources or directly get it by surveying the audiences. There are also many online survey tools available to gather data and feedback. Research instruments such as various tools, methods or techniques employed in research for gathering the information/response. One such most commonly used instrument by the entrepreneurs, is the market survey for market research.Common methods used for surveying are through personal interviews, telephonic interviews, direct mail interviews, E-mail interviews, online interviews, prepared questionnaire, field work etc. There are two types of market surveys namely (a) The census and (b) The sample. Whether census or sample survey, the objective is to produce information with required degree of accuracy, within the planned time span and keeping the expenditure to the minimum.

The basic difference between market research and *marketing research* is that market research is concerned with investigating markets (customers, consumers, distribution, etc.) while marketing research is concerned with investigating any issue related to marketing (consumer behaviour, advertising effectiveness, sales force effectiveness etc. as well as everything contained in market research).

1.12 SOURCES OF FINANCE FOR ENTREPRENEUR

The scarcest of all resources for an enterprise is the finance which is referred as "life blood of enterprise" having persistent demand. For an enterprise, finance is one of the basic foundations of all kinds of economic activities. The demand for finance is for the requirements of capital for commencement of enterprise, day to day operation, modernization activities, expansion, diversification and research and development activities. Commencement, sustenance and growth of any enterprise depend on the timely availability and optimum utilization of finance which may be arranged from the borrowed capital or debt capital and equity capital. Raising of finance is therefore the most important function of an entrepreneur. Finance can be raised from two important sources i.e. internal and external sources.

Internal sources refer to the owner's own fund which can be invested as equity. Equity is the own fund of the enterprise.

External sources refer to the funds raised from different financial institutions like commercial banks, development banks, financial corporations etc. by the entrepreneur. However, funds required by an enterprise are primarily known as fixed capital and working capital.

Fixed capital and working capital: The money invested in fixed assets like land and building, plant and machinery, furniture etc. which are fixed for long period of time is called fixed capital. Similarly, the money invested in the current assets like raw materials, finished goods etc.and are used for day-to-day operation of the business is called working capital.

Besides, on the basis of utilization of finance according to time period, funds raised by the entrepreneur can be categorized into the following types.

(i)   Short-term capital: The money whose repayment period is one year or less.

(ii) Medium-term capital: The money whose repayment period ranges from one to five years and is usually required for permanent working capital, small expansion, replacement and modification of production units.

(iii) Long-term capital: The money whose repayment period exceeds five years and is usually required for installation of plant and machinery.

1.12.1 Various Ways of Raising Finance for an Enterprise

After the total cost of establishing an enterprise has been estimated, the entrepreneur is to decide the various means of raising the funds. Generally, the means for raising the finance can be divided into two parts i.e. (i) owned funds (ii) borrowed funds. Owned funds are from the internal sources such as (a) owner's own investment such as equity (share capital) (b) personal loan from provident fund, life insurance policy etc. (c) ploughing back of profits into own enterprise. Borrowed funds are from the external sources such as (a) debentures (b) term loans (c) public deposits (d) commercial bank (e) deferred credits etc.

Share capital: The capital invested by the owner/owners in the enterprise is called the share capital. Shareholder's capital is called the equity share capital, while shareholder's loan is called debt capital. Shareholder's capital unlike loans, is a capital recorded under the equity account instead of a liability. The amount of capital invested into the business translates into shares that are distributed to the owner/owners accordingly. Debt capital involves borrowing money for a limited period from the external sources for repayment along with interest according to the terms and conditions. The equity shareholders are the owners of the enterprise. This is the permanent capital collected among the owner/owners with no liability for repayment or any obligation for paying dividends. Hence, the enterprise needs not to return equity capital except during its winding up. The equity shareholders bear the risk of ownership and also enjoy the rewards of profits.

Ploughing back of profits: If an enterprise earns ample profit, it invests some portion of such profit as a reserve source of funds instead of distributing all profits among the shareholders. Such profit invested for the enterprise for further developments, modifications, expansion etc. is termed as ploughing back of profit. Legal provisions and limits about ploughing back of profit have been made by the Government. Due to higher back of profit, an enterprise can survive even during recession time and can face easily the financial crisis. Hence, it increases company's financial soundness.

Debentures: Debentures are medium to long-term debt instrument used by enterprises/ companies to borrow money at a fixed rate of interest. The interest paid by them is a charge against profit in the company's financial statements. If an enterprise/company needs medium to long term capital and in a position to repay such capital after a fixed period, it can issue debentures. Debenture is an acknowledgement of debt, issued under the seal of the enterprise/company and containing an agreement for repayment of principal sum at a specified date and for the payment of interest at fixed rate until the principal sum is repaid. Security of assets of the company against loan may or may not be offered. Hence, a debenture is a kind of document acknowledging the money borrowed, containing terms and conditions of the loan, payment of interest and repayment of the loan.

Term loans: It is generally a long-term loan issued by a bank or any financial institution for a fixed amount and fixed repayment schedule for a fixed period with either a fixed or floating interest rate. The loan is to be repaid during 5 to 10 years in equal semi-annual instalments or as per the terms and conditions decided by the lending organization.

Public deposits: Public deposits refer to the unsecured deposits invited by enterprises/ companies from the public mainly to fulfil the finance for the working capital. A company wishing to invite public deposits makes an advertisement in the newspapers. Any member of the public can fill up the prescribed form and deposit the money with the enterprise/

company. The company in return issues a deposit receipt. This receipt is an acknowledgement of debt for the company. The rate of interest on public deposits depends on the period of deposit and reputation of the company. A company can invite public deposits for a period of six months to three years. Therefore, public deposits are primarily a source of short-term finance. However, the deposits can be renewed from time-to-time. Renewal facility enables companies to use public deposits as medium-term finance. The limit of accepting public deposits is up to 25 % of the share capital and free reserves.

Commercial banks: Commercial banks provide assistance for the requirement of working capital and other short-term needs of an enterprise. Commercial banks collect deposit from public generally for short-term and such savings are provided as financial assistance to the entrepreneurs.

Deferred credits: Many times the suppliers of machinery/equipment provide the facility of deferred credit under which payments for purchase of machinery can be made over a period of time.

1.12.2 Finance Requirement Stages

An enterprise requires its finance at various steps in order to make it a full-fledged organization for production or offering services to the people and are as follows.

1st step: This is related to seed money finance. Small amount of financing is needed to prove a concept or to create a product. Marketing is not included in this stage.

2nd step: This is related to start-up money financing. Financing for the enterprise that has already started up in the past one year. Funds are now required for marketing and development of product.

3rd step: This is related to first round financing. Additional money is required to start sales and manufacturing after a firm has spent his start-up capital.

4th step: This is related to second round financing. Finance is required to fulfil the working capital for an enterprise that is selling its products but is still losing money.

5th step· This is related to third round financing. Finance is required for the enterprise that has reached breakeven point and is planning for expansion.

6th step: This is related to fourth round financing. Finance is required for the enterprise to fulfil its bridge financing. Bridge financing refers to a short-term loan lent by the bank to the fund in order to bridge the gap between the time when the money is needed and the time that the fund expects to receive the money from investors.This makes sense considering that the bridge financing does not increase the fund's investment capacity.

1.13 FUNCTIONS OF ENTREPRENEUR

Entrepreneur is to perform a set of activities through the process of entrepreneurship by establishing the relevant enterprise as per his/her idea/planning. He/she is to identify the opportunities as per the demands of the market and to utilize the resources in an effective manner to pursue the opportunities for long term gains. His /her key objective is to create value of the resources. He/she not only perceives the business opportunities but also mobilizes the other resources **like 5 Ms,** such as man, money, machine, materials and methods. The important functions, an entrepreneur is to perform before setting up an enterprise are as follows.

(i) *Idea generation:* This is the first function to be performed by the entrepreneur before setting up enterprise. It depends on his/her education, vision, insight, observation, experience, training and exposure. Ideas can be generated mainly through market survey.

(ii) *Determination of objectives:*The next function of the entrepreneur is to determine and choke out the objectives of the business. For this, he/she is to prepare the clear blue-print on the nature and type of business. This implies whether the enterprise belongs to the category of manufacturing or service or trading business and accordingly plans and steps need to be taken for the proposed venture.

(iii) *Raising of funds:* Finance is the life-blood of an enterprise and therefore funds raising is the next important function of an entrepreneur. The responsibility lies on the entrepreneur to raise funds internally as well as externally. For this, he/she is to be quite aware of the various sources of funds and the formalities to raise funds. He/she should have sufficient information about different Government sponsored schemes, through which he/she can avail Government assistance in the form of seed capital, fixed and working capital for the business.

(iv) *Procurement of raw materials:* Another important function of the entrepreneur is to produce raw materials. He is to identify the cheap and sustainable availability of raw materials which would ultimately help the entrepreneur to manage the business successfully with respect to reduction in the cost of production and to earn more profit.

(v) *Procurement of machinery/equipment:* The entrepreneur is to procure the relevant machinery and equipment for the establishment of the proposed enterprise. For the procurement of machinery, he/she has to collect the right source of information on the details of the specifications of machinery with respect to the available technology, name of the manufacturer or supplier, whether indigenous made or foreign made, after sale services, warranty period etc.

(vi) *Market research:* The next important function of the entrepreneur is the market research and product analysis. It is the systematic collection of data regarding the product which he/she wants to start with the business. He/she is to undertake a thorough survey on the supply, demand, price of the product and the size of the customer in the proposed area for the profitability of the future venture.

*(vii) Determination of form of enterprise:*The function of the entrepreneur is to decide the form of enterprise such as may be on sole proprietorship, partnership, joint stock company and cooperative society. This can be decided based on the nature of the product, volume of investment, nature of activities, type of product, quality of product, quality of human resources etc.

(viii) Recruitment of manpower: For performing this function, entrepreneur is to choke out the activities such as estimation of the manpower need, laying down of selection procedure, devising scheme of compensation and laying down the rules of training and development

*(ix) Implementation of the venture:*Finally, the entrepreneur is to prepare the implementation schedule or action plan for the project. The project needs to be implemented in a time-bound manner. All the activities from the conception stage to the commissioning stage are to be accomplished by him/her according to implementation schedule to avoid cost and time overrun. Hence implementation of the project is also an important function of the entrepreneur.

The functions of an entrepreneur can therefore be summarised and presented in Figure 1.3 below. The above functions can precisely be put into entrepreneurial, promotional, managerial and commercial functions.The entrepreneurial function mainly includes innovation and risk taking, promotional aspect looking into discovery of idea, detailed investigation and to choke out financial proposition, managerial component emphasizes planning, organizing, staffing, facilitating effective communication among the staffs and finally the commercial part which stresses upon the production, marketing and accounts for the smooth functioning of the enterprise.

1.14 SMALL SCALE ENTERPRISES (MICRO, SMALL AND MEDIUM ENTERPRISES) (MSME)

Small Scale Enterprise is considered to be the backbone of the economy of a nation. The role played by this sector cannot be neglected. One of the most essential parts of enhancing the capabilities of Small Scale Enterprises has been identified as up-gradation of technology. The government is making efforts to improve the technology in the small scale sector through formation of appropriate policy, establishment of research institutes, poly technology centers and technology up gradation cells. The government has also made provision for availability of the requisite finance for technology up gradation. Hence, managing small scale enterprises effectively depends on competency of the entrepreneur as well as external factors such as business environment, technology, equipment, finance, human resource etc.

```
                          ┌─────────────────────────┐
                          │ Function of an entrepreneur │
                          └─────────────────────────┘
```

Entrepreneurial functions	Promotional Function	Managerial functions	Commercial Functions
Innovation	Investigation of an idea	Planning	Production
Risk-taking	Detailed Investigation	Organising	Marketing
Organisation building	Assembling the requirements	Staffing	Personnel
	Financing the proposition	Directing	Accounting
		Leadership, Communication, Motivation,	Finance
		Motivation	
		Co-ordination	
		Controlling	

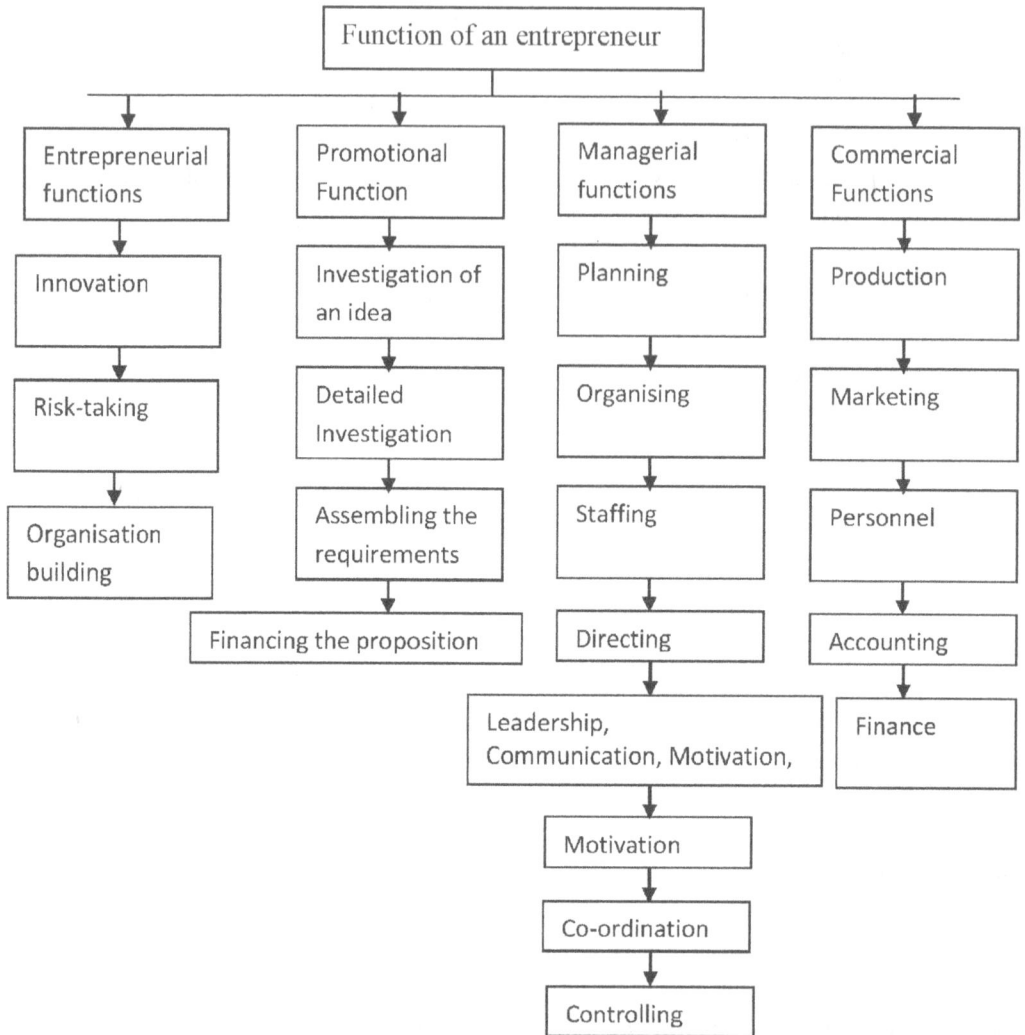

Fig. 1.3. Various functions for an entrepreneur

1.14.1 Meaning and Definitions of Small Scale Enterprise

Small-scale enterprise comprises of a variety of undertakings. The definition of small-scale enterprise varies from one country to another and from one time to another in the same country depending upon the pattern and stage of development, Government Policy and administrative set up of the particular country. As a result, there are nearly 50 different definitions of SSEs found and used in 75 countries. All these definitions either relate to capital or employment or both or any other criteria. We trace here the evolution of the legal concept of small-scale enterprise with the example of India. There can be two bases for defining small scale enterprise and these are as follows:

1.14.2 Scale of Enterprise

The size or scale of enterprise can be measured in various ways like,

(i)   Investment on plant and machinery

(ii)   Employment generation.

(iii)   Investment and employment.

(iv)   Volume and/or value of production.

(v)   Volume and/or value of sales.

Qualitative Aspects: These canbe of

(vi)   Ownership of small enterprise is in the hands of an individual or a few individuals.

(vii) Management and control of small-scale firm is with the owner or owners.

(viii) Technology adopted in small-scale unit is normally labour intensive.

(ix)   Small-scale enterprise is normally carried on in a limited or localarea.

Before Second World War, a small concern was defined as an unit having capital invested up to Rs. 30,000 and those concerns having capital in excess of that amount were classified as large- scaleunits. The definition of small-scale enterprise has undergone changes over years with the ceiling raised to take into account the rising cost of machinery as well as falling the value of rupee.

In accordance with the provision of Micro, Small and Medium Enterprises Development (MSMED) Act, 2006, the Micro, Small and Medium Enterprises are classified in two classes i.e. manufacturing enterprise and service enterprise. Under the MSMED Act, 2006, the earlier concept of "Industries" has been widened to that of "Enterprises". Enterprises have been categorised in terms of investment in plant and machinery/equipment (excluding land and building) as presented in the following Table1.7. The MSMED, Act, 2006 is the first act enacted for establishment and promotion of Micro, Small and Medium Enterprises in India.

Table 1.7: Investment in plant and machinery/equipment (excluding land and building)

Category	Manufacturing Enterprises	Service Enterprises
Micro	Does not exceed twenty five lakh rupees	Does not exceed ten lakh rupees:
Small	More than twenty five lakh rupees but does not exceed five crore rupees	More than ten lakh rupees but does not exceed two crore rupees
Medium	More than five crore rupees but does not exceed ten crore rupees	More than two crore rupees but does not exceed five core rupees

1.14.3 Objectives of Small Scale Enterprises

The small scale sector can stimulate economic activity and isentrusted with the responsibility of realizing the following objectives: -

i.   To create more employment opportunities with less investment.

ii. To remove economic backwardness of rural and less developed regions of the economy.

iii. To reduce regional imbalances.

iv. To mobilise and ensure optimum utilization of unexploited resources of the country.

v.   To improve standard of living of people.

vi. To ensure equitable distribution of income and wealth.

vii. To solve unemployment problem.

viii. To attainself-reliance.

ix. To adopt latest technology aimed at producing better quality products at lowercosts.

1.14.4 Small Enterprise as a Seedbed of Entrepreneurship

Seedbed refers to the preparing of soil for the sowing of seeds so that we may have good crop. Small Scale Enterprise is regarded as a seedbed for entrepreneurship as it provides conducive conditions for the emergence and growth of entrepreneurs. Small-scale units employ available technology and can be started with less investment. They are going to use local resources and cater mainly to local demand. These units normally limit to one individual who is called upon to perform various roles. He/she is the owner, manager and risk bearer and hence can be called an entrepreneur. The emergence, growth and success of entrepreneurs are linked with the growth of small enterprise. Governments of many countries too have given small-scale enterprises an important place in the framework of economic planning for overall development of the society. Thus setting up of more small scale units will create more opportunities for entrepreneurial development and more andmore educated unemployed will come forward for setting up their own enterprises. It will usher in an era where in enterprising persons will assume entrepreneurial career in future.

Small enterprises are called seedbed of entrepreneurship due to the following reasons:

i.   Small-scale enterprises can be started with lesser investment, which can be contributed by the promoter or arranged from friends andrelatives.

ii. Small-scale units can be taken up on a small scale and as such the element of risk is less.

iii. Small-scale units are generally based on local resources and as such there is no problem regarding their availability.

iv. Small-scale entrepreneur adopts labour intensive technology. Thus he generates employment for himself aswell as for others.

v.   Small-scale units can be located anywhere and thus help in the development of backward areas of thecountry.

vi. Small-scale units generally cater to local demand and necessary modifications can be made in the products keeping in mind the changing demand of people.

vii. Small-scale units provide ample opportunities for creativity and experimentation.

viii. Small-scale units have shorter gestation period and hence waiting period for getting return on investment is less.

ix. These units are relatively more environmental friendly.

x. Small-scale units help in building achievement motivation amongst entrepreneurs.

xi. Small-scale units are viewed favourably by the government and society because these help in equitable distribution of income and wealth.

Keeping in mind the above potentials of small scale enterprises, as a developer of entrepreneurial talent, the government has facilitated this sector by providing it with various concessions and incentives.

1.14.5 Steps to Follow for Establishment of Small Scale Enterprise

Before setting up an enterprise, an entrepreneur is to pass through a number of steps in a systematic manner in order to achieve his/her mission. However, the first and foremost decision on his/her part is to whether to become an entrepreneur or not. Likewise, choosing a viable venture ensures the proposition that "well begun is half done". Hence the steps, an entrepreneur is to follow while launching an enterprise are as discussed below.

(a) **Deciding to become self-employed:** This is the most crucial decision, an individual has to take, particularly while avoiding the mentality of wage-employment and opting for self-employment/ entrepreneurship.

(b) **Analysing strengths/ weaknesses:** After deciding to choose the entrepreneurial career, the entrepreneur is to assess his / her strengths/ weaknesses. This will enable him/her to judge the type and size of business to be most suitable for achieving the goal of entrepreneurship.

(c) **Product selection:** The next step is to decide what business to venture into, the product or range of products that shall be taken up for manufacturing and in what quantity. The level of activity will help in deciding size of business and form of ownership. One could generate a number of project ideas through environmental scanning, shortlisting a few of them, closely examine each one of these and go on to a final product.

(d) **Human resource development through training:** After finalising the business to undertake, the potential entrepreneur is to assess his/her deficiencies which can be compensated through acquiring skills by participating the relevant training programmes, provided by the training institutes meant mainly for Entrepreneurship Development Programme. The training programmes help not only to the new entrepreneur but also for skill upgradation to the existing entrepreneur and for the employees of small scale enterprise.

(e) Market survey: It is easy to manufacture an item but difficult to sell. So it is highly necessary to survey the market before going on producing the product and ascertain that the product chosen is in demand. The survey would also give the information about demand-supply lag, extent of competition, frequency of demand, pattern and design of demand, distribution policy, market pricing and distribution policy, etc.

(f) Form of ownership: A firm can be constituted as proprietorship, partnership, limited company (public or private) co-operative society, etc. This will depend upon the type, purpose and size of your business. One may also decide on the form of ownership based on resources in hand or from the point of view of investment.

(g) Location: The next step will be to decide on the place where the unit is to be established. Will it be hired or owned? The size of plot, area and the exact site will have to be decided.

(h) Technology: To manufacture any item, various processes/methods and technologies are available. Information on all these available technologies should be collected and the best one is to be identified. This will be useful in determining the machinery and equipment to be installed.

(i) Machinery and equipment: Having chosen the technology, the machinery and equipment required for manufacturing the chosen product have to be decided, suppliers to be identified and then costs have to be estimated. One may have to plan well in advance for machinery and equipment especially if it is to be procured from outside the proposed area, state or country.

(j) Preparation of project report: After finalising the form of the ownership, location, technology, machinery and equipment, the entrepreneur is to prepare detailed project report in order to assess the feasibility of the proposed venture. The economic viability and technical feasibility of the venture decided has to be established through a project report. A project report that may now be prepared will be helpful in formulating the production, marketing, financial and management plans. It will also be useful in obtaining finance, constructing shed, power and water connection, raw material quotas, etc.

(k) Finance: Finance is a major component for an enterprise. Entrepreneur has to take certain steps and follow specified procedures of the financial institutions and banks to obtain it. A number of financial agencies provide capital assistance for starting an enterprise. The detailed information regarding financial assistance and subsidy facilities available under Government schemes should be aware to the entrepreneur.

(l) Provisional registration: It is always necessary to get the unit registered with the Government. The entrepreneur is to apply for provisional registration to DIC (District Industries Centre) or Directorate of Industries. After obtaining the registration, the entrepreneur can avail the various government facilities, incentives and also financial assistance under different existing schemes.

(m) Power and water connection: The site chosen should either have adequate power and water connection or this should be arranged by following the required procedure for the purpose.

(n) Installation of machinery: Having completed the above formalities mainly finance, work shed, power and water connection etc., the next step is to procure the machinery and their installation work be taken up as per the plan layout.

(o) Recruitment of man power: Once machines are installed, man power will be required to run them. So the quantum and type (skilled, semi-skilled, unskilled, administrative, etc.) of labour have to be determined. Further, sources of getting desired labour and staff members be indented and recruited. Possibly the labour has to be trained either at the entrepreneurs' own premises or in a training establishment.

(p) Procurement of raw materials: Raw materials are the important ingredients for running the enterprise. These materials may be procured either indigenously or may have to be imported by the entrepreneur. Entrepreneur has to identify the cheap and assured sources of supply of raw materials for the smooth running of the enterprise. Government agencies can assist in case the raw materials are scarce or imported.

(q) Production: The unit established should have an organizational step up. To operate optimally, the organization is to employ its man power, machinery and methods effectively. This ensures smooth and effective running of the unit. There should not be any wastage of man power, material or machine capacity installed.

Production of the proposed items should be taken up in two stages.

(i) Trial production

(ii) Commercial production

Trial production helps knowing the problems confronted in production and marketing of the product. By the way, the chances of losses in the eventuality of mistakes can be minimized. Only after successfully launching the product at test-market stage, commercial production should be commenced.

(r) Marketing: Marketing is one of the most important activities for the products manufactured and services offered by the entrepreneur. After manufacturing the product, the great responsibility lies on the selling or providing services to the customer. This is known as marketing. Various aspects lie in the way of marketing such as how to reach the customer, distribution channels, commission structure, pricing, advertising / publicity etc. This has already been decided upon at market survey / project formulation stage. Like production, marketing should also be attempted in two stages: (i) Test marketing and (ii) Commercial marketing

Test marketing will save the enterprise in knowing the merits and demerits of products as per the opinions of the customer and accordingly, the information gives a feed back to carry out modifications, additions in design, characteristics and the other features of the product as per the preference of consumer. Having successfully test marketed the product, commercial marketing can be undertaken.

(s) Quality assurance: Before marketing the product, quality certification like ISO, BIS (Bureau of Indian Standard, earlier ISI), Agmarketc., should be obtained depending upon the product. If there are no quality standards specified for the product, the entrepreneur should evolve his/her own quality control parameters. Quality, after all, ensures long-term success by establishing reputation and credibility of the enterprise.

(t) Permanent registration: After the small scale unit goes into production and marketing, it becomes eligible to get permanent registration based on its provisional certificate obtained earlier from DIC or Directorate of Industries.

(u) Market research: Once the product or service is introduced in the market, there is the strong need for undertaking market research time to time for assessing the areas requiring modifications, upgradation and growth for sustainability of the enterprise. Most of the entrepreneurs neglects the marketing research further resulting sometimes the losses for their enterprises.

(v) Fulfilling legal formalities: For establishing or setting upon a small scale enterprise, entrepreneur is to face many situations in following the guidelines of the law. Right from the conception stage in promoting the enterprise, entrepreneur needs to be careful and also to abide by the legal formalities, procedures, policies and plans of the Government so as to make him/her free from any sort of legal complicacies in future. As Government is emphasizing more on the establishment of small scale enterprises because of their contribution to the economy of the society, the entrepreneurs should adhere to the rules and regulations of the law strictly for the safety and smooth running of the enterprises. Hence it is the duty of the entrepreneurs to obtain required no-objection certificates from the concerned Government organization to remain free from any sort of legal hassles in future.

(I) No-Objection Certificate (NOC) from local body/Panchayat/Municipality/NAC: No-Objection Certificate(NOC) from local body/NAC/Panchayat/Municipality along with permission to start a new venture need to be obtained by the entrepreneurs. This is done for the purpose of construction of industrial sheds, land utilization, pollution control measures etc., for which the enterprises would get the benefits from the Government as per the prevailing facilities existing.

(II) Registration of the unit in DIC: Small scale units are required to be registered with the District Industries Centres (DICs) which mainly function under Directorate of Industries of the state in which the industries are established.As a rule, entrepreneurs are to obtain two types of registration for their units, one is provisional and the other one is the permanent registration certificates. As the name implies, provisional registration is provisional in nature and the enterprise gets provisional registration certificate from the DIC as an incipient unit to grow for production or offering services in future normally within a period of five years. Permanent certificate is provided when the unit has gone into production. The above registration certificates are necessary to make the unit eligible to get the incentives, subsidies, exemptions as and when provided by the Government based on the benefits envisaged in the MSME policy of the Government.

1.15 ENTREPRENEURSHIP DEVELOPMENT PROGRAMMES

The entrepreneurship development is a key to achieve overall economic development through higher level of industrial activity. Every individual has qualities of an entrepreneur. If some or other qualities are developed, such individuals can become successful entrepreneurs. Entrepreneurship development plays a very important role in this direction. The economic and industrial developments of the developing countries are only due to entrepreneurship development. The more the entrepreneurship development programme, the more is the rapid growth of small scale sector, resulting into more economy development of a nation. Entrepreneurship development programme focuses on identifying entrepreneurial qualities of an individual, providing relevant training required for the entrepreneurs by delivering relevant aids/information with regard to finance, production, technology, marketing, management, infrastructure facilities etc. and trying to solve the problems and difficulties faced by the entrepreneurs by guiding them with the remedial measures. Therefore, it is in this context that an increasingly important role has been assigned to entrepreneurship development programme for promotion of an entrepreneurial culture in society.

During early sixties, the small sector was considered exclusively as the employment generating sector, but gradually this sector began to be recognized as the crucial tool for tapping latent entrepreneurial talents and now in the post-liberation period, there seems to be ample opportunities for entrepreneurship and entrepreneurial growth. In order to bring about entrepreneurial growth, the policy makers and financial institutions started thinking in terms of promoting entrepreneurship culture through training interventions. Thus entrepreneurship development programmes (EDPs) were emerged as the pioneering efforts for accelerating entrepreneurial flow among the energetic and interested persons who have the willingness and desires to achieve the goal. The purposes of the programmes are to encourage and train people to set up small scale ventures, how to manage the units effectively, to earn profit from the ventures and to undertake personal responsibility of the business. The notion i.e. the entrepreneurs are born and not made, has no longer been accepted. Ordinary persons can be turned into successful entrepreneurs through well designed training programmes conducted by the Entrepreneurship Development (ED) institutes.

1.15.1 Meaning and Definition

Entrepreneurship Development Programme (EDP) is designed with a view to help a person in strengthening and fulfilling his/her entrepreneurial motives and in acquiring skills and capabilities necessary for playing his/her entrepreneurial role effectively. It is necessary to promote his understanding of motives, motivation pattern, their impact on behaviour and entrepreneurial value.

Entrepreneurship development programmes may be defined as an action plan to create an entrepreneur with the motivation to success, who can develop business, can take strategic decisions, can cope successfully with the internal and external environment and can face the risk of investment. It is an entrepreneurial training programme.In short, Entrepreneurship Development Programme (EDP) is a continuous process of training and motivating entrepreneurs to set up profitable enterprises.

Entrepreneurship development programme is not therefore meant only a training programme, but it is a process;

(a) To increase motivational skills and knowledge of proposed entrepreneur.

(b) To create and improve entrepreneurial behaviour in day to day activities of entrepreneur and

(c) To help in developing own venture after implementing entrepreneurship.

1.15.2 Importance of EDP

EDP is an important innovative breakthrough in the strategy for developing human resource and promoting economic development of a country. It is an innovation in which those persons who possess certain identifiable qualities of entrepreneurship are counselled, motivated and trained to strengthen their self-confidence, grab a business opportunity, initiate an enterprise to become an entrepreneur instead of passively waiting for suitable wage-based employment or continue suffering from frustration in their current jobs. Under the above backdrop, the role of EDP plays an important instrument to tackle the problems of poverty and widespread issue of unemployment through rapid growth of small scale enterprises. EDP basically follows a cycle consisting of stimulatory, support and sustaining activities. An emphasis on the balance on all those three activities ensures accelerated and healthy environment for creation and proper management of enterprise on a sustainable basis. EDP therefore primarily looks into the roles such as stimulatory role (referring to all such activities that stimulate the emergence of entrepreneurship in a society), supportive role (helping the entrepreneurs in establishing and running enterprises by group support activities like registration of unit, arranging finance, providing land, shed, power, water and supply of raw materials etc.), sustaining role (activities which help the entrepreneurs in running their enterprise on a sustained basis amidst competitive market conditions for helping modernization, diversification, expansion, quality testing, additional financing etc.) and socio-economic role (guiding the promotion of the enterprises as well as enhancing the socio-economic status of the people by augmenting latent qualities of the persons, exploring unutilized resources, creating additional employment opportunities etc.).

Entrepreneurship Development Programme is therefore primarily meant for developing those first generation entrepreneurs, who cannot become successful entrepreneurs by their own. Various institutions, governments concerned department, NGOs, etc., have been established for entrepreneurship development plan and such programmes train entrepreneurs for specific objectives.

The entrepreneurship development programmes covers all aspects of business creation, risk-taking, management and the initiative for the socio-economic development and creation of a business culture among the community. These programmes focus to create entrepreneurial qualities among people who will be able to develop business activities, i.e. to set up new business and will be ready or willing to face the risk involved therein. The emphasis is more on the financial and business viability of the project that can create employment opportunities. However, in low income subsistence level communities, the Entrepreneurship Development Programmes focus to create self-employment and alleviate poverty. Entrepreneurship development programme generates motivation and develops entrepreneurial competencies among the young prospective entrepreneurs. Development of entrepreneurship incorporates four basic issues, viz;

(a) The availability of material resources,

(b) The selection of real entrepreneurs,

(c) The formation of industrial units and

(d) Policy formulation for the development of the region.

All these issues are closely interrelated. With the resources and the entrepreneurs, expected to exploit them, the focal issue that remains is that of the type of the industrial unit, particularly because of the proper utilization of raw material and the marketing of the product.

Entrepreneurial development is an organized and systemic development. It is considered as a tool of industrialization and a solution to unemployment problem. One trained entrepreneur can guide others on how to start their own enterprise and to approach various institutions. In fact, trained entrepreneurs become source of developing enterprise and economic progress.

An entrepreneur development programme is based on the belief that individuals can be developed. Their outlook can be changed and their ideas can be converted into action through an organized and systematic programme. Entrepreneurship Development Programme is therefore not merely a training programme, but a process of enhancing the motivation, knowledge and skills of the potential entrepreneurs, arousing and reforming the entrepreneurial behaviour in their day to day activities, assisting them developing their own ventures or enterprise as an outcome to the entrepreneurial action.

1.15.3 Objectives of Entrepreneurship Development Programme

The main objectives of an entrepreneurship development programme are as follows;

(a) To prepare the entrepreneur to take strategic decisions

(b) To let the entrepreneur set or reset the objectives of his/her business enterprise and work individually along with his/her team for the accomplishment of pre-decided objectives

(c) To develop a broad vision and provide visionary leadership to his/her enterprise

(d) To prepare the entrepreneur to bear the unexpected business risks for a long time after training

(e) To induce in his/her mind, the values like integrity, honesty and legal compliance

(f) To provide productive self-employment avenues to a large number of educated and low educated young men and women coming out of school and colleges

(g) To develop and strengthen the entrepreneurial competencies or characteristics among the young persons, especially success to motivation, hard work, perseverance and optimistic thinking

(h) To develop the art of communication and coordination among the entrepreneurs

(i) To enable the entrepreneur to build an effective organization to achieve enterprise goals.

(j) To make the entrepreneur analyse the relevant business environment and understand the procedures and laws involved in establishing small scale enterprise.

(k) To teach the entrepreneurs about the sources of financial and managerial assistance and their availability for small scale business units.

1.15.4 Relevance of Entrepreneurship Development Programme

The relevance of entrepreneurial resource, as the most critical input for economic development has been widely recognized by economists, policy makers, planners in recent years. Continuous emphasis has been put on the role of small scale enterprises in view of their contribution to employment generation, balanced regional development, dispersal of industries in both rural and urban areas, import substitution, export promotion and ultimately economic growth. Hence the relevance of EDP are as follows.

(a) Rapid industrial development: Entrepreneurs and entrepreneurship are the base for industrial and economic development in a given area. Entrepreneurship development programme makes entrepreneurs for creation of entrepreneurship environment. More amount is allotted now-a-days at Government level for Entrepreneurship Development. After freedom, due to Entrepreneurship development programme, the country has shown rapid industrial and economic development. Because of entrepreneurship Development programme, now-a-days entrepreneurs are on the increasing trend in engineering, chemical, electrical and agricultural industries.

(b) Reduction in regional industrial imbalance: Qualities of entrepreneurship can be developed in industrially backward zone by arranging EDP. Hence, for balanced industrial development, the Government has designed incentive plans as a part of the industrial policy. In industrial backward zone, Entrepreneurship Development Programme arranges various training programmes and makes individuals aware about government incentives. The programmes are also framed to train them for entrepreneurship development according to opportunities available. As a result, such zones can also develop in parallel with developed regions. Thus Entrepreneurship development programme helps in removing or reducing regional industries imbalance.

(c) Creation of suitable industrial environment for rapid economic development: Before arranging Entrepreneurship Development Programme, opportunities available in different zones are identified and accordingly the programme is designed. Various institutions are involved for giving their active contribution. Relevant courses of EDI, innovation centers, entrepreneurship environment and support system, strategic international programme, various publications etc. create the suitable environment both in the state and country. Entrepreneurship development programmes are arranged in rural, semi-urban and urban areas. Therefore, in every area, entrepreneurship awareness is spread. The programme plays a dominant role in creating the most suitable industrial environment for rapid economic development.

(d) Optimum utilization of available local resources: If available resources are utilized for production purpose, new opportunities for local entrepreneurship development and employment generation can be created which lead to rapid industrial and economic development. Moreover, because of creation of employment opportunities, migration of people towards urban areas can also be checked.

(e) Reduction in unemployment and poverty problems: Poverty and unemployment are the twin problems existing in any developing country. Government has designed various programmes like rural development programmes in this direction. But Entrepreneurship development programme can play a very important role in this direction. By arranging Entrepreneurship Development Programmes for educated as well as uneducated unemployed persons, they can be trained for self-employment through entrepreneurs and they can also create employment opportunities for others in future. As a result, entrepreneurship development programme helps in reducing unemployment and poverty problem by creating self-employment opportunities.

(f) Protection of entrepreneurs from industrial monopoly: Some large industrial units have created monopoly conditions in their zones. Thus government has been forced to reserve some goods and services for cottage and SSIs. By entrepreneurship development programme, entrepreneur can be trained and prepared for such reserve sector and large scale sectors cannot enter into such reserve sector. Thus due to such decentralization, small scale entrepreneurs can be protected from industrial monopoly of large scale industries.

(g) Generation of new source of income for government: For management of Entrepreneurship development programme, state and central government have to allot a large amount. But government also gets income by levying various taxes on the rise of new entrepreneurs. Some export oriented industrial units help government in earning foreign exchange also.

(h) Progress and expansion of social wealth: Government's special schemes for generation of rural wealth have been reached to rural areas especially through entrepreneurship development programmes. EDP helps to mobilize unorganized and hidden youth of various areas towards planned industrialization. Thus, EDP makes unorganized and rural area's community aware about various government incentives as well as motivating them to be entrepreneurs for taking benefits of these incentives. This improves the social standards and as a result, the progress and expansion of social wealth would be enhanced in rural areas.

(i) Development of socialistic philosophy: Economy of a nation influences both capitalism as well as socialism. Because of establishment of entrepreneurship development programme institution, philosophy of socialism has been spread at various levels through utilization of unused local resources, establishment of tiny, cottage, small and large scale industries. Entrepreneurs have accepted the concept of social responsibility. In this way, EDP helps in extending the philosophy of socialism.

(j) Instrument of social and economic revolution: EDP is designed by considering various classes of the society. In rural, semi urban and urban areas, with a view to assist women towards self-employment and freedom, special EDPs are designed for women, ST, SC as well as ex-service man and physically handicaps i.e. to say that EDPs have played an important role for bringing the poor, exploited and weaker class of our society into economic development. Thus EDP is considered an instrument for bringing social and economic revolution.

(k) Development of entrepreneurial opportunities: EDP is designed and implemented to develop entrepreneurial opportunities for potential entrepreneurs as well as to upgrade managerial skills for the existing entrepreneurs.

1.15.5 Barriers to Entrepreneurship

The factors which inhibit the growth of entrepreneurship may be classified under two categories i.e. Environmental Barriers and Personal Barriers. Environmental barriers include (a) economic (b) social (c) cultural and (d) political factors and personal barriers include (a) motivational and (b) perceptual factors

I. Environmental Category

The factors such as economic, social, cultural and political may have both positive and negative influences on the emergence and growth of entrepreneurship.

Economic factor: The factors which are responsible for economic development such as land, labour, capital,material, market etc., are equally responsible for the development of entrepreneurship. Thus, an environment, where all these factors are available to the entrepreneurs, will naturally support and promote entrepreneurship. On the other hand, if any of these factors are not available or of inadequate quality and quantity, they can become barriers to entrepreneurship.

For example: Unavailability of cash deters an entrepreneur from starting a new venture.

Social factor: Sociological factors such as caste structure, mobility of labour, customer needs, cultural heritage, respect for senior citizens, values etc. might have a far reaching impact on business. In the modern days, attitudes have been changed with respect to food and clothing as a result of industrialisation, employment of women in factories and offices, and the increased level of education. This has resulted in the growth of food processing and garment manufacturing units thus the emergence and growth of a new class of entrepreneurs.

For example: Readymade shirts, instant food, vending machines for tea and eatables.

Cultural factor: Every society has its own cultural values, beliefs and norms. If the culture of a society is conducive to creativity, risk-taking and adventurous spirit, such cultural background encourages entrepreneurship development.

For example: An entrepreneur will have to keep in mind the cultural reference of the region that he/she is going to cater to, this will enable him/her to get a quicker acceptance in that region.

Political factor: It provides the legal framework within which business is to function. The viability of business depends upon the ability with which it can meet the challenges arising out of the political environment. This environment is influenced by political organisations, stability, Government's intervention in business, constitutional provisions etc.

For example: War related stress between two countries can also stop the trade between these countries.

II. Personal Category

In a given society, a few people may take up the career of entrepreneurship. Even among the societies which are considered entrepreneurially progressive, only some selected persons venture toset-up their own enterprises.

Perceptual factor: There are certain perceptual barriers that can hamper the progress of an entrepreneur. Lack of a clear vision and misunderstanding of a situation, can result in a faulty perception. Having preconceived notions and prejudices against a particular business activity discourage the growth of entrepreneurship.

For example: One should overcome the barriers of selecting a business venture according to one's gender. There is hardly a business left where both the genders have not explored and achieved equal success.

Motivational factor: Sustained motivation is an essential input in any entrepreneurial venture. Lack of motivation is a strong barrier to entrepreneurship. Many entrepreneurs start with enthusiasm, but when they face some difficulties in the execution of their plans, they lose motivation.

For example: Failure of a venture.

1.16 DISTINCTION BETWEEN FEASIBILITY STUDY AND PLAN FOR A NEW VENTURE

Feasibility study for a new venture is defined as a controlled process for identifying problems and opportunities, determining objectives, describing situations, defining successful outcomes and assessing the range of costs and benefits associated with several alternatives for solving a problem. It basically takes the guess work prior to set up an enterprise and provides an entrepreneur with a more secure notion whether the business idea is feasible or not. A feasibility study is therefore done with calculations, analysis and estimated projections while a business plan is made up of mostly tactics and strategies to be implemented in order to grow the business. A feasibility study is not the same thing as a business plan. The feasibility study is generally completed prior to the business plan. The business plan is developed after the business opportunity is created. Before anything is invested in a new business venture, a feasibility study is carried out to know if the business venture is worth the time, effort and resources. It may seem that the feasibility study is similar in many ways to the business plan, but it is important to keep in mind that the feasibility study is developed prior to establish the venture. It's important to think of the business plan in terms of growth and sustainability and the feasibility study in terms of idea viability.

The information gathered and presented in a feasibility study help the entrepreneurs to

● List in detail, all the things they need to make the business work

● Identify logistical and other business-related problems and solutions

● Develop marketing strategies to convince a bank or investor that their business is worth considering as an investment

● Serve as a solid foundation for developing their business plans.

Hence, in order to create a feasibility study, entrepreneurs need to define dimensions of business viability including: market viability, technical viability, economic and financial viability and organisational viability.

Market feasibility: It includes a description of the enterprise, current market, anticipated future market potential, competition, sales projections, potential buyers, etc.

Technical feasibility: It includes the details on how to manufacture/deliver a product or service (i.e., materials, labour, transportation, where the business will be located, technology needed, etc.).

Financial feasibility: It includes the assessment of start-up capital, sources of capital, returns on investment, etc.

Organisational feasibility: It defines the legal and corporate structure of the business (may also include professional background information about the founders and what skills they can contribute to the business). The feasibility study has been shown in Figure 1.4.

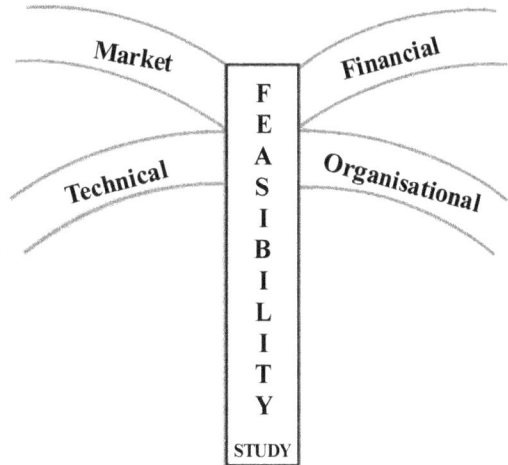

Fig. 1.4. Features of Feasibility Study

Some of the important features of feasibility study are as follows

- A feasibility study is a process in which an idea is studied to see if it is "feasible," that is, if and how it will work.
- A comprehensive feasibility study looks at the entire structure, needs, and operations of a business.
- It looks at one specific task, program, idea, or problem.
- It looks at both sides, considering advantages and disadvantages, and troubleshoots as well as potential problems.
- It is not a business plan, but serves as a foundation for developing a business plan.
- A market feasibility study is not a marketing plan, but studies markets and market potential and can be used to support or develop a marketing plan.

Preparation of a Business Plan

A Business Plan is a written summary of various elements involved in starting a new enterprise of how the business will organize its resources to meet its goals and how it will achieve progress. Hence, a business plan serves the following purposes,

a) Provides a blue print of actions to be taken in future

b) Guides the entrepreneur in raising the factors of production

c) Serves as a guide to organizing and directing the activities of the business venture

d) Helps in measuring the progress of the venture at successive stages

e) Communicates to investors, lenders, suppliers etc., initiating the programmes of the business

1.17 FORMS OF OWNERSHIP FOR AN ENTERPRISE

There are various forms of business organization in which the business entity can be organized, managed and operated. They are sole proprietorship, partnership, joint stock company and cooperative organisation (or societies).

Sole proprietorship is one of the oldest and easiest forms, which is still prevalent in the world. In this type of business, only one person owns, manages and controls the business activities. The individual who runs the business is known as a sole proprietor or sole trader. Sole proprietorship, as its name suggests, is a form of business entity in which the business is owned as well as operated by a single person. The person uses his capital, knowledge, skills and expertise to run a business solely. In addition to this, he/she has full control over the activities of the business. As this form of business is not a separate legal entity, therefore the business and its owner are inseparable. All the profits earned by the owner go to his/her pockets and the losses are also borne by him/her only. The owner of sole proprietorship business is called as the proprietor.

Partnership, on the contrary, is that form of business organization in which two or more individuals come together and agree to share profit and losses of the business in the specified ratio, which is carried on by them. The individuals who run the business are called partners. Members are separately known as partners, but jointly known as firm.

The basic differences between sole proprietorship and partnership are as follows,

(i)   The registration of sole proprietorship business is not necessary but it is the discretion of the partners that whether they want to register their firm or not

(ii) In sole proprietorship, the minimum and maximum limit of the owners are one, conversely, in partnership, there should be at least two partners and it can exceed up to 100 partners.

(iii) In sole proprietorship the liability is borne by the proprietor only, in contrast to, partnership where the liability is shared between the partners

Joint Stock Company

A joint stock company is an association of individuals, called shareholders, who join together for profit and agree to supply capital divided into shares that are transferable for carrying on a specific business. A joint stock company consists of more than twenty persons for carrying any business other than the banking business. The shareholders give a name to the company, mention the purpose for which it is formed, and state the nature and the amount of capital (shares) to be issued, etc., and submit the proposal to the Registrar of Companies. After getting the certificate from the Registrar of Company, the company starts operating. The managing body of a joint stock company is Board of Directors elected by the shareholders.

Cooperative Organization (or Societies)

A cooperative organisation is an association of persons, usually limited in number who join voluntarily to achieve a common eco-nomic end through the formation of a democratically controlled organisation, making equitable distributions to the capital required, and accepting a fair share of risk and benefits of the undertaking.It is a kind of voluntary, democratic ownership formed by some motivated individuals for obtaining necessities of everyday life at rates less than those of the market. The principle behind the cooperative is that of cooperation and self-help.

The basic differences between joint stock company and cooperative organization (society) are as follows,

(a) The primary objective of a cooperative society is to provide service, whereas a company seeks to earn profits. This does not mean that a cooperative society does not earn profits or a company does not render service to society.

(b) It simply means that all the activities of a cooperative society are guided by service motive and profits are incidental to this objective. On the other hand, the activities of a company are inspired by profit taking and services rendered to society are incidental to profit motive.

(c) The minimum number of members is ten and there is no maximum limit for cooperative organization whereas the minimum number is seven and there is no maximum limit in a joint stock company.

(d) Cooperative organization is regulated by the Co-operative Societies Act but joint stock company is regulated by the Companies Act.

1.18 RESOURCE MOBILIZATION FOR AN ENTERPRISE

A very important function of entrepreneurs is the arranging and procuring the required resources in an effective manner for smooth running of the enterprise. An economic or productive factor required to accomplish an activity, or as a means to undertake an enterprise and to achieve the desired outcome is referred as 'Resource'. Therefore, 'resource' refers to anything or means (physical tangible/non-physical-tangible) required or required to support the activities of organisation to achieve pre-determined organizational goals. These resources are the life blood of any economic activity. Their successful and timely identification, procurement and utilization ensure the success of an organisation. Human, financial, physical and knowledge are the basic factors that provide a firm, the means to perform its business processes and are referred as 'Business Resources'. The planning for effective resource mobilization includes to

(a) Evaluate and judge the need for resource.

(b) Identify the type of resource required.

(c) Locate the availability of resource.

(d) Effective Communication with the suppliers of resources.

(e) Evaluate the quality and quantity of resources required.

(f) Identify problems pertaining to mobilization of resources.

(g) Arrange funds for acquisition of resources.

(h) Plan out inventory management for the procured resources.

The various types of resources for an enterprise are (a) physical (b) financial (c) human (d) technical (e) material (f) intangible.

*Physical:*Physical resources are those that are made by human through his abilities and skills. They are available to an organisation in the form of buildings, plants, machinery etc. required for running of an enterprise.

Human: The most important asset for an enterprise is the human resources whose absence makes all other non-living resources useless. An organization's performance and resulting productivity are directly proportional to the quantity and quality of its human resource. A rich and continuing supply of qualified people/personnel is the best assurance an enterprise can have that it will flourish. 'Right man at right job at the right time'is the key for successful enterprises because it ensures: (a) benefits of specialization to the firm (b) minimizes wastages of resources (c) reduces inefficiencies (d) reduces labour turnover ratio and (e) saves cost of production.

*Financial:*Finance is one of the important prerequisites to start an enterprise. It is the availability of finance that facilities an entrepreneur to bring together, personnel, machines, materials, methods, land and convert a dream into reality. It is required to assess how much finance is needed, terms for which finance is required and sources of generating finance.

Technical: The enterprise whether involved with manufacturing products or offering services must be equipped with the updated technology/methods/processes for the success of the business. The technical experts or the trained man power are engaged for accomplishing the technical aspects of the enterprise. The latest information need to be collected time and again by the entrepreneur for skill upgradation of the technical experts by introducing the modern technology.

Material: Whether a business enterprise markets a product or offers a service, certain operations are essentially carried out combining raw materials, processing and assembling machines,tools, power etc. A deep insight into the production processes is essential for effective handling of the enterprise. To successfully convert raw materials into finished products with value addition, a wide range of arrangements is needed to be worked out by the entrepreneur. Some of the important decisions are regarding: (a) size of the unit and its installed capacity (b) identifying machinery and the technical know-how required (c) technical training involved (d) quality control systems required (e) type of technical staff required (f) maintenance cost (g) availability of spares parts and support services (after sale services) (h) wear and tear rate of assets (i) type of raw materials required (j) supplies of the raw material, their number and location.

Intangible: A much ignored resource, which is otherwise quite crucial, is the intangible resource. These resources are neither felt nor seen, far from being touched or preserved but helps immensely in providing a strong support for the enterprise. The intangible resources are good will of the associates in the enterprise, reputation from the employee, brands of the product i.e. with a strong brand, an entrepreneur can command and sustain higher margin and intellectual property viz. key commercial rights protected by patents and trademarks may be an important factor to be worked out by the entrepreneur.

1.19 CONCEPT OF RISKS IN ENTREPRENEURSHIP

Risk can be defined as a possibility of failure or loss or other adverse consequences in pursuing some activity or venture. Risk bearing and entrepreneurship are inseparable from each other. Risk, as an attribute, affects entrepreneurial behaviour. It is, among other things, the element of risk involved in entrepreneurial career and therefore many people become hesitant to become entrepreneur. Even those who take risk by joining entrepreneurship differ in the degree of risk taking ability and willingness. Depending on the degree of risk, risks can be categorized as high risk, moderate risk, and low risk. All three category of risks influence entrepreneurial behaviour differently. High risk refers to the situation when an entrepreneur starts a venture without much knowledge about it. Hence, there are more likely chances of failure or loss resulting into facing the high risk. On the contrary, moderate risk refers to the situation in which entrepreneurs do a lot of exercises and calculations to educate him/her properly before setting up an enterprise in order to reduce the chances of failure if arises. In this circumstances, the entrepreneur tries to control the risk. However, low risk refers to the situation in which some of the entrepreneurs undertake the activity or venture where they would like to achieve 100 percent of what is desired. In fact, those entrepreneurs don't take any risk. They are like fabian entrepreneurs.

Types of Risk

Business risks are of diverse in nature and arise due to innumerable factors. These risks can either be insurable or non-insurable. Insurable risks are those which can be covered through different types of insurance policies. The probability of an insurable risk can be determined. In other words, such risks can be forecasted. These are related to life and property against fire, theft, riots etc. Non-insurable risks are those whose probability cannot be determined and which cannot be insured due to lack of sufficient measurable parameters/indicators. One such example is the fluctuations in price and demand of a product. These risks may be broadly classified into two types i.e. internal and external risks, depending upon their place of origin.

Internal Risks

Internal risks are those risks which arise from the events taking place within the business enterprise. Such risk arises during the normal course of a business. It can beforecasted and the probability of its occurrence can be determined. Hence, it can be controlled by the entrepreneur to an appreciable extent. The various internal factors causing such types of risk are as follows,

Human Factors: Human factors may result from the strikes and lock-outs by trade unions, negligence and dishonesty of an employee, accidents or deaths in the industry, incompetence of the manager or other important people in the organisation etc. Also, failure of suppliers to supply the materials or goods on time or default in payment by debtors may also adversely affect the business enterprise.

Technological Factors: Technological factors are the unforeseen changes occurring in the techniques of production or distribution. They may result particularly in the technological obsolescence. One such example is that if there are some technological advancements which result in the products of higher quality, then an enterprise which is using the traditional technique of production, might face the risk of losing the market for its products of inferior quality.

Physical Factors: Physical factors refer to the loss or damage to the property of the enterprise. They include mostly the failure of machinery and equipment used in business; fire or theft in the industry; damages in transit of goods, etc. They also include losses to the firm arising from the compensation paid by the firm to the third parties on account of intentional or unintentional damages caused to them.

External risks arise due to the events occurring outside the business organisation. Such events are generally beyond the control of an entrepreneur. Hence, the resulting risks cannot be forecasted and the probability of their occurrence cannot be determined with accuracy.

External Risks

The various external factors which cause risks are as follows,

Economic factors: Economic factors are the most important causes of external risks. They result from the changes in the prevailing market conditions. They may be in the form of changes in demand for the product, price fluctuations, changes in tastes and preferences of the consumers and changes in income, output or trade cycles. The conditions like increased competition for the product, inflationary tendency in the economy, rising unemployment as well as the fluctuations in world economy may also adversely affect the business enterprise. Such risks which are caused by changes in the economy are known as 'dynamic risks'. These risks are generally less predictable because they do not appear at regular intervals.

Natural factors: Natural factors are due to the unforeseen natural calamities over which an entrepreneur has very little or no control. They result from the events like earthquake, flood, famine, cyclone, lightning, tornado, etc. Such events may cause loss of life and property to the enterprise or they may spoil its goods.

Political factors: Political factors have the great influence on the functioning of a business, both in the long and short term. They result from political changes in a country like change in the Government, communal violence or riots in the country, civil war as well as hostilities with the neighbouring countries. Besides, changes in Government policies and regulations may also affect the profitability and position of an enterprise.

Changes in taste and preference: Tastes change with the change in the dynamics in the living standard of the people. One example is that earlier there was a preference to eat most of the time at home and fast food concept was completely unknown. But nowadays, there has been an upsurge of fast food outlets all over the country.

Thus, business risk takes a variety of forms. In order to face such risks successfully, every businessman should understand the nature and causes of these risks as well as the various measures which must be taken in appropriate time in order to minimise them. Risks are present everywhere in our society; an entrepreneur should not be disappointed looking into all these risks. Every entrepreneur is to be trained accordingly to face the risks and be successful in his/her ventures.

1.20 AGRIPRENEURSHIP

Agripreneurship is the entrepreneurial process taken up in agriculture or the allied sectors. It is the process of adopting new methods, processes, techniques in agriculture or the allied sectors of agriculture, for better output and economic earnings. Agripreneurship converts agricultural activity into an entrepreneurial activity. Agriprenuer is an innovator, who drives change in rural economy, by adopting innovative ideas in agriculture and allied sectors. He/she takes risk, adopts innovation, creates new ways of doing things and taps new markets. Entrepreneurship in agriculture or agripreneurship is therefore defined as the creation of innovative economic organization for the purpose of growth or gain under conditions of risk and uncertainty in agriculture. It is a sustainable, community oriented and directly marketed agriculture with a focus to the interrelationships of social, economic and environmental processes. Agripreneurship is therefore required due to (a) increasing demand for quality food (b) competitive advantages for many primary production activities in agriculture such as tropical fruits and vegetables, livestock rearing, aquaculture, rainfed farming etc. (c) willingness of the private sector to enter into agri-business at all levels of operation (d) changing consumer demand (e) adequate scope for exporting agricultural commodities in the era of globalization (f) fast growing food processing industry (g) supportive policies of Government, NGOs and financial agencies.

1.20.1 Agripreneurship Development

Agriculture is an important and dominant sector for the economy of most of the countries because of its high share in employment and the source of livelihood for majority of population. Due to changing scenario in the socio, economic, political, environmental and cultural dimensions throughout the world, the farmers' and nations' options for survival and sustainability in the economic environments have become increasingly critical. The rapid growth of agriculture is essential not only for self-reliance but also for meeting the food and nutritional security of the people to bring about equitable distribution of income and wealth in rural areas as well as to reduce poverty and improve the quality of life. Growth in agriculture has a maximum cascading impact on other sectors, leading to the spread of benefits over the entire economy and the largest segment of population. The emergence of free market for sustaining economy has led to the development of a new spirit of enterprise i.e. 'Agripreneurship' and the increased individual need for responsibility in running their own businesses relating to the production, marketing and offering services in the field of agriculture.

1.20.2 Types of Agripreneurship

Agripreneurship is a type of entrepreneurial activity relating to agriculture and allied sectors. It deals with the cultivation, marketing and providing necessary services on commercial basis so as to make profit. There are various avenues where agripreneurship can be taken up and is broadly categorized as follows.

(a) Agricultural Equipment Business or Agricultural Machinery Business: Just like business on agricultural products, agricultural equipment/machinery business is focused on manufacturing and selling of agricultural equipment or machinery. So many commercial farmers depend on farm equipment or machinery for modernization of agricultural operations. Thus agricultural equipments are either sold to local buyers or international buyers. With the growth of agricultural sector, there is the vast scope of promotion and circulation of equipment among the exporters and importers. Agripreneurship in this direction has also great opportunity in the rural sector looking into the popularization of hand tools and implements among majority of small and marginal farmers. The selling and buying of equipment can also be taken up online or offline for widening business sphere due to recent advancement in information technology.

(b) Agricultural Services Business: This type of agricultural business provides agricultural services to the farmers. Agriculture related service is another area for the promotion of agripreneurship. These services include procurement of seeds, fertilizers, pesticides etc. and their distribution at village level, providing agri-tech equipment for pre and post-harvest operations starting from land preparation to threshing, storage and marketing and offering technical services like soil testing, plant protection, weed control, installation and maintenance of irrigation facilities, transportation and storage of agri products etc. These agricultural services are also carried out either online or offline.

(c) Agrochemicals Business: This is a type of agricultural business that is concerned with the manufacturing and selling of chemicals used in agriculture for increasing production and productivity and making profit. These chemicals are either used as fertilizers to enhance fertility of soil, herbicides to control weeds and pesticides to control pests.

(d) Agro-allied Business: This refers to a business dealing with agricultural products and agricultural services for agricultural and allied sectors. The products are called agro-allied products while their services are called agro-allied services. For the promotion of agricultural inputs, there are many potential opportunities for entrepreneurship. Agriculture process needs so many kinds of inputs like seeds, fertilizers, pesticides and farming technology. So many opportunities lie in the areas of developing and producing these inputs. There are excellent opportunities for entrepreneurial process in the areas like bio-pesticides, bio-fertilizers, vermin compost, testing and amending soil etc. The increasing focus on organic farming has also attracted more opportunities. Different varieties of species are being developed with respect to fruits, vegetables and other crops. There is a lot of scope for R&D with respect to seed development. New varieties of seeds are to be developed to improve the agri output. These varieties of seeds are expected to serve even in odd/adverse climatic conditions. By the way, per hectare agri production can be improved a lot so that the agriculturist can realize maximum revenue.

1.20.3 Agripreneurship for Small Farm Mechanization

Mechanization of small farms has been a major challenge especially for developing countries because of the involvement of majority of small and marginal farmers in agriculture, the only and chief means of their livelihood. Farm mechanization is therefore essential to sustain the interest of small farmers. However, mechanization needs to be adopted in an entrepreneurship mode among such category of farmers for their wide promotion and popularization. Environmental issues are becoming more important and crucial to sustainability in any sector. Hence, the importance of mechanization for successful implementation of conservation agriculture technologies is imperative. Many climate smart technologies can't be adopted without introduction of suitable mechanization. Robustness and affordability of the smart technologies is another important issue for the developing nations. Besides, smart agricultural technologies need to be popularized based on location and crop and can be achieved only through entrepreneurial culture in the society. Though various types of time saving, affordable and drudgery reducing small hand tools and agricultural implements have been developed in many research organization, those have not yet been popularized to a great extent due to the lack of entrepreneurship development and the traditional attitude of farmers for non-adoption to the modern technologies. The role of mechanization in agriculture has been increased as it enhances productivity through increased input use efficiency, timeliness of agricultural operations, reduced drudgery as well as the cost of cultivation. Mechanization interventions have been reported to increase the productivity by 15% and reduce the cost of production to the tune of 20%. The mechanization also facilitates conservation and sustainable agriculture while improving the livelihood opportunities, income and environmental sustainability. Agripreneurship in small farm mechanization is therefore the right option in the present context of enhancing mechanization among small and marginal farmers due to introduction of intensive agriculture and scarcity of required man-power during peak season of cultivation.

Agripreneurship in farm mechanization is therefore the need of the hour for any country to make agriculture mechanized in a more attractive and profitable venture. There is a great scope for entrepreneurship in agriculture mechanization. The potentiality of the country can be tapped only by effective, precise and timely management of agri elements such as soil, seed, water and market needs. The youth who can bear the risk and having a quest for latest knowledge in agricultural sector can prove to be the right agripreneurs. It also has a large potential to contribute to the national income while at the same time providing direct employment and income to the numerically larger and vulnerable section of the society. Agripreneurship for mechanization is not only an opportunity, but also a necessity for improving the production and profitability in agriculture by disseminating need based and location specific hand tools and implements. It is required to replace the traditional inefficient agricultural tools with efficient mechanized cultivation to facilitate timely, precise and scientific farm operations, thereby increasing farm input and labour use efficiency.

1.20.4 Need for Agripreneurship in Small Farm Mechanization

Entrepreneurship development in the farming community for the small hand tools and implements would offer the following benefits particularly to the small and marginal category of farmers who contribute their major involvement in farming.

(a) Input savings: Studies have established that there is a direct relationship between farm mechanization (farm power availability) and farm yield. Farm mechanization is said to provide a number of input savings: seeds (approximately 15-20 percent), fertilizers (approximately 15-20 percent) and increased cropping intensity (approximately 5-20 percent).

(b) Increase in efficiency: Apart from the above stated inputs, farm machinery also helps in increasing the efficiency of farm labour and reducing drudgery and workloads. It is estimated that farm mechanization can help reducing time by approximately 15-20 percent. Additionally, it helps in performing timeliness in pre-harvesting operations, reducing the post-harvest losses and improving the quality of agricultural produces. These benefits and the savings in inputs help in the reduction of production costs and allow farmers to earn more income.

(c) Social benefits: There are various social benefits of farm mechanization as well, such as helping in conversion of uncultivable land to agricultural land through advanced tilling techniques and also in shifting land used for feed and fodder cultivation for animals towards food production, decreasing workload on women as a direct consequence of the improved efficiency of labour, improving safety of farm practices, helping in encouraging the youth to join farming and attracting more people to work and live in rural areas.

(d) Only alternative to deal with increasing cost of labour: The cost of deploying labour for agriculture operation is increasing substantially. Farm mechanization is the only way to reduce labour cost, and thus cost of cultivation.

(e) Farm equipment availability for each small and marginal farmer: Tractor penetration has increased from one per 150 hectares to one per 30 hectares. However, such an increase in penetration has not been seen in other segments of farm equipment. As per-capita land holding of farmers is decreasing, small farm machinery / implements (individually operated) need to be promoted keeping in view the versatility of various crops, cropping pattern and agricultural operations. Further advanced machinery and implements developed in various research organizations can also be promoted if any interested entrepreneur desires with setting up the enterprise. Various initiatives launched by the Government can also be used to support the manufacture of inputs and farm implements currently not available in the rural areas. This would help in reducing the overall capital cost and increasing mechanization level in the society.

(f) Value addition and marketing: Keeping long term perspective to make agriculture sustainable and a lucrative activity, value addition and marketing can be strengthened at farmers' level with assured forward linkages. It would boost confidence of the farmers to invest in farm mechanization.

(g) Entrepreneurship for service providing in mechanization sector: The consumption of farm power in India stands at an average of 2.02 kW/ha in 2017-18 and comparatively poor even with the Asia-Pacific countries. A target of at least 4 kw/ha should be the aim and need to be achieved by 2022. Considering the preponderance of small and marginal holdings in the country, R and D should aim at developing and designing gender-friendly machinery so that both male and female members can be involved with the farming. Further, machinery that can suit different terrain of the geography deserves priority attention. Farm machinery can become a part of 'Farming

as a Service' (FaaS), which means that farmers should have easy access to mechanization and related services on rent in preference to owning the same. This can be facilitated by promoting (a) 'Custom Hiring Centres' (CHCs) at the rate of a minimum of 1 (one) per village (when large) and 1 (one) per Gram Panchayat comprising in cluster of small villages. These should be able to meet the demand for all basic services, and would therefore be expected to possess low duty machinery (b)'Agriculture Machinery Banks' (AMBs) at the district/sub-district level, possessing heavy duty machinery like combine harvester, laser land leveller etc. 'State/Regional Services' possessing more sophisticated and heavier machinery, that can cover larger areas to meet certain specific demands. The above types of services can be promoted by adopting in an enterprise mode or by setting up custom hiring centres.

(h) Providing service and repair facility: Enterprises may also be taken up in providing quick repair and service facilities in close proximity, so that operation and maintenance issues of farm machinery can be addressed. Mobile service centres can also be promoted to cater to minor repair demands. All those requirements would also generate more scope for enterprise creation. The services may also include offering labour, managing actual field operations in respect of not only agricultural machinery, but also other agricultural operations such as harvesting of coconut, arecanut, traditional tree climber etc.

(i) Machinery for waste management: In the Indo-Gangetic Plains (IGP), where rice-wheat is the dominant cropping system, burning of rice straw to meet the deadline for sowing of wheat is a common practice. This not only poses adverse effects ecologically but also causes a loss of opportunity to capture the value that lies in the paddy straw. This practice needs to be discouraged for exploring a viable solution for effective utilization of agro-residues. In the strategy for doubling farmers' income, gainful use of all biological products and not just the grain or fruit, it is necessary to generate additional farm incomes. Hence, agricultural mechanization in small scale can also look into the farm waste management machines and devicesand make residue management a productive activity.

1.21 ENTREPRENEURSHIP ON RENEWABLE ENERGY TECHNOLOGIES

It is an established fact that renewable energy technologies have a large potential of meeting energy demands of the society in an environment friendly manner. The technologies are fast developing and still in progress to harness the more usable power from the renewable sources of energy available abundantly in the nature. However, these technologies are yet to gain market acceptance and facing a variety of barriers (technical, financial and socio-cultural) against their large-scale dissemination. High capital cost of renewable energy systems compared to their conventional alternatives is the major barrier for wide-scale promotion. Although, in many situations, the cost of operating them by considering the long life and little maintenance of renewable energy systems into accounts, is much less than those of conventional energy options, their higher initial costs often make them out of reach for most of the users in low/middle income category and also equally discouraging commercial and industrial users from adopting the renewable energy devices for fulfilling their energy requirements. Availability of easily accessible financing is the key to overcoming the initial cost barriers. For wide scale popularization of the renewable energy technologies, Governments of many countries have introduced subsidy facilities and implemented incentives of various types for the users who are interested in renewable energy devices. The provision of loan facilities has also been included in many financial institutions and banks to meet the initial investment of the renewable energy devices for the users. In spite of these, the dissemination of renewable energy systems in the society has not been achieved as per the desired target. This may be due to the lack of awareness of the technologies among the users, non-availability of suitable skilled man-power to install the devices and deficiency in the existence of entrepreneurial spirit among the existing young generation to take up the venture either in manufacturing the devices or providing necessary services in the renewable energy sector. Hence, to make the renewable energy systems easily and quickly accessible to the society, there is the necessity of conceive of new ideas, innovative thinking and risk taking ability of starting a new venture that can be obviously expected from the entrepreneurs. For the successful and sustained adoption of new technology, the entrepreneur is to undertake an in-depth evaluation of the various feasibility aspects (resource availability, financial and economic viability, energetic feasibility, socio-cultural acceptability and environmental sustainability) of the technology, prior to its dissemination. While the satisfaction of each of these feasibility aspects is necessary for a large scale sustainable dissemination of renewable energy technologies/ systems, but in the initial phase, the financial viability may be of direct relevance for motivating the users towards adoption of a new technology. In this context, the role of an entrepreneur is very important in creating awareness and providing necessary services for the popularization of renewable energy technologies which have now gained increasing attention to derive reliable and sustainable power from them. The development of renewable energy technologies will not happen without the involvement of entrepreneurs as they are considered to be the pillars of sustainable growth. Many economists claim that entrepreneurship is an important determinant of economic growth and development. The development of entrepreneurship in renewable energy sector has enormous benefits as the adoption of renewable energies contributes to sustainable development worldwide. Entrepreneurs are the key agents in facilitating their promotion, as they improve the mix of the means of production and thus transform renewable energy technologies into viable energy systems. Evidence across the globe indicates that renewable energy (RE) is gradually substituting the mature energy systems that are often based on fossil fuels. RE currently accounts for 18.2% of global energy consumption and 26.5% of global electricity production (IEA, 2019 and Zervos and Lins 2018). While RE technologies vary, they have several major advantages with sustainability aspects over fossil fuels systems. They rely on non-perishable resources, produces very low to zero greenhouse gas emissions and reduces emissions of other pollutants that harm the environment.

1.21.1 Importance of Renewable Energy Resources and Technologies for Sustainable Development

Renewable energy is the term used to describe a wide range of naturally occurring, replenishing energy sources. Marketable energy can be produced through renewable energy technologies by converting natural resources into useful energy forms. These technologies use the energy, inherent in sunlight and its direct and indirect impacts on the Earth (photons, wind, falling water, heating effects, and plant growth), gravitational forces (the tides), and the heat of the Earth's core (geothermal) as the resources from which they produce energy. These resources represent a massive energy potential compared to fossil resources. Therefore, the magnitude of these is not a key constraint on energy production, however, they are generally diffuse and not fully accessible, some are intermittent and all have distinct regional variabilities. Despite having such difficulties and constraints, the research and development on renewable energy resources and technologies has been expanded during the past two decades because of the improvement in the collection and conversion efficiencies, lowering of the initial and maintenance costs, increase in the reliability and applicability and understanding the phenomena of renewable energy systems.

The exploitation of renewable energy resources and technologies is a key component of sustainable development which refers to the activities that meet the needs of the present without compromising with the ability of future generations. Renewable energy is also considered as a sustainable source of energy because of its availability from the resources which are continually replenished by nature such as sunlight, wind, rain, geothermal heat, biomass, waves and tides. The more specific reasons for this, are also as follows.

(a) They have much less environmental impact compared to other sources of energy since there is no any energy sources with zero environmental impact. There are a variety of choices available in practice so that a shift to renewables could provide a far cleaner energy system than would be feasible by tightening controls on conventional energy.

(b) Renewable energy resources cannot be depleted unlike fossil fuel and uranium resources. If used wisely in appropriate and efficient applications, they can provide a reliable and sustainable supply of energy almost indefinitely. In contrast, fossil fuel and uranium resources are finite and can be diminished by extraction and consumption.

(c) They favour power system decentralization and locally applicable solutions more or less independent of the national network, thus enhancing the flexibility of the system and the economic power supply to small isolated settlements. That is why, many different renewable energy technologies are potentially available for use in urban areas.

1.21.2 Parameters for Sustainability of Renewable Energy Technologies

Various parameters are essential to achieve sustainable development in a society. Some of them are as follows,

(a) *Public awareness:* This is the initial step and very crucial in making the sustainable energy program successful. This should be carried out through the media and by public and/or professional organizations.

(b) *Information:* Necessary informational input on energy utilization, environmental impacts, renewable energy resources, etc. should be provided to public through public and government channels.

(c) *Environmental education and training:* This can be implemented as a completing part of the information. Any approach which does not have an integral education and training is likely to fail. That is why, this can be considered as the significant prerequisite for a sustainable energy program. For this reason, a wide scope of specialized agencies and training facilities should be made available to the public.

(d) *Innovative energy strategies:* These should be provided for an effective sustainable energy program and, therefore, require the efficient dissemination of information, based on new methods and consisting of public relations, training and counselling.

(e) *Promoting renewable energy resources:* In order to achieve environmentally benign sustainable energy programs, renewable energy sources should be promoted in every stage. This will create a strong basis for the short- and long- term policies.

(f) *Financing:* This is a very important tool that can be used for reaching the main goal and will accelerate the implementation of renewable energy systems and technologies for sustainable energy development of the country. Some countries, e.g., Germany, has introduced the support in a different way and simply exempt the people who use such systems and technologies from some portion of their taxes

(g) *Monitoring and evaluation tools:* In order to see how successfully the program has been implemented, it is of great importance to monitor each step and evaluate the data and findings obtained. In this regard, appropriate monitoring and evaluation tools should be used.

1.21.3 Barriers to Renewable Energy Technologies Adoption

In the present context of rapidly increasing energy demand and growing concern about economic and environmental consequences, there is the need of gradual switching from conventional sources of energy to the adoption of renewable energy technologies in the society. Renewable/sustainable energy technologies have faced a number of constraints that have affected their rate of adoption. Some barriers may be specific to a technology, while some may be specific to a country or a region. The major barriers are categorized into seven dimensions (Table 1.8) based upon their nature in the adoption of renewable/

sustainable energy technologies. These are (a) economical and financial(b) market(c) awareness and information(d) technical(e) ecological and geographical(f) cultural and behavioural and(g) political as well as governmentIssues.Theseare explained in details in the following table.

Table 1.8 Dimensions of barriers for adopting renewable energy technologies

Dimensions of barriers to adopt renewable/sustainable energy technologies	Barriers to adopt renewable/sustainable energy technologies
Economical & Financial (EF)	High initial capital cost (EF1)
	Lack of financing mechanism (EF2)
	Transmission & distribution losses (EF3)
	Inefficient technology (EF4)
	Lack of subsidies (EF5)
Market (MA)	Lack of consumer awareness to technology (MA1)
	Lack of sufficient market base (MA2)
	Unable to meet electricity power demand alone (MA3)
	Lack of paying capacity (MA4)
Awareness & Information (AI)	Need for backup or storage device (AI1)
	Unavailability of solar radiation data(AI2)
	Lack of IT enablement (AI3)
Technical (TE)	Lack of awareness of technology (TE1)
	Less efficiency (TE2)
	Technology complexity (TE3)
	Lack of research & development work (TE4)
	Lack of trained people & training institutes (TE5)
	Lack of local infrastructure (TE6)
	Lack of national infrastructure (TE7)
Ecological & Geographical (EG)	Scarcity of natural & renewable resources (EG1)
	Geographic conditions (EG2)
	Ecological issues (EG3)
Cultural & Behavioral (CB)	Lack of experience (CB1)
	Rehabilitation controversies (CB2)
	Faith & Beliefs (CB3)
Political & Government Issues (PG)	Lack of political commitment (PG1)
	Lack of adequate government policies (PG2)
	Lack of public interest litigations (PG3)

For overcoming the above barriers and to promote the RE technologies sustainably, there is the necessity of innovations which would enable them to gradually become more eficient, economically viable, cost-effective and environmentally sustainable than their fossil rivals. This can be possible by focussing and encouraging entrepreneurship development on renewable energy technologies.

1.21.4 Renewable Energy Entrepreneur

Entrepreneurs are essential economic agents, responsible for providing different goods or services by combining various means of productions, usually through innovative mechanisms. RE entrepreneurs are, therefore, the agents who promote and establish RE facilities in practice and they carry the various risks associated with the process while also being its main beneficiaries. They do this by optimizing the mix of the means of production (e.g., capital, natural resources, labour force, and capabilities) and thus promoting viable RE systems while harnessing varied innovations. Unlike many other entrepreneurs in various fields (e.g.,finance, real-estate development, etc.), the uniqueness of RE entrepreneurs lies in the broad heterogeneity of their characteristics. For example, RE entrepreneurs are not motivated just by financial incentives but possibly also by energy-utilization, environmental, and social incentives. Another example relates to the fact that RE entrepreneurs are not an exclusively private sector phenomenon but also as a public sector and third sector entrepreneurs, such as state-owned enterprises and cooperatives. Hence, despite common objectives in the promotion of RE, RE entrepreneurs should not be regarded as a homogenous group. Rather, they are defined by the characteristics that distinguish them from one another and classified as external influential factors, motivations, functional features, risks, and innovations. Out of these characteristics of RE entrepreneurs, external influential factors, motivation and innovation are more important and have been discussed in the following section.

External Influential Factors

There are several external factors that influence the RE entrepreneurs and also distinguishing them from one another.

(a) RE entrepreneurs operate in different countries and regions under a range of regulatory, political, economic and physical systems that affect their promotion of RE facilities.

(b) Political and economic environments can also significantly impact RE entrepreneurs. This may include the level of democracy and economic liberalism as well as other factors such as economic growth rates, the prices of fossil fuels, solar panels or wind turbines and more. For example, RE entrepreneurs in Norway have traditionally avoided promoting solar facilities due, primarily, to the insufficient solar radiation.

(c) Physical conditions such as access to natural resources, considerably affect the activities of RE entrepreneurs and the promotion of RE across the regions. The ability of any entrepreneur to promote RE facilities depends on the existence of suitable physical conditions, mostly the availability of natural resources such as sun light, wind, or water sources. Without their sufficient availability, the entrepreneurs will not be able to promote the relevant RE facilities.

Motivations

A main inherent distinction between different entrepreneurs relates to their motivations, i.e., the incentives encouraging their involvement in RE projects, which relate to financial, energy-utilization, environmental, and social aspects. RE entrepreneurs are usually driven by more than one type of motivation and differ in the way they prioritize the various motivations. The complexity of the incentives and their different relative weights in driving RE entrepreneurs can be analytically represented in the following prototypes.

(a) *Financially-Oriented Entrepreneurs:* Financially-oriented entrepreneurs are motivated by financial incentives for promoting RE facilities. These incentives refer to profit-seeking through the sale of energy or other means such as cost savings originating in self-consumption, land lease payments for establishing RE projects and thermal energy use for industry purposes (i.e., cogeneration). For example, Danish financially- oriented entrepreneurs have made the local wind energy industry a world leader, with the support of suitable government regulation.

(b) *Energy-Oriented Entrepreneurs:* Energy-oriented entrepreneurs are motivated by energy-utilization incentives, namely, their desire or need to use the produced energy. One example of this can be found in Spain, where many communities, with the support of a new designated government regulation, have started promoting RE facilities for purposes of self-consumption.

(c) *Environmental and sustainability oriented entrepreneurs* Environmental and sustainability oriented entrepreneurs seek to be involved in RE projects in order to produce clean energy, reduce pollutant emissions, advance climate change adaptation and mitigation efforts. For example, the government of India has established several wind farms in an attempt to reduce pollutant emissions and the ecological footprint of the Indian sub-continent. When considering their environmental externalities, some entrepreneurs are aware that RE facilities can be more cost-effective than conventional electricity means.

(d) *Socially-oriented entrepreneurs:* Socially-oriented entrepreneurs are motivated by incentives that relate to aspects such as job creation, socio-economic improvements, and others. The Maranchón wind farm in Spain is an example of an RE project that has created jobs in a remote and peripheral location with limited employment opportunities, thus benefiting the community partner-entrepreneur (Del Rio and

Burguillo, 2009). Furthermore, by establishing RE projects with the involvement of different communities, community members can work together toward common goals while strengthening social ties, thus further motivating community entrepreneurs to establish RE projects. In the United Kingdom, for example, thousands of "green" jobs have been created as part of community RE projects, motivating community entrepreneurs to promote more RE projects in order to form new employment opportunities and to enjoy the accompanying financial, energy, and environmental benefits.

Innovation

Promoting RE technologies requires entrepreneurs to innovate in order to make RE more efficient, economically beneficial, cost-effective, and environmentally sustainable. Innovation can be defined as the introduction of something new in the form of devices, methods, or actions. Even though RE entrepreneurs are not necessarily the sources of the innovation, they are the main agents who identify and harness various RE innovations, be they financial, energy-utilization, environmental or social innovations in order to promote RE systems.

(a) Financial Innovations: Different innovations, related to financial issues, enable entrepreneurs to promote RE by influencing RE's cost-effectiveness and prices. Attempts to reduce the manufacturing costs of RE technologies have enabled entrepreneurs to promote such technologies by offering energy prices that are lower than fossil fuels systems. An example of this can be found in the low prices of photovoltaic panels. Many of these panels are mass produced in China due to technological innovations that simplify the production process, thus enabling RE entrepreneurs to offer cheaper solar energy prices all over the world. Moreover, innovation improves the efficiency of renewable systems, i.e., more energy is produced from the same facilities. Wind turbines are one example; their characteristics, for example, aerodynamic capabilities, have improved significantly in recent years, thus enabling RE entrepreneurs to offer more efficient energy production than in the past while also influencing international wind energy prices.

(b) Energy-Utilization Innovations: Innovation is also associated with significant improvements related to energy utilization. Innovation enables RE entrepreneurs to produce energy from various previously unexploited sources. For example, in recent years, significant progress has been made in energy production from sea waves in Greece, Spain, Italy, China, and elsewhere. Also innovation improves energy usability, influencing the quality and quantity of the energy produced by RE entrepreneurs. For example, there have been significant improvements in storage facilities, enabling RE entrepreneurs to control the timing of RE production while limiting the dependency on external factors such as wind speed and solar radiation. Furthermore, innovation enables RE entrepreneurs to reach remote areas that have poor or no connectivity

to electricity grids.For example, remote villages in developing countries, such as India, rely on off-grid RE facilities forelectricity in a way that was not possible a few years ago.

(c) Environmental Innovations: Innovation also makes RE more environmentally sustainable. First and foremost, through innovation, RE entrepreneurs can promote RE facilities with the reduced emissions of pollutants and greenhouse gases (GHGs). Because, they produce few or no GHG emissions, RE systems have emerged as popular energy alternatives for climate change mitigation. One example is the use of landfill biogas for electricity generation, which contributes to the reuse of pollutants such as methane. Moreover, innovation also enables RE entrepreneurs to promote decentralized RE infrastructures that can reduce environmental damage relative to larger centralized facilities. Since RE systems tend to be more decentralized and diversified than fossil fuels systems, relying on them also reduces the vulnerability to extreme climate change-related events and contributes to climate change adaptation strategies.

(d) Social Innovations: RE innovation can enhance social acceptability. Despite broad positive societal views on the desirability of RE infrastructure, such projects often face local opposition due, mainly, to the hazards they may cause, which can vary by their scope and level of influence, as well as by other variables, such as the characteristics of the impacted population. While some hazards, such as noise disturbance often influence only the adjacent populations, other hazards, such as landscape destruction can influence a much larger and distanced populations.Moreover, through innovative compensation mechanisms, entrepreneurs can increase the social acceptability of RE facilities by local residents. Such practices are found in various RE projects worldwide.

1.22 FORMULATION OF DPR IN RELATION TO ESTABLISHMENT OF AN ENTERPRISE

1.22.1 Importance

Detailed Project Report (DPR) is basically a written statement of what an entrepreneur proposes to take up for planning his/her business. It indicates the objectives or goals of an enterprise and states in detail how these objectives are going to be achieved at various stages of the enterprise. It is like a road map and describes the direction in which the entrepreneur should go in order to reach his/her goal. In other words, it is a kind of course of action, the entrepreneur hopes to achieve in his/her business and how he/she is going to achieve it. It is simply a document wherein all details obtained from technical analysis, financial analysis, profitability analysis, economic analysis etc. are put together. It is prepared before the execution of activities of any proposed enterprise based on certain information and factual data. Therefore, a detailed project report enables an entrepreneur to realize what he/she needs for implementing the project well in advance. It also gives a general idea of his/her various resource requirements like raw materials, man power, finance, infrastructure facilities etc. and also the means of procuring them, thus enabling an entrepreneur to foresee his/her requirements in advance and helps him/her to take suitable decisions accordingly.

It also gives an indication of likely benefits which a prospective entrepreneur can get from his/her venture. This profitability indication will help an entrepreneur to take an important investment decision. Thus, the financial rewards can be visualized in advance. Crucial decisions have to be made at various stages of production. How much to produce to achieve Break-Even-Level? and how to fix the repayment schedule? such important decisions can be taken with the help of a project report prepared well in advance. It also anticipates problems in advance so that suitable decisions can be taken then and there to solve those problems.

Further, the survival of any business depends upon the marketability of its products. The project report identifies the demand and supply position, competitor's position in the market, expected price etc. and thus ensures the survival of the business unit. This report as well highlights the practicability of a project in terms of different factors like economy, finance, technology and social desirability. It is needed by the entrepreneur for carrying out expansion or starting a venture for new production and for getting loans from banks or financial institutions. It enables the entrepreneur to understand, at the initial stage, whether the project is sound on technical, commercial, financial and economic aspects and finally helps the entrepreneurs in establishing techno-economic viability of the project.

A project report is therefore prepared with a view to plan in advance about the fulfilment of expected performance in various areas like technology, marketing, finance, personnel, production, customer satisfaction and social endowment.

1.22.2 Contents of a Detailed Project Report (DPR)

Followings are the broad contents under which complete information on relevant aspects can be included in the project report.

(i) *General Information:* A project report must provide information about the details of the enterprise to which the project belongs to. It must give information about the past experience, present status, problems and future prospects of the enterprise. It must also give information about the product to be manufactured and the reasons for selecting the product if the proposed business is a manufacturing unit. It must spell out the demand of the product in the local, national and the global market. It should clearly identify the alternatives of business and should clarify the reasons for starting the business. This section should clearly provide the information about the bio-data (name, address, qualification, experience and other capabilities) of the entrepreneur, profile of the enterprise (past performance, present status, way it is formed, problem if any etc.), constitution and organization structure of the enterprise, product (utility, range of production, design/drawing etc.).

(ii) Executive Summary: A project report must state the objectives of the business and the methods through which the business can attain success. The overall picture of the business with regard to capital, operations, methods of functioning and execution of the business must be stated in the project report. It must mention the assumptions and the risks generally involved in the business.

(iii) Organization Summary: The project report should indicate the organization structure and pattern proposed for the unit. It must state whether the ownership is based on sole proprietorship, partnership or joint stock company. It must provide information about the bio data of the promoters including financial soundness. The name, address, age, qualification and experience of the proprietors or promoters of the proposed business must be stated in the project report.

(iv) Project Description: A brief description of the project must state and give the details about the following:

Information regarding location of the site,capacity of the enterprise, raw material requirements,target of production,area required for the work shed,list of machinery and equipment, power requirements,fuel requirements,water requirements, communication system (telephone, internet, e-mail facilities etc.), transport facilities, employment requirements of skilled and unskilled labour,technology selected for the project,production process,projected production volumes, unit prices,pollution treatment plants, quality control/testing and inspection facilities need to be mentioned under this item.

If the business is service oriented, then it must state the type of services rendered to customers. It should also state the method of providing service to customers in detail.

(v) Marketing Plan: The project report must clearly state the total expected demand for the product. It must state the price at which the product can be sold in the market. It must also mention the strategies to be employed to capture the market. If any, after sale service is provided that must also be stated in the project. It must describe the mode of distribution of the product from the production unit to the market. Project report must also state the type of customers,target markets,nature of market,market segmentation,seasonality factor, future prospects of the market,sales objectives,marketing cost of the project,market share of proposed venture,demand for the product in the local, national and the global market,It must indicate potential users of products and distribution channels to be used for distributing the product.

(vi) Capital Structure and operating cost: The project report must describe the total capital requirements of the project. It must state the source of finance, it must also

indicate the extent of owners' funds and borrowed funds. Working capital requirements must be stated and the source of supply should also be indicated in the project. Estimate of total project cost that must be broken down into land, construction of buildings and civil works, plant and machinery, miscellaneous fixed assets (furniture, vehicles, tools etc.), preliminary and preoperative expenses and working capital, contingency cushion against price rise/unforeseen expenses. The proposed financial structure of venture must indicate the expected sources and terms of equity and debt financing. This section must also spell out the operating cost. The operating cost refers to cost of production in case of an enterprise manufacturing a product. The cost of production includes cost per unit of the raw materials, utilities (power, fuel, gas, water etc.), repair and maintenance, rent, insurance taxes, wages and salary, administrative expenses, sales expenses, financial expenses (interest on term loan, interest on bank borrowings for working capital etc.) and depreciation of assets.

Management Plan: The project report should state the (a) business experience of the promoters of the business (b) details about the management team (c) duties and responsibilities of team members (d) current personnel needs of the organization (e) methods of managing the business (f) plans for hiring and training of personnel (g) programmes and policies of the management.

(vii) *Financial Aspects:* In order to judge the profitability of the business, a projected profit and loss account and balance sheet must be presented in the project report. It must show the estimated sales revenue, cost of production, gross profit and net profit likely to be earned by the proposed unit. In addition to the above, a projected balance sheet, cash flow statement and funds flow statement must be prepared every year and at least for a period of 3 to 5 years. The income statement and cash flow projections should include a three-year summary, detail by month for the first year and detail by quarter for the second and third years. Break-even point and rate of return on investment must be stated in the project report. The accounting system and the inventory control system, used are generally addressed in this section of the project report. The project report must state whether the business is financially and economically viable.

(viii) *Technical Aspects:* Project report provides information about the technology and technical aspects of a project. It covers information on technology selected for the project, production process, capacity of machinery, pollution control plans etc.

(ix) *Project Implementation:* Every proposed business unit must draw a time table for the project. It must indicate the time within the activities involved in establishing the enterprise for completion. Implementation schemes show the timetable envisaged for project preparation and completion. The likely dates of the activities such as acquisition of land, registration of unit, bank loans, construction of building, power connection, ordering plant/machinery, recruitment of staff, training of the employee, starting of production etc.

(x) Social Responsibility: The proposed unit draws inputs from the society. Hence its contribution to the society in the form of employment, income, exports, local resource utilization, development of area etc. need to be described. The output of the business must be indicated in the project report.

1.22.3 SWOT Analysis

SWOT analysis is a structured planning method used to evaluate the strengths, weaknesses, opportunities and threats involved in a project or in a business venture. It enables the businesses to identify the internal and external factors needed to achieve their goal. Entrepreneur needs to follows the scientific methodology to assess the status of his/her organization in respect of probability of sustaining and succeeding. It gives the idea about strong and weak points of the organization. The most important advantage of this analysis is that it warns before hand the any likely danger for the organization, enterprise or programme. SWOT is an acronym for Strengths, Weaknesses, Opportunities and Threats in an organization. The SWOT frame work can be designed in short as presented in the following table (Table 1.9).

Table 1.9: SWOT frame work for an enterprise

Attributes of organization	Helpful in attaining objectives	Harmful in attaining objectives
Internal attributes	Strengths	Weaknesses
External attributes	Opportunities	Threats

Internal attributes: The organization has some inbuilt potentialities which need to be identified and used for the development of the organization. These are the strengths and weaknesses of the organization.

External attributes: These are the factors which affect the organization from outside. Cooperation of outside agencies can be obtained for improving the growth and progress of the organization. These are the opportunities and threats of the organization.

*Strengths:*It is the positive internal factors that contribute to accomplishing the mission, goals and objectives of an organization. Strength is the basic asset of an organization that would provide competitive advantage for its growth and development. The internal strength of the organization can be assessed by analysing the environment in which it functions. Some of the strengths of an organization are availability of necessary infrastructure, skilled manpower, adequate production capacity, good manufacturing practices, quality assurance and quality control, low cost of manufacture, adequate source of capital, efficient management, good reputation, strong system of monitoring and evaluation etc. The qualified and devoted staffs, enthusiastic management team are the assets of the organization.

Weaknesses: It is the negative internal factors that inhibit the accomplishment of the mission, goals and objectives of the organization. Weaknesses is the liability of an organization that hamper the growth and development of an organization. There may be some short-comings in the organization due to which it is difficult to progress. Some of the weaknesses of an organization generally exist are rising cost of operations, low level of motivation of staff, non-availability of raw materials, scarcity of capital, outdated technology, inadequate infrastructure, shortage of trained technicians, lack of effective coordination, inefficient management, inadequate distribution network etc. There may be older staff with outdated ideas, the poor pay scale, long process of appointing new staff, lack of facilities for the staff. In order to develop the organization, it will be necessary to remove these weaknesses.

*Opportunities:*It is positive external factors, the organization can employ to accomplish its mission, goals and objectives. Opportunity is the advantage to the organization to grow and achieve its specific objectives in a given situation. The organization has to utilize the opportunities available outside the organization. Some of the opportunities of an organization are good location, availability of appropriate technology, favourable government policies, availability of different task environment like market information, distribution outlets and media, presence of favourable cultural environment, availability of common basic infrastructural facilities etc. There should be merging with the local institutions having reputation and expertise to carry out different programmes.

Threats: It is negative external factors that inhibit the organization's ability to accomplish its mission, goals and objectives. Threats are the external situation that block the abilities of the organization to grow and develop for reaching its ultimate goal. It is the fear for the existence of the organization. If the threats are not properly attended, then they will have harmful effect on the organization. Some of the threats of an organization are shortage of power, water, fuel, rejection by the market, tough competition, fiscal policy resulting into increased taxes, duties, imports reservations, licensing etc., obsolete technology, resource crunch, shifting customer tastes and preferences, economic instability with very high ups and downs, prolonged economic depressions, frequent natural calamities, adversely changing government policies etc. Loss of identity, strengths and reputation of organization can also be a big threat. There may be risk of losing experienced staff and the dominance of the attitudes of the other organization. Threats need to be identified and attended to avoid the loss in the organization. In order to assess the competition or competitive situations, an entrepreneur should try to study the following key areas so that he can efficiently establish his enterprise (a) to prepare the profiles of customers, industry and competition (b) to identify the size of customers (c) to prepare the list of firms producing and marketing similar products (d) to find out the goodwill and the image of the product in the market (e) to examine the advantages of price, quality, warranties, service and distribution of the products (f) to examine the operational strengths and weaknesses of the products (g) to diagnose the market share, trends in sales and profitability (h) to study the frequency of demand for the product and (i) to observe the pattern of demand for the product.

Opportunity Analysis

Entrepreneur is an opportunity seeker. For establishing a new business unit, he constantly undertakes the scanning of environment. He even goes on scanning the environment until he finds the best of opportunities, out of several such opportunities for preparation of his business plan. The process by which an opportunity is identified is described as *opportunity analysis* or *opportunity–sensing and identification*. Entrepreneur analyses the opportunities taking into account his own SWOT i.e. strengths, weaknesses, opportunities and threats as described above.

In identifying an opportunity, an intending entrepreneur will have to involve themselves the following phases to reach the right venture.

Stage I: Scanning of business environment

Stage II: Shortlisting of opportunities

Stage III: Finalizing the selection of product

Scanning of business environment: To enter into the field of entrepreneurship, it is always wise on the potential entrepreneur to try and generate as many ideas as possible which will be commercially viable. This idea generation can be possible through environment scanning. Ideas can be generated by the following ways:

a) By discussion with friends, relatives, businessmen, industrialists and the persons associated with trade, commerce and industries.

b) By contacting promotional agencies like District Industries Centres (DICS), Micro, Small and Medium Enterprises (MSMEs), entrepreneurship development institutes, Small Industries Corporation, chamber of commerce and industries, industry associations, etc.

c) Generating project ideas from technical consultancy cells of commercial banks, vendor development cells of large industrial houses.

d) Generating ideas by observing products or services required by the society.

e) In the present-days, product ideas can be easily generated through internet browsing. A person can get enough ideas through internet by accessing different sites related to business and industry.

Shortlisting of opportunities/project ideas: As opportunities are unlimited, the prospective entrepreneurs generate a good many ideas. It is necessary that these opportunities should be shortlisted to three or four for closer analysis and observation. This analysis can be done on the basis of 'SWOT' (Strengths, Weaknesses, Opportunities and Threats) analysis whereby the entrepreneurs can arrive at the final decision. The strengths for them could be educational or technical qualification, skills, trade-related knowledge, family connections, etc. while the weaknesses could be the nature of the entrepreneur, financial background, lack of market, etc.

If the entrepreneurs finalize good opportunities, it is always necessary to discuss about the ideas, its pros and cons regarding the techno-economic viability with the financial institutions particularly bankers, state financial corporation and development financing institutions. If their comments and views are negative on some of the opportunities or all of them, entrepreneurs can be advised to drop such ventures and go for searching of new opportunities. This is because, if they are not convinced, the entrepreneurs may not be in a position to raise finance for such initiative. If they are convinced, the entrepreneurs should go ahead with the best of opportunities and should not hesitate to establish the enterprise.

Before finalizing an opportunity, certain product-related factors should be considered. The factors may be like stability, growth, marketability, gestation period required, etc. A product whose future may not be very stable, may be a risky venture. Such product idea or opportunity has to be dropped.

Finalization of opportunity or product: While finalizing the decision, it is essential to assess some parameters before deciding on the venture out of two or three shortlisted ones. These parameters are mainly total investment, returns on investment, return in equity, likely volume of sales and profit as well as profitability.

Now the final decision which is selected by the entrepreneurs, will, in all probability, be accepted as the most feasible one to be taken up as a venture. A part from this analysis, entrepreneur should collect a variety of data or information as given below for successful establishment and profitable management of the venture. These data are (a) total investment in plant and machinery (b) sources of raw materials for manufacturing the product (c) infrastructure facilities (d) size of the project (e) requirement and availability of skilled and unskilled manpower (f) total financial requirement (g) policy of the government (h) procedure adopted by the government to start enterprise (i) Incentives, subsidies, exemptions, if any, provided by the government.

Keeping the above things into consideration, entrepreneur should select that opportunity which will be commercially viable and technically feasible. Further, the opportunity should be capable of giving adequate returns in a sustainable manner on long term basis and can be feasible enough to be diversified in accordance with the changing needs of the time.

1.22.4 An Example of SWOT Analysis

Solar Photovoltaic (PV) water pumping device

Considering the growing demands of energy and water particularly in the context of achieving secured irrigation in a decentralized manner, it has become essential to adopt reliable, environment-friendly and water saving technologies so as to combat against the energy crisis and water stress in the future. It has been established that conventional sources of energy like oil, gas, coal etc. will not be able to provide the desired levels of energy security to mankind in foreseeable future. Hence, there is a global consensus for exploitation and utilization of different renewable energy resources. The search for new options should be eco-friendly as well as plentifully available in nature. Among the different available renewable energy resources, solar energy seems to be more promising and sustainable. Solar powered agricultural irrigation may be an attractive application of renewable source of energy for replacing fossil fuel powered irrigation devices. The use of solar photo voltaic systems may provide good solution not only for all energy related problems of the present society but also perform excellently in terms of productivity, reliability, sustainability and environmental protection ability. With the cost of solar PV system falling steadily and price of diesel/gasoline soaring, solar PV pumping may be an appropriate and economically feasible option. Utilization and popularization of solar photovoltaic system is recognized as an important part of the future energy generation, because it is non-polluting, free in its availability, and is of high reliability. These facts therefore make the PV energy resource attractive for many applications, especially in rural and remote areas of most of the developing countries like India. Solar photovoltaic (PV) water pumping has been identified as suitable for grid-isolated rural locations where there are high levels of solar radiation. Solar photovoltaic water pumping systems can also provide water for irrigation without the need for any kind of fuel or the extensive maintenance as required by diesel pump sets. They are easy to install and operate, highly reliable, durable and modular, which enables future expansion. They can be installed at the site of use, rendering unnecessary long pipelines.

Strength

(a) Solar photo voltaic powered water pumping device with micro- irrigation system can be used in decentralized manner in remote areas

(b) Cost of solar photo voltaic system is decreasing and fast approaching to grid parity

(c) Less maintenance cost and no fuel requirement

(d) The source of energy is freely and abundantly available in the nature

(e) Environment friendly technology favouring National Action Plan on Climate Change

(f) Most suitable for tropical areas of the globe where clear sun shine is received in 300-320 days in a year

(g) Technology due to its portability, affordability, enhanced water use efficiency and sustainability can solve the vagaries of diesel/electric powered pump sets in the state

(h) Device proposed can easily be adopted by individual farmer by installing shallow tube well by his own effort for irrigation in small areas without depending on Government subsidy

(i) Easy use in different wells or water sources

Weakness

(a) High initial cost of the system

(b) Variability in the output of solar PV system according to the prevailing weather condition

(c) Low efficiency of the system needs to be improved at par with diesel/electric powered pump sets

(d) Lack of technical knowhow of the users regarding connecting, disconnecting, assembling and dismantling the system when required

(e) Poor marketing facility and after sale service of the solar photo voltaic micro-irrigation system.

(f) Lack of awareness about the technology and product among the users and stakeholders

Opportunity

(a) Crop diversification from non-remunerative to remunerative crop

(b) Promotion of export oriented high value vegetable crops

(c) Increasing irrigation potential through sustainable source of energy with optimum water management practices by installing private lift irrigation points

(d) Encouraging high value crop cultivation among the farming community

(e) Encouraging farmers to install shallow tube well for private lift irrigation points

(f) Encouraging and popularizing solar energy devices under international or national level scheme

(g) Emphasis has been given at the government level to promote the use of solar energy for long term energy and ecological security by reducing carbon emission.

Threat

(a) Decline in the business transaction of pump sets manufacturing and distributing company in the country

(b) Over exploitation of ground water through the expected large number of private lift irrigation points

(c) Fear of theft and breakage of solar modules when kept outside under sun shine.

1.23 GLOSSARY FOR ENTREPRENEURSHIP

Accounting period: The time for which profits are being calculated, normally months, quarters or years.

Affiliate marketing: A retailer or service provider advertising its goods or services via a third party in return for a commission on any sales.

Amortize: It is to allocate a large expense, such as research and development or closing fees paid to lenders, over the period of time and that expense benefits the company. When a business amortizes an expense, that amortized expense is recorded on the company's income statement not when the cash expense is incurred, but rather as a series of amortization expenses spread out over the amortization period

Annual equivalent rate (AER): A quote of what interest paid on savings and investments would be. It is calculated by adding each interest payment to the original deposit, then working out the next interest payment, compounding the interest.

Annual percentage rate (APR): This is the rate of interest you agree to pay on money borrowed. The higher the amount, the more you will have to pay.

Annuity: This is a type of insurance policy. Upon retirement, a lump sum is paid into it and the insurance company then provide a regular income.

Business Incubator: Often confused with office business centres, these organizations do provide workspace, but clients have to pay for business coaching courses and support services in the early-stage of their businesses. These additional costs are normally covered by Employment Insurance benefits to get entrepreneurs to start businesses, instead of returning to the workforce.

Cofounder: An individual who starts a new business jointly with another person or persons (fellow cofounder or cofounders). (see founder)

Copyright: Copyright, a term that is more associated with writers and music producers, is a form of protection for published and unpublished literary, scientific and artistic works that a business may create to associate with or promote their product or brand.

Corporation: A corporation (basically a LLC) is a legal entity that is separate and distinct from its owners. Corporations enjoy most of the rights and responsibilities that an individual possesses. That is, a corporation has the right to enter into contracts, loan and borrow money, hire employees, own assets and pay taxes.

Capital: Money invested into a company or project by its owners.

Cash flow: The movement of cash into and out of a business

Copyright: The exclusive legal right, owned by the individual or group who created a work, or by an individual or group assigned by the originator, to use certain material and to allow others the right to use the material. It is an exclusive legal right to use and reproduce literary, artistic, or musical material. Example, the small photography business creates a detailed guide to sell to photographers in training. They are sure to copyright the book so that the materials wouldn't be reproduced by a different business.

Debt: Borrowed funds

Debtor: A person or entity who owes money to another person or entity (a debtor owes money to a creditor or lender).

Dividend: Money paid regularly by a company to its shareholders.

Due Diligence: The evaluation of an opportunity, which is often performed by investors in order to allow to them to make a more confident and prudent investment decision by evaluating risk. Entrepreneurs should also perform due diligence to verify they are working with a legitimate investor. Both parties should always know whom they are dealing with. Due diligence is an investigation of a business or person prior to signing a contract, or an act with a certain standard of care. Due diligence is done to collect other information which may influence the outcome of a transaction.

Entrepreneur: A person who organizes, operates and assumes the risk for a business venture. An Entrepreneur is someone who exercises initiative by organizing a venture to take benefit of an opportunity and, as the decision maker, decides what, how, and how much of goods or service will be produced.

Economic growth: This is the term used to describe an increase in the amount of goods and services produced by the country, known as gross domestic product (GDP).

Fiscal year: Also known as a financial year, the fiscal year is a set period used to calculate financial statements. The period used differs between countries and between businesses, although in the UK the year between 6th April and 5th April is most often used for personal taxation. The 'official' period for corporation tax runs from 1st April to 31st March, however companies can adopt any yearly period for corporation tax.

Gross: The total amount of money you have earned in a period of time before deductions such as taxes.

Gross domestic product (GDP): GDP is the sum of all goods and services produced in the country's economy. If it is up on the previous three months, the economy is growing. If GDP is down, the economy is contracting.

Gross national product (GNP): GNP is another way to measure the economy. This is the GDP plus the profits, interest and dividends received from the residents abroad and minus those profits, interest and dividends paid to those of overseas residents.

Hyperinflation: This is the inflation that is rapid or out of control. It usually only occurs during wars or during severe political instability.

Insolvency: A legal term, financial condition in which a person or company is unable to meet its near term financial obligations. If a company cannot recover from insolvency quickly, they may be forced to file for bankruptcy protection. When a company becomes unable to pay off its creditors, or its liabilities exceed its assets.

Intellectual property: The works or inventions that are original and of creative designs. The individual or company responsible for the designs will be entitled to apply for a copyright or trademark on the designs.

Investor: A person that puts money into a project or small business, with the hope of eventually receiving profit. Example, when the entrepreneurs create their eco-friendly cleaning products business, they are fortunate to secure several investors to support their growth.

Incubator: An organization or space dedicated to supporting new business ventures. The start-up incubator provides the necessary office space and professional support to grow.

Joint Venture: A joint venture is a business arrangement in which two or more parties agree to pool their resources for the purpose of accomplishing a specific task. This task can be a new project or any other business activity. It is important to note that a joint venture differs from a strategic alliance because it is a legal entity created by two or more businesses, joining together to conduct a specific business enterprise with both parties sharing profits and losses. It differs from a strategic alliance in that there is a specific legal entity created.

Leverage: It is to borrow money (use debt capital rather than equity capital) to fund a business or project, with the goal of increasing return on equity. It is the use of debt in a business, the beneficial effects of the use of debt (i.e., additional capital to fund the business without taking on more equity investment and the associated dilution).

Liquid asset: Any asset which can be easily converted into cash.

Liquidity: The ease with which a company's assets can be converted into cash.

Net: The amount of profit remaining after deductions such as taxes and insurances have been made.

Net asset value: A way of measuring investment trusts. It is to calculate the total number of its assets minus its liabilities.

Nominal interest rate: An interest rate that isn't adjusted for inflation.

Nominal values: These values do not take inflation into account.

Opportunity Cost: The value of the next-best alternative that must be given up or sacrificed to obtain something.

Overheads: Costs that do not vary regardless of the level of production and are not usually directly involved with the cost of production, such as rent.

Outsourcing: To obtain goods or resources from a source outside one's company. Example, we do all of our creative work in-house, but we outsource the production of our actual products. The act of outsourcing is the acquisition of standard operational services from another business. Outsourced services typically including accounting, payroll, telemarketing, IT support, advertising and more.

Patent: A patent is an exclusive right granted for an invention, such as a product or a process. This is a property right granted to an inventor to exclude others from making, using, offering for sale, or selling the invention for a limited time in exchange for public disclosure of the invention when the patent is granted. It is an official legal document confirming that an individual or company has the sole right to make, use or sell a particular invention. An authorization or license granted to an inventor that prevents others from making, using, or selling their invention. Example, we get a patent for our artificial limb technology so that another company wouldn't steal our invention.

Profit: A financial gain. Example, entrepreneurs take financial risks with the hopes of profiting from their business venture.

Start-Up: A newly established business.

Seed financing: Financing to fund to an early-stage company, generally provided by either angel investors and/or a venture capital firm, to fund the early stages of a company's business plan. Seed financing typically occurs before a company has commercially released its product or service, and therefore before the business is generating revenue.

Stakeholders: Any individual or party that has an interest in or may be affected by a business and/or its activities. This can include any one, from shareholders to residents of the local community.

Trade-Offs: The sacrifice of some or all of an economic goal, goods, or service to achieve a different goal, goods, or service. (See also "Opportunity Cost," above.)

Trademark: A trademark is another form of legal protection for intellectual properties. It is more about words, names, symbols, sounds or colours that distinguish brand, goods and services. Unlike patents, it has a limited time span, trademarks can be renewed forever as long as they are being used. A symbol, words, or a phrase officially registered for a company. Example, an important part of developing a company's brand is to be sure it has an official trademark.

Turnover: The total sales of a business or company during a specified period.

Venture: A risky project or business idea. Example, let one's latest venture is developing a new smart phone application. Even though it's a risky venture, he is hoping. he will find some big investors.

Venture Capital: Financial capital made available by an investor or group of investors. The venture capitalists are often willing to assume a higher level risk than banks or lending institutions in return for a greater profit.

REFERENCES

1. Grover, Indu. 2008. Handbook on empowerment and entrepreneurship. Agrotech Publishing Academy, Udaipur.

2. Viramgami, H.S. 2008. Entrepreneurship Development. APH Publishing Corporation, New Delhi.

3. Holt, David. 2006. Entrepreneurship-New venture creation. Prentice Hall of India Pvt. Ltd. New Delhi.

4. IEA. World Energy Outlook 2019; International Energy Agency: Paris, France, 2019.

5. Zervos, A and Lins, C. Renewables 2018 Global Status Report 2018. Available online: http://www.ren21.net/gsr.

6. Del Rio P and Burguillo, M. An empirical analysis of the impact of renewable energy deployment on local sustainability. Renew. Sustain. Energy Rev. 2009, 13, 1314–1325.

ENERGY RESOURCES

2.1 PERSPECTIVES ON ENERGY USES

It is established that increased uses of energy in various sectors of national development and even in personal living have improved the social and economic developments of a country. The history of life on the earth is based on the history of photosynthesis and the availability of energy i.e. solar energy resulting into biological evolution. Since industrial revolution, energy has been the driving force for technological development leading to economic and social improvement and well-being. The term usually used to express the socio-economic status of a country is human development index (HDI). It focuses on three basic dimensions of human development i.e. long and healthy life (average life expectancy), access to knowledge and standard of living. Its value ranges from 0 to 1. A country scores a higher HDI with higher life span, education level, more gross national income per capita and improved standard of living. It differentiates the country whether being developed, developing and underdeveloped. Over the years, the HDI values for the countries are also increasing (Fig. 2.1) due to more accessibility of energy in various spheres of human living, thus creating widespread amenities that promote the growth of development. The average life expectancy in India was below 35 years during 1950s and now (year 2020), it is 70 years. This is due to a ready access to energy which has promoted the development of the agricultural, industrial, and medical resources that have played a key role in increasing this life expectancy.

The two major reasons for the increase in the energy consumption at all times are the steady increase in population and the strive for better development and comfort. The world population is expected to almost double by 2050. Such an increase in the population will demand more energy to fulfil higher standard of living and higher growth in industrial, technological, manufacturing, agricultural, transport, communication sectors etc. and all productive processes in a society. It has been proved that there is a strong correlation between standard of living and per capita energy consumption for a country. Per capita energy consumption is another measure to compare the status with respect to the level of human development among the countries.

Since the beginning of the industrial revolution, the global energy consumption has increased in line with the rising standards of living. But increased energy use has not been a consequence of greater wealth and prosperity, it has been a driving force to enhance the technological and economic growth. The per capita energy consumption of United States is 13 times more than India (Table 2.1). It has been reported that with only 4.5 % of world population and the country with higher standard of livings, the United States consumes about 18 % of total global energy. On the other hand, India having population of 1.3 billion during 2020 (17 % of world population) consumes only 6.5 % of the total energy consumption of the world. The higher, a country's per capita energy consumption, the more prosperous that country is. Often, access to energy is interpreted as access to electricity. Per capita energy consumption is generally calculated by dividing total annual electrical energy consumption by the total population of a country. Electricity is a very convenient energy vector and more and more systems are now powered by electricity because of its more versatility for conversion to other forms of energy such as thermal, mechanical, illumination, sound, chemical etc., better controllability and higher efficiency.

Human Development Index

The Human Development Index (HDI) is a summary measure of key dimensions of human development: a long and healthy life, a good education, and having a decent standard of living.

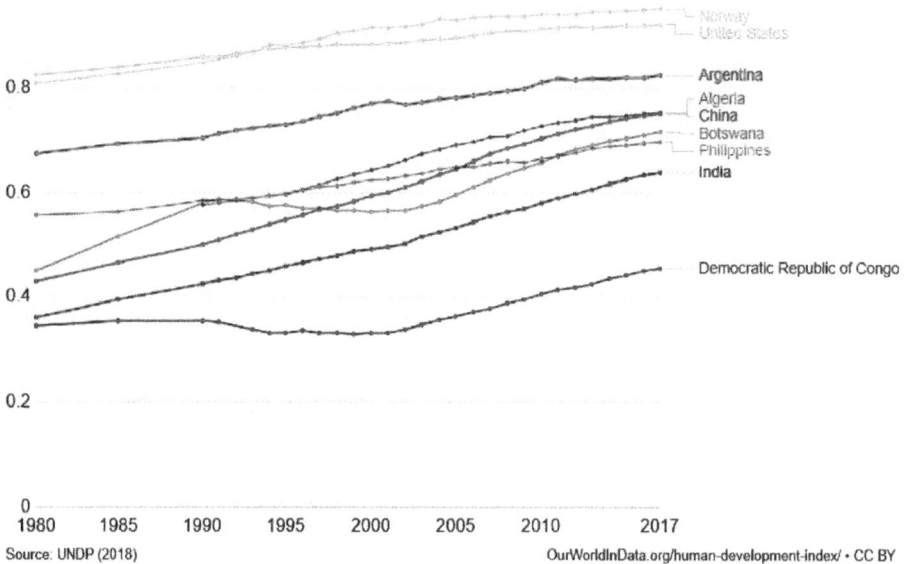

Fig. 2.1. HDI values of different countries

Hence, energy is an indispensable input for the development of a society. The energy needs of the world have increased rapidly in the recent years and, as populations grow, combined with the economic and technological growth, they are expected to increase even more. Today, energy is a continuous driving power for future social and technological developments. It is the essential ingredient for all human transactions and without it, human activity of all kinds and aspects cannot be progressive. Its increased use has also accelerated the other following areas of socio-economic progresses in a country.

Gross Domestic Product(GDP) growth: GDP growth rate is an important indicator for the economic performance of a country. It indicates the total market value or monetary value of all the finished goods and services produced within the borders of a country, during a specific time period. It thus provides information on manufacturing, savings, investments, employment, production outputs of various products for uses among the consumers and other economic variables. Enhanced production of goods and quick as well as fast provision of services rely heavily on the adequate availability of energy at the door steps. HDI indicates the quality of life whereas GDP looks solely at the general progress of the economy. It has been revealed that with the increase of the energy consumption over the years, the GDP of a country is also rising (Fig. 2.2).

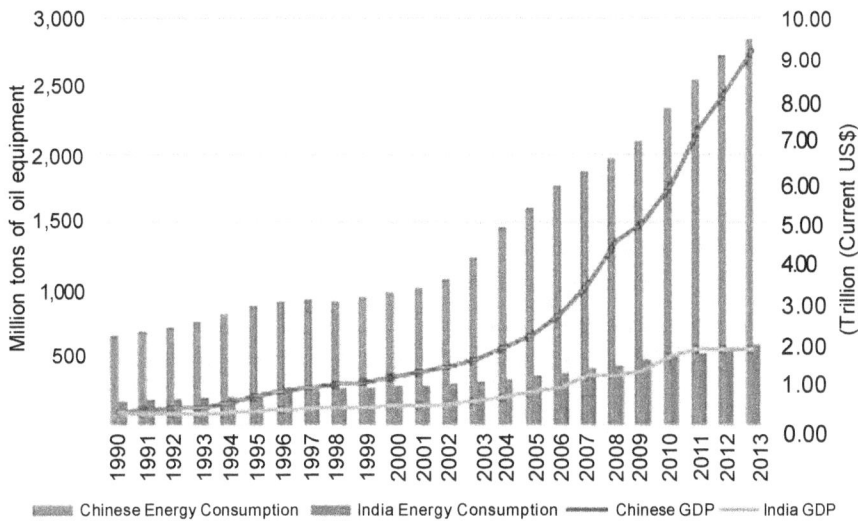

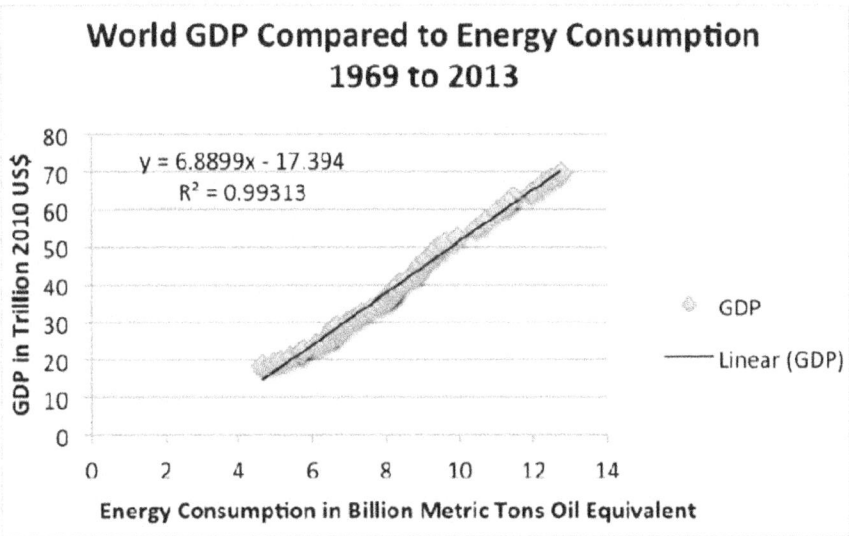

Fig. 2.2. Relation between energy consumption versus GDP

Industrial growth: Industrial sector includes manufacturing, transport, energy supply and demand, mining, construction, and related informal production activities. Increased energy use has significantly promoted the growth of industry in different segments of the production unit by facilitating the use of advanced technology, machinery, control system and safety devices leading to more employment, export potential and income generation in a country.

Agricultural productivity: During last decades, there has been substantial increase in the use of mechanical and electrical energy in Indian agriculture resulting into enhanced productivity and profitability. Increased use of mechanical energy through tractor, power tiller, oil engines etc. and electrical energy for operating irrigation pumps and stationary machinery, especially post-harvest agricultural machines have enhanced the farm power availability and productivity from 0.32 kW/ha to 2.02 kW/ha and 0.636 t/ha to 2.11 t/ha over the years from 1965-66 to 2013-14 in India (Fig. 2.3). It has also been established that food grains productivity is positively associated with unit power availability (Fig. 2.4). The relationship between food grains productivity and unit farm power availability for the period 1960-61 to 2013-14 in Indian agriculture was estimated by linear function, with highly significant value of coefficient of determination (R^2) with relationship as $Y = 0.5512 + 0.8195\ X$; $R^2 = 0.990$ where, Y = food grains productivity, t/ha, and X = power availability, kW/ha. This indicates that productivity and unit power availability is associated linearly. It is therefore evident that energy input in agriculture has to be increased further to achieve higher food grains production for the growing population of the country.

Table 2.1: Per capita energy consumption of some countries (EIA, International Energy Outlook, 2020)

Name of the country	Electricity consumption (GW-h/year) during 2019	Per capita energy consumption (kWh/person/year) during 2019
Canada	549263	14612
United States	3989566	12154
Japan	902842	7150
Russia	965156	6685
Germany	524268	6306
China	7225500	4617
United Kingdom	300520	4496
India	1547000	1055
Srilanka	13438	616
Nepal	6562	229
World	23398000 (23,398 TWh)	3081

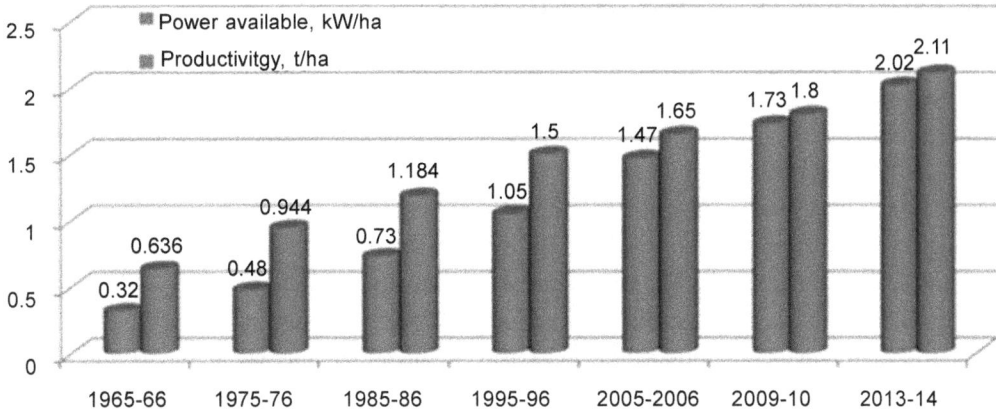

Fig. 2.3. Power Availability and Food Grain Productivity

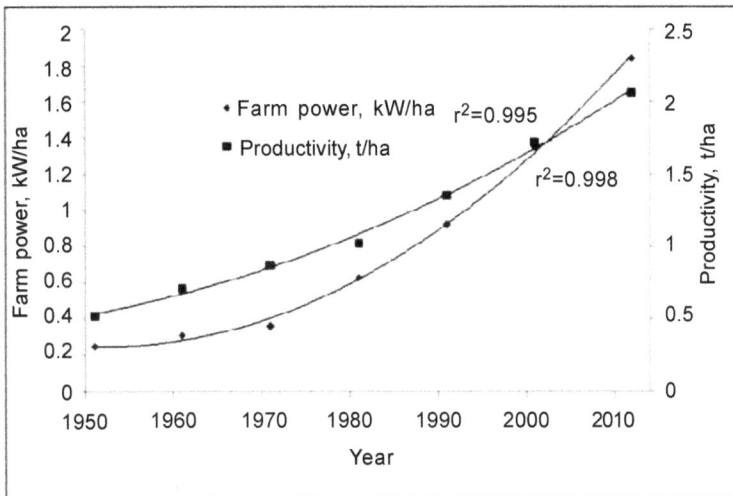

Fig. 2.4. Food grain productivity and farm Power availability

Entrepreneurship Development

Entrepreneurs are essential economic agents, responsible for providing different goods or services by combining various means of productions, usually through innovative mechanisms. They are the innovator. Their innovative ideas and entrepreneurial skills may be reflected into practical action by following the updated technologies. The present day's technologies are usually based on electrical energy as a prime mover for their operation. The technologies can effectively be used if there is the availability of energy. Increased uses of energy therefore facilitate the adoption of technologies to produce need based and location specific goods and provide services for the people. This indirectly creates the scopes for entrepreneurship development and thus employment generation and income raising opportunity in a society.

2.2 EVOLUTION OF MAN'S ENERGY DEMAND

The demand of energy has been increasing steadily for a human being for his sustenance and well-being from the period, he/she came on the earth a few million years ago. Primitive man was only using his own muscles to help him converting energy into useful work. At the beginning of human history, he had only two sources of energy. One was the radiant energy of sunlight, which warmed him and the other was the source of chemical energy from the food he ate. He derived this by eating plants or animals which he hunted. In fact, sun was the only source of these two forms of energy. The chemical energy in the food is nothing but the radiant energy of the sun stored up in the cells of plants. If there were no sunlight, the existence of earth and thus the plant was impossible and the primitive man could have not utilized his muscles to derive useful work. Subsequently, he discovered fire. The fire was discovered by the early man as a result of rubbing two stones accidentally. This was the turning point of making fire from other objects available at his surroundings. A spark, was generated while shaping a stone by rubbing it against another stone. The fire spark burnt the dry leaves and twigs laying nearby. The ancient humans subsequently learnt and repeated that practice to make fire with other biomass materials. The human being, then developed the skill to build the suitable devices to make and control fire. Afterwards, he learnt to use the heat energy obtained from the biomass for cooking as well as keeping himself warm. With the passage of time, man started to cultivate land to fulfil the demand of food. He added a new dimension to the use of energy by domesticating and training animals to work for him. With further demand for energy, man began to use the wind for sailing ships and for driving windmills, and the force of falling water to turn water wheels. Till this time, it would not be wrong to say that the sun was supplying all the energy needs of man either directly or indirectly and the ancient man was using only renewable source of energy. Firewood and techniques for using wind and hydro power provided the entire energy demand of the human beings. During that period, life was simple and unsophisticated and the environment was relatively clean.

In 1769, James Watt of Scotland laid the foundations for industrialization by developing the steam engine. The industrial revolution which began with the discovery of the steam engine brought about a great many changes in the history of energy evolution. This was the beginning of a mechanical age or the age of machine. During the end of the eighteenth century, man discovered coal and got a new source of stored chemical energy. He began to use coal in large quantities in steam engine and for other thermal applications. In the late nineteenth century, the internal combustion engine was invented and the other fossil fuels, oil and natural gas began to be used extensively. The fossil fuel era of using non-renewable sources had begun and energy was available in a concentrated form. The invention of heat engines and the use of fossil fuels made energy portable and introduced the much needed flexibility in man's movement. Gradually, the industrial revolution spread to the whole world. In 1888, Nikola Tesla, a Serbian-American engineer and physicist invented the commercial induction motor. The introduction of electrical machines along with the commercial availability of electrical power started the new electrical age. All those developments led to an increase of energy requirement by leaps and bounds.

In the beginning of the 20th century, crude oil was an important source of energy as coal. In prospecting for mineral oil, engineers found another source of chemical energy, namely natural gas, consisting mainly of methane. The natural gas was also a supplementary source of energy in the industry. Firewood lost its importance as an energy supply in the industrial nations, and large hydro-electric power stations replaced the watermills. The world energy demand rose sharply after the Great Depression of the 1930s. Natural gas entered the scene after World War II. In the 1960s, nuclear power was added to the array of conventional energy sources. The share of nuclear electricity for the primary energy demand is still relatively low. The fossil energy sources such as coal, crude oil and natural gas are now providing more than 85 per cent of the world primary energy demand.

Thus today, every country draws its energy needs from a variety of sources. We can broadly categorize these sources as commercial and non-commercial. The commercial sources include the fossil fuels (coal, oil and natural gas), hydroelectric power and nuclear power, while the non-commercial sources include wood, animal wastes and agricultural wastes. In an industrialized country like the USA, most of the energy requirements are met from commercial sources while in an industrially less developed country like India, the use of commercial and non-commercial sources is of equal importance.

2.3 ENERGY SOURCES

Any source from which energy can be extracted and utilized for mankind is called source of energy. Energy is a key input in economic growth. The growth of a nation largely depends on the availability of energy sources. Basically, there are five ultimate sources of useful energy and are as follows,

(i)   *Solar energy:* The source of solar energy is the nuclear interactions at the core of the sun, where the energy originates from the conversion of hydrogen nuclei into helium nuclei. This energy is primarily transmitted to the earth by electromagnetic waves, which can also be represented by particles (photons). In this way, solar energy is utilized as electromagnetic radiations in the earth.

(ii) *Planetary energy:* The different celestial bodies, in particular the moon, exchange mutual forces with the earth. The motion of the celestial bodies results in continuously varying forces at any specific point on the earth's surface. The tides are the most obvious manifestation of these forces. The movement of enormous water masses in the oceans creating the tides involves enormous amounts of energy. Tidal energy can be used by power plants on coasts with high tidal ranges. At high tide, water is allowed to enter into the reservoirs and is prevented from flowing back as the tide ebbs, creating a potential difference between the collected water and water outside the reservoir. The collected water is then released though turbines into the sea at low tide. The turbines drive electric generators to produce electricity. Hence, tidal energy is primarily due to the gravitational interaction of the earth and the moon.

(iii) *Geothermal energy:* Geothermal energy relates to the heat energy of the earth's interior. It is the internal heat energy available at a considerable depth below the surface of the earth. Geothermal power plants utilize this heat and convert it into electricity. This heat energy was originated during the formation of the planet and from the radioactive decay of materials.

(iv) *Nuclear energy:* The nuclear energy is released when atoms of certain unstable material split in the process of fission. Nuclear fission is the splitting of an atom's nucleus into parts by capturing a neutron. Nuclear fission produces heat because of the loss of the mass of the reactant nucleus during the process. This loss of mass is converted into the heat and electromagnetic radiation and it produces large amounts of energy that can be utilized for generation of power. Nuclear power plant uses the energy created by controlled nuclear reactions to produce electricity.

(v) *Chemical reactions from mineral sources:* These sources of energy are derived from the chemical reaction of mineral sources such as tar sand and oil shale The term 'oil shale' refers to a finely textured rock mixed with a solid organic material called kerogen. When oil shale is heated, the kerogen decomposes and yields crude oil. Similarly, tar sands are porous carbonate rocks that are intimately mixed with a very heavy, asphalt-like crude oil called bitumen. The bitumen is too viscous to be recovered by traditional petroleum recovery techniques. Tar sands contain about 10-15% bitumen; the remainder being sand or other inorganic materials. If tar sand is heated to about 80 °C, by injecting steam into the deposit in a manner analogous to that of enhanced oil recovery, the elevated temperature causes a decrease in the viscosity of the bitumen just enough to allow its pumping to the surface. Alternatively, it is sometimes easier to mine the tar sand as a solid material. When the mined tar sand is mixed with steam and hot water, the bitumen will float on the water while the sand sinks to the bottom of the container, allowing for easy separation. Heating the bitumen above 500 °C converts about 70% of it to a synthetic crude oil.

Infinite sources of energy i.e. renewable energy are derived from the sources (i), (ii) and (iii), while finite energy is derived from sources (i), i.e. fossil fuels, (iv) and (v).

2.4 ENERGY: ITS BASIC CONCEPTS AND DEFINITION

Energy is a physical quantity, which can manifest itself as heat, mechanical work, as motion and in the binding of matter by nuclear or chemical forces. The modern concept of energy was developed early in the twentieth century following the work of Albert

Einstein, whose special theory of relativity made it clear that mass itself is also a manifestation of energy. This led to the pioneering works of quantum mechanics, analysing the energy at the microscopic level. Mass is just a concentrated form of energy i.e. $E = mc^2$ where c is the speed of light and m is the mass of the object. Conversion of a small amount of mass gives a lot of energy (e.g., in an atomic or hydrogen bomb). Energy is closely related to force. When a force causes an object to move, energy is being transferred from the force to kinetic energy.

The word 'energy' itself is derived from the Greek word 'en-ergon', which means 'in-work' or 'work content'. 'En-ergon was simplified to the word 'Energy'. Energy is the cause for the motion of the particles or objects. It has the capability to produce motion, force, work, change in shape, change in form etc. The work output depends on the energy input. The capability to do work depends on the amount of energy, one can control and utilize. For instance, a piece of wood can be burnt to boil water and generate steam. The generated steam can be used to move an object (steam engine). It implies that the piece of wood possesses some energy. Likewise, an object raised to a certain height possesses energy. A hot object can do some work, therefore it possesses energy. The energy is quantitatively expressed as the product of force applied on an object and its displacement along the direction of force.

Mathematically, $W = F \times d$ where, F is the acting force in "Newton" and d is the displacement along the direction of force in "meter". A force of 1 Newton (N) acting through a distance of 1 meter in the same direction of applied force performs an amount of work equivalent to one joule (J). It is also important to note that work, the capacity for doing work and energy have the same units. A system may possess energy even when no work is being done. Since energy is measured by the total amount of work that the body can do, hence energy is expressed in the same unit of work as mentioned above. To perform the work, energy must be available. In case of a person pushing an object, the energy may be in the form of food calories. In another case, in which the force is exerted by an electrical device, the electrical energy used may be from the electrical energy generating source.

Gravitational unit of work is *m-kg.wt* or simply *m-kg*. If F = 1 *kg-wt* and d − 1 m; then work done = 1 *m-kg.wt or*, 1 *m-kg*. Hence, 1 *m-kg* is the work done by a force of one *kg-wt* when applied over a distance of 1 metre. Obviously, 1*m-kg* =9.8 *N-m* or *J* (as 1 *kgf* =9.8 N). Often it is referred to as a force of 1 kg, the word '*wt*' being omitted. To avoid confusion with mass of 1 *kg*, the force of 1 *kg* is written in engineering literature as *kgf* instead of *kg. wt.*

Example 2.1 An object has mass of 2 kg. What is its weight?

Solution: The mass of the object, $m = 2$ *kg*. Acceleration due to gravity, g = = 9.8 m/s^2. Weight (gravitational force), w = mass × acceleration of gravity ($m \times g$) = 2 kg × 9.8 m/s^2 = 19.6 N. The force experienced by a body due to gravity is called its weight. Weight = Force due to gravity = (mass) × (acceleration due to gravity).

Example 2.2 An object of mass 20 kg is raised through a height of 5 meter in 2 minutes. How much energy is used and the power utilized?

Solution: The height of raising the object, h = 5 m. The mass of the object, m = 20 *kg*. Work done = gravitational potential energy = energy spent = force × displacement = $(m\ g) \times h$ = (20 kg × 9.8 m/s^2) × 5 m = 980 J. Hence, energy used = 980 J. Power (watt)

$$\frac{\text{Energy used}}{\text{Time (second)}} = \frac{980}{2 \times 60} = 8.16\text{ watt}$$

Example 2.3 Calculate the power input of a tractor for moving a load of 1 tonne at 45 kilo meter per hour with a tractive resistance of 5 *kg*/tonne along a level track Assume the overall efficiency of tractor to be 30 percent.

Solution: In this case, force required is equal to the tractive resistance only. Force required at the rate of 5 *kg-wt*/tonne= 5 *kg-wt.* = 5 × 9.81 = 49.05 N. Distance travelled/second = 45 × 1000/3600 = 12.5 m/s. Power output of the tractor = 49.05 × 12.5 J/s = 613.12 W. Efficiency, h = 0.3, hence, power input = 613.12/0.3 = 2043 W.

There are many other units of energy that we come across in day-to-day transactions. For instance, the energy content of food is expressed in terms of calorie. The domestic electrical bill is charged in kWh. Calorie is the amount of energy required to raise the temperature of 1 g water by 1°C and British thermal unit (Btu) is the amount of energy required to raise the temperature of 1 pound of water by 1°F. The *tonne of oil equivalent* (*toe*) is a unit of energy defined as the amount of energy released by burning one tonne of crude oil. Conversion of different units of energy is mentioned in Table 2.2. Also prefixes defining multiples of any physical quantity are shown in Table 2.3.

Table 2.2: Energy units and their conversion

Energy unit	Equivalent energy unit
1 joule (or 1 J)	1 newton-meter = 1 N-m
1 kWh (kilo watt hour)	3600 kJ = 3.6 × 10^6 J = 3.6 MJ
1 calorie	4.182 J
1 BTU (British thermal unit)	1055 J
1 toe (ton of oil equivalent)	11634 kWh
1 tce (ton coal equivalent)	8141 kWh

Table 2.3: Multiple prefixes

Prefix	Multiplicative factor	Symbol	Prefix	Multiplicative factor	Symbol
Deca	10^1	*da*	deci	10^{-1}	*d*
Hecto	10^2	*h*	centi	10^{-2}	*c*

[Table Contd.

Contd. Table]

Prefix	Multiplicative factor	Symbol	Prefix	Multiplicative factor	Symbol
Kilo	10^3	k	Milli	10^{-3}	m
Mega	10^6	M	Micro	10^{-6}	μ
Giga	10^9	G	Nano	10^{-9}	n
Tera	10^{12}	T	Pico	10^{-12}	p
Pecta	10^{15}	P	Femto	10^{-15}	f
Exa	10^{18}	E	Atto	10^{-18}	a

2.5 POWER

Any physical unit of energy when divided by a unit of time becomes a unit of power. Hence, it is the rate at which energy is used or rate of doing work. Its unit is watt (W) which represents 1 joule per second. $1\ W = 1\ J/s$. If a force of F newton moves a body with a velocity of v m/s, then power $= F$ x v watt.

2.6 FORMS OF ENERGY

Energy exists in different forms. Energy may change from one form to another, but the total amount in any closed system remains constant. This is known as the conservation of energy and is very important for understanding a variety of phenomena occurring in nature. Energy can also be changed from some useful form to some other form which may not be effectively utilized for practical purposes, even though formally, energy is conserved in the process. Some of different forms of energy are listed below.

(a) Kinetic energy

Energy of an object in motion is called kinetic energy. If the object has a mass, m and

velocity v, then its kinetic energy is $KE = \dfrac{mv^2}{2}(J)$ where m is in kg and v is in meter/sec.

Example 2.4 In a swing, a boy moves to the highest point of 2.5 m and lowest point of 1 m above the ground. What is the velocity at the lowest point?

As the swing moves from the highest point (2.5 m, h_1) to lowest point (1 m, h_2) the gravitational potential energy, E_{pe} is converted into kinetic energy E_{ke}. The height difference, $h_3 = (h_1 - h_2) = 2.5-1=1.5$ m. Now, $E_{pe} = mgh_3$ and E_{ke} = ½ mv^2 where, m is the mass of the boy, g is acceleration due to gravity and v is the velocity of the moving body at

the lowest point. Equating, $E_{pe} = E_{ke}$, $v = \sqrt{2gH}\left(\dfrac{m}{s}\right) = \sqrt{2\times9.8\dfrac{m}{s^2}\times1.5m} = 5.4 m/s$. The

velocity is not affected by mass m.

(b) Potential energy

The energy that an object possesses as a result of its elevation in the earth's gravitational field is called potential energy and is expressed as $PE = mgh (J)$, where m is the mass (kg), g is the acceleration due to gravitational (m/s^2), and h is its height (m).

Example 2.5 An object of 2 *kg* mass is lifted to a height of 1.5 *m*. How much potential energy gained by the object?

Solution: Height of lift (h) = 1.5 m, mass of the object (m) = 2 kg, acceleration due to gravity (g) = 9.8 m/s^2, potential energy gained by the object = work (W) done to lift the object against gravitational force. W = force × distance = $F× h = mg × h = (2\ kg × 9.8\ m/s^2) × (1.5\ m) = 29.4$ J

(c) Chemical energy

The energy stored in the bonds of molecules is the chemical energy. It arises out of the capacity of atoms to evolve heat as they combine or separate. When certain chemical combine, energy released is usually in the form of heat. It is the chemical energy in coal, natural gas, oil, wood that heats our homes, powers cars and is used to generate electricity. The chemical energy in the food helps us to sustain our life. The chemical energy is directly converted into electrical energy in fuel cells, storage batteries etc. and into thermal energy by combustion.

Example 2.6: Calculate annual requirement of coal for a thermal power plant rated 2000 MW under following conditions.

$$\text{Annual Load factor} = \frac{\text{Energy Delivered}}{\text{Rated power ×total hours of use in a year}} = 0.5$$

$$\text{Plant Efficiency} = \frac{\text{Electrical power output}}{\text{Thermal power input}} = 0.25$$

$$\text{Utility factor of fuel} = \frac{\text{Useful energy in fuel}}{\text{Available Energy in fuel}} = 0.7$$

Heat value of coal = 14 *MJ / kg*.

Solution. Total energy delivered by the plant in a year = rated power × hours of use per year × load factor = 2000 × 8760 × 0.5 = 8760000 MWh

$$\text{Thermal energy input per year} = \frac{\text{Electrical energy output}}{\text{plant efficiency}} = \frac{8760000}{0.25} = 35040000 =$$

MWh = 3600 × 35040000 MJ

Useful Energy in coal per kg = Available energy in fuel × utility factor = (14 *MJ/kg*) × 0.7 = 9.8 *MJ/kg*

Input of coal to plant for one year

$$= \frac{\text{Thermal energy input per year}}{\text{Useful energy of coal,} \dfrac{MJ}{kg}} = \frac{35040000}{9.8 \dfrac{MJ}{kg}} = 12872 \times 10^6 \text{ kg.}$$

(d) Electrical energy

Electrical energy arises out of the capacity of moving electrons to evolve heat, electromagnetic radiation and magnetic field. It is a highly versatile form of energy and can easily converted into other forms for utilization.

Example 2.7: A hydro-electric power station supplies water to a reservoir of capacity 6 million m^3 per hour at a head of 170 m. How much energy is used by the station in one hour and power required for this? Density of water = 1000 kg/m^3. Hydraulic and electrical efficiencies are 80 and 90 percent respectively.

Solution: The weight of water (w) = 6 × 10^6 × 1000 kg wt = 6 × 10^9 × 9.81 N. Water head (h) = 170 m. Potential energy stored in the lifted water = wh = 6 × 10^9 × 9.81 × 170 J = 10^{12} J. Overall efficiency of the station = 0.8 × 0.9 = 0.72. Input energy of the power station required for the purpose = $\dfrac{10^{12}}{0.72} = 1.38 \times 10^{12} J$. This much energy is supplied by the power station in one hour. Hence, power rating of the power station

$$= \frac{1.38 \times 10^{12}}{3600} = 383 \text{ MW.}$$

Example 2.8: Estimate the rating of an induction heater to melt two tonnes (m) of zinc in one hour if it operates at an efficiency of 70 %. Specific heat of zinc = 0.1 kcal/kg ^{0}C. Latent heat of fusion (L) of zinc 26.67 kcal/kg. Melting point= 455 ^{0}C. Assume the initial temperature to be 25 ^{0}C.

Solution: Heat required to bring 2000 kg of zinc from 25 ^{0}C to the melting temperature of 455 ^{0}C = 2000 × 0.1 × (455-25) = 86000 kcal. Heat of fusion or melting = mL = 2000 × 26.67 = 53340 kcal. Total heat required = 86000 + 53,340 = 139,340 kcal. Heater input

$$= \frac{139340}{0.7} = 199{,}057 \text{ kcal. Now, } 860 \text{ kcal} = 1 \text{ kWh. Hence, heater input} = \frac{199057}{860} =$$

231.5 kWh. Power rating of heater = energy input/time $= \dfrac{231.5}{1} = 231.5$ kW.

(e) Heat energy

Heat energy is the kinetic energy of molecules. It is just the internal kinetic energy (random motion of the atoms) of a body. Heat is the energy and temperature is the potential for transfer of heat from a hot body to a cold body. It is also called the thermal energy.

Example 2.9: A paddle wheel is rotated 30 minutes by an electric motor of 10 W output power rating in a tank containing water. How much thermal energy is transferred to the water? If the tank initially contains 1 litre of water at 10 ^{0}C, then what would be the water temperature after the motor is stopped.

Solution: In this system, all the energy is eventually transferred as thermal energy of the water (assuming the heat capacity of the paddle is negligible). Therefore, the thermal energy transferred to the water can be calculated by multiplying power with time. $E = 10\ W \times (30\ minutes \times 60\ s) = 18\ kJ$. To calculate the change in the temperature of the water, specific heat of water is to be considered. The specific heat of water, $c = 4.2\ J/g\ ^0C$. But, $E = c \times m \times \Delta T$ where m is the mass of water and ΔT is the temperature rise due to use of paddle wheel. $m = 1000g$. Putting all the values, $\Delta T = 4.3\ ^0C$. The final temperature of water = $10\ ^0C + 4.3\ ^0C = 14.3\ ^0C$.

Example 2.10: A person does 30 kJ work on 2 kg of water by stirring using a paddle wheel. While stirring, around 5 kcal of heat is released from water through its container to the surface and surroundings by thermal conduction and radiation. What is the change in internal energy of the system?

Solution

Work done on the system (by the person while stirring), $W = -\ 30\ kJ = -30,000J$

Heat flowing out of the system, $Q = -5\ kcal = 5 \times 4184\ J = -\ 20920\ J$

Using First law of thermodynamics, $\Delta U = Q\text{-}W$

$\Delta U = -20,920\ J\text{-}(-30,000)\ J$

$\Delta U = -20,920\ J+30,000\ J = 9080\ J$

Here, the heat lost is less than the work done on the system, so the change in internal energy is positive.

(f) Radiant energy

Radiant energy is the energy in transit through space. It is emitted by electrons as they change orbit and by atomic nuclei during fission and fusion; on striking matter, such energy appears ultimately as heat.

Example 2.11: Assuming the sun to be a black body emitting radiation with maximum intensity at $\lambda = 0.5$ mm, calculate (i) The surface temperature of the sun, and (ii) The heat flux at surface of the sun.

Solution.

(i) Citing the relationship (Wien's Displacement Law) between the temperature of a black body and the wavelength at which the maximum value of monochromatic emissive power occurs, the product of $\lambda_{max}T = 2898\mu m\,K$

$$\therefore T = \frac{2898}{\lambda_{max}} = \frac{2898}{0.5} = 5796K.$$ Hence, surface temperature of Sun, becomes 5796

(ii) The heat flux at the surface of the sun, $(E)_{sun}$, $(E)_{sun} = \sigma T^4 = 5.67 \times 10^{-8}T^4$

$$= 5.67\left(\frac{T}{100}\right)^4 = 5.67 \times \left(\frac{5796}{100}\right)^4 = 6.39 \times 10^7 W/m^2$$ where σ is Stefan-Boltzmann

constant $\left(5.67 \times 10^{-8}\dfrac{W}{m^2K^4}\right)$.

Example 2.12: Assuming the sun (diameter = 1.4×10^9 m) as a black body having a surface temperature of 5750 and at a mean distance of 15×10^{10} m from the earth (diameter = 12.8×10^6 m), estimate (i) The total energy emitted by the sun, (ii) The emission received per m^2 just outside the atmosphere of the earth and (iii) the total energy received by the earth if no radiation is blocked by the atmosphere of the earth.

Solution: Radius of the sun, $r_s = \dfrac{1.4 \times 10^9}{2} = 0.7 \times 10^9 m$. Mean distance of the sun

from the earth, R=15×10^{10} m. Radius of the earth $r_e = \dfrac{12.8 \times 10^6}{2} = 6.4 \times 10^6 m$. Surface

temperature of the sun, T = 5750 K.

(i) The total energy emitted by the sun, E_b, $E_b = \sigma AT^4 = 5.67 \times 10^{(-8)} \times 4\pi r_s^2 \times$

$(5750)^4 = 5.67 \times 4\pi \times (0.7 \times 10^9)^2 \left(\dfrac{5750}{100}\right)^4 = 3.816 \times 10^{26}\ W$, where A is the

surface area of sun.

(ii) *The emission received per m^2:* The sun may be regarded as a point source at a distance of 15×10^{10} m from the earth. The mean area just outside the earth's atmosphere over which the radiation will fall is $4\pi R^2 = 4\pi \times (15 \times 10^{10})^2\ m^2$ $\therefore$ The emission received outside the earth's atmosphere

$$= \frac{3.816 \times 10^{26}}{4\pi \times \left(15 \times 10^{10}\right)^2} = 1349.6 W/m^2$$

(iii) *The total energy received by the earth:* Assuming the earth a spherical body, the energy received by it will be proportional to the perpendicular projected area, i.e., that of a circle $(= \pi r_e^2)$. Hence, energy received by the earth $= 1349.6 \times \pi \times (6.4 \times 10^6)^2$ $= 1.736 \times 10^{17}$ *W*.

(g) Nuclear (mass) energy

Nuclear energy is the energy stored in the nucleus of an atom. It arises out of the elimination of all or part of the mass of atomic particles. Matter can be changed into energy when larger atoms are split into smaller ones (atomic fission) or when smaller ones combine to form larger atoms (atomic fusion). As a consequence of theory of relatively devised by Albert Einstein, the amount of energy produced when mass disappears is governed by the equation, $E = \Delta m c^2$ where Δm is the reduction in the mass of one system in nuclear reaction and c equals the speed of light (3×10^8 *m/s*). The above simple equation is the basis of the energy derived when a ^{235}U nucleus fissions, as in nuclear reactor or when a deuteron and tritium (^2H and ^3H) fuse in a thermonuclear reaction.

Example 2.13: The sun converts approximately 4×10^9 kg matter into thermal energy per second. What is the power rating of the sun?

Solution: As mentioned above, energy of matter $= E = mc^2 =$ where m is the mass of the matter and c equals the speed of light (3×10^8 *m/s*). $= (4 \times 10^9$ *kg*$) \times (3 \times 10^8$ *m/s*$)^2$ *J*.

$$\text{Power,} = P = \frac{Energy}{Time} = \frac{E}{s} = \frac{\left(4 \times 10^9\right) \times \left(3 \times 10^8\right)^2}{1} = 3.6 \times 10^{26} J/s = 3.6 \times 10^{26} W. \text{ Power of}$$

sun $= 3.6 \times 10^{26}$ W. This is due to fusion reactions continuing in the sun.

Example 2.14: A nuclear power plant converts energy of matter to the electrical energy by following energy chain. Energy in matter $\rightarrow$ thermal energy $\rightarrow$ mechanical energy $\rightarrow$ electrical energy. Neglecting losses, how much matter is converted into electrical energy per day by a 100 MW power plant?

Solution: Matter converted = electrical energy delivered. Energy delivered = power $\times$ time. Electrical energy delivered, $E_e = (100$ W $\times 10^6) \times (24$ hr $\times 3600$ s$) = (10^8$ W$) \times (8.64 \times 10^4$ s$) = 8.64 \times 10^{12}$ J. Electrical energy delivered = matter converted into

$$\text{energy } = E = mc^2 \leftrightarrow m = \frac{E}{C^2} = \frac{8.64 \times 10^{12}}{\left(3 \times 10^8 m/s\right)^2} = 9.6 \times 10^5 kg.$$

Table 2.4 summarizes the various forms, source and end uses of energy. Similarly, Table 2.5 lists a variety of devices to illustrate conversions of energy from one form to another. For example, a toaster illustrates the conversion of electrical energy to thermal energy; a battery converts chemical energy into electrical energy. Mechanical energy (the kinetic energy part) of a car is converted into heat when the brakes are applied.

Table 2.4: Forms of energy

Chemical
Nuclear
Radiant
Electrical
Mechanical (kinetic, potential)

Primary source		End uses
coal		heat
oil	Chemical	light
Natural gas		motion
Uranium-nuclear		electricity
Sun-radiant/solar		Light, heat, electricity

Table 2.5: Energy conversions

	To chemical	To electrical	To heat	To light	To mechanical
From chemical	Food plants	Battery fuel cell	Fire wood	Candle phospho-rescence	Rocket, animal muscle
From electrical	Battery electrolysis/ electroplating	Transistor/ transformer	Toaster, heat lamp, spark plug	Fluorescent lamp, light-emitting diode	Electric motor
From heat	Gasification, vaporization	Thermocouple	Heat pump, heat exchanger	Fire	Turbine, gas engine, steam engine
From light	Plant (photosynthesis) camera film	Solar cell	Heat, lamp, Radiant solar	Laser	Photoelectric, door opener
From mechanical	Heat cell (-crystallization)	Generator, Alternator	Friction brake	Flint spark	Flywheel, pendulum, water wheel

2.7 ENERGY AND THERMODYNAMICS

As energy cannot be created or destroyed, only transferred, the primary objective of any energy system must therefore be the conservation and efficient use of energy. Hence, it is necessary to look into the economic use of any energy resource rather than trying for its new exploration or generation. Efficient conversion and saving of energy are thus the two important factors need to be considered for an energy system. These two aspects are based on the laws of thermodynamics specially for heat energy.

"Thermodynamics" is a branch of energy which deals with conversion of heat into work or vice versa. In most of the energy conversion processes, first law and second law of concept of thermodynamics are applicable:

● First law of thermodynamics relates to conservation of energy and throws light on concept of internal energy.

● Second law of thermodynamics indicates the limit of converting heat into work. Every energy conversion process has certain 'losses'.

2.8 ENERGY PARAMETERS

There is the necessity of monitoring the consumption of energy as well as the growth of energy requirement of a nation as a measure of its economy. In order to conserve fuel, it is imperative to adopt measures for maximizing economic development with minimum energy consumption. The energy parameters are assessed using the yardstick of gross domestic product (GDP), which indicates the value of all finished goods and services produced in a given period.

(i) **Energy intensity:** Energy intensity is a measure that is often used to assess the energy efficiency of a particular economy. The numerical value is traditionally calculated by taking the ratio of energy use (or energy supply) to gross domestic product (GDP), indicating how well the economy converts energy into monetary unit. High energy intensities indicate a high price or cost of converting energy into GDP. Low energy intensity indicates a lower price or cost of converting energy into GDP. Hence, the term 'energy intensity' is defined as energy consumption per unit of GDP.

$$\text{Energy intensity} = \frac{\text{Energy consumption}}{\text{GDP}}$$

When the per unit energy consumption for the production of energy intensive raw materials, like steel and aluminium, is reduced, the energy intensity gets reduced. Developed countries have reduced 'energy intensity', resulting in less energy consumption and at the same time achieving higher production. Factors affecting energy intensity are as follows:

a) Energy efficiency of the appliances and buildings by proper design and provision of insulation

b) Fuel economy of vehicles

c) Economical means and pattern of transportation

d) Efforts to conserve energy

(ii) Energy elasticity: It is defined as the percentage growth in energy requirement for 1% growth in GDP. The lower the value of elasticity, the higher is the overall efficiency.

$$\text{Energy elasticity} = \frac{\text{Growth in energy requirement}}{\text{GDP}}$$

The energy should contribute in increasing the GDP. The lower is the value of energy elasticity, the more is the growth of GDP. There is the requirement of improvement in the efficiency of energy consumption and energy savings in generation, distribution and utilization sectors. The value of energy elasticity for the developed countries ranges from 0.8 to 1.0.

2.9 CLASSIFICATION OF ENERGY RESOURCES

The energy resources can be classified on the basis of usability of energy, traditional use, long-term availability, commercial application, muscle power and their origin.

(i) Based on usability of energy: Based on the usability of energy, the resources can be classified as primary and secondary categories.

(a) Primary resources: These are available in the nature in raw form. These energy resources cannot be used in raw form. These resources need to be located, extracted, processed and converted into a suitable form before use by the end users. Examples of these resources are fossil fuels (coal, oil and gas), uranium, hydropower, sunlight, wind, biomass etc. These are either found or stored in the nature.

(b) Secondary resources: These resources are obtained from primary energy resources by processing. Processing helps in transformation of primary resources into the secondary or usable energy form so that it can be utilized by consumers. Examples are electrical energy, thermal energy (in the form of steam or hot water), chemical energy (in the form of hydrogen or fossil fuels) etc. The resources of primary energy and their conversion to secondary resources are shown in the fig. 2.5

(ii) Based on traditional use: Based on traditional use, the energy resources can be classified as conventional and non-conventional categories.

(a) Conventional energy resources: These energy resources have been traditionally used for many years. These resources are also widely used at present and fulfilling the energy requirements of the world. Examples are fossil fuels, nuclear and hydro power.

(b) Non-conventional energy resources: These are the alternate energy resources to the conventional energy resources. These are being considered to be used on

large scale looking into the fast depletion and environmental impacts of conventional energy resources. Examples are solar, wind, biomass etc.

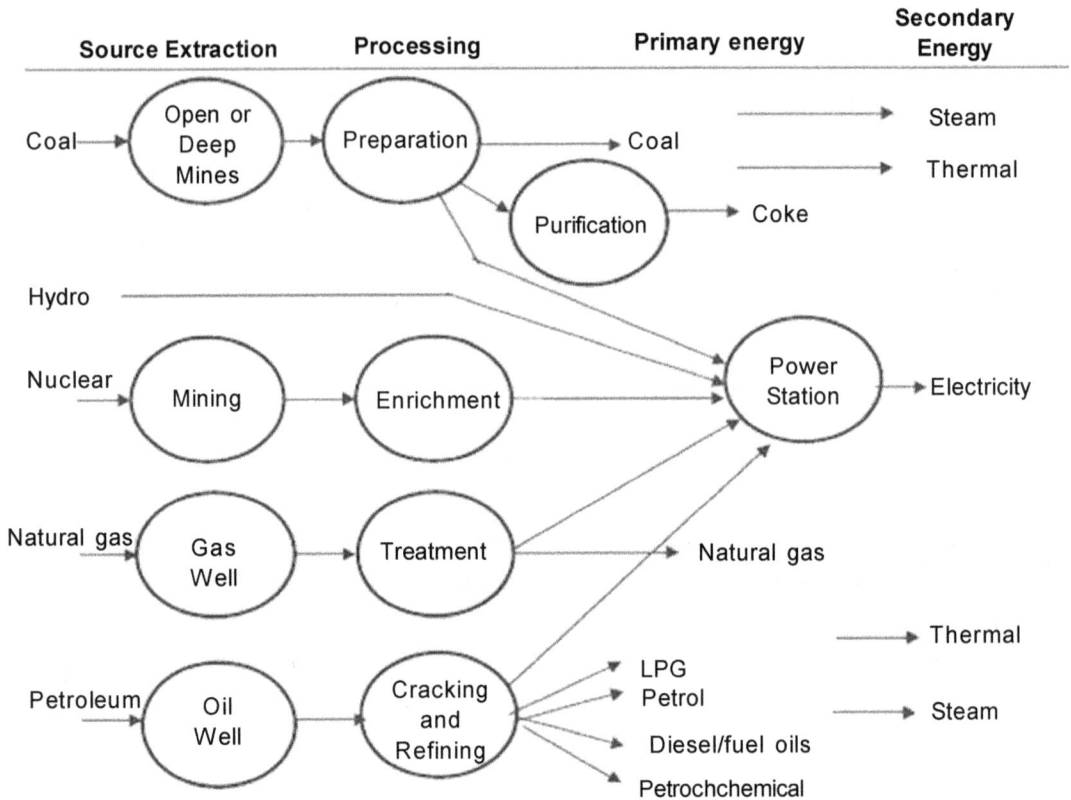

Fig. 2.5. Resources of primary energy and conversion to secondary resources

(iii) Based on long-term availability: Based on long-term availability and renewability, the energy resources can be classified as renewable and non-renewable categories.

  (a) Renewable energy resources: These energy resources can be renewed by nature again and again so that their supply is not adversely affected by the rate of their consumption. They are considered to be the infinite sources of energy and inexhaustible. Examples are solar, wind, biomass, ocean (thermal, tidal and wave), geothermal, small hydro energy etc. fossil fuels, nuclear and hydro power. These are also considered as non-conventional sources of energy.

  (b) Non-renewable energy resources: These are available in finite quantity and don't get replenished after their consumption. These are obtained from static stores of energy that remain unutilized if not released by human interaction. This is also referred as conventional sources of energy. Examples are fossil fuels (coal, oil, natural gas), uranium etc.

The basic difference between renewable and non-renewable energy resources is shown in the fig. 2.6. Table 2.6 presents the features of comparison between renewable and non-renewable energy sources.

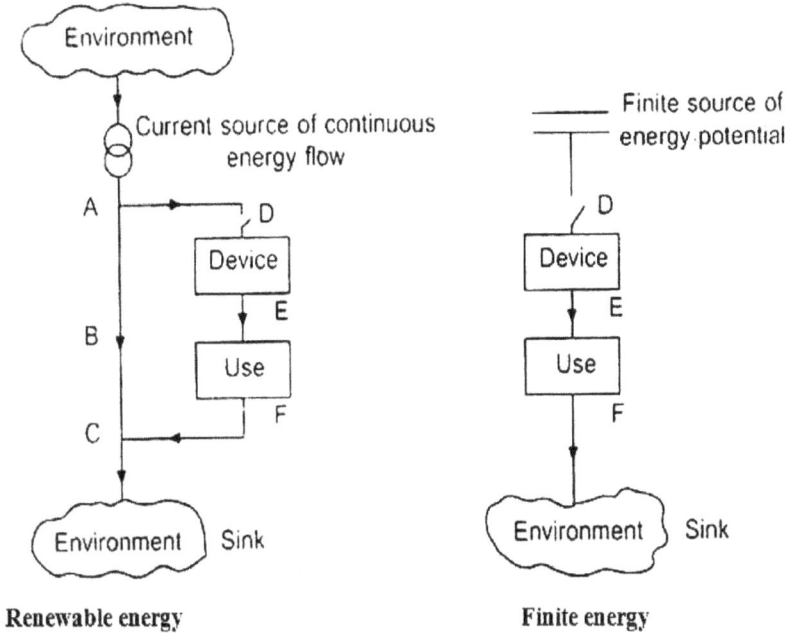

Fig. 2.6. Basic difference between renewable and non-renewable energy resources

Table 2.6. Comparative features of renewable and non-renewable energy sources

Features of comparison	Renewable energy uses	Non-renewable energy uses
Examples	Wind, solar, biomass, tidal etc.	Coal, oil, gas etc.
Source	Natural local environment	Finite and static stock
Normal state	A continuous energy flow	Static store of energy
Life time of supply	Infinite	Finite
Cost of source	Free	Increasingly expensive
Location for use	Site and society specific	General and international use
Scale	Small scale, economic, large scale may present difficulties	Increased scale often improves supply costs, large scale frequently favoured
Skills	Interdisciplinary and varied wide range of skills	Strong links with electrical and mechanical engineering areas. Narrow range of skill
Context	Rural, decentralized industry	Urban, centralized industry
Dependence	Self-sufficient system encouraged	Systems dependent on outside inputs

[Table Contd.

Contd. Table]

Features of comparison	Renewable energy uses	Non-renewable energy uses
Pollution and environmental damage	Usually little environmental harm, especially at moderate scale. Hazards from excessive wood burning, soil erosion from excessive biofuel use, large hydro reservoirs disruptive	Environmental pollution common, especially of air and water.
Safety	Less hazards in operation, usually safe when out of action	Most dangerous when faulty

(iv) Based on commercial application

(a) Commercial energy sources: The secondary usable energy resources such as electricity, CNG, LPG, petrol and diesel are essential for commercial activities. The economy of a nation highly depends on its ability to process and transform the naturally occurring raw energy sources into usable commercial energy resources. Commercial energy forms the basis of industrial, agricultural, transport and commercial development in the modern world. In the industrialized countries, commercial fuels are the predominant source not only for economic growth, but also for many household tasks of most of the population of our country

(b) Non-commercial energy sources: The sources of energy which are derived directly from the nature so as to be used without passing through any commercial outlet are known as the non-commercial sources of energy. These sources of energy are comparatively cheaper than the commercial energy sources. Examples are wood, animal dung cake, crop residues etc.

(v) Based on muscular energy

(a) Animate energy sources: These sources of energy are obtained from living organisms. These sources are generally derived from muscles.

(b) Inanimate energy sources: These sources of energy are obtained from non-living sources. It is not derived from the muscular energy. Examples are electricity, fossil fuels, solar energy etc.

(vi) Based on origin: Based on the origin, the energy resources can be classified as follows,

(a) Fossil fuel energy (b) Hydro energy (c) Nuclear energy (d) Solar energy (e) Wind energy (f) Biomass energy (g) Geothermal energy (h) Tidal energy (i) Ocean thermal energy (j) Ocean wave energy

2.10 ENERGY CHAIN OR ENERGY ROUTE

Energy chain or route signifies a sequential path for transforming primary energy source to usable form for direct use by the users. The primary energy source in the raw form cannot be used directly. It is impossible to drive a vehicle or electric motor using coal, petroleum or uranium. The energy available in primary energy sources is called raw energy. The primary energy sources have to be transformed into useful forms of energy sources which are then called secondary energy sources.

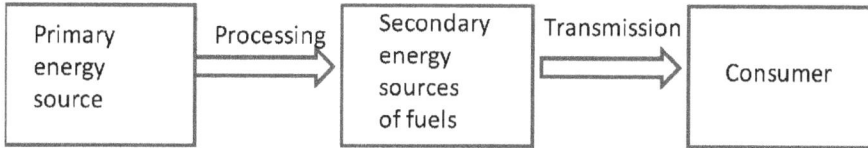

| Primary energy source | → Processing → | Secondary energy sources of fuels | → Transmission → | Consumer |

The secondary sources of energy may be of electrical or non-electrical routes.

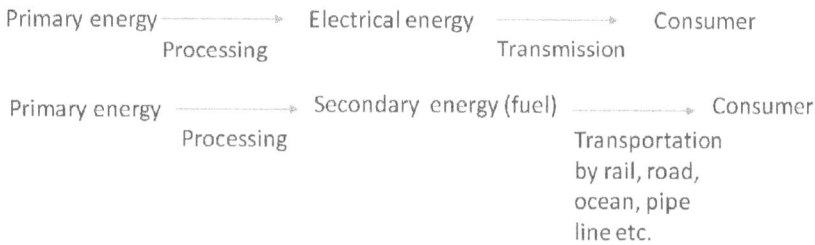

Primary energy ————→ Electrical energy ————→ Consumer
            Processing   Transmission

Primary energy ————→ Secondary energy (fuel) ————→ Consumer
            Processing   Transportation
                                                        by rail, road,
                                                        ocean, pipe
                                                        line etc.

At present about 40-50 % of world energy supply is met through the electrical energy route.

2.11 COMMON SOURCES FOR GENERATION OF ELECTRICAL ENERGY

Electrical energy is the most popular form of energy, used directly as the usable form either in thermal form (heating applications), in mechanical form (electrical motor applications in industries, in lighting purposes (illumination systems) or in transportation systems. The following common resources are used for generating electrical energy.

(i) Conventional methods

   (a) Thermal: Thermal energy (from fossil fuels) or nuclear energy are used for producing steam for turbines which drive the alternators (rotating a.c. generators).

   (b) Hydro-electric: Potential energy of stored water at higher altitudes is utilized as it is allowed to pass through water-turbines which drive the alternators.

(ii) Non-conventional methods

   (a) Solar photovoltaic cell: This device directly converts solar energy into electrical energy at atomic level of semi-conducting materials used in the solar cells. It operates based on the photo-voltaic effect, which develops an emf on absorption of ionizing radiation from the sun.

   (b) Wind power: Speeds of wind in suitable magnitude are utilized in driving wind turbines coupled to the alternators. Wind energy is a renewable source of energy. It is available plentifully at certain places. It suffers from the disadvantages of

its availability being uncertain (since dependent on nature) and control being complex (since wind speed has wide range of variation, as an input and the output required is at constant voltage and constant frequency).

(c) *Fuel cell:* This device enables direct conversion of energy, chemically into electrical form. This is an up-coming technology and has a special merit of being pollution-free and noise-free. It is yet to become popular for bulk-power generation.

2.12 HEATING VALUES OF VARIOUS FUELS

The heating value of a fuel is the amount of heat released during its combustion. Also referred to as energy or calorific value. Heating value is a measure of a fuel's energy density, and is expressed in energy (joules) per specified amount (*e.g.* kilograms). The heating values of some fuels are given in the table 2.7.

Table 2.7: Heating values of various fuels

Primary resource	Heating value
Coal: Anthracite	32-34 MJ/kg
Bituminous	26-30 MJ/kg
Coke	29 MJ/kg
Brown coal: Lignite (old)	16-24 MJ/kg
Lignite (New)	10-14 MJ/kg
Peat	8-9 MJ/kg
Crude petroleum	45 MJ/kg
Petrol	51-52 MJ/kg
Diesel	45-46 MJ/kg
Natural gas	50 MJ/kg, (42MJ/m^3)
Methane (85% CH_4)	45 MJ/kg, (38 MJ/m^3)
Propane	50 MJ/kg, (45 MJ/m^3)
Hydrogen	142 MJ/kg, (12 MJ/m^3)
Wood	10-11 MJ/kg
Natural uranium	$0.26\text{-}0.3 \times 10^6$ MJ/kg
Enriched uranium	$2.6\text{-}3.0 \times 10^6$ MJ/kg
U233	83×10^6 MJ/kg
U235	82×10^6 MJ/kg
Pu 239	81×10^6 MJ/kg

2.13 CONSUMPTION TREND OF PRIMARY ENERGY RESOURCES

Primary energy resources correspond to those that exist prior to any human-induced modification. This includes fuels extracted from the ground (coal, crude oil, or natural gas) or energy captured from or stored in natural sources (solar radiation, wind, biomass, etc.). Primary energy resources are either found or stored in nature. Secondary energy resources are obtained from the transformation of primary sources. Gasoline or diesel fuel from crude oil and charcoal from wood are examples of secondary sources. The basic difference between these two can be explained through the fig. 2.7 below.

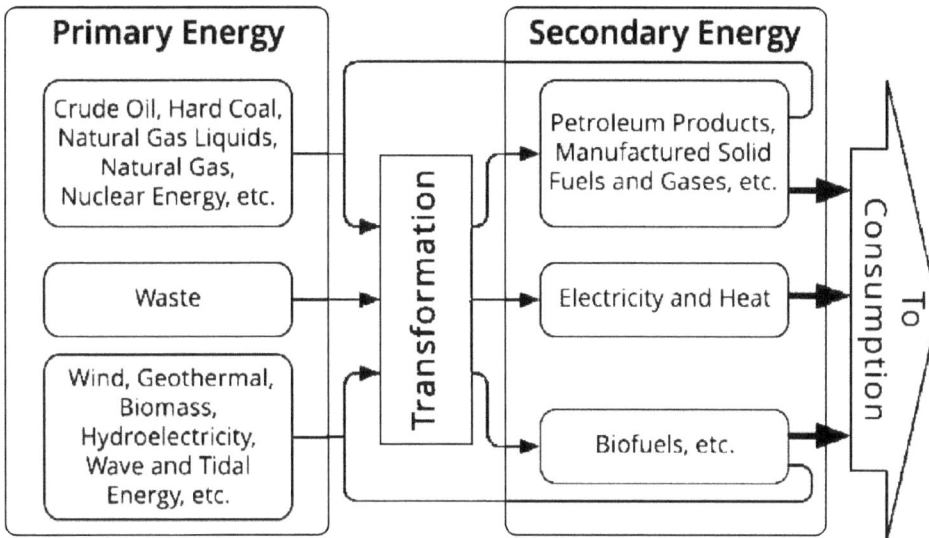

Fig. 2.7. Basic difference between primary and secondary energy resources

The primary energy resources fulfilling the world's energy requirement are from coal, crude oil, natural gas, nuclear, hydro and renewable energy sources. Hence, the primary energy sources can further be classified as (i) conventional energy sources, examples are fossil fuels (coal, crude oil and natural gas), nuclear energy and hydro energy and (ii) non-conventional energy sources (examples are solar energy, biomass energy, wind energy etc.). The world's primary energy consumption (in Exa-joules) for different countries during 2019 is presented in Table 2.8.

Table 2.8: World's primary energy consumption (Exa-joules) by fuels (2019)

Fuels	Oil	Natural gas	Coal	Nuclear	Hydro	Renew-ables	Total	% of total
North America	44.78	38.07	12.41	8.59	6.03	6.70	116.58	19.96
South & Central America	11.86	5.95	1.48	0.22	6.37	2.73	28.61	4.89
Europe	30.40	19.95	11.35	8.28	5.66	8.18	83.82	14.35
Middle East	17.80	20.10	0.40	0.06	0.30	0.12	38.78	6.64
Africa	8.28	5.40	4.47	0.13	1.98	0.41	20.67	3.53
India	10.24	2.15	18.62	0.40	1.44	1.21	34.06	5.83

[Table Contd.

Contd. Table]

Fuels	Oil	Natural gas	Coal	Nuclear	Hydro	Renew-ables	Total	% of total
China	27.91	11.06	81.67	3.11	11.32	6.63	141.70	24.26
Japan	7.53	3.89	4.91	0.59	0.66	1.10	18.68	3.21
Australia	2.14	1.93	1.78	——	0.13	0.42	6.41	1.09
Others	32.09	32.95	20.77	3.54	3.77	1.48	94.6	16.20
Total World	**193.03**	**141.45**	**157.86**	**24.92**	**37.66**	**28.98**	**583.90**	**100**
Percentage of world's total (%)	**33.05**	**24.22**	**27.03**	**4.26**	**6.44**	**4.96**		

Ref: BP statistical review of world energy 2020

The average percentage consumption trend of various primary energy resources of the world during 2017 is indicated in fig. 2.8.

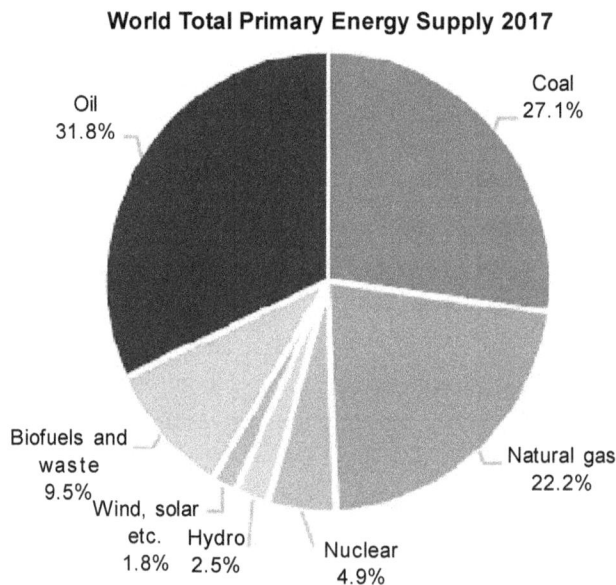

World Total Primary Energy Supply 2017

Fig. 2.8. Percentage consumption of various primary energy resources (year 2017 data)
Source: International Energy Agency, 2019.

Looking at this figure, it is clear that there is the heavy dependence of energy on fossil fuels. About 84 % of the world's energy supply comes mainly from fossil fuels. The share of fossil fuel is about 91 % in India (table 2.8). The global primary energy consumption is more than doubled, from 270.5 EJ in 1978 to 583.90 EJ in 2019. The rise in the total global primary energy consumption within the past 40 years was mainly covered by fossil fuels. On account of this, the share of fossil fuels in global primary energy consumption was only eight percentage less in 2019 (84 %) than in 1978 (92 %). Around 84% of the total primary energy consumption was still covered by coal, oil and gas in 2019. It is also clear that United States with about 4 % of the world's population consumes around 23 % of the total world's primary energy resources whereas India with about 18 % of the world's population consumes around 5 % of the total world's primary energy resources.

2.14 CONVENTIONAL ENERGY RESOURCES (WORLD STATUS)

Conventional sources of energy can be described as those which have been used since a long time in order to fulfil most of the daily energy requirements such as cooking, lighting, transportation etc. Conventional sources of energy are used extensively by mankind and the magnitude of usage is so high that the reserves have got depleted to a great extent. Examples are coal, petroleum, natural gas, hydro power and nuclear energy.

At present (2019), the annual primary energy consumption of the world is 583.90 exajoules (equivalent to 162.2×10^{12} kWh of energy or average power of 1.85×10^{7} MW). Fossil fuels roughly provide about 85% of this energy and will continue to provide more than 80% of the total energy demand well into the future. Approximately 30 % of this energy is consumed in transportation sector and the remaining 70% by industries, domestic, agriculture and social consumers. Between 1971 and 2017, total final consumption (TFC) was increased by 2.3 times. The share of energy uses in commerce and services or industry for instance has been stable as shown in fig. 2.9. However, energy use in transport sector has significantly been increased, from 23% of TFC in 1971 to 29% in 2017. Nevertheless, in 2017, industry remained the largest consuming sector, only one percentage lower than in 1971 (37%). The residential sector ranked third in 2017 (21%).

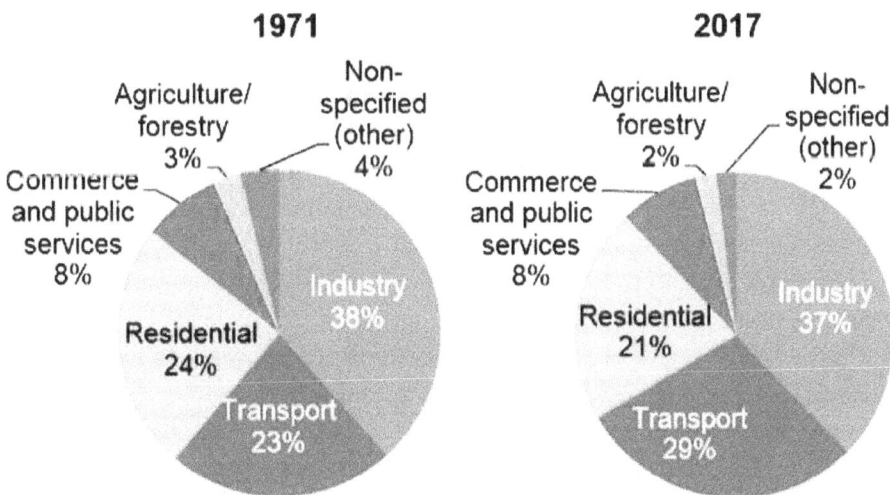

Fig. 2.9. World total energy consumption by sector **Source:** IEA

The energy demand has grown very fast in recent years, with primary energy demand increasing by more than 20% since 2009 (Table 2.9). The average annual growth rate from 2009 to 2019 was 2.09 %. Over 70% of this growth was from the developing countries.

2.15 AVAILABILITY OF CONVENTIONAL ENERGY RESOURCES AND FUTURE TRENDS (WORLD STATUS)

As mentioned above, fossil fuels, hydro energy and nuclear energy are the prevailing sources of conventional energy to meet the present demand of the energy for the mankind. The energy requirement of the world is increasing day by day and this growing demand puts a lot of pressure on the conventional sources of energy. Fossil fuels such as crude oil, coal and natural gas are the main resources presently fulfilling the requirement of world's energy.

Fossil fuels: Fossil fuels have been a major source of energy since about 1850, the start of the industrial era. The dictionary meaning of the word 'fossil' is the remains of an animal or plant which have hardened into rock (naturally occurring solid mass or aggregate of minerals). In case of fossil fuels, the word fossil is used to indicate the fuels, derived from plants and animals that had lived millions of year ago. Fossil fuels are so called because these are in fact the fossils of old biological life that once existed on the surface of the earth. These were formed in several parts of the earth at varying depth during several millions of years by slow decomposition and chemical actions of buried organic matter under favourable pressure, heat and bacterial action. The fossil fuels include coal, crude oil and natural gas.

Millions of years ago, the dead plants and animals were slowly decomposed into organic materials and formed as fossil fuels. Different types of fossil fuels were formed depending on what combination of animal and plant debris was involved, how long the material was buried, and what conditions of temperature and pressure existed under which they had been decomposed. For example, oil and natural gas were created from organisms that lived in the water and were buried under ocean or river sediments. Long after the shifting of seas or drying of the rivers, the combined action of heat and pressure compressed and boiled the organic material under the layers of silt. In most areas, a thick liquid which is now called crude oil was formed as a result of this slow boiling of the buried materials over thousands of years. In deeper and hotter regions, the boiling process continued beyond the stage of crude oil formation until natural gas was formed. Over the time, some of this oil and natural gas began to move upward through the earth's crust until they ran into rock formations called cap rocks (harder and more resistant rocks) that are dense enough to prevent them from seeping to the surface. It is from under these cap rocks that most oil and natural gas are extracted today.

Table 2.9: World's primary energy consumption (Exajoules)

Year	2009	2010	2011	2012	2013	2014	2015	2016	2017	2018	2019	Average annual Growth rate (%)
North America	109.76	113.29	113.35	110.86	113.72	114.78	113.83	113.74	114.34	117.79	116.58	0.62
South & Central America	24.82	26.6	27.26	27.93	28.53	28.76	28.80	28.50	28.61	28.53	28.61	1.52
Europe	85.55	88.69	88.66	86.32	85.43	82.10	82.77	83.90	84.76	84.76	83.82	-0.20
Middle East	28.22	29.74	30.86	32.12	33.06	34.05	35.04	36.23	36.83	37.61	38.78	3.71
Africa	15.57	16.07	16.13	16.69	17.14	17.66	17.91	18.38	18.79	19.39	19.87	2.76
India	21.52	22.55	23.88	25.11	26.08	27.86	28.77	30.07	31.33	33.30	34.06	5.82
China	97.52	104.28	112.54	117.06	121.37	124.20	125.38	126.96	130.83	135.77	141.70	4.53
Japan	19.83	21.13	20.06	19.92	19.75	19.24	18.97	18.65	18.89	18.84	18.76	-0.53
Australia	5.48	5.50	5.70	5.63	5.67	5.75	5.84	5.88	5.87	6.00	6.41	1.69
Others	74.55	78.17	79.87	83.34	84.16	84.85	85.86	88.29	90.17	94.24	95.31	2.78
Total World	**482.82**	**506.02**	**518.31**	**524.98**	**534.91**	**539.25**	**543.17**	**550.60**	**560.42**	**576.23**	**583.90**	**2.09**

The same types of geologic conditions which created oil and natural gas also created coal, but with a few differences. Coal was formed from the dead remains of trees, leaves, branches and other plants that lived millions of years ago. In some areas, such as portions of what is now the eastern United States, coal was formed from swamps covered by sea water. The sea water contained a large amount of sulphur, and as the seas dried up, the sulphur was left behind in the coal. When this coal burns, the sulphur is converted to SO_x which causes air pollution. Some coal deposits, however, were formed from freshwater swamps which had very little sulphur in them. These coal deposits, located largely in the western part of the United States, have much less sulphur in them. It is estimated that about a ten feet deep layer of prehistoric debris was needed to make each one-foot-deep layer of coal.

Fossil fuels are generally considered as non-renewable resources because they take millions of years to form, and the known reserves are being depleted much faster than the new ones being made available. Presently, we are passing through the peak period of the fossil age. Many estimates have been made regarding the amount of fossil fuels still available in the earth. Before estimating these, it will be useful to distinguish between the terms 'resources' and 'reserves'. The term 'resources' implies an estimate of the total quantity available that may eventually be successfully exploited and used by man. The term 'reserves' generally refers to that portion of the resources which has been proved and can be economically recovered with available techniques.

Reserve-to-production ratios (RPR) may be calculated for individual countries or globally for specific resources. Oil, coal, and gas are the three most important fuels for powering the modern world. These resources are not uniformly distributed over the earth, so some countries have larger reserves than others. Due to the uncertainty of the reserves numbers, estimates of RPR vary widely. RPR indicates the life of reserves that would last if production would continue at the present rate. It is calculated by dividing the production of the resource in a particular year to the reserves remaining at the end of that year and is usually expressed in years. RPR is most commonly applied to fossil fuels, particularly petroleum and natural gas. The *reserve* portion (numerator) of the ratio is the amount of a resource known to exist in an area and to be economically recoverable (proven reserves). The *production* portion (denominator) of the ratio is the amount of resource produced in one year at the current rate.

RPR = (amount of known resource) / (amount used per year)

If a country has 10 million barrels of proven oil reserves, for example, and is producing 250,000 barrels a year, then the RPR, or life of the reserves, is 10,000,000 / 250,000 = 40 years.

As per an estimate, if the world continues to consume fossil fuels at year 2019 rates, the reserves of coal, oil and gas would last about 180,50 and 55 years respectively (Table 2.11). This gives only an indication and not a very realistic figure. It is generally accepted that the rate of production of an economic commodity of which a finite quantity exists is governed by the laws of supply and demand. As the amount, available depletes, the commodity becomes costlier, and its use gradually declines. Also, new reserves are continuously being discovered and new technologies are being invented for those resources which were not considered economical earlier. The locations and estimates of the world's main fossil fuel reserves are indicated in Table 2.10.

From the Table 2.10, it is evident that coal constitutes approximately 60 % of the fossil fuel reserves in the world, with the remaining 40 % being oil and gas. The size and location of reserves of oil and gas are limited in the Middle East, coal remains abundant and broadly distributed around the world. Economically recoverable reserves of coal are available in more than 70 countries worldwide. Hence, coal reserves are not limited to mainly one location, as against oil and gas mostly available in the Middle East. The coal reserves have therefore the potential to be the dominant fossil fuel in the future. The increase in fossil fuel resources is due to the availability of improved data, as well as technological improvements.

After the industrial revolution, our energy demands have increased tremendously which results in the rate of consumption of fossil fuels at a much faster rate than their formation. As a result, the fossil fuel reserves of the world have become items of limited quantity while our demands of these resources are unlimited, clearly indicating a situation of imbalance. This imbalance implies that our activities on the earth (at current arte) cannot be sustained forever; at the most it can last only a century or two more with ever-increasing consumption of fossil fuels. Also, there is an imbalance in the distribution of fossil fuel reserves across the planet. For instance, the USA contains about 25% of coal reserves and five countries in the Middle East contain about 60% of the oil reserves compared to the total reserves of the corresponding resources of the world. This results in energy insecurity for the countries that are devoid of fossil fuels and could be a potential cause of political crisis. Let us have a look at these fossil fuels and their reserves. World hydrocarbons reserves, production rates and depletion years assuming no future discoveries are shown in Table 2.11. The depletion years may shift ahead after new discoveries.

Table 2.11 gives an idea of total worldwide reserve of fossil fuel sources. The estimation of reserves varies as it depends on various factors used in estimation. The estimation depends on the advances in technologies for detecting the presence of fossil fuels. It depends on the advances in technology for extracting fossil fuels in an economical way. Based on the current reserves, the number of years these reserves are going to last can be estimated. This can be obtained by considering the amount of current reserves and dividing it by the current rate of production/consumption of these reserves. According to the current rate of consumption (Table 2.11), the coal, oil and natural gas can last only for about 180, 50 and 55 years respectively.

Table 2.10: World's fossil fuel reserves (Exajoules) (as in 2019)

Region	Fossil fuel reserves (Exa-joules)				Fossil fuel reserves (Percent)			
	Oil	Natural gas	Coal	Sum	Oil (%)	Natural gas (%)	Coal (%)	Sum (%)
North America	1495.72	630.42	7205.24	9331.38	3.05	1.28	14.73	19.06
South & Central America	1983.49	336.34	383.29	2703.12	4.05	0.68	0.78	5.51
Europe	88.12	142.80	3783.05	4013.97	0.18	0.29	7.73	8.20
Middle East	5102.85	3175.2	449.12	10122.25	10.43	6.49	0.918	20.67
Africa	769.28	625.80			1.57	1.27		
India	28.76	54.61	2966.06	3049.43	0.05	0.11	6.06	6.22
China	160.34	352.8	3964.66	4477.80	0.32	0.72	8.10	9.14
Japan	1.83	—	—	1.83	0.003	—	—	0.003
Australia	14.68	100.8	4174.21	4289.69	0.030	0.20	8.53	8.76
Others	966.39	2931.05	7027.01	10924.45	1.97	5.99	14.36	22.32
Total World	10611.46	8349.82	29952.64	48913.92	21.69	17.07	61.23	100

Table 2.11: Global fossil fuel statistics based on proved reserved (as in 2019)

Fuels	Total reserves	Production per year	Reserves to production ratio (availability no. of years)
Coal	29952.64 (exa-joules)	167.58 (exa-joules)	180
Oil	10611.46 (exa-joules)	0.5 ton (209.41 exa-joules)	51
Natural gas	8349.82 (exa-joules)	3989.3 billion m³ (153.30 exa-joules)	55

Source: BP Statistical Review of World Energy, 2020.

Coal

Coal is a product of natural process of decomposition of organic materials (plants) buried in swamps and has been out of the contact with oxygen. It is formed when plant material is converted into peat, which is then converted into lignite, then sub-bituminous coal, after that bituminous coal and finally anthracite. Peat constitutes the young organic material and is the first stage in the formation of coal from the biomass. It is made of plant material which has not been transformed into coal. During the formation of coal, it passes through different stages involving biological and geological processes that take place over geological time. The plant materials buried in the soil undergo a series of complex transformations which occur in two steps. The first step is a biochemical degradation assisted by organisms (bacteria and fungi). The second step is a physico-chemical decomposition, a "coalification" process occurring because of the burial environment. The combination of pressure and temperature in this environment produces the thermal cracking of the buried materials. Water is squeezed out and carbon dioxide is released. Later, hydrogen rich volatile compounds escape. Above a temperature of the order of 110 ° C, this thermal cracking produces coal.

The harder forms (e.g. anthracite coal) are regarded as metamorphic rock due to their exposure to high temperatures and pressures. Coal is primarily composed of carbon, along with variable quantities of other elements, mainly hydrogen, sulphur, oxygen, and nitrogen.

During different times in the geologic past, the earth had quite dense forests in wet land areas. These forests were buried under the soil due to a number of natural processes including flooding. They were compressed further as more soil was deposited over them. As the dead plant materials moved deeper in the earth due to the further deposition of soil over them causing more rise in the temperature and pressure. It is considered that the mud or acidic conditions preserved the deeply buried dead plant material from biological degradation and oxidation. Eventually, the dead material was converted to coal by a slow process under high pressures and high temperatures. This conversion is called carbonization because coal contains primarily carbon.

Factors such as surrounding environment and pH of the water available have an impact on the characteristics of coal. The greater the degree of coalification, the better the quality of the coal. The degree of coalification is specified by the "rank" of the coal, an indicator of its maturity. This is an important factor for determining its quality. Peat has the lowest rank and anthracite has the highest. The carbon content of coal varies about 55% for lignite and 90% for anthracite. The hydrogen content in coal is about 4%, around half of that in crude oil. Coal also contains volatile matter which is liberated by heating at high temperature in the absence of air. This volatile matter includes hydrocarbons, aromatic hydrocarbons, and some sulphur. Larger carbon content is equivalent to higher energy content and high energy content is considered high quality coal. The ranks of coal during the different stages of its formation are summarized in Figure 2.10. Figure 2.11 summarizes the varying energy content of different types of coal.

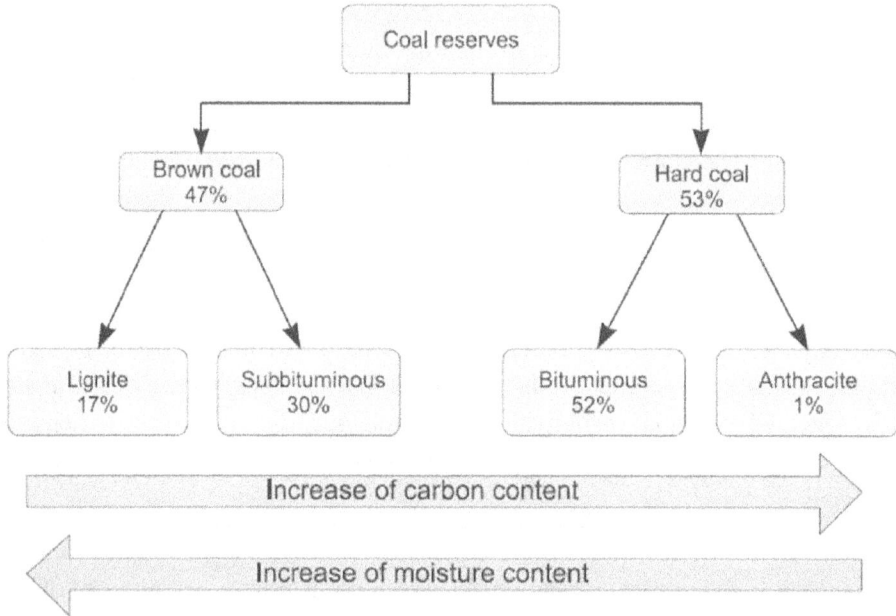

Fig. 2.10. Various forms of coal

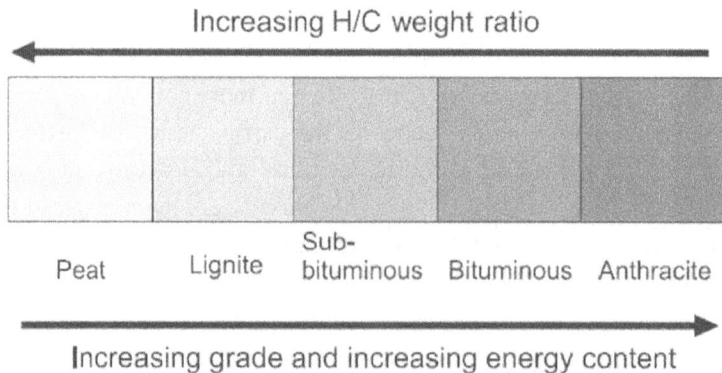

Fig. 2.11. Various forms of coal having varying energy content and H/C weight ratio.

(From peat to anthracite, the energy content increases and the hydrogen - to - carbon (H/C) ratio decreases. This ratio, which is typically around 1 for peat falls below 0.5 for anthracite)

Oil

Oil or the crude oil occurs in the form of liquid. It is a complex mixture of hydrocarbon and some amount of inorganic elements like sulphur, oxygen and nitrogen. The crude oil in itself is not useful for consumption in appliances. Crude oil is refined to get various products like petrol, diesel, kerosene and some other products like fertilizers, pesticides, pharmaceuticals, plastics, nylon, paints etc. Petroleum is also called as crude oil or just oil, is a naturally occurring, yellowish-black liquid found in geological formations beneath the earth's surface. However, petroleum broadly includes both crude oil and petroleum products. The terms oil and petroleum are sometimes used interchangeably. Petroleum products are materials such as fuel oils, propane, butane, lubricants, paraffins etc. derived from crude oil when it is processed in oil refineries. Oil is formed by the anaerobic decomposition of organic material including phytoplankton and zooplankton, which were settled in large quantities in deposition basins, such as a seabed or lakebed, over geological time. This organic matter, after being mixed with mud, remained buried under increasing layers of sediment over millions of years. As a result, the organic matter and mud mixture were exposed to high pressures and temperatures over millions of years causing the organic matter to chemically change, first into kerogen (a waxy material, which is found in oil shales), and then, with more exposure to further heat, into liquid and gaseous hydrocarbons in a process known as catagenesis. Catagenesis is a term used in petroleum geology to describe the cracking process which results in the conversion of organic kerogens into hydrocarbons. Throughout all these processes, the energy density of hydrocarbons increases, however, the origin of the oil is mostly from the deceased sea organisms (zooplankton, phytoplankton, shellfish, algae, animals, and so on) buried in the sediments of sand and mud. Hence, petroleum or oil is a fossil fuel formed by decomposition and bacterial anaerobic reactions on the buried vegetation and animal masses under favourable temperatures and pressures and marine surroundings over several millions of years.

Natural Gas

Natural gas is one of the fossil fuels and is formed by the decomposition of the remains of the dead animals and plants buried under the earth. Natural gas is a hydrocarbon gas mixture mainly consisting of methane (CH_4) with varying amounts of other higher alkanes and sometimes a small portion of non-hydrocarbons such as carbon dioxide, nitrogen, hydrogen sulphide, or helium, largely of inorganic origin. Natural gas is formed when the phytoplankton and zooplankton sink to the bottom of the ocean and are exposed to very high temperatures and pressures under the surface of the Earth over millions of years. It is often referred to as "gas" when compared to other energy forms, such as oil or coal and is a non-renewable fossil fuel resource that is used as a source of energy for heating, cooking, electricity generation, as a fuel for vehicles and as a chemical feedstock for manufacturing fertilizer, plastics, paints, and other chemicals.

It is classified into two categories such as (i) associated gas and (ii) non-associated gas. Associated gas is a natural gas originating from the fields producing both liquid and gaseous hydrocarbons simultaneously. On the other hand, non-associated gas is a natural gas which is obtained independently. It is generally found in the space above an oil reservoir. Hence, it is mainly found along with crude oil but there are some reserves where it is obtained in the absence of crude oil. Typically, the major constituents of the associated natural gas are methane (about 50 percent by volume), ethane (about 20 percent) and propane (about 10 percent). The major constituent of non-associated natural gas is methane (more than 90 percent by volume).

Like crude oil, the raw natural gas cannot be used in the form as obtained from the well. It must be processed and refined to make a cleaner-burning and purified methane. It is refined to eliminate hydrogen sulphide, carbon dioxide, nitrogen, water etc. before its transportation. The materials particularly associated hydrocarbons such as ethane, propane and butane are also removed from natural gas so it can be safely transported and processed (including liquefaction and compression). Vehicles running on compressed natural gas (CNG) are increasing because they generate much lesser NO_x and SO_x than vehicles running on petrol or diesel.

As mentioned earlier, natural gas is primarily composed of methane. After the release of methane to the atmosphere, it is removed by oxidation to CO_2 and water gradually in the troposphere or stratosphere. The life time of atmospheric methane is relatively short compared to that of CO_2, but methane, therefore natural gas, has higher global warming potential when compared to CO_2. When natural gas is burned, it produces more water than CO_2 by mole, compared to coal, which obviously produces primarily CO_2. Natural gas produces only about half the CO_2 per kilowatt-hour that coal produces. Also, natural gas produces much lower amounts of SO_2, NOx, and PM emissions when compared to other fossil fuels, hence it is the cleanest fossil fuel. Natural gas releases almost 30% less CO_2 than oil and 43% less than coal.

2.16 UNCONVENTIONAL FOSSIL FUELS

As the supplies and reserves of oil and gas are depleting, attention is increasingly being focussed in other naturally occurring static sources of fuels, known as unconventional fossil fuels. Coal, crude oil and natural gas are categorized under conventional sources of fossil fuels. Tar sand, oil shale, shale gas, coal bed methane and methane hydrate are categorized under unconventional fossil fuels. These sources of fuels are basically hydrocarbons. Crude oil can be obtained from tar sand and oil shale whereas natural gas can be derived from shale gas, coal bed methane and methane hydrate. The availability of oil and gas would become less in the coming years and this can only be substituted with production of fuels from unconventional sources. The difference between conventional and unconventional fossil fuels is not so much about their chemical compositions, but with their geological formations around the resources and how those resources are extracted beneath the earth surface. Generally, conventional oil and gas are easy to extract, however, unconventional ones are more difficult to mine. Unconventional fuels differ from conventional hydrocarbons in their production, transportation or processing. As such, the costs of unconventional production have traditionally been higher, because of the complex technologies that their exploitation requires. Extraction of conventional fossil fuels from their geological formations can be done with standard methods that can be used to economically remove the fuels from the deposit. Hence, their extraction is easier and less expensive because they require no specialized technologies and can utilize common methods. Unconventional fossil fuels could contribute to a reserve if they could be extracted economically. It is a well-known fact that it is more labour and resource intensive to produce fuels from the unconventional sources, which obviously require additional energy to refine (i.e., higher production costs). Unconventional resources are trapped in reservoirs with poor permeability and porosity, meaning that it is extremely difficult for them to flow through the pores and into a standard well. Geological formations in the deposit of conventional and unconventional fuels have been shown in the fig. 2.12. Extraction of fuels from these difficult reservoirs requires specialized techniques and tools and are not economical compared to the conventional ones. However, the present days' advanced technologies could potentially be able to exploit the unconventional fuels economically and contribute to the reserves for possible future fuels.

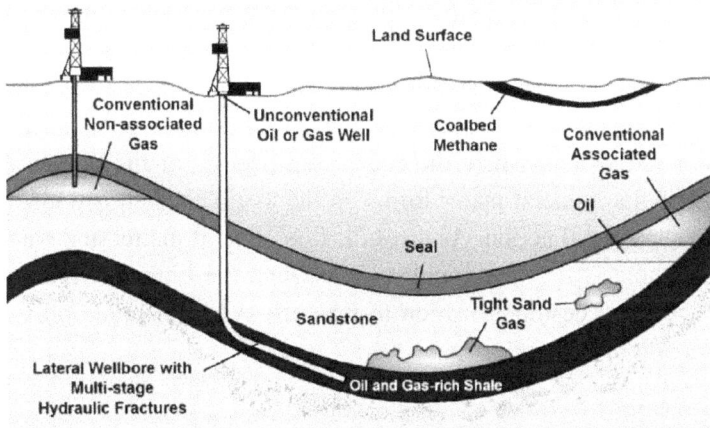

Fig. 2.12: Geology of conventional and unconventional fossil fuels

The potential of unconventional oil and gas has been recognized. According to the latest estimate, the recoverable amount of global unconventional oil resources is 6200 ton, roughly the same as conventional oil resources; the recoverable amount of unconventional gas resources is about 4000 m^3, roughly eight times that of conventional gas resources, and it is mainly concentrated in four regions with rich unconventional oil and gas resources, i.e. North America (34%), Asia- Pacific (23%), South America (14%) and Russia (13%), (International Energy Agency, 2015).

In the past 10 years, major breakthroughs have been made in the exploration and development of unconventional oil and gas in North America. The rapid development of unconventional oil such as tight oil and tar sands oil rapidly promoted oil production increase in North America by 31.0% and it became a major growth point in global oil production. The rapid development of unconventional gas such as shale gas and tight gas rapidly promoted the increase in natural gas production in the United States by 38.4% and set off a wave of unconventional oil and gas development around the globe.

Tar sands

Tar sands, also called bituminous sands, oil sands, or extra heavy oil, are a mixture of sand or clay, water, and extremely heavy crude oil. It is a porous sandstone deposit impregnated with bitumen. Bitumen is a black and highly viscous mixture of hydrocarbons available naturally in semi-solid form. In order to transport the tar sands, the natural bitumen is processed or diluted. Tar sands are found in many locations, notably in Canada, the United States, Russia, and some Middle Eastern countries. If a tar sand deposit is close to the surface, it can be exploited by open pit mining techniques. After transportation to the extraction plant, bitumen is separated from the sand using hot water. Roughly three quarters of the bitumen contained in the tar sand is recovered. About 1 barrel of oil may be obtained from about 2 tons of tar sand. If the tar sands are not near the surface, in situ techniques can be used to extract bitumen from tar sands. A schematic diagram indicating the processes involved in producing synthetic crude oil from tar sands is shown in Fig. 2.13. The sands obtained from surface mining are first passed through a conditioning drum where water, steam and caustic soda are added and slurry is formed. The slurry passes into a separation tank where the coarse sand settles at the bottom and a froth of bitumen, water and fine mineral mater forms on the top. The froth is diluted with naphtha and subjected to centrifugal action. As a result, fine mineral matter and water is removed. After this, the naphtha is recovered and recycled, and the bitumen obtained is subjected to hydro-processing and desulphurization to produce synthetic crude oil.

Oil shale

Oil shale is a fine grained sedimentary rock containing a large amount of kerogen. Actually the name is misleading since oil shale does not contain oil and it is usually not shale. Shale is a fine-grained, sedimentary rock, composed of mud that is a mix of flakes of clay minerals and tiny fragments of other minerals, especially quartz and calcite. Shale from which oil is extracted contains a large amount of kerogen, and is known as oil shale. Kerogen is fossilized organic material within a sedimentary rock. When oil shale is heated, the kerogen decomposes and yields crude oil. Hence, there is no oil in the oil shale; however, it contains a unique material called kerogen, which is a precursor of crude oil. The distillation of crude oil from oil shale is not economical so far. Deposits of oil shale exist in the United States, Russia, Brazil, and China. The United States has a little more than 60% of these reserves and three countries i.e. the United States, Russia, and Brazil have 86% of the recoverable reserves. The largest known deposit of oil shale is in the western United States in the Green River Formation in Wyoming, Utah, and Colorado. It is estimated that there is a deposit of about 1.8 trillion barrels of oil shale. It is also estimated that oil shale deposits could provide about 40 liters of oil per ton of rock. There are two main technologies for extraction of oil shale deposits. In the first, the rocks are fractured in place and heated to recover gases and liquids. In the second, the oil shale is mined and transported to another location where it is heated to 450 ^{0}C and hydrogen is added. Both processes need large quantities of water and also large amount of energy, which often makes the exploitation of oil shale uncompetitive at present. There is a need to make this resource competitive on a large scale due to the sustained high price of oil at present. A schematic diagram indicating the processes involved is shown in fig. 2.14.

One major problem associated with the use of both tar sands and oil shale is the environmental degradation associated with surface mining and with the disposal of large amounts of sand and spent shale rock which remain after the crude oil is obtained. This problem would need careful attention if either of these energy alternatives is to be used on large scale in the future.

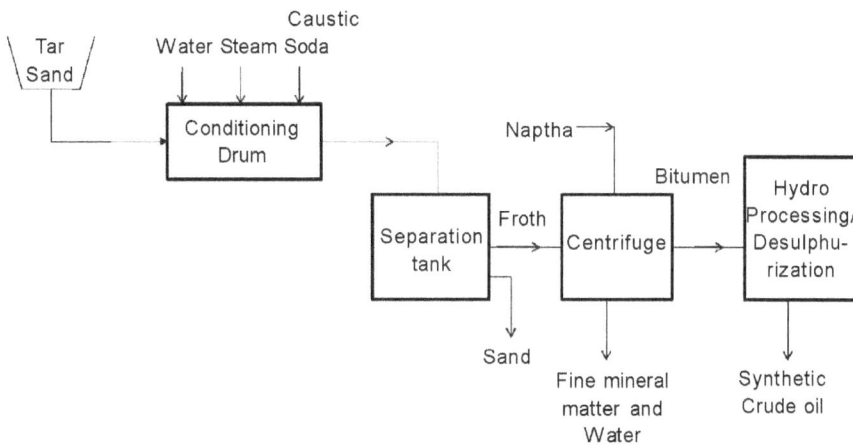

Fig. 2.13. Production of synthetic crude oil form tar sands

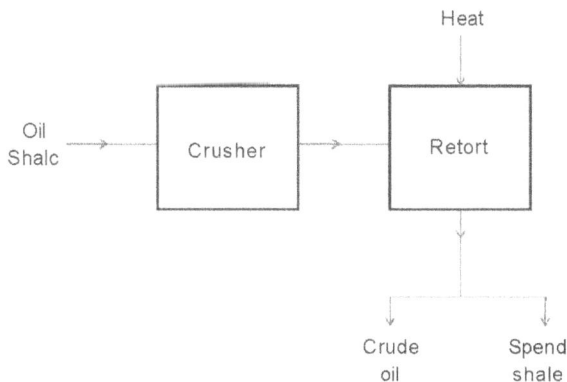

Fig. 2.14. Production of crude oil from oil shale

Shale gas

Shale gas is a natural gas that is found trapped within shale formations. Shale gas has become an increasingly important source of natural gas in the United States and Canada since 2000. Shales ordinarily have insufficient permeability to allow significant fluid flow to a wellbore. Many of the shale formations are not commercial sources of natural gas. Shale gas is one of a number of unconventional sources of natural gas and the other one is coalbed methane. Shale formations have low permeability, thus fluids do not flow easily through the formation and generally require extra effort to access gas or oil trapped within them. Because of this, special techniques such as fracturing must be used to produce shale gas. Since accessing shale gas is difficult and requires special techniques, shale gas is considered an unconventional resource. After the success in the United States, shale gas exploration has begun in other countries, such as China, Poland, and South Africa. Due to this increase in shale production, the United States is now a leading one in producing natural gas in the world. Unconventional natural gas resources have a great prospect in natural gas markets because of limited resources of conventional natural gas.

Coalbed methane

Coalbed methane, coalbed gas, coal seam gas, or coal mine methane is a form of natural gas extracted from the coal beds. Coal bed methane is the methane found in coal seams. A *coal seam* is a dark brown or black banded deposit of *coal* that is visible within layers of rock. When natural gas is obtained from a *coal seam*, it is known as *coal seam* gas or *coal* bed methane. This gas bonds to the surface of underground *coal seams*, which are generally filled with water. It is formed by either biological processes (microbial action) or thermal processes occurring at large depths where the temperature is high. Not all coal seams contain coal bed methane. The term refers to methane adsorbed into the solid matrix of the coal. Coalbed methane is distinct from a typical sandstone or other conventional gas reservoir, as the methane is stored within the coal by a process called adsorption which refers to the adhesion of atoms, ions or molecules from the gas to a surface. Coal bed methane is held in the coal by the surrounding water pressure. Because of that, there are coal seam aquifers. Coal bed methane travels with groundwater in the coal seams and is extracted when water is pumped from the seam in order to decrease the pressure and liberate the methane. Coal bed methane is used in the same way as traditional natural gas. It is estimated that at least 1.5–2 m^3 of gas can be exploited per ton of coal. In recent decades, it has become an important source of energy in United States, Canada, Australia, and other countries.

Methane Hydrates

Methane hydrates is another source of unconventional natural gas. It is a white crystalline solid. Also called methane clathrate or methane ice, it is a material in which methane molecules are trapped inside a crystalline cage structure (hence the name "clathrate"), similar to ice. On the average, methane hydrates contains about 13% methane, corresponding to one molecule of methane for about six molecules of water. The average density is 0.9 g/cm^3. Methane hydrates is stable at low temperatures and high pressures. Gas hydrates can occur only in polar or high-altitude permafrost regions, or in oceanic sediments or deep inland seas where the water temperature is close to 0 ^{0}C and the water depth exceeds 300 m. Under normal conditions of temperature and pressure, methane hydrates melts and the methane that is released can easily be ignited with a lighted match. Harnessing methane from methane hydrates is not a simple task. Only a small part of the deposit is expected to be recoverable and currently no accepted industrial method exists for that recovery.

2.17 HYDRO RESOURCES

Hydro resource is usually restricted to the generation of shaft power from the falling water. The power is then used for direct mechanical purposes or more frequently for generating electricity. Other sources of hydro power are waves and tides. Water flowing from higher level to lower level has kinetic energy which can be converted to mechanical energy and then electrical energy. Let a volume of water falling per second Q (m^3/s) down a slope. The density of water is ρ (*kg/m^3*) and the mass of water falling per second is ρQ (*kg/s*). The potential energy lost by the falling water per second is $\rho Q g H$ *(watt)*, where '*g*' is acceleration due to gravity and '*H*' is the vertical height of water falling. Thirty-six (36) liters of water (36 kg. of water) falling from a height of 10 m carries only 1 Wh. The energy routes of hydro power generating system are

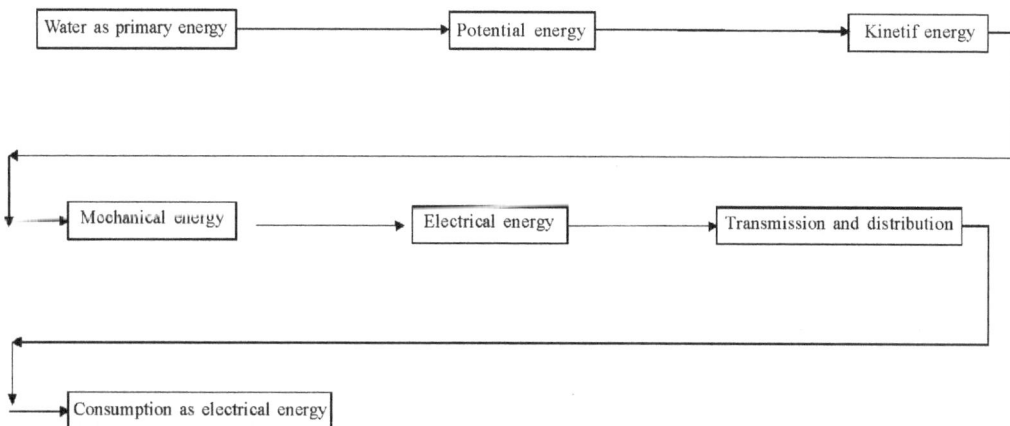

Hydro power is the most advanced and flexible source of power. It is a well-developed and established source of electric power. The early generation of electricity from the year of about 1880, was often derived from hydro turbines. A number of large and medium-sized hydro scheme have been developed. Due to the requirement of huge capital investment and strong environmental concerns about large power plants, only about one-third of the realistic potential has been tapped so far. Hydro installations and plants are long lasting (turbine life is about 50 years). This is due to continuous steady operation without high temperature or other stresses. Therefore, it often produces electricity at low cost with consequent economic benefits.

The global gross electricity production from the hydro sources is about 16.2 % to the total electricity generation during the year 2018. Likewise, hydro source accounts for about 6.47 % of the world's primary energy supply. (Fig. 2.15 and Table 2.12).

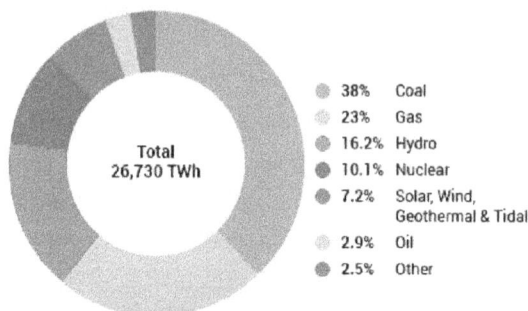

Total 26,730 TWh	38% Coal
	23% Gas
	16.2% Hydro
	10.1% Nuclear
	7.2% Solar, Wind, Geothermal & Tidal
	2.9% Oil
	2.5% Other

Fig. 2.15. World's gross electricity production (2018) **Source:** IEA

Table 2.12: World's primary energy consumption (TWh) during 2018

Sources	Quantity (TWh)	Percentage
Oil	53030	33.22
Natural gas	38400	24.06
Coal	43980	27.55
Nuclear	6690	4.19
Hydro	10340	6.47
Renewables	7150	4.48
Total	**159590**	**100**

Source: BP Statistical Review of World Energy, 2020.

Industrialized counties account broadly for two-thirds and developing countries for one-third of the present hydro power production. Five countries make up more than half of the world's hydro power production as per the data (BP statistical review of world energy 2020) during 2018 i.e. China (1200 TWh), Brazil (390 TWh), Canada (385 TWh), US (290 TWh) and Russia (190 TWh). The worldwide installed capacities of commissioned hydropower plants are respectively 1267 GW, 1291 GW and 1308 GW during the year 2017, 2018 and 2019. The electricity generation from hydropower was 4300 TWh during 2018 (Table 2.13). Norway derives about 90 % of its required electric power from hydro resources. The world's biggest hydroelectric power station is located at China, built over the river Yangtze, capacity 22,500 MW and flooded area 1084 km². The second world's highest hydroelectric power station is at Itaipu, Brazil, built over the river Parana, capacity 14,000 MW and flooded area 1350 km².

Table 2.13: World's gross electricity production and consumption (during 2018)

Production from sources	Percentage	Consumption of electricity by sectors	TWh	% of electricity consumption by sectors	World's total electricity production (TWh)
Coal	38.0	Industry	9362	41.95	26,730
Oil	2.90	Transport	390	1.75	
Natural gas	23.0	Residential	6008	26.92	
Hydro	16.20	Commercial and public services	4799	21.50	
Nuclear	10.10	Others	1757	7.87	
Renewable energy	9.80				
Total	**100**		**Total**	**22316**	**100**

Source: International Energy Agency (IEA), 2019.

2.18 NUCLEAR RESOURCES

Nuclear resources refer to the conversion of the energy content of a nucleus in an atom to electrical energy. In general, electrical energy can be generated from (i) burning of mined and refined energy sources such as coal, natural gas, oil, and nuclear fuel and (ii) harnessing energy sources such as hydro, biomass, wind, geothermal, solar, and wave resources. Today, the main sources for electrical energy generation are (i) thermal power, primarily using coal and secondarily natural gas (ii) "large" hydro power plant and (iii) nuclear power from nuclear power plants. The balance of the energy sources is from using solar, wind, biomass and geothermal resources whose utilization are increasing day by day in some countries.

The process that is used currently to generate nuclear energy is called fission reaction in which, a nucleus splits into smaller nucleus and results in release of energy. Currently, the reactors that are used for fission reactions are called 'Burner reactors' which use Uranium isotope U^{235}, which is less than one per cent of the naturally occurring uranium. U^{235}, U^{233} (isotopes of uranium) and Pu ^{239}(plutonium) are used as nuclear fuels in nuclear reactors (thermal reactors) and are known as fissile (or fissionable) materials. Out of these, only U^{235} occurs in nature. U^{233} and Pu^{239} are produced from Th^{232} (Thorium) and U^{238} respectively in Fast Breeder Reactors (FBRs). Th^{232} and U^{238} are known as fertile materials. Fissionable (fissile) materials are the materials that can undergo fission reaction with high energy (fast) neutrons. Fertile materials are the materials that can be transformed into fissile materials by the bombardment of neutrons inside a reactor. Fertile is a term used to describe an isotope that is not itself fissile (it cannot simply undergo fission by thermal neutrons), but can be converted into a fissile material through bombardment with neutrons in a nuclear reactor. Naturally occurring uranium contains three isotopes, U^{234}, U^{235} and U^{238}. The relative percentages of these isotopes are U^{234} (0.006 per cent), U^{235} (0.711 per cent) and U^{238} (99.283 per cent). Of these isotopes, only U^{235} undergoes spontaneous fission when subjected to bombardment by slow neutrons. It is the only naturally occurring fissile material. It is estimated that the complete fission of 1 kg of uranium^{235} releases about 6.73×10^{10} kJ of heat, which is more than the heat released when 3000 tons of coal are burned. Therefore, for the same amount of fuel, a nuclear fission reaction releases several million times more energy than a chemical reaction. The safe disposal of used nuclear fuel, however, remains a concern.

Though conversion of nuclear energy into electricity does not pollute the environment, it has some other negative effects, one of them is how to dispose of the nuclear waste. The nuclear waste has some radioactive elements and cannot be left open. Also there are possibilities of accidents associated with the nuclear power plant, where the radioactive elements are released into the environment, which are harmful for human beings.

The nuclear option

Under the nuclear option, the two alternatives are (i) The breeder reactor and (ii) Nuclear fusion.

The breeder Reactor

In order to understand the working of a breeder reactor, it is necessary to describe the fission reactions in a little detail. The fission reaction which occurs is shown in Fig. 2.16. The break-up of U^{235} when subjected to neutron bombardment yields fission products, neutrons and the release of a large amount of energy as heat (6.73×10^7 kJ per g of U^{235}). The neutrons are slowed down by a moderator, and are used to bombard the U^{235} nucleus again, thereby setting up a controlled chain reaction. A nuclear reactor working on the above reaction and utilizing only U^{235} is called a burner reactor. All reactors working in present day's nuclear power plants are essentially burner reactors and thus of no use for the abundant U^{238} isotope present in naturally occurring uranium.

Although U^{238} is not a fissile material, it is a fertile material, i.e. it can be converted by neutron bombardment into a fissile material, plutonium^{239}. Similarly, naturally occurring thorium^{232} is also a fertile material. It can be converted into U^{233} which is a fissile material Both these reactions are shown in Fig. 2.17.

$U^{235} \longrightarrow$ (fussion Products) + (neutrons) + (energy released as heat)

Moderator to slow
down neutrons

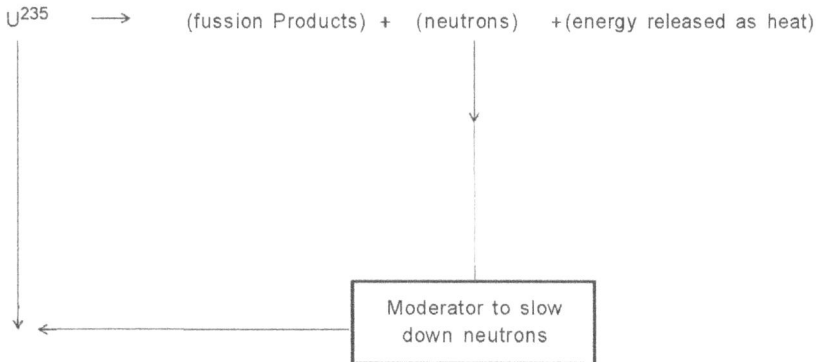

Fig. 2.16. Fission reaction of U^{235}

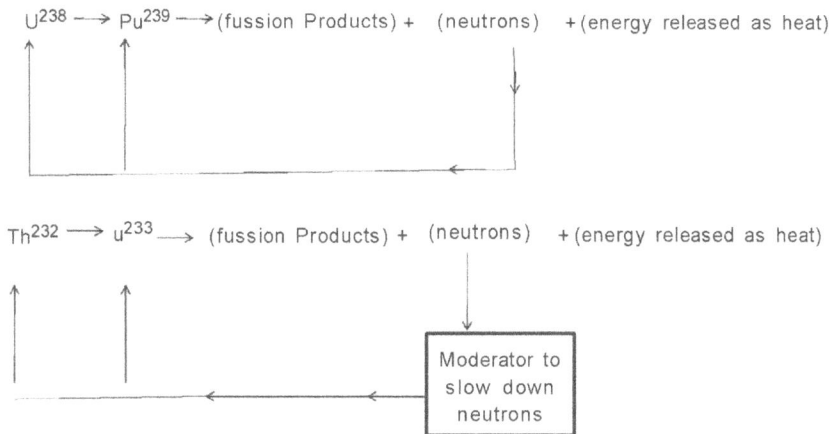

$U^{238} \dashrightarrow Pu^{239} \dashrightarrow$ (fussion Products) + (neutrons) + (energy released as heat)

$Th^{232} \longrightarrow U^{233} \longrightarrow$ (fussion Products) + (neutrons) + (energy released as heat)

Moderator to
slow down
neutrons

Fig. 2.17. Breeder reactions for U^{238} and Th^{232}

The neutrons generated by the fission reaction serve two purposes. They help in converting a fertile material to a fissile material and also sustain the fission reaction for the fissile material formed. The above reactions are called breeder reactions if they produce more fissile material than they consume, and the nuclear reactor in which they are caused to occur is called a breeder reactor. Breeding is achieved by having both fissile and fertile materials in the reactor core under the conditions which provide enough neutrons to propagate a chain reaction in the fissile material as well as to convert more fertile material into fissile material than was originally present. Thus, a breeder reactor working on the U^{238} to Pu^{239} cycle utilizes naturally occurring uranium almost completely. Similarly, a breeder reactor working on the Th^{232} to U^{233} cycle helps in utilizing the vast thorium resources of the world.

Reactor working on various breeder cycles have been built. However, the major effort has been on cooling on liquid-metal and the fast breeder reactors working on the U^{238} to Pu^{239} cycle. In the seventies, it appeared that breeder reactors would be in commercial operation by the turn of the century. However, fears of nuclear accidents, difficulties associated with radioactive waste disposal and the possibility of plutonium being misused for weapons have caused the breeder development programme to be closed or to be slowed down in some countries. Breeder reactor technology is not well developed and not currently used for producing significant amount of electricity.

Nuclear Fusion

In nuclear fusion, energy is released by joining very light atoms. If the current research on controlled fusion is eventually successful, and fusion reactors are built, they could provide the ultimate solution to the world's energy problem.

The reactions of interest involve the fusing of the heavy isotopes of hydrogen (deuterium D and tritium T) into the next heavier element, viz, helium. They are as follows

$$D + D \rightarrow T + p + 4.0 \text{ Mev} \tag{I}$$

$$D + D \rightarrow {}^3He + n + 3.2 \text{ Mev} \tag{II}$$

$$D + T \rightarrow {}^4He + n + 17.6 \text{ MeV} \tag{III}$$

$$\underline{D + {}^3He \rightarrow {}^4He + p + 18.3 \text{ MeV}} \tag{IV}$$

$$6D \rightarrow 2\ {}^4He + 2p + 2n + 43.1 \text{ Mev} \tag{V}$$

Equations (I) and (II) show that two nuclei of deuterium can fuse in two ways. Both ways are equally probable. In the first, tritium and one proton are formed, while in the second, helium-3 and one neutron are formed. The energy released by the fusion reaction is as indicated. Tritium is unstable and combines with deuterium to form helium-4 and one neutron (Eq. III), while helium-3 combines with deuterium to form helium-4 and one proton (Eq. IV). The net result is the addition of all the four reactions (Eq. V). It indicates that six deuterium nuclei are converted to two helium-4 nuclei, two protons and two neutrons with an energy release of 43.1 MeV. Eqs (I) and (II) are referred to as the D-D reactions, Eq. (III) as the D-T reactions and Eq. (IV) as the D- ${}^3He$ reaction.

Deuterium occurs naturally in sea water and it is estimated that the fusion of all the deuterium in just one cubic metre of sea water would yield an energy of 12×10^9 kJ. Thus the fusion of deuterium represents an essentially inexhaustible source of energy.

Fusion reaction are much more difficult to achieve in practice because of the strong repulsion between the positively charged nuclei, called the coulomb repulsion. To overcome this repulsive force and to enable the two nuclei to fuse together, the energy level of the nuclei must be raised by heating them to about 100 million ${}^0C$. High temperature is essential to force the lighter nuclei to fuse with each other and the fusion process to occur. But such high temperatures are found only in the sun/stars or in exploding atomic bombs (the A-bomb). In fact, the uncontrolled fusion reaction in a hydrogen bomb (the H-bomb) is initiated by a small atomic bomb. The uncontrolled fusion reaction was achieved in the early 1930s, but all the efforts since then to achieve controlled fusion by massive lasers, powerful magnetic fields, and electric currents to generate power have failed. Hence, the key problems in the development of a nuclear fusion reactor are the attainment of the required high temperature by initially heating the fuel charge and the confinement of the heated fuel for a long enough time for the reaction to become self-sustaining.

Of the reactions given in Eqs. (I)-(IV), the D-T reaction takes place at the lowest temperature, at about 10^7 K. It is therefore the least demanding in terms of its heating and confinement requirements. Attention has been focussed on using this fact. It has been suggested that since tritium does not occur naturally, the D-T reaction should be supplemented by one using, lithium as follows

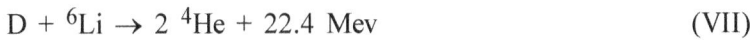

$$D + \ T \ \rightarrow \ ^4He \ + n + 17.6 \ Mev \qquad \text{(III)}$$

$$n + \ ^6Li \rightarrow \ ^4He \ + T + 4.8 \ Mev \qquad \text{(VI)}$$

$$D + \ ^6Li \rightarrow 2 \ ^4He + 22.4 \ Mev \qquad \text{(VII)}$$

Thus it appears likely that the D-T reaction in association with lithium will be exploited first for achieving controlled nuclear fusion. However, its use will be limited by the availability of lithium. Intensive research in nuclear fusion is in progress in all the technologically advanced nations. Although no research group has as yet come close to achieving the conditions mentioned.

Nuclear power is based on low-emission technology that can provide base load power. During 2019, there are around 455 nuclear power plants, operational in the world, operating in 31 counties and generating 2586000 GWh of electricity, which is about 10% of the world's electricity. France produces about 71% of its total electrical power by nuclear means. In the European Union as a whole, nuclear energy provides 30% of the electricity. The commercial reactors are usually thermal reactors. Fast breeder Reactors (FBR$_s$) utilize fast neutrons and generate more fissile material than they consume. They generate energy as well as convert fertile material (U^{238}, Th^{232}) into fissile material (Pu^{239} and U^{233} respectively). The breeder technology is not yet commercially developed, the main problem being their slow breeding rate and, therefore, long doubling time (the time required by an FBR to produce sufficient fissile material, to fuel a second identical reactor) of around 25 years. With continuing R and D efforts in this direction, it is hoped that by 2050, FBRs will be the main source of power after overcoming the present difficulties. Nuclear fusion reaction has a lot more potential and vast resources are available, however, controlled fusion reaction has not been achieved yet.

2.19 NON-CONVENTIONAL ENERGY SOURCES

Non-conventional sources of energy are the ones which are not being usually or traditionally used by the human beings in the present context of energy usage. Contrary to these, conventional sources of energy are the ones which the human beings have been using so far to meet most of their energy requirements. These conventional sources of energy have been traditionally used from many years. These are the natural energy resources which are present in a limited quantity and are being used for a long time. They are called non-renewable sources as once they are depleted, they cannot be generated at the speed which can sustain its consumption rate. They have been formed from the decaying matters over hundreds of millions of years. The non-conventional energy is also considered to be renewable source and alternate to the conventional energy. However, non-conventional sources of energy are the energy sources which are continuously replenished by natural processes and these are inexhaustible. These are continually replenished by nature and derived directly from the sun (such as thermal and photo-electric), indirectly from the sun (such as wind, hydropower, and photosynthetic energy stored in biomass), or from other natural movements and mechanisms of the environment (such as geothermal and tidal energy). Fig. 2.18 shows an overview of renewable energy resources. These do not include energy resources derived from fossil fuels, waste products from fossil sources, or waste products from inorganic sources. Technologies for the use of non-conventional sources of energy are presently under the development stage and its share in the energy mix is very small, but increasing day by day. The conventional sources of energy are further classified as commercial and non-commercial sources of energy (fig. 2.19). Coal, electricity, natural gas and petroleum are known as commercial energy since the consumer needs to pay its price to buy them. However, non-commercial sources of energy are generally available cheaply or freely. The examples are such as agro-residues, straw, dried dung, firewood etc.

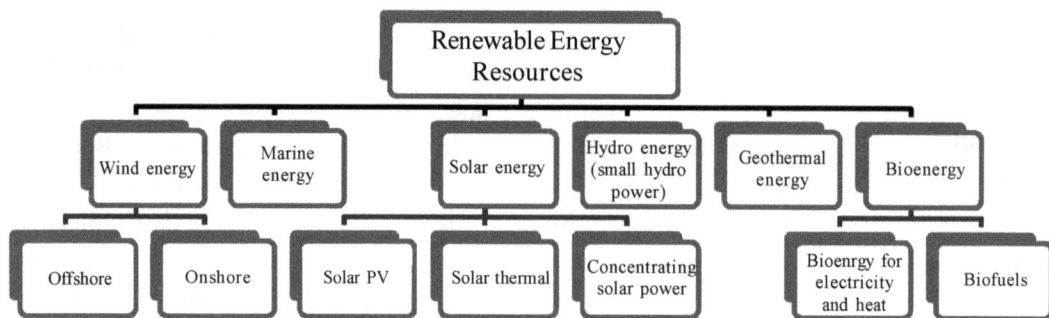

Fig. 2.18. Overview of renewable energy resources

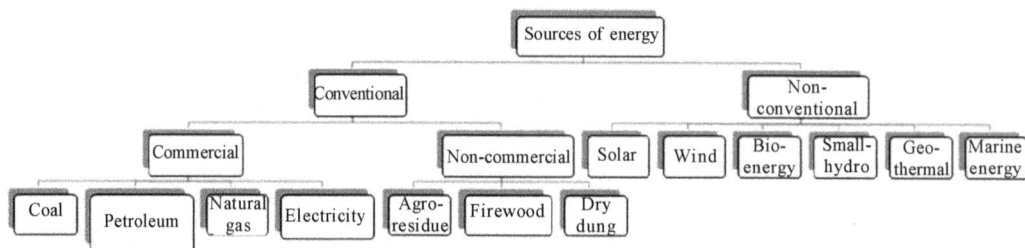

Fig. 2.19. Overview of sources of energy

Based on the above information, some of the key differences between conventional and non-conventional sources of energy are as follows.

Table 2.14: Comparison of conventional and non-conventional energy resources

Conventional resources	Non-conventional resources
Traditional	Non-traditional
These have been in use for many years	These are not in routine use at present
These resources can be easily converted into mechanical energy	These resources require some costly method to be converted into mechanical energy
These are likely to be depleted, that is, these have limited availability	These are non-depletable or may be available in vast quantities
These pollute the environment by emitting harmful gases and also contribute to global warming	These are environment-friendly and don't pollute the environment
Coal, petrol, diesel, nuclear fuels CNG and LPG are conventional energy resources	Solar, wind, biomass, small hydro, geothermal and marine energy are non-conventional energy resources

Solar Energy

Solar energy refers to the energy that comes from the light of the sun in the form of electromagnetic radiations. The sun has played a dominant role since time immemorial for different natural activities in the universe at large and in the earth in particular for the formation of fossil and renewable energy sources. Almost all the renewable energy sources originate directly and indirectly from the sun. Solar energy is referred to as renewable and/or *sustainable energy* because it will be available as long as the sun continues to shine. The earth continuously intercepts the solar power of 1.73×10^{17} W, which is equivalent to 1.5×10^{18} kWh annually. Taking into the account the annual global primary energy consumption to be 584 exa-joules (163×10^{12} kWh during the year 2019), the solar energy received by the earth is about 10,000 times greater than that is consumed on the earth annually. Likewise, considering this energy to be uniformly spread all over the world's surface (radius of earth, 0.63×10^{7} m), the amount of solar radiation incident on the outside of earth's atmosphere is about 1360 W/m^2. Taking into the account of reflection and absorption of incoming solar radiation in the earth's atmosphere, the solar radiation received on the surface of the earth on a bright sunny day at noon is approximately 1 kW/m^2 in the tropical regions.

Solar energy is the prime and free source of inexhaustible energy available to all. Power from it, is generated either in photovoltaics (PV) or in the concentrated solar power (CSP) systems. In PV, sunlight is directly converted into electricity when it is incident on the solar panels. The generated power from the panels can be used either in decentralised or centralised manner for wide scale applicability and one of the most important modern renewable energy technologies. CSPs are the application of solar thermal energy and use mirrors to collect solar rays which heat fluid, and the resulting steam moves a turbine to produce electricity. The solar thermal (STE) power plants are technically similar to conventional coal-based power plants while the concentrating photovoltaics (CPV) systems concentrate sunlight onto photovoltaic surfaces for power generation. STE technology collects the solar radiation by reflection and thereafter harvests the heat energy of rays by the collector and the convection devices. Based on the scale of attaining temperatures, three broad categories of collectors are recognised, (i) low temperature collector, designed as flat-plate, such collectors are mostly used for drying or heating water in homes, commercial buildings and swimming pools (ii) medium temperature collectors, comprising of a group of advanced flat-plate collectors or evacuated tube collectors, such devices acquire temperature above $100\ ^{0}C$ for large commercial complexes or even residential area with higher energy demands and (iii) high temperature collectors, capable of concentrating large volume of radiations, hence suitable for solar power plants. In addition, solar energy does not pollute and harm the environment while it could provide energy security, since sun light is available everywhere in the earth. Even though the sun generates an infinite amount of energy, the global power generated from solar PV during 2019 is about 2.6 percent (720 TWh solar PV against total electricity production in the world as 27000 TWh). The reason for that is related to its cost, since photovoltaic panel installation still is expensive, despite the new technological developments and even though the cost has decreased a lot in the last decade. At present (during the year 2019), the capital cost of a solar PV system is about Rs. 75/W as against Rs. 60/W for coal-fired thermal plant.

Wind Energy

Wind energy is an indirect form of solar energy. Solar irradiation causes temperature differences on the earth and these are the origin of winds. Temperature differences between different locations on the surface of the earth and between different altitudes create pressure differentials which cause the airflows thus creating wind. These temperature differences result from differential solar heating of the earth, which varies from region to region and with the time of day. Heated air has a tendency to rise and be replaced by colder air. This leads to wind, which carries energy. Around 40 percent of the kinetic energy possessed by the moving air can be tapped to do useful work. It has been estimated that the total solar power received by the earth is approximately 1.73×10^{17} W. Of this solar input, only 2% (i.e. 3.5×10^{15} W) is converted into wind energy and about 40 % of wind energy (1.2×10^{16} kWh) is dissipated within 1000 m altitude of the earth's surface. This much energy is about 70 times more than the annual global primary energy consumption of 163×10^{12} kWh as per the data during the year 2019. The installation cost of wind power plant is Rs. 105/W (which is comparable to that of conventional thermal plants) (IRENA 2019). There has been remarkable growth of wind-power installation in the world. Wind-power generation is the fastest growing energy source. Wind-power installation capacity in the world is 590 GW during 2018, which is about 1% of the world's electrical power generation capacity (5910 GW). Wind (1290 TWh) supplies about 5% of worldwide electrical generation (27000 TWh) as per the data during the year 2018. It accounts for approximately 23 % of electricity production in China, 20 % in US, 8 % in Germany, 3.8 % in UK, 3.75 % in Spain and 3.5 % in India from the total electricity generation of the wind energy source. China is the world leader in wind power with an installed capacity of 236 GW.

Bioenergy

Biomass energy or bioenergy refers to that form of energy which is produced from the conversion of biomass, that is originated by the sunlight through the photosynthesis process. The process is sunlight + CO_2 + H_2O → CH_2O (carbohydrate) +O_2. In this process, carbohydrates are produced. Oxygen, required for breathing by all living animals, is also produced in the photosynthetic process. For each mole of CH_2O produced (30 g), this reaction absorbs 500 kJ of energy (energy stored ≈ 16 MJ/kg, the heat value of dry biomass). Biomass is therefore an organic material which stores solar energy in the form of chemical energy and can be considered as natural solar cell. More specifically, food crops, residues from agriculture and municipal or industrial wastes, grassy or woody plants, and even methane fume from landfills can be used for biomass energy production. Biomass is therefore a living or dead biologic material which can be used as a source of energy. It is a source of carbon supplying energy when burnt. Synthetic fuels can be made from the biomass as a substitute of fossil fuels. The biomass inventory is dominated by plant matter which accounts for about 90% the total but also includes animal matter and biodegradable wastes. Combustible wastes are included in the biomass inventory provided they are derived from living or dead biologic materials. This is the case for wood and crop wastes, animal materials and wastes, and even black liquor (the alkaline spent liquor from the digesters used in paper manufacturing). Also included are the municipal wastes coming from residential, commercial, and public service sources. The latter encompasses hospital wastes, which are biodegradable and renewable resource because of their continuous availability from the human beings as long as they exist on the earth. Biomass material may be transformed by chemical or biological processes to produce intermediate bio-fuels such as biogas (methane), producer has, ethanol, biodiesel and charcoal. Hence, the bioenergy can be obtained from the following sources:

Crops: corn, sugarcane, potatoes, beets, wheat, sorghum

Oilseed crops: largest source of fats and oils are cottonseed, soybean, rapeseed (canola), palm oil; minor sources are sunflower, peanuts, flax, safflower, sesame, jatropha, chinese tallow, castor

Agriculture residues: bagasse from sugarcane, corn fiber, rice straw and hulls, nutshells

Major research area: production of ethanol from cellulose

Wood: sawdust, timber slash, mill scrap, paper trash, fast-growing trees like poplars and willows

Municipal solid waste (MSW)

Hospital solid wastes

Animal wastes

Grasses: fast-growing biomass like switch grass, elephant grass, and prairie bluestem

Methane: landfills, municipal wastewater treatment, manure, lagoons from confined animal feeding

Biomass is not only useful for producing energy, it is also valuable as a raw feedstock for building materials, paper, biodegradable plastics, and so on. It can as well be used to produce basic chemical compounds for organic chemistry applications and development of new materials. As commonly used, the term *'biomass'* does not include organic fossil fuels, coal, oil, and gas, which were formed hundreds of millions of years

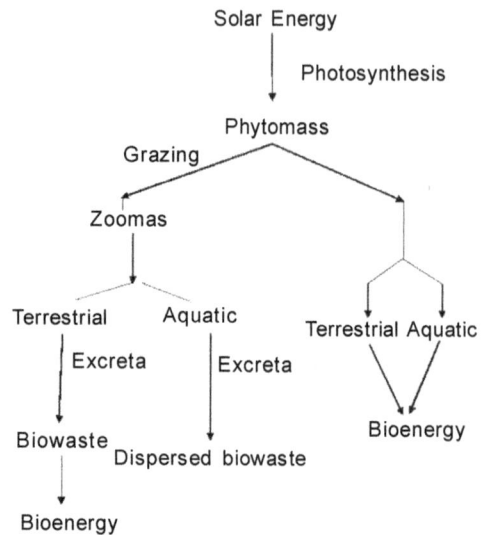

Fig. 2.20: The solar energy and biomass energy pathways in earth

ago. Biomass is a renewable source because the timescale for harnessing and producing again through plantation is within the timescale of a human life. This is not the case for fossil fuels, which require millions of years to be created not within a year or short period.

Biomass is the general term used to include phytomass or plant biomass and zoomass or animal biomass. The biomass energy availability in the earth is shown in the fig. 2.20.

Out of 1.2×10^{17} W, solar power incident on the earth's surface, 0.1 % (1.2×10^{14} W) i.e. 120 terra watt power is utilized for photosynthesis purposes. The biomass production resulting from photosynthesis is basically carbohydrate with energy content of 3744 cal/g ($\approx$ 16 MJ/kg). Hence, biomass production per year in earth's surface =

$1.2 \times 10^{14} \frac{J}{s} \times \frac{1\,cal}{4.18J} \times 3.15 \times 10^7 \frac{s}{year} \times \frac{1g}{3744\,cal} = 240 \times 10^9 \frac{t}{year}$ and incorporating about

100×10^9 t/year of atmospheric carbon. Out of the available solar power for photosynthesis purposes, around 50 % is absorbed by the plants, hence the solar energy absorbed by plants (through the photosynthesis process) in the biosphere is estimated to be about 2 $\times$

10^{21} J/year $\left(\approx 0.5 \times 1.2 \times 10^{14} \frac{J}{s} \times 3.15 \times 10^7 \frac{s}{year} \right)$. Of this, about 0.5 % by weight is the

biomass used for human food. Taking into the account the annual global primary energy consumption to be 584 exa-joules (during the year 2019), the biomass produced in the earth is about 4 (2×10^{21} J/584 exa-joules) times greater than the present day world'sannual energy consumption.Around 9.3 % of the world's total energy supply by fuels is from the biofuels (Table 2.15).

Table 2.15: Total primary energy supply by fuels (as per year 2018 data)

S.N.	Fuels	Percentage
1	Coal	26.9
2	Oil	31.5
3	Natural gas	22.8
4	Nuclear	4.9
5	Hydro	2.5
6	Biofuels	9.3
7	Other renewable energy	2.1

Likewise, the contribution of biofuels towards world's electricity production is 637 TWh (about 2.5 % of total world's electricity production) as per data during 2018.

Geothermal Energy

Geothermal energy is the energy available in the interior of the earth. This energy can be extracted from the earth's interior in the form of heat. Billions of years before the evolution of life began on the earth, the earth was a ball of fire. Then gradually, the outer surface (crust) started cooling sufficiently to allow the life to begin. But in the earth's interior, very vast quantities of heat was trapped and stored. This stored heat energy in the earth's interior, when extracted is called the geothermal energy. This vast reservoir of energy consists of about 10 % of the remaining heat, trapped in the interior during the formation of earth and the major part (90%) results from the energy released in the decay of radioactive elements such as uranium, thorium, and potassium which are contained in the earth. A geothermal resource can be simply defined as a reservoir inside the earth from which heat can be extracted economically (cost-wise less expensive than or comparable with other conventional sources of energy such as hydroelectric power or fossil fuels) and utilized for generating electric power or any other suitable industrial, agricultural or domestic applications. Volcanoes, geysers, hot springs and boiling mud pots are the visible evidences of the great reservoirs of heat that lie beneath the earth. Although the amount of thermal energy within the earth is very large, the useful geothermal energy can be extracted only at certain suitable sites. It is not feasible to access and extract heat economically from very deep locations. Where it is available near the surface and is relatively more concentrated, its extraction and use may be considered feasible. These sites are known as the geothermal fields. Because of non-homogeneities in the earth's crust, the earth's surface is not flat everywhere, there are hills and valleys. Due to geological changes, molten rocks formed in the deeper hot regions of earth's crust are pushed upward and trapped in certain regions called 'hot spots'. Due to this, there are some places where the hot spot is just beneath the surface where temperature is in fact much higher than the average value expected. At such places, when ground water comes in contact with the hot rocks, either wet or dry steam are formed. A well drilled to these locations causes the dry or wet steam to emerge at the surface where its energy can be utilized either for generating electricity or for space heating. Geothermal energy is considered to be a renewable source of energy because the heat extraction is small compared with the earth's heat content. Further, the energy can be extracted without burning fossil fuels such as coal, gas or oil, resulting in less harmful emissions. Hence, huge quantity of heat energy in the earth's interior plus the heat generated by the radioactive decay of radioactive isotopes such as K_{40}, Th_{232}, and U_{235} which are abundant in the earth's interior are the forms of geothermal energy and considered to be a renewable energy source.

The geothermal heat therefore occurs from a combination of two sources i.e. (i) the original heat produced during the formation of earth by gravitational collapse (contraction due to gravitational force tending to draw matter inward towards the centre of gravity) and (ii) the heat produced by the radioactive decay of various isotopes. As the thermal conductivity of rocks is very low, it is taking many billions of years for the earth to cool. The geothermal gradient is not the only important factor which determines accessibility of geothermal energy. The permeability of the rocks is also an important parameter because it determines the rate at which heat can be conducted to the surface. The average energy flux for conduction of geothermal heat to the surface is 0.06 W/m^2, which is negligible as compared to power incident to the earth surface, particularly for solar energy (~ 1 kW/m^2). We are fortunate that the geothermal heat flux is not much larger otherwise the earth's surface would be much hotter. The earth's core is surrounded by a region known as 'mantle', which consists of a semi-fluid material called the 'magma'. The mantle is finally covered by the outermost layer known as 'crust', which has an average thickness of about 30 km. The temperature in the crust increases with the depth at a rate of 30 $^0C/km$. The temperature at the base of the crust is about 1000 ^0C and then increases slowly into the core of the earth. The temperatures of the core of the earth are as high as 4000 ^0C. In this true furnace, the thermal energy is continued to be produced by the decay of radioactive materials within the interior. For this reason, some scientists refer to the geothermal energy as a form of 'fossil nuclear energy'. A section through the earth is shown in fig. 2.21. Though the general distribution of layers is as shown in the fig. 2.21, at certain locations, anomalies exist in the composition and structure of these layers. There are regions in which the hot molten rock (magma) of the mantle is pushed up through faults and cracks towards the surface. In an active volcano, the magma actually reaches the surface, but more often 'hot spots' occur at moderate depths (within 2 to 3 km), where the heat of the magma is conducted through an overlaying rock layer. We see the evidence of such 'hot spots' in volcanic eruptions, geysers and bubbling mud holes. In fact, the zone of likely geothermal sites corresponds roughly to the regions of earthquake and volcanic activities. These regions are at the junctions of *tectonic plates* that make up the earth's crust. These plates are in in a state of constant relative motion (at rates of several centimetres per year). Where they collide or grind, there are very strong forces that cause earthquakes. It is near the junction of these plates that heat travels most rapidly from the interior via subsurface magma to surface volcanoes. Geothermal energy potential of these regions is great, owing to increased temperature gradients of about 100 ^{0}C/km. Moderate increase in temperature gradient to about 50 ^{0}C/km occur in localized region away from plate boundaries, owing to anomalies in the composition and structure of crust.

Each year more than 100,000 TWh of heat energy is conducted from inside the earth to the surface. The total outward flow is around 30 TW. The geothermal flux is however only about one-thousandth of the incoming flux of solar energy incident on the earth (1.2 × 10^{17} W). The Earth's heat content is about 1×10^{19} TJ (2.8 ×10^{15} TWh) (Wikipedia 2019). This is a huge amount of energy, enough to supply our energy needs at current rates

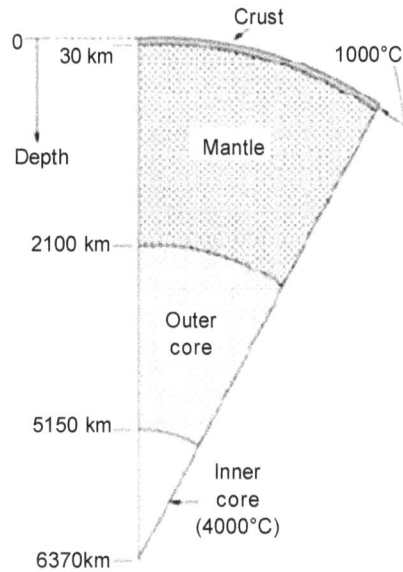

Fig. 2.21. Cross section of the earth

for 3,50,000 years and hence considered to be inexhaustible. This heat naturally flows to the surface by conduction at a rate of 44.2 TW and is replenished by radioactive decay at a rate of 30 TW. These power rates are more than double humanity's current energy consumption from primary sources, but most of this power is too diffuse (approximately 0.1 W/m^2 on average) to be recoverable. The Earth's crust effectively acts as a thick insulating blanket which must be pierced by fluid conduits (of magma, water or other) to release the heat underneath. Because of low grade thermal energy, its economic recovery is not feasible everywhere on the surface of the earth. Practically, it is not the size of the resource that limits its use but the availability of technology that can tap the resource in an economic manner. Its overall contribution in total energy requirement is negligible. However, it is very important resource locally. At the end of 2019, the world's total installed electrical power-generating capacity from geothermal resources was about 15,406 MW$_e$ and direct thermal-use installed capacity was 107,727 MW$_t$. Globally, use of geothermal power is growing annually at a rate of about 3.8 % electrical and 10 % thermal. The island of Hawaii procures 30 % of its electricity from geothermal resources. Likewise, geothermal electrical energy production in El Salvador is 25 % of the country's total installed electricity-generation capacity. The oldest geothermal power generation is located at Lordarello in Italy, commissioned in 1904 and presently (during 2019) producing 944 MW$_e$ of power.

Ocean Energy

The oceans in the world constitute a vast natural reservoir for receiving and storing the energy of the sun incident on the earth. Oceans in the globe cover around an area of 361×10^6 km^2, occupying about 71% of the surface of the earth and containing 97% of the total water available on the earth. The average depth of the ocean is 3800 m, much larger than the average height of the land above sea level, 850 m. The total volume of seawater is around 1335×10^6 km^3. Hence, oceans are the huge reservoirs of various useful and renewable energy sources. The various types of energy resources which ocean can provide are as (i) the tides of the ocean can be used to generate electricity; (ii) the wind produces large waves in the ocean having high kinetic energy which can be converted into electric power and (iii) the temperature gradient from the surface of ocean to the greater depth inside the ocean can be used to provide thermal energy to generate electricity. Tides and waves produce mechanical energy whereas temperature difference produces thermal energy. The technologies available as on today for harnessing power from the tidal energy is relatively more developed compared to the other two, which are still under evaluation and in the initial development stages. The main disadvantages common to all of them are (i) low energy density and (ii) their occurrence at distances remote from the consumption centre.

(i) *Tidal energy*

Tidal energy is a form of hydro power that converts energy of ocean tides into electricity or other useful forms of power. Tides are produced by gravitational attraction of the moon and the sun acting on the rotating earth. Tides are periodic rise and fall of the water level in the oceans due to various positions of the sun and rotating moon. The ocean level difference caused due to tides contains large amount of potential energy. The highest level of tidal water is known as flood tide or high tide. The lowest level is known as low tide or ebb. The level difference between high and low tide is known as tidal range. The tidal range varies greatly with the locations. Only sites with large tidal ranges (about 5 m or more) are considered suitable for power generation. Tides, the daily rise and fall of ocean levels, are the result of the gravitational force of the moon and sun as well as the revolution of the earth, the moon and the sun both exert a gravitational force of attraction on the earth. The magnitude of the gravitational attraction of an objects is dependent upon the mass of an object and its distance. The moon exerts a larger gravitational force on the earth (about 70 per cent of the tide-producing force), because, although it is much smaller in mass, it is very closer to earth's surface than the sun. The effect of sun is about 2.2 times less than the moon. The rise and fall of the ocean water level are used in energy conversion process. when water rises, it is captured in an artificial basin, which is then allowed to flow back during the low tide. The escaping water is used to drive water turbines which in turn drive electrical generators. The basic techniques are to block estuaries with a barrier, forcing the water to flow through turbines in order to generate electricity. An alternative is to couple a dam and a storage basin for water. When the tide rises, the dam is open and seawater fills the basin. The dam is closed when the tide goes down and the water which is contained in the basin can be released to flow through a turbine and produce electricity. The principle of tidal power generation is shown in Figure 2.22.

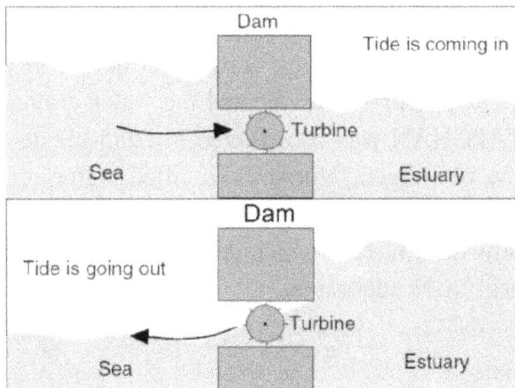

Fig. 2.22. Principle of tidal power generation with sea separated from estuary by a dam

It is in the developing stage and although not yet widely used, tidal power has potential for future electricity generation. Tides are more predictable than wind energy and solar power. There are at present only a few operational tidal power plants. The first and the biggest, a 240-MW tidal power plant was built in 1966 in France at the mouth of the La Rance river, near St. Malo on the Brittany coast. A 20-MW tidal plant is located at Nova Scotia, Canada, and a 400-kW capacity plant is located at Kislaya Guba, near Murmansk, Russia, on the Barents Sea. Many sites have been identified in USA, Argentina, Europe, India and China for development of tidal power.

(ii) *Wave energy*

The ocean waves are caused by wind, which in turn is caused by uneven heating of land masses and water bodies and the rotation of the earth. Energy from the surface wind is transferred to the ocean to form waves. The rate of energy transfer depends upon the wind speed and the distance over which it interacts with water. The energy flux in waves is more than that available from solar, wind and other renewable sources. The wave energy depends upon amplitude of the wave and the period of motion. Wave power is commonly expressed in kW per meter, that is the rate of energy transfer across a line of 1 meter length parallel to the wave front. Wave energy consists of kinetic energy resulting from the propagation of waves and potential energy resulting from lifting of water mass with respect to mean sea level. Waves after formation continue to travel even when wind dies down. This is the reason why a long swell of wave can be seen even when sea is calm. Waves are caused by surface winds, whereas tides are caused by the gravitational forces of moon and sun on ocean water.

Wave motion is primarily horizontal, but the motion of the water particles is primarily vertical. Mechanical power is obtained by floats making use of the motion of water. The concept visualizes a large float that is driven up and down by the water within a relatively stationary guide. This reciprocating motion is converted to mechanical and consequently the electrical power (fig. 2.23). Another option is when a moving wave is constricted, a surge is produced raising its amplitude. Such a device is known as tapered channel device (TAPCHAN). TAPCHAN is a wave energy conversion device in which a funnel directs the incoming wave into a temporary reservoir and the water drains out through a turbine house (fig. 2.24). The TAPCHAN was constructed in 1985 for demonstration purpose at Toftestallen (40 km N.V. of Bergen, Norway). It lifted water up into a reservoir 3 m above the sea level. At the back side of the reservoir, the water returned to the ocean through a conventional low head turbine which driving a 350 kW generator. The generator was connected to the grid in Nordhorland.

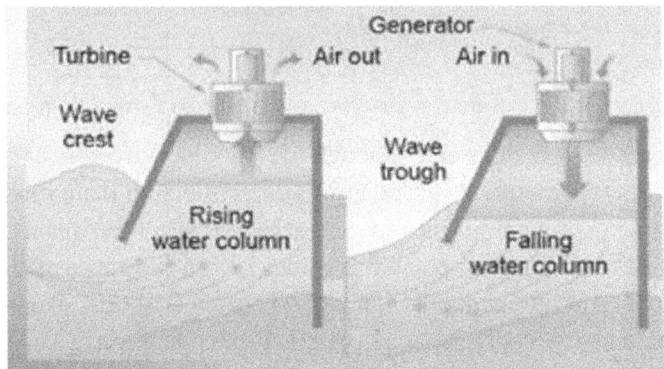

Fig. 2.23. Oscillating water column

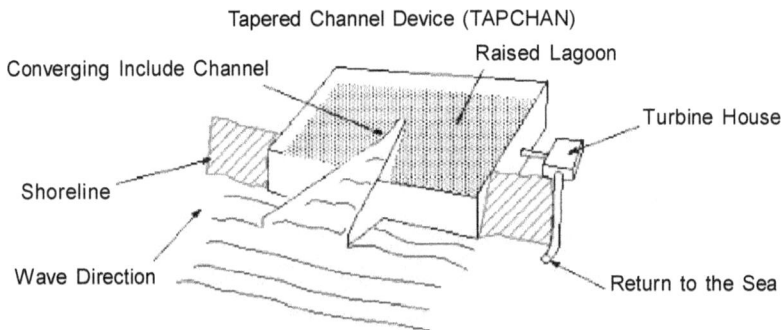

Fig. 2.24. Tapered channel device

Wave power refers to the energy of ocean surface waves and the capture of that energy to do useful work. Good wave power locations have a flux of about 50 kilowatts per metre of shoreline. As per an estimate, the potential for shoreline-based wave power generation is about 50,000 MW. Deep-water wave-power resources are truly enormous, but perhaps impractical to capture. Some wave plants are recently deployed at few places. The world's first 2250-MW, commercial wave farm is based in Portugal. Other plans for wave farms include a 3 MW plant in the Orkneys, off northern Scotland, and the 20MW wave-hub development off the north coast of Cornwall, England.

(iii) *Ocean thermal energy conversion (OTEC)*

Ocean thermal energy exists in the form of temperature difference between the warm surface water and the colder deep water. A heat engine generates power utilizing a well-established thermodynamic principle, where heat flow from a high temperature source to a low-temperature sink though an engine, converting a part of the heat into work. In this conversion system, the surface water is used as a heat source and the deep water as a heat sink to convert part of heat to mechanical energy and then into electrical energy. A minimum temperature difference of around 20 ^{0}C is required for practical energy conversion. This resource potential is expected to be very high. In this conversion system, a low boiling point liquid such as ammonia, propane or freon can be vaporised into high pressure vapour using the heat of warm water available at the ocean surface into a boiler as shown in fig. 2.25. The liquid vapour is then used to run a turbine coupled with a generator to produce electricity. After expansion in the turbine, the liquid vapour is condensed into liquid in the condenser using cold water from the deep layer of ocean. The condensed liquid is pumped back to the boiler so as to be heated by warm water from the ocean surface. This cycle is repeated. The advantages of OTEC are as (i) power generation is continuous throughout the year (ii) energy is available from nature at no cost (iii) the resource supplies steady power without fluctuation and independent of vagaries of weather and (iv) the availability hardly varies from season to season. The disadvantages are as (i) it has a small temperature gradient which gives a small thermodynamic efficiency and (ii) capital cost in high due to necessity of heat exchanger; boiler and condenser. OTEC technology is still in its infant stages. Conceptual designs of small OTEC plants have been finalized. Their commercial prospects are quite uncertain.

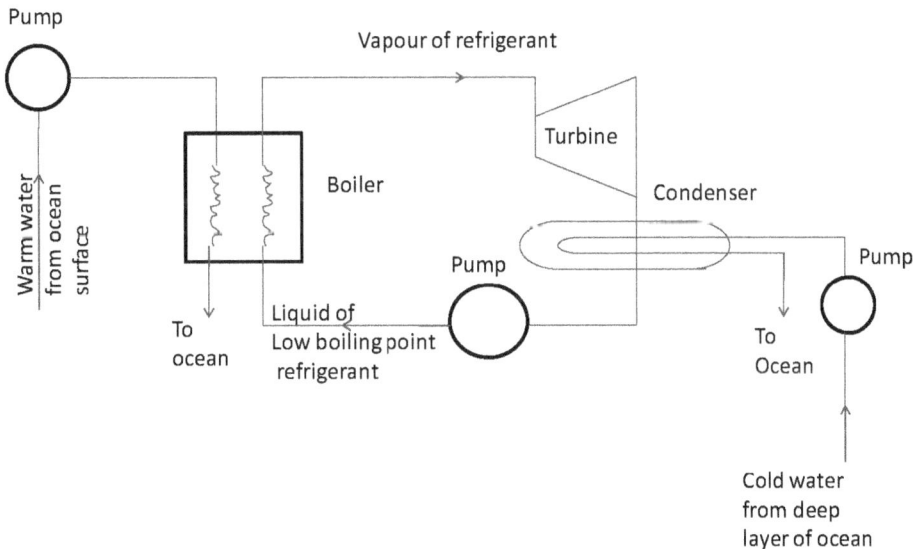

Fig. 2.25. Line diagram of OTEC power plant

Table 2.16: Advantages and disadvantages of different renewable energy resources

S.N.	Energy sources	Advantages	Disadvantages
1	Solar energy	• Potentially infinite energy supply • Causes no air or water pollution	• May not be cost effective • Storage and backup are necessary • Reliability depends on availability of sunlight
2	Wind energy	• Is a free source of energy • Produces no water or air pollution • Wind farms are relatively inexpensive to build • Land around wind farms can have other uses	• Requires constant and significant amounts of • Requires constant and significant amounts of • Wind farms require significant amounts of land • Can have a significant visual impact on landscapes • Need better ways to store energy
3	Biomass energy	• Abundant and renewable • Can be used to burn waste products	• Burning biomass can result in air pollution • May not be cost effective
4	Hydropower	• Abundant, clean, and safe • Easily stored in reservoirs • Relatively inexpensive way to produce electricity • Offers recreational benefits like boating, fishing, etc.	• Can cause the flooding of surrounding communities and landscapes. • Dams have major ecological impacts on local hydrology. Can have a significant environmental impact • Can be used only where there is a water supply • Best sites for dams have already been developed
5	Geothermal energy	• Provides an unlimited supply of energy • Produces no air or water pollution	• Start-up/development costs can be expensive • Maintenance costs, due to corrosion, can be a problem
6	Marine energy	• Ideal for an island country • Captures energy that would otherwise not be collected	• Construction can be costly • Opposed by some environmental groups as having a negative impact on wildlife • Takes up lots of space and difficult for shipping to move around

Table 2.17: Some negative environmental impacts of different renewable energy resources.

S.N.	Energy sources	Potential negative impacts on the environment
1	Solar	Soil erosion, landscape change, hazardous waste
2	Wind	Noises in the area, landscape change, soil erosion, killing of birds by blades
3	Biomass	May not be CO_2 natural, may release global warming gases like methane during the production of biofuels, landscape change, deterioration of soil productivity, hazardous waste
4	Hydropower	Change in local eco-systems, change in weather conditions, social and cultural impacts
5	Geothermal	subsidence, landscape change, polluting waterways, air emissions
6	Marine energy	Landscape change, reduction in water motion or circulation, killing of fish by blades, changes in sea eco-system

2.20 ENERGY AND ENVIRONMENT

The conversion of energy from one form to another often affects the environment and the air we breathe in many ways. Thus, the study of energy is not complete without considering its impact on the environment. Fossil fuels such as coal, oil and natural gas have been powering the industrial development and the amenities of modern life that we are enjoying since 1700s, but these have been creating many undesirable side effects. From the soil we farm, the water we drink and the air we breathe, the environment has been paying a heavy toll for it. Pollutants emitted during the combustion of fossil fuels are causing smog, acid rain, global warming and climate change. The environmental pollution has now reached such high levels that it becomes a serious threat to vegetation, wild life and human health. Air pollution has been the cause of numerous health problems. Hundreds of elements and compound such as benzene and formaldehyde are known to be emitted during the combustion of coal, oil, natural gas in the engines of vehicles, furnaces and even fireplaces. The largest source of air pollution is the motor vehicles and the pollutants released by them are usually grouped as hydrocarbons (HC), nitrogen oxides (NO_x) and carbon monoxide (CO). The HC emissions constitute a large components of volatile organic compounds (VOC) emissions and these two terms are generally used interchangeably for motor vehicle emissions. Hence, environmental problems are continuously growing with respect to emissions of pollutants, hazards and degradation of ecosystem over wider areas. The major areas of environmental problems include (i) water pollution (ii) hazardous air pollutants (iii) ambient air quality (iv) maritime pollution (v) solid waste disposal (vi) land use and siting impact (vii) acid rain (viii) stratospheric ozone depletion (ix) global climate change (greenhouse effects) etc. Among these environmental issues, the internationally most vital problems are the acid precipitation, the stratospheric ozone depletion and the global climate change. In conjunction with this, we will focus on these three concerns in detail.

Acid Rain

Fossil fuels are mixtures of various chemicals, including small amounts of sulphur. The sulphur in the fuel reacts with oxygen to form sulphur dioxide (SO_2), which is an air pollutant. The main source of SO_2 is from the coal based thermal power plant where high-sulphur containing coal is burnt to produce electrical energy. Motor vehicles also contribute to SO_2 emissions since gasoline and diesel fuel also contain small amounts of sulphur. Volcanic eruptions and hot springs also release sulphur oxides. The sulphur oxides and nitric oxides react with water vapour and other chemicals, high in the atmosphere in the presence of sunlight to form sulphuric and nitric acids (fig. 2.26). The acids formed usually dissolve in the suspended water droplets in clouds or fogs. These acid-laden droplets, which can be as acidic as lemon juice, are washed from the air on to the soil by rain or snow. This is known as acid rain. The soil is capable of neutralizing a certain amount of acid, but the amount produced by the power plants using high-sulphur containing coal has exceeded this capability and as a result, many lakes and rivers in industrial areas have become too acidic for fish to grow. Forests in those areas also experience a slow death due to absorption of acids through the leaves, stems and roots of the trees. The major evidences observed in the damages of acid precipitation are such as (i) acidification of lakes, streams and ground waters (ii) toxicity to plants from excessive acid concentration (iii) resulting in the damage to fish and aquatic life (iv) damage to forests and agricultural crops (v) deterioration of materials, e.g., buildings, metal structure, fabrics etc. The magnitude of the problem was not recognized until the early 1970s and the serious measures are taken since then, to reduce the sulphur dioxide emissions drastically by installing scrubbers in power plants and by desulphurizing coal before combustion.

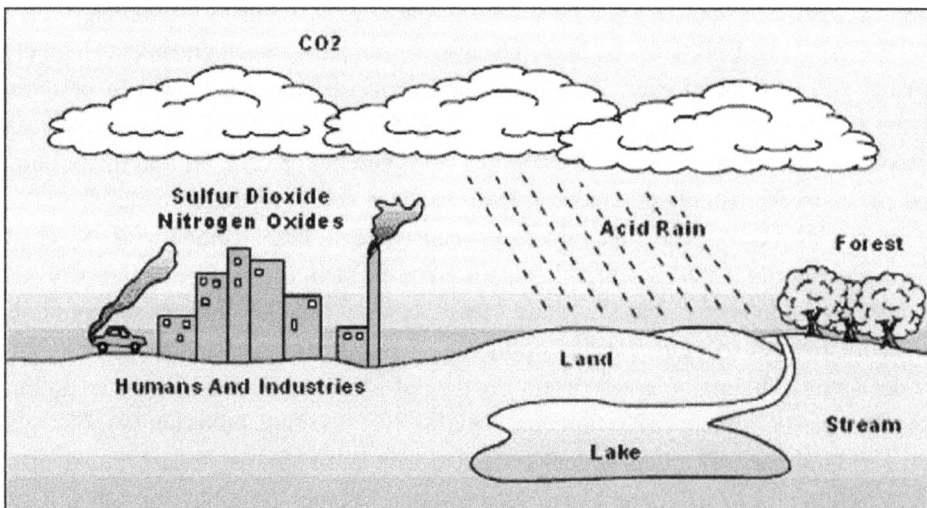

Fig. 2.26 Formation of sulphuric acid and nitric acid when sulphur oxides and nitric oxides react with water vapour and other chemicals in atmosphere

Stratospheric Ozone Depletion

It is well known that the ozone, present in the stratosphere, roughly between the altitudes of 12 to 25 km, plays a natural and equilibrium-maintaining role for the earth through the absorption of ultraviolet (UV) radiation (240-320 nm) (1 nm = 10^{-3} μm or 10^{-9} m) and absorption of infrared radiation. A global environmental problem is due to the regional depletion of the stratospheric ozone layer caused by the emissions of NO_x and CFCs (chlorofluorocarbons) etc. (figure 2.27). Ozone depletion in the stratosphere can lead to the increased levels of damaging ultraviolet radiation reaching the ground, causing increased rates of skin cancer, eye damage and other harms to many biological species. CFCs are the refrigerants, which are used in air conditioning and refrigerating equipment and NO_x emissions are produced by fossil fuels. Hence the major pollutant, NO_x emission, needs to be minimised to prevent-stratospheric ozone depletion.

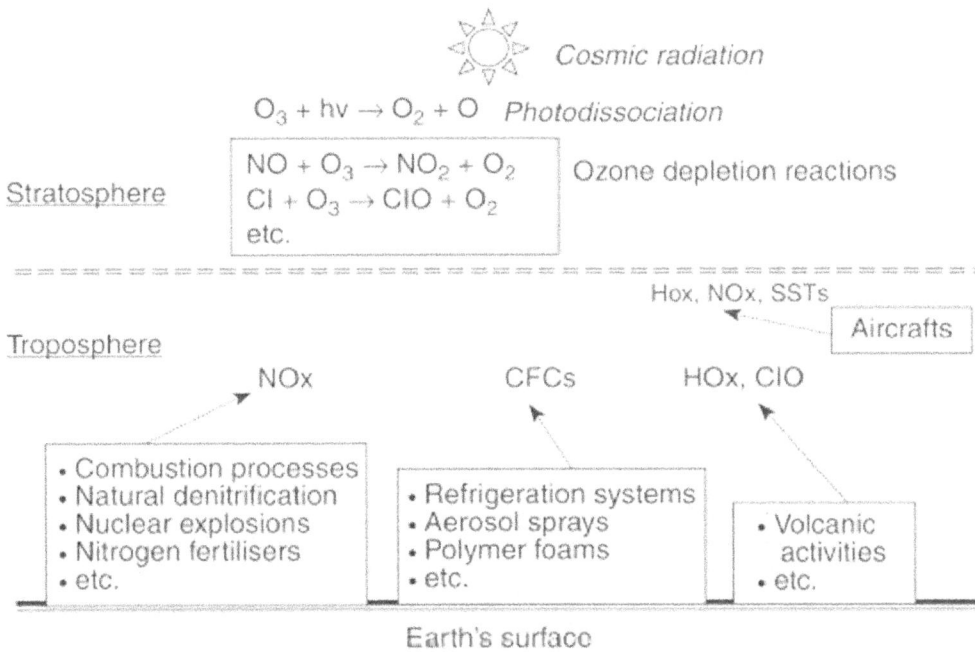

Cosmic radiation

$O_3 + hv \rightarrow O_2 + O$ Photodissociation

Stratosphere

$NO + O_3 \rightarrow NO_2 + O_2$
$Cl + O_3 \rightarrow ClO + O_2$
etc.

Ozone depletion reactions

Hox, NOx, SSTs

Aircrafts

Troposphere

NOx

CFCs

HOx, ClO

• Combustion processes
• Natural denitrification
• Nuclear explosions
• Nitrogen fertilisers
• etc.

• Refrigeration systems
• Aerosol sprays
• Polymer foams
• etc.

• Volcanic
  activities
• etc.

Earth's surface

Fig. 2.27. A schematic representation of sources of natural and anthropogenic ozone depletion

Global Warming and Climate Change (Greenhouse Effect)

The greenhouse effect is a process by which the radiative energy leaving a planetary surface is absorbed by some atmospheric gases called greenhouse gases. They transfer this energy to the other components of the atmosphere and it is reradiated in all directions, including back down to the surface. This transfers energy to the surface and lower atmosphere, so the temperature there, is higher than it would be if direct heating by solar radiation were the only warming mechanism. One practical example is that when one enters his car parked under direct sunlight on a sunny day, the interior of the car gets much warmer than the air outside, and one may have wondered why the car acts like a heat trap. This is because the glass in practice transmits over 90 percent of radiation in the visible range and is practically opaque (non-transparent) to radiation in the longer wavelength infrared regions. Therefore, glass allows the solar radiation to enter freely but locks the infrared radiation emitted by the interior surfaces. This causes a rise in the interior temperature as a result of the energy build-up in the car. This heating effects is known as the greenhouse effects, since it is utilized primarily in greenhouses. The greenhouse effect is also experienced on a larger scale on the earth, which warms up during the day as result of the absorption of solar energy, cools down at night by radiating part of its energy into deep space as infrared radiation (long wavelength). Carbon dioxide (CO_2), water vapour and trace amounts of some other gases such as methane and nitrogen oxides in the atmosphere act like a blanket and keep the earth warm at night by blocking the heat radiated from the earth (fig. 2.28). Therefore, they are called 'greenhouse gases'. In this case, the CO_2 is the primary component. Water vapour is usually taken out of this list, since it comes down as rain or snow as a part of the water cycle and man's activities in producing water (such as burning of fossil fuels) do not make much difference on its concentration in the atmosphere. It is mostly due to evaporation from rivers, lakes, oceans etc. The CO_2 is different due to the man's activities that do make a difference in CO_2 concentration in the atmosphere.

The greenhouse effects make life on earth possible by keeping the earth warm (about 30 ^{0}C). However, excessive amounts of these gases emitted by the human beings disturb the delicate balance by trapping too much energy, which causes the average temperature of the earth to rise and the climate generally changes at some localities. Theses undesirable consequences of the greenhouse effect are generally referred to as **Global warming** or **climate change.**

The global climate change is due to the excessive use of fossil fuels such as coal, petroleum products, and natural gas in electric power generation, transportation and manufacturing processes. The current concentration of CO_2 in the atmosphere is about 410 ppm (or 0.41 percent) and is also in the rising trend. This represents a 47 percent increase since the beginning of the industrial age. It is projected to increase over 700 ppm by the year 2100. Under normal conditions, vegetation consumes CO_2 and releases O_2 during the photosynthesis process, and thus keeps the CO_2 concentration in the atmosphere in check. A mature growing tree consumes about 12 kg of CO_2 a year and exhales enough oxygen to support a family of four. However, deforestation and huge increase in the CO_2 production through the use of fossil fuels in recent decades disturbed this balance. Also a major source of greenhouse gas emissions is transportation. Also, a major source of greenhouse gas emission is in the transportation sector. Each litre of gasoline burned by a vehicle produces about 2.5 kg of CO_2 and therefore a car emits about 6000 kg of CO_2 to the atmosphere in a year, which is nearly 4 times the weight of the car.

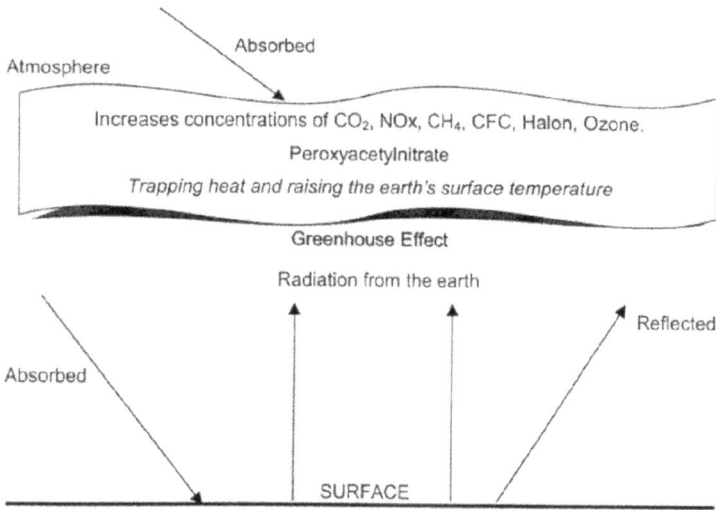

Fig. 2.28. Greenhouse effect on earth

2.21 POTENTIAL SOLUTION OF ENVIRONMENTAL PROBLEMS

Efforts need to be made to find ways to replace fossil fuels with more environmental friendly alternatives as potential solutions to the current environmental problems associated with the harmful pollutant emissions. The use of renewable sources of energy should be encouraged worldwide, with incentives, as necessary, to make the earth a better place to live in. The advancements in thermodynamics in recent time have contributed greatly in improving conversion efficiencies of systems and thus to reduce pollution for clean environment. As individuals, we can follow practising the energy conservation measures and by making energy efficiency a high priority in our purchases. Some of the potential solutions to environmental problems are (i) renewable energy technologies (ii) energy conservation (efficient energy utilization) (iii) alternative energy dimensions in transportation (iv) energy sources switching from fossil fuel to environmentally benign energy forms (v) energy storage technologies for better uses (vi) coal cleaning technologies (vii) recycling (viii) encouraging forestation (ix) changing life style (x) increasing public awareness (xi) education and training on use of low carbon technologies. Among the potential solutions mentioned above, emphasis need to be focussed on the use and popularization of renewable energy technologies.

2.22 GREEN ENERGY

Green energy is loosely defined as the form and utilization of energy that has no or minimal negative environmental, economic and societal impact. The term 'green energy' is used to describe the sources of energy which are considered environmental friendly, non-polluting and therefore may provide a remedy to the systematic effects of certain forms of pollution and global warming. This is, in fact, the renewable energy sourced from the sun, wind, falling water, biomass and wastes. Green energy can be utilized to reduce the negative effects of hydrocarbon energy resources and their emissions, especially greenhouse gases. Green energy provides an important option for meeting clean energy demands for both industrial and non-industrial applications and consequently is a major factor in future sustainable development and world stability. Green energy can contribute to energy security, sustainable development, and social, technological and industrial innovations in a country. Increasing the green energy utilization of a country often positively impacts economic growth and social development. The supply and utilization of low-priced green energy is particularly significant for global stability, since energy plays a vital role in industrial and technological development and living standards around the world. The promotion of green energy sources and technologies for sustainability and global stability has become one of the primary goals of energy policy makers in many countries. Policy makers increasingly assign high priority to promoting green energy-based technologies because they can help mitigate climate change pollution, increase fossil fuel source reserves, decrease dependence on imported energy, increase employment, and support remote and rural communities. Furthermore, green energy-based technologies can increase energy supply diversity, improve the national balance of trade and increase security.

2.23 RENEWABLE ENERGY TECHNOLOGIES FOR SUSTAINABLE DEVELOPMENT

A secure supply of energy resources is generally agreed to be a necessary but not sufficient requirement for the development within a society. Furthermore, sustainable development demands a sustainable supply of energy source that, in long term, is readily and sustainably available at reasonable cost and can be utilized for all required tasks without causing negative societal and environmental impacts. Supplies of energy resources like fossil fuels (coal, oil and natural gas) and uranium are generally acknowledged to be finite and other energy sources such as sunlight, wind and falling water (hydro) are generally considered renewable and therefore sustainable over the relative long term. Wastes (convertible to useful energy forms through, for example, waste-to-energy incineration facilities) and biomass fuels are also usually viewed as sustainable energy sources. In general, the implications of these statements are numerous and depend on how the word 'sustainable' is defined.

Environmental concerns are an important factor in sustainable development. For a variety of reasons, activities which continually degrade the environment are not sustainable overtime i.e., the cumulative impact of such activities on the environment often leads over time to a variety of health, ecological and other problems. A large portion of the environmental impact in a society is associated with its utilization of energy resources. Ideally, a society seeking sustainable development utilizes only energy resources which cause no or less environmental impact (e.g., which release no or low emissions to the environment). However, since all energy resources lead to some environmental impact, it is reasonable to suggest that some (not all) of the concerns regarding the limitations imposed on sustainable development by environmental emissions and their negative impacts, can be in part overcome through increased energy efficiency. Clearly, a strong relation exists between energy efficiency and environmental impact, since, for the same services or products, less resource utilization and pollution is normally associated with increased energy efficiency.

Hence, the exploitation of the renewable energy resources and technologies is a key component of sustainable development. There are three significant reasons for it and are as follows

(i) They have much less environmental impact compared to other sources of energy since there is no any energy source with zero environmental impact. There are a variety of choices available in practice that shift to renewable energy option for providing a far cleaner energy system than would be feasible by tightening controls on conventional energy

(ii) Renewable energy sources cannot be depleted unlike fossil fuel and uranium resources. If used wisely in appropriate and efficient applications, they can provide a reliable and sustainable supply of energy almost indefinitely. In contrast, fossil fuel and nuclear energy sources are finite and can be diminished by extraction and consumption

(iii) They favour power system decentralization and locally applicable solutions more or less independent of the national network, thus enhancing the flexibility of the system and the economic power supply to small isolated settlements. That is why, various renewable energy technologies are potentially available for use at present in urban areas.

2.24 BARRIERS TO RENEWABLE/SUSTAINABLE ENERGY TECHNOLOGIES ADOPTION

There are several barriers which restrict to the effective deployment and large scale diffusion of renewable energy technologies among the users. These may include financial barriers, technical barriers, and market barriers such as inconsistent pricing structures, institutional, political and regulatory barriers and social and environmental barriers. Some barriers may be specific to a technology, while some may be specific to a country or a region. Barriers for the adoption of renewable or sustainable energy technologies in the Indian context have been identified through the extensive literatures review.

High initial capital cost

Technology imported, from highly industrialized and technologically advanced countries, is more efficient but at the same time it may be more expensive than technology manufactured locally and requiring higher initial investment costs. Initial costs for renewable/sustainable energy technologies tend to be high and uncompetitive which may prohibit the consumers from adopting them so easily. Many consumers give more importance on the lower initial cost rather than on the less operating cost.

Lack of financing mechanism

Many of the renewable energy technologies in India are still in the development stage. As in other developing countries, economic and financial issues appear to be crucial for the development of renewable/sustainable energy technologies in India. There is a lack of sufficient government incentives schemes or financing mechanisms to promote the adoption of renewable/sustainable energy technologies in businesses and industries. Poor credit availability for the promotion of renewable energy technologies is a major barrier in their adoption in India. Small and medium scale enterprises (SMEs) above all face 'lack the finances' towards adoption of cleaner technologies.

Lack of subsidies

Lack of financial incentives of different forms (subsidy, tax exemption, low interest loan, long-term credit, specific funds for grid connected projects in rural/mountainous areas, etc.) may potentially hinder the adoption of renewable energy projects.

Lack of consumer awareness to technology

Market imperfection is the most important barrier, indicating relatively poor access to information compared to the conventional energy technologies. Newspapers and magazines are the primary source of their information. Lack of adequate awareness about costs and benefits of the renewable/sustainable energy technologies among the stakeholders may result in a lack of interest and commitment to promote them.

Lack of sufficient market base

Renewable energy based establishments and companies are generally small in India. Thus they have fewer resources than large generation companies or integrated utilities. These small companies are less able to communicate directly with large numbers of customers. In India, the size of sustainable electricity market is not big enough. Due to lack of sufficient market base, private investment is discouraged and also lack of policies that attract sufficient private investments and developmental assistance in sustainable energy technologies adoption.

Lack of paying capacity

Renewable/sustainable technologies are primarily targeted at rural areas or poor customers, who have limited capacity to adopt these technologies. High first costs and investments associated with mass manufacturing remain as barriers both for the users and the manufacturers.

Need for backup or storage device

Due to weather conditions and the fact that daylight hours are limited, along with an uneven geographic distribution of solar resources, solar power is intermittent. To supply un-interrupted and continuous power supply, backup or storage devices are required. Introduction of the storage device adds cost to the system and in addition, disposal of battery (storage device) is also another major environmental issue.

Lack of information technology resources

Poor 'information flow and communication' may be one of the greatest barriers to technology transfer. Information support is necessary for developing linkages to achieve efficient renewable energy technologies.

Technology complexity

Most of the renewable energy/sustainable technologies are complex in nature. Wind energy is generated by complex mechanisms involving the rotation of the earth, heat energy from the sun, the cooling effects of the oceans and polar ice caps, temperature gradients between land and sea and the physical effects of mountains and other obstacles. Bioenergy production also poses a complex multi-phase location/technology/capacity/route analysis problem.

Lack of research and development work

Renewable energy technologies in India are still in the development stage. Lack of appropriate and need based Research & Development (R&D) hampers the adoption of renewable energy technologies among the users in a large scale. Large investment required for R&D work is also a major barrier in taking up the activities of renewable energy technologies in a sustainable manner.

Lack of trained people and training institutes

In India and much of South East Asia, there is a need for technically trained people and people with strong management skills to take up the renewable technologies on an entrepreneurship basis. For developing skilled man power, there is the requirement of locally availability of training institutes which are also lacking in India. There is also the lack of adequate guidance and technical support for the operators that prevent the efficient exploitation of renewable/sustainable resources on long term basis.

Lack of local infrastructure

Infrastructure here broadly refers to not only the physical facilities of transmission and distribution networks, but also necessary equipment and services for power companies. There is no clear division of authority between units functioning at the state level and local authorities for exploiting and developing renewable energy resources for electricity production. Also there are no institutional mechanisms to provide after-sale support to the technologies. Limited private sector participation and target linked programs have not been able to infuse motivation to the existing institutional mechanisms, to be capable of catering to the new markets.

Lack of national infrastructure

Renewable Energy Technologies such as harnessing energy from the wind energy may need strong infrastructure development. Lack of infrastructure is another aspect of institutional barriers and this is related to the availability of infrastructure such as roads, connectivity to grid, communications and other logistics. The regulatory barriers arise in land acquisition and problems in getting clearances for setting up the power plants based on renewable energy sources.

Scarcity of natural and renewable resources

Unprecedented growth in human population has been observed in India. This growth indirectly accelerates the exploitation of natural resources due to their increasing consumption and destruction by the human beings. Scarcity of natural resources and increase in price of non-renewable resources will place renewable and recycled resources progressively on the business agenda of industries.

2.25 ENTREPRENEURSHIP ON RENEWABLE ENERGY TECHNOLOGIES

Entrepreneurship in any domain depends on the presence of opportunities that entrepreneurs discover and transform into realized products and services. This process consists of a number of distinct steps. The steps are research, demonstration and commercialization. Research includes feasibility study and assessment. Demonstration consists of planning and development of prototypes. The last step is the commercialization which includes activities related to attaining full-scale production, self-sufficiency and business maturity.

Entrepreneurs must be aware of both technological potential and market need. They have to identify a match between potential and need and must develop business models to capitalize on that match. The existence of renewable energy technologies and market opportunities, therefore, are necessary conditions for renewable energy technology entrepreneurship to emerge. However, such supportive conditions do not exist everywhere to the same extent. In particular, local policy and social conditions can create important differences in the degree to which there are technological and market opportunities across geographic and political boundaries. The most advanced technologies are of little use if potential entrepreneurs are unable to identify customers willing to pay for novel products and services. In the absence of potential markets, entrepreneurs can neither create nor capture value from the provision of new technologies. The barriers for diffusion of renewable energy technologies as mentioned above may be overcome if entrepreneurs would come up locally to take the venture in a small scale looking into the potential and requirement of the society. They would become the prime agent to solve the barriers such as developing skilled man-power, simplification of technologies, creating awareness among the users, providing services at the door steps, establishing liaison with the Government and financial institutions, creating a market base, providing relevant information to policy makers and planners for successful implementation of technologies, popularizing need based renewable energy technologies looking into the availability of resources locally etc. In sum, the positive effect of technological opportunities and market opportunities on renewable energy technology entrepreneurship will be augmented when there is greater local attention to renewable energy.

2.26 CARBON CREDIT

Carbon credit is a general term for any tradable certificate or permit representing the right to emit one tonne of carbon dioxide or the equivalent amount of a different greenhouse gas (tCO_2e). *One carbon credit is equal to reduction in emission of one tonne of carbon dioxide or carbon dioxide equivalent gases.* Since GHG mitigation projects generate credits, this approach can be used to finance carbon reduction schemes between trading partners and around the world.

Carbon trading is the buying and selling of a new and artificially-created commodity, the right to emit carbon dioxide. Unlike trading in other commodities like crude oil or vegetables, carbon trading is not a voluntary exchange between producers and those who want to consume or sell on the goods. Instead, it results from the action by governments to create this new commodity and the right to emit carbon but then to limit its emissions for the sustenance of environment. Carbon trading allows the industrialized nations to meet their greenhouse gases (GHG) emission compulsions by buying GHG reduction credits from other countries which make profit by trading the carbon credits. It means that if a country cannot meet its greenhouse gas reduction target, it can buy credits from other countries that have credits in excess. As a result, carbon has become a commodity, which like other commodities, is traded in open market, called carbon market. On the one hand, the idea of carbon trade seems like a win-win situation i.e. greenhouse gas emissions may be reduced while some countries reap economic benefit. Carbon credits can therefore be viewed as a means of empowering the market to care for the Environment. Carbon credits can be bought and sold in international markets at the prevailing market price. Carbon credits create market for reducing greenhouse emissions by giving a monetary value to the cost of polluting the air.

The Carbon credits (often called a carbon offset) are certificates issued to countries that have successfully reduced emissions of GHG which cause global warming. Carbon credits (or) certified emission reductions are a certificate just like a stock. This can be used by governments or industry to balance the damaging carbon emissions that are polluting the environment day by day. Each carbon credit represents one tonne of CO_2 either removed from the atmospheres or saved from being emitted. Carbon credits can be created in many ways but there are 2 broad types.

(i) Sequestration (retaining or capturing CO_2 from the atmosphere) such as afforestation and reforestation activities.

(ii) CO_2 saving projects such as the use of renewable energies (wind power, solar energy, biomass power, hydel power etc.).

The carbon trade is an idea that came about in response to the Kyoto Protocol. The Kyoto Protocol is an agreement under which industrialized countries will reduce their greenhouse gas emissions. The current price of carbon trading in the international market is $20.81 USD during 2020.

DECENTRALIZED ROOFTOP SOLAR PHOTOVOLTAIC (PV) POWER PLANT

INTRODUCTION

A standalone or off-grid PV system refers to the sole source of power for fulfilment of electric load of a set-up, residential building, factory etc. It does not depend on the grid electricity (central electricity supply system) or any other source of electric power supply. In this system, the solar panels supply DC electricity to the batteries. Inverter converts the stored energy in the battery into AC electricity. The batteries allow the system to provide power when the sun is not shining. Such an arrangement is suitable where grid connectivity is not possible especially in the rural areas. This system is more viable for houses or commercial buildings in the remote places where the availability of grid electricity is erratic, uncertain and accessible for short time. The storage device i.e. batteries is an important component for this system. All kinds of load like fan, computer, TV, tube light, CFL, LED lamps and other electrically run home appliances can be operated using this type of system configuration which mostly consists of solar panels, charge controller, batteries, inverter and loads. The common configurations of standalone PV system are shown below. The generation of electricity through solar PV system (SPVS) is considered to be eco-friendly and reliable. The propagation of such device even in small capacity can create jobs for the local community and lower down the cost of the generation of power. However, local communities particularly the rural dwellers need to be sensitized with the benefits of a PV power plant. Therefore, a holistic approach involving the public, non-government, Government, academia, media and entrepreneurs need to be established for strengthening social acceptance of solar power generation. Recently, strong initiatives have been taken throughout the globe for awareness in utilizing renewable sources of energy to combat against global warming and climate change. Due to the development and remarkable reduction in the cost of solar PV technology in recent years, solar PV electricity has been proven to be a secured source of energy for the places where sun shine is available abundantly throughout the year. When energy security comes, the "4A theory" always comes into context. The **four A's** of energy security are availability, affordability, accessibility and acceptability. It has also been established that the "4A theory" is applicable to solar PV system because of uninterrupted availability of energy sources at an affordable price. Hence, entrepreneurship for providing services on installation of roof top solar power plant has now-a-days a great demand to fulfil the growing demands of energy in the society. Moreover, the installation of SPVS is more feasible and favourable in the shade free space of the concrete roof top due to the increasing construction of pucca houses at present in rural areas. The various configurations of standalone solar PV system are shown in Fig. 3.1

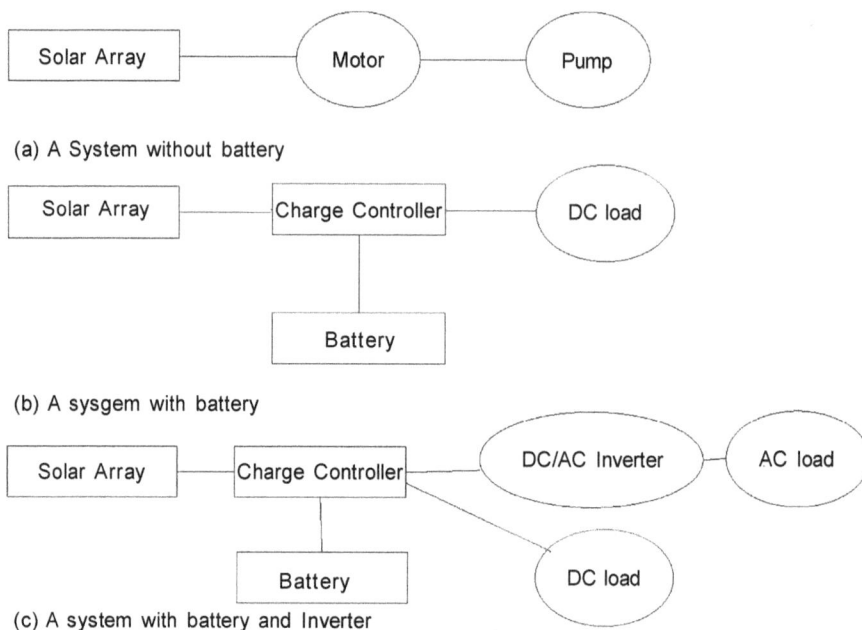

(a) A System without battery

(b) A sysgem with battery

(c) A system with battery and Inverter

Fig. 3.1. Common configurations of standalone solar PV system

3.1 EXECUTIVE SUMMARY

Renewable energy is a viable alternative to meet the growing energy demand of the society. Realizing this fact, Government of India has recently focussed on achieving 100 GW of solar electricity by 2022. Out of which, 40 % is being expected through decentralized and roof top based solar project. Due to the impressive growth in the construction of pucca houses in rural areas through Indira Gandhi Awas Yojana of Govt. of India and already existing pucca houses in the urban areas, the installation of roof top solar power plant is a viable option to generate own power in a decentralised manner at individual level. It is also feasible where grid connectivity is difficult due to inaccessible locations. The unutilized space of the roof top can successfully be used for the installation of SPVS both in rural and urban areas due to constraints in the availability of land on the ground surface. Initiatives have as well been taken by Government and non-Government organisations to install solar PV power generation system over the roof top of some of the public offices, buildings, private houses and institutions to tackle recent power crisis.

Subsidy facilities have also been introduced for the users through Jawaharlal Nehru National Solar Mission, Ministry of New and Renewable Energy, Govt. of India. The importance of SPV plant is going to increase with the rising electricity tariff. Hence, a techno-economic feasibility and environmental impact study of installing a standalone roof top based solar power plant have been assessed for a residential building comprising 5-7 members in the city of Bhubaneswar, Odisha, India. In this project report, all information and data (primary and secondary) required for setting up of a solar power plant have been collected for its viability in small scale applications. The step by step procedure presented in the report would become a ready reckoner for the entrepreneur interested to provide services for roof top solar power plant.

3.2 RATIONALE OF THE PROJECT

India is one of the top consumers of coal and its consumption in 2015 was 10% of the world's coal demand which is expected to get doubled by 2035 (Energy Outlook, 2017). As on July 2016, India is generating 304.76 GW of electricity and the major share i.e. 60 % of it was from coal based thermal power plant (Berwal et al, 2017). The generation of electricity from coal based thermal power plant is at present a main source for emission of greenhouse gases for the country. Heavily dependent on the imported fuel as well as coal and escalating demands of the electricity, the economic growth of the country has become unstable due to the increasing cost of fuels. To maintain a steady economic growth, it has to increase the electricity supply, but the sources of electricity need to be eco-friendly, secure and reliable, looking into one aspect of reducing greenhouse gas emissions. Renewable energy is one of such suitable options which can relieve India from the above concern of power availability, especially through the use of solar energy as the country lies in the sunny belt of the northern hemisphere between the Tropics of Cancer and the Equator and most parts of it experience about 300 clear sunny days in a year and the daily average solar energy incident over the country varies 4-7 kWh/m^2 (Bandopadhaya et al 2011), thus offering a great potential for utilizing solar energy. Solar energy in India therefore, easily qualifies for the above mentioned four 'A' theory and it is expected that solar resources and PV technology are likely to play a key role in decarbonizing India's electricity sector (Jain et al 2018). During post-independence era, Government of India has implemented a wide range of policies by creating an efficient regulatory mechanism to support the growth of solar energy. In order to explore its huge potential of solar energy, the central government has also set a target of achieving 100 GW of solar power till 2022 under Jawaharlal Nehru National Solar Mission (JNNSM). Up to March 2016, the total cumulative installed capacity of solar power projects in India is 8118 MW only. But, during 2014–16, a capacity addition of 14.30 GW has been reported under grid connected renewable power, of which, 5.8 GW is from solar power (Anon, 2017). Though, Government of India has made constant efforts to improve the energy accessibility for its citizens, still the per capita energy consumption (1075 kWh/year) stands low as against the world average (3126 kWh/year). Furthermore, 580 million people lack access to electricity in India (Energy Outlook, 2017). As discussed earlier, due to an abundant availability of sunshine throughout the country and maturity of solar energy conversion technologies, solar energy has become quite popular in India and need to be popularized among the people through the comprehensive program covering R & D, demonstration and utilization, commercialization and awareness creation. Dissemination of roof top solar power plant among the people through entrepreneurship venture is one of the viable and affordable options for promotion of solar PV electricity in a large scale.

Objectives

The objectives of the above initiative are as follows,

- To promote decentralized roof top based solar power plant on standalone or off-grid way both in rural and urban areas for small scale applications
- To create awareness and to demonstrate effective use of solar PV systems for individual, community, institutional and industrial applications.
- To encourage entrepreneurship development on providing services for installation of roof top solar power plant
- To support consultancy services, capacity building and human resource development on solar PV system.
- To assess the techno-economic viability of small scale solar roof top power plant
- To sensitize the replacement of kerosene and diesel, wherever possible.

3.3 EARLIER WORKS DONE BY OTHERS

Most of the work on the feasibility of standalone and grid connected roof top solar power plant has been undertaken by the researchers in various climatic conditions, may be referred before taking up the entrepreneurship of providing services on the above venture. The principle of working and performance of standalone and on-grid solar PV system are same, the only difference lies in the inclusion of connectivity of electrical energy to the grid supply structure in case of on-grid system. The role of grid connectivity for small power household system is to supply the excess of the generated energy to the grid if more energy is generated by the PV system than required by the load. The inclusion of batteries is optional in on-grid system, but the use of batteries is compulsory in standalone system. Hence feasibility study and performance analysis of on-grid roof top solar PV system may also be referred for off-grid system and same for both the cases.

In India, Sharma and Chandel (2013) conducted a performance study of 190 kW_p grid interactive solar photovoltaic power plant in Punjab and observed that the final yield (ratio of net annual AC energy output to the array rated power), reference yield (ratio of total daily in-plane solar insolation to the reference radiation) and performance ratio (PR) (ratio of final yield to reference yield) varied from 1.45 to 2.84 kWh/kW_p-day, 2.29 to 3.53 kWh/kW_p-day and 58% to 83% respectively. In another study conducted by Mondol et al. (2006), it was found that, for a 13 kWp roof mounted grid connected PV system, the monthly average daily PV array, PV system and inverter efficiencies varied from 4.5% to 9.2%, 3.6% to 7.8% and 50% to 87% respectively while performance ratio ranged from 0.29 to 0.66. A techno economic study of a 1 kWp grid connected PV system conducted by Tarigan and Kartikasari (2015) in Indonesia revealed that the system could technically meet the basic electrical needs of a household. This system injected about 1 MWh/year to the grid with a PR of 72% and reduced 1296 kg CO_2. The payback period of this system was estimated as 17.6 years after which the system would make profit. Pundir et al. (2016) made a comparative study of the performance of a grid connected solar PV power system in IIT Roorkee and found that the generation cost of electricity from the system was 8.50 INR per kWh without subsidy with a performance ratio of 63.68% having capacity factor of 8.77%. The payback period of this system was found to be 7.5 years and this system is eco-friendly and good for environment as it reduced 2464 ton of CO_2 per annum. In another study, Peerapong and Limmeechokchai (2014) made a comparative study of three types of grid connected solar photovoltaic power plant namely solar residential roof top (11.04 kW_p), integrated ground mounted roof top (330 kW_p) and utility scale (38.5 MW_p) in Thailand and found that the lowest cost of electricity of \$0.27/kWh was with the utility scale solar system. The cost of electricity with residential solar roof top and ground mounted roof top system were \$ 0.46/kWh and \$0.29/kWh respectively. They concluded that the new feed in tariff schemes of Govt. for residential, integrated ground mounted and utility scale with installed capacity larger than 1 MW are reasonable for investment and make the investor feasibly profitable for the whole life time of a project. Sharma and Goel (2017) studied the performance of 11.2 kWp grid connected solar photovoltaic system, installed on the roof top of a constituent institute of Siksha 'O'Anusandhan University, Bhubaneswar, Odisha and found that the total energy generated per annum was 14.960 MWh and the PV module efficiency, inverter efficiency and performance ratio were calculated to be 13.42%, 89.83% and 0.78 respectively. Similarly, Berwal et al. 2017 studied the technical and financial viability of 50 kW_p capacity roof top solar power plant installed at Saraswati library building of Deenbandhu Chhotu Ram University of Science & Technology, Sonipat, Haryana and found that the PV plant generated more than 5200 kWh/month of electricity and reducing 4070 kg/month of GHG emissions. Further, this plant was installed with govt. subsidy and found that the subsidy amount made it lucrative by reducing pay back duration to 5.7 years. But even without subsidy, the pay back duration came out to be 10.3 years along with its financial viability.

From the above referred works, it can be concluded that the roof top solar power plant is technically and financially viable as well as environmentally sustainable venture for its adoption either on small or large scale applications. The annual average final yield of power varies from 2.5-3.5 kWh/kW_p per day. The promotion of entrepreneurship for the above project has also a great prospect for the entrepreneurs looking into the present context of wide dissemination of solar PV electricity in the society.

3.4 CURRENT STATUS OF STANDALONE SOLAR ROOFTOP PV POWER PLANT

The progress of solar rooftop PV sector in India is very slow in recent years, even after the announcement of Jawaharlal Nehru National Solar Mission (JNNSM) with a target of installing a cumulative capacity of solar power generation of 100 GW by 2022, out of which 40 GW would be from off- grid/decentralized solar power and 60 GW is allotted for the large solar utility power plants. As of 2016, the total rooftop installed capacity in India was less than 2% of the total cumulative solar PV installation in the country, whereas countries like Germany, USA, and Japan are leading the world in terms of rooftop solar installations. Germany is leading the world in solar rooftop installation having more than 70% PV installations in the form of the solar rooftop (MNRE 2017). Similarly, in Japan, the solar program has grown primarily through decentralized solar installations which include standalone domestic, standalone non-domestic and grid connected distributed ways. Likewise, India has to develop its off-grid decentralized solar installations for residential and commercial purposes in order to provide sustainable and clean energy and energy security to every citizen, which is only possible through appropriate Government policies and support systems. According to the International Energy Agency, India has only 1 watt/capita solar PV whereas countries like Germany, Italy, and Belgium have 491, 308 & 287 respectively (Brunisholz, M 2016). The solar resource is abundant in India. Hence, there are various factors for which India is lagging behind in decentralized solar rooftop PV segment for power generation.

3.4.1 Business Model for Solar Rooftop Installations

Business models play a very important role in introducing new technology to the market. Innovative business models play an important role in motivating the end consumer for the use of solar power technology. Followings are the rooftop solar business models in India.

(a) *Owned by the customer (CAPEX):* In this model, the rooftop owner installs the solar system on his rooftop. The operation and maintenance cost is bearable by the owner and the power generated is used by the owner. In some cases, a third party takes the responsibility of O&M and the cost of which is bearable by the rooftop owner. This is the most popular business model in India and more than 75% of the total solar rooftop PV in India are installed using this model.

(b) *Third party owned (RESCO):* This type of business model is really popular in USA and in European countries. In this model the rooftop owner may or may not invest in the rooftop solar PV according to the mutual agreement between them. The third party takes the responsibility of operation and maintenance and sold the power to the rooftop owner or feed it to the grid and gives the rent of solar PV installation to the rooftop owner. This model is not much popular in India because long term rooftop availability remains a key risk for private developers and consumer does not want to indulge in legal formalities for leasing their rooftop for a long span of 20-25 years. Therefore, less than 25% of the total solar rooftop PV installations in India are developed under this business model.

(c) *Utility-owned:* In this type of business model, the utility itself owns, operates and maintains a rooftop solar PV installation or it may appoint a 3rd party to install a solar facility. The power generated is then utilized to fulfil the renewable purchase obligation. In this model, the utility has to sign a lease agreement with the rooftop owner for a period of about 20-25 years.

In India, the promotion of the grid connected SPV rooftop and small SPV power generating plants among residential, community, institutional, industrial, and commercial establishments is being carried out under JNNSM with the minimum capacity installation of 1 kW to a maximum capacity installation of 500 kWp through various agencies such as Solar Energy Corporation of India (SECI), State Nodal Agencies (SNA), Channel partners, Financial Institutions etc. India has also deployed net metering and /or gross metering facilities for development of rooftop solar under JNNSM.

Gross metering: In this facility, all the power generated from the solar rooftop is fed into the grid without any internal use at the rooftop installed facility. All the energy exported to the grid is recorded using a feed-in meter at a predetermined feed-in-tariff approved by the regulator through a power purchase agreement (PPA).

Net metering: Under this facility, the power produced is first utilized by the rooftop owner to meet internal loads and the excess power is then fed to the grid. In this facility, a single meter is used which shows net power consumed by the consumer from the grid after exporting excess solar power to the grid.

3.4.2 Barriers for Solar Photovoltaic Rooftop

The barriers usually encountered for solar rooftop segment in India are in the form of economic barriers, financial barriers, policy and regulatory barriers, education barriers, technological barriers etc.

(a) *Lack of awareness among end customer:* The lack of awareness, information and knowledge of solar energy among the people of India are the major barrier to the growth and development of solar PV rooftop. Unfortunately, because of lack of awareness about solar power systems, lack of education and poor training and development mechanism, people are not able to take benefits of capital subsidy and other Government support. The Government has to prepare a robust mechanism to educate, aware and train people about the Government support and about the benefits of solar power technology.

(b) Lack of clarity in rooftop policies: Some states in India have announced policies for the solar rooftop, but there is a significant lack of clarity in these policies. There is a very slow progress on net metering policy across the country. It has been found that the average time taken in solar rooftop connection is as high as 90 to 100 days in some states. The local implementing authorities or state nodal agencies are facing lack of appropriate training and process protocol, as a result, they are unable to deliver proper services of installing solar rooftop systems to the customers. There is no clear road map of permit mechanism, commercial agreements, and technical requirements and therefore this solar segment is still largely unexploited. There should be a proper coordination between the state nodal agencies, state electricity regulatory commission and the DISCOM.

(c) Lack of financial institutes: Like any other nation, commercial banks constitute a major source of financing for infrastructure projects, including renewable energy in India. Since the solar power technology is new in India and doesn't have any proven track record as like conventional power projects, thus lack of awareness and familiarity, shorter history and lengthy payback period makes it next to impossible to obtain any financial help from banks.

(d) Poor technical or commercial skills: The diffusion of any new technology in the market requires good technical and commercial skill workforce. Similar is the case with the solar power, it also requires a large number of skilled and semi-skilled labour to effectively implement the policies on ground level. But in some specific markets, a skilled workforce, which can design, install, operate and maintain the solar power establishment is not available in large numbers. India is also, lacking in large numbers of the skilled and semi-skilled workforce, which is required to achieve the solar rooftop target of 40 GW by 2022. As per the report released by Council on Energy, Environment and Water (CEEW) and the Natural Resources Defence Council (NRDC), 2016, in order to achieve the 40 GW of solar rooftop target, India requires 210800 skilled plant design and site engineers and approximately 624600 semi-skilled and low-skilled technicians for construction purposes.

3.4.3 Remedial Measures

There are various critical barriers as discussed above, for the growth and development of solar rooftop in India, prevailing at the grass root level like lack of awareness about the technology, lack of transparency in implementing the policies, limited number of institutions for financial support and the shortage of skilled workforce. As Government has prepared a roadmap for achieving a target of 40 GW solar rooftop installations by the year 2022, in the similar context, there is a need to prepare a blueprint to pacify these critical barriers. These barriers cannot be pacified in short span of time. Therefore, the Government has to develop such policies which will effectively improve the implementation process at the grass root level. The followings remedial measures, therefore, need to be looked into in order to boost the growth of the solar rooftop sector in India with a view to provide an effective tool for sustainable energy security.

(a) A key element to develop the market of the solar rooftop is to increase awareness among people and to demonstrate the benefits of the decentralized solar rooftop system on a regular basis.

(b) The state nodal agencies can publicize the technology through rural social marketing, street plays, site visits and word of mouth.

(c) Entrepreneurial spirits among the unemployed youths need to be promoted for taking up the solar rooftop as a service providing venture by strengthening their skills/capacity through conducting various entrepreneurship development programmes.

(d) Apart from the capital subsidy, the Government must think upon providing easier loans at very low interest rates and tax incentives for the end consumers. For this purpose, the Government must empower the banks for providing easier loans in minimum time and having a minimum rate of interest.

(e) There must be sufficient institutional arrangements in every district of the country for providing regular training to the common people regarding technical, financial, policies and other benefits aspects of the solar rooftop system.

(f) Certificates courses, short-term courses, workshops and etc. must be introduced in universities and colleges in order to build a skilled workforce for the growth and development of decentralized solar rooftop sector.

(g) The Government should take steps to improve the research and development and must invest in setting up integrated manufacturing facilities across the country in order to further reduce the system cost.

3.5 SITE DETAILS AND METEOROLOGICAL DATA

The off-grid roof top solar PV power plant is proposed to be installed in a residential building located at Bhubaneswar, Odisha. The latitude and longitude of the place is 20.24 ^{0}N and 80.85 ^{0}E with an altitude of 45 m above MSL. The global horizontal solar irradiance data can be collected from the local weather station where all meteorological data are available. If relevant data are not available for a place, those data can also be collected from a secondary source i.e. NASA research centre by mentioning the latitude and longitude of a place. The climate of the place is warm and humid. The annual average ambient temperature is 26.6 ^{0}C and varies from 10.5 ^{0}C to 37.8^{0}C. The annual mean wind speed is 3.5 m/s with the maximum value of 11 m/s. The annual average relative humidity is 76.9 % and varies from 29.6 to 96.4 %. The global horizontal irradiance of the place varies from 4.22 to 5.4 kWh/m^2/day as per solar resource map (Fig. 3.2) of India.

The mean annual sunshine hours are 6.6 hours, varies from 3.5 to 8.0 hours.

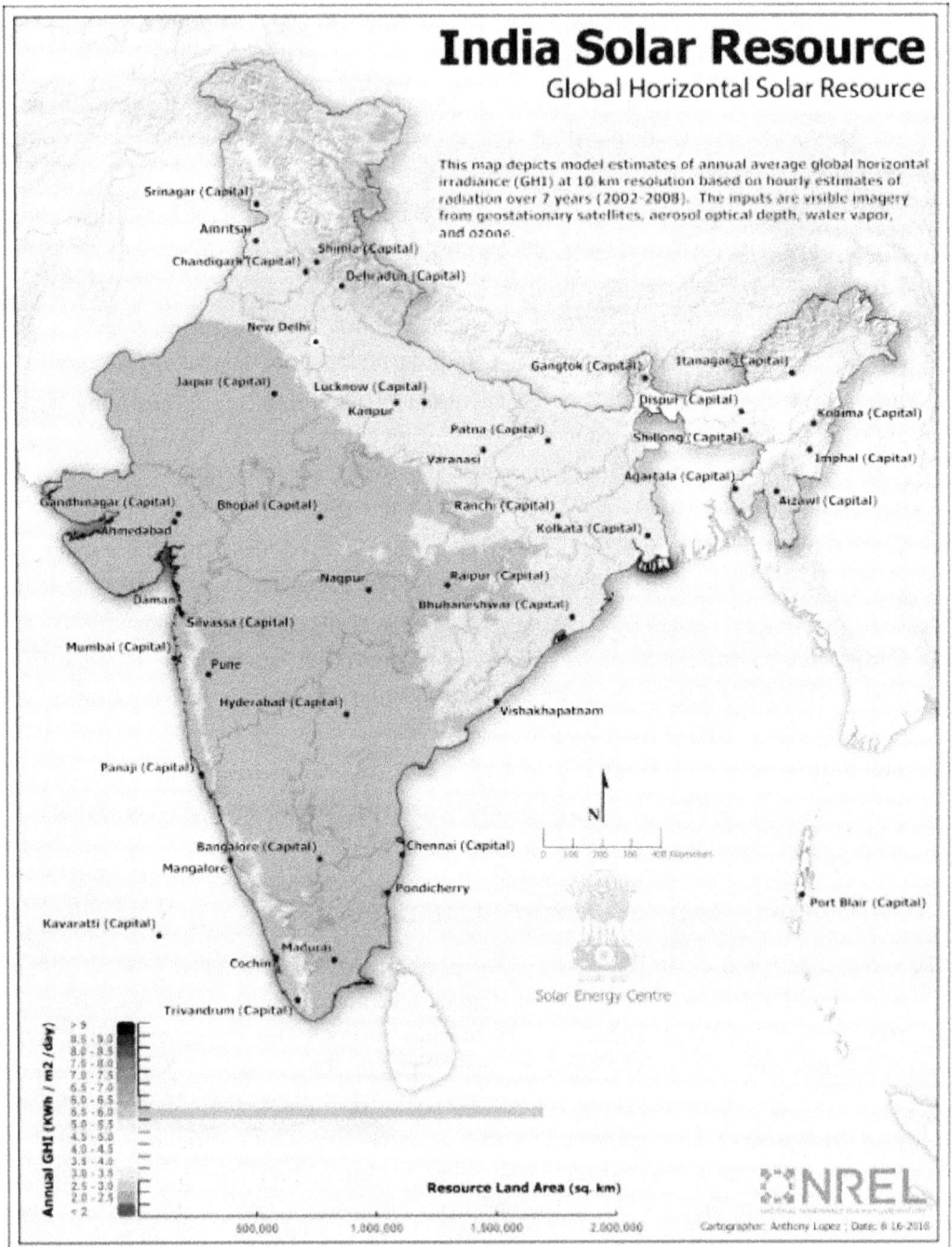

Fig. 3.2. Solar resource map of India

3.6 ESTIMATION OF INSTALLED CAPACITY OF SOLAR PV SYSTEM FOR THE ROOF

The installed capacity of solar PV system, C_R, in kW$_p$ that could fit for the roof can be estimated by the equation, $C_R = \left(\dfrac{C_M}{1000}\right) \times \left(\dfrac{RCR \times A_R}{A_M}\right)$ where A_R is the total roof area available for installation of solar modules in m^2, C_M is the individual module rated capacity in W$_p$, A_M is the area of one module in m^2 and RCR is the roof cover ratio, which is the fraction of roof area that the modules will cover. The roof area (A_R) should be the total area after deducting the space occupied by any obstacle on the roof such as water tanks, utility rooms, air conditioning systems etc. A typical value for the roof cover ratio (RCR) is 0.85, which would allow for 15% of the roof to be free for spacing between modules and away from obstructions. As a thumb rule, for 1 W$_p$ sizing of a module, the area of module is 9.02×10^{-3} m^2. The W$_p$ size of module may be selected as per the availability in the market.

3.7 ESTIMATED ENERGY (kWh/ANNUM) GENERATED BY ROOF TOP SOLAR POWER PLANT

The energy, E (kWh/annum) generated by the PV array can be estimated by the following ways.

(i) E = C_R (kWp)× *peak sun shine hours per day×clear sunny days in a year* × *derating factor*

Peak sun shine hours refer to the number of equivalent sun shine hours in a day for the solar radiation of 1000 W/m^2, the standard radiation for which the panel is characterized.

Solar radiation data are available from the meteorological department on daily, monthly or yearly basis. The daily radiation data (kWh/m^2/day) can be considered as the load is estimated on the daily basis. The daily solar radiation in India generally varies between 4-7 kWh/m^2/day. For the location under study, the solar radiation for example is 5 kWh/m^2/day. The solar PV modules are rated for solar radiation intensity of 1000 W/m^2 or 1 kW/m^2 (standard test condition or STC). Therefore, this solar radiation of 5 kWh/m^2/day can be written as

$$5kW \times \dfrac{h}{m^2} \times \dfrac{1}{day} = 5\dfrac{h}{day} \times 1\dfrac{kW}{m^2} = \text{equivalent daily peak sun shine hours} \times \text{STC}$$

power density

In this way, for 5 kWh/m^2/day, the solar radiation is equal to 5 hours of 1 kW/m^2 or equivalent daily peak sunshine hours is 5. Though the day length may be longer, light equivalent to number of peak sunshine hours is considered for sizing of solar module.

The derating factor indicates the fraction of direct current (DC) being converted to alternating current (AC). The derating factor generally ranges between 0.6 and 0.8 (Honda et al. 2012). A typical derating factor of 0.75 can be used for calculation.

(ii) E (kWh/annum) = Insolation(kWh/m^2/year) × system efficiency (combined efficiency of panel, batteries, inverter) × area of array (m^2)

Annual average insolation (kWh/m^2/year) = Average radiation on the optimum tilted surface (kW/m^2) peak sun shine hours/day number of clear sunny days/year.

The information of average radiation on the horizontal surface, peak sun shine hours and number of clear sunny days are available in Meteorological Department. The data of horizontal solar radiation can be converted for the inclined surface by taking the tilt angle of the panel.

(iii) E (kWh/year) = Average daily final yield (h/day) × number of clear sunny days/year × installed capacity of solar PV array, C_R *(kWp)*.

The average daily final yield is defined as the ratio of net daily AC energy output to the rated power of the installed PV array. $Y_{Fd} = \dfrac{E_{AC,d}}{P_{pv,rated}}$ (h/day), where $E_{AC,d}$ is the net daily AC output (kWh/day) and $P_{(pv,rated)}$ is the rated installed capacity of array (kW). The final daily yield varies between 3.5 to 4.5 h/day in India depending upon the system efficiency and climatic conditions of the place. However, it provides a rough estimate of energy generated by the PV system based on the data available from the literatures for a particular location or nearer to it.

3.8 MOUNTING ARRANGEMENT OF ROOFTOP SOLAR POWER PLANT

A rooftop solar PV installation comprises of PV panels assembled in arrays, mounting frames to support the panels to the roof, wiring, inverters, and other components depending on the type of installation. The rooftop must be able to accommodate all of these components. Generally, a flat, concrete roof will normally have the strength to accommodate the additional weight of the panels and supporting structures. Inclined roofs of storage sheds and residential buildings, made of metal sheet, tiles, or similar materials need to be examined whether the trusses can support additional weight or not. The mounting structures must also be capable of withstanding maximum wind loads.

Static load on rooftop due to PV array and mounting structure: Total static load on the rooftop of the building due to various components of the power plant need to be estimated in kg/m^2 and to assess the safety measures as per building guidelines. As per thumb rule, a crystalline PV system will place a dead load of about 20-25 kg/m^2 on the roof (California Energy Commission 2001). Most solar panels are certified to withstand winds of up to 2,400 Pascal (240 kg/m^2), equivalent to approximately 140 mile-per-hour (MPH) (224 kmph) winds (NABCEP 2005).

Shading analysis: The shading analysis is done to keep the spacing between modules and array such that they would receive uninterrupted sunlight from 9.00 am to 4.00 pm throughout the year and to decrease the inter-module and inter-array shading for preventing related losses because of it. The modules mounted on the structure are to provide 100 mm grounding spacing for adequate rear ventilation and 10 mm module to module spacing to accommodate thermal expansion. The PV modules/panels are mounted on metal frames supported by concrete pillars. It is advisable to clean the surface of the module manually with water at an interval of 15 days in order to eliminate the soiling loss. The inter-row distance of PV array is considered using a thumb rule that the minimum spacing between the rows is equal to three times the height of the module (Chakraborty et al. 2015). Sufficient distance between arrays and around the periphery wall of rooftop is also allowed for facilitating maintenance purposes and leaving space for rain water exhaust pipe, water tank and rooftop entrance etc. The figure below (4.3) shows the arrangement of modules to prevent shading.

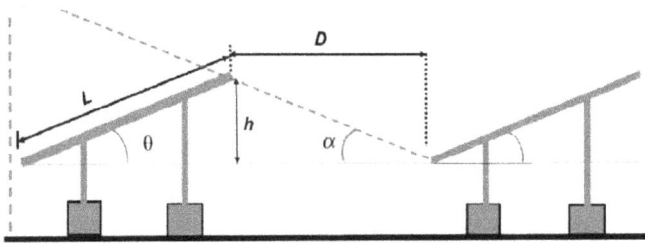

Fig. 3.3. Mounting of module to prevent shading

α= Sun elevation angle, θ = Solar module tilt angle, *D* = Distance between the two rows (m), h = Height of the solar module (m), L = Length of the solar module (m). *D* ≈ 3*h*.

Tilt and orientation: In order to harness optimum radiation, the tilt angle of the module is taken as the latitude of the place and its orientation is considered to be 0^0 azimuth facing due south (charles, 2015).

3.9 GROUTING OF STRUCTURE

The structure is grouted on the rooftop by digging 30 cm×30 cm square hole to the depth of concrete structure and the surface area of the hole is cemented by using water proofing compound of IS-2649. The stainless steel nut bolt assembly, 304-M6 size is grounded in the pit chamber with concrete mix of M-20/ M-25 grade. The schematic of the grouted nut bolt assembly is shown below.

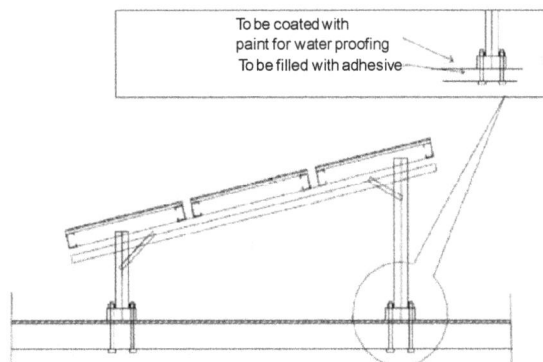

To be coated with paint for water proofing
To be filled with adhesive

3.10 CABLE SIZE

The selection of a proper cable size is another important and crucial parameter in PV system after selecting appropriate array, batteries and inverter. In solar PV applications, one part of the system circuit is working with DC electricity and another part may be working with AC electricity. The sizes of cable are different in both the cases due to difference in their amount of current and voltage. Hence, appropriate choice of cable will reduce the electrical losses (voltage drop or power loss) in it. Appropriate choice of cable is also important not only from the perspective of better system performance but also for avoiding shock hazards, fire hazards and improving reliability of PV system. The easy and fast way to determine the size of cable (diameter or cross sectional area) is to calculate voltage drop index(VDI) of the system by using the following equation.

$$\text{VDI} = = \frac{\begin{array}{c}25\,\%\text{ more of the actual current to flow in the system (ampere)}\ \times\\ \text{one way cable length in the circuit (feet)}\end{array}}{\%\text{ allowable voltage drop in the circuit} \times \text{system operating voltage (V)}}$$

After calculating VDI, the standard wire gauge (SWG) of the cable is obtained by using VDI table for a particular material like copper or aluminium. Copper is more preferable and commonly used as it is durable than aluminium. After knowing SWG, standard wire gauge dimensions' table is referred to find out the diameter or cross sectional area of the cable to be used for the system.

3.11 SYSTEM PERFORMANCE ASSESSMENT:

In order to analyse the energy related performance of a PV system, some important parameters are computed using data collected during its operation in a given location. The International standard IEC 61724 published by the International Electro Technical Commission (IEC) in 1998 describes few parameters for evaluating the performance of the photovoltaic systems (Photovoltaic, 2010). This standard has been accorded by Bureau of Indian Standard (BIS) in 1998. The parameters to assess the performance are array yield (Y_A), final yield (Y_F), reference yield (Y_R), system efficiency (η_{sys}), performance ratio (*PR*) and capacity utilization factor (*CUF*). These indicators provide a basis under which PV system can be compared under various operating conditions.

Array yield (Y_A): It is defined as the ratio of energy output from a PV array over a defined period (day, month or year) to its rated power and is given by Ayompre et al.

(2011); $Y_A = \dfrac{E_{DC}}{P_{pv,rated}}$ (h/day) where array energy output per day, $E_{DC} = I_{DC} \times V_{DC} \times$

t(kwh). $I_{DC} = DC$ current (*A*), $V_{DC} = DC$ voltage (*V*), $t =$ sun shine hours in a day generating PV energy. It is also equal to the time (h/day) which the PV array is to operate with rated power to generate DC energy in a day.

Final yield (Y_F): It is defined as the ratio of annual, monthly or daily net AC energy output of the system to its rated power of the installed PV array at standard test conditions (STC) of 1 kW/m^2 solar radiation and 25 ^{0}C cell temperature and is given by

$Y_F = \dfrac{E_{AC,day}}{P_{pv,rated}}$ (h/day). This is a representative data which indicates the comparison of

similar PV systems in a specific geographic location. It is dependent on the type of mounting, roof structure, cables used and shading losses.

Reference yield (Y_R): It is defined as the ratio of total daily in-plane solar irradiation,

H_t, (kWh/m^2) to the reference irradiance (1kW/m^2) and is given by $Y_R = \dfrac{H_t\left(\dfrac{kwh}{m^2}\right)}{1\,kw/m^2}$

(h/day). This yield represents the number of peak sun shine hours per day. It indicates the solar radiation resource for the PV system. It is a function of PV array orientation, location and month as well as year wise weather variability.

System efficiency: The system efficiency depends upon the efficiencies of the module, inverter and batteries if exist. The combined efficiencies all the three components represent the system efficiency which enables one to compare with the other PV systems.

a. *PV module efficiency:* The instantaneous PV module efficiency as given by Ayompe

et al. (2011) is $\eta_{PV} = \left(\dfrac{P_{DC}}{G_t \times A_m}\right) \times 100$ where P_{DC} is the *DC* power generated by

the module, G_t = total in-plane solar radiation (W/m^2) and A_m is the area of module (m^2). As a function of temperature, it can also be represented (Sharma and Goel,

201?) as $\eta_{PV} = \eta_{T,ref}[1 - \beta_{ref}(T - T_{ref})]$ where $\eta_{T,ref}$ = efficiency of PV module

at reference temperature, β_{ref} = temperature coefficient of power, T_{ref} = reference temperature (25 ^{0}C) and T = the cell temperature which is given by

$T = T_{amb} + \dfrac{(NOCT - 20)G_t}{800}$ where T_{amb} is the ambient temperature, NOCT is nominal

operating cell temperature of module declared by the manufacturer. One can reach nominal in nominal incident condition under an irradiance of 800 W/m^2 and ambient temperature of 20 ^{0}C (Mulcué-Nieto, L.F. and Mora-López, L., 2014).

b. Inverter efficiency: The instantaneous PV module efficiency as given by Ayompe et al. (2011) $\eta_{inv} = \dfrac{P_{AC}}{P_{DC}}$ is and daily inverter efficiency $\left(\eta_{inv,daily}\right) = \left(\dfrac{E_{AC,daily}}{E_{DC,daily}}\right) \times 100$

c. System efficiency: The instantaneous PV module efficiency as given by Ayompe et al. (2011) is $\eta_{sys} = \eta_{PV} \times \eta_{inv}$

The system efficiency can also be expressed as $\eta_{sys,daily} = \dfrac{E_{AC,daily}}{H_t \times A_m}$ where $E_{(Ac,daily)}$ is the daily AC energy output.

Performance ratio: Performance ratio (PR) indicates the overall effect of losses on the rated output due to PV module temperature, incomplete utilization of incident solar radiation, system components inefficiencies or failure, wiring mismatch, soiling etc. It is a dimensionless quantity. Generally, PR values are greater in winter than in summer because of increased losses due to more rise of PV module temperature. Normally PR varies from 0.6 to 0.8 depending on the location, solar irradiance and climatic conditions. It does not represent the amount of energy produced because a system with low PR in high solar irradiation area may produce more energy than a system with high PR in a low solar irradiation location. The PR indicates how close a PV system approaches to ideal performance during real operation and allows comparison of PV systems independent of location, tilt angle, orientation and their nominal rated power capacity. Performance ratio is defined as the ratio of the energy fed to the load (final yield) to the energy that the system could have produced at DC rated power for the number of peak sun hours per day (reference yield) and expressed as $PR = \dfrac{Y_F}{Y_R}$

Also $PR = \dfrac{Net\ energy\ available\ at\ the\ load\ for\ period\ of\ evaluation}{Calculated\ rated\ energy\ output\ for\ period\ of\ evaluation}$

Capacity Utilization Factor (CUF): It is defined as the ratio of actual annual energy generated by the PV system $E_{(AC,annual)}$ to the amount of energy the PV system would generate if it is operated at full rated power for 24 h per day for a year and is expressed as $CUF = \dfrac{E_{AC,annual}}{P_{PV,rated} \times 24 \times 365} \times 100$ (Ayompe et al. 2011). The capacity utilization factor for PV system is also expressed by $CUF = \dfrac{\frac{h}{day}\ of\ peak\ sun}{24\,h/day}$

If a system delivers full rated power continuously, its CUF would be unity i.e. 100%. CUF is dependent on the location of the PV system. The higher the capacity factor, the better the PV system. The capacity utilisation factor of all roof top solar PV system in India is 16%–17%. CUF is location specific.

Energy losses: There exists a variety of sources through which energy losses occur in PV system. These losses affect the performance of PV system. Hence, it is necessary to evaluate these losses in order to minimise them for enhancing the performance of PV system. The losses are array capture loss, system loss, soiling and degradation. Soiling and degradation losses are more difficult to evaluate because of their small effects and are generally ignored.

Array capture loss (L_C): Array capture loss includes thermal capture loss (L_{CT}) and miscellaneous capture loss (L_{CM}). The thermal capture loss is caused by the increase in the cell temperature higher than 25 ^{0}C in actual condition. The electrical behaviour of PV system has strong influence on cell temperature thereby affecting the final yield. $L_{CT} = Y_R - Y_T$ where $Y_{(T)} = Y_R[1 - \gamma(T_C - 25)]$ where Y_T is the corrected reference yield, Y_R is the reference yield and γ is the temperature coefficient for solar cell.

Miscellaneous capture loss (L_{CM}) includes cable loss, shading effect, mismatch loss, degradation loss, maximum power tracking errors etc. $L_{CM} = Y_T - Y_A$ where Y_A is the array yield. Hence, $L_C = Y_R - Y_A$

System loss(L_S): This loss is due to conversion of DC into AC by inverter. The loss in inverter is calculated by $L_{inv} = 1 - \dfrac{E_{AC}}{E_{DC}}$. Therefore, $L_S = Y_A - Y_F$

Total loss $L_T = L_C + L_S = Y_R - Y_F$ All yield terms mentioned in the calculation of losses are expressed in h/day.

3.12 SIZING OF SOLAR PV SYSTEM COMPONENTS

The components for sizing of standalone solar PV system consists of PV modules, MPPT charge controller, batteries, inverter and cables. The arrangement of the components of standalone solar PV system is shown in fig. 3.4.

Determination of designed electrical load: The electrical load for a residential building of about 7 members has been estimated and presented in Table 3.1.

The electrical load for the residential building = 18974 Wh/day. The designed load for the PV system should be 30 % higher than the actual load to account for the losses in PV system components. The designed load is therefore determined to be 24,667 Wh/day (18974 × 1.3) = 24.667 kWh/day.

SOLAR PANEL

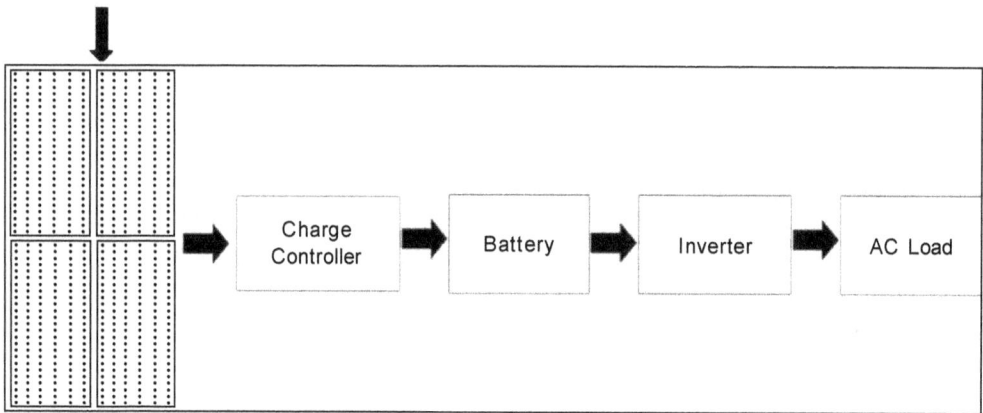

Fig. 3.4. Arrangement of components in standalone solar PV system

Determination of number of peak sun shine hours for the location: The solar radiation data on daily/ monthly/annual basis for the location under study are collected from the meteorological department. The daily average data of solar insolation at the tilted surface is computed in $kWh/m^2/day$. The daily average solar insolation in India varies from 4-7 $kWh/m^2/day$. Suppose, for the location under study, the insolation is 5.0 $kWh/m^2/day$.

Determination of PV panel size (W_P): PV panel size required (W_P) = {Designed load(Wh/day)}/ (peak sun shine hours /day) = 24667/5 = 4933.4 (W_P). The PV module peak watt is selected as per the availability in the market. The common operating voltages of PV array and batteries are 12/24/48 V

$$Number\ of\ modules\ required = \frac{Total\ watt\ peak\ rating}{PV\ module\ peak\ rated\ output} = \frac{4933.4}{175} = 28.19 \approx 28.$$

Determination of MPPT (Maximum Power Point Tracking) Charge Controller Capacity: MPPT is an electronic device whose function is to extract maximum available power from the modules under any given condition (less radiation, more radiation, high temperature etc.). Its presence between the source (PV module) and load ensures the maximum extraction of available power from the module. Charge controller protects the battery from over charging and over discharging. Though the functions of MPPT and charge controller are different, many manufacturers combine the function of charge controller and MPPT in one single device which is called MPPT charge controller. MPPT charge controller is rated by the output current, it can handle and not by the input current from the solar panel/array. The solar PV panel is connected to the input side of MPPT and the battery is connected to the output side. Therefore, load for the MPPT is the battery. The specification of MPPT is given in terms of output voltage and current it can handle. Therefore, the current requirement of the MPPT for its load is determined. Let

Table 3.1: Estimation of Electrical Load for a Residential Building of about 7 Members

S.N.	Electrical appliances	Rated power (W)	Number	Total power (W)	Daily electrical loads		
					(working hours/day)/ Summer season load (Wh)	(working hours/day)/ Rainy season load (Wh)	(working hours/day)/ Winter season load (Wh)
1	CFL	36	6	216	(14)/ (3024)	(12)/ (2592)	(14)/ (3024)
2	Ceiling Fan	60	6	360	(20)/(7200)	(16)/(5760)	—
3	TV	100	1	100	(10)/ (1000)	(10)/ (1000)	(10)/ (1000)
4	Refrigerator	250	1	250	(24)/ (6000)	(24)/ (6000)	—
5	Laptop	60	2	120	(10)/(1200)	(10)/(1200)	(10)/(1200)
6	Submersible pump	1000	1	1000	(0.5)/ (500)	(0.5)/ (500)	(0.5)/ (500)
7	Mobile charger	5	5	25	(2)/(50)	(2)/(50)	(2)/(50)
	Total daily load (Wh)				**18974***	**18002**	**5774**

* Maximum daily load is considered for sizing of PV panel.

the system operating voltage be 48 V. MPPT charge controller capacity (A)

$$= \frac{PV \ array \ size \ (W_P)}{Battery \ bank \ design \ voltage \ or \ system \ voltage \ (V)} = \frac{4933.4}{48} = 102.77 \ ampere.$$ So, a 48

V and 103 A MPPT charge controller suits well for this system. Taking current and voltage into consideration, MPPT charge controller of just higher capacity can be chosen as per the availability in the market.

Determination of Inverter size (W): The input rating of the inverter should never be lower than the PV array rating. The maximum continuous input rating of the inverter (W) should be about 10 % higher than the PV array size to allow safe and efficient operation of PV power system. The inverter converts DC power to AC power to operate AC loads.

Inverter capacity $(W) = 1.1 \times$ PV array size $(W_p) = 1.1 \times 4933.4 = 5426.74 \approx 5.5 \ kW$

Battery sizing: Off-grid PV system needs a storage medium and in case of PV power plant, batteries are the most common storage medium. The capacity of batteries should be sufficient to meet the energy requirements of the load considering the number of days of autonomy, depth of discharge and efficiency of battery. Taking days of autonomy as 2, depth of discharge as 80 % and efficiency of battery as 90 %, the total energy need

to be stored in the batteries $= \dfrac{18974}{0.8 \times 0.9} = 52,705 \ Wh.$ For sizing of battery, if the system

voltage (V_{sys}) is 48 V, then ampere-hour requirement of the system (C_{sys}) or battery bank $= 52705/48 = 1098 \approx 1100$ Ah. Specifications of the available battery is 12 V $(V_{battery})$ and 220 Ah $(C_{battery})$.

Therefore, the number of batteries (N_s) to connect in series to get $V_{sys} = \dfrac{V_{sys}}{V_{battery}} = \dfrac{48}{12} = 4$

Similarly, the number of batteries (N_p) to connect in parallel to get $C_{sys} = \dfrac{C_{sys}}{C_{battery}} = \dfrac{1100}{220} = 5$

Hence, total number of battery = 4 × 5 =20

Four batteries are to connect in series to get a string and five strings are to connect in parallel to get the required energy for the load from the batteries.

Alternatively,

Determination of battery bank size (Ah)

$$= \frac{Total \ Wh \ required/day \ or \ electrical \ load \ (Wh/day) \times days \ of \ autonomy}{Each \ battery \ voltage \ (V) \times depth \ of \ discharge \times battery \ efficency}$$

$$= \frac{18974 \times 2}{12 \times 0.8 \times 0.9} = 4392.12 \ Ah$$

The depth of discharge indicates how much percent of total charge of the battery can be used.

$$\text{No. of batteries} = \frac{Battery\ capacity\ (Ah)}{Ampere\ hour\ capacity\ of\ each\ battery} = \frac{4392.12}{220} = 19.96 \approx 20$$

Twenty number of batteries, each of 12 V and 220 Ah are required for the above energy requirement.

Module circuit: The module circuit indicates the number of module/panel to be connected in series i.e. the size of an array and voltage input to the inverter, total number of arrays in the solar field and number of array to connect in parallel. The size of an array depends on the inverter maximum V_{OC} and V_{OC} of the module used.

$$\text{Size of an array} = \frac{Open\ circuit\ voltage\ of\ inverter}{Open\ circuit\ voltage\ of\ each\ PV\ module} = \frac{600}{21.52} = 27.88 \approx 28 \text{ modules}$$

Maximum voltage input to the inverter = maximum voltage of module × number of modules in series = 17.3 × 28 = 485V, which is less than the maximum DC voltage of inverter and also in the range of MPPT voltage.

$$\text{Total number of arrays in the solar field} = \frac{No.\ of\ modules}{No.\ of\ modules\ in\ an\ array} = \frac{28}{28} = 1$$

Twenty-eight number of modules may be connected in series by arranging them in 4 rows instead of one row for better incidence of solar radiation.

$$\text{Number of strings in parallel} = \frac{Total\ number\ of\ PV\ modules\ in\ the\ system}{No.of\ modules\ in\ a\ string} = \frac{28}{28} = 1$$

Total current of strings in parallel = current in one string (I_{sc}) × *total number of strings in parallel* = 10.28 A × 1 = 10.28 *A* which is less than maximum DC current of Inverter.

The specifications of battery, PV module and Inverter suitable for the PV power plant are presented in Table 3.2.

Sizing of cable: The amount of current flowing both in DC and AC circuit and system operating voltage are required to calculate voltage drop index(VDI) of the system which is an easy way to determine the size of cable (diameter or cross sectional area). The expression for VDI is as follows.

$$\text{VDI} = \frac{25\%\ more\ of\ the\ actual\ current\ to\ flow\ in\ the\ system\ (ampere)\ \times\ one\ way\ cable\ length\ in\ the\ circuit\ (feet)}{\%\ allowable\ voltage\ drop\ in\ the\ circuit\ \times\ system\ operating\ voltage\ (V)}$$

Table 3.2: Specifications of battery, PV module and Inverter

Battery specifications	
Nominal voltage	12 V
Depth of discharge	80 %
Battery capacity	220 Ah
Battery efficiency	90 %
Life of battery	8 years
System voltage	48 V
Days of autonomy	2 days

Module specifications	
Type	Polycrystalline silicon
W_P	175
I_{sc}	10.28 A
V_{oc}	21.52 V
I_{max}	9.12 A
V_{max}	17.3 V
Module efficiency	15 %
$L \times W \times T$	1490 × 990 × 40 mm
Temperature coefficient of Power (P_m), γ (%/°C)	-0.3845

Inverter specifications	
Rated power	5.5 kW
Max. DC input Voltage	600 V
Max. DC input current	42 A
MPPT voltage range	400-550V
Max output AC current	24 A
Output AC voltage (single phase)	220-260 V
Max. efficiency	98.3

Type of circuit	Max. wattage (W)	System voltage (V)	Max. current (A)
DC	4933	48	102
AC	5500	230	23.91

After calculating VDI with 2 % allowable voltage drop, the standard wire gauge (SWG) of the cable is obtained by using VDI table for a particular material like copper or aluminium. After knowing SWG, standard wire gauge dimensions' table is referred to find out the diameter or cross sectional area of the cable to be used for the system.

3.13 TECHNO-ECONOMIC ENVIRONMENTAL ASSESSMENT OF PV POWER PLANT

Techno-economic environmental assessment of solar PV power plant comprises of life cycle assessment, financial evaluation and environmental benefits.

I. **Life cycle assessment:** Life cycle assessment is a method for assessing different energy aspects related to the development of a system and its potential impacts throughout its life. The assessment of any renewable energy system includes the amount of energy consumed by the system components for the materials, their manufacturing and transportation i.e. embodied energy of the system, amount of energy generated by the plant, its energy payback time (EPBT), its life cycle conversion efficiency (LCCE) and capacity utilization factor (CUF).

a. *Energy payback time of the plant:* It is defined as how long does a PV system need to operate to recover the energy that has been spent in making the system

and is given by: $EPBT = \dfrac{\left(E_m + E_{mf} + E_t + E_i + E_{mg}\right)}{E_g}$ where, E_m is the primary

energy demand to produce materials comprising PV system, E_{mf} is the primary energy demand to manufacture PV system, E_t is the primary energy demand to transport materials used during life cycle, E_i is the primary energy demand to install the system, E_{mg} is the primary energy demand for management during life cycle, E_g annual electricity generation in primary energy terms.

The value for the total energy consumed in materials, manufacturing, transport, installation and management for each m^2 area of module was computed by Tiwari et al. (2009) as $(E_m + E_{mf} + E_t + E_i + E_{mg})$ to be 1516.59 kWh/m^2 of module.

Total energy of modules = No. of modules × (length × width) of each module = 28 × 1.49 × 0.99 = 41.30 m^2. Hence, total embodied energy = 1516.59 × 41.30 = 62,640 kWh.

Annual electricity generated (E_g) = Electrical load (Wh/day) × no. of clear sunny days in a year = 18974 × 300 = 5692 kW/year.

Energy Payback Time (EPBT)

$$= \frac{\textit{Total embodied energy of the modules used in the system (kWh)}}{\textit{Annual electricity generated from the system (kWh/year)}}$$

$$= \frac{62640}{5692} = 11 \text{ years.}$$

b. *Electricity production factor (EPF):* It is the ratio of annual energy generated by the system to the embodied energy. It predicts the overall performance of the PV module. It is the reciprocal of EPBT and is equal to $\frac{5692}{62640} = 0.09$

c. *Capacity utilization factor (CUF):* It is the ratio of actual energy generated by solar PV plant in a year to the energy output at its rated capacity (peak watt) over the yearly period. $CUF = \frac{5692000}{4933 \times 8760} = 0.13$

d. *Life cycle conversion efficiency (LCCE):* It is the net energy productivity of the PV system with respect to the solar energy (insolation) over the lifetime of the system. $LCCE = \dfrac{E_g \times L_s - E_{em}}{\textit{Solar insolation per year} \times L_s} = \dfrac{5692 \times 30 - 62640}{5 \times 300 \times 41.30 \times 30} = 0.058$

where L_s is the total life of the system and E_{em} is the embodied energy.

II. **Financial Evaluation of SPV power plant:** Financial evaluation of any system is important for its sustained growth in this competitive world. The realistic values or current market prices of the components associated with the project need to be taken for its convincing financial assessment. This would ultimately provide an insight regarding the economic viability of a proposed venture. The table below presents the cost break-ups of standalone roof top SPV power plant. The financial evaluation of the power plant has been assessed taking into account all the relevant present and future costs as well as revenues involved with it. The useful lives of PV system, battery bank and inverter have been assumed to be respectively 30, 10 and 10 years respectively. The break-up costs of solar PV power plant are presented in Table 3.3.

The line diagram of various cash flows at different intervals of time during the useful life of the set-up is shown in the figure 3.5 below. The life cycle cost assessment of the set up includes the following items.

Table 3.3: Break-up cost of solar PV power plant

Name of PV component	Number	Sizing	Approximate cost range (Rs.)	Cost per unit (Rs.)	Cost (Rs.)
Solar module (total 4900 W_p)	28	175 W_p	32-35/ W_p *	35/W_p	1,71,500
Charge controller	1	103 ampere × 12 Volt	50-60/Amp #	60/Amp	6,180
Batteries	20	220 Ah x 12 V	40-50/Ah #	45/Ah	1,98,000
Inverter	1	5.5 kW	7-12/ W_p*	10/ W_p	55,000
Structure			2-5 / W_p *	3/ W_p	14,700
Cables			1-2 / W_p *	1.5/ W_p	7350
Peripherals (Junction box, DC protection system, earthing, lightning arrestor)			1.5-2.5 W_p *	2.0/ W_p	9,800
Supervision and installation			3-4/ W_p *	3/ W_p	14,700
Total					**4,77,230**

* https://www.solarmango.com/2016/cost-break ups-rooftop-solar power plants -India -2016.

https://ezysolare.com/elite

Cash Flow Diagram

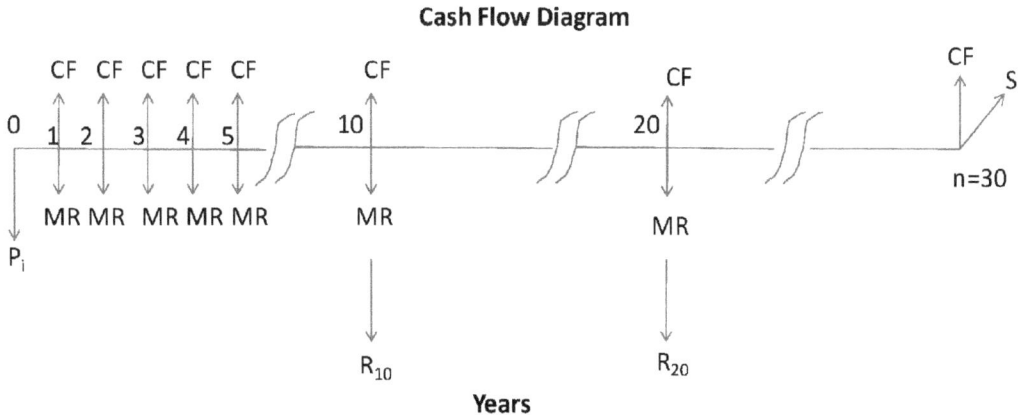

Years

Fig 3.5. Cash flow diagram for life cycle costs and revenues of the experimental set-up

Capital cost (P$_i$): Capital cost is the total cost involved for all the components in the set-up. The capital cost is mentioned in the Table 4.3 above i.e. Rs.4,77,230.

Maintenance and repair cost (P$_{mr}$): It is the cost, incurred during the operation of the power plant on annual basis. It includes the costs for maintenance of the components involved and repair of structure, peripherals and replacement of cables as and when required. It is assumed to be 0.5 % of the total capital cost *(P$_i$)* in 1st year. The inflation in the cost of maintenance and repair (MR) has been considered at the rate of 0.5 % per annum. The net maintenance and repair cost in terms of present value (P_{mr}) is calculated by using the formula for present worth of geometric gradient series as per the following equation.

$$(P_{mr}) = \frac{A_1}{(i-g)}\left[1-\left(\frac{1+g}{1+i}\right)^n\right]$$

Where A_1 is the 1st year maintenance and repair cost (0.5 % of P_i), i = interest rate (4 %), g = inflation rate (constant percentage increase rate in each year) and n is the useful life of the set-up (30 years). The inflation rate (g) in annual maintenance and repair cost is assumed to be 0.5 %. The annual interest rate has been considered as 4 %, a subsidized interest rate offered by Government of India to promote the use of renewable energy applications in India (MNRE, 2016).

$$P_{mr} = \text{Rs. } 43,760$$

Replacement cost (P_r): It is the cost incurred during replacement of battery bank and inverter at every 10 years interval. If R_{10} and R_{20} are the replacement costs after 10 and 20 years respectively, hence the net replacement cost in terms of present value is

$$(P_r) = R_{10} \times \left[\frac{1}{(1+i)^{10}}\right] + R_{20} \times \left[\frac{1}{(1+i)^{20}}\right]$$

Where (P_r) is the replacement cost in terms of present value. However, inflation rate per annum for the equipment, need to be replaced is considered as 0.5 % of their current price as per the present market survey. Hence, $R_{10} = R \times (1+0.005)^{10}$ and $R_{20} = R \times (1+0.005)^{20}$ where R is the capital cost included for battery bank and inverter. P_r = Rs.306977.

Salvage value (P_s): It is the cost incurred in demolition and disposal of the system. If S is the salvage value at the end of the useful life of the set up, then the salvage value in terms of present value is

$$(P_s) = S \times \left[\frac{1}{(1+i)^n}\right]$$

Salvage value is considered to be 20 % of capital cost (P_i), the depreciation of balance of system was considered equivalent to the rate of escalation in the price of the structural components, mainly steel and aluminium per kg. P_s = 29458.

Hence the overall life cycle cost of the proposed set-up in terms of present value is expressed as $P_{net} = P_i + P_{mr} + P_r - P_s = 4{,}77{,}230 + 43{,}760 + 306977 - 29458$ = Rs. 798509

Annualized uniform cost (C_A): Annualized uniform cost is defined as the product of net present value of the system and Capital Recovery Factor (CRF) and can be written as

$$C_A = P_{net} \times \left[\frac{\left(i \times (1+i) \right)^n}{\left((1+i)^n - 1 \right)} \right] = \text{Rs. } 45515$$

Cost per unit electricity (C_u): Cost per unit electricity (C_u) = (Annualized uniform cost) / (Annual energy output) $= \frac{C_A}{E_{out}} (Rs/kWh) = 45515/5692 = \text{Rs. } 8.00$

Payback Period (n_{pp}): Payback period is the time (generally in years) required to recover the investment costs. Let us assume that net cash flow (CF) is same for each year, then the net present value (P_{net}) in terms of payback period can be expressed as (Tiwari and Tiwari 2007)

$$P_{net} = CF \times \left[\frac{(1+i)^{n_{pp}} - 1}{i(1+i)^{n_{pp}}} \right] \rightarrow n_{pp} = \frac{\ln \left[\dfrac{CF}{CF - iP_{net}} \right]}{\left[\ln(1+i) \right]}$$

The net cash flow per year (CF) includes the annual income from the power plant in saving of electrical energy (5692 kWh/year) from the grid electrical system. The current electrical tariff is Rs. 7.00/kWh

The cash inflow in first year $5692 \times 7.00 = \text{Rs. } 39844$

The annual escalation of electric tariff as per current trends is about 2 %. The net cash inflow in savings of energy cost in terms of present value (P_{ci}) during its life time is calculated by using the formula for present worth of geometric gradient series as per the following equation.

$$(P_{ci}) = \frac{A_1}{(i-g)} \left[1 - \left(\frac{1+g}{1+i} \right)^n \right] = \text{Rs.} 879602$$

Where A_1 is the 1st year cash inflow (Rs. 39844), i = interest rate (4 %), g = inflation rate (constant percentage (2 %) increase rate in each year) and *n* is the useful life of the set-up (30 years).

Annualized uniform cash inflow (CF) is the product of net present value of cash inflow during the life period of the system and Capital Recovery Factor (CRF) and can be written as

$$CF = P_{ci} \times \left[\frac{\left(1 \times (1+i)^n\right)}{\left((1+i)^n - 1\right)} \right] = \text{Rs. 50137 and } n_{pp} \text{ (Pay back period)} = 25 \text{ years}$$

III. Financial Evaluation of Power Plant with 30 % subsidy provided by Govt.

Capital cost (P_i): Capital cost is the total cost involved for all the components in the set-up. The capital cot is mentioned in the Table above i.e. Rs. 3,34.061 (70% of total cost)

Maintenance and repair cost (P_{mr}):

$$\left(P_{mr}\right) = \frac{A_1}{(i-g)} \left[1 - \left(\frac{1+g}{1+i}\right)^n \right] = \text{Rs. 30628}$$

Replacement cost (P_r): If R is the capital cost included for battery bank and inverter i.e. Rs. 2,53,000, then P_r = Rs.306977.

Salvage value (P_s):

$$\left(P_s\right) = S \times \left[\frac{1}{(1+i)^n} \right] \text{ where } S = 20 \text{ % of capital cost } (P_i) \text{ i.e Rs. 477230 and } P_s =$$

29458.

Hence the overall life cycle cost of the proposed set-up in terms of present value is expressed as $P_{net} = P_i + P_{mr} + P_r - P_s$ = 334061 + 30628 + 306977 – 29458 = Rs. 642208

Annualized uniform cost (C_A)

$$C_A = P_{net} \times \left[\frac{\left(i \times (1+i)^n\right)}{\left((1+i)^n - 1\right)} \right] = \text{Rs. 36605}$$

Cost per unit electricity (C_u): Cost per unit electricity (C_u) = (Annualized uniform cost) / (Annual energy output) = $= \frac{C_A}{E_{out}} (Rs/kWh)$ = 36605/5692 = Rs. 6.43

Payback Period (n_{pp}):

$$P_{net} = CF \times \left[\frac{(1+i)^{n_{pp}} - 1}{i(1+i)^{n_{pp}}} \right] \rightarrow n_{pp} = \frac{\ln\left[\frac{CF}{CF - i P_{net}} \right]}{\left[\ln(1+i) \right]}$$

The annual escalation of electric tariff as per current trends is about 2 %. The net cash inflow in savings of energy cost in terms of present value (P_{ci}) during its life time is calculated by using the formula for present worth of geometric gradient series as per the following equation.

$$(P_{ci}) = \frac{A_1}{(i-g)}\left[1 - \left(\frac{1+g}{1+i}\right)^n\right] = \text{Rs.879602}$$

Where A_1 is the 1st year cash inflow (Rs. 39844), i = interest rate (4 %), g = inflation rate (constant percentage (2 %) increase rate in each year) and n is the useful life of the set-up (30 years).

Annualized uniform cash inflow (CF) is the product of net present value of cash inflow during the life period of the system and Capital Recovery Factor (CRF) and can be written as

$$CF = P_{ci} \times \left[\frac{\left(i \times (1+i)^n\right)}{\left((1+i)^n - 1\right)}\right] = \text{Rs.50,137}$$

The net cash flow per year (CF) includes the annual income from the power plant in saving of electrical energy (5692 kWh/year) from the grid electrical system. The current electrical tariff is Rs. 7.00/kWh.

The cash inflow in first year 5692 = Rs. 39844

$$n_{pp} \text{ (Pay back period)} = 18 \text{ years}$$

IV. **Environmental Benefits:** Power generated from solar PV system or any other renewable resource does not release any harmful emissions. Coal based thermal power plant in the present context of generating and fulfilling the major shares for the power demands of the country releases huge amount of greenhouse gases like carbon dioxide (CO_2), nitrogen oxide (NO_x), sulphur dioxide (SO_2) and ash. The use of solar power plant reduces the emissions of pollutants to the environment as the source for its operation i.e. sun is a clean source of energy and it requires no fuel and releases no greenhouse gas emission during the service periods. The reduction in greenhouse gases emissions through the use of solar PV power plant (4.9 kW) has been assessed. The reduction in the emission of greenhouse gases has been presented in table 3.4.

Table 3.4: Greenhouse gases (GHG) reduction by SPV system

GHG from coal fired thermal power plant	Emission per kWh of electricity (g/kWh)	Total annual reduction for 5692 kWh (kg)	Reference
CO_2	980	5578	Vasisht et al (2016)
SO_2	1.24	7.05	Tarigan and Kartikasari
NO_x	2.59	14.75	(2015)
Ash	68	387	

Carbon Dioxide (CO_2) Emission, Mitigation and Carbon Credit

With the growing significance of environmental issues, clean energy generation has become increasingly important. Solar PV system is a favourable and viable option for a clean source of energy and is environment friendly. As the space in the roof top is generally not utilized properly, it can be effectively used by installing PV power plant as a standalone or grid-tied manner.

CO_2 Emission

The average carbon dioxide emission for electricity generation from coal based thermal power plant is approximately 0.98 kg of CO_2 per kWh at the source. If the transmission and distribution losses for Indian condition are taken as 40% and poor inefficient electric equipment losses are around 20%, then the figure 0.98 can be taken as 1.58 kg per kWh (Sharma and Tiwari 2013).

CO_2 Mitigation

The CO_2 mitigation is the amount of CO_2 emission reduction by generating the energy from the PV plant that would otherwise be released by the thermal power plant in case of India.

Yearly CO_2 mitigation = yearly electricity generation (5692 kWh) × 1.58 = 8.88 tonnes of CO_2

Carbon Credit Earned

One carbon credit represents the mitigation of 1 ton of CO_2 emission. Carbon credit is awarded against reduction in greenhouse gases emissions mostly CO_2. It can be traded in the international market at the current market price. The emissions of CO_2 due to various energy utilizations and generations pose a major contributing factor to the greenhouse effects. International treaties for mitigating emissions of greenhouse gases across the globe have therefore introduced carbon credit trading which is an administrative approach for controlling pollution by providing economic incentives for the efforts to reduce the emissions of pollutants. Carbon credits are therefore a tradable permit scheme prevailing among the countries included in the strategic planning of controlling the emissions of greenhouse gases. The amount of carbon credit earned due to the use of this proposed PV system can be ca lculated by the following equation:

Carbon credit earned = € 27/ton × CO_2 emission mitigated by the set up (tons/life). € = Rs 83. The amount € 27/ton represents the monetary value of mitigating one ton of CO_2 emission. The price of carbon credit is as per European Climate Exchange (www.ecx.eu). The summary of techno-economic environmental parameters of the existing set-up is presented in the table 3.5 below.

Table 3.5: Estimated techno-economic environmental parameters of the study

S.N.	Parameters	Unit	Calculated values (without subsidy)	Calculated values (with subsidy of 30 %)
1	Capacity of solar PV power plant	kW_p	4.9	4.9
2	Capital cost of set-up	Rs.	4,77,230	3.34,061
3	Life of set-up	year	30	30
4	Maintenance and repair cost in terms of present value	Rs.	43,760	30,628
5	Replacement cost in terms of present value (Replacement of battery and inverter after each 10 years)	Rs.	3,06,977	3,06,977
6	Salvage value in terms of present value	Rs.	29458	29458
7	Net present value of set-up in terms of present value	Rs.	7,98,509	6.42,208
8	Annualized uniform cost of set-up in terms of present value	Rs.	45,515	36,605
9	Electricity generated from the set-up/year	kWh	5692	5692
10	Cost of unit electricity from PV system	Rs.	8.00	6.43
11	Cash inflow/year (without using grid electricity)	Rs.	39,844	39,844
12	Payback period	year	25	18
13	Total reduction of CO_2 emission from the proposed set-up during its life	ton	267	267
14	Total reduction of SO_2 emission from the proposed set-up during its life	ton	0.25	0.25
15	Total reduction of NO_x emission from the proposed set-up during its life	ton	0.45	0.45
16	Total reduction of ash emission from the proposed set-up during its life	ton	12	12
17	Carbon credit earned from the proposed set-up during its life	Rs.	5,98,347	5,98,347

3.14 SIZING STANDALONE SOLAR PV SYSTEM CONSIDERING LOADS SEPARATELY ON RESISTIVE AND INDUCTIVE BASIS

The nature of electrical loads for the appliances used in various sectors are different. The electrical load may be resistive, inductive, capacitive or some of the combination between them. The classifications of electrical load are shown in the figure (3.6) below.

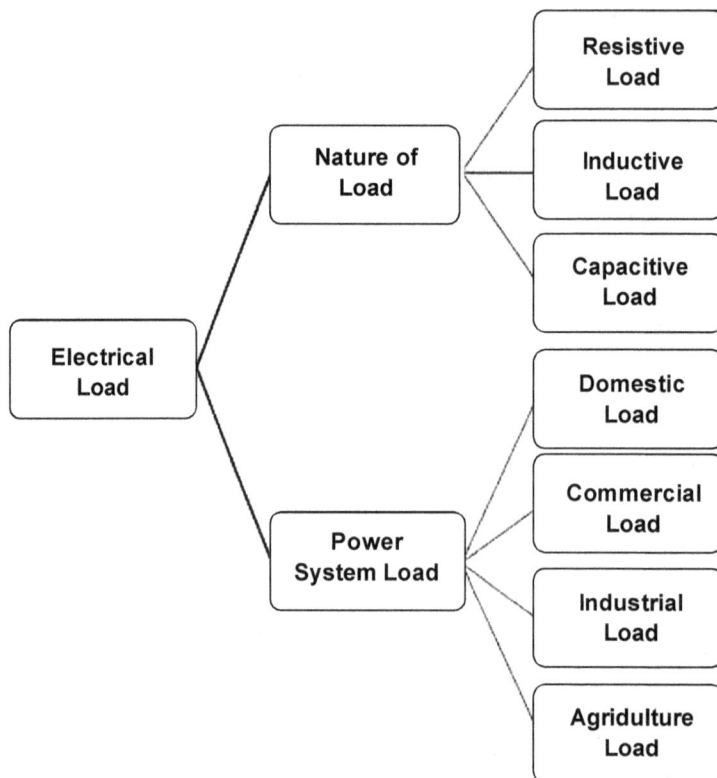

Fig. 3.6. Various types of electrical load for different applications

Resistive Load

The resistive load obstructs the flow of electrical energy in the circuit and converts it into thermal energy, due to which, the energy drop occurs in the circuit. The lamp and the heater are the examples of the resistive load. The resistive loads take power in such a way, that, the current and the voltage wave remain in the same phase. Thus, the power factor of the resistive load remains in unity. The examples are electrical heater, candescent bulb. mobile, laptop, TV, LED light, toaster, oven etc.

Inductive load

The inductive loads use the magnetic field for doing the work. The transformers, generators, motor are some of the examples. The inductive load has a coil which stores magnetic energy when the current pass through it. The current wave of the inductive load is lagging behind the voltage wave and the power factor of the inductive load is also lagging. This type of load pulls a large amount of current when first energized. After a few cycles or seconds, the current settles down to the full-load running current. The appliances operating under this type of load have much higher start-up power than the rated power. (3-5 times).

Capacitive Load

In the capacitive load, the voltage wave is leading the current wave. The examples of capacitive loads are capacitor bank, three phase induction motor starting circuit etc.

Resistive loads do not necessitate any significant surge current when energized. Like light bulb, electric heater etc. are resistive loads. On the other hand, inductive load requires a large amount of surge current when first energized which is about three times the normal energy requirement. Fan, electric motor, air-conditioner etc. are inductive load. Hence, the type of electrical load needs to be identified separately for precise sizing of PV system. The electrical loads for the appliances considered in the above section for a residential building as per Table 3.1, can also be taken for sizing of PV system separating the resistive and inductive loads and are presented in Table 3.6.

Most of the PV applications require operation of the load during non-sunshine hours and smooth operation of the load during sunshine hours irrespective of the variation in radiation intensity. This necessitates the use of battery in a standalone PV system. Whenever, a battery is used, a charge controller is also used to ensure long battery life. Most of the loads are of AC types, which require an inverter to be used in the system to convert DC into AC. An MPPT circuit is usually used in the system to optimize the PV source utilization. All the components included in a standalone PV system are shown in the figure (3.7) below.

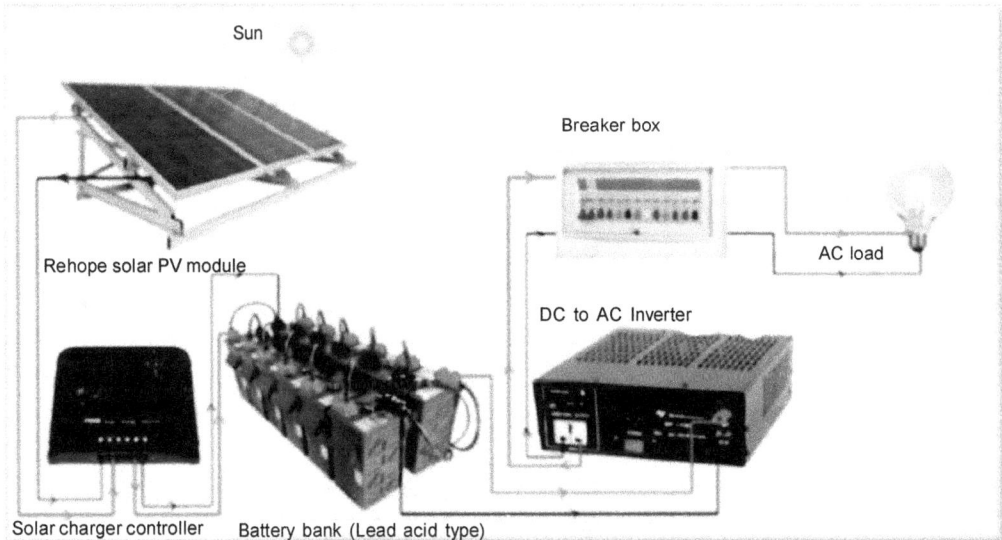

Fig. 3.7. Components included in a standalone PV system

During sunshine hours, energy flows from PV source to the battery through MPPT based charge controller. When the load is operating, the energy flows from the battery to the load through inverter in case of AC load. While designing the system, it is required to proceed in the reverse direction i.e. from load to the inverter, then to the battery, the charge controller and finally for the PV panel.

Table 3.6: Electrical loads for the residential building under consideration

S.N.	Electrical appliances	Resistive/ Inductive	Rated power (W)	No.	Total power (W)	(working hours/ day/Summer season load (Wh)	Daily electrical loads	
							(working hours/day) /Rainy season load (Wh)	(working hours/ day)/Winter season load (Wh)
1	CFL	Resistive	36	6	216	(14)/ (3024)	(12)/ (2592)	(14)/ (3024)
2	Ceiling Fan	Inductive	60	6	360	(20)/(7200)	(16)/(5760)	———
3	TV	Resistive	100	1	100	(10)/ (1000)	(10)/ (1000)	(10)/ (1000)
4	Refrigerator	Inductive	250	1	250	(24)/ (6000)	(24)/ (6000)	———
5	Laptop	Resistive	60	2	120	(10)/(1200)	(10)/(1200)	(10)/(1200)
6	Submersible pump	Inductive	1000	1	1000	(0.5)/ (500)	(0.5)/ (500)	(0.5)/ (500)
7	Mobile charger	Resistive	5	5	25	(2)/(50)	(2)/(50)	(2)/(50)
	Total daily load (Wh)					18974*	18002	5774
Total resistive load (Wh) (based on summer period)/ total resistive wattage (watt)						5274/461		
Total inductive load (Wh) (based on summer period)/ total inductive wattage (watt)						13700/1610		

* Maximum daily load is considered for sizing of PV panel.

3.15 SIZING OF PV SYSTEM

The sizing of PV system involves four steps which are to be proceeded in the following sequence.

(a) Load estimation

(b) Determining capacity of inverter and system DC voltage

(c) Determining capacity of battery bank

(d) Determining the specification of MPPT based charge controller

(e) Sizing of PV array

Load Estimation

The daily resistive and inductive loads, as estimated above are respectively 5274 Wh and 13700 Wh and their corresponding wattages are 461 W and 1610 W.

Determining Capacity of Inverter and System DC Voltage

An inverter is rated by its output power (kVA) and DC input voltage (V_{dc}). Power rating of the inverter should not be less than the total power consumed in different loads. On the other hand, it should have the same nominal voltage of battery bank that is charged by solar PV module. *Typically, capacity of the inverter is taken to be the sum of all the loads running simultaneously and 3.5 times the total power of the inductive loads to take care of surge protection* (Pal et al, 2015). Further, the obtained value is to be multiplied by 1.25 to get the requirement, if an option of 25% extra is kept for a reasonable future load expansion. As this is not mandatory, so in this analysis, the consideration of 25 % extra wattage has been neglected. Hence, the power ($P_{inv, \, delivered}$) that should be delivered by the inverter is as follows,

$$P_{inverter, \, delivered} = (TP + 3.5 \times P_{inductive}) = 2071 + 3.5(1610) = 7706 \text{ watt} \qquad ...(1)$$

Total power rating of all loads (TP) $= P_{resistive} + P_{inductive} = 461 + 1610 = 2071 \; W \quad ...(2)$

However, this is an ideal situation. This power calculation has to be corrected for power factor of inverter. Power rating of inverter (P_{output}) is related to the real power that is delivered by inverter as output and it is given by the following expression of power

factor (PF), $PF = \dfrac{Deliverable \; real \; power}{Power \; rating \; or \; inverter}$. Here, 'Real power' is the power that is

consumed for work on the load ($P_{inverter, \, delivered}$ in this case) and it is as calculated from Equation (1). Value of PF is generally taken as 0.8 for most of the inverters. So, the power rating of inverter (kVA) (kilo volt ampere) is expressed as

$$P_{kVA} = \frac{P_{inverter, \, delivered}}{PF} = \frac{7706}{0.8} = 9.6 \; kVA \approx 10kVA \qquad ... (3)$$

Inverter converts DC power to AC power. But this conversion is not 100% efficient. So, efficiency ($\eta_{inverter}$) of inverter is an important parameter which has to be taken care of. Continuous AC power load, which is the total power (*TP* as obtained above) needed when all the appliances are running at steady state condition and has to come from a DC power source, such as battery. Therefore, the continuous power load to the inverter,

$$P_{continuous\ power\ load,\ inverter} = \frac{TP}{\eta_{inverter}} = \frac{2071}{0.9} = 2300\ W \qquad ...(4)$$

Now the continuous (DC) input current ($I_{dc,\ inverter}$) to an inverter from PV modules can be determined, if the system DC voltage (V_{dc}) is specified.

Deciding System Voltage

The voltage and current from inverter to the load are of AC in nature and from PV array to the inverter, they are DC in nature. The output voltage of the inverter is decided by the voltage at which the load operates. But the input voltage to the inverter (also referred as system voltage, which is also the battery bank's terminal voltage, MPPT charge controller voltage and PV array terminal voltage) is decided mainly from the data sheet of the inverter selected as per its kVA rating for the loads under consideration. Higher PV system voltage may be chosen to minimize the current carried by the cables resulting into reduction of the power losses and voltage drop in the cables. Usually, the nominal or system voltage for the PV system may be taken as 12 V, 24 V, 48 V or 72 V or any multiple of 12. Hence, for this case, let us consider 276 V (23 × 12) as PV system voltage.

Daily Energy to be Supplied to Inverter

In this example, daily energy used by the load is 18,974 Wh. This energy is to be supplied by the battery through the inverter. In the inverter, energy conversion takes place from DC to AC. The process also needs its own efficiency for conversion. Therefore, the energy to be supplied to the inverter input should be more than the energy used by the load. In most cases, the inverter efficiency is reasonably high, at least 90 %.

$$\text{Input energy of inverter} = E_{input\ to\ Inverter} = \frac{E_{daily}}{\eta_{Inverter}} = \frac{18974}{0.9} = 21,082\ Wh. \qquad ...(5)$$

$$I_{dc,\ inverter} = = \frac{P_{continuous\ power\ load,\ inverter}}{V_{dc}} = \frac{2300}{276} = 8.33\ A \qquad ...(6)$$

Determining Capacity of Batteries

The terminal voltage of the battery bank, same as system voltage is 276 V in this case. The input energy to the inverter is same as the output energy of the battery bank. The charge storage capacity, which is essentially the energy storage capacity of the battery bank is determined by the daily energy requirement, depth of discharge, number of days of autonomy and battery bank terminal voltage.

$$\text{Required capacity of batteries} = \frac{E_{(input\ to\ inverter)} \times days\ of\ autonomy}{V_{dc} \times DoD} = \frac{21082 \times 2}{276 \times 0.8} = 191\ Ah$$

$$\text{... (7)}$$

Let us consider the battery of 12 V and 100 Ah, as per availability in the market

$$\text{Number of battery in series} = \frac{V_{dc}}{voltage\ of\ a\ single\ battery} = \frac{276}{12} = 23 \qquad \text{... (8)}$$

$$\text{Number of battery in parallel} = \frac{Capacity\ of\ battery\ bank}{Ah\ capacity\ of\ single\ battery} = \frac{191}{100} = 1.91 \approx 2 \text{ ...(9)}$$

Total number of batteries = 23 × 2 = 46 ... (10)

If we take battery efficiency (η_{Bat}) to be about 85% typically for lead acid battery, then energy required ($E_{Battery\ bank}$) from solar MPPT charge controller to charge the battery bank is given by the following equation

$$E_{battery\ bank} = \frac{V_{dc} \times Battery\ bank\ capacity}{\eta_{Bat}} = \frac{276 \times 191}{0.85} = 62\ kWh \qquad \text{... (11)}$$

Charge Controller Specification

The voltage handling capability of the MPPT based charge controller is same as the system voltage and 276 V in this case. Similarly, the current handling capability should also be same as the total current of the PV array. The energy to the input terminal of the battery bank is supplied through controller electronics (charge controller and MPPT). The efficiency of the controller circuit is generally quite high. Let the controller circuit efficiency be 90 %. Hence, the energy that should be supplied by the PV array at the input of the controller circuit be $\frac{62}{0.9} = 68kWh$. Thus about 68 *kWh* should be generated by the PV array every day. From the design point of view, the controller circuit should be able to handle the current flowing from the PV array at its input and from the controller to the battery bank.

Sizing of PV array

In order to charge battery bank, the PV array needs to supply the energy to the battery bank at 276 V. Therefore, the total Ah generated by the PV array should be 68000/72 = 246 Ah. The equivalent peak sun shine hours for the location under study is 5. The current requirement from PV array would be 246/5 = 49 A

Hence, $V_{PV\ array}$ = 176 V and $I_{PV\ array}$ = 49 A.

As per the availability from the market, let the watt peak of one module be 350 W_p (WAAREE make) whose $I_{max\ (module)}$ = 9.5 A and $V_{max\ (module)}$ = 37.50 V.

The number modules to be connected in parallel = 49/9.5 = 5.15 ≈ 5

The number of modules to be connected in series = 276/37.5 = 7.36 ≈ 7

Hence, the total number of modules = 5×7 =35.

Each row containing 7 modules would be in series connection and there would be 5 such rows in parallel. The complete sizing information of the PV system is presented in Table 3.7

The PV array operating voltage and current should lay within the allowable limits of inverter's maximum input DC voltage and input DC current in all operating conditions. The break-up costs of different components of PV system under consideration are presented in Table 3.8.

3.16 LIFE CYCLE COSTING (LCC)

The LCC is the cost of using a system during its lifetime. The cost of owning and running a system or equipment over its lifetime can be divided into three components:

(i) Capital cost (paid while purchasing the system /equipment).

(ii) Operation and maintenance cost (cost incurred for running the equipment, could be regular maintenance cost, or it could be fuel cost etc.)

(iii) Replacement cost (cost incurred if any component of system or equipment needs replacement before the life of the system is over).

Thus, the LCC includes all associated costs of the system that can be incurred over its lifetime. The LCC can be used to evaluate different designs of the set-up (having same functionality) of PV systems. The design with lowest LCC is chosen.

Present Worth of Future One-time Investments

The present worth of a product is defined as the amount of money that needs to be invested today with the discount rate, 'd', and the inflation rate, 'r', such that we would

Table 3.7: Complete sizing information of the PV system

Battery specifications	
Nominal battery voltage	12 V
Depth of discharge	80 %
Battery capacity	100 Ah
Battery efficiency	85 %
Life of battery	8 years
System voltage	276V
Days of autonomy	2 days
No. of battery in series	23
No. of battery in parallel	2
Total no. of battery	46

Module specifications	
Type	Polycrystalline silicon
W_P	350
I_{sc}	9.8 A
V_{OC}	46.4 V
I_{max}	9.5 A
V_{max}	37.5 V
Module efficiency	16 %
No. of modules in series	7
No. of modules in parallel	5
Total no. of modules	35
Solar cells/module	72
L × W × T	1960 × 990 × 40 mm
Temperature coefficient of Power (P_m), γ (%/°C)	-0.3845

Inverter specifications	
Rated power	10 kVA
Max. DC input Voltage	600 V
Max. DC input current	50 A
MPPT voltage range	180-400V
Max output AC current	24 A
Output AC voltage (single phase)	220-260 V
Max. efficiency	90

MPPT based charge controller

276 V and 49 A

Table 3.8: Break-up costs of solar PV power plant

Name of PV component	Number	Sizing	Approximate cost range (Rs.)	Cost per unit (Rs.)	Cost (Rs.)
Solar module (total 12250 W_p)	35	350 W_p	30-35/ W_p	30/W_p	3,67,500
Charge controller	1	49 ampere x 276 Volt	50-60/Amp	50/Amp	2450
Batteries	46	100 Ah x 12 V, each	40-50/Ah	40/Ah	1,84,000
Inverter	1	10 kVA	7-12/VA	10/VA	1,00,000
Structure			2-5 / W_p	3/W_p	36,750
Cables			1-2 / W_p	1.5/W_p	18,375
Peripherals (Junction box, DC protection system, earthing, lightning arrestor)			1.5-2.5 W_p	2.0/ W_p	24,500
Supervision and installation			3-4/ W_p	3/W_p	36,750
Total					**7,70,325 ≈ 7,70,000**

be able to purchase the product in the future, 'n' years later. Suppose, we have with us $ 'X' today and the cost of a product (that if we buy today) is also $ 'X'. But we need to buy this product in future as a replacement component (one-time investment). Because of the inflation, the cost of the product will increase and because of the discount rate (supposedly higher than the inflation rate), the value of money will also increase. Since we are assuming that the discount rate is higher than the inflation rate, the cost of the product in the future (in n years) will decrease due to the effective increase in the purchasing power of money and it will decrease by a factor, $F_{PW\text{-}one\ time}$, given by the following equation. Thus, the present worth of the investment that needs to be made 'n' years later would be lower by the factor given by $F_{PW\text{-}one\ time}$.

$$F_{PW-one\,time} = \frac{Future\ \cos t}{Future\ value} = \frac{X(1+r)^n}{X(1+d)^n} = \left(\frac{1+r}{1+d}\right)^n \qquad \dots (12)$$

Where $F_{PW\text{-}one\ time}$ can be read as 'present worth factor for future one-time investments'. Now if the present cost of a product is C_0, then its present worth (PW) for the actual one-time investment after 'n' years would be given as:

$$PW_{replacement} = C_0 F_{PW-one\,time} = C_0 \left(\frac{1+r}{1+d}\right)^n \qquad \dots (13)$$

Present Worth of Future Recurring Investments

There are requirements when the investment is required on recurring basis to run the system. in such cases the present worth of the future investments can be obtained by using Eq. (13), where recurring investment is considered as one-time investment for each year of operation. Thus, present worth of recurring expenses PW_{rec} can be obtained by summing up the present worth of one-time investments for the entire life of the system as given in Eq. (14).

$PW_{rec\text{-}beg}$: When the investment is made in the beginning of the year, (for instance buying diesel for diesel generator) or in Eq. (15), for $PW_{rec\text{-}end}$, when investment is made at the end of the year, (for instance buying grease for the diesel generator):

$$PW_{rec-beg} = C_0 + C_0\left(\frac{1+r}{1+d}\right)^1 + C_0\left(\frac{1+r}{1+d}\right)^2 + \dots + C_0\left(\frac{1+r}{1+d}\right)^{n-1} \qquad \dots (14)$$

$$PW_{rec-end} = C_0\left(\frac{1+r}{1+d}\right)^1 + C_0\left(\frac{1+r}{1+d}\right)^2 + \dots + \left(\frac{1+r}{1+d}\right)^n \qquad \dots (15)$$

Note that in either case [Eq.(14) or Eq. (15)], there are n number of purchases made in the lifetime on the system. The above equation can be simplified if we replace with the following term i.e.

$$k = \frac{1+r}{1+d} \qquad \dots (16)$$

The simplified version of Eqs. (14) and (15) in terms of parameter k can be written as:

$$PW_{rec-beg} = C_0 \left(\frac{1-k^n}{1-k} \right) = C_0 F_{PW-rec-beg} \text{ where } F_{PW-rec-beg} = \left(\frac{1-k^n}{1-k} \right) \qquad \dots (17)$$

$$PW_{rec-end} = C_0 k \left(\frac{1-k^n}{1-k} \right) = C_0 F_{PW-rec-end} \text{ where } F_{PW-rec-end} = k \left(\frac{1-k^n}{1-k} \right) \qquad \dots (18)$$

Here, $F_{(PW\text{-}rec\text{-}beg)}$ is the present worth factor for the recurring investments which are made in the beginning of each year and $F_{(PW\text{-}rec\text{-}end)}$ is the present worth factor of the recurring investment made at the end of each year. However, for financial evaluation, all calculations are done usually at the end of the year.

Normally, the value of 'k' is less than 1 and varies in the range of 0.95 to 1.0. This gives nearly the same value of $F_{(PW\text{-}rec\text{-}beg)}$ and $F_{(PW\text{-}rec\text{-}end)}$. Also, in many cases the investment may not be made exactly at the end or at the beginning, but may be in the middle of the year, or throughout the year (as purchase of fuel). Therefore, it is difficult to decide whether to use Eq. (3.17) or Eq. (3.18). In practice, either of the equation can be used to get a good estimate of cumulative recurring cost. The solution for the equations (3.17) and (3.18) can be done by using the sum of finite geometric series like $1 + k + k^2 + k^3 +$

$$\dots + k^n = \left(\frac{1-k^n}{1-k} \right) \text{ if } k \neq 1$$

Life Cycle Cost

In order to calculate the LCC of a system, first the present worth factors, as discussed in the above two paragraphs ($F_{(PW\text{-}one)}$ and $F_{(PW\text{-}rec\text{-}end)}$, are calculated. They are multiplied by the appropriate present cost components and added together to find out the total cost of operating a given system over its lifetime. For instance, a system has a lifetime of 10 years and has a capital cost C_1 (in terms of present cost). It requires replacement of a critical component in 5 years, having a cost C_2 (in terms of present cost). The system also has an annual maintenance cost of C_3 (in terms of present cost). Then the LCC of the system would be given by the following equation:

$$LCC = C_1 + F_{PW-one}\left(5\ years\right) \times C_2 + F_{PW-rec-end}\left(10\ years\right) \times C_3 \qquad \ldots (19)$$

If there are other cost components (such as fuel cost, electricity cost, etc.) associated with the system operation, then they should be added in Eq. (19) with appropriate present worth factor.

The LCC analysis can be used to compare the cost of owing and running different systems for a given period. For instance, it can help us in choosing a PV system design among different possible designs with a given type of electrical load.

Annualized LCC (ALCC)

It is sometimes to compare the LCC of a system on an annual basis. In this case, dividing the LCC by the expected life of the system will not give us the annualized LCC or ALCC of the system. Because it is assumed that the annual cost of operation of the system will not be the same throughout its lifetime; therefore, in order to find out the annual cost in terms of current value of money, it is necessary to divide the LCC by the present worth factor $F_{PW-rec-end}$ or $F_{PW-rec-beg}$.

Unit cost of generated electricity

In several situations, a given requirement of electrical energy (lighting, irrigation, etc.) can be fulfilled by several sources. Possible alternatives for electricity generations are such as use of solar PV modules, use of diesel generator, use of wind turbines or simply use of grid electricity (if available). Among these available options, which option should be chosen? The answer to this question can be obtained by finding the cost of generated electricity from each of these options over a period of time (e.g., 10 or 20 years). Then, from the economic point of view, the source which provides the lowest unit cost of electricity generated should be chosen. This analysis can be done by finding out ALCC of a given source and then dividing it by the annual number of electrical units generated by the given source. This would provide the generated cost of electricity from a source over its period of operation.

Discounted payback period (considering inflation and discount rate)

As a modification of the simple payback period (which fails to consider the time value of money), the costs and benefits may be adjusted to account for the changing value of money and their inflation over time. This leads to the estimation of a discounted payback period which is the length of time required for the project's equivalent receipts to be equal to the project's equivalent capital outlays. The length of time is usually expressed in years, required to recover the first cost of an asset or project.

There are two types of payback analysis as determined by the required return.

No return: d=0%: Also called simple payback, this is the recovery of only the initial investment.

Discounted payback: d > 0%: The time value of money is considered by using both inflation (decline in value of money over time) and discount rate (increase in value of money over time). For example, 10% per year must be realized in addition to recovering the initial investment.

If $C_{initial\ investment}$, capital investment, n_p, the simple payback period, n_{dp}, the discounted payback period, net cash flow, NCF = cash inflows - cash outflows (annual benefit-annual cost of operation) and 'd' is the discount rate and r is the inflation rate, the following situations may arise

(i) No return, *d=0% and r=0%*, NCF, varies annually,

$$-C_{Initial\ investment} + \sum_{n=1}^{n_p} NCF_n = 0 \qquad \ldots (20)$$

(ii) No return, *d=0% and r=0%*, annual uniform NCF, $n_p = \dfrac{C_{Initial\ investment}}{NCF}$ ... (21)

(iii) Discounted, *d>0% and r>0%*, NCF, varies annually,

$$-C_{Initial\ investment} + \sum_{n=1}^{n_{dp}} NCF_n \left(\frac{(1+r)^n}{(1+d)^n} \right) = 0 \qquad \ldots (22)$$

(iv) Discounted, *d>0% and r>0%*, annual uniform NCF,

$$-C_{Initial\ investment} + NCF \left(\frac{(1+r)^n}{(1+d)^n} \right) = 0 \qquad \ldots (23)$$

Therefore, the discounted payback period, n_{dp} is the smallest 'n' that satisfies

$$-C_{initial\ investment} + \sum_{n=1}^{n_{dp}} (B_n - C_n) \left(\frac{(1+r)^n}{(1+d)^n} \right) = 0 \qquad \ldots (24)$$

Where B is the annual benefit and C is the annual cost of operation and $k = \left(\dfrac{1+r}{1+d} \right)$.

Calculation of Life Cycle Cost (LCC)

Using the Eq. (19), the LCC for the present PV system can be calculated and as follows.

$LCC = C_{(initial\ investment)} + F_{(PW\text{-}one\ time)}$ *(10 years)* $\times C_{(present\ cost\ of\ battery\ and\ inverter)}$ + $F_{(PW\text{-}one\ time)}$ *(20 years)* $\times C_{(present\ cost\ of\ battery\ and\ inverter)}$ + $F_{(PW\text{-}rec\text{-}end)}$ *(30 years)* $\times C_{(Present\ operation\ and\ maintenance\ cost)}$ ... (25)

Initial investment = Rs. 7,70,000, present cost of batteries and inverter = Rs. 2,84,000,

Inflation rate per annum for the equipment, need to be replaced is considered as 0.5 % of their current price as per the present market survey and discount rate = 4%. The discount rate has been considered as 4 %, a subsidized discount rate offered by Government of India to promote the use of renewable energy applications in India (MNRE, 2016).

Annual recurring cost, (operation and maintenance cost), =0.25 % of initial investment for 1st year i.e. Rs. 1925 (0.25 % of Rs. 7,70,000). After that, the inflation rate for recurring cost is 0.25% per annum and discount rate = 4%, $k'_{recurring\ cost}$ = 0.96.

$$PW_{replacement} = C_{(present\ cost\ of\ battery\ and\ inverter)} \times F_{(PW\text{-}one\ time)} =$$

$$C_{present\ cost\ of\ battery\ and\ inverter} \times \left(\frac{1+r}{1+d}\right)^n \text{ and } F_{PW-red-end} = k\left(\frac{1-k^n}{1-k}\right) \text{where}$$

$$k = \left(\frac{1+r}{1+d}\right) = \frac{1+0.0025}{1+0.04} = 0.96$$

$$LCC = 7,70,000 + 2,84,000 \times \frac{(1+0.005)^{10}}{(1+0.04)^{10}} + 2,84,000 \times \frac{(1+0.005)^{20}}{(1+0.04)^{20}}$$

$$+1925 \times 0.96\left(\frac{1-0.96^{30}}{1-0.96}\right) = 7,70,000 + 2,01,671+1,43,209+32802=11,47,682$$

$ALCC$ = 11,47,682/17.04 =67,352

Unit cost of generated electricity = ALCC/electrical units generated per year =Rs67352/ 6270 = Rs. 10.74/unit of electricity

With subsidy

$$LCC_{with\ subsidy} = 5,39,000 + 2,84,000 \times \frac{(1+0.005)^{10}}{(1+0.04)^{10}} + 2,84,000 \times \frac{(1+0.005)^{20}}{(1+0.01)^{20}}$$

$$+1925 \times 0.96\left(\frac{1-0.96^{30}}{1-0.96}\right) = 5,39,000 + 2,01,671+1,43,209+32802=9,16,682$$

$ALCC_{(with\ subsidy)}$ = 9,16,682/17.04 =53,795

Unit cost of generated electricity = ALCC/electrical units generated per year = Rs. 53795/6270 = Rs. 8.5/unit of electricity.

Annual benefit is the amount saved due to use of PV electricity from the PV system and the amount in this case is (19 kWh/day × 330 $\dfrac{days}{year}$ × Rs.7.00/kWh)=Rs.43,890.

Inflation rate for electrical unit is 2% per annum and this is required to calculate the variations in annual amount saved due to use of solar PV system. $K_{annual\ benefit}$ = 0.98. Estimations for the payback period of the system under consideration are presented without and with subsidy in the following Tables 3.9 and 3.10 respectively.

Environmental Benefits

GHG from coal fired thermal power plant	Emission per kWh of electricity (g/kWh)	Total annual reduction of emissions for 6270 kWh of electricity generated from PV system (kg)	Reference
CO_2	980	6144	Vasisht et al (2016)
SO_2	1.24	7.77	Tarigan and
NO_x	2.59	16.23	Kartikasari (2015)
Ash	68	426	

CO_2 mitigation

The CO_2 mitigation is the amount of CO_2 emission reduction by generating the energy from the PV plant that would otherwise be released by the thermal power plant in case of India.

Yearly CO_2 mitigation = yearly electricity generation (6270 kWh) × 1.58 = 9.90 tonnes of CO_2

Carbon credit earned

Carbon credit earned = € 27/ton × CO_2 emission mitigated by the set up (tons/life). €= Rs 83. Hence carbon credit earned = Rs. 665577. The summary of techno-economic environmental parameters of the existing set-up is presented in the table 3.11 below.

3.17 CONCLUSION

The above discussion provides a complete analysis of a proposed standalone rooftop solar PV power plant and the possible future scope of this venture as service based

Table 3.9: Estimation of discounted payback period considering both inflation and discount rate

Year	Initial investment	Annual Benefit	Present worth factor (k)	Present worth	Annual cost	Present worth factor (k')	Present worth	Net cash flow	Cumulative cash flow
0	-7,70,000	0	1	0	0	1	0	-7,70,000	-7,70,000
1	0	43890	0.98	43012	1925	0.96	1848	41,164	-7,28,836
2	0	43890	0.96	42,134	1925	0.92	1771	40,363	-688473
3	0	43890	0.94	41256	1925	0.88	1694	39562	-648911
4	0	43890	0.92	40378	1925	0.84	1617	38761	-610150
5	0	43890	0.90	39501	1925	0.81	1559	37942	-572208
6	0	43890	0.88	38623	1925	0.78	1501	37122	-535086
7	0	43890	0.86	37745	1925	0.75	1443	36302	-4,98,784
8	0	43890	0.85	37306	1925	0.72	1386	35920	-462864
9	0	43890	0.83	36428	1925	0.69	1328	35100	-4,27,764
10	0	43890	0.81	35550	1925	0.66	1270	34280	-393484
11	0	43890	0.80	35112	1925	0.63	1212	33900	-359584
12	0	43890	0.78	34234	1925	0.61	1174	33060	-326524
13	0	43890	0.76	33356	1925	0.58	1116	32240	-294284
14	0	43890	0.75	32917	1925	0.56	1078	31839	-262445
15	0	43890	0.73	32039	1925	0.54	1039	31000	-231445
16	0	43890	0.72	31600	1925	0.52	1001	30599	-200846
17	0	43890	0.70	30723	1925	0.49	943	29780	-171066
18	0	43890	0.69	30284	1925	0.47	904	29380	-141686
19	0	43890	0.68	29845	1925	0.46	885	28960	-112726
20	0	43890	0.66	28967	1925	0.44	847	28120	-84606
21	0	43890	0.65	28528	1925	0.42	808	27720	-56886
22	0	43890	0.64	28089	1925	0.40	770	27319	-29567
23	0	43890	0.62	27211	1925	0.39	750	26461	-3106
24	0	43890	0.61	26772	1925	0.37	712	26060	**22954**
25	0	43890	0.60	26334	1925	0.36	693	25641	

Payback period=23 years and 2 months

Table 3.10: Estimation of discounted payback period considering both inflation and discount rate with availing 30 % Govt. subsidy on capital investment.

Year	Initial investment	Annual Benefit	Present worth factor (k)	Present worth	Annual cost	Present worth factor (k')	Present worth	Net cash flow	Cumulative cash flow
0	-5,39,000	0	1	0	0	1	0	-5,39,000	-5,39,000
1	0	43890	0.98	43012	1925	0.96	1848	41,164	-4,97,836
2	0	43890	0.96	42,134	1925	0.92	1771	40,363	-4,57,473
3	0	43890	0.94	41256	1925	0.88	1694	39562	-417911
4	0	43890	0.92	40378	1925	0.84	1617	38761	-379150
5	0	43890	0.90	39501	1925	0.81	1559	37942	-341208
6	0	43890	0.88	38623	1925	0.78	1501	37122	-304086
7	0	43890	0.86	37745	1925	0.75	1443	36302	-267784
8	0	43890	0.85	37306	1925	0.72	1386	35920	-231864
9	0	43890	0.83	36428	1925	0.69	1328	35100	-196764
10	0	43890	0.81	35550	1925	0.66	1270	34280	-162484
11	0	43890	0.80	35112	1925	0.63	1212	33900	-128584
12	0	43890	0.78	34234	1925	0.61	1174	33060	-95524
13	0	43890	0.76	33356	1925	0.58	1116	32240	-63284
14	0	43890	0.75	32917	1925	0.56	1078	31839	-31445
15	0	43890	0.73	32039	1925	0.54	1039	31000	-445
16	0	43890	0.72	31600	1925	0.52	1001	30599	**30154**

Payback period=15 years

Table 3.11: Estimated techno-economic environmental parameters of the study

S.N.	Parameters	Unit	Calculated values (without subsidy)	Calculated values (with subsidy of 30 %)
1	Capacity of solar PV power plant	kW_p	12	12
2	Capital cost of set-up	Rs.	7,70,000	5,39,000
3	Life of set-up	year	30	30
4	Life Cycle Cost (LCC)	Rs.	11,47,682	9.16.682
5	Annualized Life Cycle Cost (ALCC)	Rs.	67,352	53,795
6	Electricity generated from the set-up/year	kWh	6270	6270
7	Cost of unit electricity from PV system	Rs.	10	8
8	Cash inflow/year (without using grid electricity)	Rs.	39,844	39,844
9	Payback period	year	23	15
10	Total reduction of CO_2 emission from the proposed set-up during its life	ton	297	297
11	Total reduction of SO_2 emission from the proposed set-up during its life	ton	0.23	0.23
12	Total reduction of NO_x emission from the proposed set-up during its life	ton	0.48	0.48
13	Total reduction of ash emission from the proposed set-up during its life	ton	12.78	12.78
14	Carbon credit earned from the proposed set-up during its life	Rs.	6,65,577	6,65,577

entrepreneurship. As the use of solar PV electricity is increasing day by day due to the reduction in the price of solar module, advancement in the technology of enhancing the efficiency of solar cell, an environment friendly power source and provision of subsidy facility at Government level for its wide popularization, this would become a better and most suitable option for the users both at individual and community level. Roof top solar power plant is most feasible at domestic level to obtain a secured and reliable source of energy particularly in off-grid areas. However, this practice can also be followed in urban areas where the number of pucca building is more and shadow free space in the roof top is plentifully available for installation of power plant. Hence the future scope for entrepreneurship option in providing service for solar power plant both in rural and urban areas is high. In comparison to off-grid PV system, the on-grid system is more preferable due to generation of more revenue by supplying surplus electricity to the grid and eliminating the high cost of storage component i.e. battery, a major device in standalone system. This ultimately results into high payback period of the system. The Government is also now focussing on the promotion of more number of grid connected solar PV power plant. Nevertheless, standalone system is to be used where grid electricity is not available. Dissemination of the practice will eventually help in developing a sustainable environment and improving policies for the enhanced use of solar energy. Rooftop mounted solar PV system is always advantageous from the economic performance point of view than similar ground mounted system due to exclusion of the cost of land. Grid connected roof top solar PV plant is also more preferable to standalone system. Solar power plant is not only used for reducing the consumption of fossil fuel but also it can be a continuous source of energy for critical areas where uninterrupted supply is demanded. Reduction in CO_2, SO_2, NO_x and ash emissions from the energy generated with solar energy which could otherwise be generated with highly polluting coal based thermal power plant has also been analysed. The life of the plant, the current discount rate, inflation rate, escalation in energy cost, Govt. subsidy etc. have been considered for detailed analysis of the plant. Earned carbon credit for the proposed set-up has been evaluated. The estimated data presented, would definitely be helpful on the entrepreneurship development of PV system for manufacturers, designers, installers and end users in their efforts to predict the long term effect.

REFERENCES

Anonymous. 2017. Bridge to India. SOLAR Including the 2017. Available: https://bridgetoindia.com/report/india-solar-map-september2017.

BP 2017 Energy Outlook. Available: https://safety4sea.com/ wp-content/uploads/2017/01/BP-Energy-Outlook-2017_01.pdf.

Abhik Milan Pal, Subhra Das and N.B.Raju. 2015. Designing of a Standalone Photovoltaic System for a Residential Building in Gurgaon, India. *Sustainable Energy, 2015, Vol. 3, No. 1, 14-24*

Bandyopadhyay B, Gupta MK, Vashishtha RD. Solar Radiation Assessment over India. Book on Ä Solar future for India. Edited by G M Pillai, WISE 2011, isbn:81-902925-2-8:349-363.

Berwala, Anil K., Kumarb, Sanjay, Kumaria, Nisha, Kumara, Virender and Haleemc, Abid. 2017. Design and analysis of rooftop grid tied 50 kW capacity Solar Photovoltaic (SPV) power plant Renewable and Sustainable Energy Reviews.

Brunisholz, M. (Ed.). (2016). Snapshot of global photovoltaic markets (Publication No. 601 IEA PVPS T1-29:2016). International Energy Agency. Retrieved August 24, 2016, from 602 http://www.iea-pvps.org.

California Energy Commission. 2001. *A Guide to Photovoltaic (PV) System Design and Installation*. June. http://www. energy.ca.gov/reports/2001-09-04_500-01-020.PDF

Charles R. 2015. Optimum Tilt of Solar Panels. Available: http://www.solarpaneltilt.com

Chakraborty S, Sadhu P K and Pal N. 2015. Technical mapping of solar PV for ISM-an approach toward green campus. Energy Sci. Eng. 3: 196–206.

Honda, S., A. Lechner, S. Raju, and I. Tolich. 2012. *Solar PV System Performance Assessment Guideline*. San Jose, CA:San Jose State University.

Jain S, Jain N K and Vaughn W J 2018 Challenges in meeting all of India's electricity from solar: An energetic approach. Renew. Sustain. Energy Rev. 82: 1006–1013.

Mulcué-Nieto, L.F., Mora-López, L., 2014. A new model to predict the energy generated by a photovoltaic system connected to the grid in low latitude countries. Sol. Energy 107, 423–442.

Mondol, J.D., Yohanis, Y., Smyth, M., Norton, B., 2006. Long term performance analysis of a grid connected photovoltaic system in Northern Ireland. Energy Convers. Manage. 47 (18), 2925–2947.

Peerapong, P., Limmeechokchai, B., 2014. Investment incentive of grid connected solar photovoltaic power plant under proposed feed-in tariffs framework in Thailand. Energy Procedia 52, 179–189.

Pundir, K.S.S., Varshney, N., Singh, G.K., 2016. Comparative study of performance of grid connected solar photovoltaic power system in IIT Roorkee campus. In: Paper of International Conference on Innovative Trends in Science, Engineering and Management held at New Delhi, India, pp. 422–431.

Sharma, V., Chandel, S.S., 2013. Performance analysis of a 190 kWp grid interactive solar photovoltaic power plant in India. Energy 55, 476–485.

Sharma, Renu and Goel, Sonali. 2017. Performance analysis of a 11.2 kWp roof top grid-connected PV system in Eastern India. Energy Reports 3 (2017) 76–84.

Tarigan, E., Kartikasari, F.D., 2015. Techno-economic simulation of a grid-connected PV system design as specifically applied to residential in Surabaya, Indonesia. Energy Procedia 65, 90–99.

North American Board of Certified Energy Practitioners (NABCEP). 2005. *Study Guide for Photovoltaic System Installers and Sample Examination Questions*. August. http://www.brooksolar.com/files/NABCEP_Study Guide-Revised_Version_3_-_08_05-FINAL.pdf

Tiwari Arvind, Barnal P, Sandhu G.S and Sodha M.S. 2009. Energy metrices analysis of hybrid-photovoltaic (PV) modules. Appl. Energy, 86, 2615-2625.

Tiwari GN, Tiwari AK. Solar Distillation Practice for Water Desalination Systems. Anamaya Publishers. New Delhi: India, 2007.

Tarigan, E., Kartikasari, F.D., 2015. Techno-economic simulation of a grid-connected PV system design as specifically applied to residential in Surabaya, Indonesia. Energy Procedia 65, 90–99.

Sharma Rakhi and Tiwari G.N. 2013. Life cycle assessment of stand-alone photovoltaic(SAPV)system under on-field conditions of New Delhi, India. Energy Policy63(2013)272–282.

GRID-CONNECTED ROOFTOP SOLAR PHOTOVOLTAIC POWER PLANT

INTRODUCTION

Rooftop solar photovoltaic power plant is an electricity generating system (on or off grid), installed in the rooftop of any residential, institutional, social, Government, commercial, industrial buildings etc. in order to use the shadow free and unutilized space where the sunlight is incident plentifully throughout the day. The solar photovoltaic system used for generating electricity, is reliable, sustainable, environment friendly and maintenance free. It has the great potential to become a major source of electricity in future due to consistent increase in prices and limited energy resources of fossil fuels. India's rooftop solar installation expanded recently due to consumers' awareness and incentives provided by Jawaharlal Nehru National Solar Mission of Ministry of New and Renewable Energy, Govt of India, with a projection of 100 GW solar installation by 2022. As a result, in many urban areas, rooftop solar systems have been installed and the power generated is utilised for own use and the surplus power is fed to the grid. Installation of solar photovoltaic system on the surface is not always possible due to the constraints of land. Hence, the roof top area can be utilised successfully for generation of electricity by installing solar PV system. Initiatives have been taken by Government and non-Government organisations to install solar power generation system over the roof top of all the public office buildings, private houses and institutions to tackle recent power crisis. The rooftop solar power generation system will not only resolve the power crisis, but also reduce the harmful effects of greenhouse gases which are produced by fossil fuel based energy generation system. A significant part of the solar energy can be utilised for promoting grid connected solar photovoltaic power systems on the rooftop based on the varying sizes as per the need and availability of the free space of rooftop.

India being a tropical country lies between 8^0 and 37^0 north latitude, has average annual temperature ranging between 25 ^{0}C and 27.5 ^{0}C, with about 300 clear sunny days in a year and daily average solar energy incident over India varies from 4-7 kWh/m^2/day, offers great potential for utilizing solar energy because of the availability of about 5000 trillion (10^{12}) kWh of solar energy per year over its land area. With such a vast source of solar energy, the Ministry of New and Renewable Energy (MNRE), Government of India, launched the Jawaharlal Nehru Solar Mission (JNNSM) on 11th January 2010, as mentioned above, with the objective of focussing to create an environment for the dissemination of solar PV technology widely in the country. A small fraction of the total incident solar energy, if captured effectively, can meet the country's entire power requirements. The aim of the programme is also to reduce the cost of solar power generation in the country through (i) long term policy (ii) large scale deployment goals (iii) strengthening R & D and (iv) domestic production of critical raw materials, components and products. With a strong commitment to increase the generation of energy from renewable sources to a capacity of 175 GW by 2022, India has a target to install 100 GW of solar energy capacity. Of this, 40 GW would be the share of grid connected solar PV rooftop. One hundred seventy-five (175 GW) of the proposed installed capacity includes 60 GW from wind power, 100 GW from solar power, 10 GW from biomass power and 5 GW from small hydropower. Currently, solar power, wind power, biomass power and small hydropower together contribute to about 60 GW or 18% to India's total power capacity (Figure 4.1) with solar alone accounting for 24% of the total renewable electricity mix. However, in order to meet India's renewable energy goals, wind power installations need to be doubled from their current capacity, while the solar power capacity has to be enhanced by more than seven times in the next five years. If India succeeds in achieving these targets, the dependence on the coal for generating power will be decreased to a great extent after 2022 (CEA 2016).

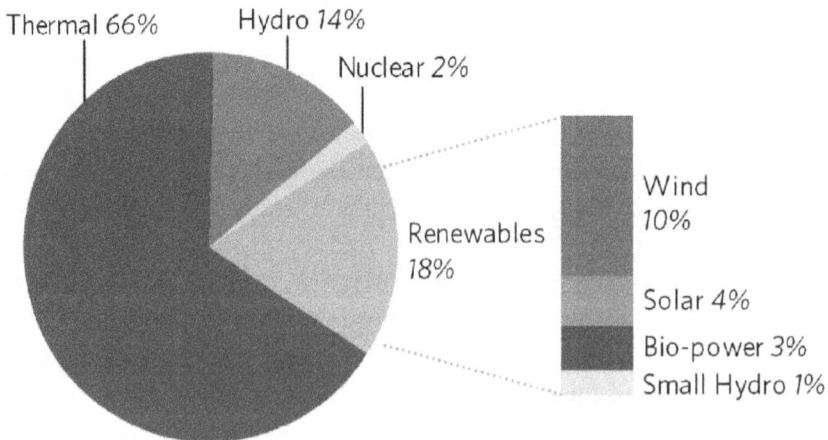

Fig. 4.1. Power capacity in India by various sources (as of 2017/11/30, % share)

From 2013-2016, installed rooftop solar capacity has grown from 117 MW to 1,250 MW (BTI 2017), indicating that the Indian rooftop solar capacity has increased tenfold in just three years. The accelerating growth in rooftop solar systems is driven by the fact that the technology is already price-competitive in many Indian states and sectors (CPI 2016). Recognizing the potential of solar technologies, the National Solar Mission of India has earmarked 40 GW of its 100 GW by 2022 solar power target for rooftop solar. Despite India's considerable progress in bringing down the cost of rooftop solar installations, the deployment rate is still insufficient to achieve the national target of 40 GW by 2022. Expected shares of rooftop solar capacity by various sectors in India to achieve 2022 target (38.75 GW) are shown in figure-4.2.

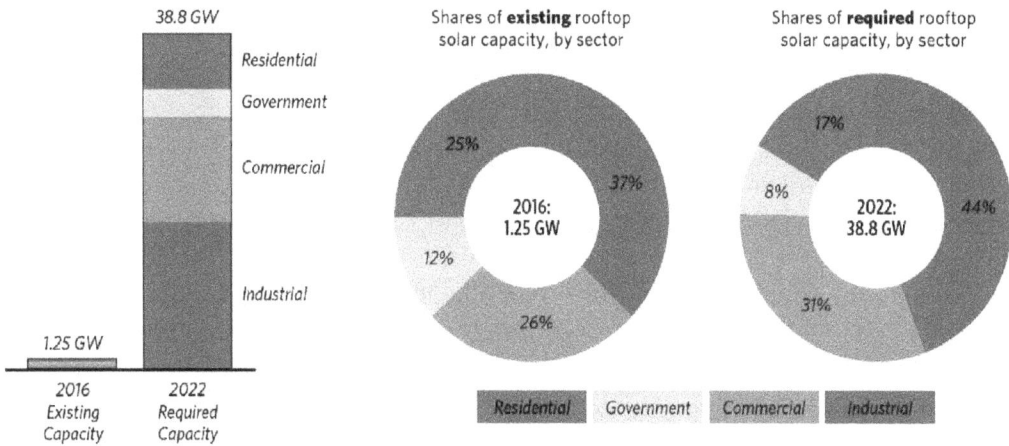

Fig. 4.2. Expected shares of rooftop solar capacity by various sectors in India by 2022
Source: BTI 2017, Niti Aayog 2015.

Solar photovoltaic rooftop has therefore emerged as a potential green technology to address climate change issues by reducing reliance on conventional fossil fuel based energy and to fulfil the increasing demands of the energy due to fast rise of population throughout the world. During operation, a solar plant contributes to a significant reduction of CO_2 emissions without consuming any fuel. In India, the practice of solar photovoltaic electricity has already been followed in a decentralized manner, a new thrust is through grid connected solar electricity and solar rooftop would thus emerge as an achievable goal for residential, commercial and industrial sectors.

4.1 HISTORICAL DEVELOPMENTS OF SOLAR PV IN INDIA

Solar Photovoltaic (SPV) Program of India was conceived in 1970s in response to the oil crisis occurred in the world during the period. This was considered as one of the largest national programs throughout the world. The SPV research & development in the country began in late 1970s and a programme for energy development through solar energy was launched in early 1980s with three main objectives such as (i) research on solar cell materials (ii) development of production and manufacturing capabilities of SPV module and (iii) promotional measures and incentives for installation of SPV system. The manufacturing base was strengthened and over 300,000 smaller systems aggregated to 22MW have been installed until 1995 making India third largest solar PV user. India's production of solar module increased from 9.5 MW in 1998–1999 to about 40 MW in 2004–2005 corresponding to 4.2- fold rise in 6 years. Although the growth is impressive, it is low compared with the global growth in recent years. The PV production in the world increased from 287.7 MW in 2000 to 1759 MW in 2005 corresponding to a 6.1-fold rise in five years (Bhattacharya and Jana 2009). Global demand for PV modules has been driving up their export from India. The percentage of export from the cumulative production was 35% in 2001, 43% in 2002, 50% in 2003, 55% in 2004 and more than 60% in the year 2005 (Goel, 2016). The rest was utilized in telecommunication towers, street lighting, agricultural water pumping and others. Solar Home Systems (SHS) were also encouraged. The first major PV plant of 1MW capacity, connected to the grid was set up in Jamuria under Asansol district in West Bengal during 2009-10. In 2010 Jawaharlal Nehru National Solar Mission (JNNSM) was introduced as a part of National Action Plan on Climate Change 2008, with a target to install 20 GW solar capacity by 2022. The mission was to be implemented in three phases namely Phase I (2010-12), II (2013-17) and III (2017-22). Under the phase I of JNNSM, one of the components related to solar rooftop was 'Rooftop PV and Small Scale Generation Programme' (RPSSGP) for encouraging development of rooftop or ground mounted solar systems with a maximum capacity of 2 MW. A total of 100 MW was targeted under this scheme. Projects under the RPSSGP scheme were mostly ground-mounted with negligible share of solar rooftop. The capacity of grid connected solar power plants was assessed as 45.5 MW in July 2011. Being a tropical country, India is endowed with abundant availability of solar energy and having 300 sunny days in a year. The country also experiences higher solar irradiance compared to many other countries with the potential ranging from 4 to 7kWh per sq. m per day in many parts of it. Subsequently, Government of India has revised Solar Mission Programme in 2014 with a target of 100 GW installed capacity of solar electricity by 2022. Out of which, 40 GW is now projected to come through grid connected rooftop solar systems. Both centralized grid connected and standalone solar energy development strategy have now been emphasized for achieving energy security of the nation in providing '24×7 power to all'. States and Union Territories in the country have identified their solar energy potential (Fig 4.3).

With such a vast potential of solar energy in the country, there is a great scope for taking up service providing entrepreneurship for the installation of grid connected rooftop / ground mounted solar photovoltaic power plant both in private and public sector. Policies for growth of rooftop solar in India have also been framed for wide installation of solar PV power plant on government buildings, airports, railways network, educational institutions, residential sector and commercial complexes. Steps have as well been taken to link the promotion of solar PV power plant through 'Make in India', 'Smart city mission' and 'Digital India' programmes of Govt. of India.

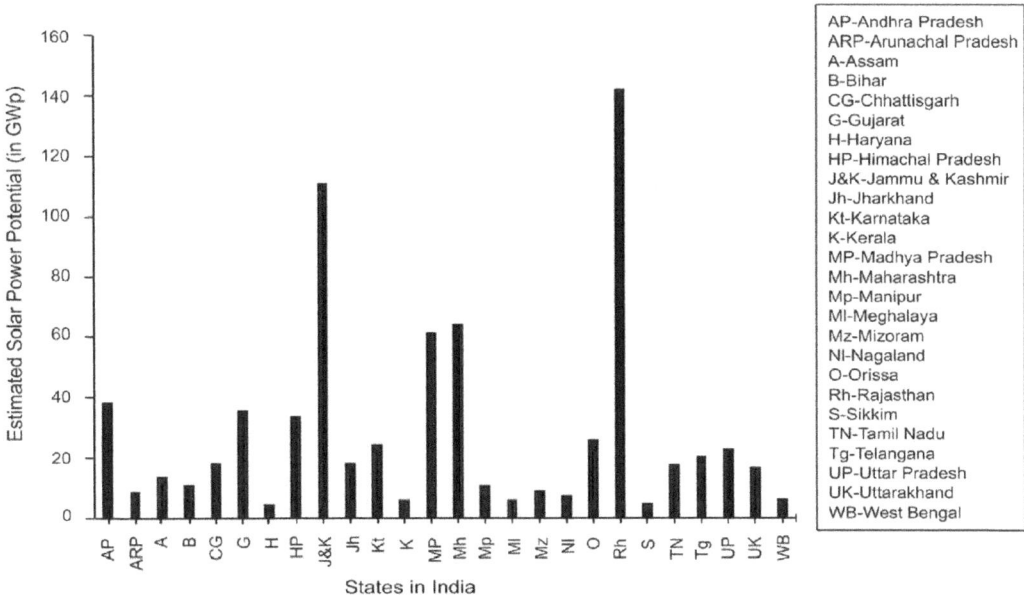

AP-Andhra Pradesh
ARP-Arunachal Pradesh
A-Assam
B-Bihar
CG-Chhattisgarh
G-Gujarat
H-Haryana
HP-Himachal Pradesh
J&K-Jammu & Kashmir
Jh-Jharkhand
Kt-Karnataka
K-Kerala
MP-Madhya Pradesh
Mh-Maharashtra
Mp-Manipur
Ml-Meghalaya
Mz-Mizoram
Nl-Nagaland
O-Orissa
Rh-Rajasthan
S-Sikkim
TN-Tamil Nadu
Tg-Telangana
UP-Uttar Pradesh
UK-Uttarakhand
WB-West Bengal

Fig. 4.3. Solar energy potential of different states in India (MNRE Report 2014.)

4.2. RATIONALE OF GRID-CONNECTED SOLAR ROOFTOP POWER PLANT FOR RESIDENTIAL BUILDING

In grid connected rooftop or small SPV system, the DC power, generated from SPV panel is converted to AC power using power conditioning unit/Inverter and is fed to the grid either for 440/220 Volt, three/single phase line or for 33 kV/11 kV, three phase lines depending on the capacity of the system installed at the residential, institutional or commercial establishment and the regulatory framework specified for the respective States. Grid-connected solar PV systems supplies solar energy directly to the loads of the building without the provision of battery storage arrangement. Surplus energy if generated is exported to Discom (Distribution company) grid and in case of shortfall, the required energy is imported from the grid to meet the load. These guidelines are followed only in grid-connected rooftop solar PV systems. These systems generate power during the day time which is utilized by powering captive loads and feed the excess power to the grid as long as grid is available. In case, solar power is not sufficient during the period due to cloud cover or for other reasons, the captive loads are met by drawing the required power from the grid. The grid connected rooftop system generally works on net metering basis wherein the beneficiary pays the bill to the utility on the basis of net meter reading only. A 1 kW rooftop system generally requires 10 sq. metres of shadow-free area. Actual area, however, depends on local factors of solar radiation and weather conditions, efficiency, shape and type of solar module, shape of the roof etc. The figure for the grid connected set-up is shown below (Fig. 4.4).

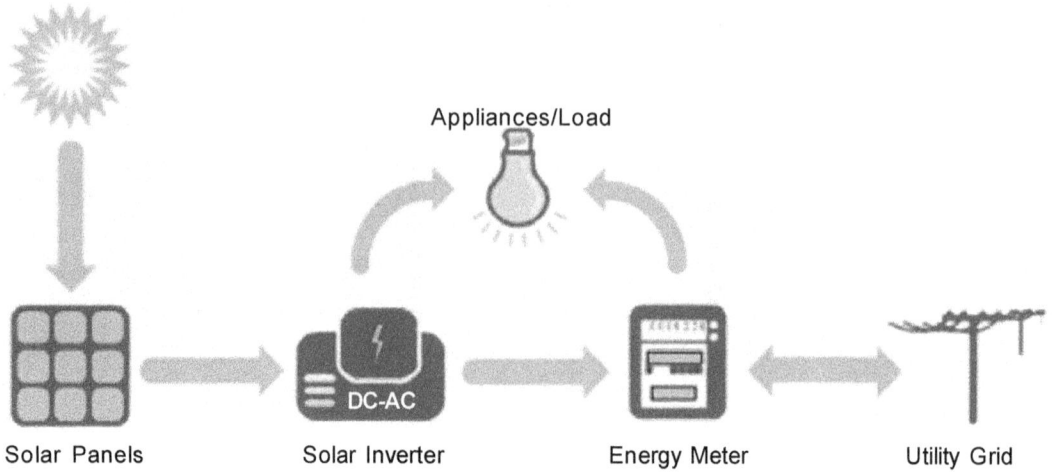

Fig. 4.4. A set up showing grid connected rooftop SPV arrangement

4.3 NET METERING

The Distribution Companies (DISCOM) agree to allow grid connectivity and purchase the electricity on feed-in-tariff or through net metering arrangement. If a building/facility consumes more power than being produced by SPV system during a billing month, the consumer is to pay for the net amount of power, drawn from the grid and in case the consumption is less than the power produced by SPV system, the excess power generated will be credited to the billing account of consumer and carried over to the next month. For example, if a Solar Plant produces 700 units over a billing month, but 1000 units are consumed by the loads in a building, the consumer is to pay the energy charges on the net 300 units only. Thus, net metering systems are primarily aimed at providing an opportunity to the consumers to offset their electricity bills, where, a single meter records both import of conventional energy from distribution grid and export of solar energy into distribution grid. The arrangement of net metering system is shown below (Fig. 4.5).

4.4 GROSS METERING

In gross metering system, the owner of rooftop or the investor sells the generated solar PV energy to the distribution company (DISCOM) without consuming for own use. All the energy generated from the PV system is exported to the grid and is separately recorded through a different 'feed-in-meter'. The developer exports the solar energy to the utility at a pre-determined feed-in-tariff as per long term power purchase agreement with the utility. The consumer uses the conventional energy only from the utility. The arrangement for gross metering system is shown below (Fig. 4.6).

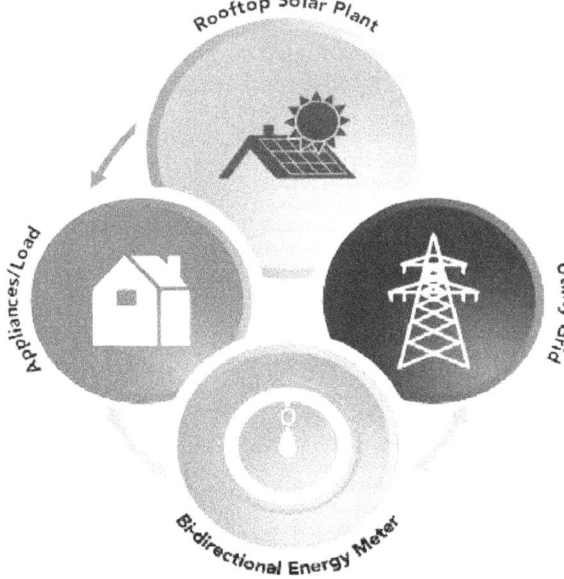

Fig. 4.5. Grid-tied net metering arrangement

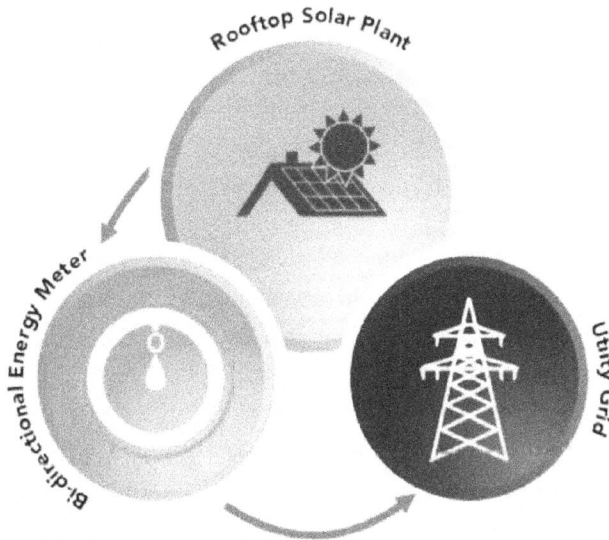

Fig. 4.6. Grid-tied gross metering arrangement

Energy metering consists of three energy meters such as PV meter, load meter and net meter, used in a PV system. The metering configuration used depends on the application. PV meter is used to determine the amount of energy produced by solar PV system. It is connected in series with the inverter output to monitor PV generated energy. The load energy meter connected in series with the load is used to monitor energy consumed by load. The net energy meter is used to monitor net energy import from or export to the grid. Minimum two energy meters are required to determine overall energy flow. The PV and load energy meters are uni-directional and register the flow of energy in forward direction. The net energy meter is bi-directional and registers the flow of energy both in forward and reverse directions as shown in fig. 4.7.

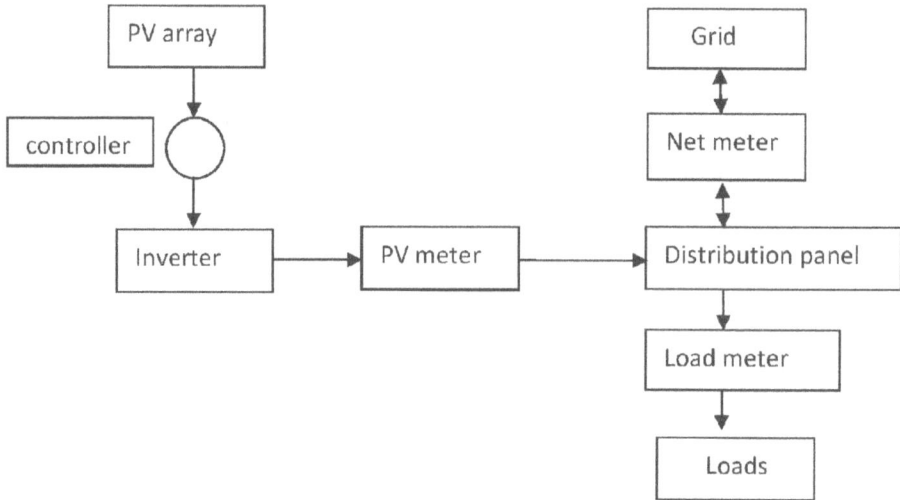

Fig. 4.7. Energy metering arrangement in grid-connected PV system

When the PV energy generated is less the energy consumed by the load, the net energy meter runs in forward direction (positive reading). If PV generated energy is greater than energy consumed by load, then energy is exported to grid. In this case, the net meter runs in reverse direction (negative reading). Thus, net meter can run in forward as well as reverse direction. During certain period if the more energy is consumed by the load (than produced by PV), then the net meter reading will be positive and if more energy is produced by the PV modules than consumed by the load, then net meter reading will be negative. Negative net meter reading implies that energy is fed to grid and positive net meter reading implies that energy is imported from the grid and the energy flow in the net meter is in the forward direction.

During positive net meter reading, Energy consumed by load = PV generated energy + Energy drawn from grid

Negative net meter reading occurs if PV generated energy is greater than energy consumed by load and energy is exported to the grid. The energy flow in the net meter is in the reverse direction. In this case, PV generated energy = Energy consumed by load + energy fed to grid. Net meter reading is negative when energy is fed to grid.

Example 4.1: If PV meter reads 10 kWh and load meter reads 8 kWh, what will be the Net energy meter reading?

Solution: PV meter reading = 10 kWh and Load meter reading = 8 kWh

Since PV generated energy is greater than energy consumed by load, it indicates that energy is fed to the grid.

PV generated energy = Energy consumed by load + Energy fed to grid

Energy fed to grid = PV generated energy – Energy consumed by load = 10 kWh – 8 kWh = 2kWh

Net energy flow is 2kWh into the grid and hence the net energy meter would read – (minus) 2 units. The negative sign indicates the energy is fed to the grid.

Example 4.2: If net energy meter reads + 4kWh and load meter reads 10 kWh, what is the PV generated energy?

Solution: From the definition, net energy meter reading is positive, indicating that more energy is consumed by the load than that produced by PV. Net energy meter reading, +4 kWh and Load meter reading: 10kWh

Energy consumed by load = PV generated energy + Energy drawn from grid.

PV generated energy = Energy consumed by load – Energy drawn from grid = 10 kWh – 4kWh = 6kWh. Therefore, PV generated energy is 6 kWh.

4.5 VIABILITY OF ROOFTOP SOLAR POWER PLANT (5 kW$_p$) FOR A 3-BED ROOM HOUSE

As per 2011 census, it is estimated that 330 million residential houses are present in India. Out of these, 166 million houses are electrified, 76 million houses use kerosene for lighting and 1.08 million houses are using solar energy for lighting. There are 140 million houses having proper roof (concrete or asbestos / metal sheet). More than two rooms are present in 130 million houses. On an average, a house can accommodate 1-3 kW$_p$ of solar PV system. The large commercial roofs can accommodate larger capacities of solar plant. With an estimate from the lower side, about 26000 MW capacity power plant can be accommodated on the roofs of buildings having more than 2 rooms by considering only the 20 % of 130 million houses, each of 1 kW$_p$ capacity roof top solar power plant.

4.5.1 Advantages of Grid Connected Rooftop Solar PV Plant

- Electricity generation at the consumption point and hence savings in transmission and distribution losses
- Low gestation time
- No requirement of additional land and utilization of unused space of rooftop
- Local employment generation
- Benchmark cost of rooftop solar system of about Rs. 70,000/- per kW
- Central financial assistance of Rs. 21,000/- per kW by availing 30 % subsidy
- Net cost to customer of Rs. 49,000/- per kW
- Generally, a 1 kW system generates about 1200- 1500 units per year

● Savings per annum of Rs. 7200 to Rs. 9000 (considering average tariff of Rs. 6 per unit)

● Payback period of the setup of about 5.5 - 7 years

4.5.2 The Loads for a 3 Bed-room Building

The expected loads for a 3 bed-room building is presented in Table 4.1

Table 4.1: The loads for a 3 bed-room building

Name of appliance	Wattage	Number	Hours of use per day	Total Wh per day
Air conditioner	2000 W	1	4	8000
Ceiling fan	100 W	6	10	6000
CFL bulb	20 W	10	10	2000
Refrigerator	250 W	1	20	5000
LED TV	100 W	1	6	600
Computer	100 W	1	4	400
Total load				**22000 Wh (22 units)**

Daily electricity generation from 5 kW_p plant @ 4-5 units generated per kW capacity solar power plant = 20 units

Monthly generation = 600 units

Average monthly electricity consumption = 800 units (approx.) of a 3-bedroom house

Electric tariff = Rs. 6.00 per unit

After adjusting the net consumption and net generated units of power, the average electricity bill during the month = 200×6 = Rs. 1, 200 and saving = Rs. 3,600.

4.5.3 Project Viability Calculation

The capital cost of power plant = Rs. 5,00,000 for 5 kW_p plant

Subsidy @ 30% = Rs.1,50,000

Beneficiary share = Rs. 3,50,000

Annual savings for payment towards electric energy @ Rs. 3600 per month = Rs. 43,200

Simple payback period = 3,50,000/43,200 = 8 years

The life of the set-up is about 25 years and hence the project is economically viable. The entrepreneurship option for providing services in installation of the set-up is also profitable and sustainable. The venture for both the entrepreneurs and users is therefore economically, socially and environmentally viable. The technology needs to be promoted in a wide scale for achieving energy security for the society.

4.6 GRID-CONNECTED SOLAR PV SYSTEM FOR SMALL POWER APPLICATIONS

Solar PV system can broadly be divided into two categories i.e. one is standalone system which uses battery for energy storage and the other is grid-connected system which does not use battery. The grid connected system can also be broadly divided into two categories i.e. (i) grid-connected PV system for small power applications especially in households and (ii) grid-connected PV system for large power applications mainly in solar power plant. Main components of a grid-connected solar PV systems are (i) solar PV modules (ii) a power condition unit/inverter (iii) load and (iv) grid. An electric grid can be considered as a large sink of energy wherein energy generated by solar PV can be supplied or taken from the grid whenever required. The idea of connecting PV system to the grid is to use the grid as energy storage medium so that the use of battery can be avoided and the cost of the system can be reduced. The main purpose of the grid-connected system that are used for small power household applications is to generate energy and consume it within the household itself. Typically, the power rating of this type of PV system can be in the range of 1 kW to several 10s of kWs. The role of grid connectivity in case of small power generation in household system is to supply excess generated energy to the grid (if more energy is generated by the PV system than the required load) or take energy from the grid if there is any shortage. Inverter plays a very important role in the small grid-connected PV power system.

The grid-connected PV systems for large power applications are mainly designed to act as power plants and the chief purpose of these power plants is to generate power and supply it to the grid. In this way, they are equivalent to coal or hydro based power plants. Since, the key purpose of this type of grid-connected PV system is to supply the power to the grid, their power ratings are normally in the range of 1 MW to several 10s of MW. It is assumed that a grid is a large sink of energy and a large amount of energy generated by PV plants can be fed into the grid and no battery storage is included in such plants.

In both grid connected small or large solar PV systems, the components included are (i) solar PV array (ii) array combiner box (iii) DC cabling (iv) DC distribution box (v) Inverter (vi) AC cabling and (vii) AC distribution box.

4.7 SIZING OF GRID-CONNECTED PV SYSTEM FOR A HOUSEHOLD

Some basic principles, followed when designing the quality aspects of grid-connected PV power plant for small applications are as follows:

i. Select a packaged system that meets the owner's needs. Customer criteria for a system may include meeting annual energy requirements, reduction in monthly electricity bill, environmental benefits, desire for back-up power, initial budget constraints, etc.

ii. Size and orient the PV array to provide the expected electrical power and energy.

iii. Ensure the roof area or other installation site, capable of handling the desired system size.

iv. Locate the array to minimize shading and interference with obstructions.

v. Ensure that suitable cable ducts are available to lay DC cables from array to the inverter.

vi. Ensuring the availability of indoor control room to house inverter as well as DC and AC distribution boxes.

vii. Specify sunlight and weather resistant materials for all outdoor equipment.

viii. Design the system with a minimum of electrical losses due to wiring, fuses, switches, and inverters.

ix. Design the system in compliance with all applicable building and electrical codes.

x. Ensure that the design meets local utility interconnection requirements.

4.7.1 Steps of System Design

We need to design the system to meet annual energy consumption requirements. The energy flow happens from PV array to the load and the PV system design should process in the reverse direction of energy flow, i.e., from load to PV array. In the reverse path, we should account for all losses in PV system. In the grid-connected PV system, since power capacity is normally large as compared to the small standalone system, and hence power losses in DC as well as AC cables must also be considered.

A simple block diagram of grid-connected PV system is shown in Figure 5.8, below. In this case, the PV system is assumed to be connected with the grid. The PV design analysis is presented assuming that all the power required by the load is to be supplied by the PV array alone. However, since the grid is connected to the PV system, any additional power generated by the PV array (than required by the load) will be supplied to the grid, and any deficit of power required by the load will be drawn from the grid. In principle, a grid-connected PV system can also be designed to take some percentage of power from the PV array. The grid connected PV system also considers the rooftop area available for the installation of PV module as a limiting factor. If the whole rooftop area is covered by PV modules, the power generated may be surplus or in shortage, depending on the case, power can either be fed to the grid or power can be drawn from the grid. In this way, there are a number of possibilities of a grid-connected PV system design and design depends on the constraints which may come from load side, grid side or physical infrastructure side.

Step 1 *Estimate annual energy usage*

In this step, the estimation of energy consumption by the load is done. As an example, let us assume that energy consumed by the load is 22 kWh or 22 units per day as estimated above.

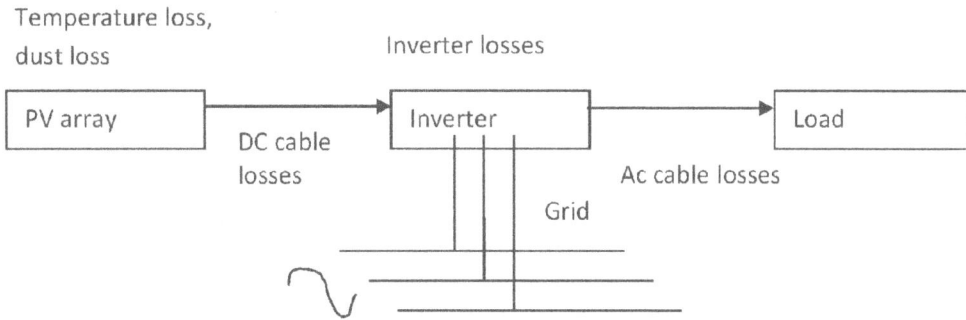

Fig. 4.8. Block diagram of grid-connected PV system indicating various losses in PV system

Based on the daily energy requirements, monthly and yearly energy load can also be estimated.

Step 2 *To calculate the average daily solar radiations in terms of equivalent sunshine hours*

Average daily solar radiation for a given location, where solar PV plant is to be installed, is normally given in terms of kWh/m^2/day. Normally in India, the daily solar radiation varies between 4 to 7 kWh/m^2/day. Now, since the solar PV modules are rated for standard test condition (STC) which is equal to sunlight power density of 1000 W/m^2 or 1kW/m^2 and temperature of 25 ^{0}C. It means that a 100 W$_p$ PV module will provide output power of 100 W$_p$ only when the condition corresponding to STC is matched.

Suppose at a given location, average daily solar radiation is 5 kWh/m^2/day, this can be written as 1 kW/m^2 × 5 hour/day. In this way, we can say that daily solar radiation of 5 kWh/m^2/day is equal to solar power density of 1 kW/m^2 for 5 hours per day. Although the sunshine hours are usually longer than that in a day, and intensity changes from morning to evening, but in this case, 5 hours/day is referred as equivalent peak sunshine hours. The equivalent sunshine hour is a very useful concept because PV modules are rated for STC condition.

Step 3 *Estimate the AC power supplied to the load*

AC power output (in terms of kW) can be obtained by dividing the total energy supplied to the load (given in terms of kWh) by the equivalent daily hours of peak sunshine (given in terms of hours or 'h'). In this example, daily energy consumed is 22 kWh and h is taken as 5 hours.

$$\text{AC power to load (in kW)} = \frac{daily\ energy\ to\ load\ (in\ kWh)}{equivalent\ daily\ sunshine\ hours\ (in\ h)} = \frac{22\,kWh}{5h} = 4.4\,kW$$

Step 4 *Estimate the AC power output of inverter or PV plant output power*

The AC output power losses consist of transformer and AC cable losses. The AC output power losses are typically of the order of 2% to 5%. In this example, let us say this to be 5%.

$$\text{AC power output of inverter (in kW)} = \frac{Power\ fed\ to\ load\,(kW)}{(1 - AC\ power\ lossess)}$$

$$\text{Thus, inverter AC output power} = \frac{4.4}{1-(5/100)} = \frac{4.4}{1-0.05} = \frac{4.4}{0.95} = 4.63\,kW$$

Step 5 *Estimate the DC power input to inverter*

The input to inverter will be DC power. The input DC power will be more than the inverter output AC power due to losses in the inverter. The inverter power losses consist of MPPT tracking and DC to AC conversion losses. The inverters are becoming very efficient and power losses in inverter are in the range of 2% to 5%. Let us assume that inverter losses are 3% in this example.

$$\text{Inverter DC power input (in kW)} = \frac{Inverter\ AC\ power\ output\ (kW)}{(1 - Inverter\ power\ lossess)}$$

$$\text{Thus, inverter DC input power} = \frac{4.63}{1-(2/100)} = \frac{4.63}{1-0.03} = \frac{4.63}{0.97} = 4.77\,kW.$$

Step 6 *Estimate DC power output of PV array*

Between the PV array and the inverter input, there are DC cables. The DC cables also have power losses. The DC cable losses depend on the cable thickness and length. Typical systems are designed for maximum DC cable losses of 3%. Due to the cable losses, the PV array will have to produce more power. Assuming DC cable losses of 3%:

$$\text{PV array DC power output (in kW)} = \frac{Inverter\ DC\ power\ input\ (kW)}{(1 - DC\ cable\ losses)}$$

$$\text{Thus, PV array DC output power} = \frac{4.77}{1-(3/100)} = \frac{4.77}{1-0.03} = \frac{4.77}{0.97} = 4.91\,kW.$$

In this way, the estimated size of PV array is 4.91 kW. Remember that this power rating of the PV modules is power rating under STC condition, which means it is peak power rating with subscript 'P', i.e., 4.91 kW_p or almost equal to 5.0 kW_p.

Step 7 Take operating losses in account

In practice, while PV modules operate in field, there are several losses that can occur. Some losses are small and some are large. The module or PV array output gets reduced by the following way:

i.   Manufacturing tolerance/module mismatch loss

ii. Module temperature loss

iii. Module soiling loss

Module mismatch losses: The typical manufacturing tolerance of PV module output power is +/- 3%. The module mismatch loss is, therefore, 3%

Module temperature losses: In order to determine module temperature losses, module average operating temperature needs to be determined. The normal operating cell temperature (NOCT) is defined as module cell temperature under STC of 25^0C. Typical PV module NOCT is specified at 45^0C. The average operating temperature of the module cell depends on the average ambient temperature conditions. For average ambient temperature of 30^0C, the module average operating temperature is taken as about 50^0C·

The temperature coefficient for output power of PV module is in the range of 0.25%/^{0}C to 0.5%/^{0}C. This power temperature coefficient gives an indication of loss of power for a given module for every degree centigrade rise in temperature. The crystalline silicon modules have power temperature co-efficient of about 0.45%/^{0}C while thin film modules have lower temperature coefficient ranging from 0.2%/^{0}C to 0.3%/^{0}C. Normally, the power temperature coefficient of PV modules is given in the datasheet of manufactures. It mostly depends on material of PV modules but to some extent also depends on the way of manufacturing as well. It is to be noted that the temperature coefficient is given in terms of per degree centigrade. Thus, if PV module is characterized at STC condition of 25^0C and if average operating temperature is 50^0C, then the loss in power output of module or array due to increase in temperature is as follows.

- The percentage power losses for crystalline silicon modules or PV array due to

$$\text{temperature} = 25°C \times \frac{0.45\%}{0°C} = 11.25\%$$

- The percentage power losses for thin film PV modules due to temperature, assuming power temperature coefficient of 0.3%/^{0}C:

$$\text{Module temperature loss} = 25°C \times \frac{0.3\%}{0°C} = 7.5\%$$

Module soiling losses: The modules are typically soiled by dust, dirt, bird droppings etc. The periodic cleaning of modules is recommended to minimize soiling loss. Typical module soiling loss is estimated at 5%. Thus, the total PV module losses while in operation are expressed as, the total module loss = mismatch loss + temperature loss + soiling loss =3% + 11.25% + 5% = 19.25%

Step 8 *Estimate the final required PV array capacity*

The PV array must supply the DC power required as estimated in step 6, but the array must also supply the operating losses as estimated in step 7. Considering this, the total PV array required would be:

$$\text{Final PV array power output (in kW)} = \frac{PV\ array\ DC\ power\ output\ (kW)}{1 - operating\ losses}$$

$$= \frac{5.00}{1 - (19.25/100)} = \frac{5.00}{1 - 0.192} = \frac{5.00}{0.808} = 6.18\ \text{kW}_\text{p}$$

Thus, the final PV array capacity required is 6.18 kW$_\text{p}$. The subscript 'P' is added in power, i.e., W$_\text{p}$ to indicate that this power rating is for STC condition.

Step 9 *Estimate required number PV modules*

For this purpose, one can choose the modules that are available in the market, their making and power rating. PV modules are available in the capacity of 30W$_\text{p}$ to 300 W$_\text{p}$. Normally, for applications needing PV modules in several kW$_\text{p}$, we should choose modules of higher wattage rating. In higher wattage rating, crystalline Si PV modules of 230 W$_\text{p}$, 235 W$_\text{p}$ and 240 W$_\text{p}$ are easily available. For example, if one chooses to use PV modules of 240 W$_\text{p}$, which is now-a-days available, then the total number of PV modules required would be:

$$\frac{6.18\,kW}{240} = \frac{6180}{240} = 25.75$$

Or, taking the next integer value, we can say 26 PV module will be required.

The connection of these PV modules would depend on the inverter power rating and input DC voltage the inverter can take.

Step 10 *Estimate inverter power rating*

In step 6, it is estimated that the DC power input to inverter is 4.91 kW, it means that an inverter should be able to process this much power. Normally, for the safety purpose little higher power rating is chosen. This is required because reflection from clouds or other objects may temporarily increase the sunlight falling on module, resulting in higher power generation. About 10% higher inverter capacity should be chosen. Also, we have considered about 3% mismatch losses, in case there is no mismatch in modules. This extra power would appear at inverter input, we should take safety for this purpose as well.

Overall, our inverter capacity should be 10% to 15% higher than input power value estimated in Step 6.

$$\text{Inverter power rating (in kW)} = \frac{DC~input~power~to~inverter~(kW)}{1 - saftey~factor}$$

$$= \frac{4.91}{1-(15/100)} = \frac{4.91}{1-0.15} = \frac{4.91}{0.85} = 5.77\,kW$$

Note that this is the input power rating of inverter. Due to losses within inverter, the output power rating of the inverter will be less.

Step 11 *PV array configuration*

The PV array configuration or the number of PV modules to be connected in series or in string and the number of parallel strings depend on the inverter voltage ratings. The PV string V_{oc} and V_{mp} must be within allowable limits of inverter ratings. Maximum I_{sc} of the PV array should also be within acceptable limit to the inverter.

Final array configuration is selected based on inverter input DC voltage window of operation. Inverter manufacture specifies a range of V_{mp} values for its MPPT operation. Based on module tolerance and its minimum and maximum temperature of operation, a range of module V_{mp} output is determined. For a number of modules, connected in a series string, a range of string V_{mp} is calculated. The number of modules in a series string is determined such that its range of V_{mp} output falls well within the permitted range of V_{mp} values of the inverter. This is demonstrated in Figure below (fig. 5.9).

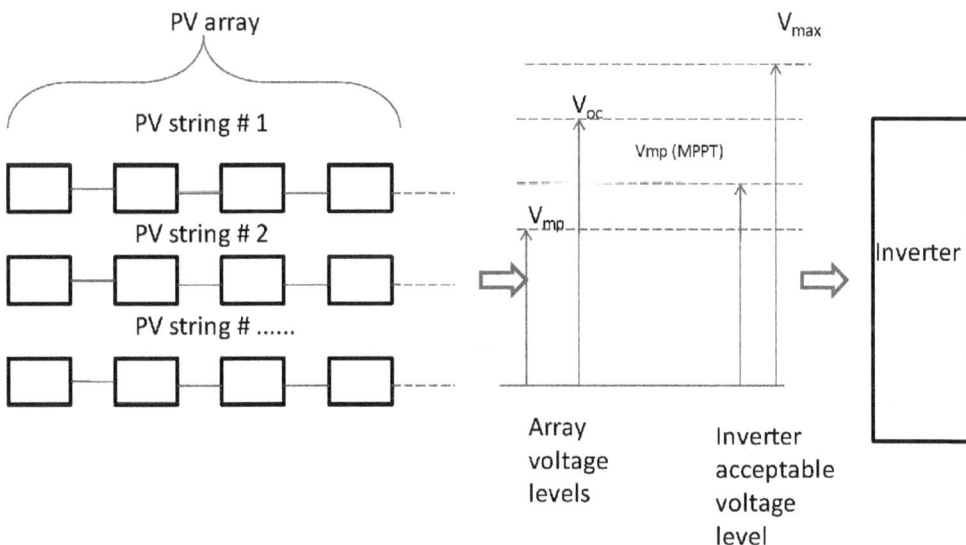

Fig. 4.9. Array voltage level and voltage levels acceptable at the input of the inverter

Inverter manufacture also specifies maximum input DC voltage rating. Based on module tolerance and its minimum temperature of operation, maximum module V_{oc} output is determined. For a number of modules connected in a series string, maximum string V_{oc} output is equal to the maximum module V_{oc} multiplied by the number of modules. The maximum array V_{oc} is equal to the maximum string V_{oc} for a number of strings connected in parallel to make an array. It is then verified that the maximum array V_{oc} is less than maximum input DC voltage rating of the inverter. This is demonstrated in figure 4.9.

It is also verified that the maximum array I_{sc} is within the maximum input DC current rating of the inverter. The maximum module I_{sc} is determined based on module tolerance, maximum available solar radiation and maximum temperature of operation. The maximum string I_{sc} is equal to the maximum module I_{sc} for a series connected string, and the maximum array I_{sc} is maximum string I_{sc} multiplied by the number of strings in an array.

Let us now go to find out the number of modules, we should be connecting in a string. This is decided by considering the allowable V_{mp} and V_{max} at the input of inverter and V_{mp} and V_{oc} of the PV module selected for installation.

Table (4.2) below gives the list of commercially available PV modules with their parameters. These modules can be used for grid-connected PV system applications. Using these PV modules, the PV systems of capacity from few kW to MW level plants can be installed. Normally, for MW level plants, higher wattage rating PV modules are preferred. Table 4.3 gives the typical parameter of commercially available grid-connected inverter. The main parameters of these inverters are maximum DC input voltage, maximum power point tracking voltage range, maximum DC input current, efficiency of inverter and its AC power rating. Referring to Figure 4.9, and by using Table 4.2 and Table 4.3, we can estimate the number of panels that we can connect in PV string and the number of such strings that will be required to make parallel connections for the given power applications.

Table 4.2: Typical parameters of commercial PV modules

S.N.	Model X	P_m (W$_p$)	V_{oc} (V)	I_{sc} (A)	V_m (V)	I_m (A)
1	Module 1	125	22.10	7.9	18.2	6.9
2	Module 2	135	22.00	8.38	17.7	7.63
3	Module 3	145	22.40	8.5	18.35	7.9
4	Module 4	200	32.50	8.42	26.1	7.66
5	Module 5	210	41.59	7.13	33.81	6.21
6	Module 6	230	36.00	8.45	28.6	8.05
7	Module 7	240	37.38	8.45	30.60	7.85

Table 4.3 Typical parameters of commercially available grid-connected inverters

AC power output (kVA)	Model	Maximum DC input voltage (V)	Maximum power point (MPP) voltage tracking range (V)	Maximum DC current (A)	Peak efficiency (%)
5	Inverter 1	600	80-120	50	97.0
6	Inverter 2	600	90-120	60	97.3
10	Inverter 3	600	100-160	75	97.5
17	Inverter 4	600	400-800	35	98.2
250	Inverter 5	600	330-600	800	97.5
500	Inverter 6	600	330-600	1470	98.6
1000	Inverter 7	1000	460-875	2100	98.6

Number of PV modules required

In this example, total PV array capacity required is 6.18 or about 6 kW_p (from step 8). Suppose we choose Module 3 (from table 4.2) for this installation. The rated power capacity of Module 3 is 145 W_p, therefore, the number of Module '3' required for 6 kW_p system would be:

$$\text{Total no. of modules required } = \frac{Total\ PV\ plant\ capacity}{Rated\ power\ of\ on\ module} = \frac{6kW_p}{145W_p}$$

$$= \frac{6000}{145} = 41.37$$

To round off, we need about 42 PV modules of 145 W_p for this PV system.

Number of PV modules in a string

In order to determine the number of PV modules in the array, we should look at the acceptable voltage rating of inverter and try to match it with the voltage ratings of strings as per the presentation given in Figure 4.9. The V_{oc} of PV Module '3' is 22.4 V, V_{mp} is 18.35 V and I_m is 7.9 A. Now, from step 10, the inverter power rating is 5.77 kVA. Let us take 6 kVA commercial inverter available in the market. Referring to inverter of 6 kVA, its Maximum Power Point (MPP) voltage tracking range is 90 V-120V and maximum input DC current is 60A.

When we connect PV modules in series, the voltage gets added, therefore, we need to connect Module '3' in series to get the required voltage in the range of Inverter 2. If we divide V_{mp} range of Inverter 2 with the V_{mp} of the Module '3', we will get number of PV module to be connected in series to form one string is:

$$\text{No. of modules in string} = \frac{V_{mp} \text{ of selected inverter}}{V_{mp} \text{ of selected PV module}}$$

Since the inverter V_{mp} tracking range is 90V to 120 V, let us have an average of it, that is, 105 V. This 105 V will be the V_{mp} of the PV module string. Note that any other voltage level within the inverter tracking range can also be chosen. We have to only ensure that the array V_{mp} point should lay within the inverter V_{mp} range in all operating conditions, $\frac{105}{18.35} = 5.72$.

The above calculation shows that we will have to connect 5.72 PV modules in series, or to round it off, we get 6 modules. The V_{mp} of 6 module series will be $18.35 \times 6 = 111.9$.

It is still in the range of V_{mp} tracking of inverter and should be acceptable. Also, remember that if the temperature of cells in module is higher than 25^0C, the actual module V_{mp} in the field will be lower. It means that the V_{mp} of 6 nos. of PV modules in series are within the acceptable limit.

Number of PV strings in parallel

In total, we need to connect 42 PV modules in this grid-connected PV system. Each string needs 6 nos. of PV modules. Therefore, the number of strings to be connected in parallel will be, No. of strings in parallel $= \dfrac{Total\ no.\ of\ PV\ modules\ in\ system}{No.\ of\ modules\ in\ a\ sting} = \dfrac{42}{6} = 7$

Thus, we need to connect 7 strings in parallel and each string having 6 nos. of modules in series. Note, that when we connect strings in parallel, the current of each strings gets added. The total current of the parallel combination will be the sum of current of each string. We have to make sure that the total current of the PV array is within the limit of the inverter. In this example, the maximum DC current limit of Inverter 2 is 60 A. In this case, the Module '3' I_{sc} is 8.5 A, and 7 strings are connected in parallel, therefore, the total current of all the strings will be, total current of strings in parallel = Current of one string ′ Total number of strings = $8.5 \times 7 = 59.5$ A

From this calculation, we can see that the current of 7 strings put together is less than the maximum allowable DC current of the inverter. Hence, the Inverter 2 is suitable for this application.

Step 12 Select Balance of system (BoS components)

DC BoS components: Various Balance of system (BoS) components are required to integrate in the system. PV array support structure, DC cables, string and array combiner boxes and DC distribution boxes are used on the DC system side.

PV array support structure is of three types such as fixed, manual tracking and auto tracking. The structure is usually custom designed to suit the type of modules used, string and array design and the type and the slope of terrain where the array is installed. The module mounting frames are typically fabricated using galvanized iron or aluminium sections and are designed to withstand outdoor temperature, humidity, wind and other environmental factors. The modules are either bolted onto the frames or held with clips. The frames are secured onto the roof structure or on the ground using suitable support members. The support members are typically fabricated using metal sections matched with module frame sections or using treated wooden sections. The support structures are typically grouted into concrete blocks, secured onto terrace or firm ground. The support structures can also be rammed into soft ground.

The string/array combiner box is typically double insulated and rated for IP65. The surge protection devices are typically rated for 600 V to 1000 V DC. The double pole DC disconnect switch is rated for minimum of 1.25 times PV array short circuit current. The DC disconnect switch inside array combiner box is used to isolate PV array from the DC cables in the building and the inverter. The Ground Fault Detector Interrupter (GFDI) device is typically designed to detect 5% of PV array maximum output current as ground fault current and interrupt DC input circuit.

The DC disconnecting switch and surge protection devices are also housed in the DC distribution box located near the inverter. Alternatively, they are integrated with the inverter DC input circuit. The DC disconnecting switch inside DC distribution box is used to isolate DC circuit from the inverter.

If the DC cables are rated for a minimum of 1.25 times the string short circuit current at each location and then string fuses are optional. The string blocking diodes are also optional for small size power plants using less than four strings. The DC cables are either rated for outdoor environment or are housed inside conduits which, in turn, are rated for outdoor use.

The total length of string DC cables is two times (positive and negative cables), the distance between modules and the string combiner box. The total length of array DC cables (positive and negative cables) is two times the distance between array combiner box and inverter. The DC cable size is selected such that the total power loss over the entire length of DC cable is within 3% of the rated PV array output power.

AC BoS components

The BoS components on the AC system side are AC circuit breakers, AC overvoltage protection devices, AC output transformers, AC cables, AC distribution boxes and AC energy meters.

The AC circuit breakers and AC overvoltage protection devices are typically integrated with the inverter output AC circuit. Alternatively, they can be housed inside AC distribution box to facilitate the isolation of inverter output near AC output transformers or loads. The AC circuit breakers are typically rated for 1.2 times the maximum AC output current of the inverter.

The AC voltage protection devices are typically rated for 400 V AC for 230 V single phase circuits and 600 AC for 415 V three phases circuits.

The AC output transformer is normally built within the inverter for LV circuits and is rated for 1.2 times the maximum AC output power of the inverter. For MV circuits, an external three phase step-up transformer is used to supply AC power to MV grid. The MV transformer is rated for 1.05 to 1.1 times the maximum AC output power of the inverter. The AC input and output circuit breakers and AC voltage protection devices are used with AC output transformer. The transformer losses are typically in the range of 2% to 5% of its rated output power.

The AC cables are typically rated for 1.2 times the maximum AC current carried. The AC cables from inverter output to transformer housed inside control room are rated for indoor environment. The AC cables from transformer output to load housed inside building are also rated for indoor specifications. The AC cables from transformer output to grid are rated for outdoor environment. The total length of LV cable is twice the distance between inverter and transformer or load. The total length of MV AC cables is twice the distance between transformer and grid distribution centre. The AC cable size is selected such that the total power loss over the entire length of AC cables is within 2 % of the rated inverter AC output power.

Step 13 *Select energy meter*

The energy meters are used to monitor energy generated by the PV power plant, energy consumed by load and energy supplied to grid by the plant

In PV power plant, for small power applications, net energy meter is used to monitor net energy drawn by the load from the grid or net energy supplied by inverter to grid. The net energy meter is rated for Low Tension (LT) grid voltage and PV power plant maximum output AC current or maximum load AC current whichever is higher. The PV generated AC energy and the AC energy consumed by the load can be monitored by separate energy meters. Alternatively, DC energy meter can be used at the output of PV array to monitor PV generated DC energy. In PV power plant, for large power applications, three phases AC energy meters are used to monitor energy produced by PV power plant and supplied to MV (medium voltage) grid. These energy meters are rated for MV, AC voltage. Additionally, inverter also monitors AC energy produced by PV power plant.

4.8 GRID-CONNECTED PV SYSTEM DESIGN FOR POWER PLANTS (1 MW$_p$)

The PV systems which are designed for power plant applications have primary function to feed power to the grid only, and not to any local load like in captive power plants. When we design a captive power plant, the basic criterion is to design PV system to meet energy demands of the local load. Here, the grid connectivity helps to supply the load if there is additional energy demand or excess generated energy is fed to the grid. On the other hand, when we design a solar PV system for power plants, the basic concept is to fix the size or capacity of power plant and then design the plant to get the best yield or best electricity generation from the plant. For instance, under Jawaharlal Nehru National Solar Mission, the government is awarding the contracts to install solar PV power plants in capacity of 5 MW$_p$. There are other schemes in which the PV plants can be installed in capacity of 1 MW$_p$, in such situation, the PV plant installer tries to design the plant in such a way that the losses in the plants would be minimum and therefore, energy generation would be maximum. Mostly, the electricity generated in PV power plants is sold to the Government at predefined rates. In the PV plant design, we can do the design in two parts:

(i) Part 1: Estimation of energy output of PV plant

(ii) Part 2: Determining the configuration of PV plant (string size, inverter rating, etc.).

In small grid-connected PV plants or in standalone PV system design, the design or sizing of components is done in the reverse order of energy flow. But in PV power plants, the design or sizing of the components is done in the direction of energy flow. This is done because, as discussed, the PV plant capacity is fixed in advance. In this section, let us see how we can design MW scale power plants and how much energy can be generated from these plants.

4.8.1 Estimation of Energy Output of PV Plant

Step 1. Fixing the capacity of PV power plant

The first step is to fix the size or capacity of power plant. This is normally fixed by the Government scheme under which the PV power plant is being installed, as the main purpose of PV power plants is to generate electricity and to sell it to the Government. Typical power rating of these plants is 1 MW_p, 2 MW_p or 5 MW_p. Some Governments are also installing PV power plants of very large capacities, ranging up to 500 MW_p. The capacity of the overall plant is normally in the multiple of 1 MW_p. In this case, let us design the power plant for 1 MW_p capacity. Figure below (4.10) shows the arrangement of a large capacity PV power plant with associated losses.

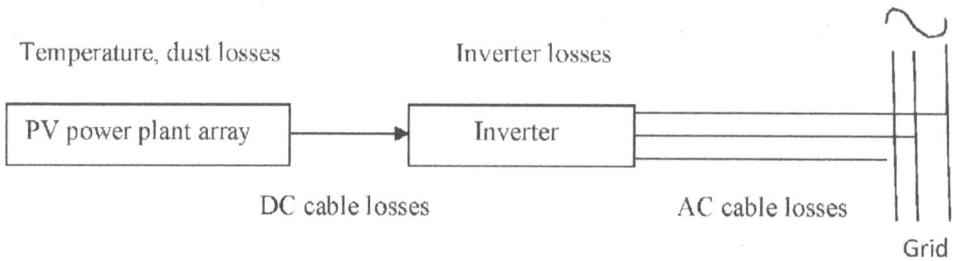

Fig. 4.10. Arrangements of a large capacity PV power plants with associated losses.

Step 2 Average daily solar radiation and equivalent peak sunshine hours

As discussed earlier (steps of system design), we should get daily solar radiation data of a location where PV plant is being installed. Using these data, we can estimate the amount of electricity that can be generated using power plant of given capacity. The solar radiation for a given location, where solar PV plants is to be installed, is normally given in terms of $kWh/m^2/day$. Normally, in India, the daily solar radiation varies between 4 to 7 $kWh/m^2/day$.

Since the solar intensities change on day to day and month to month basis, therefore, average daily solar radiation falling on a given location also changes. If we want more accurate estimation of possible energy production from 1 MW plant, we should try to get monthly average data for the site and therefore, we should estimate the possible electricity generation for each month. Also it is known that solar radiation data are available on horizontal surface. In order to collect more solar radiation, solar PV modules are installed in a tilted manner facing south. The tilt angle varies from one site to the other site. Therefore, it is important that the solar radiation data are to be obtained for tilted plane. Normally, the solar radiation data are always higher on tilted surface as compared to solar radiation data on horizontal surface. As an example for some site, these solar radiation data is given in table below (Table 4.4).

Let us assume that we need to design 1 MW_p plant using solar radiation data given in table 4.4 above. In this section, we will do the calculation for one month, and it is expected that the readers will do the calculation for other months on their own.

4.8.2 Equivalent Sunshine Hours or Peak Sun Shine Hours

Equivalent or peak sun shine hours refer to number of daily sunshine hours equivalent to 1000 W/m^2 for which the solar module is characterized. The solar PV module's power capacity is measured at an input solar radiation of 1000 W/m^2. During a day, from sunrise to sunset, the solar radiation intensities vary significantly. Normally, the number of daily sunshine hours equivalent to 1000 W/m^2 is estimated for the location at which PV system needs to be installed.

Table 4.4. Example of solar radiation data of one location, giving average daily and monthly solar radiation data (the radiation data vary from one site to other site)

Month	Radiation on horizontal surface (kWh/m^2)	Radiation on tilted surface (kWh/m^2)	Equivalent daily sunshine hours (*Hours*)	No. of days	Monthly radiation on tilted surface (kWh/m^2)
January	4.95	5.50	5.50	31	171
February	5.50	6.12	6.12	28	171
March	5.8	6.45	6.45	31	200
April	5.9	6.66	6.66	30	200
May	5.93	6.59	6.59	31	204
June	5.58	6.20	6.20	30	186
July	5.49	6.10	6.10	31	189
August	4.63	5.15	5.15	31	160
September	4.68	5.20	5.20	30	156
October	4.70	5.23	5.23	31	162
November	4.84	5.38	5.38	30	161
December	4.87	5.42	5.42	31	168

Annual solar radiation on tilted surface (kWh/m^2) 2128

The conversion of radiation falling on a module to electricity not only depends on its tilt but also the intensity of the insolation and its variation over the day. The intensity, in turn, depends on the latitude and the atmospheric conditions of the site. In space, beyond earth's atmosphere, the solar radiation is approximately 1.35 kW/m^2. While passing through the atmosphere, the radiation loses some of its power and reduces to 1 kW/m^2 on the horizontal surface of the earth. This peak value reduces to zero during morning and evening hours and is highest in the mid of the day, mostly at solar noon when the light has to travel through a thinner atmosphere. In addition to this, there is always a seasonal variation when the declination angle changes. This makes the variation of the radiation intensity a regular feature for a system designer to worry about for every site of installation.

A typical variation of incident solar radiation on a horizontally placed surface on earth is shown in Figure below (4.11). In this case, the sun rises at about 7 hours (7 a.m.), while the intensity rises from zero to a peak, more than 800 W/m^2 at 12 noon, then it reduces to zero again at 18 hours (6 pm) in the evening. On integration, the area under the curve gives the energy received on the surface from dawn to dusk. The insolation for the figure below has been 6.0 *kWh/m^2* per day. The Equivalent Hours of Peak Sunshine is a convenient way to express the effective time of solar radiation on a unit surface if the solar radiation was effectively constant at its peak value of 1 *kW/m^2*. In this case, it means that if the sunlight was assumed to be constant at 1 *kW/m^2*, then the equivalent hours of peak sunshine is 6.0 *kWh/m^2/day* divided by 1 *kW/m^2* = 6.0 h.

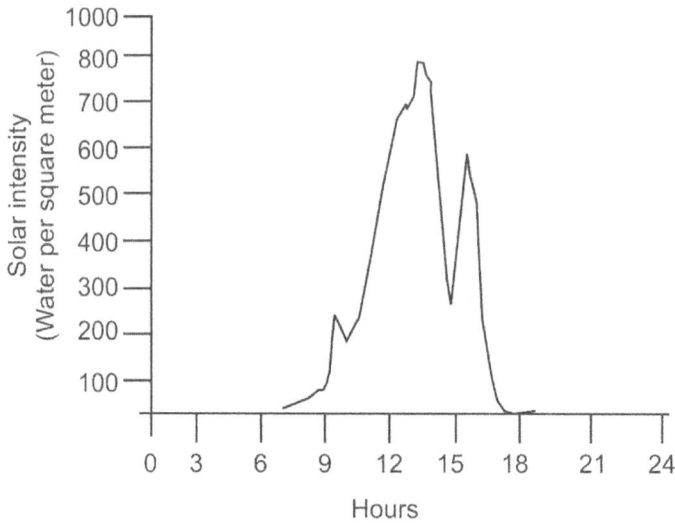

Fig. 4.11. Variations of solar radiation with the time of the day on horizontal surface

In India, the peak equivalent sunshine hours vary between 4-7 hours corresponding to 4000 Wh/m^2-day and 7000 Wh/m^2-day. The sunshine hours and equivalent sunshine hours in a day have also been explained with the help of the following figure (4.12). The peak equivalent sunshine hour is not to be mistaken for the actual hours of sunlight.

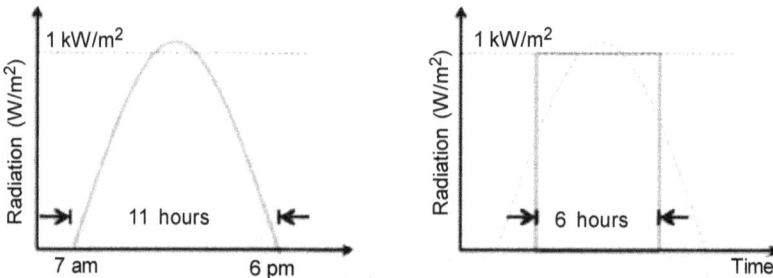

Fig. 4.12. Showing sunshine and equivalent sunshine hours in a day

The integration of this bell-shape curve yields the total net solar energy i.e. insolation for one day in kWh/m^2/day. The same value is also equivalent to the number of hours of irradiance if the sun would be shining at a constant irradiance of 1 $kW/$ m^2. This value (i.e. number of hours) is known as "Peak Sunshine Hours" or "Equivalent sunshine hours". For example, it is seen in the figure above (left) that the Sun shines from 7 am to 6 pm and as a result yielded an insolation of 6 kWh/m^2 for that day (determined by integrating the bell curve). This insolation is equivalent to that resulting from the Sun shining at an intensity of 1 $kW/$m^2 for 6 hours. Hence, it can be said that the insolation received was for "6 Peak Sun Hours".

Let us take the month of January for the calculation. For the site under consideration, the average daily solar radiation on the tilted surface for the month of January is 5.5 $kWh/$

m^2/day (mentioned in the table above). The average daily solar radiation of 5.5 kWh/m^2/day is equal to solar power density of 1 kW/m^2 for 5.5 hours per day. Thus, in this case, equivalent sunshine hours are 5.5 hours per day and monthly equivalent sunshine hours is 5.5 × 31 = 170.5 or 171 hours per month (for the month of January). These hours are for the radiation on the plane of array, i.e., for tilted surface.

Step 3 Estimating the energy production over a given period

In this example, we will do calculation for the month of January. For the month of January, the average daily solar radiation is 5.5 kWh/m^2/day and the corresponding average daily sunshine hours is 5.5 hours. The estimation of possible energy from the plant can be obtained by simply multiplying the PV plant capacity with the daily solar radiation data on the plane of array or plane of modules. In this example, the PV plant capacity is 1 MW_p. Therefore, the maximum possible energy produced by the plant in the month of January is, Monthly energy production = PV plant capacity × Average sunshine hours × Number of day in month

= 1 MW_p × 5.5 hours × 31 = 1000,000 × 5.5 × 31 = 170500000 Wh

On dividing by 1000, we will get electricity in terms of kWh. Therefore, $\frac{170500000}{1000}Wh = 170500 kWh.$ This much amount of the electricity is being produced by the plant in the month of January. Similarly, we can also calculate the electricity produced by 1 MW_p PV plant in other months as well. Calculation for each month is given in Table 4.5. We can see from this table that a total of 2.12 Million kWh electricity can be produced by this 1 MW_p PV power plant.

Here, we have not taken any losses in the PV plant into account. In the following calculation, we need to consider all kind of losses to estimate the actual energy that can be delivered to the grid by this 1 MW_p PV plant.

Step 4 consider losses at PV array level

We can see from table (4.5) above that the total electricity production potential for site under consideration (for which solar radiation data are taken) and for the month of January is 170500 kWh. This electricity production is for ideal condition given by standard test condition (STC) wherein PV module temperature (more precisely temperature of cells in modules) of 25^{0}C is considered. In practice, due to non-ideal conditions, several losses occur at PV module level which reduces the PV module output. These losses include:

i.   Module temperature loss

ii. Module soiling loss due to dust

Table 4.5: Monthly and annual electricity production potential of 1 MW_p solar PV plant for a location under consideration

Month	Power plant capacity	Monthly radiation on tilted surface (kWh/m^2)	Equivalent monthly sunshine hours (hours)	Monthly energy production (kWh)
January	1 MW_p	171	171	170500
February	1 MW_p	171	171	171360
March	1 MW_p	200	200	199950
April	1 MW_p	200	200	199800
May	1 MW_p	204	204	204395
June	1 MW_p	186	186	186000
July	1 MW_p	189	189	189100
August	1 MW_p	160	160	159650
September	1 MW_p	156	156	156000
October	1 MW_p	162	162	162130
November	1 MW_p	161	161	161400
December	1 MW_p	168	168	168020

Total electricity production potential by 1 MW_p power plant in one year without considering any losses = 2,128,305 kWh

2.12 million kWh

iii. Module mismatch loss

iv. DC cable loss

v. Solar radiation loss

Module temperature losses

This is also discussed earlier where the design of grid connected plant for small power application is considered. The loss of power due to temperature is given by temperature coefficient of power. The crystalline silicon modules have power temperature coefficient of about 0.45%/^{0}C while thin film modules have lower temperature coefficient ranging from 0.2%/^{0}C to 0.3%/^{0}C.

Also, it should be noted that the cells in PV modules are at 20^0 C to 25^0C higher temperature than the ambient temperature and it is the cell temperature which results in power losses. Suppose, for the site for which this plant is designed will have 35^0C ambient temperature in the month of January, and the cell temperature is 20^0C higher than the ambient temperature, i.e., the cell temperature is 35+20 = 55^0C. Then, the percentage of power losses for crystalline silicon modules or PV array due to temperature is

$$\text{Module temperature loss} = (55^0\text{C} - 25^0\text{C}) \times \frac{0.45\%}{^\circ\text{C}} = 13.5\%$$

Module soiling losses

This is the loss due to dust settlement on the PV module which hinders solar radiation entering inside and thus reduces power output. Normally, even after regular cleaning of PV modules, the losses are in the range of 1% to 2%. For this example, let us take it to be 1.5%.

Module mismatch losses

The typical manufacturing tolerance of PV module output power is +/- 3%. Let us take these losses to be about 1.5%.

DC cable loss

This is the resistive losses that happens in DC cables which connect modules. Typically, cable thickness is chosen to keep the losses within 2%. For this example, let us assume these losses to be about 1.2%.

Solar radiation loss

This is the loss due to the reflection of PV module from the glass surface. This should normally within 2% to 4%. Here, let us consider this as 2.5%.

In this way, the total PV module losses will be the sum of all the losses mentioned above. Among all losses, the loss of temperature is most dominant which depends on the selection of a site. Normally, the sites which have higher prevailing ambient temperature also have higher solar radiation. The advantage of higher solar radiation is normally more than the loss of power due to higher temperature.

Thus, total module losses are = 13.5% + 1.5% + 1.5% +1.2% + 2.5% = 20.2%

Step 5 Energy generation potential after considering module losses

So far, we have estimated the energy generation potential of 1 MW_p plant for a given site under ideal condition. In step 4, the estimated losses at module level is about 20.2% for the month of January. Ideal energy generation for the month of January in the power plant is 170500 kWh. After considering the losses, the energy output of 1 MW_p PV plant would be:

Energy generated without losses × (1-PV module level losses)

$$= 170500 \text{ kWh} \times \left(1 - \frac{20.2}{100}\right) = 170500 kWh \times (1-0.202) = 170500 \text{ kWh} \times (0.798)$$

= 136059 kWh (in the month of January)

Similar to this, calculation can also be done for other months.

Step 6 Available energy at the output of inverter

Electricity gets generated by PV modules and then it flows at the input of inverter in the DC form. This DC power is converted to AC power by inverter. Other than converting DC power to AC power, the inverter also performs the function of maximum power point tracking (MPPT).

In this PV plant design, we are so far doing the calculation in terms of energy at the inverter level; the energy output of the inverter will be less than the energy that gets into the inverter due to losses in inverter. The inverter power losses consist of MPPT tracking and DC to AC conversion losses. The high power inverters (range of several 100 kW to 1 MW) are becoming very efficient and power losses in inverter are in the range of 1.5% to 3%. Let us assume that inverter losses are 2% in this example. The energy coming at the input of inverter in this example for the month of January is 136059 kWh. Then, the available energy at the output of inverter will be:

Energy at output of inverter (kWh) = Energy at input of inverter (kWh) × (1-losses

at inverter) $=136059 \ kWh \times \left(1-\dfrac{2}{100}\right) = 136059 \ kWh \times (1\text{-}0.02) = 136059 \ kWh \times (1\text{-}$

$0.98) = 133337.8 \ kWh$ (in the month of January)

Step 7 Energy fed to the grid

Inverter is connected to the grid through AC cables and isolation transformer, if it is already not a part of inverter. There are some losses that occur in AC cables too. This loss is normally very small and usually less than 1%. In this example, let us say this loss is about 0.5%. The energy available at the output of the inverter is 133337.8 kWh is eventually the energy fed to the grid will be:

Energy fed to the grid (kWh) = energy at output of inverter (kWh) × (1- losses in AC cables)

$$= 133337.8 \ kWh \times \left(1-\dfrac{0.5}{100}\right) = 133337.8 \ kWh \times (1\text{-}0.005) = 133337.8 \ kWh \times (0.995)$$

$= 132671.1$ kWh (in the month of January)

Thus, the electricity fed to the grid after considering all the losses in the month of January (from 1 MW_p plant and for this particular location) is 132671.1 kWh.

Similar calculations can also be done for other months. The summary of calculation for each month is presented in table 4.6.

Table 4.6 Monthly electricity generation that can be fed to the Grid from 1 MW_p plant

Month	Power plant capacity	Equivalent monthly sunshine hours (hours)	Monthly energy production (without any losses) (kWh)	Monthly energy production (considering all losses) (kWh)
January	1 MW_p	171	170500	132666
February	1 MW_p	171	1711360	133335
March	1 MW_p	200	199950	155581
April	1 MW_p	200	199800	155464
May	1 MW_p	204	204395	159040
June	1 MW_p	186	186000	144727
July	1 MW_p	189	189100	147139
August	1 MW_p	160	159650	124224
September	1 MW_p	156	156000	121384
October	1 MW_p	162	162130	126153
November	1 MW_p	161	161400	125585
December	1 MW_p	168	168020	130736
Total annual electricity production (kWh)			2.12 million kWh	1.65 million kWh

PV plant performance ratio

A PV power plant performance ratio is the ratio of actual electricity fed to the grid to that of electricity that would be produced by given plant capacity in the absence of any losses. It can be given as:

$$\text{PV plant performance ratio} = \frac{Electricity\ fed\ to\ grid}{Electricity\ generated\ by\ plant\ in\ absence\ of\ any\ losses}$$

In this case, the electricity generated by 1 MW_p PV module in whole year is 2.12 Million kWh and factor considering all these losses, the electricity that can be fed to the grid is 1.65 Million kWh. Therefore, the performance ratio of plant is:

$$\text{Performance ratio} = \frac{1.65}{2.12} = 0.778 \text{ or } 77.8\%$$

Sankey diagram is shown (Fig. 4.13) for overview of energy losses in PV power plants.

It has been discussed that the loss of electricity occurs in many ways between the generation of electricity at the module level to feeding of the electricity to the grid. The losses due to temperature, dust, DC cable losses, AC cable losses, inverter losses, etc. can be presented in the form of Sankey diagram (Fig. 4.13) to get good overview of overall energy flow in PV power plants. An example of energy flow in PV plant of 1 MW with various losses occurring at different stages is shown in Figure 4.13.

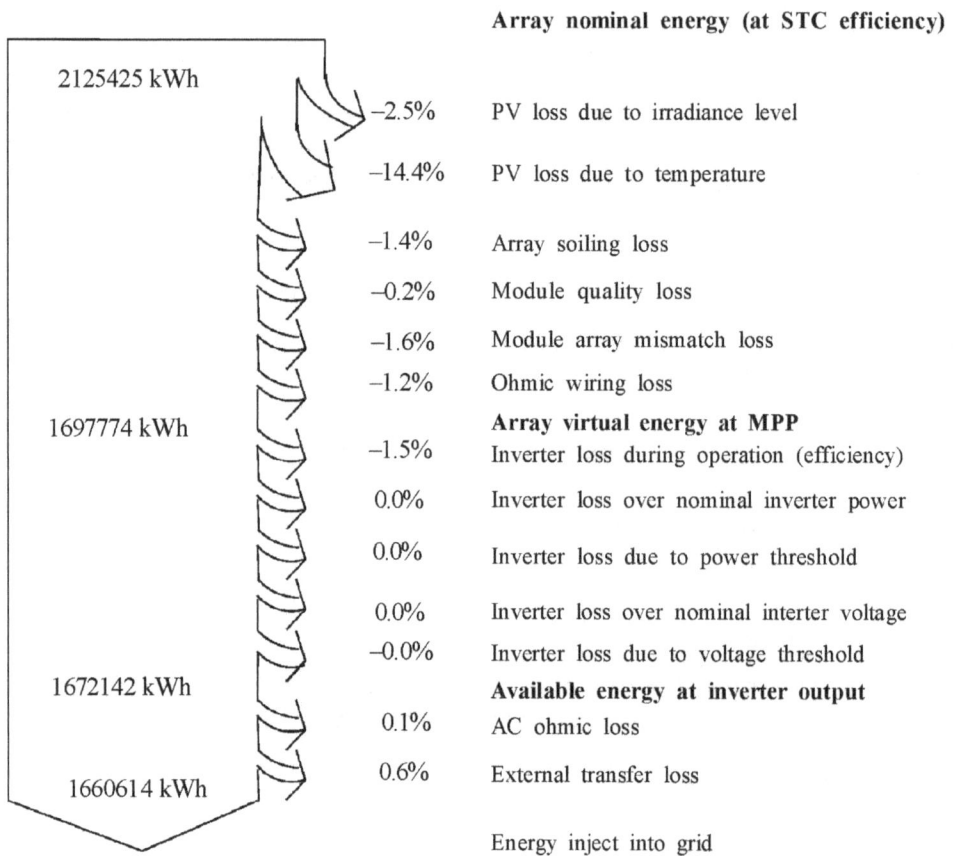

Fig. 4.13 Sankey diagram of PV power plant showing various losses at different stages of conversion

Determining configuration of PV plant

So far, we have studied at the electricity generation in $1\,MW_p$ PV plant and have considered various losses and estimated the electricity fed to the grid for the month of January. We have not considered the number of PV modules which will be required in $1\,MW_p$ PV plant. A detailed description of how to select a PV module and how to select an inverter is discussed earlier. The details about series and parallel connection of PV modules and corresponding addition of their currents and voltages are to be determined.

Select PV module and inverter

Let us first make a choice of PV modules for use in this $1\,MW_p$ PV plant. Normally, for large scale power plants, PV modules of higher wattage ratings are chosen. Looking at the PV module parameters given in Table 4.1, let us choose a PV module '7' having rated wattage of 240 W_p, the other parameters of PV Module '7' are as follows:

i.   Rated power of module No. '7' = 240 W_p

ii. Open circuit voltage (V_{oc}) = 37.38 V

iii. Voltage at maximum power point (V_{mp}) = 30.60 V

iv. Short circuit current (I_{sc})= 8.45 A

v.   Current at maximum power point (I_{mp}) = 7.85 A

Now, let us select an inverter. After looking at table 4.2, we can choose Inverter No. '6' whose power rating (1000 kVA) matches with the need. The other parameters of inverter No. '6' are as follows:

i.   AC power output = 1000 kVA

ii. Max DC input voltage range = 1000 V

iii. MPP tracking range voltage = 460-875 V

iv. Max DC current = 2100 A DC

v.   Peak efficiency = 98.6%

Many times, one can also use small capacity inverters for a large power application. For instance, in order to design 1 MW_p plant, instead of using one 1000 kVA inverter, one can also use to inverters of 500 kVA capacity. In the extreme case, one can use one inverter for each string or even one inverter for each module as well.

Number of PV modules required

In this example, total PV array capacity required is 1 MW_p and we have chosen 240 W_p PV module for installation. Therefore, the number of modules required for this plant would be:

$$\text{Total no. of modules required} = \frac{Total\ PV\ plant\ capacity}{Rated\ power\ of\ one module}$$

$$= \frac{1MW_p}{240W_p} = \frac{1,000,000}{240} = 4166.6 \text{ modules}$$

To round off, we need about 4167 PV modules of 240 W_p for this 1 MW_p PV power plant.

Number of PV modules in a string

In order to determine the number of PV modules in a string, we should look at the acceptable voltage rating of inverter and try to match it with the voltage ratings of strings. The V_{oc} of PV module No. '7' is 37.38 V, V_{mp} is 30.60 V and I_m is 7.85 A. Also, for Inverter No. '6' of 1000 kVA, its maximum power point (MPP) voltage tracking range is 460 V-875 V and maximum input DC current is 2100 A. While deciding the number of PV modules in a string, we have to ensure that V_{mp} of the string is within the V_{mp} range of inverter and V_{oc} of the string is less than the maximum DC voltage of Inverter.

When we connect PV modules in series the voltage gets added. Therefore, we need to connect many Module No. '7' in series to get the required voltage in the range of inverter '6'. If we divide V_{mp} range of inverter '6' with the V_{mp} of the Module '7', we will get the number of PV modules to be connected in series to form one string as:

$$\text{No. of modules in string} = \frac{V_{mp} \text{ voltage of selected inverter}}{V_{mp} \text{ of selected PV module}}$$

Since the inverter V_{mp} tracking range is 460 V to 875 V, let us have an average of it. That is, about 667 V. This 667 V will be the V_{mp} of the PV module string. Note that any other voltage levels within the inverter tracking range can also be chosen, but we have to only ensure that the array V_{mp} point should lay within the inverter V_{mp} range in all operating conditions.

$$\text{Number of 240 } W_p \text{ modules in strings} = \frac{V_{mp} \text{ of inverter}}{V_{mp} \text{ module}} = \frac{667}{30.60} = 21.79$$

The above calculation shows that we will have to connect 21.79 PV modules in series, or to round it off, we get 22 modules. While designing the number of PV modules to be connected in series (as string), we have to take care that V_{mp} of string should lay within V_{mp} tracking range of inverter under all possible operating conditions.

Number of PV strings in parallel

In total, we need to connect 4167 PV modules for this 1 MW_p PV power plant. One string of PV modules will have 22 PV modules. Therefore, the number of strings to be connected in parallel will be as follows:

$$\text{No. of strings in parallel} = \frac{Total \ no. \ of \ PV \ modules}{No. \ of \ modules \ in \ a \ string} = \frac{4167}{22} = 189.4 \text{ parallel strings}$$

To round off we will have to connect 190 parallel strings (each having 22 PV modules in series) in this 1 MW$_p$ power plant.

Note, that when we connect strings in parallel, the current of each string gets added. The total current of the parallel combination will be the sum of current of each string. We have to make sure that the total current of the strings is within the limit of the inverter. In this example, the maximum DC current limit of inverter No. '6' is 2100A, and I_{sc} of Module '7' is 8.45 A. The maximum current of a PV module will be the maximum current of a string. Therefore, the total current of 190 strings connected in parallel will be:

Total current of strings in parallel = current of one string × total number of strings = 8.45 × 190 = 1605.5 A

From this calculation, we can see that the current of 190 strings when put together is less than the maximum DC current limit of inverter (which is 2100 A). Therefore, this design is acceptable and can be implemented.

This completes the design of 1 MW_p solar PV power plant. In summary, the details of 1 MW_p plant is presented in Table 4.7. Due to rounding off that we have done in design, the PV plant parameters are slightly different from targeted value.

Table 4.7: Summary of parameters of designed 1 MW_p PV power plant

Power rating of selected PV module	240 W_p
Number of PV modules in a string	22
Number of strings in parallel	190
Total number of modules used in plant	22 × 190 = 4180
Total rated peak power of plant	4180 × 240 = 1003200 W_p
V_{mp} voltage of PV array	22 × 30.60 = 673.2 V
I_{mp} current of PV array	190 × 7.85 = 1491.5 A
Expected annual energy generation of plant	1.65 million kWh

4.9 DISTANCE BETWEEN PV MODULES IN AN ARRAY TO PREVENT SHADING

In a string of array, the modules are to be placed in such a manner that the module will not cast shadow over its nearby module. Therefore, an adequate distance should be kept between the first and the second and between the second and third and so on, such that under the worst condition, when the sun is over the tropic of Capricorn, all the modules would be exposed to light. At the winter solstice in the northern hemisphere, around December 21, the Sun is directly over the tropic of Capricorn. During winter period, the elevation of sun in the sky is low compared to the summer period and hence the length of shadow is highest. The minimum distance between two modules in an array is therefore considered when sun is directly over the tropic of Capricorn.

While designing a power plant, shading analysis is a crucial step in order to avoid any losses due to shades from nearby structures, objects, trees and nearby modules. Shading reduces the amount of irradiation to be actually received by the modules. The length of the shadow casted on PV modules increases during the "low-sun" days or during winter season. Therefore, the worst-case shadow condition causing the longest shadow occurs during these days. This results in the lowest access to sunlight radiation during the period. The winter solstice, generally falls on 21st or 22nd of December, which gives the longest shadow length. For calculating the minimum distance between the two modules, shadow length during 8am and 4 pm in 22nd December is generally considered. The local latitude angle for all the places is taken as the tilt angle of the module in order to harness optimum radiation. The calculation is done as per the following diagram (Fig. 4.14).

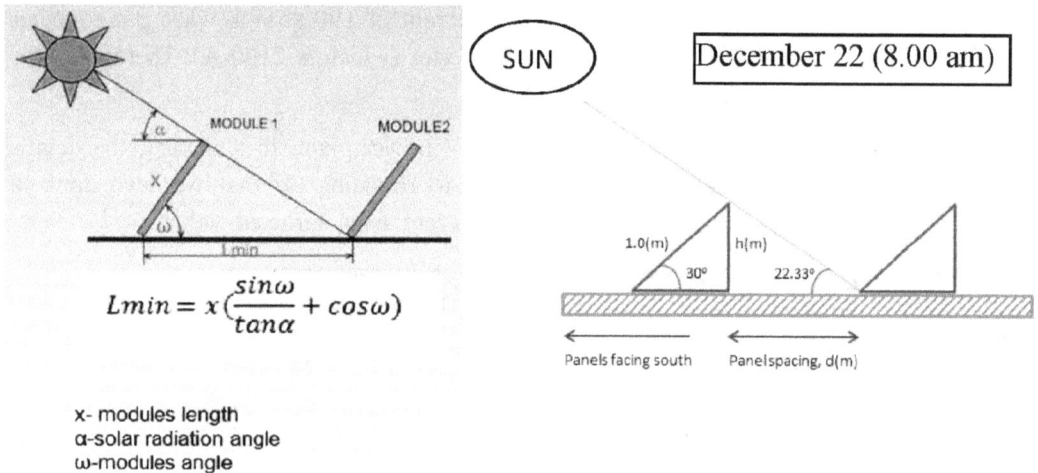

$$Lmin = x\left(\frac{sin\omega}{tan\alpha} + cos\omega\right)$$

x- modules length
α-solar radiation angle
ω-modules angle

Fig. 4.14. Arrangement of panel spacing with respect to shading

The inter-module distance is considered using a simple thumb rule where the minimum spacing between the modules is equal to three times the height of module (Das and Das 2019), i.e., *(d= 3×h)* as per the following diagram

4.10 WIRE SIZING IN PV SYSTEMS

In solar PV system, appropriate dimensions (diameter and length) of the wires or cables for interconnection of modules, batteries and loads should be used. The size of the wires should be chosen to avoid excessive voltage drops in a line (wire or cable connecting two points electrically). Normally, the voltage drop in the line connecting modules to batteries should not be more than 5% of the line voltage. Similarly, voltage drop in the line connecting batteries and the load should not be more than 5% of the line voltage. Also, it should be ensured that the maximum current passing through the cables should be within the current handling capacity of the cables.

A schematic representation of line A-B connecting a battery of voltage V and load R_L is shown in fig. 4.15. The voltage drop (ΔV) in the line due to current flow is the voltage that is used to overcome the resistance of the line. As per the Ohm's law, if I is the current flowing in a line having a resistance of R ohm, then the voltage drop (across the line) will be IR. Due to this voltage drop in the line, actual voltage available to the load will be $V-\Delta V$, which is less than the supply or battery voltage. Such drop in the line voltage is not desirable and should be restricted to a limit, in many cases within 5%. For instance, if drop in the line voltage connecting PV modules and battery is high, then the available voltage may not be enough to charge the batteries resulting into the inefficiency of the whole purpose.

The voltage drop for a given cable of specified material resistivity (r), length (L) and cross-sectional area (A) can be estimated as:

$$V_d = 2 \times \Delta V = 2 \times I \times \frac{\rho L}{A} \qquad \qquad \dots (4.1)$$

The factor of 2 in the above equation (4.1) is introduced due to the fact that the length of the cable used is actually double (for taking current to and fro) the physical distance between PV module and the battery or battery and the load. As mentioned earlier, the V_d should be within 5% limit of the line voltage, i.e., if the line voltage is 12 V, the voltage drop should not be more than 0.6 V. It is clear from the above equation that the cable having more length and smaller cross-sectional area will have larger voltage drop. Also, resistivity of the cable material (Cu or Al) plays an important role. The resistivity of Cu and Al is 1.678×10^{-8} Ω-m and 2.82×10^{-8} Ω-m, respectively. For a required length of the cable, proper diameter and material should be chosen to limit the voltage drops. Due to lower resistivity, Cu cables are generally used in solar PV systems.

Typically, the diameter of cables used in DC electrical system is higher than the diameter of cables used in AC electrical systems. In the solar PV systems, the DC voltage levels are normally in the range of 12 V to 48 V (as against 220 V in AC systems). Due to this, the loads of similar power rating need to be supplied with higher current in DC systems compared to AC systems. For instance, a 100 W load supplied with 220 V AC supply would require 0.45 A (rms), while the same load will draw 8.33 A when supplied by a 12 V DC supply. In this case, if the cable of the same diameter (same length and material as well) is used in both AC and DC systems, the voltage drop in the DC systems would be nearly 18.5 times more than in AC system as per the above equation. Therefore, to limit the voltage drop in DC system, cables of larger diameter are used.

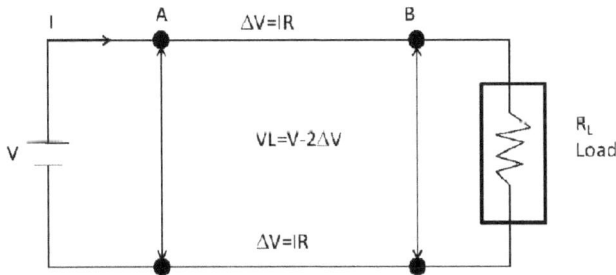

Fig. 4.15: Schematic diagram showing drop in the line A-B due to resistance R and current flow I

Example 4.3. In a PV system the distance between the solar PV module and battery is 10m. The system voltage is chosen to be 12 V DC. The PV module and battery are connected by Cu cable of 2.5 mm^2 cross-sectional area. Estimate the voltage drop in the cable if it is carrying 3.5 A current. What would happen if the cable of 1.5 mm^2 cross-sectional area is used?

Solution: Equation for sizing of cable as mentioned above can be used to estimate the voltage drop. It is given that $V = 12\ V$, $L = 10\ m$, $A = 2.5\ mm^2$, $\rho(Cu) = 1.678 \times 10^{-8}\ \Omega\text{-m}$ and

$$I = 3.5A.\ V_d = 2 \times I \times \frac{\rho L}{A} = 2 \times 3.5 \times \frac{1.678 \times 10^{-8} \times 10}{2.5 \times 10^{-6}} = 0.47V$$

It is 3.9% of the 12 V. Since V_d is within 5% of the line voltage (12 V), therefore it is acceptable. It also indicates that an appropriate cable is chosen for the given requirement.

Now if a cable of only 1.5 mm^2 cross-sectional area is chosen, then the estimated voltage drop is:

$$\frac{2 \times 3.5 \times 1.678 \times 10^{-8} \times 10}{1.5 \times 10^{-6}} = 0.783V$$

In this case, the voltage drop is 6.5% of the 12 V, which is beyond the acceptable limit.

4.11 ECONOMIC ANALYSIS OF PV SYSTEM

After investing some amount for any business enterprise, it is required to assess its financial viability to make the venture sustainable and profitable. It is also desirable to find out the time period during which, the invested amount would be recovered. The period within which the invested money can be recovered or saved is known as payback period. The estimation of payback period can be done in two ways, (i) by estimating the simple payback period and (ii) by estimating the lifecycle cost (LCC) of the system. Among these, the estimation of simple payback period is relatively easy. The estimation of life cycle cost is comparatively complicated. The simple payback period is the amount of time that is obtained by dividing the initial investment in a PV system by the cost of annual energy savings due to the use of PV system in which money is invested. Thus, if the initial investment cost is 'X' and the cost of annual energy saving is 'Y', then, simple payback

$$\text{period} = \frac{X}{Y}\ \text{years}$$

However, the simple payback period has limitations. It does not consider many important parameters in estimation such as time-value of money, inflation rate, lifetime of system and operation and maintenance cost. These parameters are taken into consideration in LCC. Due to this reason, the simple payback period is only useful for making a rough estimate of how long, during which, the initial investment can be recovered.

4.11.1 Life Cycle Costing (LCC)

The LCC is the cost of using a system during its lifetime. The cost of owning and running a system or equipment over its lifetime can be divided into three components:

(i) Capital cost (paid while purchasing the system /equipment).

(ii) Operation and maintenance cost (cost incurred for running the equipment, could be regular maintenance cost, or it could be fuel cost etc.)

(iii) Replacement cost (cost incurred if any component of system or equipment needs replacement before the life of the system is over).

Thus, the LCC includes all associated costs of the system that can be incurred over its lifetime. The LCC is calculated before making a purchase or taking up a venture. The LCC can be used to evaluate different designs (having same functionality) of PV systems. Of course, the design with lowest LCC should be chosen. In general, this approach of calculating the LCC can be used to take a decision on purchase of any other system or appliances. For instance, it can help us to decide whether to buy an expensive refrigerator with high efficiency or a cheap refrigerator with low efficiency.

In order to calculate the LCC, three cost components as mentioned above should be known at the time when investment decision is made. The component of capital cost is known as the time of investment, but the other two cost components, operation and maintenance cost and the replacement cost, occur in future. For this, cost estimation (in terms of current value of money or present worth) of future investment is required. It therefore requires an understanding of the time value of money.

Time value of money

Since the cost is calculated for the lifetime of the system (which could be a long period, up to 30 years in the case of PV systems), it is important to consider the time value of money, i.e., it is necessary to consider the increase or decrease in the value of money over the systems' lifetime. There are two ways in which the value of money over time changes, due to inflation and due to interest (more appropriately discount rate).

Inflation rate, r: The inflation rate, r, is a measure of the decline in the value of money over time. Thus, if inflation rate is 5%, the value of money will decrease by 5% after one year and as compared to today, a product would cost 5% more in the next year. If the cost of a product today is C_0, then its cost, a year later, will be $C_0 \times (1+r)$. Two years later, the cost of the same product would be $C_0 \times (1+r) \times (1+r)$. Here '$r$' should be inserted in decimal and not in percentage. Extending this logic to n years, the future cost of the product in n years, $C(n)$, due to inflation rate, r, can be given as:

$$C(n) = C_0 (1+r)^n \qquad \qquad \text{... (4.2)}$$

Discount rate, d: The value of money over time increases due to the interest it can earn. It gives the rate at which value of saved money increases. If we have 'X' amount of money today, it will have some purchasing power. The purchasing power of this 'X' amount will be higher next year due to the interest that it will earn. Thus, money has increased its value with time. In order to represent the true increase in the value of money over time, discount rate, d, (instead of interest rate) is used. The discount rate gives the increase in value of money with respect to future value of money. While the interest rate gives the increase in value of money with respect to current value of money as given below:

$$\text{Discount rate } (d) = \frac{Future\ value - present\ value}{Future\ value} \qquad \ldots (4.3)$$

$$\text{Interest rate } (i) = \frac{Future\ value - present\ value}{Present\ value} \qquad \ldots (4.4)$$

Similar to inflation rate, we can also write the expression for increase in value of money over 'n' years, $M(n)$, due to the discount rate d. It can be written as

$$M(n) = M_0(1+d)^n \qquad \ldots (4.5)$$

Where M_0 is the amount of money at this moment and 'd' is presented in decimal and not in percentage.

One should take a note that the value of money over time will increase only when the discount rate is higher than the inflation rate. If the reverse is true, i.e., inflation rate is higher than the discount rate, the value or the purchasing power of money actually decreases. One should also note that the discount rate and inflation rate depend on the future, and therefore, by nature they are unpredictable. While doing the calculations, reasonable numbers for the discount rate and inflation rate are taken as per the prevailing conditions.

With the above discussion on discount and inflation rates, we are now in a position to find out the present worth of future investments, useful for calculation of the LCC. The future investments could be on replacement of components (which are one-time investments, e.g., batteries in a PV system) or the future investments could be on the operation and maintenance (which are recurring in nature, e.g., diesel in a generator set).

Present worth of future one-time investments

The present worth of a product is defined as the amount of money that needs to be invested today with the discount rate, 'd', and the inflation rate, 'r', such that we would be able to purchase the product in the future, 'n' years later. Suppose, we have with us Rs. 'X' today and the cost of a product (that if we buy today) is also Rs. 'X'. But we need to buy this product in future as a replacement component (one-time investment). Because of the inflation, the cost of the product will increase as given by Eq. (4.2) and because of the discount rate (supposedly higher than the inflation rate), the value of money will increase as given by Eq. (4.5). Since we are assuming that the discount rate is higher than the inflation rate, the cost of the product in the future (in n years) will decrease due to the effective increase in the purchasing power of money and it will decrease by a factor, $F_{PW-one\ time}$, given by the ratio of Eq. (4.2) to Eq. (4.5). Thus, the present worth of the investment that needs to be made 'n' years later would be lower by the factor given by $F_{PW-one\ time}$.

$$F_{PW-one\ time} = \frac{Future\ cost}{Future\ value} = \frac{X(1+r)^n}{X(1+d)^n} = \left(\frac{1+r}{1+d}\right)^n \qquad \dots (4.6)$$

Where $FP_{W-one\ time}$ can be read as 'present worth factor for future one-time investments'. Now if the present cost of a product is C_0, then its present worth (PW) for the actual one-time investment after 'n' years would be given as:

$$PW_{replacement} = C_0 F_{PW-one\ time} = C_0 \left(\frac{1+r}{1+d}\right)^n \qquad \dots (4.7)$$

Present worth of future recurring investments

There are requirements when the investment is required on recurring basis to run the system. In such cases, the present worth of the future investments can be obtained by using Eq. (4.7), where recurring investment is considered as one-time investment for each year of operation. Thus, present worth of recurring expenses PW_{rec} can be obtained by summing up the present worth of one-time investments for the entire life of the system as given in Eq. (4.8).

$PW_{rec-beg}$: When the investment is made in the beginning of the year, (for instance buying diesel for diesel generator) or in Eq. (4.9), for $PW_{rec-end}$, when investment is made at the end of the year, (for instance buying grease for the diesel generator):

$$PW_{rec-beg} = C_0 + C_0\left(\frac{1+r}{1+d}\right)^1 + C_0\left(\frac{1+r}{1+d}\right)^2 + \dots + C_0\left(\frac{1+r}{1+d}\right)^{n-1} \qquad \dots (4.8)$$

$$PW_{rec-end} = C_0\left(\frac{1+r}{1+d}\right)^1 + C_0\left(\frac{1+r}{1+d}\right)^2 + \dots + C_0\left(\frac{1+r}{1+d}\right)^n \qquad \dots (4.9)$$

Note that in either case [Eq.(4.8) or Eq. (4.9)], there are n number of purchases made in the lifetime on the system. The above equation can be simplified if we replace with the following term i.e.

$$k = \frac{1+r}{1+d} \qquad \dots (4.10)$$

The simplified version of Eqs. (4.8) and (4.9) in terms of parameter k can be written as:

$$PW_{rec-beg} = C_0 \left(\frac{1-k^n}{1-k} \right) = C_0 F_{PW-rec-beg} \text{ where } F_{PW-rec-beg} = \left(\frac{1-k^n}{1-k} \right) \quad \text{... (4.11)}$$

$$PW_{rec-end} = C_0 k \left(\frac{1-k^n}{1-k} \right) = C_0 F_{PW-rec-end} \text{ where } F_{PW-rec-end} = k \left(\frac{1-k^n}{1-k} \right) \quad \text{... (4.12)}$$

Where $F_{PW-rec-beg}$ is the present worth factor for the recurring investments which are made in the beginning of each year and $F_{PW-rec-end}$ is the present worth factor of the recurring investment made at the end of each year. However, for financial evaluation, all calculations are done usually at the end of the year.

Normally, the value of 'k' is less than 1 and varies in the range of 0.95 to 1.0. This gives nearly the same value of $F_{PW-rec-beg}$ and $F_{PW-rec-end}$. Also, in many cases the investment may not be made exactly at the end or at the beginning, but may be in the middle of the year, or throughout the year (as purchase of fuel). Therefore, it is difficult to decide whether to use Eq. (4.11) or Eq. (4.12). In practice, either of the equation can be used to get a good estimate of cumulative recurring cost. The solution of equations (4.8) and (4.9) can be made by following the sum of finite geometric series like

$$1 + k + k^2 + k^3 + \cdots + k^n = \left(\frac{1-k^n}{1-k} \right) \text{ if } k \neq 1$$

4.11.1.1 *Life Cycle Cost*

In order to calculate the LCC of a system, first the present worth factors, as discussed in the above two paragraphs (F_{PW-one} and $F_{PW-rec-end}$), are calculated. They are multiplied by the appropriate present cost components and added together to find out the total cost of operating a given system over its lifetime. For instance, a system has a lifetime of 10 years and has a capital cost C_1 (in terms of present cost). It requires replacement of a critical component in 5 years, having a cost C_2 (in terms of present cost). The system also has an annual maintenance cost of C_3 (in terms of present cost). Then the LCC of the system would be given by the following equation:

$$LCC = C_1 + F_{(PW-one)} \text{ (5 years)} \times C_2 + F_{(PW-rec-end)} \text{ (10 years)} \times C_3 \quad \text{... (4.13)}$$

If there are other cost components (such as fuel cost, electricity cost, etc.) associated with the system operation then they should also be added in Eq. (4.13) with appropriate present worth factor.

The LCC analysis can be used to compare the cost of owing and running different systems for a given period. For instance, it can help us in choosing a PV system design among different possible designs, for a given type of electrical load. Let us consider an example.

Example 4.4: Compare the use of fluorescent lamp with incandescent lamp on the basis of LCC.

Solution: The fluorescent lights have the characteristics of being expensive, long life and efficient and the incandescent lamp sources have the characteristics of being inexpensive, shorter life and less efficient. Roughly, the cost of fluorescent lamp could be about 10 to 20 times higher for same wattage rating, but they are about 4 to 5 times more efficient and they last about 4 to 5 times longer. Let us consider an example wherein an industry has to decide whether to use incandescent lamp or the fluorescent lamp. The time span of study is 5 years. It turns out that for a given light requirement, industry needs to use 20 pieces of incandescent lamp (each costing $1). The lifetime of incandescent lamp is 1 year and needs 4 replacements during 5 years of study. It is given that the cost of electricity is $0.1/unit in the region where industry is located. Total energy consumption of the incandescent lamp would be about 15 units/day (as per requirement). The same number of fluorescent light would cost about $400 (20 times more) but they would consume 4 times less energy per day (3.75 units/day) and if we assume the life of the fluorescent to be 5 years, then there would not be any replacement during the study time of 5 years. Which type of light source should be chosen?

The LCC provides the basis of comparison. The LCC of the two cases is calculated and given in table 4.8. It is considered that the inflation rate is 3% and the discount rate is 10%. The present worth factor for replacement ($F_{PW\text{-one time}}$) of the incandescent light at the end of the 1st year, 2nd year, 3rd year and 4th year will be 0.94 (1.03/1.1), 0.88 $\{(1.03)^2 / (1.1)^2\}$, 0.82 and 0.77, respectively. The electricity expenses can be considered as the recurring cost. The present worth ($F_{PW\text{-rec-end}}$) factor for recurring cost for 5 years is 4.40. The annual electricity bill in the case of fluorescent lamp and incandescent lamp would be $135 and $540, respectively considering 360 days/year.

It can be seen from the table that the LCC of the fluorescent lamp for 5 years of operation is much lower than that of the incandescent lamp; therefore, fluorescent lamp should be the preferred choice.

Annualized LCC (ALCC)

It is sometimes to compare the LCC of a system on an annual basis. In this case, dividing the LCC by the expected life of the system will not give us the annualized LCC or ALCC of the system. Because it is assumed that the annual cost of operation of the system will not be the same throughout its lifetime; therefore, in order to find out the annual cost in terms of current value of money, it is necessary to divide the LCC by the present worth factor F_{PW}-rec-end or $F_{PW\text{-rec-beg}}$. In example 4.4 of the use of incandescent lamp or fluorescent lamp, the $F_{PW\text{-rec-end}}$ is 4.4. Therefore, the ALCC for incandescent lamp ($ALCC_{IL}$) or fluorescent lamp ($ALCC_{FL}$) can be given as:

$$ALCC_{IL} = \frac{LCC}{F_{PW-rec-beg}} = \frac{2465.6}{4.4} = \$560.3 \text{ and } ALCC_{FL} = \frac{LCC}{F_{PW-rec-beg}} = \frac{794.4}{4.4} = \$180.5$$

Thus, the annualized cost of operation (example 5.4) is much higher in the case of the use of incandescent lamp as compared to the use of fluorescent lamp.

Table. 4.8: Comparison of the LCC of incandescent lamp and Fluorescent lamp for the operation of 5 years. All costs are in US$

Lamp cost and replacement	Incandescent lamp			Fluorescent lamp		
	Present cost	Present worth factor	Present worth	Present cost	Present worth factor	Present worth
1st year	20	1	20	200	1	200
2nd year	20	0.94	18.7	-		0
3rd year	20	0.88	17.5	-		0
4th year	20	0.82	16.4	-		0
5th year	20	0.77	15.4	-		0
Annual Electricity cost	540	4.40	2377.6	135	4.40	594.4
Life cycle Cost (LCC)				2465.6		794.4

Unit cost of generated Electricity

In several situations, a given requirement of electrical energy (lighting, irrigation, etc.) can be fulfilled by several sources. Possible alternatives for electricity generations are the use of solar PV modules, use of diesel generator, use of wind turbines or simply the use of grid electricity (if available). Among these available options, which option should be chosen? The answer to this question can be obtained by finding the cost of generated electricity from each of these options over a period of time (e.g., 10 or 20 years). Then, from the economic point of view, the source which provides the lowest unit cost of electricity generated should be chosen. This analysis can be done by finding out ALCC of a given source and then dividing it by the annual number of electrical units generated by the given source. This would provide the generated cost of electricity from a source over its period of operation.

Example 4.5 A remote hospital where no grid electricity is available has a load of 8 kWh/day. The power should be available to the load 24 h a day. For this hospital, the use of solar PV module or a diesel generator is being planned. Find out and compare the unit cost of electricity generated from a solar PV system and a diesel generator, if the load to be operated for 20 years.

Solution: For the above per day electrical energy requirement (8 kWh/day), a solar PV system or a diesel generator can be used. We need to choose a solution which will cost lowest among the two. The sizing of various components of the system is given in table 4.9 with zero-day autonomy for the battery.

Table 4.9: PV system components sizing for supplying 8 kWh/day Load

Component	Value	Remarks
Inverter	1 *kVA* (closest to 863 *W*)	70% DoD
Battery	6 × 12 *V* × 180 *Ah*	2 in series in a string
		3 strings in parallel
Solar panels	32 × 75 *W* (2400 *W*)	75% operating efficiency
		Assumed solar irradiance = 6 h
		16 panels in parallel, 2 strings in series
Operation and maintenance		2% of the system cost

Similarly, for supplying the 8 kWh/day load, the diesel generator of the following specification (table 4.10) can be chosen.

Table 4.10: Diesel generator specifications for supplying 8 kWh/day load

Component	Value	Remarks
Diesel genset	2 *kVA*	Assuming 40% efficiency
Diesel consumption per day	3.2 litre/day (1170 litres/year)	2.5 *kWh*/litre
Operation and maintenance		De-carbonization engine overhaul

Let us now calculate the LCC for the PV system as well as for the diesel generator. LCC of solar PV system is first calculated. The life of the system is given as 20 years. In order to calculate the LCC, apart from the current investment, it requires to calculate the present worth factor of future investment in batteries (to be replaced every 5 years) and diesel (to be bought every year). Table 4.11 gives the present worth of PV system (and present worth factor), including the battery replacement cost and maintenance cost. Batteries would be replaced in every 5th year, i.e., 10th and 15th year. For this, the present worth factor, $F_{PW\text{-one time}}$, for the 5th, 10th and 15th year should be used to estimate the present worth. Some other components may have to be replaced once in the 10th year. Operation and maintenance expenses are recurring costs that occur every year end.

Therefore, we need to use $F_{PW\text{-rec-end}}$ present worth factor in calculating the present worth. Adding all the present worth cost, we get LCC. The annualized LCC analysis measures the cost of the respective systems that may occur every year. It is the annual expenses incurred by the systems.

Table 4.11: LCC of solar PV system having 20 years lifetime (inflation rate 3% and discount rate 10%)

Solar PV component	Initial cost	Present factor	Present worth
PV array	$10,666.67	1	$10,666.67
Battery	$2533.33	1	$2533.33
Battery 5th year		0.72	$1823.54
Battery 10th year		0.52	$1312.62
Battery 15th year		0.37	$944.84
Other components(inverters, wires, etc.)	$2640.00	1	$2640.00
Other components (inverter, wire) replacement every 10th year		0.48	$1267.2
O & M recurring at the end of the year	$348.89	11.10	$3875.35
LCC			$25063.55
Annualized LCC			$2257.97

Let us now calculate the LCC for the diesel generator. Here, the lifetime of the diesel generator is considered to be 6 years. Thus, it needs to be replaced after every 6th, 12th and 18th year. The consumption of diesel and operating and maintenance cost are considered as the recurring costs. The calculation of present worth of various costs for diesel generator is given in Table 4.12. The LCC and ALCC are also given.

Table 4.12: LCC of diesel generator system having 20 years' lifetime (inflation rate 6 % and discount rate 10 %)

Component	Initial	Present worth factor	Present worth
DG set	$ 1222.22	1	$ 1222.22
DG set 6th year		0.80	$ 978.65
DG set 12th year		0.64	$ 783.63
DG set 18th year		0.51	$ 627.46
Diesel expenses recurring at end of year	$ 1092.00	14.39	$15,715.00
O & M recurring at the end of the year	$1111.11	14.39	$ 15988.87
LCC			$ 35315.83
Annualized LCC			$ 2454.19

Now in order to compare the suitability of solar PV and diesel generator system based on economics, we can calculate the per unit cost of electricity generated from each system. The cost of per unit energy from both the systems can be calculated by dividing the ALCC of the system by the annual energy generated per year. In this example, the daily energy generation is 8 kWh; therefore, the annual energy generation is 2920 kWh per year. Thus, per unit cost is obtained by dividing ALCC with 2920 kWh/year. From tables 4.10 and 4.11, the per unit cost of electricity from solar PV is $0.77/kWh and from diesel generator is $0.84/kWh. Clearly, the cost of electricity generated per unit is less for solar PV, therefore, solar PV system should be chosen for the hospital.

4.12 CONCLUSION

The rooftop solar photovoltaic (PV) segment is one of the fastest growing clean energy segments across the globe due to its ability to provide reliable power for both rural and urban customers, to scale up investments through entry of multiple investors and to help the power utilities in minimizing transmission and distribution losses as occurred normally in conventional grid system. In fact, Indian Government is keen to enhance solar capacity within coming five years by promoting decentralized and grid-connected rooftop solar PV systems. India has taken the challenge of developing 40 GW of rooftop solar power capacity as a part of its Green Commitments before United Nations Framework Convention on Climate Change (UNFCCC), as India has tremendous scope of generating solar energy. The reason being the geographical location and it receives solar radiation almost throughout the year, which amounts to 3000 h of sunshine. This is equal to more than 5000 trillion kWh. Almost all parts of India receive 4–7 kWh of solar radiation per square meters in a day. Considering the clear advantages of rooftop solar power (minimal distribution losses, no need of land or dedicated transmission, etc.), the Ministry of New and Renewable Energy (MNRE) is pursuing development of proactive ecosystem for the fast development of this segment.

Rooftop solar projects, like other renewable energy projects (wind, large solar and small hydro), are more cost-effective over the long run, especially when compared to the increasing costs of conventional energy generation. But, the implementation of these projects is challenging as these projects have a high initial cost and may be more expensive in the short term as compared to the existing cost of conventional sources of power generation. However, the government of India provides 30% subsidy on the capital investment made by investors in solar PV installations up to 250 kW_p (SECI, 2014) in order to disseminate the solar PV system widely in the country. Hence, there is the great scope for grid-connected roof top solar power plant both for the users and the entrepreneurs providing services for the installation of the system in rural as well as in urban areas.

Some Related Terminologies

Alternative Current: Electric current in which the current changes polarity 120 times per second (in the U.S.) and is commonly referred to as 60 Hertz (cycles per second) AC. Many other countries including India use 50 Hertz as a standard.

Alternative Energy: A term coined for "non-conventional" energy systems usually on a smaller scale and includes solar electric systems, thermal heat devices, wind generator systems, and small hydro-electric systems.

Alternator: A electric generator producing AC (alternating current) rather than DC (direct current). Alternators are more efficient than DC generators and in automobiles the alternator output is converted to 12 volts DC using rectifier diodes built into the alternator.

Amorphous Semiconductor: A non-crystalline semiconductor material for PV cells also called thin film. Thin film solar cells are easier to make and cheaper than crystalline semiconductors for solar cells, but also less efficient.

Ampere: One Ampere (Amp) is the amount of current that flows in a circuit driven by an electromotive force (voltage) of one Volt through a resistance of one ohm.

Amp-hour: A measure of current over a period of time. One amp used or generated for one hour equals to one Amp-hour.

Angle of Incidence: The angle at which direct sunlight strikes the surface of the solar panel relative to the normal. Sunlight incident on the solar panel at an angle of 90 degrees produces electric energy more efficiently.

Angle of Inclination: The angle, at which a solar panel is placed above the horizontal. (90 degrees would be vertical). A general rule is to set the angle of a solar panel to the latitude with plus or minus 15 degrees.

Anti-reflection Coating: A thin coating of a material applied to a PV cell that reduces the amount of reflection of light striking on its surface.

AEER: (Appliance Energy Efficiency Ratings) Operating efficiency of appliances as set by the Department of Energy guidelines.

Array (solar): Any number of solar photovoltaic modules or panels electrically connected together to provide a single electrical output.

Asynchronous Generator: An electric generator that produces alternating current that matches an existing electric power source or conventional electric grid.

Autonomous System: A stand-alone power system with battery backup, without electric grid connection. These systems are designed for 3 to 5 autonomous days of uninterrupted power supply from the batteries without recharging from a solar panel or standby diesel generator set.

AWG: The abbreviation for American Wire Gauge, a standard system for designating the size of the electrical wires.

Backup Energy System: Back-up electric power systems using batteries and inverters and diesel AC generator, or both.

Battery: Number of electrolytic cells connected in series or in parallel that store electrical energy in chemical energy form.

Blocking Diode: A blocking diode allows the flow of current from a solar panel to the battery but prevents/blocks the flow of current from battery to solar panel thereby preventing the battery from discharging. A semiconductor device connected in electric circuits to stop the flow of current from one direction, but allow it in the opposite direction.

Bus (electrical): An electrical conductor that serves as a common connection means for multiple electrical connections.

Bypass Diode: A semiconductor diode connected across a PV cell in a photovoltaic module, so that the diode will conduct if the cell becomes reverse biased due to shading or the failure of other cells.

Cell (battery): An electrochemical device that produces an electromotive force. Several cells interconnected is called battery. A lead cell produces about 2.12 volts and a 12 volt battery uses 6 of these cells and fully charged measures about 12.72 volts.

Cell (solar): A solar cell converts light when incident on it into electric energy. A single solar cell produces about 0.5V DC. It is the basic unit of a photovoltaic solar panels. A 12-volt solar panel consists of 36 individual cells; a 24-volt solar panel uses 72 cells.

Charge Controller: An electronic device which regulates the voltage from a PV array and ensures maximum transfer of energy and prevent overcharging of the battery bank.

Circuit: Interconnection of individual electronic components or devices in which the flow of electrical current performs useful work or functions.

Circuit Breaker: A safety device used to cut off the flow of electricity in an electric device or circuit to prevent damage or fire when an overload condition occurs.

Combiner Box: A solar array junction box where multiple solar modules are electrically connected together and contains electric fusing devices.

Conductor: Generally, metals that conduct electricity through them are called conductors. Gold, silver, copper, and aluminium are the good conductors in the order mentioned.

Conventional Power: Power generation from hydroelectricity power stations, petroleum, natural gas, coal, or nuclear power plants.

Conversion Efficiency (solar panel): The ratio of the energy produced by a solar panel to the total light energy incident on it expressed in percentage. Solar panels are generally 9 to 14 percent efficient.

Converter (DC): An electronic device for changing AC volts to the other levels of DC voltage.

Crystalline Silicon Photovoltaic Cell: Photovoltaic cell made from a single crystal or a polycrystalline silicon. Several cells are joined together to form a solar module.

Current (Electrical): The rate of flow of charges in an electrical circuit, measured in amperes (amps).

Cycle (AC): In alternating current (AC), the current goes from zero to maximum in one direction then zero to maximum in the other direction, is called a cycle, then repeats. Standard AC cycle in India 50 cycles per second (Hertz)

Deep Discharge: Discharging a battery down to 20 percent or less of its full charge condition.

Diffuse Solar Radiation: Sunlight scattered by atmospheric particles and arriving at the earth's surface from all directions.

Dimmer (switch/control): An electronic device that allows light intensities to be adjusted from dim to full bright.

Diode: A semiconductor device that allows current to flow in one direction only.

Direct Current (DC): Electric current in which the flow of electrons is in one direction only, unlike Alternating Current (AC).

Discharge Rate: The rate at which the energy is supplied from a battery, usually expressed in Amp-hours.

DOD (Depth of Discharge): The percentage that a battery is discharged from a fully charged condition.

Electrical Grid: Large integrated system of electricity distribution net work from centralized locations to individual homes and businesses.

Electrolyte: A liquid or gel type electrolytic conductor that carries current by the movement of ions (instead of electrons) between the plates of the batteries.

Electromagnetic Energy: Energy generated from an oscillating electromagnetic field produced by a magnet or an electric current flowing through a conductor.

Electromagnetic Field (EMF): The electrical and magnetic fields produced by any device through which electricity flows.

Electron: A fundamental particle of an atom with a negative electrical charge and a mass of 1/1837 of a proton. Electrons surround the positively charged nucleus (protons and neutrons) of an atom and the number of electrons is equal to number of protons which is called the atomic number of an element. Elements that loose electrons easily make good conductors of electricity and elements whose electrons are tightly bound together make good insulators. It is the movement of charges which constitutes the electric current in electrical circuits.

Energy Efficiency Ratio (EER): A fractional measure of the energy efficiency of room air conditioners based on the Department of Energy guidelines.

Equinox: Two times of the year when the sun crosses the equator, when night and day are of equal length, occurring on March 21st (spring equinox) and September 23 (fall equinox) in the northern hemisphere.

Filament (tungsten): A coil of tungsten wire inside an incandescent bulb in a vacuum or inert gas surrounding. When electricity flows through the tungsten wire "filament" it radiates energy in the form of heat and light. These incandescent bulbs are less energy efficient.

Fluorescent Tube Light: An electric lamp using a phosphor coated glass tube that glows when ions in the tube from anode to cathode with high voltage electricity. These are more efficient than incandescent type lamps.

Foot Candle: A unit of luminance equal to one lumen per square foot.

Fossil Fuels: Fuels such as petroleum, natural gas, and coal are called fossil fuels which are stored sun energy in the ground from the decayed remains of the plants and animals over millions and millions of years.

Frequency: The number of cycles through which an alternating current changes direction twice each second. In India, the standard frequency for electricity is 50 cycles per second (50 Hertz). In the U.S. the standard frequency for electricity is 60 cycles per second (60 Hertz).

Fuel cell: An electrochemical device that converts chemical energy directly into the electricity.

Full Sun: The amount of sunlight energy striking the earth's surface at noon on a clear day (about 1,000 Watts per square meter).

Fuse: A safety device which cut off the flow of electricity to prevent damage and fire to the electrical devices and circuits under overload conditions. Also refer to 'circuit breaker'.

Generator (AC): An electric generator driven by an internal combustion-engine to produce electricity as stand by electricity for stand-alone systems and for back-up electrical power. The fuel used may be gasoline, diesel, or propane (LPG)

Giga-watt (GW): A large unit of power: 1,000,000,000 watts, 1 million kilowatts, or 1,000 megawatts.

Green Power: A popular term referred to the energy produced from clean, renewable energy resources such as wind, solar, or hydroelectric systems.

Grid: A common term referring to electricity transmission and distribution network system run by electricity generating companies.

Grid Tie System: A solar powered electrical system that is connected to the electricity grid so that power can be supplied by the grid when needed and fed back into the grid during excess power production.

Ground: The electrical reference potential where voltage is zero or at the minimum. Connecting one side of the electric system and metallic surfaces of electric systems to the ground potential prevents shocks by stabilizing the voltage to safe level.

Hertz: The unit for cycles per second. It is the frequency of an AC electric system and is 60 hertz in the U.S.A and 50 hertz in India.

Hybrid System: A renewable or alternative electric energy system that uses two different sources for power, such as wind generators or and diesel generators and solar photovoltaic arrays together to supply electricity.

Hydroelectric Power Plant: A power plant that converts potential energy stored in the water body at higher elevations into electricity by allowing gravity fall of the water column on to the turbines.

Incandescent (lamp): An electric lamp that uses a tungsten wire filament placed inside a glass bulb in a vacuum or inert gas surrounding, that radiates heat and light when electricity flows through the filament. This is an inefficient electric device radiating light.

Infrared radiation: Infrared, sometimes called infrared light, is electromagnetic radiation with wavelengths longer than those of visible light. It is therefore generally invisible to the human eye; it can be detected as a sensation of warmth on the skin.

Inverter (AC): An electronic device that converts DC electricity into AC electricity. In the PV electric systems, 12,24,48, or higher volts DC power from the batteries or solar panels are converted to 240/440 volts AC to operate normal appliances and to connect to grid respectively.

Joule: S.I unit of energy is joule. One joule of energy is defined as displacing an object with 1 Newton of force by one meter. 1 Joule = 1 Newton × 1 meter (J = N.m)

Kilowatt: Standard unit of electrical power which equals to 1000 watts.

Kilowatt-hour (kWh): 1000 watts produced or consumed for a period of 1 hour. It is one unit of electric energy, on the basis of which electrical consumers are billed.

Lead Acid Battery: Battery made from grouping electrochemical cells that use lead and lead oxide for electrodes and sulphuric acid for the electrolyte.

Line Loss: Energy lost due to the resistive loss in the wire that causes inefficiencies in an electrical transmission and distribution system.

Load: The devices and appliances that draw electric power from an electrical supply system.

Long Wave Radiation: Infrared radiation or heat wave energy.

Luminance: The measure of the intensity of light emitted from a luminous body, measured in lumens.

Maximum Power Point (MPP): The point on the current-voltage (I-V) curve of a solar module, where the product of the current times the voltage equals to maximum wattage.

Maximum Power Point Tracking (MPPT): An electronic charge controller technique that attempts to supply maximum power to the batteries by tracking the maximum power point (MPP) at all times achieving a 15% to 35% increase over other types of battery charging techniques.

Megawatt: Large unit of power: 1 million watts or 1 thousand kilowatts (1,000,000 watts)

Megawatt-hour: Large unit of electric power 1 million watts produced or used for a period of 1 hour.

Module (solar): Several solar cells electrically connected together in an environmentally protected housing producing a standard output voltage and power. Multiple modules or panels can be assembled into a PV array for increased power and or voltage.

Mono-crystalline: A single crystal (silicon) produced using complicated crystal growth process. Long silicon rods are produced which are cut into slices of 0.2 to 0.4 mm thick discs or wafers which are then processed into single cells. This is the basic building unit of the crystalline silicon modules.

Multi-crystalline: A semiconductor material composed of randomly oriented, single crystals, also referred to as polycrystalline or semi-crystalline.

Name Plate: An identifying plate usually located near the AC cord of an electrical appliance that contains information such as model number, serial number, operating voltage, and power consumption.

National Electrical Code (NEC): The NEC is a set of regulations and standards that most electrical equipment installations must follow that makes the electrical system in the United States, one of the safest in the world.

Net Metering: Using a single meter to measure usage and generation of electricity by customers with a wind or PV power energy system. The net energy used or produced is either purchased from or sold to the electric grid.

Off-Peak (demand): The times during a 24-hour period of low electricity demand, unlike peak-demand.

Ohm: The S.I unit of resistance.

Ohm's Law: The various formulas that define the relationship between resistance (R), voltage, (V), and current (I) as in V = IR, I = V/R, R = V/I etc.

One-Axis Tracking: A sun tracking system that moves in only one direction or axis generally following the sun in its arc across the sky from east to west.

Open-Circuit Voltage (V_{oc}): The maximum possible voltage across a PV cell or module in sunlight when no current is flowing

Panel (solar): A solar photovoltaic device consisting of groups of individual solar cells connected in series, in parallel, or in series-parallel combinations to produce a standard output.

Parallel: A wiring technique where all the positive leads coming from a device are connected to one terminal and all the negative leads to another terminal. In case of batteries such groupings increase the current output but the voltage remains the same.

Payback Period: The time required before the savings resulting from an energy producing system equals to the cost of the system. This time period has been steadily decreasing as alternate energy systems are developing fast.

Peak Sun Hours: The equivalent number of hours per day when solar radiation averages to one kilo-Watt per square meter. That is 4 peak sun hours means that the energy received had the solar radiation for 4 hours being 1 kilo Watt per square meter.

Photon: A light particle that acts as an individual unit of energy.

Photovoltaic Array: A group of solar photovoltaic modules connected together to increase voltage and or power to the level required for a given system design.

Photovoltaic cell: Specially made semiconductor materials like silicon, cadmium sulphide, cadmium telluride and gallium arsenide that convert sunlight directly into electricity. Three common types of cell are mono-crystalline, poly-crystalline and amorphous or thin film.

Photovoltaic Conversion Efficiency: The ratio of the energy produced by a photovoltaic module or panel to the sunlight energy incident on it expressed as a percentage. Solar panels are generally 9 to 15 % efficient.

Photovoltaic module or panel: Vide module or solar module.

Poly-crystalline: A semiconductor material composed of randomly oriented, small individual crystals, capable of converting light energy into electric energy.

Power: Inherent capacity of a product or built in a product that can utilize (consume) certain amount of energy to perform work. Earlier, it was measured in horsepower. For electrical products, the power is rated in Watts and is equal to the Voltage (V) times the Current (I) that flows through the device (that is used or consumed). Power = VI.

Radiation: The transfer of heat energy through space or vacuum in the form of electromagnetic waves.

Reflective Glass: A window glass that is coated with a reflective film to reduce unwanted solar heat gain by the modules during the summer.

Renewable Energy: Energy from the sources that are available in the nature such as moving water (hydro, tidal and wave power), biomass, geothermal energy, solar energy, wind energy and energy from solid waste treatment plants.

Resistance (electrical): The inherent property of a material to inhibit the flow of electrons through it, producing heat in conductors, devices or components and its S.I unit in Ohms (W)

Resistor: An electronic device that resists the flow of charges, usually connected in electric or electronic circuits.

Ribbon Cells (photovoltaic): A type of solar photovoltaic cell made by pulling the material from its molten bath, such as silicon, to form a thin continuous sheet of cell material.

Safety Disconnect: A switch which disconnects one circuit from another circuit to isolate power generation or storage equipment from each other.

Self-Discharge (rate): The rate at which a battery without being used will loose its charge over a time period.

Semiconductor: It is neither a good conductor nor a good insulator. It has a limited capacity for conducting an electric current. This characteristic allows it to be built into units of products such as diodes, transistors, and integrated circuit packages, which perform controlled operations similar to earlier vacuum tube devices. Semiconductor material can be densely packed and layered through photographic process and has given rise to the modern world of electronics. Certain semiconductors, including silicon, gallium arsenide, copper indium diselenide and cadmium telluride are generally used in the photovoltaic conversion of light energy into electricity process.

Series: A wiring technique where multiple devices are wired connecting the positive lead of the device to the negative terminal of the next device, then to the positive terminal of the third device and so on. Solar cells together in this type of arrangement result in the increase in voltage but current remains the same.

Shallow Cycling: Allowing a battery bank to only loose 20% of its full state of charge when being used then recharging back to full state of charge. With a large backup battery system this technique can supply sufficient power between charge cycles and will greatly increase the life and performance of the batteries.

Short Circuit: Generally, the maximum current that flows freely through an external circuit that has no load or resistance, (usually due to the failure of a device or component).

Silicon: An element of atomic number 14, that is semi-metallic, and an excellent semiconductor material due to the electrons in the outer shell are neither tightly bound nor loosely bound to the nucleus. Silicon is the most abundant material in the earth's crust.

Sine Wave: The wave generated by alternating current generators and sine wave solid-state inverters. The ideal sine wave is the graph generated by plotting a sine function against time intervals.

Single –Crystal Material: Refer to mono-crystalline.

Sizing: The process of designing a solar electric system, to increase or decrease the number of components to meet the required operating loads based on the total wattage of all appliances in the system.

Solar Array: Refer to photovoltaic Array

Solar Cell: The building block of a photovoltaic solar panel. A single solar cell produces about 0.5 Volt. Configuration of solar cells, for example, a 12-volt solar panel has 36 individual cells, a 24-volt solar panel uses 72 cells.

Solar Constant: The average amount of solar irradiation that reaches the earth's outermost atmosphere on a surface normal to the sun's rays; which is equal to 1363 watts per square meter.

Solar energy: Electromagnetic energy emanating from the sun (solar radiation). The amount of the solar energy that reaches the earth is equal to one billionth of total solar energy generated, or the equivalent of about 420 trillion kilowatt-hours.

Solar Module (Panel): Several solar cells electrically connected together in an environmentally protected case, producing a standard output voltage and power. Multiple modules or panels can be assembled into a PV array for increased power and or voltage.

Solar Noon: The time of the day, at a given location when the sun reaches its highest zenith in the sky.

Solar radiation: The various wavelengths of electromagnetic radiation that is emitted by the sun including all wavelengths of radiation and also the visible light.

Solstice: The two times of the year when the sun is apparently farthest north and south of the earth's equator occurring on or around June 21 (summer solstice) and December 21 (winter solstice) in the northern hemisphere.

Square Wave Inverter: A type of inverter that produces square wave output. The square wave inverter is the simplest and the least expensive, produces less quality of power.

Stand-Alone Inverter: An inverter that operates independent of any electric grid or electric transmission and distribution network.

Stand-Alone System: A system that operates independent of any electric grid or electric transmission and distribution network; generally, contains battery backup.

State of Charge (SOC): The remaining charge available in a battery system, expressed as a percentage of the battery when fully charged.

Synchronous Inverter: An inverter that produces AC electricity from direct current electricity and uses another AC source, such as the electric power grid or a generator to synchronize its output voltage and frequency to the external power source.

Thin-Film: A thin layer of semiconductor material of few microns or less thickness, used to make PV cells.

Tilt angle (of a solar array): The angle at which a solar array is set to face the sun relative to a horizontal base line and is usually adjusted seasonally due to the changing declination of the sun.

Tracking Solar Array: A solar PV array that follows the path of the sun during the day to maximize the solar radiation it receives. A single axis tracked device tracks the sun east to west and two axis tracker device tracks the daily east to west movement of the sun and the seasonal declination movement of the sun. The tracking devices are complex and expensive weighted against overall performance of the system.

Transformer: An electromagnetic device that changes the voltage of an AC supply voltage up or down, consists of an induction coil with an iron core and a primary and secondary winding, the ratio of which determines output voltage against input voltage.

Trickle Charge: The small charging voltage that is required to maintain a battery in a fully charged condition after it has been charged. Trickle charging is used to recharge a battery for losses from self-discharge as well as to restore the energy discharged during intermittent use of the battery.

Ultraviolet Radiation: Electromagnetic radiation that ranges from 10 and 400 nanometres in wavelength.

Utility Company: Generally, a commercial venture that supplies electricity.

Visible radiation: The visible radiation portion of the electromagnetic spectrum ranges from 0.4 to 0.76 microns in wavelength.

Volt: A unit of electrical potential difference measurement. One ampere of current flows through a resistance of one ohm, across a potential difference of one volt.

Voltage: The difference in electrical potential across two points in a electric circuit measured in volts.

Wafer: A thin cross section of semiconductor material made by slicing it from a single crystal or rod.

Watt: The unit of electric power: one watt equals to one ampere under an electrical pressure of one volt.

Watt-hour: One watt produced or consumed for a period of one hour.

Wattmeter: A device that measures power production or usage displayed in watts.

REFERENCES

BTI, Bridge to India (2017). Indian Solar Handbook 2017. Accessed from http://www.bridgetoindia.com/reports/india-solar-handbook-2017/ on June 4, 2017.

Bhattacharya S.C and Jana Chinmoy. 2009. Renewable energy in India: Historical developments and prospects. Energy 34 (2009) 981–991.

CEA, Central Energy Agency (2016). Draft National Electricity Plan 2017-2022. Accessed from http://www.cea.nic.in/reports/committee/nep/nep_dec.pdf on April 12, 2017.

CPI, Climate Policy Initiative (2016). The Drivers and Challenges of Third Party Financing for Rooftop Solar Power in India. Accessed from https://climatepolicyinitiative.org.

Das Samar and Das Dulal. 2019. Feasibility study of installation of MW level grid connected solar photovoltaic power plant for north-eastern region of India. Sadhana (2019) 44:207.

Goel Malti. 2016. Solar Rooftop in India: Policies, Challenges and Outlook. Green Energy and Environment.

SECI. Solar Energy Corporation of India, pilot scheme on grid-connected rooftop PV systems, 14th Feb 2014. http://seci.gov.in/content/innerpage/publication-and-resources.php, 2014.

WIND-SOLAR PHOTOVOLTAIC HYBRID POWER SYSTEM

INTRODUCTION

The decentralized power generation using renewable energy sources is one of the viable options for urban, rural and remote locations. Renewable energy resources such as solar and wind energy are vast, omnipresent and unlike fossil fuels, they are very well distributed all over the world. They have shown increasing growth for power generation in recent years as these are freely available, environmental friendly sources for electrical power generation. They have enough potential to become important sources for power generation in the future because of their environmental, social and economic benefits in addition to public support and government incentives. The growing concerns of global warming and depleting fossil fuel reserves have compelled the planners and policy makers across the world to look for alternative source of power generation, especially from the renewable energy resources. Countries across the globe are now showing an increasing inclination towards harnessing renewable energy resources. This change is not only desirable but is also the need of the hour.

It is a fact that neither a standalone solar system nor a wind energy system can provide a continuous supply of energy due to their seasonal and periodical variations. Therefore, in order to satisfy the continuous load demand in remote locations, hybrid energy systems are being implemented on combined solar and wind energy conversion units with battery storage facilities. The term hybrid in itself means a mixture of two different components with about similar results for a specific purpose, for example, a hybrid solar-micro wind renewable energy source is to supply reliable electricity to the rural community. Hybrid energy system, uses two or more energy sources, allows improving the system efficiency and power reliability and to reduce the energy storage requirements for stand-alone applications. The hybrid solar–wind systems are becoming popular in remote area for power generation applications due to advancements in renewable energy technologies and substantial rise in prices of petroleum products.

Hybrid stand-alone renewable energy systems based on wind–solar resources are considered to be economically viable and reliable than stand-alone system with a single source. The integration of two or more resources in a proper combination to form a hybrid system uses the strengths of one source to overcome the weaknesses of the other. For certain locations, the hybrid solar–wind power generation systems with storage banks offer a highly reliable source of power, which is suitable for electrical loads that need higher reliability. The hybrid energy system may operate in two modes i.e. simultaneous and sequential mode. In simultaneous mode, the solar and wind energy system produce energy concurrently while in sequential mode they produce electricity alternatively. The significant characteristics of the hybrid system are to combine two or more power generation technologies to make proper use of their operating characteristics and to obtain efficiencies higher than that could be obtained from a single power source.

Of course, with the increased complexity in comparison with single energy system, the optimum design of a hybrid system becomes complicated through uncertain renewable energy supplies and load demand, non-linear characteristics of the components, high number of variables and parameters that have to be considered for their optimum design. This complexity makes the hybrid systems more difficult to be designed and analysed. In order to efficiently and economically utilize the renewable energy resources, optimum sizing of the components is necessary. The optimum sizing method can help to guarantee the lowest investment with full use of the PV array, wind turbine and battery bank, so that the hybrid system can work at the optimum conditions in terms of investment and system power reliability. This type of optimization also includes economic aspects and it requires the assessment of the system's long-term performance in order to reach the best compromise for both reliability and cost.

Hybrid renewable energy system offers sustainable application in the rooftop of the residential buildings in order to minimize conventional and fossil fuel based energy consumption and CO_2 emissions. The roof mounted wind turbines with high elevation are exposed to higher wind speeds in comparison to ground based ones as such wind based hybrid systems have the potential to make a significant impact on rooftop electricity generation which needs to be popularised. The large wind power generation is mainly focused in coastal or high windy regions, but presently roof mounted micro wind turbines are being used in low/medium windy locations as single or hybrid system with other sources like solar photovoltaic unit. Also, the energy produced by large wind farms has the impacts on the environment like bird mortality, noise pollution, interference to transmission of microwave signals of TV and communication etc. The large wind turbine farms occupy a larger area with generating much noise than roof mounted micro wind turbines. The technical and economic feasibility of a renewable energy based hybrid system have to be established based on wind and solar resource potential for the location of interest. Currently, there is a widespread adoption of hybrid energy systems in remote locations as well as in the residential sector in the developing countries like India. The main goal of using hybrid system is therefore to improve electrical power production, to minimize cost, to reduce negative effects associated with burning fossil fuels and to improve the overall system efficiency. The integrated solar energy system that is considered in hybrid system consists of a PV module array, hybrid charge controller and inverter (which is shared with the wind turbine) (Fig. 5.1).

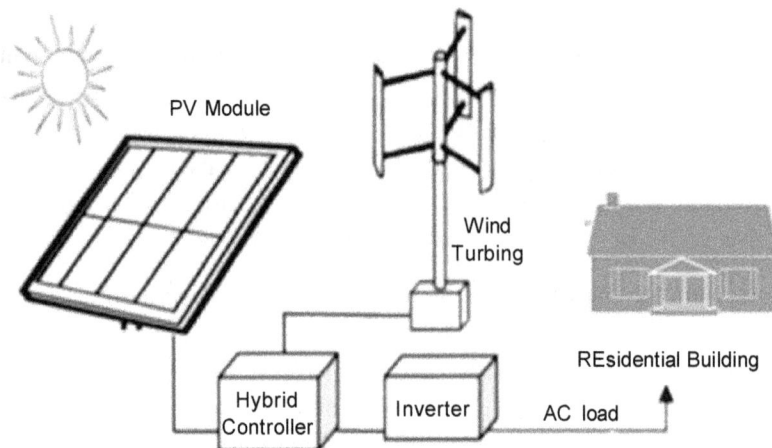

Fig. 5.1. Schematic of wind-PV hybrid energy system

Apart from rooftop solar panel for residential power generation, a vertical axis wind turbine can also be used to supplement the domestic power demand due to its low-wind operational capability. Conceptual design of power system having two renewable energy sources for off-grid as well as on-grid is shown in Fig. 5.2.

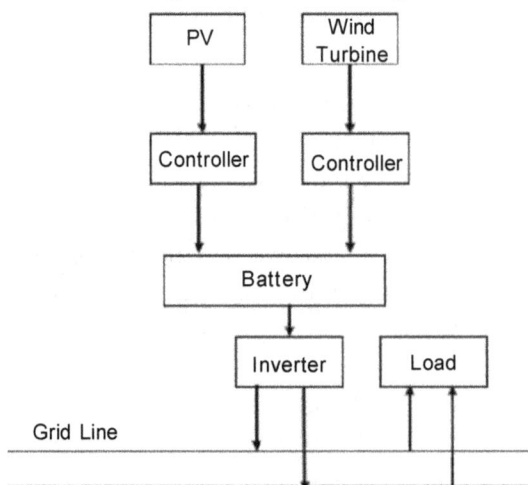

Fig. 5.2. Conceptual design of hybrid wind-solar renewable energy sources

Micro wind turbines can accelerate the growth of power generation in low windy but good solar potential regions if integrated with photovoltaic systems. Most of the wind turbines available in the market have higher cut-in and rated speeds as present focus is on large wind power generation only. Thus the incentive based government policies need to be formulated for promoting solar micro-wind based hybrid systems in urban, rural and remote locations.

5.1 SOLAR ENERGY RESOURCES

Solar energy is one of the non-depletable, site-dependent, non-polluting energy sources and is available in abundance. It is a potential source of alternative/renewable energy and utilization of solar radiation for power generation reduces the dependence on fossil fuel. Solar energy is received in the earth in the form of electro-magnetic radiations and is harnessed by human beings by using various devices. The convenient way of deriving electrical energy from solar energy is through the photovoltaic (PV) technology. Solar photovoltaic (PV) technology is used to convert sunlight into direct current electricity. About 95% of solar radiation reaches earth with the wavelength intervals of 0.3–2.4 ìm only. The radiation consisting of photons travels through the space at the speed of 3.0×10^8 m/s and each photon carries different amount of energy, measured in electron volts. Currently, the world's population consumes about 15 terawatts ($1 \text{tera} = 10^{12}$) of power. Hence the human beings could sustain their power requirement by using only 0.01% of the incident solar power (1.7×10^{17} watt) in the earth without burning fossil fuels. The goal of solar PV technology is to achieve higher performance with lower costs. The process of electricity generation in a PV system is illustrated in Fig. 5.3. PV cells generate direct current electricity by utilising particular characteristics of semi-conducting materials when they are exposed to sunlight. Photons of the sunlight come into contact with the solar absorbent semi-conducting material, which excites its electrons, and then generates current through an external load.

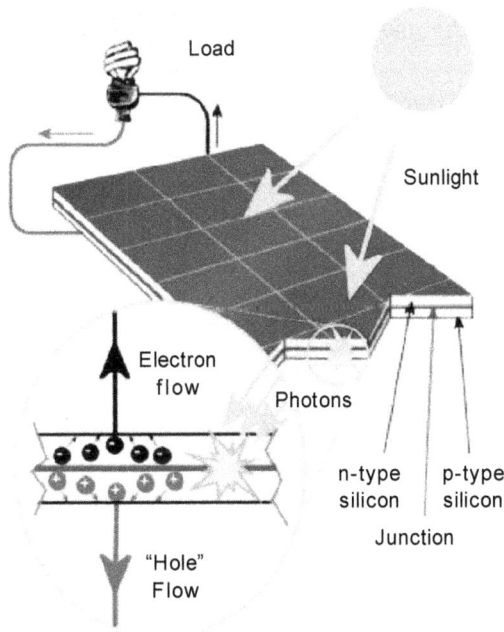

Fig. 5.3. Electricity generation process from solar PV cell

The current produced is proportional to the intensity of sunlight, striking the surface as well as the efficiency and size of the cell. The performance of PV modules and array are basically rated according to their maximum power output (W), tested under Standard Test Conditions (STC). Standard Test Conditions are specified as a cell (or a module) operating temperature of 25°C (77°F) and incident solar irradiance level of 1000 W/m^2. The countries existing in tropical and sub-tropical zones have sufficient availability of sunshine. Hence, there is a great prospect for expansion of solar PV power system in those countries along with future scope for this renewable source of energy through 'Grid Parity'. Many of the places in the globe are still facing power shortages, though have good solar power potential that is not being used efficiently. With the increasing shortage in power supply and energy in those places, it is highly necessary to exploit the potential of available energy sources through micro-generation technologies that can meet the energy needs under the distributed generation in a standalone way to provide long-term solutions.

5.2 MICRO-WIND ENERGY RESOURCES

Wind energy, being renewable, provides a great opportunity to generate energy due to its abundance. Wind is atmospheric air in motion. The energy from the flowing wind is harnessed through the device called wind turbine. Wind turbines are those which convert the kinetic energy present in the wind to mechanical energy and eventually into electricity. Power that is obtained from wind, flowing at a certain speed may be calculated by assuming that a parcel of air is moving towards a wind turbine at a velocity of v as shown in Fig 5.4.

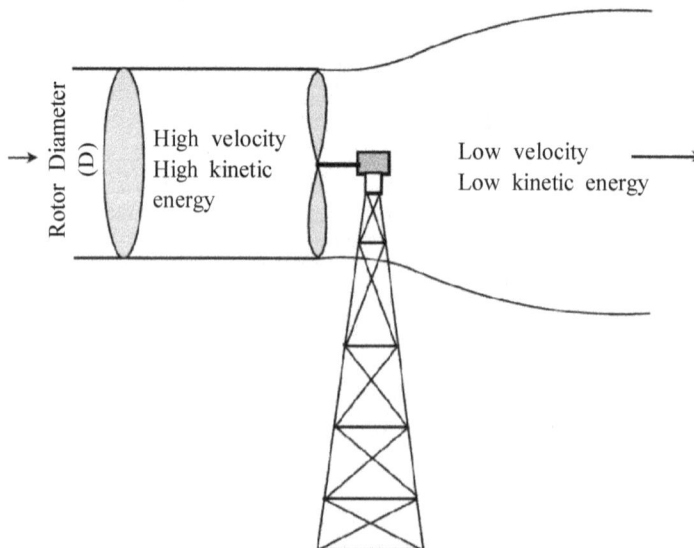

Fig. 5.4. Interaction of wind with the turbine

About 1–2% of 174,423,000,000,000,000 kWh (1.74×10^{17} Wh) of energy that the sun radiates to the earth per hour are converted into wind energy. This much amount of energy is about 50–100 times more than the energy converted into biomass by all plants on the earth and 100-200 times more than the current energy of 15 terawatt-hour (15×10^{12} Wh), consumed by the world's population. Various countries, in order to meet their growing power demand, are installing large scale wind farms both on-shore and off-shore. The large scale wind farms are not a sustainable and viable option for renewable power production due to various reasons as mentioned above. The better choice, available is therefore by installing the decentralized grid system i.e., by using small scale wind turbines. Small scale wind turbines produce power of around 10 kW which is sufficient for our domestic needs. This energy can be effectively utilized, so that the energy extracted from the conventional resources could be saved for a larger period of time. Hence, there arises the need to understand the characteristics of small scale wind turbines. The majority of work on small scale wind turbines has been done over the past few years. The disadvantages of small wind turbines are their effective placement, wind fluctuation, changes in wind direction and also initial cost per watt power and the unit cost per kWh compared to the medium and large scale wind turbines. If the cost factors will be at an affordable rate, small scale wind turbines would become a potential source for power generation. The small scale wind turbines need to be therefore cost effective. Hence, it is very much necessary to select the right types of small scale wind turbines available at present to make the system cost effective. The micro wind turbines have low cut-in and low rated speeds, indicating that they can be effective in generating small amounts of usable power from the winds of lower speeds or higher speeds and shorter durations. Wind energy is not as easy to capture as solar energy because it keeps flowing, moving and changing direction. Moreover, the complex physical conditions and structure of residential areas make it even harder to predict.

Wind turbines can be divided into a number of broad categories in view of their rated capacities i.e. micro, small, medium, large and ultra-large wind turbines. The classification of wind turbines is as follows.

Table 5.1: Classification of wind turbines

S.N	Types of wind turbine	Range of power ratings (kW)
1	Micro	0.004-0.25
2	Mini	0.25-1.4
3	Household	1.4-16
4	Small	16-100
5	Medium	100-1000
6	Large	1000-3000
7	Ultra-large	> 3000

Micro wind turbines are especially suitable in locations where the electrical grid is unavailable. They can be used for street lighting, water pumping, and residents at remote areas, particularly in developing countries. Because micro wind turbines need relatively low cut-in speeds at start-up and operate in moderate wind speeds, they can be extensively installed in most areas around the world for fully utilizing wind resources and greatly enhancing the availability of wind power generation. Small wind turbines usually refer to the turbines with the output power less than 100 kW (Tummala et al 2016). Small wind turbines have been extensively used at residential houses, farms, and other individual remote applications such as water pumping stations, telecom sites, etc. in rural regions. Distributed small wind turbines can increase electricity supply in the regions while delaying or avoiding the need to increase the capacity of transmission lines. The most common wind turbines have medium sizes with power ratings from 100 kW to 1 MW. This type of wind turbines can be used either on-grid or off-grid systems for village power, hybrid systems, distributed power, wind power plants, etc. Megawatt wind turbines up to 3 MW may be classified as large wind turbines. In recent years, multi-megawatt wind turbines have become the mainstream of the international wind power market. Most wind farms presently use megawatt wind turbines, especially in the offshore wind farms. Ultra-large wind turbines are referred to wind turbines with the capacity more than 3 MW. This type of wind turbine is still in the earlier stages of research and development.

There are two types of wind turbines (WT), vertical axis wind turbines (VAWT) and horizontal axis wind turbines (HAWT). Betz, a German physicist, in 1919, theoretically determined that a wind turbine can only convert 16/27 (59.3%) of the kinetic energy of the wind into mechanical energy by turning a rotor. This is known as the Betz Limit or the Betz's Law. The maximum efficiency (C_p), also known as coefficient of performance or power coefficient of wind turbine is therefore 0.593, which is known as Betz limit and the power of wind turbine can be calculated as given below.

$$P_{WT} = C_p \times \left(\eta_b \times \eta_g \right) \times \frac{1}{2} \times \rho \times A \times v^3$$ where P_{WT} is the maximum power obtained

from the wind turbine, ρ is air density, A is swept area of wind turbine and v is wind speed, η_b is the efficiency of the gearbox/bearing and η_g is the efficiency of the generator. However, the actual power obtained from the wind turbine is at lower efficiency than given by the Betz limit, due to various losses (drag on the blade, swirl imparted to airflow by the rotor etc.). The practical value of C_p is in the range of 0.35–0.40. A value of greater than 0.80 is possible for η_g if a permanent magnet generator or grid connected induction generator is used. The efficiency of gearbox and bearings can be greater than 95%. One approach to maximising performance is to develop wind turbines with lower cut-in speed to deal with the slow average wind speed in rooftop of residential buildings. Also, wind direction and wind speed at a selected location need to be continuously measured to obtain correct data for analysing the potential of wind energy to generate electricity.

It is interesting, to note that, among the renewable energy sources, wind power is the fastest growing in terms of global annual and cumulative installed capacity. Wind energy is almost available everywhere around the world. But the wind speed strength varies depending on the particular area. Wind energy can be harnessed during the day and night times or throughout the day, unlike other renewable.

$$\text{Wind energy production} = \text{Power} \times \text{Time}$$

The wind turbines are most commonly classified by their rated power at a certain wind speed. Annual energy output is the most important measure for evaluating a wind turbine. In order to calculate the expected energy output, the capacity factor of the turbine should be known. A reasonable power coefficient would be between 0.25 and 0.30. A very good power coefficient would be 0.40. Power coefficient is very sensitive to the average wind speed. When using the power coefficient to calculate the estimated annual energy output, it is extremely important to know it at the average wind speed of the intended site. By multiplying the rated power output by the power coefficient and the number of hours in a year, (8,760 h), an estimate of annual energy production can be obtained as follows.

$$100 \text{ kW } (RP) \times 0.20 \text{ (CP)} \times 8760 \text{ (h)} = 175200 \text{ kWh}$$

where, *RP* is the rated power, *CP* is the power coefficient and h is hour. For accurate estimate of energy production, the wind distribution (speed and frequency) of the site should be known either collecting the data from the local meteorological department or predicting its value.

5.3 ORIGIN AND NATURE OF WIND

Wind energy is a converted form of solar energy due to the differential heating of land masses and water bodies in the earth's surface and spatial air pressure differentials, formed in the atmosphere by the effects of uneven solar radiation. Wind results from the movement of air due to atmospheric pressure gradients. Wind flows from regions of higher pressure to regions of lower pressure. The larger the atmospheric pressure gradient, the higher the wind speed and thus, the greater the wind power that can be captured from the wind by means of wind energy-conversion devices. The generation and movement of wind are complicated due to a number of factors. Among them, the most important factors are uneven solar heating, the Coriolis effect due to the earth's self-rotation and local geographical conditions. The energy contained in the wind can be converted into the useful energy through windmill. Wind energy can be converted into mechanical energy for grinding the grains or water pumping. Alternatively, the wind energy can be converted into electrical energy (the most desired form of energy) by the use of generators that convert mechanical energy to electrical energy. A wind turbine is similar to a fan but works in reverse direction i.e. a fan converts electrical energy to mechanical energy for the rotation of the blade thus produces airflow, on the other hand, a wind turbine converts airflow into mechanical energy and then produces electrical energy. Typically, wind turbines are much larger in size compared to a fan. A device for producing mechanical work directly by the flowing of atmospheric air is called a windmill or just a wind turbine. If electricity is produced, the combination of turbine and generator may be called as wind generator or an aero-generator. Because of the confusion of these terms, the acronym 'WECS' (wind energy conversion system) is increasingly used now-a-days.

5.3.1 Origin of Wind

The sources from which the wind is originated through natural phenomenon are classified as (i) global winds (or planetary winds) and (ii) local winds. The origin of wind is a complex phenomenon and depends on macro and micro level factors. The origin of winds may be traced basically to uneven heating of the earth's surface due to solar radiation. This may lead to circulation of widespread winds on a global basis, producing planetary winds or may have limited influence in a similar manner to cause local winds. Global winds are caused by the daily rotation of earth around its polar axis and unequal temperatures existing between polar and equatorial regions. The strength and direction of these global winds change with the season as the solar input energy varies. Similarly, local winds are caused by the unequal heating and cooling of earth's land surface and ocean/lake surface during day and night.

5.3.1.1 Global winds due to uneven solar heating

Among all factors affecting the wind generation, the uneven solar radiation on the earth's surface is the most important and critical one. The unevenness of the solar radiation can be attributed to four reasons.

(i) First, the earth is elliptical in shape, revolving around the sun in the same plane as its equator. Because the surface of the earth is perpendicular to the path of the sunrays at the equator but parallel to the sunrays at the poles, the equator receives the greatest amount of energy per unit area, with energy decreasing towards the poles. Due to the spatial uneven heating on the earth, it forms a temperature gradient from the equator to the poles and a pressure gradient exists from the poles to the equator. Thus, hot air with lower air density at the equator rises up to the high atmosphere and moves towards the poles and cold air with higher density flows from the poles towards the equator along the earth's surface. The rising air at equator moves southwards (southern hemisphere) and northwards (northern hemisphere) as shown in fig. 5.5. This motion of air stops when air cools down at about 30^0 North and 30^0 South latitude. At these latitudes, air begins to sink down and flows towards the equator. In this way, the air completes one circle. The motion of earth around its axis (west to east) has an effect on the direction of the wind flow. Without considering the earth's self-rotation and the rotation-induced Coriolis force, the air circulation at each hemisphere forms a single cell, defined as the meridional circulation.

(ii) Second, the earth's self-rotating axis has a tilt of about 23.5° with respect to its ecliptic plane. It is the tilt of the earth's axis during the revolution around the sun that results in cyclic uneven heating, causing the yearly cycle of seasonal weather changes.

(iii) Third, the earth's surface is covered with different types of materials such as vegetation, rock, sand, water, ice/snow, etc. Each of these materials has different reflecting and absorbing rates to solar radiation, leading to high temperature on some areas (e.g. deserts) and low temperature on others (e.g. iced lakes), even at the same latitudes.

(iv) The fourth reason for uneven heating of solar radiation is due to the earth's topographic surface. There are a large number of mountains, valleys, hills, etc. on the earth, resulting in different solar radiation on the sunny and shady sides.

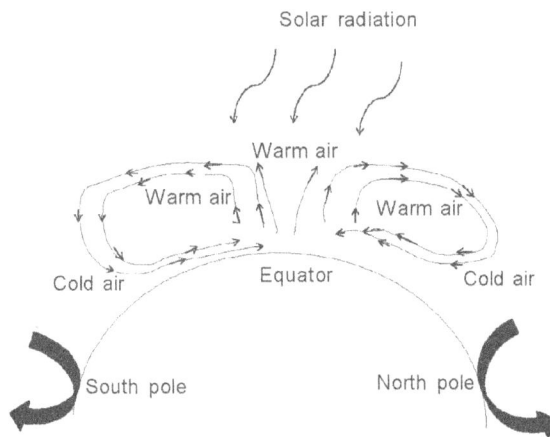

Fig. 5.5. Atmospheric air circulation between equator and 30 ^{0}N and 30 ^{0} S latitude

5.3.1.2 Coriolis force

The earth's self-rotation is another important factor to affect wind direction and speed. The Coriolis force, which is generated from the earth's self-rotation, deflects the direction of the movement of atmospheric air. In the north atmosphere, wind is deflected to the right and in the south atmosphere to the left. The Coriolis force depends on the earth's latitude; it is zero at the equator and reaches maximum values at the poles. In addition, the amount of deflection on wind also depends on the wind speed; slowly blowing wind is deflected only a small amount, while stronger wind is deflected more. The demonstration of Coriolis force on the rotation of earth about its own axis is shown in Fig. 5.6.

Between 30 ^{0} North and 30 ^{0} South latitude, the rising heated air gets cooled at higher altitude and sinks down to the equator again. Due to rotation of earth, the rising air gets deflected towards east and the return air gets deflected towards west. This is known as *Hadley circulation*. Due to Coriolis force, these winds deviate towards west. These air currents are also known as trade winds because of their use in sailing ship for trades in the past. There is little wind near the equator (5^0 around it), as the air slowly rises upwards rather than moving westwards. Between 30^0 N(S) and 70^0 N(S), predominantly western winds are found. These winds form a wavelike circulation, transferring cold air southward and warm air northward (in the northern hemisphere and vice versa in the southern hemisphere). This pattern is called *Rossby circulation*. This is shown in Fig. 5.7.

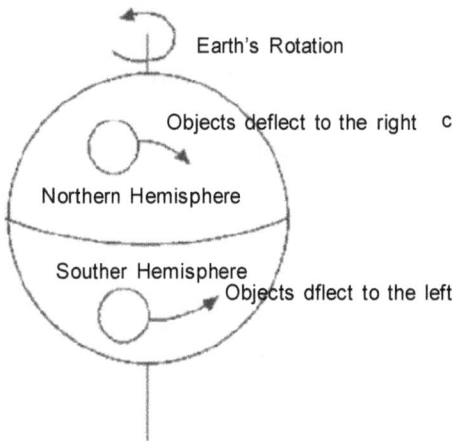

Fig. 5.6. A demonstration of Coriolis force   Fig. 5.7. Global winds and their circulation

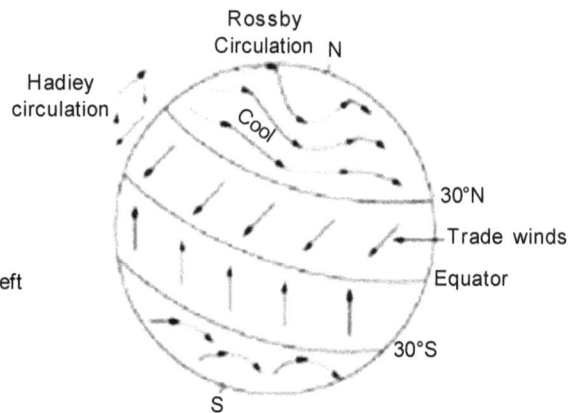

5.3.1.3 Local winds due to local geography

Localized uneven heating causes the origin of local winds. This occurs due to the following three mechanisms.

(i)   The first mechanism is the differential heating of land surface and water bodies due to solar radiation. During the day, solar energy is readily converted to sensible thermal energy on the land surface and this quickly increases its temperature, whereas on the water bodies, it is partly consumed in evaporating water and the rest is absorbed to cause an increase in temperature. The land mass thus becomes hotter than water bodies which causes differential heating of air above them. As a result, cool, heavier air blows from the water towards land. At night, the direction of wind is reversed as the land mass cools to sky more rapidly than the water. This is the mechanism of shore breeze.

(ii) The second mechanism of the local winds is the differential heating of slopes on the hillsides and that of low lands. The slope heats up during the day and cools down during the night more rapidly than the low land. This causes the heated air to rise along the slope during the day and relatively cool air to flow down at night.

(iii) The roughness on the earth's surface is a result of both natural geography and manmade structures. Frictional drag and obstructions near the earth's surface generally retard the wind speed and induce a phenomenon known as wind shear. The rate at which wind speed increases with height varies on the basis of local conditions of the topography, terrain, and climate, with the greatest rates of increase observed over the roughest terrain. A reliable approximation is that wind speed increases about 10% with each doubling of height. In addition, some special geographic structures can strongly enhance the wind intensity. For instance, wind that blows through the mountain-channel can form mountain jets with high speeds.

5.3.2 Nature of winds

Before installation of a wind turbine, it is essential to have full knowledge of the behaviour and nature of wind. The winds vary from place to place. The nature of wind at a site depends upon general climate of the region, physical geometry of the locality and the terrain around the site. All countries have national meteorological services that record and publish weather related data, including wind speeds and directions. These data, obtained from the nearest meteorological station are only useful to provide first order estimates for predicting the power from the wind for a specific site, but are not sufficient for detailed planning. Usually careful measurements around the nominated site are needed at several locations and heights for several months or years. These detailed measurements can then be related to the standard meteorological data and provide a long-term base for comparison. These data provide a good approximation for predicting the power extraction from the wind at a specific site.

In meteorology, the *Beaufort scale* is often applied to give the wind force. This scale allows an approximate estimation of wind speed without complicated measurement systems; however, this scale is less useful for technical purposes. It was initially designed for sailors to describe the sea state, but has subsequently been modified to include the effects of wind on land. This has now been expressed as wind speed, in m/s as per the SI unit. Table 5.2 compares the Beaufort classes with the corresponding wind speed values.

Table 5.2: Beaufort scale for wind description

Beaufort number	Wind speed		Observable effects	Wind Description
	m/s	Km/h		
0	0.0-0.4	0.0-1.6	Smoke rises vertically	Calm
1	0.4-1.8	1.6-6	Smoke drifts but vanes unaffected	Light
2	1.8-3.6	6-13	Leaves move slightly but vanes unaffected	Light
3	3.6-5.8	13-21	Leaves in motion, flags begin to extend	Light
4	5.8-8.5	21-31	Small branches move, dust raised, pages of book loosen	Moderate

[Table Contd.

Contd. Table]

Beaufort number	Wind speed		Observable effects	Wind Description
	m/s	Km/h		
5	8.5-11	31-40	Small trees sway, wind noticeable	Fresh
6	11-14	40-51	Large branches sway, telephone lines whistle	Strong
7	14-17	51-63	Whole tree in motion	Strong
8	17-21	63-76	Twigs break off, walking difficult	Gale
9	21-25	76-88	Slight structural (e.g., chimneys) damage	Gale
10	25-29	88-103	Trees uprooted, much structural damage	Strong gale
11	29-34	103-121	Widespread damage	Strong gale
12	>34	>121	Disastrous conditions, countryside devastated, only occurs in tropical cyclones	Hurricane

5.3.2.1 Information of wind (speed)

Data from the nearby meteorological department provide relevant information regarding the speed and direction of the wind for a particular place. Wind speed is measured using an instrument, called anemometer and wind direction is measured by a wind vane attached to a direction indicator. Anemometers work, based on, one of the following principles.

(i) The oldest and simplest anemometer is a swinging plate hung vertically and hinged along its top edge. Wind speed is indicated by the angle of deflection of the plate with respect to the vertical.

(ii) A cup anemometer consists of three or four cups mounted symmetrically about a vertical axis. The speed of rotation indicates wind speed.

(iii) Wind speed can also be recorded by measuring the wind pressure on a flat plate. When flat plate stops the wind, the wind pressure at the plate increases. The measured wind pressure is an indicator of wind speed.

(iv) A hot-wire anemometer measures the wind speed by recording cooling effect of the wind on a hot wire. The heat is produced by passing an electric current through the wire. When wind passes on a hot wire, the wire gets cooled depending upon the wind speeds. By recording the cooling effect of the wind on hot wire, the wind speed can be determined.

(v) An anemometer can also be based on sonic effect. Sound travels through still air at a known speed, however, if the air is moving, the speed decreases or increases accordingly.

(vi) The other novel techniques include the laser anemometer, the ultrasonic anemometer and the SODAR Doppler anemometer.

A typical anemograph consisting of wind speeds recorded at three heights i.e. at 10 m, 50 m and 150 m during strong winds is shown in figure 5.8. The important conclusions, which can be drawn from this anemograph are as follows,

• Wind speeds increase with heights

• Wind speeds at all heights fluctuate or change with time

• The fluctuation is spread over a broad range of frequencies

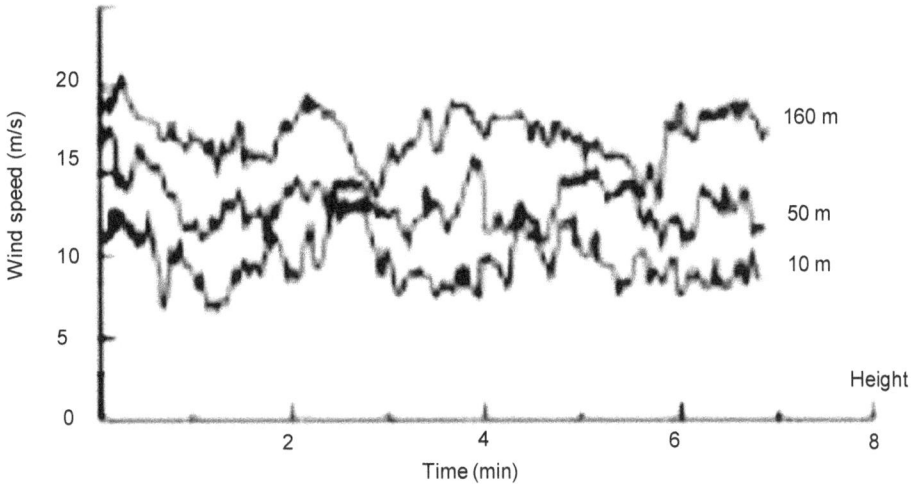

Fig. 5.8. Anemograph of wind speed at three heights

5.3.2.2 Wind direction

Wind direction is one of the wind characteristics. Statistical data of wind directions over a long period of time is very important in the selection of site and layout of wind energy

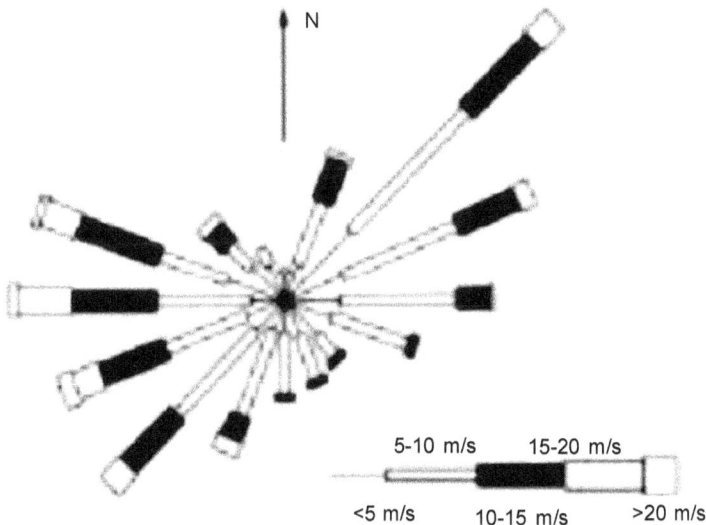

Fig. 5.9. Wind rose describing mean wind speed and direction

conversion system. The wind rose diagram is a useful tool of analysing wind data that are related to wind directions at a particular location over a specific period of time (year, season, month, week, etc.). It is used by the meteorologists to have a clear view of how wind speed and direction are typically distributed at a particular location. The mean wind speed, and its direction are depicted on a single diagram as shown in Fig. 5.9. This diagram displays the relative frequency of wind directions in 8 or 16 principal directions. As an example shown in Fig. 5.9, there are 16 radial lines in the wind rose diagram, with 22.5° apart from each other. The length of each line is proportional to the frequency of wind direction. A wind rose was also, before the use of magnetic compasses, a guide on mariners' charts to show the directions of wind in the eight principal directions.

5.3.2.3 Wind speed variations with height

At earth's surface, the speed of wind is almost zero. It increases as its height increases from the ground. The speed of wind near the earth's surface is retarded by the surface roughness.

Wind shear: The rate of change of wind speed with height is called wind shear. The lower air layers moving slowly tends to retard air layer above them, resulting in the change in mean wind speed with height until the shear forces are reduced to zero.

Gradient height: The wind speed increases as height increases because shear force reduces with height. At a certain height, the shear force reduces to zero and the wind speed does not change above this height. This height is called the gradient height.

Free atmosphere: The gradient height is generally at about 2000 m from the ground. Above this gradient height, any change in the ground conditions does not affect the wind speed; that is, wind speed is uniform above the gradient height. This atmosphere with uniform wind speed is called the free atmosphere.

Planetary boundary layer: The layer of air from ground to the gradient height is called the planetary boundary layer

Surface layer: It is the air layer considered from the height of local obstruction to a height of about 100 m.

Ekman layer: It is the air layer from surface layer (100 m) that extends up to the gradient height. The variation of shear stress can be neglected in this layer, and the mean wind speed with the height can be given by Prandtl logarithmic law. The division of various atmospheric layer above the earth's surface is shown in Fig. 5.10 (i) and variations of wind speed with height (wind shear) in Fig. 5.10(ii).

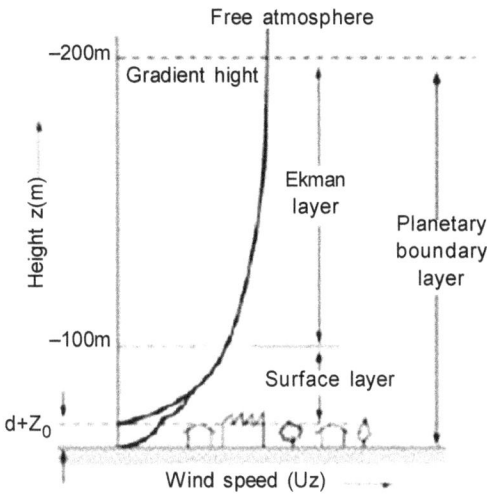

Fig. 5.10 (i) Divisions of atmospheric layers above earth's surface

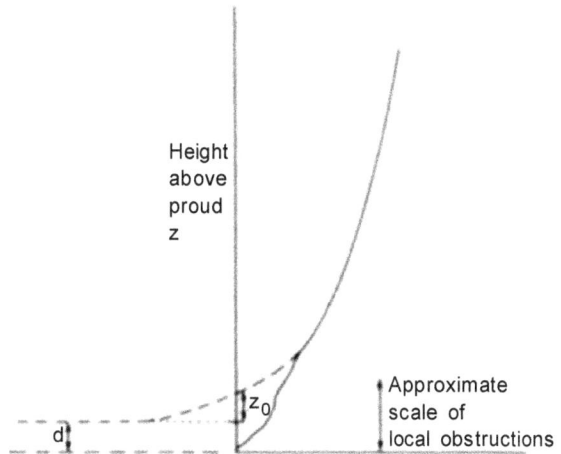

Fig. 5.10 (ii) Variations of wind speed with heights (wind shear)

Within the height of local obstructions, wind speed increases erratically and violent directional fluctuations can occur in strong winds. Above this erratic region, the height/ wind speed profile is given by expressions (Prandtl logarithmic law),

$$v(h_2) = v(h_1) \times \frac{\ln\left(\dfrac{h_z - d}{z_0}\right)}{\ln\left(\dfrac{h_1 - d}{z_0}\right)} \qquad \ldots (5.1)$$

The wind speed $v(h_2)$ at height h_2 can be calculated directly with the *roughness length*, z_0 of the ground cover and the wind speed $v(h_1)$ at height h_1. Obstacles can cause a displacement of the boundary layer from the ground. This displacement can be considered by the parameter 'd'., known as zero plane displacement. For widely scattered obstacles, parameter 'd' is zero. In other cases, 'd' can be estimated as 70 per cent of the obstacle height. The *roughness length* z_0 describes the height at which the wind is slowed to zero. (In other words, surfaces with a large roughness length have a large effect on the wind. Table 5.3 shows the classification of various ground classes depending on the roughness length. The following example shows the influence of the ground cover. The wind speed $v(h_1) = 10$ m/s at a height $h_1 = 50$ m above different ground classes is assumed. Equation above (5.1) is applied to calculate the wind speed $v(h_2)$ at a height of $h_2 = 10$ m. The displacement 'd' for the boundary layer from the ground must be considered for higher obstacles in ground classes 6 to 8. Table 5.4 shows the calculated results. The wind speed decreases significantly with rising roughness lengths z_0; thus, it does not make any sense to install wind power plants in built-up areas or large forests.

The wind speed also increases significantly with height. For instance, the wind speed at a height of 50 m is 30 per cent higher than at 10 m for ground class 4. This must be considered for the installation of large wind turbines. The usable wind speed at the top of large wind towers is much higher than at the common measurement height of 10 m. Wind turbines of the megawatt class come with hub heights of between 50 and 70 m for coastal areas (ground class 1 to 3) and even higher for inland areas with higher roughness lengths. This example should not give the impression that the wind speed is already independent of the ground at a height of 50 m. The wind speed usually becomes independent of the height, where the wind becomes known as *geostrophic wind*, at heights significantly exceeding 100 m from the ground. As seen in the diagram, near the line of local obstructions, the average wind speed does not follow the above equation i.e. up to the dotted line in the figure 5.10 (i), but deviates from it and becomes highly erratic. It is very important then to place the wind turbine well above the height of the local obstructions so that turbine disk receives a uniform wind flux around its area without erratic fluctuations. Equation (5.1) may be used for the estimation of wind speed from the line of local obstruction to the surface layer (100 m height).

Table 5.3: Rouglutess length z_0 for different ground classes

Ground class	Roughness length z_0 in m	Description
1. Sea	0.0002	Open sea
2. Smooth	0.006	Mud rials
3. Open	0.03	Open flat terrain, pasture
4. Open to rough	0.1	Agricultural land with a low population
5. Rough	0.25	Agricultural land with a high population
6. Very rough	0.5	Park landscape with bushes and trees
7. Closed	1	Regular obstacles (woods, village, suburb)
8. Inner city	2	Centres of big cities with low and high buktngs

Source: Christoffer and Ulbricht-Eissing, 1989

Table 5.4: Example of the Decrease in Wind Speed $v(h_2)$ at Height $h_2 = 10$ m as a function of the Ground Class for $v(h_1) = 10$ m/s at $h_1 = 50$ m

Ground class	z_0	d	$v(h_2)$ at h_2 = 10m	Ground class	z_0	d	$v(h_2)$ ar h_2 = 10 m
1	0 0002 m	0 m	8.71 m/s	5	0.25 m	0m	6 96 m/s
2	0.005 m	0 m	8.25 m/s	6	0.5 m	3 m	5 81 m/s
3	0 03 m	0 m	7.83 m/s	7	1 m	5 m	4.23 m/s
4	0.1 m	0 m	7.41 m/s	8	2 m	6 m	2.24 m/s

As the speed is measured at a height of 10 m from the ground and also wind turbines are operated at a height more than 10 m, a simpler empirical relation, called (1/7) power law, based on the data from several locations for the sites of low surface roughness, can be used to determine the mean wind velocity at other heights.

With $z = \sqrt{(h_1 \times h_2)}$ and $a = \dfrac{1}{\ln\dfrac{z}{z_0}}$, equation (5.1) becomes $\dfrac{v(h_2)}{v(h_1)} = \left(\dfrac{h_2}{h_1}\right)^a$ ... (5.2)

Where $v(h_1)$ is the wind speed at the reference height h_1. 'a' is the parameter that depends upon the surface roughness and range of height(z) being considered. For $z = 10$ m and $z_0 = 0.01$ m, the parameter 'a' is about (1/7), this equation is then called a (1/7) power law. However, this power law is only valid if the displacement d of the boundary layer from the ground is equal to zero. The value of 'a' is often taken as 0.14 (1/7) for open sites. is called the power index. It varies with the season and time of the day. Great care should be taken in using this formula, especially, $z > 50$ m. Good sites should have low value of 'a'. The value of 'a', power index for a particular site can be obtained from the measured wind speeds at two heights by using the equation 5.2.

5.3.2.4 Wind speed variations with time and its frequency distribution

Implementation of wind power requires the knowledge of future wind speed at the turbine sites. Such information is essential for the design of the machines and the energy systems, and for their economics. As wind speed fluctuates with time, it is more important to know about the continuity of supply, rather than the total amount of energy available in a year. For electric power generation, the minimum average speed required is 5 m/s. A site is not considered favourable for wind power generation if average wind speed remains less than this, for prolonged period, as there will be no generation of power during these periods. Also no generation is possible if wind speed is very high i.e. 25 m/s and above. The sites may be considered good where favourable winds (wind speeds 5-25 m/s) are available for most of the time (typically 70 to 80 per cent of time).

The amount of energy available in the wind (the wind energy resource) at a site is the average amount of power available in the wind over a specified period of time, commonly one year. If the wind speed is 20 m/s, the available power is very large at that instant, but if it only blows at that speed for 10 h per year and the rest of the time the wind speed is very low, the resource for the year is small. Therefore, the site wind speed distribution (the relative frequency of occurrence for each wind speed), or the wind speed probability density function (pdf) is very important in determining the resource.

A wind speed frequency distribution gives much better information about the wind conditions of a certain site than the mean wind speed. The frequency distribution can be presented in a tabular form with wind speed intervals or as statistical functions. The most common statistical functions that are used for wind power calculations are the Weibull and the Rayleigh distributions. If the actual wind speed probability density distribution is not available, it is commonly approximated with the generalized two-parameter Weibull distribution, given by

$$f_{weibull}(v) = \frac{k}{a} \times \left(\frac{v}{a}\right)^{k-1} \times \exp\left(-\left(\frac{v}{a}\right)^k\right) \qquad \dots (5.3)$$

Where $f_{weibull}(v)$ is the frequency of occurrence of wind speed (v) and 'a' is the Weibull scale factor (m/s) and 'k' is the Weibull shape factor (unit less). The shape and scale factor depend on the site. The unit of wind speed in Eq. (5.3) is m/s. The Weibull k value (k) is a measure of the distribution of wind speeds over the year. It is assumed to be 2 by default because this has been shown to represent most wind regimes fairly accurately. Lower k values correspond to broader wind speed distributions, meaning that the wind speeds vary over a wide range. Wind regimes where the wind speed tends to vary over a narrower range (like tropical trade wind environments) have higher k values. The mean wind speed can be estimated approximately from the Weibull parameters (Molly, 1990):

$$\bar{v} = a \times \left(0.568 + \frac{0.434}{k}\right)^{\frac{1}{k}} \qquad \dots (5.4)$$

The parameter a for $k = 2$ can be obtained from the mean wind speed,

$$a_{k=2} = \frac{\bar{v}}{0.886} \approx \frac{2}{\sqrt{\pi}} \times \bar{v} \qquad \dots (5.5)$$

Substituting 'a' in the Weibull distribution and using 'k' = 2 results in the *Rayleigh distribution* which is

$$F_{Rayleigh}(v) = \frac{\pi}{2} \times \frac{v}{\bar{v}^2} \times \exp\left(-\frac{\pi}{4} \times \frac{v^2}{\bar{v}^2}\right) \qquad \dots (5.6)$$

The Rayleigh distribution needs only the average wind speed as a parameter. Fig. 5.11 shows Rayleigh distributions for different mean wind speeds.

5.3.2.5 Wind power density

Wind power density is the available wind power in airflow through a perpendicular cross-sectional unit area in a unit time period. It is a comprehensive index in evaluating the wind resource at a particular site. Wind can reach much higher *power densities* than solar irradiance i.e. 10 kW/m^2 during a violent storm and over 25 kW/m^2 during a hurricane, compared with the maximum terrestrial solar irradiance of about 1 kW/m^2. However, a gentle breeze of 5 m/s (18 km/h) has a power density of only 0.075 kW/m^2. The classes of wind power density at two standard wind measurement heights are listed in Table 5.5.

This is based on the range of wind speed and wind power density that describe the energy contained in the wind. Wind power density is used to compare wind resources independent of wind turbine size and is the quantitative basis for the standard classification of wind resource for a site.

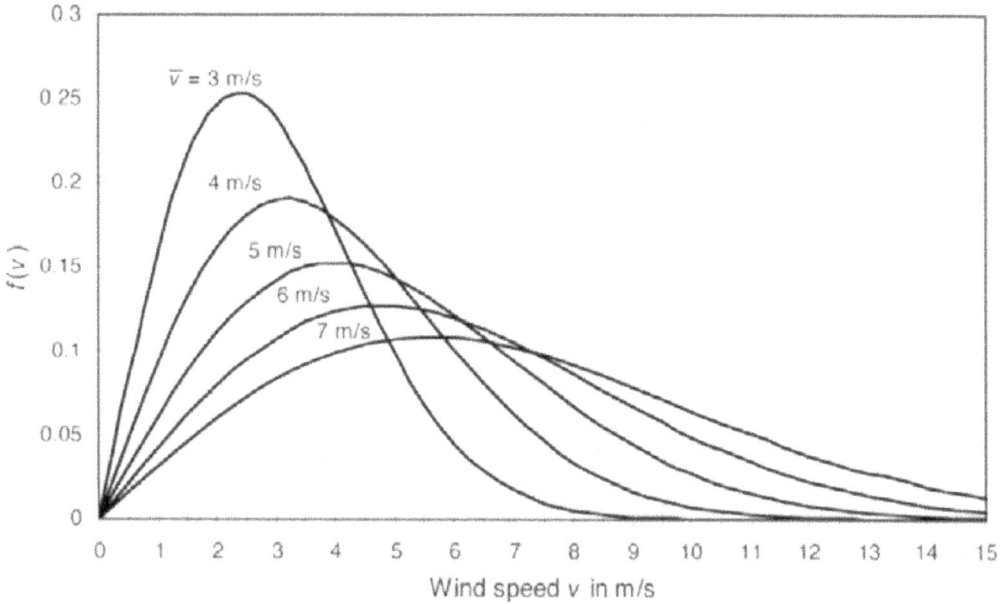

Fig. 5.11. Rayleigh distributions for different mean wind speeds

Table 5.5: Classes of wind power density

Wind power class	10 m height		50 m height	
	Wind power density (W/m^2)	Mean wind speed (m/s)	Wind power density (W/in^2)	Mean wind speed (m/s)
1	<100	<4.4	<200	<5.6
2	100-150	4.4-5.1	200-300	5.6 6.4
3	150-200	5.1-5.6	300-400	6.4-7.0
4	200-250	5.6-6.0	400-500	7.0-7.5
5	250-300	6.0-6.4	500-600	7.5-8.0
6	300-350	6.4-7.0	600-800	8.0-8.8
7	>400	>7.0	>800	>8.8

Source: American wind energy association. Basic principles of wind resource evaluation, http://www.awea.org/faq/basicwr.html

5.3.2.6 Capacity factor for a wind turbine

The capacity factor of a wind turbine is the actual energy output for the year divided by the energy output if the turbine is operated at its rated power output for the entire year. The output from a wind turbine depends on the wind speed through the rotor. In other way, capacity factor can also be defined as the portion of the year, a wind turbine produces power equivalent to its maximum rated power.

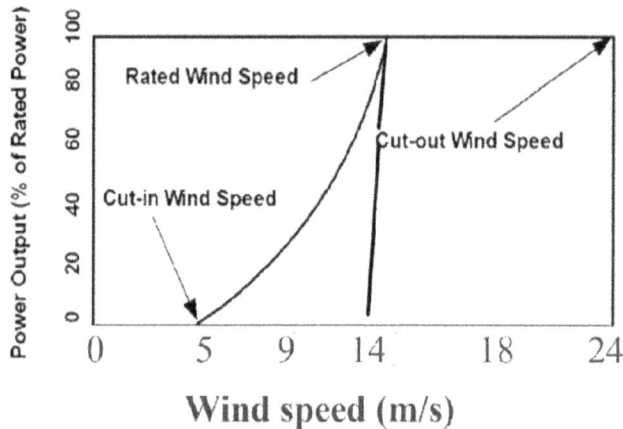

Fig. 5.12. Dependence of power output of a turbine on wind speed

The relationship between wind speed and rated power, called a *power curve*, is shown graphically in Fig. 5.12. The turbine starts to produce power only when a certain wind speed is reached (called cut-in wind speed). As the wind speed increases, the power output increases sharply. Similarly, at lower wind speeds, the power output drops off sharply. However, if the wind speed is above a certain value, the wind turbine is forced to remain idle. This is known as cut-out wind speed. The "rated wind speed" is the wind speed at which the "rated power (RF)" is achieved. There are two main options if the wind speed is above the rated wind speed. In one option, the power output above the rated wind speed is mechanically or electrically maintained at a constant level using an advanced control system. However, this is rather costly as the rotation of blades is hard to control. In the other option, the wind turbine is cut off from the power production. Using the power curve, it is possible to determine roughly how much power will be produced at the average or mean wind speed prevalent at a site. The power curve shown in Fig. 5.12 indicates that the turbine would produce about 20% of its rated power at an average wind speed of 7 m/s.

5.3.2.7 Wind energy production

While wind turbines are most commonly classified by their rated power at a certain wind speed, annual energy output is the most important measure for evaluating a wind turbine. The energy produced by a 100 kW rated power wind machine (rated wind speed, 14 m/

s, rotor area, 161 m^2, and coefficient of performance 0.37) in a year is calculated by multiplying its rated power with the number of hours (8760 h) in a year, i.e. 100 kW × 8760 = 876000 kWh. But, this is the energy when wind will flow at rated wind speed throughout the year, which is never a case. Therefore, in order to get the realistic energy output, we have to multiply the above numbers by the capacity factor. Assume that the capacity factor is 30%, then annual energy production would become 100 × 87600 × 0.30 = 262800 kWh.

When using the capacity factor to calculate estimated annual energy output, it is extremely important to know the capacity factor at the average wind speed of the intended site. The power curve can also be used to find the predicted power output at the average wind speed at the wind turbine site. For accurate estimate of energy production, the wind distribution of the site should be known. If such data are not available, there are two common wind distributions functions that can be used to make energy calculations for the wind turbines i.e. the Weibull distribution and a variant of the Weibull distribution, called the Rayleigh distribution that is thought to be more accurate at sites with high average wind speeds. Energy output is also influenced by the wind turbine design features, including cut-in and cut-out speeds. In most commercial operations, the turbine is shut down at the cut-out-speed to protect the rotor and drive train machinery from damage. Therefore, the wind turbine must be designed based on the characteristics of the site. The increased capacity factor will lead to higher reliability and availability of the wind power. A typical capacity factor for a coal-based plant is more than 80%, implying that a coal power plant of the same rated power (100 kW) will produce more than double energy than a 100 kW rated power wind turbine. Table 5.6 below presents an example of calculating annual energy output of a wind turbine.

Table 5.6: Calculation of annual energy output for a given wind speed distribution. The wind turbine having characteristics such as rated power, 100 kW, for rated wind speed, 14 metre/sec, cut-in wind speed, 3 metre/sec, rotor area, 161 m^2, coefficient of performance, 0.37.

Wind speed (metre/sec)	Hours per year	Power o/p (kW) ($=C_p$ × 0.6AU^3)	Energy o/p (kWh)
1	200	0	0
2	250	0	0
3	350	0	0
4	500	2	1163
5	600	5	2725
6	700	8	5494
7	800	12	9971
8	900	19	16744
9	1000	26	26490

[Table Contd.

Contd. Table]

Wind speed (metre/sec)	Hours per year	Power o/p (kW) (=C_p × $0.6AU^3$)	Energy o/p (kWh)
10	950	36	34521
11	850	48	40000
12	550	63	34535
13	400	80	31934
14	350	100	34899
15	250	100	25000
16	110	100	11000
17	0	100	0
	8760		274926
	(total hours per year)		(total energy produced per year)

Thus, as shown in the above table, the annual energy produced by a wind machine at a given location can be estimated if we know the annual distribution of number of hours versus wind speed.

5.3.2.8 Estimation of wind power potential

A moving object possesses kinetic energy due to its motion. Similarly, flow of air in the atmosphere possesses kinetic energy which is called wind energy. Considering a laminar flow of air perpendicular to the cross section (A) of a cylinder and moving at velocity U, the kinetic energy

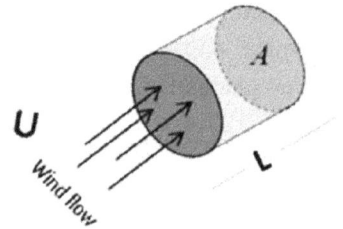

possessed by the flowing air is given by $E = \frac{1}{2}mU^2$, where

'm' is the mass of air in the cylinder. The power in the flowing wind is the energy per

unit time. Hence, $P = \frac{1}{2}\frac{dm}{dt}U^2$, where $\frac{dm}{dt}$ represents the rate of air flow through the

cylinder. However, the mass in the cylinder is equal to the density of air times the volume of the cylinder ($\rho \, A \, L$). Here ρ the density of air. As length (L) of the cylinder divided

by unit time is equal to the velocity V, thus the rate of air flow $\left(\frac{dm}{dt}\right)$ becomes $\rho \, A \, U$.

Substituting $\rho \, A \, V$ for $\frac{dm}{dt}$ in the expression for power $P = \frac{1}{2}\rho AU^3$. Wind power,

therefore, depends upon the density of the air, which varies under different conditions.

Under standard temperature (15^0C) and pressure (760 mm of Hg), the air density is consideed as 1.22 kg/m^3.

Calculation for air density

Air density ρ varies directly with air pressure. Its value is inversely proportional to air temperature, expressed in Kelvin as $\rho = \dfrac{P_r}{RT}$ where 'P_r' is the air pressure in Pa, 'T' is the air temperature in Kelvin and 'R' is the gas constant (287 J/kg.K). The standard value of air pressure = 1.01325×10^5 Pa (at 1 atmosphere) and at 15 ^0C. Therefore

$\rho = \dfrac{1.01325 \times 10^5}{287 \times 288} = 1.22 \, \text{kg/m}^3$. Air density is maximum at sea level and reduces gradually at higher altitudes.

Numerical 5.1

A wind energy generator generates 1600 W at rated speed of 7 m/s, at atmospheric pressure and temperature of 20^0 C. Calculate the power generated and the change in output if the wind generator is operated at an altitude of 1750 m, temperature 11^0C, wind speed 8.5 m/s and air pressure 0.9 atmosphere.

Solution. Given: P = 1600 W; U_w (wind speed) = 7 m/s; P_{atm} = 1.0132×10^5 Pa; T_{alt} = 11+273 =284 K; $P_{air (alt.)}$ = $0.9 \times 1.0132 \times 10^5$ = 0.912×10^5; Pa; $(U_w)_{alt.}$ = 8.5 m/s

Power generated and change in output:

Power generated at an altitude of 1750 m, P_{alt}.

Air density at 1750 m, $\rho_{air(alt)} = \dfrac{P_{air(alt.)}}{RT} = \dfrac{0.912 \times 10^5}{287 \times 284} = 1.12 \, \text{kg/m}^3$

(where R = gas constant = 287 J/kg K)

Now, $P = \dfrac{1}{2}\rho A_{bl} U_w^3$

Or, $1600 = \dfrac{1}{2} \times 1.205 \times A_{bl} \times (7)^3$ where A_{bl} is the turbine rotor area

Or, $A_{bl} = \dfrac{1600 \times 2}{1.205 \times (7)^3} = 7.74 m^2 \left(\because P_{atm.} = \dfrac{P}{RT} = \dfrac{1.0132 \times 10^5}{287 \times 293} = 1.205 \dfrac{kg}{m^3} \right)$

$$\therefore P_{alt} = \frac{1}{2}\rho_{air.} \times A_{bl} \times (U_w)^3_{alt.} \quad \text{i.e.} \quad P_{alt} = \frac{1}{2} \times 1.12 \times 7.74 \times (8.5)^3 = 2661.9W$$

Hence, change in output $P_{alt.} - P = 2661.9 - 1600 = 1061.9$ W (increase).

Energy derived from a wind turbine

Wind data (air speed versus number of hours per year) in the form of histogram are used extensively for wind power potential and energy output of wind turbines. By the help of histogram, computation of the output of wind energy can be done even in hand calculations. In order to calculate the energy output (*kWh*), one has to multiply power by the number of hours. But the power depends on the wind speed. Therefore, one should know the wind speed distribution in terms of number of hours in a year (Fig. 5.13). The sum of power produced by a given wind speed multiplied by the number of hours per year of that wind speed gives the total energy produced by a wind machine in a year (Table 5.6) and is calculated as given below;

Annual energy produced $= \Sigma P_i h_i \ (Wh)$

Where P_i is the power corresponding to a given wind speed i and h_i is the number of probable hours in a year of that wind speed. The subscript i takes values from minimum to maximum possible wind speed, in the interval of 0.5 hour or 1 hour.

Typically, a wind turbine will produce an average output power which will be some proportion of the rated output power (maximum output power for which wind turbine is designed). It is depicted in Fig. 5.12. The ratio of the average actual power to rated power (capacity factor) is typically within 30%. The capacity factor of a wind turbine is defined as the portion of the year; a wind turbine produces power equivalent to its maximum rated power. It implies that the capacity factor for wind turbines is low. For most locations, the capacity factor lies between 15 to 30%, but it can be different. Note that the wind speed specified for this graph is the wind speed at the rotor hub (axis of rotor).

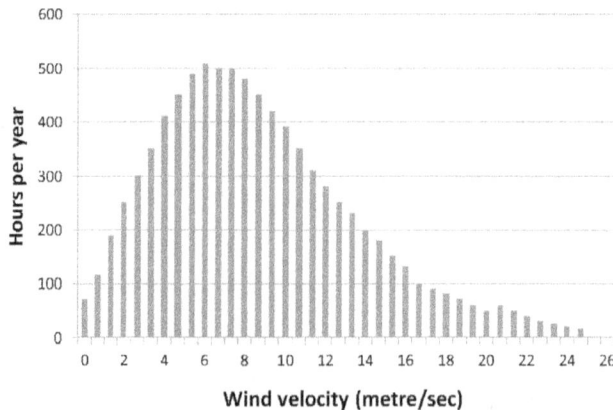

Fig. 5.13. Wind speed distribution with the number of hours in a year

5.3.2.9 Estimation of required wind turbine power rating

Let the annual energy requirement of an industry is 20,000 kWh. What should be the size of wind turbine that is required to be installed to meet the energy requirement? Following assumptions are taken into account for estimation purpose i.e. (i) annual energy requirement is 20,000 kWh (ii) propeller type (horizontal axis) wind machine is used (iii) coefficient of performance is 0.40 (iv) wind speed at 15 metre height is 5 m/s (if the turbine hub is placed at the height other than 15 metre, the wind speed should be estimated as discussed in 'wind speed variations with height' section (v) density of air, 1 kg/m^3 (vi) capacity factor is 0.30 (i.e. 30% of the time, wind machine is producing energy at rated power) and (vii) number of hours in a year, 8760 hours

Step 1: To find out power density of wind (power per unit area)

Power density of turbine considering hub height as 15 metre for ideal condition $= \frac{1}{2} \times$ air density $\times$ (velocity)^3 = 0.5 × 1 × (5×5×5) = 62.5 watt/m^2

Step 2: To find out useful power and energy density by considering various losses

Actual power density that will be converted to useful energy = C_p × transmission loss × generator loss. Overall loss factor by considering (i) transmission losses (rotor to generator) as 0.90 (ii) generator losses as 0.90 and (iii) coefficient of performance as 0.40 can be calculated as 0.40 × 0.90 × 0.90 =0.324

Actual power density= ideal power density × overall loss factor = 62.5 × 0.324 = 20.25 W/m^2.

Annual energy density (useful) = power density × number of hours per year =20.25 × 8760 = 177.39 kWh/m^2

The real annual energy density will be less as the wind of rated speed. So, real annual energy density= annual energy density (useful) × capacity factor = 177.39 × 0.30 − 53.2 kWh/m^2

Step 3: To find out the rotor size and turbine power rating, the area of the turbine can be estimated from the real energy density

$$\text{Area of the rotor } = \frac{Total\ annual\ energy\ required}{Real\ annual\ energy\ density} = \frac{20000kWh}{53.2kWh/m^2} = 345.8m^2$$

Radius of the rotor blade, (R) is calculated from πR^2 = 375.8 → R = 10.9 metre. Now, we can estimate the actual power rating of the turbine. It is obtained by multiplying the actual power density with area of the rotor. Power rating of turbine = Actual power density × area of rotor = 20.25 × 375.8 = 7.6 kW ≈ 8 kW. The cost of a wind turbine on per kW basis runs between Rs 30000 and Rs 50000. For the above calculation, an 8 kW wind turbine would cost about 8 × 40000= Rs 3,20,000.

5.3.2.9.1 Power versus speed characteristics of wind turbine generating unit

This section discusses power versus characteristics of the wind and wind turbine generator unit. The power in the wind stream passing through the swept area of turbine propeller (blades) is proportional to the cube of the wind speed (V^3) i.e. $p \propto V^3$ Fig. 5.14 shows a steady and idealised characteristics curve of wind stream.

The idealised characteristics of a variable speed wind turbine is shown as curve A-B-C-D in Fig. 5.15. Wind turbine develops less power than the wind stream power due to spillage, pitch angle, friction etc. Hence, curve BC is below curve '1' of wind stream. For operational and design purposes, the wind turbine is provided with the following limiting speeds;

Cut- in speed (V_A) is the speed (about 5 m/s) at which the wind turbine is allowed to start rotating. *Cut-out speed (V_E)* is the speed (about 24 m/s) at which the wind turbine is not allowed to deliver power due to safety precaution. As the wind speed reaches the cut-in speed V_A, the turbine comes into motion. The operating point moves along the characteristic A-B-C-D. During the segment B-C, the power of the turbines rises in tune with the rising wind speed. After (C), the blade pitch control keeps the power of wind turbine along constant power line C-D. At D, maximum velocity is reached and the turbine is stopped. Fig. 5.16 shows the further modification in the power-speed characteristics, required in practice. The following limiting speeds are (i) to start at (V_A) (cut-in speed), (ii) to stop at high speed (V_E) (cut-out speed), (iii) to restart at reduced speed V_F. ($V_F<V_E$) and (iv) to stop at lowest speed V_H ($V_H < V_A$).

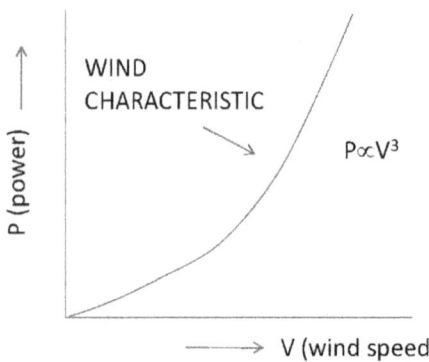

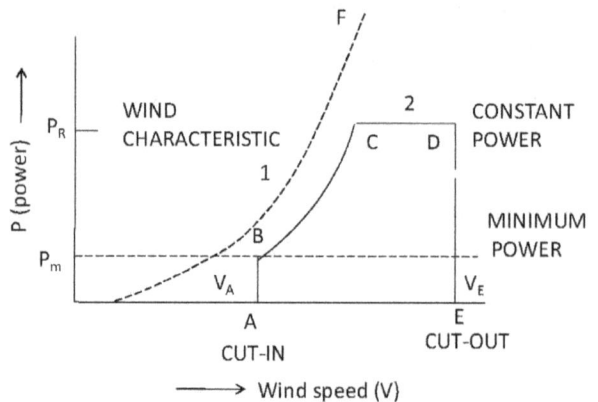

Fig. 5.14 Power-speed characteristics of a wind stream

Fig. 5.15. Power-speed characteristics (2) of a wind turbine with reference to wind stream characteristics (1)

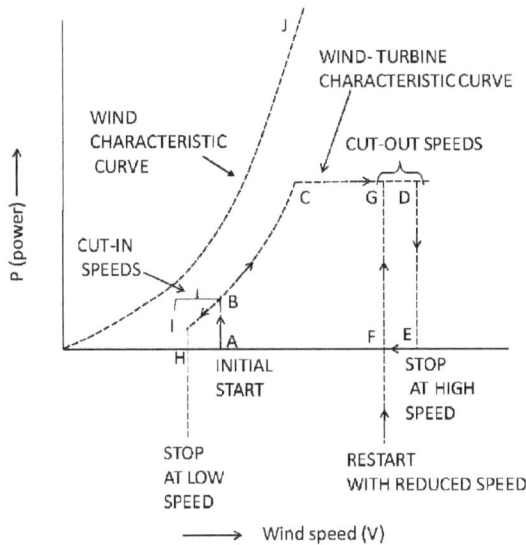

Fig. 5.16. Practical speeds of a wind turbine

Hysteresis in cut-in and cut-out speeds

As a general term, hysteresis means a lag between input and output in a system upon a change in direction. In practice, the starting of wind turbine at low speeds and stopping wind turbine at low speed are arranged at different speeds. Likewise, stopping at high speeds and restarts at reduced high speed occur at different speeds. Such a hysteresis is arranged by means of speed control system.

To prevent frequent shutdowns and restarts, which contribute to fatigue loading of the turbine, hysteresis is often applied, so that the wind turbine starts up only when the average wind speed reaches a value lower than the shutdown wind speed, i.e. the turbine will only restart after the wind speed has fallen below a level, less than the cut-out speed. This is not a problem where wind speeds rarely exceed the cut-out speed, but can be problematic in areas of high wind speeds, dominated by short gusts of wind above the cut-

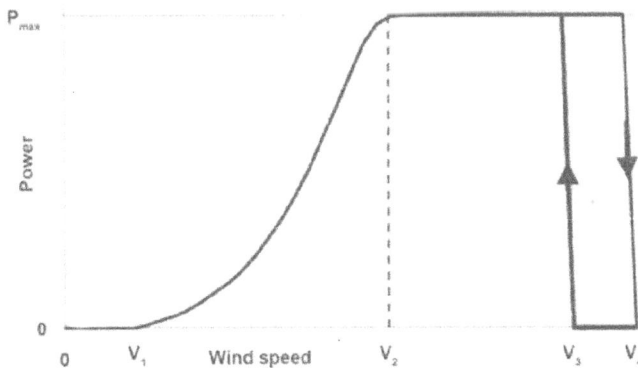

Fig. 5.17. Effect of high wind speed shutdown hysteresis

out speed and high turbulence. The net result is a loss of power production as the turbine switches in and out of operation. The effect of hysteresis on a typical power curve of a modern wind turbine is shown in Fig. 5.17. The wind turbine will shut down when the average wind speed reaches a certain value denoted V_4 in the figure. The typical shutdown wind speed is 24 m/s. When the average wind speed drops below the shutdown value to value V_3, the wind turbine starts again. Energy production is lost in the transition between V_4 and V_3.

5.4 BASICS OF FLUID FLOW IN WIND ENERGY CONVERSION SYSTEM

5.4.1 Terms and Definitions of Wind Energy

A wind turbine converts the energy of the wind into electrical energy. A wind turbine is similar to a fan but works in reverse direction i.e. a fan converts electrical energy to mechanical motion which gives us air flow, on the other hand, a wind turbine converts air flow into mechanical motion and gives us the electrical power. Typically, wind turbines are much larger in size as compared to a fan. Hence, the fundamentals of fluid (air) flow striking the rotating blades are important to assess the performance of wind turbine conversion system. The rotor blade of turbine changes frequently due to its rotation around the hub and various forces which are acting on it need to be analysed for estimating the extraction of power from the flowing wind.

Blade: Part of the rotor in a wind turbine that catches the wind

Blade element: Incremental cross section of the rotor blade as shown in Fig. 5.18 (a).

Root of blade: The area of a blade nearest to the hub.

Tip of blade: The end of a wind turbine blade farthest from the hub.

Tip speed ratio: It is the ratio of speed of the outer blade tip to the undisturbed natural wind speed. It indicates how fast the blade tips are moving compared to the undisturbed natural wind speed.

Airfoil or aerofoil: A streamlined curved surface designed for air to flow around it in order to produce low drag and high lift forces

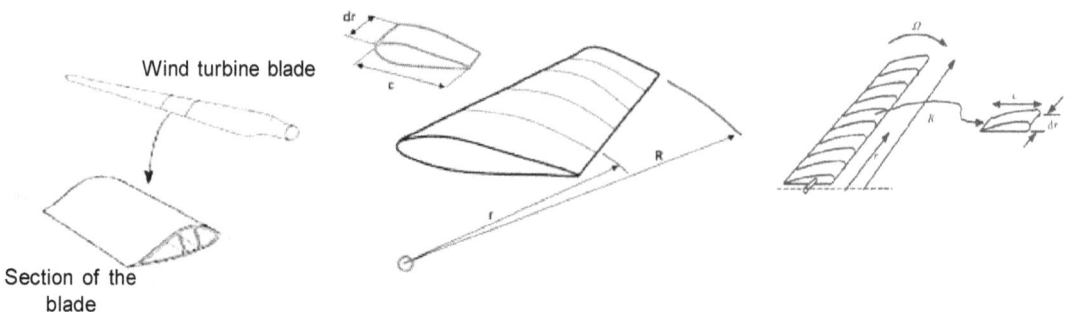

Fig. 5.18 (a). Rotor blade and its cross section

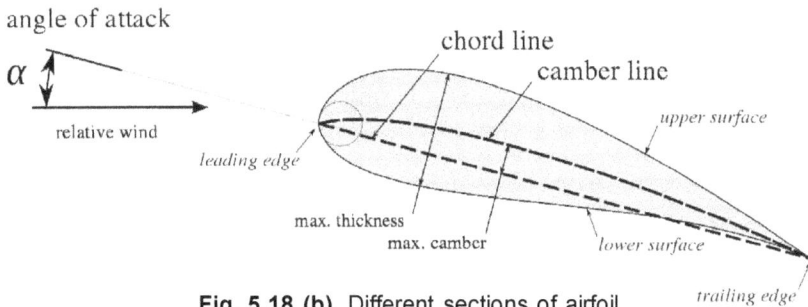

Fig. 5.18 (b). Different sections of airfoil

Angle of attack: The angle between relative air flow to the chord of the airfoil (fig. 5.18 (b)).

Heading edge: The edge of the blade that faces towards the direction of motion of wind or it is the front edge of the airfoil (fig. 5.18 (b)).

Trailing edge: The edge of the blade that faces away from the direction of motion of wind or it is the rear edge of the airfoil (fig. 5.18 (b)).

Chord line: It is the line joining the heading edge and trailing edge (fig. 5.18 (b))

Mean line: The line which is exactly equidistant from the upper and lower surface of the airfoil (fig. 5.18 (b)). It is also called camber line.

Camber: The maximum distant between the mean line and the chord line (fig. 5.18 (b)). It is a measure of the curvature of the air foil

Rotor: It is the prime part of wind turbine that extracts energy from the wind. It constitutes the blade and hub assembly (fig. 5.18 (c)).

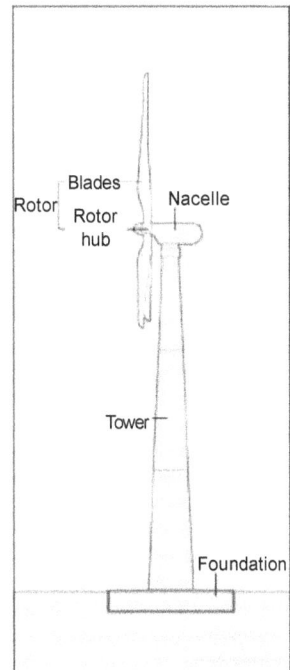

Fig. 5.18 (c) Blade hub assembly of wind turbine

Hub: Blades are fixed to a hub which is a central solid part of the turbine (fig. 5.18 (c)).

Propeller: Propeller or a wind turbine comprises essentially a hub and blades. Wind turbines and propellers are very similar from the aerodynamics point of view, the former extracting energy from the wind, the latter putting energy into the fluid to create a thrust. We consider rotors with blades similar to airplane wings, sometimes misleadingly called 'propeller type' for horizontal axis wind turbine.

Vortices and eddies: Rotational movement of the air occurs as the airstream flows off the blade. This may be apparent as distinct vortices and eddies (whirlpools of air) created near the surface. Vortex shedding occurs as these rotating masses of air break free from the surface and move away

Turbulence: The air is disturbed by the blade movement and by wind gusts, and the flow becomes erratic and perturbed. This turbulence occurs before and after the rotating blades, so each individual blade may often be moving in the turbulence created by other blades

Nacelle: The nacelle houses the generator, the gear box, the hydraulic system and the yawing mechanism.

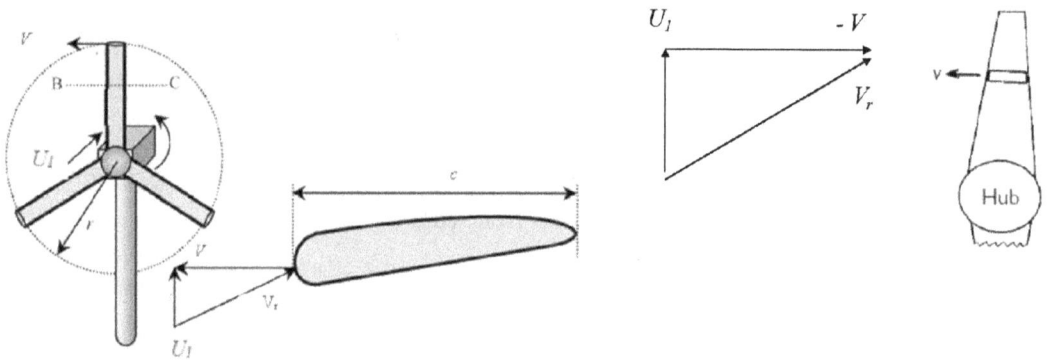

Fig. 5.18 (d). Blade element velocities

Wind velocity (u₀): Velocity of free air in the neighbourhood of a wind turbine (at a distance where the disturbances due to the rotation of a turbine does not reach) (fig. 5.18 (d)).

Incident wind velocity (u₁): Velocity of air passing through the rotor, i.e., the velocity at which the wind strikes the blade. It is slightly less than u_0 (fig. 5.18 (d)).

Blade element linear velocity (v): Linear circumferential velocity of the blade element due to rotation of blade (fig. 5.18 (d)).

Relative wind velocity (v_r): Velocity of air relative to the blade element as the both the air and blade element move (fig. 5.18 (d)).

Angular Speed (w): Angular speed of rotor in rad/s

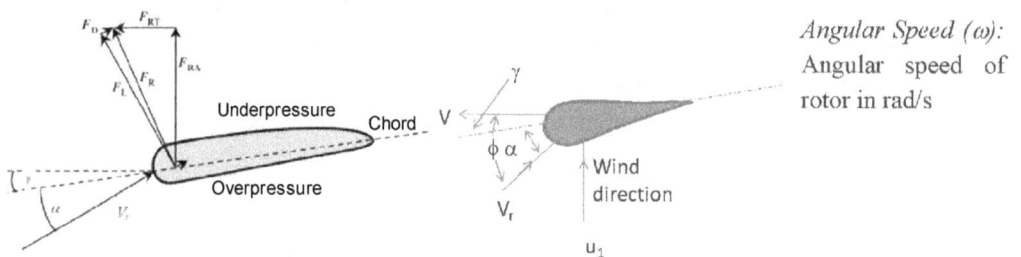

Angular Speed (ω): Angular speed of rotor in rad/s

Fig. 5.18 (e). Forces on blade element

Vector addition of drag (F_D) and lift (F_L) forces provides the resultant force, $F_R = F_D + F_L$. The resultant force can also be subdivided into an axial component F_{RA} and a tangential component F_{RT}. The tangential component F_{RT} of the resultant force causes the rotor to turn.

Angle of incidence (or angle of attack), α: Angle between central line of the blade element and relative wind velocity, v_r and normal to the plane blade. The airflow remains attached to the aerofoil for small angle of attack. The airflow is separated from the aerofoil for large angles of attack.

Pitch angle or blade setting angle (γ): It is the angle between the centreline of the blade element and the direction of linear motion of the blade element or angle between the relative wind velocity, v_r and normal to the plane of blade. The output of a turbine is greatly influenced by blade pitch angle. The blade pitch control is a very effective way to control the output power, speed or torque of the turbine (fig. 5.18 (e)).

Drag Force, DF_D: Incremental force acting on the blade element in the direction of relative velocity of wind (fig. 5.18 (e)).

Lift Force, ΔF_L: Incremental force acting on the blade element in the direction perpendicular to the relative velocity of wind (fig. 5.18 (e)).

Axial Force, ΔF_A: Incremental force acting on the blade element along the axis of rotation of the blade (fig. 5.18 (e)).

Tangential force, ΔF_T: Incremental force acting on the blade element tangential to a circular path of rotation (fig. 5.18 (e)).

Solidity: It is defined as the ratio of the projected area of the rotor blades on the rotor plane to the swept area of the rotor.

Wind turbine and wind generator: A device for direct mechanical work is often called a windmill or just a wind turbine. If electricity is produced, the combination of turbine and generator may be called a wind generator or an aero generator. Because of the confusion of these terms, the acronym WECS (wind energy conversion system) is increasingly used.

Horizontal and vertical axis wind turbine: If the axis of rotation is parallel or perpendicular to airstream, the former is a horizontal axis and the latter is a vertical axis turbine.

Pitch control: It is the control of pitch angle by turning the blades or blade tips [Fig 5.18 (e1)].

Yaw control: It is the control for orienting (steering) the axis of wind turbine in the direction of wind [Fig 5.18 (e2)].

Teetering: It is the action of smoothening forces on the rotating blades and to decrease the mechanical stress by teetering. It is required to hinge the blade independently against a spring or blades are to hinged together (teetered). The plane of wind turbine wheel is swung in inclined position at higher wind speeds by teetering control [Fig 5.18 (e2)].

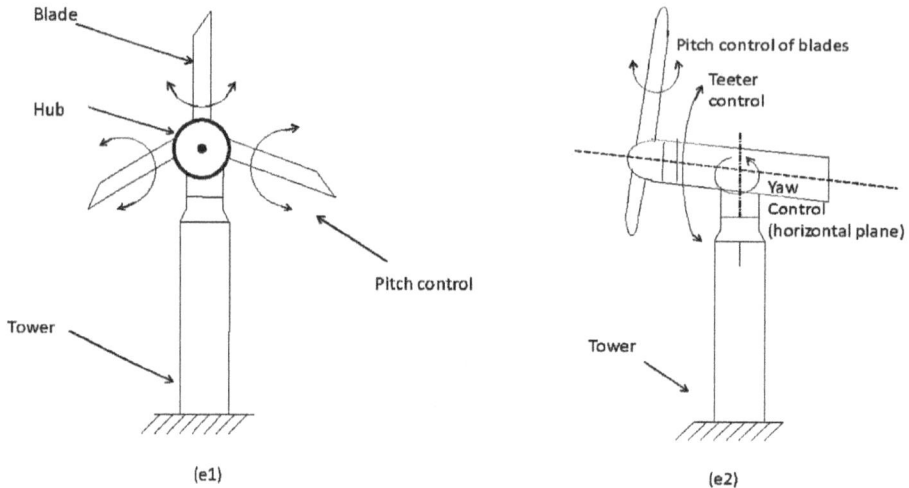

Fig. 5.18 (e1 & e2). Controls in wind turbines

5.4.2 Aerodynamics of Wind Turbine (Horizontal Axis turbine)

Aerodynamics deals with the motion of air or other gaseous fluids and the forces acting on objects as a result of relative motion between air and object. In wind turbines, aerodynamics deals with the relative motion between air and airfoil (blade element). It is the shape designed to create lift forces when air flows over it. When wind passes over both the surfaces of the airfoil shaped blade, it passes more rapidly over longer (upper) side of the airfoil in comparison to its shorter (lower) side as shown in fig.5.18 (f). The Bernoulli's equation can be used as a guide in identifying the high and low pressure region. Pressure is low at location where the flow velocity is high and pressure is high at locations where flow velocity is low. Therefore, low pressure is created in upper surface of the airfoil and high pressure in its lower surface. The pressure difference between top and bottom surface of the airfoil results in a force called aerodynamic lift as air moves from high pressure region to low pressure region. The upward force due to aerodynamic lift pushes the blades to move up. Since the blades of the wind turbine are constrained to move up with the hub at its centre, the lift force causes the rotation of blade about the hub. Air flowing smoothly over an airfoil produces two forces; lift force which acts perpendicular to the flow and drag which acts in the direction of flow. In wind turbine, drag force perpendicular to lift force also acts on the blade causing the impediment of rotor rotation. The prime objective in wind turbine design is for the blade (airfoil shaped) to have a relatively high lift to drag ratio. This ratio can be varied along the length of the blade to optimize the output energy of the turbine at various wind speed. Hence in aerodynamic analysis of wind turbine, both lift and drag forces are important for their optimization in its efficient design.

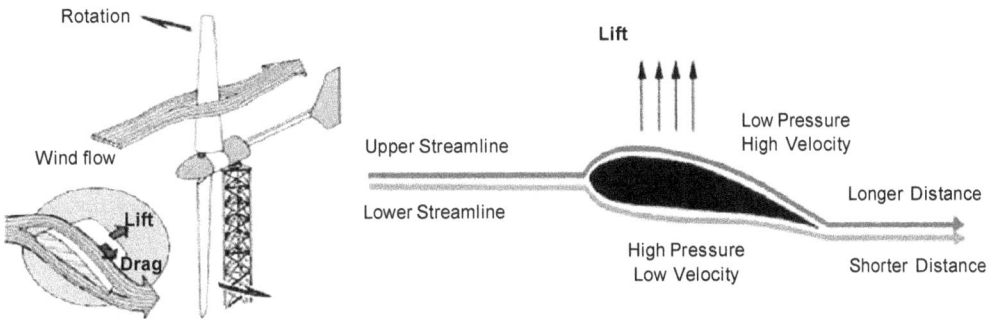

Fig. 5.18 (f). Aerodynamic lift force on airfoil section of wind turbine

The basic concepts of lift and drag forces have been discussed in the next section.

Lift and Drag force

Fluid or air flow over solid bodies frequently occurs in practice and it is responsible for numerous physical phenomena such as drag force acting on objects like trees, electric towers, the lift force developed by airplane wings, the lift force experienced by dust particles in a wind storm and the blade motion developed by a turbine. Either the fluid moves over a stationary body or a body moves through a standstill fluid, aerodynamically, both processes are equivalent to each other. The approach is to study the relative motion between the fluid and the body. Such motions are referred to as flow over bodies or external flow.

Drag

It is the common experience that a body meets some resistances when it is forced to move through a fluid especially a liquid. It is noticed and felt that it is very difficult to walk in water because of much greater resistance offered by water to motion compared to air. Also it is observed that high wind knocks down the tree, power lines and even trailer. It is felt that the strong push is exerted by the wind on the human body. The force, a flowing fluid exerts on a body in the flow direction is called drag force. Drag is usually an undesirable effect, like friction and is minimized as far as practicable. Reduction of drag is closely associated with the reduction of fuel consumption in automobile, submarines, aircraft etc. But in some cases, drag produces a very beneficial effect and is maximized. Friction for example is a 'life saver' in the brakes of automobile. Likewise, it is the drag that helps the people to parachute for safe landing, pollens to fly to distant locations etc.

Lift

When a body is immersed in a standstill fluid, only normal pressure force is exerted on it. A flowing fluid in addition exerts tangential shear forces on the surface because of the no-slip condition caused by viscous effects. Both of these forces in general have two components, one is the drag in the direction of flow and the other is perpendicular to the fluid flow, called 'lift'. Thus drag force is due to the combined effects of pressure and wall shear forces in the flow direction. The components of pressure and wall shear forces in the direction normal to the fluid flow are called lift. The relative magnitudes of drag and lift forces depend completely on the shape of the object. Streamline objects experience a smaller drag force than experienced by the blunt objects. Airfoils of a wind turbine are especially shaped to produce lift force on coming in contact with the moving air.

A good airfoil has high lift/drag ratio (LDR); in some cases, it can generate lift forces perpendicular to air stream direction, 30 times as great as the drag force parallel to the flow. The lift increases as the angle formed at the junction of the airfoil and the air stream (the angle of attack) becomes less and less acute, up to the point where the angle of the air flow on low pressure side becomes excessive. When this happens, the air flow breaks away from the low pressure side, a lot of the turbulence occurs, the lift decreases and the drag increases quite substantially, causing the phenomenon i.e. stalling.

Drag force (ΔF_D)

The total force due to pressure force and shear force exerted on the aerofoil can be resolved in the direction parallel to airflow and perpendicular to the airflow. The component of total force parallel to the direction of air flow is called drag force. The drag force always opposes the relative motion between the body and the air. It is given by the following equation:

$$Drag = \Delta F_D = C_D \times \frac{1}{2}\rho u_1^2 . A, ,$$ where C_D is the drag coefficient and 'A' is the

projected area of the body aerofoil perpendicular the direction of airflow (Fig. 5.18 (g)).

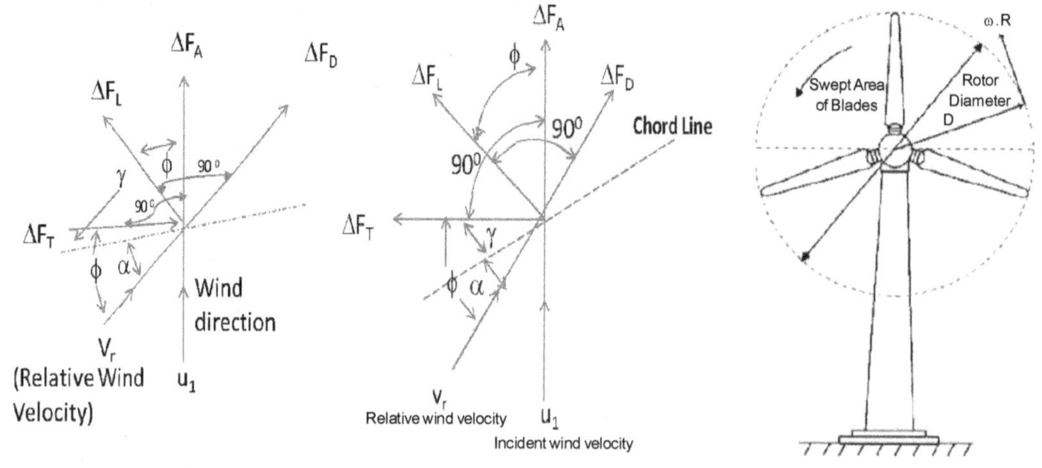

Fig. 5.18 (g). Forces acting on airfoil section of wind turbine

Lift force (ΔF_L)

The component of the total force (pressure force and shear force) on the body in the direction perpendicular to airflow is called lift force. As the name suggests, this force tries to lift the body. It is given by the following equation:

$$Lift = \Delta F_L = C_L \times \frac{1}{2}\rho u_1^2.A, \text{ where } C_L \text{ is the lift coefficient.}$$

Axial force (ΔF_A)

The total force (pressure force and shear force) can also be resolved along the axis of rotation of blade and perpendicular to it (tangential force on the blade). The component of total force acting on the blade along the axis of rotation of the blade is called the axial force. The axial force does not contribute to the rotation of the blade. It is also called thrust force and has to be balanced by a suitable reaction force generated by any thrust bearing provided on a rotor. *The axial force contributes to waste energy which cannot be extracted as an useful component from wind energy.* Hence, axial force should be as less as possible. The axial force can be given as follows:

$$\Delta F_A = \Delta F_L \cos\phi + \Delta F_D \sin\phi, \text{ where } \phi = \alpha + \gamma$$

Tangential force (ΔF_r)

It is the component of total force on the blade acting tangential to its circular path of rotation. *This is the force which contributes mainly to the useful energy extracted from the wind energy.* It should be as high as possible. It is given by

$$\Delta F_T = \Delta F_L \sin\phi - \Delta F_D \cos\phi$$

Tip speed ratio (λ)

The tip speed ratio is defined as the ratio of the speed of tip of the rotor blade to the speed of oncoming air. Hence, tip speed ratio $\lambda = \dfrac{\omega \times R}{u_0}$ is where R is the diameter of rotor.

For a particular wind speed, there exists an optimum turbine tip speed to produce the maximum output.

Numerical 5.2 Find the tip-speed ratio if a 6 m diameter rotor has rotation of 20 rpm and the wind speed is 4 m/s. What is the implication of tip speed ratio?

$$\text{Tip speed ratio} = \frac{w.R}{u_0}, \text{tip speed radio} = \frac{\dfrac{\pi R N}{60}}{4} = \frac{\pi \times 6 \times 20}{60 \times 4} = 1.6$$

The windmill rotating fast has tip speed ratio greater than 1. Two or three-bladed rotors rotate faster, thereby having tip speed ratio ranging from 3 to 10. More bladed rotors rotate move slowly, thereby having tip speed ratio between 1 and 2.

Numerical 5.2 (i) Consider a wind turbine of rotor diameter 20m, rotating at a speed of one rotation per second. Wind speed is 15 m/s. Calculate the tip speed ratio (λ) for this turbine.

Given, f, frequency of rotation (Hz), Sec^{-1}, rotor radius = 10 m, angular velocity (radian/s) is given by $2\pi f = 2\pi$ rad/s. Velocity of rotor tip $= w.R = 20\pi$ m/s. Tip speed

ratio $= \dfrac{w.R}{u_0} = 20\pi /15 = 4.18$.

Solidity

Solidity σ is defined as the ratio of the blade area to the circumference of the rotor. Solidity determines the quantity of blade material required to intercept a certain wind area. Hence, $\sigma = \dfrac{Nb}{2\pi R}$ where N is the number of blades, b is the blade width and R is radius of circle described by a blade. For example, if a 3-metre radius rotor has 24 blades, each 0.35 m wide, the solidity is

$$\sigma = \dfrac{24 \times 0.35}{2\pi \times 3} \times 100 = 44.6\%$$

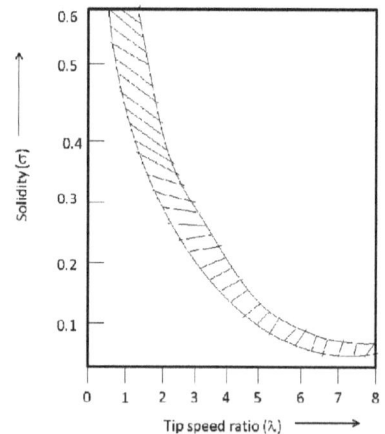

Fig. 5.18 (h). Variation of solidity(σ) with tip speed ratio (λ)

Solidity represents the fraction of the swept area of the rotor which is covered with metal. Variation of solidity σ with tip speed ratio λ is shown in Figure 5.18 (h).

The following observations can be made from Fig. 5.18 (h).

(i) A two or three-bladed wind turbine has a low solidity and so needs to rotate faster to intercept and capture wind energy with aerofoil blades like aircraft. Otherwise, the major part of wind energy would be lost through the large gaps between the blades. High speed wind turbines have a low starting torque.

(ii) Rotor having a high value of solidity, like the multi-bladed wind water pump turbine operate at low tip speed ratio. Such rotors need a high starting torque.

(iii) As tip speed ratio increases, the number of blades decreases

A high solidity rotor rotates slowly and uses the drag force while a low solidity rotor uses lift forces.

Fluid flow concept (attached and separated) around a body

The nature of flow around a body at a given velocity depends upon the relative magnitudes of the inertia and the viscous forces. The ratio of these forces is known as the Reynolds number.

In a very slow-moving (creeping) fluid, there is no relative motion at the body wall and the drag on the body may be directly attributed to viscous, frictional shear stresses set up in the fluid, the flow remains attached to the surface of the body as shown in Fig. 5.18 (i).

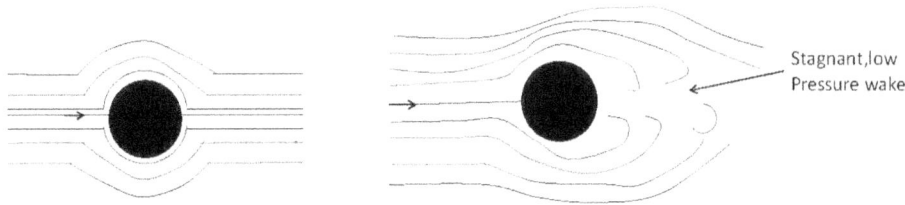

Stagnant,low
Pressure wake

Fig. 5.18 (i). Creeping flow past a cylinder   **Fig. 5.18 (j).** Separated flow past a cylinder

In practical situations, when the viscosity is low and the velocity is relatively high, the drag force that exists is due primarily to an unsymmetrical pressure distribution, fore and aft. This is caused by the fact that the fluid does not follow the boundary of the body, but separates from it leaving a low pressure, stagnant fluid in the wake as shown in Fig.5.18 (j). On the upstream side, where the flow remains attached, the pressure is high, because the flow has been slowed down and pushed aside by the body obstructing its path.

An airfoil is a streamlined body tapered gently in the aft region and has a sharp trailing edge. For an airfoil, the flow remains attached to the body throughout. Because the trailing edge is sharp, the separation leaves no low-pressure wake. The drag is very low, largely due to skin friction only, rather than the pressure difference. The drag does not depend on the frontal area but on the surface area and surface finish. This situation is shown in Fig.5.18 (k).

Consider an airfoil at a small angle of incidence 'α' to oncoming flow as shown in Fig.5.18 (l). The flow remains attached and the drag is small. The particles above the airfoil travel faster causing static pressure there to reduce, whereas, those beneath it, move slowly causing the static pressure to increase in this region. This produces a normal force on the airfoil, known as lift force.

When the incidence angle 'α' is increased beyond a certain critical value (10 to 16^0, depending on the Reynolds number), the separation of the boundary layer on the upper surface takes place as shown in Fig. 5.18 (m). This causes a wake to form above the airfoil, reducing the lift and increasing the drag substantially. This is known as stalled state of the airfoil. A well-rounded leading edge and proper thickness of the airfoil improves stalling behaviour of the airfoil.

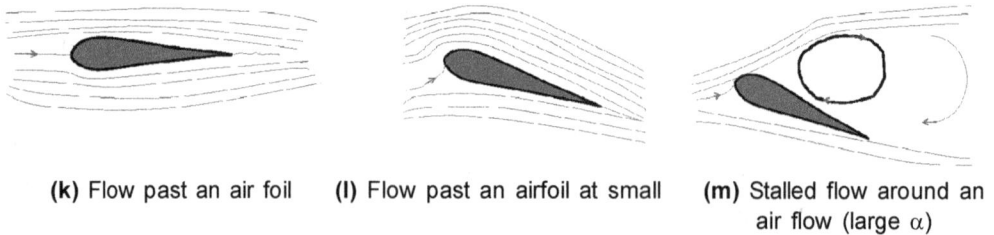

(k) Flow past an air foil   **(l)** Flow past an airfoil at small   **(m)** Stalled flow around an air flow (large α)

Fig. 5.18. Fluid flow around an airfoil

5.5 WIND POWER EXTRACTION

The wind turbine is used to extract useful power from wind. The energy or power can be extracted by partially decelerating (reduction of velocity) and expanding (reduction of pressure) using wind turbine. The rotor of wind turbine collects wind from the whole area swept by the rotor. The area swept can be considered as airstream tube which is continuously expanding as shown in Fig. 5.19. This airstream tube model is also called Betz model of expanding air. As air mass flow rate remains same everywhere within the stream tube according to the Law of Continuity, the wind speed therefore decreases as air expands. As shown, airstream tube has area of A_0 at the upstream, area of A_1 while passing through rotor blade (airfoil) and area A_2 at the downstream. In order to compute the mathematical relationships, the following assumptions are made.

i.   The flow of wind is 'compressible' and hence air stream diverges as it passes through the turbines

ii. The mass flow rate of wind remains constant at far upstream, at the rotor and at far downstream.

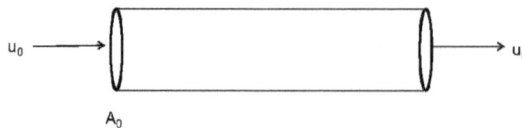

Fig. 5.19 (a). Unperturbed wind stream tube in absence of turbine

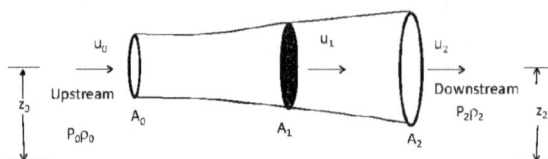

Fig. 5.19 (b). Wind stream tube in presence of turbine

As the free wind stream interacts with the turbine rotor, the wind transfers part of its energy into the rotor and the speed of the wind decreases to a minimum value leaving a trail of disturbed wind called 'wake', fig. 5.20 (a). The variation in velocity is considered to be smooth from far upstream to far downstream. However, the fall in static wind pressure is sharp as depicted in fig. 5.20 (b). The wind leaving the rotor is below the atmospheric pressure (in wake region), but at far downstream, it regains its value to reach the atmospheric level. The rise in static pressure is at the cost of kinetic energy, consequently further decreasing the wind speed. The air-mass flow rate remains same throughout the stream tube i.e. $\dot{m} = \rho A_0 u_0 = \rho A_1 u_1 = \rho A_2 u_2$ where u_0 is wind speed is upstream, which reduces to u_1 while passing through the rotor and becomes u_2 at the downstream and stream tube area of constant air mass is A_0 upstream, which expands to A_1 while passing through the rotor and becomes A_2 downstream, 'ρ' the density of the flowing air. A_0 and A_2 can be located experimentally for wind speed determination, but wind speed measurement at A_1 is not possible because of the rotating blades.

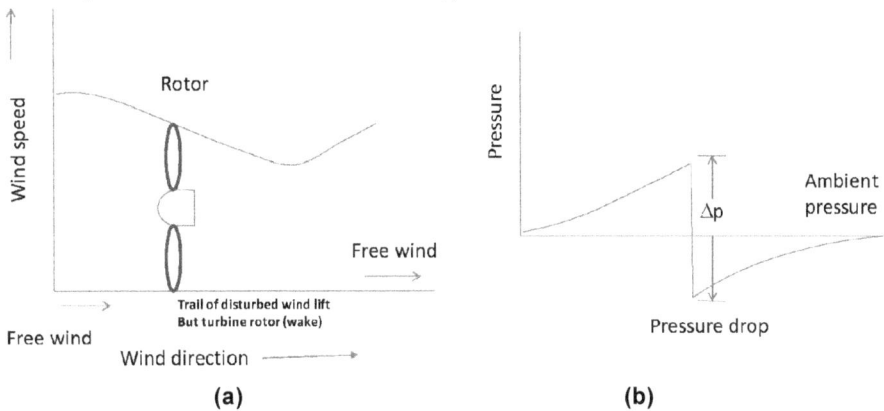

Fig. 5.20. Change in wind speed and pressure in traversing the turbine rotor

If u_0 and u_2 are the wind speeds at upstream and downstream respectively, the force or thrust on the rotor is equal to the reduction in momentum per unit time from the air mass flow rate . Hence thrust on turbine is given by

$$F = \dot{m}\, u_0 - \dot{m}\, u_2 \qquad \qquad \text{... (5.1)}$$

The power extracted by the turbine is

$$P_T = F u_1 = (\dot{m}\, u_0 - \dot{m}\, u_2) u_1 \qquad \qquad \text{... (5.2)}$$

The loss in kinetic energy per unit time of wind flow is

$$P_w = \frac{1}{2} m \left(u_0^2 - u_2^2 \right) \qquad \qquad \text{... (5.3)}$$

Equations (5.2) and (5.3) are equal and hence by equating, one gets

$$\left(u_0 - u_2 \right) u_1 = \frac{1}{2} \left(u_0^2 - u_2^2 \right) = \frac{1}{2} (u_0 - u_2)(u_0 + u_2) \qquad \qquad \text{... (5.4)}$$

Or, $$u_1 = \frac{u_0 + u_2}{2} \qquad \qquad \dots (5.5)$$

As an extreme case, considering u_2 to be zero (which is not practical as downstream air must have some kinetic energy to leave the turbine region), $u_1 = u_0 / 2$. Thus, according to this linear momentum theory, the air speed through the turbine can always be more than half the speed of upstream air.

The mass of air flowing through the rotor per unit time is given by

$$\dot{m} = \rho A_1 u_1 \qquad \qquad (5.6)$$

Substituting Eq, (5.6) in Eq (5.2), $P_T = \rho A_1 u_1^2 (u_0 - u_2) \qquad \qquad \dots (5.7)$

Substituting the value of u_2 from Eq. (5.5) in eq. (5.7), one gets

$$P_T = \rho A_1 u_1^2 \{u_0 - (2u_1 - u_0)\} = 2\rho A_1 u_1^2 (u_0 - u_1) \qquad \qquad \dots (5.8)$$

The interference factor or perturbation factor (a) is defined as the fractional decrease of wind speed at the turbine, Thus,

$$a = \left(\frac{u_0 - u_1}{u_0} \right) \Rightarrow u_1 = (1 - a)u_0 \qquad \qquad \dots (5.9)$$

Now, using Eq (5.5) in Eq. (5.9), $a = \dfrac{u_0 - u_2}{2u_0} \qquad \qquad \dots (5.10)$

Substituting Eq (5.9) in Eq. (5.8), $P_T = 2\rho A_1 (1 - a)^2 u_0^2 [u_0 - (1 - a)u_0]$

Or, $$P_T = \frac{1}{2} \rho A_1 u_0^3 \left\{ 4a(1 - a)^2 \right\} \qquad \qquad \dots (5.11)$$

Or, $$P_T = C_P P_0 \qquad \qquad \dots (5.12)$$

Where, $$C_P = 4a(1 - a)^2 \qquad \qquad \dots (5.13)$$

and P_0 is the power of free (unperturbed) wind $\left(P_0 = \dfrac{1}{2} \rho A_1 u_0^3 \right)$ and C_p the fraction of power extracted, called the power coefficient.

Sometimes, $b = \dfrac{u_2}{u_0}$ is also referred as interference factor. In order to get maximum value of C_p, it is to be differentiated with respect to 'a' and equating to zero, i.e.

$$\frac{dC_p}{da} = 1 + 3a^2 - 4a = 0, \text{ Thus } a = 1 \text{ or } 1/3, \text{ However, } a = 1 \text{ as } C_p = 0 \text{ when } a = 1.$$

so, $a = \frac{1}{3}$, putting the value of $a = \frac{1}{3}$ in Eq. (5.13), one gets $C_{p\max} = \frac{16}{27} = 0.59$

The following conditions may result from the expression of power coefficient (C_p),

(i) When $a = 0$, then $u_1 = u_0$, $u_2 = u_0$ and no power generation takes place.

(ii) When $a = \frac{1}{3}$, then $u_1 = \frac{2}{3}u_0, u_2 = \frac{1}{3}u_0$ and maximum power generation takes place.

(iii) When $a = \frac{1}{2}$, then $u_1 = \frac{1}{2}u_0, u_2 = 0$ and only turbulence occurs at the downstream.

(iv) When $a = 1$, then $u_1 = 0$ and it results in the stalling of turbine

The variation of the power coefficient C_p with the interference factor 'a' is shown graphically in fig. 5.21. The physical significance of the curve is discussed below:

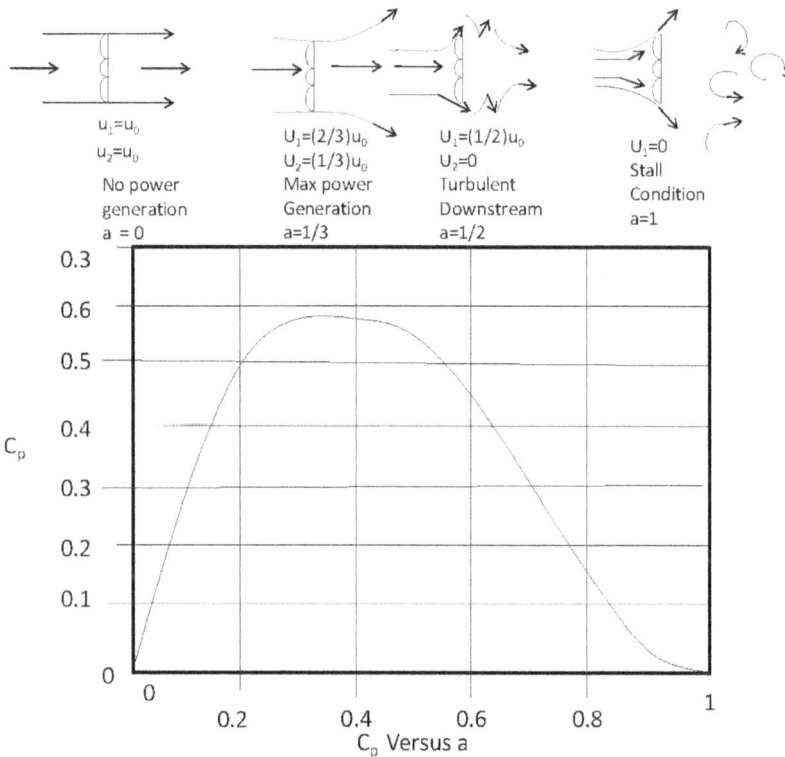

Fig. 5.21. Graph showing power coefficient versus perturbation factor (a)

When no load is coupled to the turbine, the blades act just as freewheel. There is no reduction of wind speed at the turbine, therefore, $u_1 = u_0$ and the value of 'a' is zero. The turbine does not generate any power and $C_p = 0$. The air just passes through the turbine without any reduction of speed.

Now as load is applied, power is extracted, so C_p increase as u_1 decreases. Maximum value of C_p (i.e., $C_{P\ max} = 16/27 = 0.593$) occurs at '$a$' = 1/3. At this condition,

$$u_1 = \frac{2}{3}u_0 \text{ and } u_2 = \frac{1}{3}u_0$$

That means, at maximum power extraction condition, the upstream wind speed is reduced to two-third at the turbine and further reduced to one third at the downstream. The criterion for maximum power extraction, i.e., $C_{P\ max} = 16/27$ is called the Betz criterion. This applies to an ideal case. For a commercial wind turbine, however, maximum power coefficient is less than ideal value.

When $u_2 = 0$, 'a' = 1/2 and the simple model breaks down as no wind is predicted to be leaving downstream. In practice, this is equivalent to the onset of a turbulence downstream. Power extraction decreases due to mismatch of rotational frequency and wind speed and partial stalling begins. The turbine blades will still be turning, causing extensive turbulence in the air stream, leading to more losses. When the wind speed at the turbine is reduced to zero (i.e., $u_1 = 0$), 'a' becomes unity and no power is extracted. This state is known as (complete) stall state of blades.

The Betz Limit

The maximum achievable value of the power coefficient is known as the Betz limit after Albert Betz, the German aerodynamicist. (1919). In practice, all of the kinetic energy in the wind cannot be converted to shaft power since the air must be able to flow away from the rotor area. The Betz limit, derived using the principles of conservation of momentum and conservation of energy, suggests a maximum possible turbine efficiency, (or power coefficient) of 59%. Considering the rotor efficiency to be about 70 %, and bearing, vibrations, friction losses and generator efficiency of about 90 %, the available power coefficient would be about 60 % of C_p i.e. in practice, power coefficients of 30 -35 % ($0.7 \times 0.9 \times 0.59$) are more common.

5.5.1 Axial thrust on turbine, F_A

The Betz model of expanding airstream tube is shown in fig. 5.20. In case of no energy extraction and on applying the Bernoulli's equation at upstream and downstream of the tube, we have $p_0/\rho_0 + gz_0 + u_0^2/2 = p_2/\rho_2 + gz_2 + u_2^2/2$, where p is the static pressure.

As $z_0 = z_2$ and variation in air density is negligible compared to other terms, considering ρ as average air density, the static pressure difference across the turbine may be written as

$$\Delta p = p_0 - p_2 = \left(u_0^2 - u_2^2\right)\rho / 2$$

The maximum value of static pressure difference occurs when u_2 approaches zero (which will be the situation for solid disk). If $= u_2 = 0$, $\Delta p = \dfrac{1}{2}\rho \times u_0^2$

$\Delta p_{max} = \rho u_0^2 / 2$ and maximum thrust on the disk is $F_{A\,max} = A_1\rho u_0^2 / 2$ ... (5.14)

On a horizontal machine, this thrust acts along the turbine axis and therefore is known as axial thrust F_A. This axial thrust must be equal to loss of momentum of the air stream as given by $F_A = \dot{m}u_0 - \dot{m}u_0 - \dot{m}u_2$. Using Eqs. $\left(P_0 = \dfrac{1}{2}\rho A_1 u_0^3\right)$, 5.5 and 5.9, we can write

$$F_A = 4a(1-a)(A_1)\rho u_0^2 / 2 \qquad \text{... (5.15)}$$

$$F_A = C_F F_{A\,max} \qquad \text{... (5.16)}$$

Where $C_F = 4a\,(1\text{-}a)$, the coefficient of axial thrust   ... (5.17)

Maximum axial thrust occurs when $C_F = 1$, which is achieved when $a = 0.5$, equivalent to $u_2 = 0$. Maximum power extraction by the Betz criterion occurs when $a = 1/3$, corresponding to $C_F = 8/9$.

As per Eq. (5.14), $F_{A\,max} = \dfrac{A_1\rho u_0^2}{2}$, *Hence, by doubling the wind speed, the maximum axial thrust on the wind turbine will be increased by 4 times and power generation to increase by 8 times.*

5.5.2 Torque Developed by the Wind Turbine (T)

The maximum conceivable torque T_M on an ideal turbine rotor would occur if maximum circumferential force acts at the tip of the blade with radius R. thus,

$$T_m = F_{(cir,\ max)}R \qquad \text{...(5.18)}$$

$$T_m = \frac{P_0}{u_0}R \text{ (maximum torque } = \frac{\frac{1}{2}\rho A u_0^3}{u_0} \times R = \frac{1}{2}\rho A_1 u_0^2 R) \qquad \text{...(5.19)}$$

Now, if the tip-speed ratio is 'λ' defined as $\lambda = \dfrac{Speed\ of\ tip\ of\ the\ rotor\ blade}{Speed\ of\ on\ coming\ air} = \dfrac{R\omega}{u_0}$

$$... (5.20)$$

Then T_m, can be written as $T_m = \dfrac{P_0 \lambda}{\omega}$ ... (5.21)

For a practical machine, where circumferential force is not concentrated at the tip but spread throughout the length of the blade, less shaft torque will be produced than what is given by Eq. 5.21. Thus, the shaft torque, T_{sh} is given as

$$T_{sh} = C_T\ T_M \qquad ... (5.22)$$

where, C_T is known as torque coefficient. As the product of shaft torque and angular speed equals power developed by the turbine, or $T_{sh}\omega = P_T$ or $C_T T_M \omega = C_P P_0$

Substituting for T_m from Eq. 5.21, we get, $C_T P_0 \lambda = C_P P_0$ or, $C_T = \dfrac{C_P}{\lambda}$ ... (5.23)

Both C_T and C_p are function of tip-speed ratio, λ. As per Betz limit, the maximum value of C_p can be 0.593, therefore,

$$C_{T\,max} = \frac{C_{P\,max}}{\lambda} \qquad ... (5.24)$$

Thus, machines with higher speeds (high tip speed ratios) have low value of $C_{T\,max}$ or low starting torque. The torque coefficient becomes more at low value of tip speed ratio, for high solidity turbine (multi-blades). The multi-blades turbine can give power with slow speed. However, the low solidity turbine (single or double) operates with high tip speed ratio but gives lower torque coefficient in comparison with high solidity turbine.

5.5.3 Operational Characteristics for Wind Turbine

The performance of a wind turbine can be characterized and assessed by three parameters such as power, torque and thrust and their variations with the wind speed. The power determines the amount of energy captured by the rotor, the torque developed determines the size of the gear box and must be matched by the generator, being driven by the rotor. The rotor thrust has great influence on the structural design of the tower. It is usually convenient to express the performance by means of non-dimensional characteristic performance curves from which the actual performance can be assessed regardless of how the turbine is operated i.e. at constant rotational speed or variable rotor speed. Assuming that the aerodynamic performance of the rotor blades does not deteriorate, the non-dimensional aerodynamic performance of the rotor will depend upon the tip speed ratio. It is usual, therefore, to display the power, torque and thrust coefficients as functions of tip speed ratio. Tip speed ratio is the function of three most important variables; blade swept radius, wind speed and rotor frequency. Being dimensionless, it becomes an essential scaling factor in design and analysis.

Power performance

The power curve for a wind turbine indicates the net electricity energy output from a wind turbine as a function of the wind velocity at hub height. The curve is drawn either by theoretical calculation or by field tests. A power curve for a grid connected stall-regulated wind turbine is shown in fig. 5.22. This curve indicates the performance of turbine while operating for cut-in, rated and cut-out speeds.

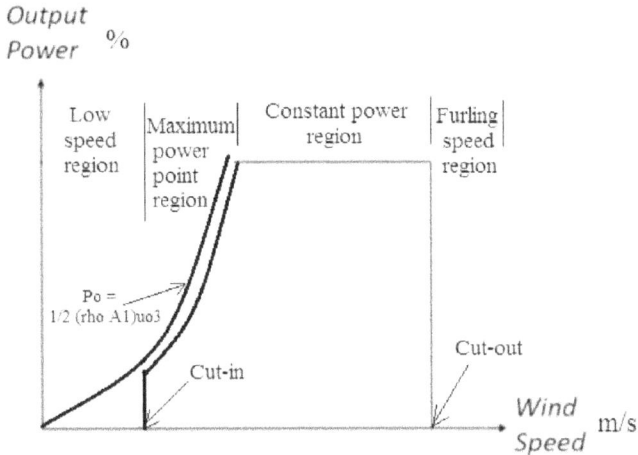

Fig. 5.22. Wind turbine operating regions and its performance

Dynamic Matching

Dynamic matching is important to create the required rotational frequency of the turbine to particular wind speed for optimum efficiency. After passing through the turbine, air moves downstream with some energy. As per the Betz limit, the accepted standard of maximum extractable power is 59%. In the derivation of Betz limit, the dynamic rotational state of the turbine has not been taken into consideration. Therefore, power extraction efficiency decreases from an optimum value as shown in fig. 5.23. The power extraction will decrease, if

(i) The blades are so close together or rotating so rapidly that a blade moves into the turbulence due to the air created by a preceding blade (fig. 5.23a).

(ii) The blades are so far apart or rotating so slowly that much of the air passes through the cross-section of the device without interfering with a blade (fig. 5.23c).

Thus, for a particular wind speed, there exists an optimum turbine speed to produce maximum output. Therefore, to obtain optimum efficiency, it is important to match the rotational frequency of the turbine to the corresponding wind speed.

Let t_b be the time taken by a blade to move into the position previously occupied by the preceding blade and t_w be the time for the disturbed wind moving past that position and normal air stream becoming re-established. For an n-bladed turbine rotating at angular velocity of $\omega, t_b = \dfrac{2\pi}{n\omega}$. A disturbance at the turbine disk created by a blade into which

the following blade moves, will last for a time, $t_w = \dfrac{d}{u_0}$, where, d is the length of the wind

strongly perturbed by the rotating blades (fig. 5.23b). Maximum power extraction occurs when $t_w = t_b$, at blade tips, where maximum incremental area is swept by the blades.

Hence, $\dfrac{2\pi}{n\omega} = \dfrac{d}{u_0} \Rightarrow \dfrac{2\pi}{d} = \dfrac{n\omega}{u_0}$. Multiplying R in both sides, $\dfrac{2\pi R}{d} = \dfrac{n\omega R}{u_0}$. But tip seed ratio,

$\lambda = \dfrac{R\omega}{u_0}$. Substituting the value of λ, the tip speed ratio for optimum power extraction

becomes $\lambda_0 = \dfrac{2\pi}{n}\left(\dfrac{R}{d}\right)$. If it is the assumed $d = kR$, then $d \to R$ when $k = 1$. In that

case, the tip speed ratio for maximum power extraction is $\lambda_0 = \dfrac{2\pi}{kn}$. Practical results

show that $k \approx \frac{1}{2}$, (d $\approx$ (1/2) R, so for n-bladed turbine, $\lambda_0 \approx \dfrac{4\pi}{n}$. Hence, for a two- bladed

turbine ($n = 2$), the maximum power extraction ($C_{p\ max}$) occurs at $\lambda_0 \approx 2\ \pi$ and that for a four-bladed turbine at $\lambda_0 \approx \pi$. Fig. (5.24) shows the relationships of C_p and λ for a variety of wind turbine types (to be discussed in the later section).

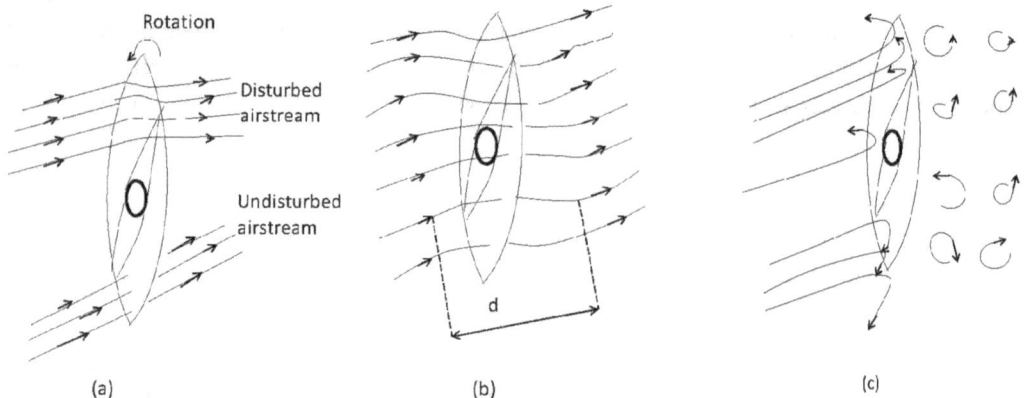

Fig. 5.23. Turbine frequency and power output. (a) rotational frequency too low, some wind passes unperturbed through the rotor swept area (b) rotational frequency optimum, whole air stream affected, 'd' is the length of the wind strongly perturbed by the rotating blades (c) rotational frequency too high, energy is dissipated in turbulent motion and vortex shedding

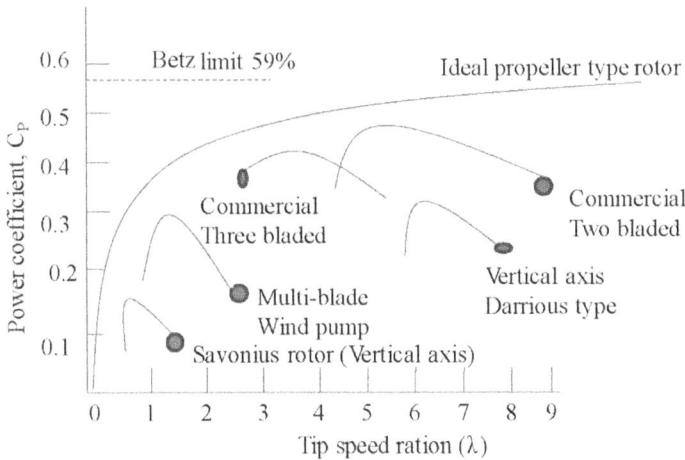

Fig. 5.24. Power coefficient C_p versus tip speed ratio λ

From the figure above, it is seen that solidity (total blade area divided by the swept area) is very important in the performance of the wind turbine. The maximum practically obtainable value of the power coefficient is approximately 0.5. A wind turbine, achieving a value of power coefficient as 0.4 or above is considered to have good performance. Low solidity (2-bladed turbine) produces a broad flat curve which means C_p changes very little over a wide tip speed ratio whereas high solidity (multi-bladed turbine) produces a narrow performance curve with a sharp peak, making the turbine very sensitive to the changes in the tip speed ratio. The optimum solidity appears to be achieved with three bladed turbines but two-bladed might also be an acceptable alternative because there occur little changes in the maximum C_p between them.

5.6 Parameters to consider for selecting wind turbine

While selecting a wind turbine, the parameters like pitch angle, solidity, tip speed ratio, torque, loads to be used etc. are to be taken into consideration for obtaining greater output from it.

(i) Blade pitch angle: The blades of a rotor are curved so that they can deflect the wind to create lift. The created lift force causes the rotor to rotate. To generate the maximum amount of lift, the blades are to be set at an appropriate angle to the wind direction., which is called the pitch angle. The tips of the blades move faster than other points near the axis. Hence, the pitch angle, λ, varies along the length of the blade. The pitch angle should be large without taking any risk of stalling the rotor. To make the pitch angle large all the way along the blade, it needs to be twisted. For rotor rotating fast as in two or three-bladed wind turbines, the blades are to be given smaller pitch. Certain rotors are provided with the mechanism to control the pitch depending upon the wind conditions. The effect of blade pitch on performance coefficient and tip speed ratio characteristics is shown in fig. 5.25 (a).

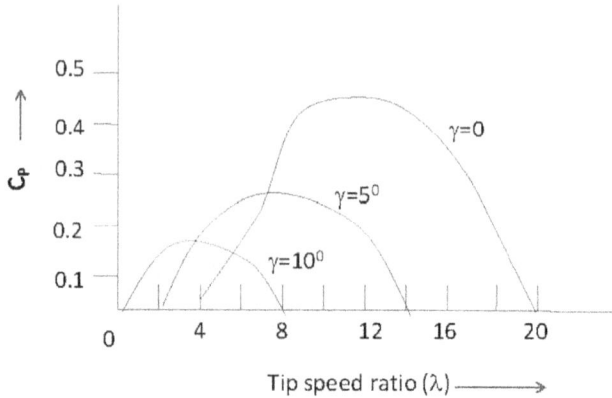

Fig. 5.25 (a). Effect of pitch angle on performance coefficient

(ii) Solidity: The greater the solidity of a rotor due to the presence of multiple blades, the slower it needs to intercept the wind with the help of rotation. The wind turbine with low solidity such as two or three-bladed rotors have to run rapidly to intercept the wind and to avoid any loss of wind energy through the large gaps existing between the blades.

(iii) Tip speed ratio (λ): Faster rotating wind turbines have tip speed ratios of more than 1, while slower rotating wind turbines have tip speed ratios of less than 1. Rotors rotating with the help of drag force have lesser tip speed ratios (tip speed is less than wind speeds). Savonius and Darrieus rotors have low tip speed ratios. On the other hand, two or three-bladed wind turbines rotate very fast and these have high tip speed ratios ranging from 3 to 10. Multi-bladed rotors have tip speed ratios between 1 and 2 and these are suitable for only wind pump applications. Each rotor has a certain optimum tip speed ratio at which it can produce maximum output. The range of tip speed ratios for optimum output or performance coefficient with rotors having different solidities is shown in fig. 5.25 (b).

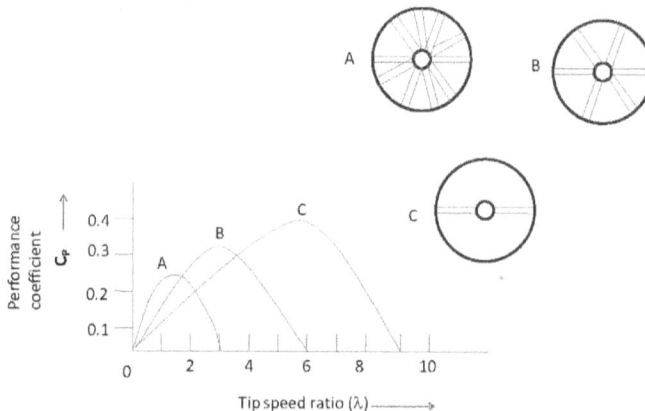

Fig. 5.25 (b). Variation of performance with solidity and tip speed ratios

(iv) Torque: Torque is generated by tangential or turning force acting on the blades of a rotor. It depends on solidity and tip speed ratio of the rotor. High solidity rotors with low tip speed ratios (as in multi-bladed rotors) produce much more torque compared to low solidity rotors as shown in fig. 5.25 (c). The high-speed wind turbines have higher performance coefficient (C_p), but has low starting torque coefficient (On the other hand, high solidity rotors produce high starting torque, but these have low performance coefficients.

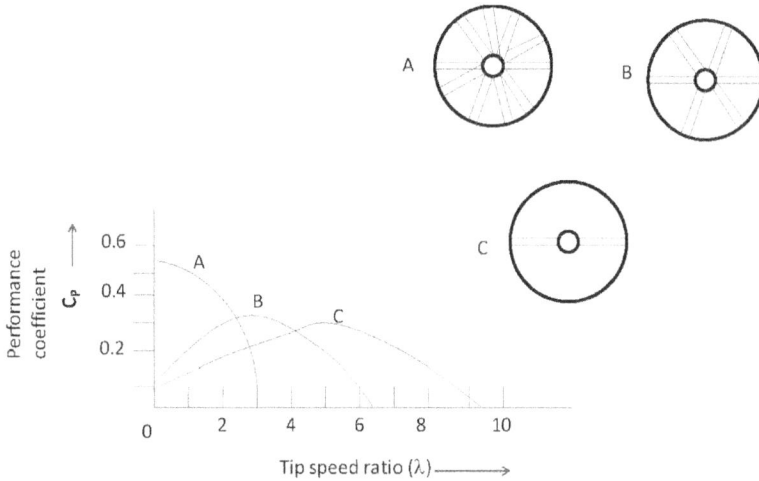

Fig. 5.25 (c). Variance of torque coefficient with solidity and tip speed ratios

(v) Load characteristics: The choice of rotor depends on the load characteristics. The loads requiring high torque (such as for piston pumps) have to be provided with rotors with either high solidity or some arrangement to detach the rotors from the load so that low starting torque wind turbines can be used. However, wind turbines used to generate electricity need low torque and high speed which are met with wind rotors of low solidity and high tip speed ratios. In case of wind pumps, these require three times more torques than what are required for normal working. Hence, these wind turbines require low wind speeds for running but a gust of higher wind speeds to start them initially. High speed turbines are unsuitable for running directly any reciprocating pumps unless these produce electricity by generators.

Solved Examples

Example 5.3 Find the maximum power output of a wind turbine if wind speed u_0 = 8 m/s, air density ρ = 1.226 kg/m^3, and rotor diameter (D) = 60 m.

Solution: $A_1 = \dfrac{\pi}{4}D^2 = 2826 m^2$. Power of wind = $P_w = \dfrac{1}{2}\rho A_1 u_0^3 = 887 kW$

Maximum power from wind turbine, = $P_T = \dfrac{16}{27} \times P_w = 524 kW$

Example 5.4 Wind speed is 10m/s at the standard atmospheric pressure. Calculate (i) The total power density in wind stream and (ii) The actual power produced by a wind turbine of 100 m diameter with an efficiency of 40%, assume air density = 1.226 kg/m^3.

Solution: u_0 = 10 m/s, air density ρ =1.226 kg/m^3, and rotor diameter (D) = 100 m.

$A_1 = \dfrac{\pi}{4}D^2 = 7850m^2$. Power of wind = $P_W = \dfrac{1}{2}\rho A_1 u_0^3 = 4812\,kW$,, Total power density

= 613 W/m^2. Actual power produced = Efficiency $\times \dfrac{1}{2}\rho A_1 u_0^3 = 0.4 \times 4812 = 1925\,kW$

Example 5.5 Wind at one standard atmospheric pressure and 15^{0}C has a speed of 10 m/s. A 10 m diameter wind turbine is operating at 5 rpm with maximum efficiency of 40%. Calculate (i) the total power density in wind stream (ii) the maximum power density (iii) the actual power density (iv) the power output of the turbine and (v) the axial thrust on the turbine structure.

Solution: u_0 = 10 m/s, air density ρ =1.226 kg/m^3, and rotor diameter (D) = 10 m.

Total power density of wind $= \dfrac{1}{2}\rho u_0^3 = 613W/m^2$. Maximum power density of wind =

(16/27) × 613= 363 W/m^2. Actual power density = Efficiency × Total power density =0.4 × 613 = 245.2 W/m^2. Power output from turbine, P_T = Actual power density

$\times A_1 = 245.2 \times \dfrac{\pi}{4} \times 10^2 = 19248W$. Axial thrust $F_{A\max} = A_1 \rho u_0^2 /2 = 4812N$.

Example 5.6 A wind turbine operates at an atmospheric pressure of 1 atm and temperature 18^{0}C. The effective diameter of the rotor is 100 m and wind blows at a speed of 8 m/s. The operating speed of the rotor being 50 rpm. Evaluate (i) the total power density of the wind stream and (ii) torque and axial thrust for maximum efficiency.

Solution: u_0 = 8 m/s, air density ρ =1.226 kg/m^3, radius of rotor (R) = 50 m

(i) Total power density $\dfrac{1}{2}\rho u_0^3 = \dfrac{1}{2} \times 1.226 \times (8)^3 = 313.85W/m^2$

(ii) Torque $T_{\max} = \dfrac{1}{2}\rho A_1 u_0^2 R = \dfrac{1}{2} \times 1.226 \times \dfrac{\pi}{4} \times (100)^2 \times 8^2 \times 50 = 15398560\,N.m$

(iii) Axial Thrust, $F_{x(\max)} = A_1 \rho u_0^2 /2 = \dfrac{\pi}{8} \times (100)^2 \times 1.226 \times 8^2 = 3079712\,N$

Example 5.7 Design the rotor for a multi-blade wind turbine that operates in a wind speed of 36 kmph to pump water at a rate of 6m^3/h with a lift of 6m and also calculate the angular velocity of the rotor.

Solution: Given, water density = 1000 kg/m³, '*g*' (acceleration due to gravity) =9.8 m/s², water pump efficiency = 50%, efficiency of rotor to pump = 80%, power coefficient C_p= 0.3, tip speed ratio, λ = 1.0 and air density = 1.226 kg/m³

$$\text{Power required to pump water} = = \frac{6 \times 1000}{3600} \times 9.8 \times 6 = 98W$$

Power required at rotor

$$= \frac{98W}{0.5 \times 0.8} = 245W, \ P_{Total} = \frac{1}{2}\rho A_1 u_0^3 = \frac{1}{2}\rho \pi R^2 u_0^3 = \frac{1}{2} \times 1.226 \times \pi R^2 \left(\frac{36 \times 1000}{3600}\right)^3. \ \text{But}$$

$$C_p \times P_{total} = P_{\max} \to 0.3\left[\frac{1}{2} \times 1.2 \times \pi R^2 \left(\frac{36 \times 1000}{3600}\right)^3\right] = 245 \to R = 0.66m, \ \text{where } R \text{ is the}$$

radius of the turbine rotor. As $\lambda = 1$, the number of blades in a multi-blade turbine varies from 8 to 18. The angular velocity of rotor, $\omega = \dfrac{\lambda u_0}{R} = 1.0 \times 36 \times \dfrac{1000}{3600} \times \dfrac{1}{0.66} = 15.1$ rad/s = 144

rpm (1 rpm = $2\dfrac{\pi}{60}rad/s$)

Example 5.8 Find the required diameter of a wind turbine to generate 4 kW at a wind speed of 7 m/s and a rotor speed of 120 rpm. Assume power coefficient = 0.4, efficiency of mechanical transmission = 0.9 and efficiency of generator = 0.95

Solution: Data given, generated power, P = 4 kW, wind speed, u_0 = 7 m/s, power coefficient, C_p = 0.4, efficiency of mechanical transmission, η_m= 0.9, efficiency of generator, η_e= 0.95, air density = 1.226 kg/m³. The generated power is given by

$$P = \frac{1}{8}\pi\rho D^2 u_0^3 \eta_e \eta_m C_p = 4(kw) = \frac{1}{8} \times \pi \times 1.226 \times D^2 \ (7)^3 \times 0.95 \times 0.9 \times 0.4 \to D =$$

$$\sqrt{\frac{8 \times 4}{\pi \times 1.226 \times (7)^3 \times 0.95 \times 0.9 \times 0.4}} = 0.26m$$

Example 5.9 Diameter of wind turbine rotor for water pumping application.

The three main parameters that are needed for water pumping applications are the total pumped head *H* (m), the pumped volume flow rate (m³/s) and the expected mean wind speed u_0 (m/s). The actual delivered power of the rotor must be equal to the required

hydraulic power. So, $C_p\left(\dfrac{1}{2}\rho A, u_0^3\right) = \rho_w gHQ$, where C_p is the power coefficient of wind turbine, ρ is the density of air (kg/m³), A_1 is the swept area (m²), u_0 is the mean velocity of air (m/s), ρ_w =1000 kg/m^3 (the density of water) and g = 9.81 m/s^2 (acceleration due to gravity). Rearranging, the above expression, $A_1 = \dfrac{1000 \times 9.81 HQ}{0.6 C_p u_0^3}$ and rotor diameter,

$$D = \sqrt{\dfrac{4A_1}{\pi}}$$

Example 5.10 A propeller-type wind turbine has the following data:

Speed of free wind at a height of 10 m = 12 m/s, air density = 1.226 kg/m^3, a = 0.14, height of tower = 100 m, diameter of rotor = 80 m, wind velocity at the turbine reduces by 20%, generator efficiency = 85%. Find (i) Total power available in wind, (ii) Power extracted by the turbine, (iii) Electrical power generated, (iv) Axial thrust on the turbine, (v) Maximum axial thrust on the turbine

Solution: From given data, u_H = 12 m/s, H = 10 m, z = 100 m, ρ = 1.226 kg/m^3, α (power index for wind speed variation with height) = 0.14, D = 80 m, A_1= = 5026.55 m^2, u_1 = 0.8 u_0, η_{Gen}=0.85. From Eq. 5.2, u_z = 16.565 m/s = u_0 and u_1 = 0.8 × 16.565 = 13.252 m/s. (i) From Eq. $P_0 = \left(\dfrac{1}{2}\rho A_1 u_0^3\right)$, P_0 = 14 MW (ii) From the expression of interference factor, $a = (u_0 - u_1)/u_0$, the interference factor, 'a' = 0.2. From C_p= 4 a(1 - a)^2, the power coefficient C_p = 0.512. From P_T=$C_p P_0$, power extracted by the turbine P_T= 7.168 MW. (iii) Electrical Power generated = 0.85 × 7.168 = 6.09 MW. (iv) F_A = 4 a(1 - a) (A_1)ρu_0^2/2, axial thrust on the turbine, F_A= 5.4 × 10^5 N. (v) Maximum axial thrust occurs when 'a' = 0.5 and C_F = 1

Example 5.11. The following data were measured for a horizontal axis wind turbine, Speed of wind = 20 m/s at 1 atm and 27 ^{0}C, Diameter of rotor = 80 m, Speed of rotor = 40 rpm. Calculate the torque produced at the shaft for maximum output of the turbine.

Solution: Given: u_0 = 20 m/s, P = 1 atm = 1.01325 × 10^5 Pa, T= 273 + 27 = 300 K, radius of rotor R = 40 m, Speed of rotor ω = 40 rpm = (2π) × (40/60) = 4.1888 rad/s, ρ = P / ($R_{gas\ constant}$ T) = 1.01325 × 10^5 / (287 × 300) = 1.177 kg/m^3 (where gas constant = 287 J/kg K). Area of rotor A_1 = π R^2 = 5026.548 m^2, for maximum output, 'a' = 1/3, $C_{p\ max}$ = 0.593

Tip-speed ratio, λ = 40 × 4.1888/20 = 8.378, from the equation of available power in wind i.e. $P_0 = \dfrac{1}{2}(\rho A_1)u_0^3 = 23.665\,MW$, $T_M = \dfrac{P_0}{u_0}R = 23.665 \times 40/20 = 47.33N$

$$C_{T\,max} = \dfrac{C_{p\,max}}{\lambda} = \dfrac{0.593}{8.378} = 0.07078$$

Torque produced at the shaft at maximum output, $T_{(sh\ max)}$ = 47.33 ×0.07078 = 3.35 N

Example 5.12. A horizontal axis wind turbine is installed at a location having free wind velocity of 15 m/s, 80-m, the diameter of rotor having three blades attached to the hub. Find the rotational speed of the turbine for optimal energy extraction.

Solution: Given, rotor diameter = 80 m, $R = 40$ m, $u_0 = 15$ m/s, $n = 3$. The tip-speed ratio for optimum output, $\lambda_0 = \dfrac{4\pi}{n} \to 4.188$. The tip-speed ratio is given by

$$\lambda_0 = \frac{4\pi}{n} \to 4.188 = \frac{40 \times \omega}{15} \to \omega = 1.57.$$ If N is rotor speed in rpm, $\omega = 2\pi N/60, \to N = 15$ rpm

Therefore, for optimum energy extraction, rotor speed should be maintained at 15 rpm.

5.7 Types of wind turbines

The wind turbines, mainly the rotor of the wind energy conversion systems (which converts wind energy to mechanical motion) can be divided into various categories based on their geometry and the way wind passes over the airfoils or blades. Wind turbines operate by slowing down the wind and extracting a part of its energy in the process. The wind turbines have blades, sails or buckets fixed to a central shaft. The extracted energy causes the shaft to rotate. This rotating shaft is used to drive a pump, to grind grains or to generate electric power.

(i) *Lift and drag type wind turbines:* These types of turbines work based on aerodynamic lift and drag. Wind can rotate the rotor of a wind turbine either by lifting (lift force) the blades or by simply passing against the blades (drag force). Earlier type of wind turbines was mainly based on the drag force. The wind machines that use drag force provide high torque but rotate at low rounds per minute (rpm) speed. Due to this characteristics, these machines are suitable for water pumping and grinding type of applications. These machines are also relatively inefficient, their coefficient of performance lies between 0.15 and 0.30, i.e. their efficiencies are between 15 and 30%. The most recent machines are based on the lift aerodynamic force for wind energy to electrical energy conversion. The characteristics of lift force-based machine is low to medium torque, high rounds per minute (rpm) speed. The coefficient of performance of these machines range between 0.30 and 0.40 (30 to 40% efficient). These types of wind turbines are mainly used for electricity generation. Low speed devices are mainly driven by the drag-forces acting on the rotor. They generally move slower than the wind and their motion reduces rather than enhancing the power extraction. The torque at the rotor shaft is relatively high. High speed turbines on the other hand rely on lift forces to move the blades of wind turbine and the linear speed of the blades is usually several times faster than the wind speed. The torque is low as compared to the drag type. With the same swept area, the power extracted by a wind turbine relying on lift forces is generally many times greater than the power from a turbine relying on drag forces. For the generation of electricity, it is usually desirable that the driving shaft of the generator needs to operate at a considerable speed. Therefore, lift type of devices is preferable to the drag type of devices for electricity generation.

(ii) *Horizontal and Vertical Axis Wind Turbines:* Wind turbines are classified into horizontal and vertical axis type, based on the axis around which their rotor revolves. When the axis of rotation is parallel to the air stream (i.e. horizontal), the turbine is said to be a horizontal axis wind turbine and when it is perpendicular to the air stream (i.e. vertical), it is said to be a vertical axis wind turbine. Over 90% of the wind turbines in the present days is of horizontal axis type. They are mostly two or three bladed propeller type of machines. They work mainly on lift force. While the Savonious and Darreius wind turbines are vertical axis type. The Darreius turbines work on lift force while the Savonious turbines work on drag force. Horizontal axis wind machines require special infrastructure to keep generator, gearbox, etc. at the height of the hub, which in case of vertical axis machines can be kept on the ground. But horizontal axis turbines take the advantage of higher hub height, i.e. as the height from the ground increases, the wind speed increases and higher power can be generated using horizontal axis wind turbines.

(iii) *Constant speed or variable speed turbines:* Turbines can be classified on the basis of its movement at variable speed or constant speed. Wind speed keeps on varying all the time which also results in variation in the rotor speed. As a result, the frequency of the generated power also varies. It requires extra efforts to convert variable frequency power to constant, 50 Hz frequency power. It is the question of design whether to allow rotor speed to vary or not. In case of small application of water pumping and battery charging, it is advisable to allow variation in speed (as more power will be generated as compared to constant speed, since power is proportional to the cube of the wind speed). But for the large generator, the rotor is allowed to rotate at constant speed. In this case, simple generator can be used and the produced electricity can directly be fed to the grid. If variable rotor speed is allowed, then the frequency of generated power also varies, in that case, suitable power conditioning needs to be required to change the frequency to the grid frequency, so that the power can be fed to the grid.

5.7.1 Horizontal Axis Wind Turbine

Horizontal axis wind turbines are now-a-days widely used all over the world. These are mainly used for generation of conventional electrical power. Their theoretical aspects have been well researched and sufficient field experience is available with them.

A. Components

The important components of horizontal axis wind turbine are shown in Fig. 5.26

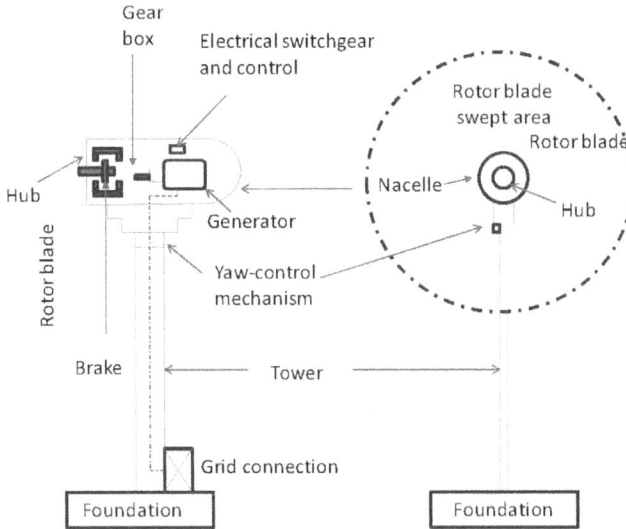

Fig. 5.26. Horizontal axis wind turbine

Turbine blades

Turbine blades have aerofoil type cross section to extract energy from wind. These blades are made of high-density materials such as wood, glass fibre, carbon fibre reinforced plastic, aluminium and epoxy composites. The requirement of the blade material is of low weight, high stiffness and high fatigue strength. The glass reinforced plastic is suitable from the material point of view and is used commonly.

The blades are twisted from tip to root to maintain pitch angle. Most of wind turbines have two-or three blades, also known as propeller type wind turbines because of similarity to the propeller of an old aeroplane, but blades of a wind turbine rotate very slowly compared to that of an aeroplane. A two bladed rotors give much smoother power output compared to three-bladed rotors. A three bladed rotor generates little more power output (more than 5%), but additional blade incorporation adds to substantial additional weight to the windmill (about 50% extra). A two-bladed rotor is also simpler to be constructed and erected on the ground.

Hub

The central solid portion of a rotor is called hub. All blades are attached to it. Pitch angle control mechanism is provided inside the hub. The hub of the rotor is attached to the low speed shaft of a wind turbine. The low speed shaft of a wind turbine connects the rotor hub to the gearbox. The shaft contains pipes for the hydraulics system to enable the aerodynamic brakes to operate

Nacelle

The term 'nacelle' is derived from the name for housing containing the engines of an aircraft. The rotor is attached to nacelle which is mounted at the top of a tower. It houses gearbox, generator, controls and brakes. The purpose of gearbox is to regulate the output rotation from the rotor with the speed of the generator. The recent design uses "direct-drive" generators that operate at lower rotational speeds and do not need gear boxes. Brakes are provided for automatic application of brakes if the wind speeds exceed the designed speed and also to stop the rotor when power generation is not desired. Protection and control functions are provided by switchgear and control block. The generated electrical power is conducted to the ground terminals through a cable.

Yaw control system

The yaw control system is provided at the base of the nacelle. The system is provided to adjust the nacelle around the vertical axis so that rotor blades are always facing the wind stream. In small wind turbine, a tail vane is used as passive yaw control. The yaw system is also used to cut off the rotor from the wind in case of very strong wind. This is done from the point of view of protecting the machine.

Generator

A generator converts mechanical motion of the rotor to electrical energy. It works on the principle of electromagnetic induction. When a conducting coil rotates in a magnetic field, voltage is generated across the coil terminals, which becomes a source of power. The generator of a grid-connected wind machine is required to produce the frequency which is same as grid frequency (50 Hz in India). Generator can be synchronous generator or asynchronous generator. A synchronous generator provides constant frequency output power but does not allow rotor speed variation. On the other hand, an induction generator can supply constant frequency output power while allowing some variations in the rotor speed. Practically, as wind speed varies and thus variation in rotor speed is expected, induction generators are normally used. Induction generation has brushless and rugged construction. It is also available at economical cost. However, both of these generators are efficient in energy conversion only for small range of blade tip speed (as the generator has to match the frequency of the grid). In order to overcome this limitation, variable speed wind turbines are used. In these machines, the generated power is first converted to DC and then back to AC power of desired frequency. In this case, any generator can be used as in the first stage (AC to DC) as there is no limitation on the frequency of generation. However, it requires extra circuitry (first AC to DC and then DC to AC conversion).

Tower

Tower is provided to support nacelle and rotor. The tower height should be sufficient so that enough wind speed can be intercepted by the rotor. For medium and large-sized wind turbines, the tower is slightly taller than the rotor diameter, while in small sized wind turbines, the tower is much larger than the rotor diameter as the air is erratic at lower heights. There should not be any obstruction in the way of wind stream in its approach to the rotor. Tower can be made of materials such as steel or concrete. The construction of tower may be tubular or lattice types.

B. Types of rotors

The rotors of horizontal axis wind turbine can be of (i) single blade rotor, (ii) two blades rotors, (iii) three blades rotors, (iv) sail wing rotor (v) chalk multi-blades rotor, (vi) multi-blades rotor and (vii) dutch-type rotor, depending on the number of blades, wind speed and nature of applications as shown in fig. 5.27.

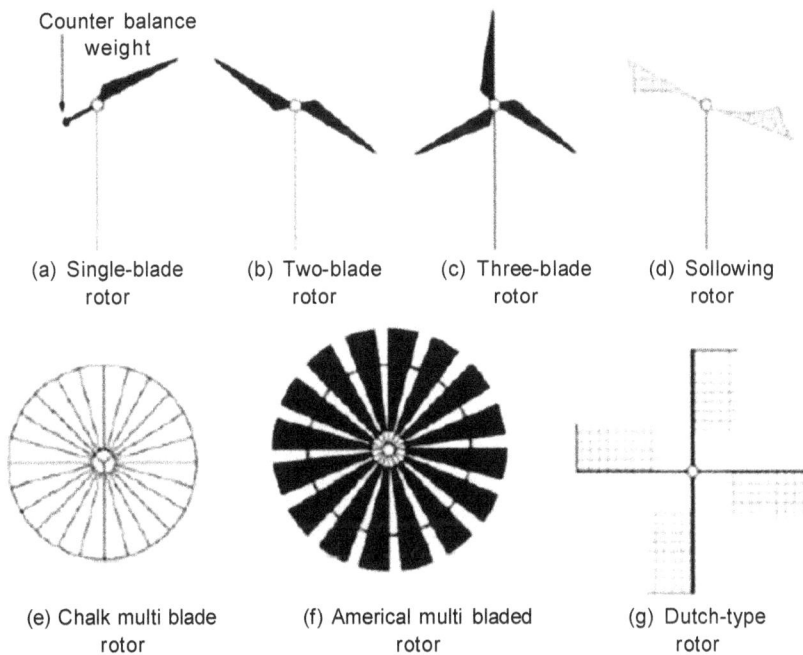

Fig. 5.27. Various types of rotors for horizontal axis wind turbine

The single, two-bladed, three-bladed, sail wing and chalk multi-blade rotors are relatively high-speed machines and these machines are, therefore, suitable for applications such as electric power generation. However, horizontal axis wind turbines are commonly fabricated with two and three-bladed rotors. A single-bladed rotor with balancing counterweight has simple construction and less cost, but it makes more noise during operation. It is used where small power is required. The multi-bladed and Dutch-type rotors are used where low speeds are required. Hence, these rotors are suitable for applications such as piston pumps where high starting torque is needed. As these rotors have high solidity, these can operate even when slow winds are present.

C. Upwind and downwind turbines

Based on the configuration of the wind rotor with respect to the wind flowing direction, the horizontal-axis wind turbines can further be classified as upwind and downwind turbines (fig. 5.28). The majority of horizontal-axis wind turbines being used today are upwind turbines, in which the wind rotors face the wind. The main advantage of upwind designs is to avoid the distortion of the flow stream as the wind passes though the wind rotor and nacelle. For a downwind turbine, wind blows first through the nacelle and tower and then the rotor blades. This configuration enables the rotor blades to be made more flexible without considering tower strike. However, because of the influence of the distorted unstable wakes behind the tower and nacelle, the wind power output generated from a downwind turbine fluctuates greatly. In addition, the unstable flow stream may result in more aerodynamic losses and introduce more fatigue loads on the turbine. Downwind turbines suffer from wind shadow effects of the tower on the blades as they pass through the tower's wake in region of separated flow. Furthermore, the blades in a downwind turbine may produce higher impulsive noise. However, an upwind machine produces higher power because of the elimination of tower shadow on the blades. This also causes lower noise, low blade fatigue and smoother power output.

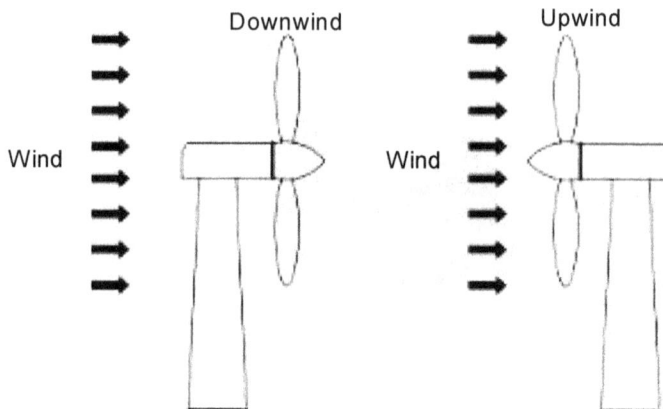

Fig. 5.28. Upwind and downwind aero-turbine

D. Teetering of rotors

Teetering (shaking) of rotor is done by providing a teeter hinge (a pivot within the hub) that allows a see-saw motion to take place out of the plane of rotation (i.e. vertical plane). Due to rise of wind speed with the increasing height, the axial force on the blade when it attains the upper position is quite high compared to that when it is at a lower position. For one and two-bladed rotors, this causes cyclic (sinusoidal) load on the rigid hub resulting into fatigue. Hence, there is the requirement of teetering action for the hub. The rotor leans backward to accommodate the extra force as shown in Fig. 5.29.

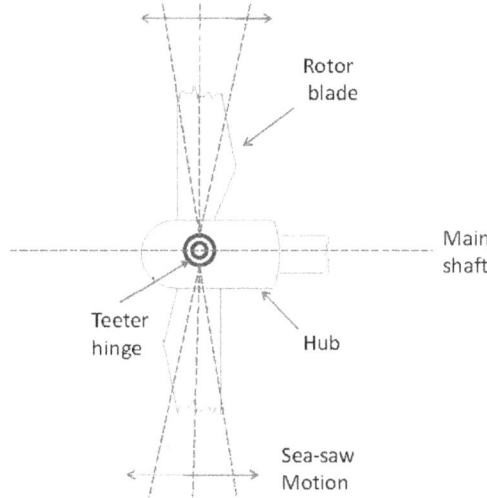

Fig. 5.29. Teetering of rotor

This also reduces blade loads near the root by approximately 49 %. The use of a third blade has approximately the same effect as a teeter hinge on the hub moments since the polar symmetry of the rotor averages out the applied sinusoidal loads. Therefore, teetering is not required when the number of blades is three or more.

E. Yaw-control system

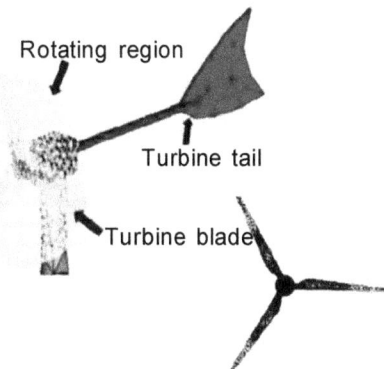

Fig. 5.30. Passive yawing or wind turbine rotor by turbine tail

Wind turbine control systems continue to play important roles for ensuring wind turbine reliable and safe operation and to optimize wind energy capture. The main control systems in a modern wind turbine include yaw control (passive and active), pitch control, stall control (passive and active) and others. Under high wind speed conditions, the power output from a wind turbine may exceed its rated value. Thus, power control is required to regulate the power output within allowable fluctuations for avoiding turbine damage and stabilizing the power output. The horizontal axis turbines have yaw that turns the rotor according to the wind direction. The yaw-control system continuously orients the rotor in the direction of wind. For localities with a prevailing wind in one direction only, the rotor can be in a fixed orientation. Such a turbine is said to be yaw fixed. Most wind turbines however are yaw active. Small wind turbines can use wind vanes for passive yawing (fig. 5.30). The wind vane moves the rotor of an upwind turbine always to a position perpendicular to the wind. In large turbines however, an active yaw control with power steering and wind direction sensor is used to maintain the orientation.

F. Pitch-control system

The pitch control system is a vital part of the modern wind turbine. This is because the pitch control system not only continually regulates the wind turbine's blade pitch angle to enhance the efficiency of wind energy conversion and power generation stability, but also serves as the security system in case of high wind speeds or emergency situations. The pitch of a blade is controlled by rotating it from its root, where it is connected to the hub as shown in fig. 5.31. The pitch control mechanism is provided through the hub using a hydraulic jack (fig. 5.32) in the nacelle.

In modern wind turbine machines, pitch control is incorporated by controlling only the outer 20 % length of the blade (i.e. tip), keeping the remaining part of the blade as fixed.

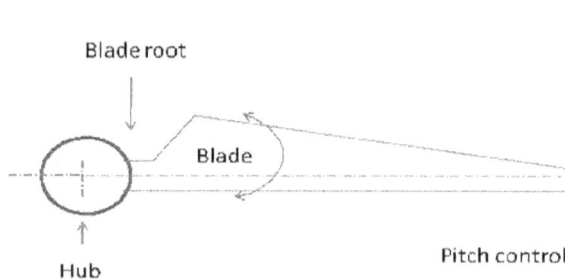

Fig. 5.31. Pitch control

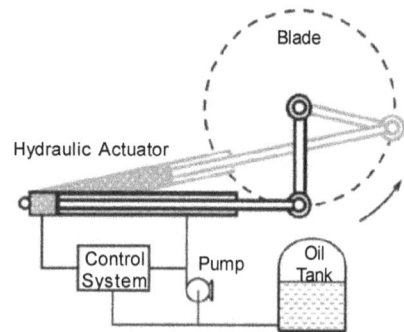

Fig. 5.32. Hydraulic pitch control system

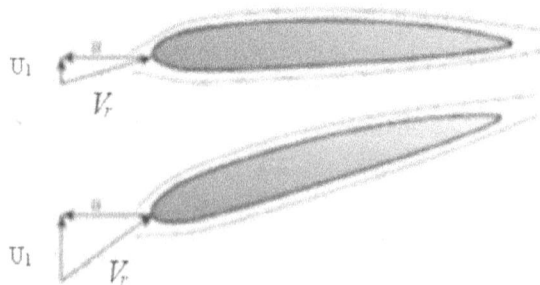

Fig. 5.33. Rotor Blade Positions for Different Wind Speeds for a Pitch-controlled System

G. Stall-control system

Besides pitch control, stall control is another approach for controlling and protecting wind turbines. The concept of stall control is that the power is regulated through stalling the blades after rated speed is achieved.

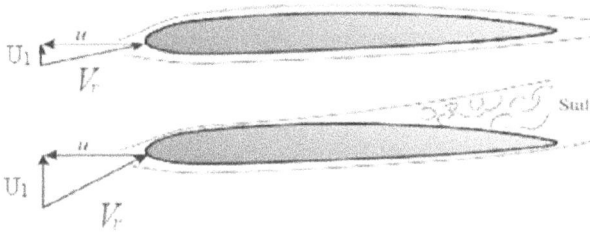

Stall control is achieved by increasing the angle of attack at higher wind speeds. The rotor blades do not pitch, i.e. the pitch angle remains constant; stall control can be realized by construction measures without advanced technical requirements.

5.7.2 Vertical Axis Wind Turbine

Wind wheels and windmills with vertical axes are the oldest systems to exploit the wind. For more than 1000 years back, drag devices with vertical axes have been constructed. Today, there are some modern wind generator concepts that also have vertical axes. Wind energy conversion system with vertical axes have some advantages. Their structure and their assembly are relatively simple. The electric generator and the gear as well as all electronic components can be placed on the ground. This simplifies the maintenance compared to rotors with horizontal axes. Rotors with vertical axes need not be oriented into the wind; therefore, they are perfectly suited for regions with very fast changing of wind direction. However, these advantages have not resulted in a breakthrough for wind generators with vertical axes. Today, almost all wind power plants use rotors with horizontal axes; systems with vertical axes are only used for very special applications. The poorer efficiency and higher material demand of systems with vertical axes have been the deciding factors for the market dominance achieved by horizontal axis turbines to date

A. Components

The construction details of a vertical axis wind turbine (Darrieus type rotor) are shown in fig. 5.34.

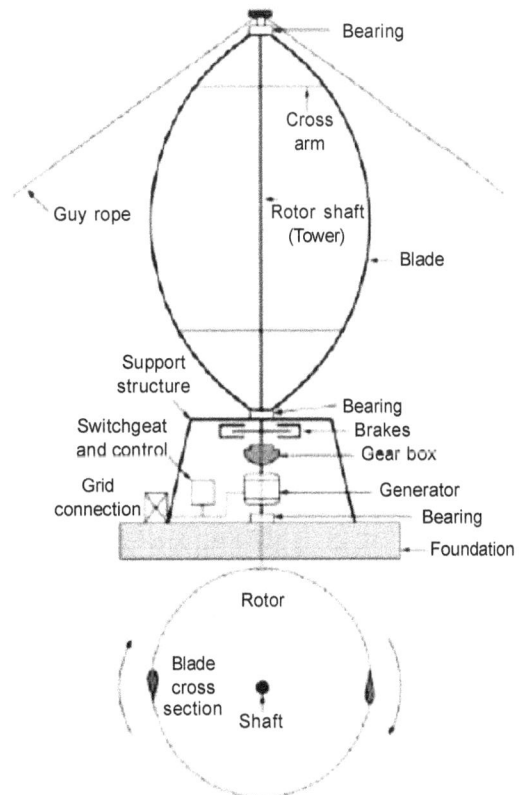

Fig. 5.34. Vertical axis (Darrieus) turbine

Tower (rotor shaft)

The tower consists of a hollow vertical shaft which can rotate about its vertical axis between its bearing at top and bottom. It is provided with a support structure at the bottom and at upper end, it is supported by guy ropes. Guy ropes are anchored to the ground at the end. The height of the tower for a large turbine is about 100 m.

Blades

This type of wind turbine has two or three blades which are thin and curved shaped similar to an "eggbeater". The blades are curved in such a way that minimum bending stress are produced on rotation due to the centrifugal forces. The blades are designed in such a way that they offer aerofoil type cross section to wind stream. The pitch of the blades cannot be changed. The diameter of the rotor is slightly less than the tower height.

Support structure

The support structure is provided at the ground to support the weight of the rotor. It houses gearbox, generator, brakes, electrical switch gear and controls.

B. Rotor of vertical axis wind turbine

The vertical axis wind turbines can have various types of rotors which include Cup type rotor, Savonious rotor, Darrieus rotor, Musgrove rotor and Evans rotor as shown in fig. 5.35. The simplest of all rotors is cup type rotor, which consists of three or four cup type structures attached symmetrically to a vertical shaft. It works on the principle that drag force on a large concave surface is more than that on a convex surface of the cup when wind stream strikes these surfaces, and so the rotor consisting of cups starts rotating. However, such rotors cannot extract enough energy from winds and therefore are not used for power generation. These devices are mainly used for wind speed measuring instruments such as the cup anemometer. The Savonius rotor (S-rotor), invented by S.J. Savonious of Finland in the year 1920, is formed with two half-cylinders attached to a vertical axis, but facing in opposite direction. S-rotor can produce high starting torque at low wind speed. it has also low efficiency of conversion of wind energy. As it can extract power even from low-speed winds, it can operate and deliver power throughout the day. Hence, such rotor is used for applications where low power is required, such as in wind pumping. Darrieus rotor, invented by a French engineer G.M. Darrieus in 1925, consists of two or three curved blades attached to a rotor shaft similar to an eggbeater. It has good power coefficient and is used for large power generation. It also has a large tip speed ratio, thereby developing large bending stress in the blades due to centrifugal forces formed. The main drawback of this rotor is that it is not-self-starting due to lower starting torque developed by it. It has generally to be run by using its electrical generator as motor.

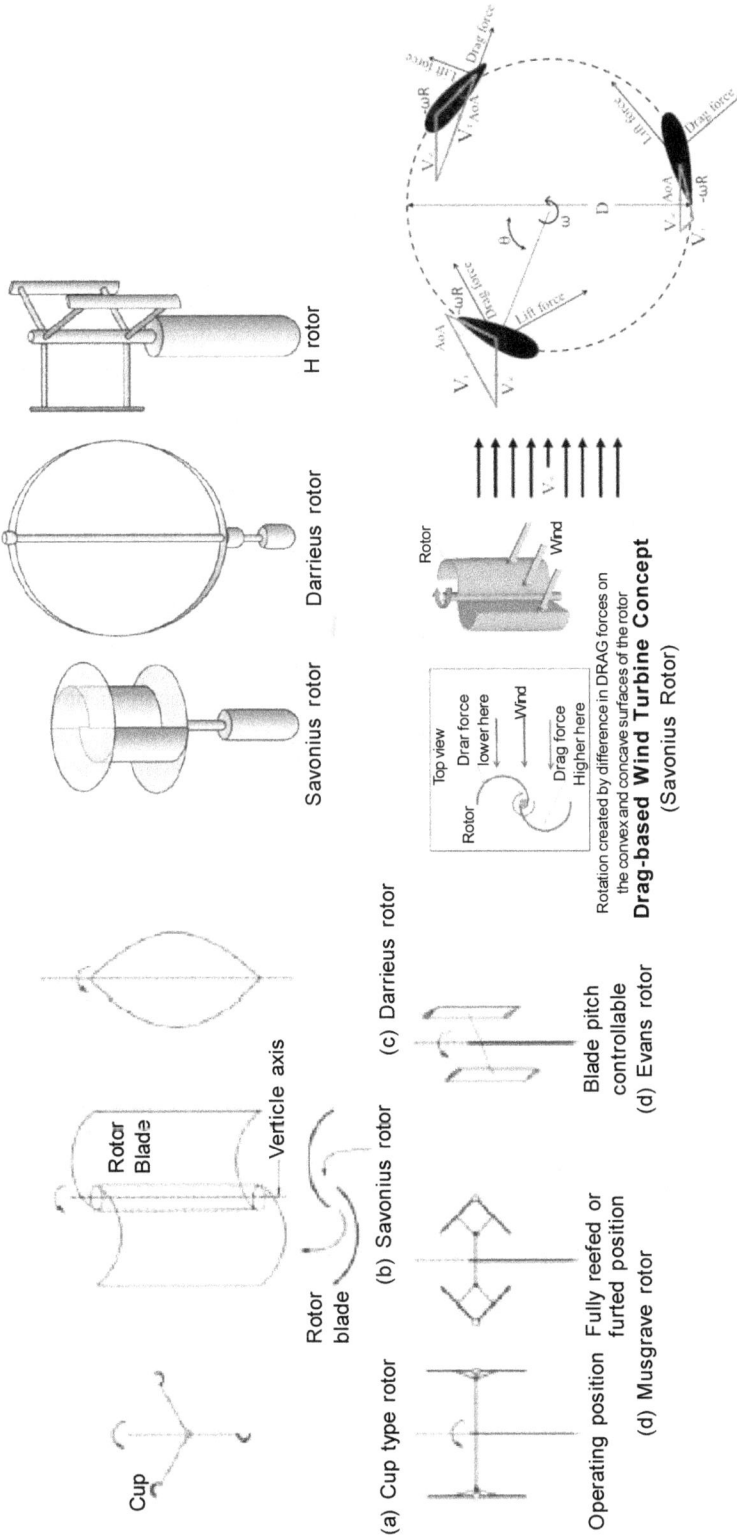

Fig 5.35. Various types of rotors of vertical axis wind turbine

The blades also have fixed pitch which cannot be changed. This results in unmanageable output at high wind speeds. Darrieus type rotors are lift devices, characterized by curved blades with airfoils. They have relatively low solidity and low starting torque but high tip to wind speed and therefore comparatively high outputs per given rotor weight and cost. Musgrove suggested the use of an H-shaped rotor where blades with a fixed pitch are attached vertically to a horizontal cross arm. Power control is achieved by controlled folding of the blades. Inclining the blades to the vertical, provides an effective means of altering the angle of attack of the blades and hence controlling the power output.

The Evans rotor has blades which are hinged on a vertical rotor shaft and the blade pitch is varied cyclically during rotation to regulate the power output. It is a self-starting rotor. The types and characteristics of various types of rotors are described in table 5.7.

Table 5.7. Characteristics of various types of wind turbine rotors

SN	Rotor type	Tip speed ratio	RPM	Torque	Typical load
1	Propeller (1-3blades) (Lift)	6-20	High	Low	Electric power generation
2	Sail wing (Lift)	4	Moderate	Moderate	Electric power generation or pumps
3	Chalk multi-blade (Lift)	3-4	Moderate	Moderate	Electric power generation or pumps
4	American multi-blade (Drag)	1	Low	High	Pumps
5	Dutch type (Drag)	2-3	Low	High	Pumps
6	Savonious (Drag)	Less than 1	Low	High	Pumps
7	Darrieus (lift)	5-6	High	Low	Electric power generation
8	Musgrove and Evan (lift)	3-4	Moderate	Moderate	Electric power generation or pump

5.8 COMPARISON OF HORIZONTAL AND VERTICAL AXIS WIND TURBINES

The comparison of horizontal and vertical axis wind turbines is given in Table 5.8.

Savonious Rotor

Advantages

i. It has vertical rotating rotor shaft which eliminates the need of any expensive transmission system from the rotor to generator.

ii. It produces power effectively in slow wind speeds (as low as 8 km/h)

iii. The generator can be located at ground level which helps in easy maintenance.

Table 5.8: Comparison of horizontal and vertical axis wind turbines

Horizontal axis wind turbine	Vertical axis wind turbine
Axis of rotation is parallel to the airstream	Axis of rotation is perpendicular to the airstream
These are commonly used and almost fully developed	These are under development stage
The rotor has to face wind stream. It is provided with yaw mechanism to keep it facing wind stream	The rotor can accept wind stream from any direction. There is no need of yaw mechanism
Nacelle carrying gear train, controls and generator has to be mounted on top of the tower	Gear train, controls and generator can be located at ground level
Tower has to be strong and designed properly	Tower is simple in construction and installation
Inspection and maintenance of windmill is difficult	Inspection and maintenance of windmill is easy
Costly	Less costly
Less noisy	More noisy
Extract more power from wind	Extract less power from wind
Technology is fully developed	Technology is under development
Less fatigue to parts due to wind action	More fatigue to parts due to wind action
Designed to use lift force	Designed to use drag force
More efficient	Less efficient
Smooth output	Fluctuating output
Produces lower starting torque	Produces high starting torque
Operates properly in moderate wind speeds	Can operate even in low wind speeds
Pitch of blade can be controlled	Pitch of blade cannot be controlled

iv. The tower is simple in construction and installation due to lower loads.

v. The cost of construction and installation is low.

vi. Yaw and pitch control mechanisms are not required.

Disadvantages

i. It utilises drag force which results in its lower efficiency.

ii. Power output is very low.

iii. The power output is low as the pitch control of the blade is impossible.

iv. It is unsuitable where tall installation is necessary due to wind conditions.

v. It has low power to weight ratio.

Darrieus Rotor

Advantages

i. The rotor blade can accept the wind from any direction.

ii. It does not require any yaw control mechanism.

iii. The blades have constant pitch. There is no need of pitch control mechanism.

iv. The gear train and generator are mounted at the ground level.

v. The tower is simple in construction.

vi. The cost of construction and installation is low.

vii. It has higher power coefficient and tip speed ratio compared to S-rotor.

Disadvantages

i. It is not self-starting machine

ii. It works on lift force and its efficiency to convert wind energy is low compared to horizontal axis rotors.

iii. It can have limited height. It cannot utilize high wind speeds available at higher level.

iv. It cannot be yawed out of the wind; special high torque braking system is required during the occurrence of high wind speeds.

5.9 WIND ELECTRIC CONVERSION SYSTEMS

A wind-electric conversion system (WECS) converts wind energy into electrical energy. Medium and large scale WECS are designed to operate in parallel with a public or local A.C. grid. This is known as a grid-connected system. A small system, isolated from the grid, feeding only to a local load is known as autonomous, remote, decentralized, stand-alone or isolated power system. The block diagram of wind-electric generating system is shown in fig. 5.36.

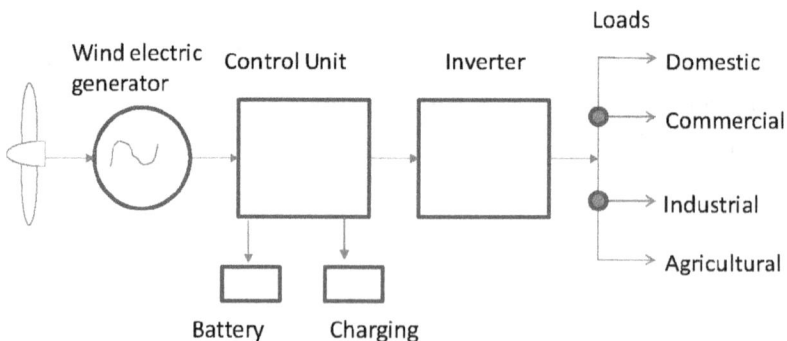

Fig. 5.36. Standalone wind electric generator

The generated A.C. power is first converted to D.C. power and to D.C. power of constant frequency through converter and then back to A.C. power of desired frequency. Hence, it requires extra circuitry (first A.C. to D.C. and then D.C. to A.C. conversion). The choice of electrical system for a wind turbine is guided by three factors,

(i) Type of electrical output: D.C., Variable frequency A.C. and constant frequency A.C.

(ii) Wind turbine rotor speed: Constant speed with pitch control, Nearly constant speed with simpler pitch control, Variable speed with fixed pitch blades

(iii) Utilization of electrical energy output: Stand alone or grid-connected load, Energy storage requirements for unfavourable wind periods.

Choices of electrical output: The output from the wind-electric conversion systems depends on the use of types of generators and their suitability in wind power generation.

(i) D.C. generator: Conventional D.C. generators are not favoured due to their high cost, weight and maintenance problems of the commutator. However, permanent-magnet (brushless and commutator-less) D.C. machines are considered in small-rating (below hundred kW) isolated systems.

(ii) A.C. generator:

(a) Constant speed constant frequency system: Constant frequency A.C. form is obtained generally from synchronous generators. A synchronous generator delivers electric power at constant frequency of voltage and current when the rotor is driven at constant speed (called the synchronous speed). If synchronous generator is connected to the grid, it has a tendency to remain in synchronism with the grid frequency. However, the wind turbine should drive the rotor at one constant speed. This requires variable pitch blade control and gears. The cost of mechanical system increases the wind turbine unit as against the lower cost of electrical generator and control. Initially, large scale wind electric generation was of constant speed, constant frequency A.C. output, obtained from synchronous generators. Constant Frequency systems (CF) are essential for modern wind farms as the output is either grid connected or delivered to consumers requiring constant frequency supply. Fig. 5.37 gives a block-diagram of simple CF system using a (constant speed) synchronous generator. Such a system is called constant Speed Constant Frequency System (CSCF). Earlier large wind-turbine generator units employed CSCF systems. Synchronization of a wind-driven generator with the power grid also poses the problems especially during gusty winds. The main advantage is that it generates both active as well as reactive powers.

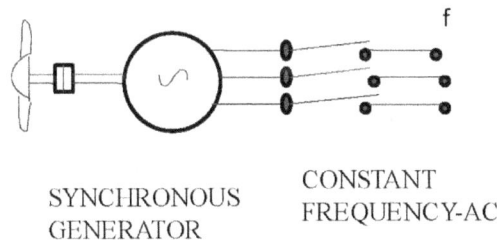

SYNCHRONOUS GENERATOR CONSTANT FREQUENCY-AC

Fig. 5.37. Constant speed constant frequency wind turbine system

(b) Variable speed constant frequency system: Recent advances in thyristor convertor technology has paved a way for variable speed constant frequency systems (VSCF). Fig. 5.38 shows a block diagram for the wind turbine with variable speed. The rotor speed is allowed to vary optimally with the wind speed. The generator produces variable frequency output ($F \propto N$). The Rectifier-Inverter combination delivers constant frequency electrical output which can be delivered to the load or the grid. VSCF requires simpler wind turbine and its controls. There is no need of pitch control. The wind energy is optimally utilized. The turbine always works with maximum efficiency (constant tip speed/ wind speed ratio). Due to variable speed, the stresses on wind turbine rotor, blades etc. are reduced greatly. These advantages of VSCF systems are with the additional cost of electrical (Rectifier-Inverter, controls) units

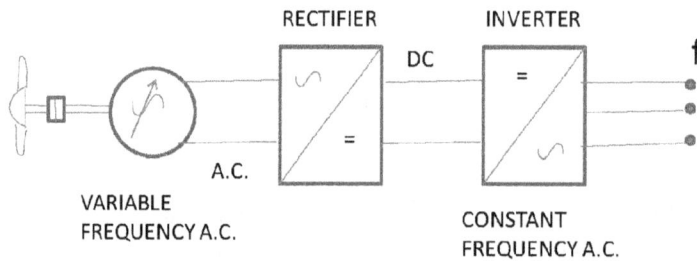

Fig. 5.38. Variable speed constant frequency wind turbine system

(c) Nearly Constant Speed and constant frequency of Grid. The small and medium generator units rated 100 kW, 200 kW, 300 kW etc. are more popular in India and other developing countries Such machines can have Induction Generators (Asynchronous generator), fig. 5.39. The Induction Generator needs grid connection and is not suitable for isolated load. The induction generator rotor speed can vary by 10% around the reference synchronous speed without problem. The primary advantages of an induction machine are the rugged, brushless construction, no need of separate D.C. field power and tolerance of slight variation of shaft speed (±10%) as these variations are absorbed in the slip. Compared to D.C. and synchronous machines, they have low capital cost, low maintenance and better transient performance. For these reasons, induction generators are extensively used in wind and micro-hydroelectric plants. The machine is available from very low to several megawatt ratings. The induction machine requires A.C. excitation current, which is mainly reactive. In case of a grid-connected system, the excitation current is drawn from the grid and therefore, the network must be capable of supplying this reactive power. The voltage and frequency are determined by the grid. In a standalone system, the induction generator is self-excited by shunt capacitors.

Hence, generator can be a synchronous generator or asynchronous generator. A synchronous generator provides constant frequency output power but does not allow rotor speed variation. On the other hand, an induction generator can supply constant frequency output power while allowing some variations in the rotor speed. Practically, since wind speed varies and thus variation in rotor speed is expected and therefore, induction generators are normally used.

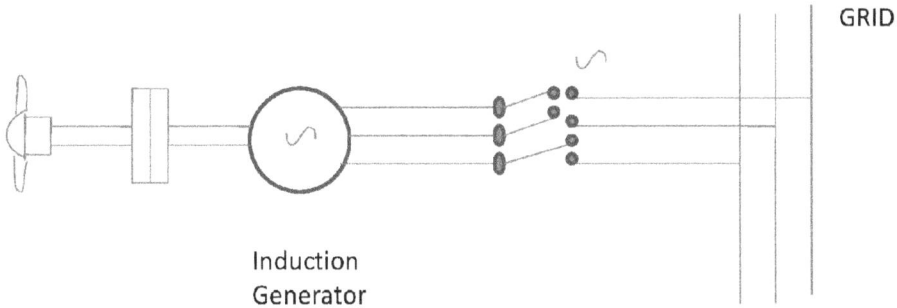

Induction
Generator

Fig. 5.39. Near constant speed grid frequency wind turbine

Comparison among the different generators used in wind turbine generating (WTG) system is given in Table 5.9.

Table 5.9: Comparison of various generators used in wind turbine generating system

Synchronous generator with WTG unit	Constant speed	Wind turbine needs fine pitch control
	Synchronised with the grid frequency	
Induction generator	Near constant speed with 10% variations	Wind turbine with simpler pitch control
	No need of synchronising with grid frequency	
Variable frequency generator (0 to 50 Hz)	Needs grid connection	Wind turbine simplest
	Variable speed	
	Needs electronic frequency convertor in output stage for connection to load/grid	

5.10 OTHER OPTIONS

Several other electrical options are in use to obtain constant frequency output from variable speed drives.

(i) Wind turbine generation unit with battery storage facility

(ii) Grid connection

(iii) Solar-wind hybrid system

5.10.1 Wind Turbine Generation Unit with Battery Storage Facility

Fig. 5.40 shows the schematic block diagram for storage of energy from the WTG to the batteries for a variable speed constant frequency system. The rectifier (2) converts A.C. to D.C. The storage battery (4) stores electrical energy in D.C. form. The inverter (3) converts D.C. to A.C. The power transformer (5) steps up the voltage from inverter to the required distribution voltage of bus bar (6). The bus bar receives energy from the WTG unit (1) via the electric circuit (2,3,5) during normal operation of the wind turbine generator. The excess electrical energy is stored in the battery (4). During unfavourable winds, the stored energy is released to the load.

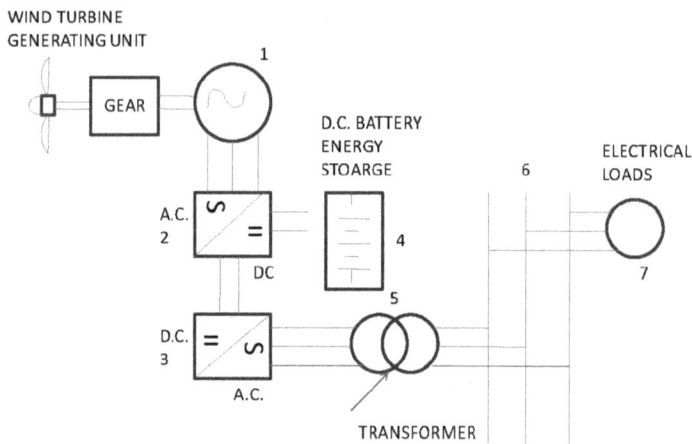

Fig. 5.40. Wind turbine generation unit with battery storage facility

5.10.2 Grid Connection

Fig. 5.41 represents a simplified schematic diagram for wind turbine generation (WTG) unit to the grid connection. The generator voltage is stepped up to the distribution voltage by means of the transformer. The transformer secondary terminals are connected to the three phase bus bars. The connections are given to the local load and to the distribution systems. The difference between the power generated by the WTG and the load is fed to the grid. During lower power generation by WTG, the load receives power from the grid supply.

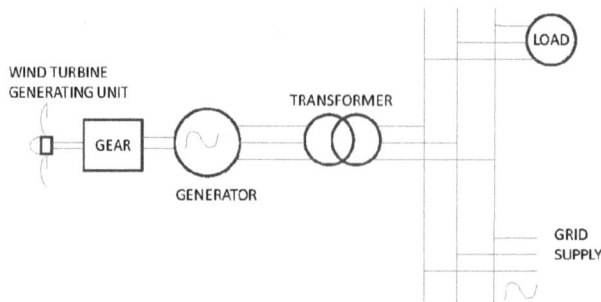

Fig. 5.41. Simplified wind turbine generating unit with grid connection

5.10.3 Solar-wind Hybrid System

The solar and wind energy systems are located in vast flat open terrains away from forests and tall buildings. Some locations may have both sun and wind favourable for several months of the year. Fig. 5.42 shows a schematic diagram of a solar-wind hybrid energy system. During favourable wind periods, the wind turbine generates energy. The excess energy is stored by the battery system. During the day time and favourable sun, the solar photovoltaic system produces electrical energy. The convertor provides D.C. to A.C. inversion, conditioning and control. The entire system may be connected to the grid and/or isolated (standalone) load.

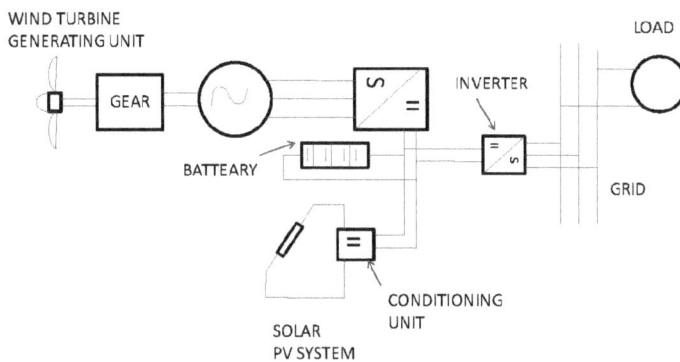

Fig. 5.42. Hybrid wind-solar energy system

5.11 POWER EXTRACTION FROM VERTICAL AXIS WIND TURBINE (DRAG TYPE)

The vertical axis wind turbine of savonius type works on the drag force. The Savonius turbine uses drag to push the curved blades to generate a torque that makes the rotor to turn and rotate. Aerodynamically, it is the simplest wind turbine to design and build which reduces its cost compared to the aerofoil blade designs of the other vertical and horizontal axis wind turbines. The turbine rotates because of the difference of the drag force acting on the concave and convex parts of its blades. Figure 5.43 illustrates the working principle of a savonius rotor.

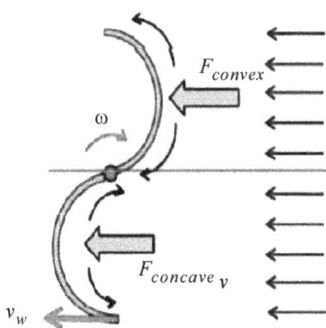

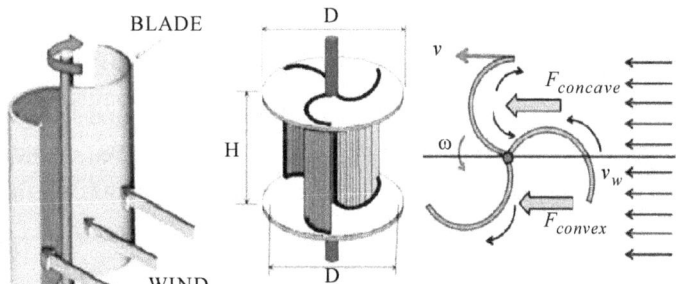

Fig. 5.43. Working principle of a savonius rotor   **Fig. 5.44.** Three-bladed savonius rotor

The air is trapped in the concave part and pushes the turbine. The flow that hits the convex part does produce a drag that is lower than the one on the concave part. It is the differential of the drag force that causes this turbine to rotate. This lowers the efficiency of the turbine as some of the wind's power is used in pushing the convex part and is hence wasted fig. 5.45.

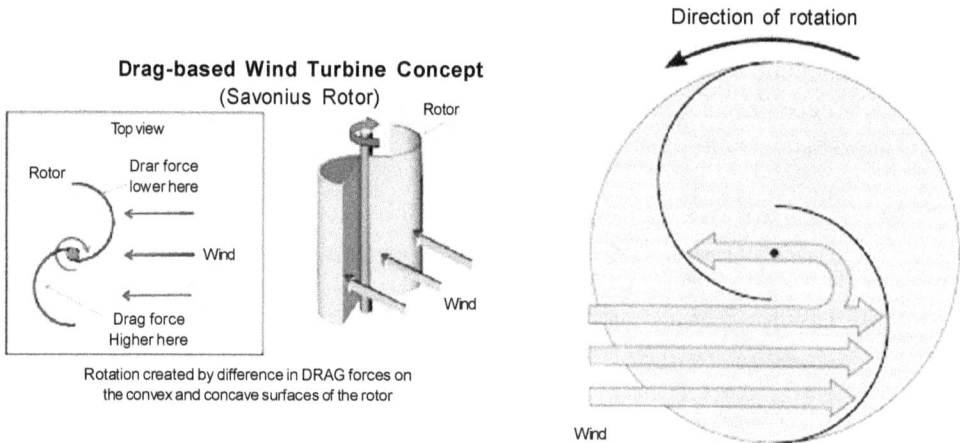

Fig. 5.45. Wastage of wind energy

More blades can be added to the S shape design, and the same principle causes it to spin as shown in Figure 5.44. A Savonius rotor requires 30 times more surface for the same power as a conventional rotor blade wind-turbine. Therefore, it is only useful and economical for small power requirements. This makes Savonius ideal for small applications with low wind speeds. Savonius rotors are hence desirable for their reliability as they are able to work at several magnitudes of wind speed. Experimental studies have confirmed that savonius wind turbines do perform well at low wind speeds (cut in speed at around 2.5 m/s). According to the same study, two blades perform better than three blades as more drag is wasted in the three blades versions. The power coefficient of the two blade design is therefore higher than that of the three blade design.

5.12 POWER EXTRACTION FROM WIND TURBINE (DRAG TYPE)

If an object is set up perpendicularly to the wind, the wind exerts a force F_D on the object. The wind speed v_w, the effective object area A and the drag coefficient C_D, which depends on the object shape is used for estimation of the drag force:

$$F_D = C_D \cdot \frac{1}{2} \cdot \rho . A . v_w^2 \qquad \qquad \text{... (5.25)}$$

Figure 6.46 shows drag coefficients for various shapes. With $P_D = F_D \times v_w$, the power to counteract the force becomes:

$$P_D = C_D \cdot \frac{1}{2} \cdot \rho . A . v_w^3 \qquad \qquad \text{... (5.26)}$$

If an object moves with speed v by the influence of the wind in the same direction as the wind, the drag force is:

$$F_D = C_D \cdot \frac{1}{2} \cdot \rho . A . (v_w - v)^2 \qquad \qquad \text{... (5.27)}$$

and the power used is:

$$P_T = C_D \cdot \frac{1}{2} \cdot \rho . A . (v_w - v)^2 . v \qquad \qquad \text{... (5.28)}$$

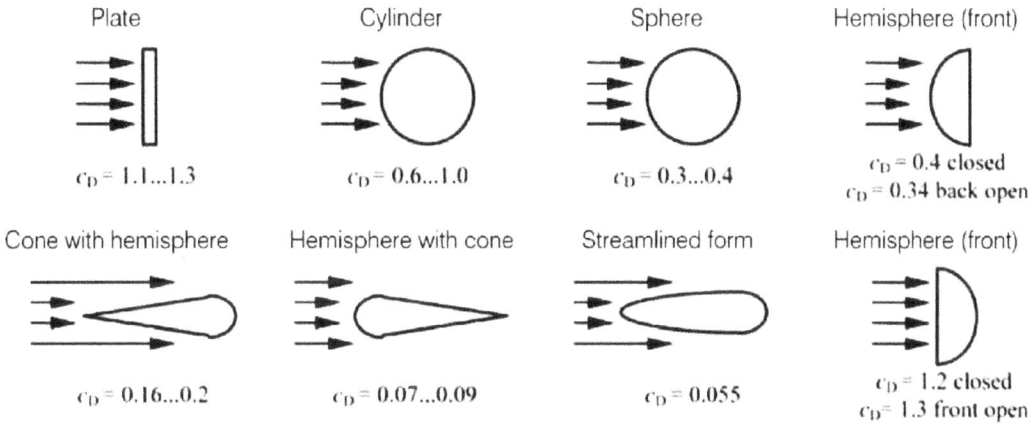

Plate	Cylinder	Sphere	Hemisphere (front)
$c_D = 1.1...1.3$	$c_D = 0.6...1.0$	$c_D = 0.3...0.4$	$c_D = 0.4$ closed $c_D = 0.34$ back open

Cone with hemisphere	Hemisphere with cone	Streamlined form	Hemisphere (front)
$c_D = 0.16...0.2$	$c_D = 0.07...0.09$	$c_D = 0.055$	$c_D = 1.2$ closed $c_D = 1.3$ front open

Fig. 5.46. Drag coefficients for various shapes

The following example calculates approximately the used power of a savonius wind rotor. It consists of two open hemispherical cups that rotate around a common axis. The wind impacts the front of the first cup and the back of the second cup (figure 5.47). The resulting force F consists of a driving and a decelerating component (Gasch and Twele, 2002):

$$F = C_{D1} \cdot \frac{1}{2} \cdot \rho . A . (v_w - v)^2 - C_{D2} \cdot \frac{1}{2} \cdot \rho A . (v_w = v)^2 \qquad \text{... (5.29)}$$

The used power is:

$$P_T = \frac{1}{2} \cdot \rho . A . \{ C_{D1} . (v_w - v)^2 - C_{D2} . (v_w + v)^2 \} . v \qquad \text{... (5.30)}$$

The ratio of the circumferential speed to the wind speed is called the tip speed ratio λ :

$$\lambda = \frac{v}{v_w} \qquad \qquad \text{... (5.31)}$$

The tip speed ratio of drag device is always smaller than one. Using the tip speed ratio, the power is:

$$P_T = \frac{1}{2}.\rho.A.v_w^3 (\lambda \left(C_{D1}(1-\lambda)^2 - C_{D2}(1+\lambda)^2 \right) \qquad \text{... (5.32)}$$

Hence, the power coefficient of the savonius wind rotor becomes:

$$C_P = \frac{P_T}{P_0} = \frac{P_T}{\frac{1}{2}.\rho.A.v_w^3} = \lambda.(C_{D1}.(1-\lambda)^2 - C_{D2}.(1+\lambda)^2) \qquad \text{... (5.33)}$$

The maximum value of the power coefficient of the cup anemometer is about 0.073. This is much below the ideal Betz power coefficient of 0.593. From the findings of the experimental studies by the researchers, the savonius rotor reaches its maximum power coefficient at a tip speed ratio of about 0.16, when the wind speed v_w is about six times higher than the circumferential speed v.

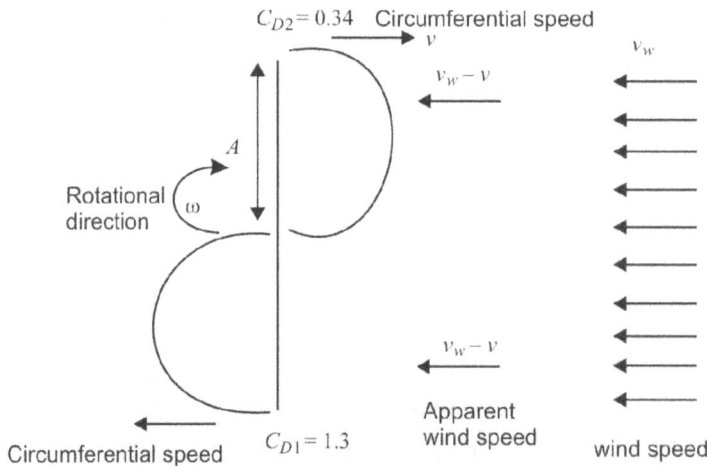

Fig. 5.47. Model of savonius wind rotor for power calculation

The optimal power coefficient $C_{p,opt,D}$ of a drag device can be calculated using

$$C_P = \frac{P_T}{P_0} = \frac{\frac{1}{2}.\rho.A.C_D.(v_w - v)^2.v}{\frac{1}{2}.\rho.A.v_w^3} = C_D.\left(1 - \frac{v}{v_w}\right)^2.\frac{v}{v_w} \qquad \text{... (5.34)}$$

Using $v/v_w = 1/3$ as well as the maximum drag coefficient of $C_{D,max} = 1.3$, this gives

$$C_{P,opt,D} = 0.193 \qquad \text{... (5.35)}$$

This value is also much below the ideal value of 0.593. Therefore, modern wind turbines are lift devices, rather than drag devices, as these achieve much better power coefficients.

5.13 SOLAR PHOTOVOLTAIC-MICRO WIND BASED HYBRID POWER SYSTEM FOR CROP IRRIGATION

The decentralized power generation using renewable energy sources is one of the viable options for urban, rural and remote locations. Solar and wind energy systems have shown increasing growth for power generation in recent years as these are freely available and environmental friendly sources for electrical power generation (Sinha and Chandel 2015). Wind energy conversion systems and solar photovoltaic systems are clean energy systems, do not emit greenhouse gas emissions like fossil fuel based power plants. The hybrid system using both solar and wind resources are more advantageous than either a solar or wind based system as it improves system efficiency, power reliability and energy storage requirements. In order to utilize solar and wind resources efficiently and economically, optimum sizing of the hybrid system with lowest cost of energy has to be carried out. The large wind power generation is mainly focused in coastal or high windy regions, but presently roof mounted micro wind turbines are being promoted and popularized in low/medium windy locations as single or hybrid system with other sources mainly the solar photovoltaic system. The technical and economic feasibility of a renewable energy based hybrid system has to be established based on wind and solar resource potential for the location of interest. Currently, emphasis is being given for a widespread adoption of hybrid energy systems in remote locations as well as in the roof tops of the residential building in developing countries like India (Bhattacharjee and Acharya 2015). Efforts need to be also focussed on the use of the same hybrid power system in agricultural sector for creating irrigation facilities at individual level in small farms.

In India, majority of the cultivable land is being irrigated from the underground water source. For lifting of water, the devices i.e. water pumps are mostly operated with the electric power and diesel/gasoline fuels which are not only polluting the environment but also becoming costly day by day due to the rising price of the petroleum fuels and electric power tariff and beyond the capacity of the majority of marginal and small farmers of the country for use in the irrigation purposes. The dependence of energy both through fossil fuels and grid connected electricity, may therefore, not be a viable solution on long term basis for increasing mechanized water lifting devices to sustain assured irrigation in crop cultivation. To combat this challenge of over-reliance on the conventional grid and motorized pumps for irrigation, solar PV micro wind hybrid power system may be a promising alternative for power reliability in agriculture in a decentralised manner. This is because the country has an equatorial climate, characterized with abundant sunshine and moderate wind speeds.

The erratic and non-uniform availability of solar and wind energy resources are the main drawbacks in restricting the use of any particular renewable energy based power system as a standalone unit, which is effective in operation either during some time of the day or a particular season, leading to increase in the unnecessary operational and lifecycle costs, as well as the oversizing of the components. These limitations can however be overcome by combining two or more renewable energy resources in the form of a hybrid system, such as a photovoltaic system and a wind turbine. A hybrid system has the advantage of improved reliability and gives better energy service when compared to a standalone supply system. Designing of renewable energy hybrid systems involves the sizing and selecting the best components to provide affordable, efficient and effective power generating unit for applications in remote areas. Extension of utility grid lines to most remote and non-electrified areas experiences problems like high capital investment, high lead time, low load factor, poor voltage regulation, and frequent power supply interruptions. Therefore, a hybrid power system, being cost-effective, non- polluting and reliable is suggested to mitigate the scarcity of electricity and its use nearer to the point of loads. This may also be a viable proposition for supplying power to the irrigation devices. However, there is the scanty of scientific information on the utilization of solar-micro wind hybrid systems to meet irrigation energy requirement in India. The annual average data for solar radiation and wind speed are required for the feasibility and designing of the system. Those data can also be collected from the nearby metrological station. The hybrid solar-micro wind system has been considered as a case study in this section for application in irrigating land of 1 hectare areas, suitable for a small farm. The

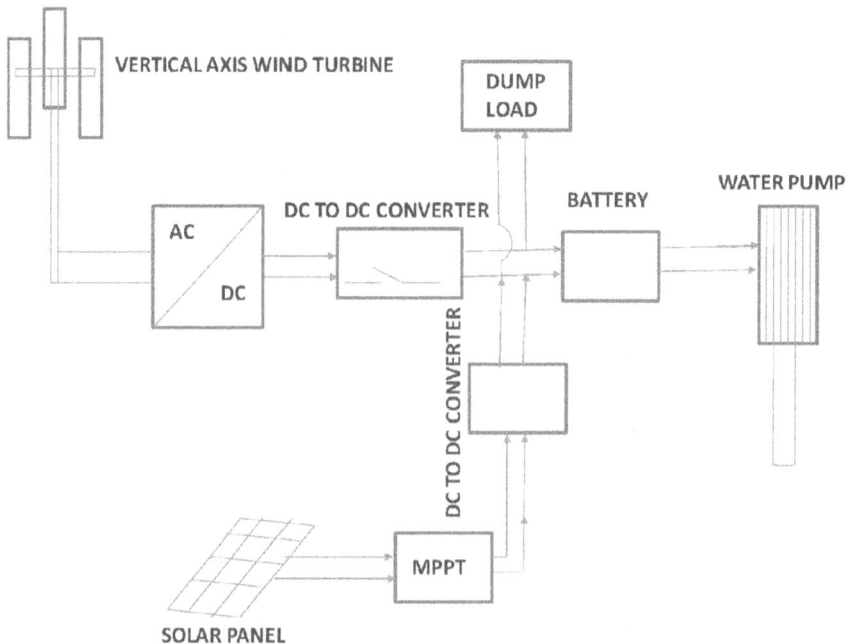

Fig. 5.48. Solar PV micro wind hybrid power

line diagram of the system is shown in fig. 5.48. The PV system produces DC voltage which is stored in a battery bank after passing through a MPPT and then in a DC to DC converter to fulfil the voltage requirement of battery. The wind turbine produces AC power, which is converted into DC and is stored in the battery. A dump load is also connected with the system to divert excess charge when the battery bank is fully charged. A permanent magnet brushless based induction generator is used in the wind turbine. The induction generator requires AC excitation current which is supplied from a shunt capacitor, mostly in a standalone system. Rice has been considered as it is one of the most water consuming field crops and is widely cultivated in India.

5.13.1 Hydraulic Energy Requirement for Water Pump in Irrigating 1 ha Land for Rice

The method of irrigation to follow is surface irrigation. The water requirement for a medium duration (120 days' duration) rice crop = 720 mm. The volume of water in a growing season for 1 ha land = 10,000 m^2 × 0.72 m = 7200 m^3 = 7,200,000 litres. Dividing 1 ha land into 4 segments, the land to cover per segment for irrigation per day = 2500 m^2/day. Assuming irrigation interval of 10 days, no. of irrigation required = 120 days /10 = 12. The depth of water per irrigation per day = 720 mm/12 = 60 mm. For area of 2500 m^2, the volume water to pump/day = 2500 × 60/1000 = 150 m^3 = 150,000 litres/day. The discharge (Q) required from the pump, taking the duration of operation as 10 hours/day (*6 hours during day and 4 hours during night time*) = (150,000) / (10 × 3600) = 4.16 litres /second. Considering the depth of water lifting as 150 feet and 9 feet extra for frictional and head losses, the total head (H) of water lifting = 53 meter. The power requirement for the pump (hp) = (γ × Q × H) /75 = where γ is the density of water (1 g/cm^3), (*lit/s*), H (m). Power = 2.93 hp ≈ 3 hp = 2.23 kW.

5.13.2 Sizing of Solar PV-micro Wind Hybrid Power System for Load of 2.23 kW

Due to higher potential of solar energy and low to medium wind energy resources for coastal zone, the shares of solar PV and wind turbine to fulfil the above load may be assumed to be 70 % and 30 % respectively. Low annual average wind speed is favourable for vertical axis wind turbine (VAWT) than horizontal axis wind turbine (HAWT). VAWT is quieter in operation, lighter in weight, capturing wind from all directions, effective at low cut-in speed, requiring less space, easy installation at roof top, simple structure, placed nearer to the load with less transmission and distribution losses but poor in aerodynamic performance. Integration of two types rotors (savonius and darrieus) with the same axis, one above the other) in VAWT improves the aerodynamic performance at par with HAWT and it has been reported by the researchers that the combined rotors (savonius and darrieus) in the vertical axis would achieve the power coefficient in the range of 0.35-0.45 (Kumar et al 2018). The wattage for solar PV will be about 2 hp (1500 watt) and for wind turbine of around 1 hp ($\approx$ 730 watt). The hybrid system is to run 10 hours a day (*6 hours during day and 4 hours during night time*). The batteries would be charged from solar system for 6 hours in day time during sun shine hours but 24 hours in a day from the wind turbine. Batteries connected with both solar and wind system are to supply the required power to operate the water pump. The output from the batteries = 22300 *Wh*/day. The input energy to the batteries by considering 95 % battery efficiency = 22300/0.95 = 23473 $\approx$ 23500 *Wh*/day. Fulfilling 70 % and 30 % load of batteries from solar PV and wind turbine respectively, the loads from the respective sub-system become 16450*Wh*/day and 7050 *Wh*/day. The DC submersible water pump of 3 hp (2.23 *kW*) and 72 volts would be used.

Sizing of batteries

Battery bank size (Ah)

$$= \frac{Total\,Wh\,required/day\,or\,electrical\,load\,(Wh/day) \times days\,of\,autonomy}{System\,voltage\,(V) \times depth\,of\,discharge \times battery\,efficency}$$

$$= \frac{23500 \times 2}{72 \times 0.85 \times 0.9} = 853 \approx 860\,Ah.$$

Specifications of the available battery is 12 V ($V_{battery}$) and 200 Ah ($C_{battery}$).

Therefore, the number of batteries (N_s) to connect in series to get

$$V_{sys} = \frac{V_{sys}}{V_{battery}} = \frac{72}{12} = 6$$

Similarly, the number of batteries (N_p) to connect in parallel to get

$$C_{sys} = \frac{C_{sys}}{C_{battery}} = \frac{860}{200} = 4.3 \approx 5.$$. Hence, total number of battery = 6 × 5 = 30 (each of 12 V

and 200 Ah). Six batteries are to connect in series to get a string and five strings are to connect in parallel to get the required energy to be stored in the batteries and to meet the load requirement.

Sizing of solar PV system (16450 Wh/day)

The electrical load requirement = 16450 Wh/day. The designed load for the PV system should be 20 % higher than the actual load to account for the losses in PV system components. The designed load is therefore determined to be 19740 Wh/day (16450 × 1.2).

$$PV\,array\,size\,required\,\left(W_p\right) = \frac{Designed\,load\,\left(\dfrac{Wh}{day}\right)}{peak\,sunshine\,hours\,per\,day} = 19740/5.5 = 3589$$

$$Number\ of\ PV\ modules\ (N)\ =\ \frac{PV\ array\ size\ (W_p)}{peak\ watt\ of\ PV\ module(W_p)} = 3589/175 = 20.50 \approx 20$$

Determination of MPPT (Maximum Power Point Tracking) Charge Controller Capacity: Let the system operating voltage be 72 V. MPPT charge controller capacity

$$(A)\ =\ \frac{PV\ array\ size\ (W_P)}{Battery\ bank\ design\ voltage\ or\ system\ voltage\ (v)} = \frac{3589}{72} = 49.84 \approx 50\quad ampere.$$

So, a 72 V and 50 A MPPT charge controller suits well for this system. Taking current and voltage into consideration, one MPPT charge controller or more than one unit of just higher capacity can be chosen as per the availability in the market.

Sizing of wind turbine

The energy to supply to the batteries from the wind energy conversion system = 7050 *Wh/day*.

The average energy to be generated by the wind turbine

$$E_{load} = \frac{Energy\ to\ be\ sup\ plied\ by\ battery\ to\ the\ pump}{\eta_{controller} \times \eta_{inverter}} \times 1.1$$

$$= \frac{7050}{0.95 \times 0.90} \times 1.1 = 9070 Wh/day$$

The annual energy required to be generated $= \dfrac{9070 Wh/day}{1000} \times 365\ days = 3310 kWh$

The size of the wind turbine that is required to be installed to meet the energy demand can be determined based on the following assumptions.

Coefficient of performance = 0.4, density of air = 1.225 kgm^{-3}, capacity factor (C_p) of 0.35 (that is to say 35% of the time wind machine is producing energy at rated power), transmission losses of rotor to generator = 0.9, generator losses of 0.9, number of operational hours per year 8760 h.

Power density of the wind (power/unit area)

The power density of wind (WPD) of moving air is given by $WPD = \dfrac{1}{2}\rho v^3$ where ρ is the density of the flowing air (kg/m^3) and v is the wind speed (m/s). The mean annual speed of the coastal areas may be taken 4.5 m/s at a height of 10 m. Hence, WPD = 56 W/m^2.

Actual wind power density to be converted into the useful energy $= WPD \times C_p \times$ transmission loss generator loss $= 56 \times 0.4 \times 0.9 \times 0.9 = 18 \; W/m^2$.

Annual useful energy density = Actual energy density $\times$ number of hours per year $= 18 \times 8760 = 158 \; kWh/m^2$.

Rotor size

The rotor size is determined from the swept area of the turbine.

$$\text{Swept area} = \frac{Total\,annual\,energy\,required}{Annual\,useful\,energy\,density} = \frac{3310kWh}{158kWh/m^2} = 20.94 \approx 21m^2$$

Taking combined savonius-darrieus vertical axis wind turbine (fig. 6.49), the swept area becomes $\frac{1}{2} \times 21 = 10.5m^2$.

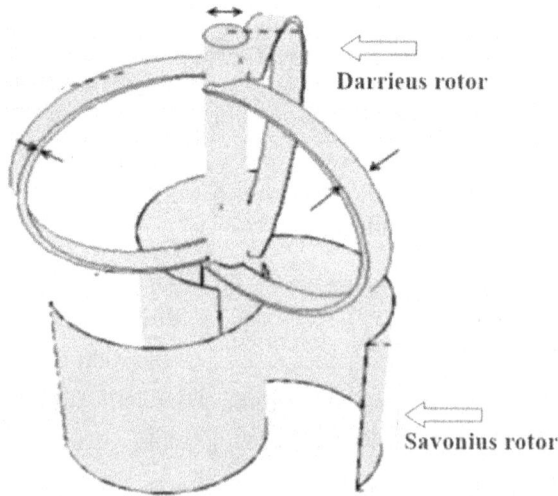

Fig. 5.49. Combined Savonius-Darrieus vertical axis wind turbine

The swept area of savonius and darrieus wind turbine can also be obtained from the relationship i.e. swept area $(m^2) = D \times h$ where h is the height and D is the diameter of the rotor. Using an aspect ratio (AR) of 2 (for turbines with more than two blades),

$AR = \frac{h}{R} = 2 \to h = 2R$ where R is the radius of the rotor.

Swept area for each rotor $= 10.5 \; m^2 = D \times h = 4R^2 \to R = 1.6 \; m$ and $h=3.2 \; m$

Power rating of turbine

The output of a wind turbine depends on the turbine's size and the wind speed through the rotor. The turbine rated power = Actual wind power density × area of rotor = 18 × 21 = 0.378 kW.

Actual rated power of the turbine rating

The capacity factor for wind energy conversion system may be assumed to be 0.35 (that is to say, 35% of time wind machine produces energy at the rated power).

$$\text{Rated power} = \frac{Power\ rating}{Capacity\ factor} = \frac{0.378}{0.35} = 1.08 \approx 1kW$$

Thus, a 1-*kW* rated power wind turbine would provide desired annual energy under the given situation.

5.14 BREAK-UP COST OF SOLAR PV MICRO WIND HYBRID POWER SYSTEM

Name of PV component	Number	Sizing	Approximate cost range (Rs.)	Cost per unit (Rs.)	Cost (Rs.)
Solar panel (total 3545 W_p)	20	175 W_p	32-35/ W_p	35/W_p	1,22,500
Charge controller (Inbuilt MPPT)	1	40 ampere × 48 Volt	50-60/Amp	60/Amp	2400
DC to DC converter	2	72 Volt	70-80/ volt	75/ volt	10,800
Batteries AC to	'30	200 Ah 12 V	40-50/Ah	45/Ah	2,70,000
DC converter	1	1 kW	7-12/ W_p	10/ W_p	10,000
Vertical axis wind turbine	1	1 kW	120-150 /W	150/ W	1,50,000
Bore well	1	200 feet	300-400 /feet	350/feet	70,000
Pump house	1	(10 ft. 10 ft)	1200-1500/ Sq. feet	1500/Sq. feet	1,50,000
DC submersible water pump	1	3 hp			1,50,000
Cables (solar PV system)			1-2 / W_p	1.5/ W_p	5320
Peripherals (Junction box, DC protection system, earthing, lightning arrestor)			1.5-2.5 W_p	2.0/ W_p	7100
Water Conveying PVC pipes					30,000
Supervision and installation (solar PV and wind system)			5-6/ W_p	5/ W_p	22725
Total					**10,00,845 ≈ 10,00,000**

Life Cycle Costing (LCC)

The LCC is the cost of using a system during its lifetime and it generally includes (i) capital cost (paid while establishing a system), (ii) operation and maintenance cost (cost incurred for running the system, could be regular maintenance cost) and (iii) replacement cost (cost incurred if any component of the system needs replacement before the life of the system is over).

Present worth of future one-time investments

The present worth of a product is defined as the amount of money that needs to be invested today with the discount rate, 'd', and the inflation rate, 'r', such that we would be able to

purchase the product in the future, 'n' years later. $F_{PW-one\ time} = \dfrac{Future\ cost}{Future\ value} = \dfrac{X(1+r)^n}{X(1+d)^n}$

$= \left(\dfrac{1+r}{1+d}\right)^n$, where $F_{(PW\text{-}one\ time)}$ can be read as 'present worth factor for future one-time

investments'. Now if the present cost of a product is C_0, then its present worth (PW) for the actual one-time investment after 'n' years would be given as:

$$PW_{replacement} = C_0 F_{PW-one\ time} = C_0\left(\dfrac{1+r}{1+d}\right)^n = C_0(k)^n \text{ where } k = \left(\dfrac{1+r}{1+d}\right)$$

Present worth of future recurring investments

There are requirements when the investment is required on recurring basis to run the system. In such cases, the present worth of the future investments can be obtained by using the above equation, where recurring investment is considered as one-time investment for each year of operation. Thus, present worth of recurring expenses PW_{rec} can be obtained by summing up the present worth of one-time investments for the entire life of the system as given in following equation.

$$PW_{rec} = C_0\left(\dfrac{1+r}{1+d}\right)^1 + C_0\left(\dfrac{1+r}{1+d}\right)^2 + ... + C_0\left(\dfrac{1+r}{1+d}\right)^n$$

There are n number of purchases made in the lifetime on the system. The above equation can be simplified if we replace with the following term i.e. $k = \dfrac{1+r}{1+d}$. The simplified version of the equation becomes $PW_{rec} = C_0 k \left(\dfrac{1-k^n}{1-k} \right) = C_0 F_{PW-rec}$ where

$F_{PW-rec} = k \left(\dfrac{1-k^n}{1-k} \right)$. Here, F_{PW-rec} is the present worth factor of the recurring investment

made at the end of each year.

Life Cycle Cost

In order to calculate the LCC of a system on present worth basis, the present worth factors, as discussed i.e. $F_{(PW-one)}$ and $F_{(PW-rec)}$, are to be calculated in the beginning. They are multiplied by the appropriate present cost components and added together to find out the total cost of operating a given system over its lifetime. In the proposed set up, the system has a lifetime of 20 years and has a capital cost C_1 (in terms of present cost). It requires replacement of critical components (batteries, MPPT, AC to DC converter, DC to DC converter and pump) in 8 years, having a cost C_2 (in terms of present cost). The system also has an annual maintenance cost of C_3 (in terms of present cost). Then the LCC of the system would be given by the following equation:

$LCC = C_1 + F_{(PW-one)}$ (8 years) $\times C_2 + F_{(PW-one)}$ (16 years) $\times C_2 + F_{(PW-rec)}$ (20 years) $\times C_3$

C_1 = Rs. 10,00,000, C_2 = AC to DC converter (Rs. 10,000) + DC to DC converter, 2 nos. (Rs. 10,800) + MPPT (Rs. 2400) + Batteries (Rs. 2,70,000) + Water pump (Rs. 1,50,000) = Rs.4,43,200.

Inflation rate per annum for the equipment and devices, need to be replaced is considered as 1 % of their current price as per the present market survey and discount rate = 10%. $k_{PW-one\ time}$ = 0.91.

Annual recurring cost, (operation and maintenance cost), $C_{(Present\ operation\ and\ maintenance\ cost)}$ = 0.5 % of initial investment for 1st year i.e. Rs. 5000 (0.5 % of Rs. 10,00,000) = C_3. After 1st year, the inflation rate for recurring cost (operation and maintenance cost) is 0.5% per annum and discount rate = 10 %, $k'_{recurring\ cost}$ = 0.91 and for n =20,

$F_{PW-rec} = k' \left(\dfrac{1-(k')^n}{k'} \right) = 8.6$

$LCC = 10,00,000 + (k_{PW-one\ time})^8 \times 4,43,200 + (k_{PW-one\ time})^{16} \times 4,43,200 + F_{(PW-rec)}$ (for 20 years) $\times 5000 = 10,00,000 + 2,08,304 + 97504 + 43,000 = $ Rs. 13,48,808

Annualized uniform cost (C_A)

Annualized uniform cost is defined as the product of net present value of the system and Capital Recovery Factor (CRF) and can be written as

$$C_A = LCC \times \left[\frac{\left(d \times (1+d)^n \right)}{\left((1+d)^n - 1 \right)} \right] = \text{Rs. } 158544 \approx \text{Rs.} 1,60,000$$

Expected income from crop cultivation of 1 ha land (kharif, rabi and summer season)

S.N.	Crop	Season of cultivation	Yield/ha (ton)	Average cost of cultivation/ ha (Rs)	Selling price of harvested crop (Rs/ton)	Profit (Rs/season)
1	Rice	Kharif	4	50,000	18000	22,000
2	French beans	Rabi	15	80,000	40,000	5,20,000
3	Okra	Summer	10	1,00,000	40,000	3,00,000
Total income expected to be generated per year						**Rs. 8,42,000**

Cost-benefit expected from the proposed system

S.N.	Item	Rs.
1	Cost of the system (solar wind hybrid system)	10,00,000
2	Annualized uniform cost of operating the system considering inflation and discount rates	1,60,000
3	Total income generated per year	8,42,000
4	Benefit/year	6,82,000
5	Benefit: cost	4.25:1
6	Payback period (year) = (Initial investment/net income per year)	1.5

Cost per unit electricity (C_u)

The electrical output from solar PV sub-system = 16450 Wh/day

Annual energy output from solar PV sub-system considering 300 sunny days/year = 4935 kWh/annum

Annual useful energy density from wind subsystem = 158 kWh/m² and for 21 m² rotor area, the energy output from wind subsystem = 158 × 21 = 3318 kWh/annum

E_{out} = The annual energy output from solar PV micro wind hybrid power system = 4935 + 3318 = 8253 kWh/annum

Cost per unit electricity (C_u) = (Annualized uniform cost) / (annual electricity generated,

kWh/year) = $\frac{C_A}{E_{out}} (Rs/kWh)$ = *1,60,000/8253* = Rs. 19.00

REFERENCES

Abhishiktha Tummala, Ratna Kishore Velamati, Dipankur KumarSinha, V.Indraja, V. HariKrishna. A review on small scale wind turbines. Renewable and Sustainable Energy Reviews 56(2016) 1351–1371

Bhattacharjee S, Acharya S. PV–wind hybrid power option for a low wind topography. Energy Convers Management 2015; 89:942–54.

Christoffer, J.; Ulbricht-Eissing, M. (1989) Die bodennahen Windverhältnisse in der Bundesrepublik Deutschland. Offenbach, Deutscher Wetterdienst, Ber.Nr.147.

Molly, J.-P. (1990) Windenergie. Karlsruhe, C.F. Müller.

Gasch, R.; Twele, J. (Eds) (2002) Wind Power Plants. London, James & James

Sunanda Sinha, S.S. Chandel. Prospects of solar photovoltaic–micro-wind based hybrid power systems in western Himalayan state of Himachal Pradesh in India. Energy Conversion and Management 105 (2015) 1340–1351.

Essalaimeh S, Al-Salaymeh A, Abdullat Y. Electrical production for domestic and industrial applications using hybrid PV–wind system. Energy Convers Manage 2013; 65:736–43.

Adaramola MS, Agelin-Chaab M, Paul SS. Analysis of hybrid energy systems for application in southern Ghana. Energy Convers Manage 2014; 88:284–95.

Rakesh Kumar, Kaamran Raahemifar, Alan S. Fung. 2018. A critical review of vertical axis wind turbines for urban applications. Renewable and Sustainable Energy Reviews 89 (2018) 281–291.

SOLAR DRYER
(CONCEPTS AND APPLICATIONS)

INTRODUCTION

Drying is a traditional method of preserving agricultural produces. Preservation deals with the processes for preventing decay or spoilage of an agricultural produce or food material thus, allowing it to be stored afterwards in a congenial condition for future use. Storage on the other hand, refers to keeping the materials in safe condition after performing preservation processes on them for minimum deterioration to occur and uses in future. Preservation therefore mainly prevents the growth of bacteria, fungi, and other micro-organisms in a produce required to be stored. Moisture present in a produce is the key component favouring the growth of harmful micro-organisms, causing the decay of the material. Hence, drying is one of the ways of preserving an agricultural produce for the removal of moisture from it. It is the process of moisture removal from a produce, involving both heat and mass transfer. Sometimes, the term 'dehydration' is also used for the same purpose. However, both the terms 'drying' and 'dehydration' mean the removal of water. But the former term is generally used for drying under the influence of non-conventional energy sources like sun and wind whereas dehydration means the process of removal of moisture by the application of artificial heat under controlled conditions of temperature, humidity and air flow. Hence, drying is a cheaper process compared to the dehydration. The removal of moisture prevents the growth and reproduction of micro-organisms like bacteria, yeasts and moulds causing decay and minimizes many of the moisture-related deteriorative reactions. It brings about substantial reduction in weight and volume, minimizing packing, storage, and transportation costs and enables storability of the product under ambient conditions. Drying is therefore an important operation in terms of improving and extending the shelf-life of agricultural produces.

When heat (naturally or artificially) is supplied to the wet material by any means, its outer surface gets warmer and heat further enters into the material generally through conduction. The moisture is transported from inner portion to outer surface of the material by various moisture diffusion processes such as liquid diffusion or vapour diffusion (movement of moisture/water vapour from higher to lower concentration). The moisture available at the outer surface of the material is evaporated and is moved away in the flowing air through natural or forced convection. As a result, the temperature and humidity of the surrounding air flowing over the material to be dried, change. Hence, in terms of physics, the exchange of heat and humidity between the air and the product to be dried is seen through the phenomena such as heating of the grain, accompanied by a cooling of the drying air and reduction in the moisture content of the grain, accompanied by an increase in the relative humidity of the drying air. However, this process does not take place uniformly. The water present in the outer layers of grain evaporates much faster and more easily than that of the internal layers. Thus, it is much harder to lower the moisture content of a product for example from 25% to 15% than from 35% to 25%. It is therefore, a misnomer to think that this difficulty can be overcome by rapid drying at high temperature. In fact, such drying conditions create internal tensions, producing tiny cracks that can lead to the rupture of the grain during subsequent treatments.

6.1 CAUSES OF SPOILAGE OF AGRICULTURAL PRODUCES

The preservation of agricultural produces and food materials refers to the practices followed in which these materials may be retained over a period of time without being contaminated by pathogenic organisms or chemicals and without losing its colour, texture, flavour and nutritional values. Therefore, in order to prevent or reduce the spoilage, its causes are to be identified and understood. All types of agricultural produces/foods contain proteins, carbohydrates, fats (lipids), vitamins and minerals. The proteins and minerals like iron, calcium and phosphorus helps for tissue building and growth of the body. The carbohydrates and fats provide energy to the body. The vitamins and minerals are essential to safeguard the body against diseases. The destruction of any one of the above mentioned components causes the spoilage of food. The spoilage period depends upon the type of food. The perishable foods like meat, fish, milk and many fruits and vegetables begin to deteriorate immediately unless properly preserved. The semi-perishable foods like eggs, onions and potatoes can be kept for several weeks in a cool dry place. The non-perishable foods like cereals, pulses and nuts can be stored for long periods of time. The spoilage of food comes in the form of bad odour, uncommon colour, bad taste and physical appearance.

There are chiefly three agencies that cause spoilage in agricultural produces i.e. (i) microorganisms (ii) enzymes and chemical changes and (iii) man, animals, rodents, insects, etc. For effective preservation, we have to check the growth and activity of these spoilage causing agencies. The spoilage of the agricultural produce is basically due to the physical and chemical changes taking place in it, as discussed below,

(A) Spoilage due to physical changes

One of the causes of the spoilage of unpackaged fresh foods, such as meat, poultry, fish, fruit, vegetables, cheese, eggs etc. is the loss of moisture from the surface of the product by evaporation into the surrounding air. This process is known as desiccation or dehydration.

In fruits and vegetables, desiccation is accompanied by a considerable loss in both weight and vitamin content. The loss of weight also effects the taste of food. In meat, cheese etc., desiccation causes discoloration and shrinkage. Eggs lose moisture through the porous shell, with a resulting loss of weight and poor grading of the egg. The spoilage of food, particularly fruits and vegetables, due to impact, brushing and squeezing is very common in packing.

(B) Spoilage due to chemical changes

The spoilage of agricultural produce/food is caused by a series of complex chemical change in the food. These chemical changes are brought about by both internal and external spoiling agents. The former are the natural enzymes, whereas later are the micro-organisms. These two types of spoiling agents responsible for the spoilage of food are discussed, in detail, as below:

(a) Enzymes. The enzymes are inherent in all organic substances such as fruits, vegetables and animals. They are organic catalysts produced by cells. The life of every cell of plant or animal tissue depends upon the chemical reactions activated by these organic catalysts. Chemically, enzymes are proteins in nature and hence may be denatured by heat. There are various types of enzymes and specific enzymes act on some specific foods starting the chemical action which is responsible for the spoilage of food. Following table shows the different types of enzymes and their chemical action with the substance.

Table 6.1 Different types of enzymes and their chemical action with the substance.

S.N.	Type of enzymes	Chemical action of enzymes
1.	Ptyalin	Converts cooked starches into soluble sugars-maltose
2.	Pepsin	Converts proteins to peptones.
3.	Rennin	Converts caseinogens to casein.
4.	Amylase	Converts sugars and starches to maltose.
5.	Lipase	Reduces fats to glycerine and fatty acids.
6.	Trypsin	Reduces proteins and peptones to polypeptides and amino acids.
7.	Erepsin	Reduces all protein substances to amino acids.

The moisture (water) is necessary for the proper activity of the enzymes. The optimum temperature at which most enzymes act rapidly is about 37^0C, but they are destroyed by heating the foods or their activity can be reduced by freezing the foods.

(b) Micro-organisms. The micro-organisms are undetectable living organisms present in the surroundings. They grow in and on the surface of the food. The various micro-organisms responsible for the spoilage of food are discussed, in detail, as below:

(i) *Bacteria.* The bacteria are single-celled organisms found in soil, water, dust and air. Many bacteria are useful in preserving the food and their presence is necessary in some specific foods such as those which ferment apple juice to produce cider. Some bacteria cause the spoilage of foods. Since bacteria are capable of withstanding extreme temperature, therefore they may be classified according to their temperature ranges into three general groups, as shown in the following table:

Table 6.2: Types of bacteria.

S.N.	Type of bacteria	Temperature in ^{0}C		
		Minimum	Optimum	Maximum
1.	Psychrophilic	0	15-20	30
2.	Mesophilic	15-30	25-40	50
3.	Thermophilic	25-45	45-55	55-85

The psychrophilic bacteria are those organisms which play an important part in the spoilage of food in the refrigerator and in cold storages. The bread left in the refrigerator shows grey or black specks due to the activity of psychrophilic bacteria. The thermophilic bacteria are those organisms which are capable of withstanding high temperatures. The food and canning industry and milk processing plants are generally affected by thermophilic bacteria. Since bacteria may be aerobic or anaerobic, therefore they are likely to flourish anywhere and everywhere. Some of them may cause spoilage of food while others may cause food poisoning and diseases borne activities through the food. The bacteria may be destroyed be sunlight, ultra-violet rays, extreme heat and by the use of certain chemical substances.

(ii) *Yeast.* It is another micro-organism which is also responsible for food spoilage. They require water and a source of energy (usually sugar) for their growth. The growth is most rapid at temperature between 25 ^{0}C to 30 ^{0}C. Since sugar serves as a source of energy to yeast, therefore they are generally found in places where sugar is available. The yeasts find their way into the ground when they are washed or blown from the surface of fruits, particularly grapes. The yeast cells which are always present in the atmosphere may contaminate food and cause its spoilage. They produce pigments and undesirable chemical products during their metabolism. The yeast may cause spoilage of fruit juices, syrups, molasses, honey, jellies and other foods converting their sugar into alcohol and carbon dioxide. Generally, all types of yeast will be destroyed when subjected to a temperature of 100 ^{0}C and their activity will be stopped under low temperatures, but they remain in food for a considerably long period as their cells are hard.

(iii) Moulds (Fungus). The moulds are multi-cellular, filamentous fungi that contain sporangium. The spores in the sporangium spread through the air and start new mould plants. When these spores find a favourable environment, they germinate and produce a fluffy growth. They are found in different colours such as white, grey, blue, green, red, orange or some other colour depending upon the variety of the mould. Most of the moulds grow between 25^0C to 30^0C in warm and damp places. Some moulds also grow even at refrigerator temperature. The growth of mould is rapid on acidic foods such as lemon and on foods having high sugar content such as jams and jellies. They also grow on mature fruits and vegetables, neutral foods such as bread and other starchy foods which are spoiled by the rhizopus (commonly known as black mould). A small proportion of moulds found on food stuffs is capable of producing toxic materials known as mycotoxins. The best known of these are aflatoxins produced by moulds growing on peanuts, ragi, wheat and millet.

(C) Spoilage due to external agents other than micro-organisms

Man, animal, rodents, insects etc. may also cause only physical loss to the agricultural produces and can be easily minimized by proper packing and handling. The damaged food by animals, man, insects, rodents etc. may later on give way for the initiation of microbial and self-decomposition.

Overall, from the food preservation point of view, the causes of the spoilage can be considered in the decreasing order of importance and emphasis. Highest emphasis is given on the control of microbial decomposition followed by self-decomposition, ultimately followed by damages caused by man, animals, insects, rodents etc.

In a nutshell, spoilage of fruits, vegetables and their products may also be categorized of two types, (i) abiotic and (ii) biotic spoilage. Abiotic spoilage causes less health hazards to consumers and occurs due to action of enzymes, oxidation of fats (rancidity), purification of proteins, browning reactions between proteins and sugars, physical changes (weight loss, shrivelling, colour fading etc.), absorption of moisture (hygroscopic foods) etc. The biotic spoilage causes more health hazards and is associated with micro-organisms such as bacteria, yeasts, moulds etc. It may also be caused by insects, animals, rodents etc. The microbial spoilage is also of two types i.e. spoilage by plant pathogens which appear on living tissues and spoilage by saprophytes, appearing on dead and decaying tissues. The factors affecting spoilage of agricultural produces are mainly *moisture* (higher the moisture more the spoilage and microbial spoilage is reduced at moisture below 10%), *temperature* (higher the temperature (within limits) more the spoilage), p^H (lower the p^H lesser the spoilage and vice versa and bacteria cannot grow at low p^H) and the composition of fruits and vegetables (more the carbohydrates, higher are the chances of spoilage and fats generally are not spoiled by microorganisms).

6.2 METHODS OF PRESERVATION

All the methods of food preservation must provide such an environment in and around the preserved food so that unfavourable conditions may be created to restrict the continued activity of the spoilage agents. When the product is to be preserved for a long time, the unfavourable conditions must be of sufficient severity to eliminate the spoilage agents entirely or at least make them ineffective. Followings are the various methods used for food preservation

(i) *Application of heat.* All types of spoilage agents (i.e., enzymes, bacteria and moulds) are destroyed when subjected to high temperatures over a period of time. The temperature of the product is raised to a level, fatal to all spoilage agents and is maintained at this level until they are all destroyed. The product is then sealed in sterilized and air tight containers. A product so processed will remain in a preserved state for long time.

(ii) *Drying/Dehydration.* The process of removing the moisture from the product is called dehydration (i.e., drying). It is one of the oldest methods of preserving foods and still it is widely used. Since both enzymes and micro-organisms require moisture for their growth, therefore it is necessary to stop their growth completely by dehydrating the foods. A variety of dehydrated products are available in the market. They include dried milk, dehydrated soups, pre-cooked peas and cereals. A very common method for dehydration is sun-drying. However, this method cannot preserve the taste of foods

(iii) *Chemical preservation.* This method may employ high concentrations of salt, sugar and acids. Salt as a preservative is used for preserving vegetables and fruits like tamarind, raw mango, amla, fish and meat. It may be used in dry or brine from. The presence of a high concentration of salt prevents the water from being available for bacterial growth. This is because the concentration of salt in the water is higher than that is the bacterial cells. Thus the water be absorbed by the cellular membrane of the bacteria. The principle of sugar as preservative is same as that of salt. However, preservation with salt is a cold process while in case of sugar, the mixture is heated. Sugar acts as a preservative because the high concentration of sugar solution withdraws water from the micro-organisms, thereby preventing their growth. The moulds will, however, grow on the surface of jams, jellies, if proper sterility is not maintained. When the medium, in which, food is preserved, is strongly acidic, then most of the micro-organisms cannot survive. The use of vinegar (acetic acid) and lemon juice (citric acid) is common in the domestic methods of pickling. The benzoic acid as a preservative is used up to a concentration of 0.1% for all coloured fruits and vegetables. The pesticides are also sprayed over fruits and vegetables and food grains to prevent spoilage.

(iv) *Oils and spices.* The oils and spices along with salt and sugar provides a medium that resists the activity of the micro-organisms in food. Moreover, they improve the flavour of the food being preserved. Spices such as chillis, fenugreek, mustard and pepper are used in pickling. When oil is used in pickling, the top layer of oil prevents the micro-organisms in the air from coming into contact with the food.

(v) *Canning.* It is a method of preserving food from spoilage by storing it in containers that are hermetically sealed and then sterilized by heat. During this process, some of the micro-organisms are destroyed and the rest are rendered inactive. The enzymes are also inactivated.

(vi) *Pasteurization.* This method is generally used on large scale to protect milk against bacterial infection. The milk used for the preparation of milk products like cheese, butter and ice cream is pasteurized. The pasteurization may be brought about by the holding process or high temperature short time method. In the holding process, the milk is heated to at least 62 ^{0}C and kept at the temperature for at least 30 minutes. In the high temperature short time method, the milk is heated to 70 ^{0}C and kept at that temperature for at least 15 seconds. The milk may be sterilized either by boiling for a period of time or by the application of heat as in the preparation of evaporated milk. The sterilization deepens the colour of milk and gives it a slightly heated flavour, while pasteurization does not change the colour or flavour of milk.

(vii) *Refrigeration:* The refrigeration is the only means of preserving food in its original freshness. It may be noted that when food is to be preserved by refrigeration, the refrigerating process must begin very soon after harvesting of crop or killing of micro-organisms and must be continuous until the food is finally consumed.

The preservation of perishable foods by refrigeration involves the use of low temperature as a means of eliminating or retarding the activity of spoilage agents. The low temperatures are not as effective as high temperatures in bringing about the destruction of spoilage agents. The storage of perishable foods at low temperatures provides a practical means of preserving perishable foods in their original fresh state for longer periods of time. The degree of low temperature required for adequate preservation varies with the type of product stored and the length of time the product is to be kept in storage. The application of refrigeration for preserving foods is common in domestic refrigerators, commercial refrigerators and cold storage facilities.

6.3 IMPORTANCE OF STORAGE

Storage of the agricultural produces is taken up after performing proper preservation processes on them. The objectives of storage are to (i) slow down the biological activity i.e. respiration rate, ethylene (an important natural plant hormone) evolution (ii) to reduce the drying of product and loss of moisture (shrivelling), (iii) to reduce the growth of micro-organisms and to reduce the pathogenic damage (rotting, diseases etc.), (iv) to avoid the physiological disorders (freezing injury, chilling injury, heat injury, storage disorder etc.) and (v) to reduce the physical damage due to rodents, animals etc. Temperature and relative humidity are the two important storage considerations for the agricultural produces. The rate of deteriorative reactions is doubled for each 10 ^{0}C rise in temperature. Therefore, the storage temperature should be within the permissible limit. Higher temperature causes increase in respiration and lower temperature may cause chilling or freezing injury. Likewise, relative humidity should be within the permissible limit, though it varies for different products. Lower humidity level may cause shrinkage and weight loss. Generally, for most of the fruits and vegetables, the relative humidity should be in the range of 85-95 % except some commodities i.e. onion, garlic, potato etc. where humidity levels of 65-70 % are required. Higher humidity in such commodities may cause rotting. Composition of gases in the surrounding environment also affects the storage life of any commodity in a particular set of conditions of temperature and humidity. More the availability of oxygen, higher would be the respiration rates (up to a specific level). Absence of oxygen may initiate anaerobic respiration, leading to the formation of alcohol. Presence of ethylene in the storage atmosphere also favours respiration. Based on the shelf life, the agricultural produces/food materials can be classified as non-perishable, semi-perishable and perishable. *Non-perishable* products can be stored under room conditions for months together. These contain very low moisture content generally < 8-10% e.g. sugar, flour, dry beans etc. *Semi-perishable* products can be stored under room conditions from few days to about a month or two e.g. potato, nuts, almond etc. and *Perishable* products that cannot be kept in safe and sound condition at room temperature for more than a day or two. These contain very high concentrations of moisture generally more than 70-80% e.g. milk, meat, fish, fruits, leafy vegetables etc.

6.3.1 Dried and Storage Conditions of Some Agri-produces

The initial moisture content, the final moisture content and the maximum temperature at which product may be dried are very important and the values for a variety of products are given in Table 6.3.

Table 6.3: Maximum temperature allowable for drying and the initial and final moisture contents of various products

Products	Moisture content %		Maximum temperature allowable for the drying (^{0}C)
	Initial	Final	
Paddy, raw	22-24	11	50
Paddy, par-boiled	30-35	13	50
Maize	35	15	60
Wheat	20	16	45
Millet	21	14	-

[Table Contd.

Contd. Table]

Products	Moisture content %		Maximum temperature allowable
	Initial	Final	for the drying (^{0}C)
Corn	24	14	50
Rice	24	11	50
Green peas	80	5	65
Cauliflower	80	6	65
Carrots	70	5	75
Green beans	70	5	75
Onions	80	4	55
Garlic	80	4	55
Cabbage	80	4	55
Sweet potato	75	7	75
Potatoes	75	13	75
Spinach	80	10	-
Cassava	62	17	-
Cassava leaves	80	10	-
Chillies	80	5	65
Fish, raw	75	15	30
Fish, water	75	15	50
Onion rings	80	10	55
Prunes	85	15	55
Apples	80	24	70
Apricots	85	18	65
Peaches	85	18	65
Grapes	80	15-20	70
Bananas	80	15	70
Guavas	80	7	65
Mulberries	80	10	65
Figs	80	24	-
Opra	80	20	65
Pineapple	80	10	65
Yams	80	10	65
Nutmeg	80	20	65
Sorrel	80	20	65
Coffee	50	11	-
Coffee beans	55	12	-
Cocoa beans	50	7	-

[Table Contd.

Contd. Table]

Products	Moisture content %		Maximum temperature allowable
	Initial	Final	for the drying (^{0}C)
Cotton	50	9	75
Cotton seed	50	8	75
Copra	30	5	-
Groundnuts	40	9	50
Silk cocoons	68-70	10-12	80
Timber			
Foliage trees	25-35	17-20	50
Conifers	30-40	10-15	50
Soaked trees	60	12	50
Mahogany	35	11	-
Leather	50	18	35
Fabrics	50	8	75

The recommended storage conditions of some agri-produces are given in Table 6.4 below.

Table 6.4: Recommended storage conditions of some vegetables and fruits

Fruit/vegetable	Temperature (^{0}C)	Relative humidity (%)	Approx. storage life
Asparagus	0	95	3-4 weeks
Beans green	4-7.2	90-95	1-2 weeks
Bitter gourd	0.6-1.7	85-90	1 month
Brinjal	8-12	90-95	1-3 weeks
Broccoli	0	95	1.5-2 weeks
Cabbage	0-1.7	92-98	3-6 weeks
Carrot	0	95	20-24 weeks
Cauliflower	0-1.7	85-98	3-7 weeks
Chilli	7	90-95	2-3 weeks
Colocasia	11.1-12.8	85-90	21 weeks
Cucumber	10-13	92-95	1.5-2 weeks
Garlic	0	65-70	6-9 months
Ginger	7.2-13	65-75	4-6months
Lettuce leaves	0	95-98	1-3 weeks
Mushroom	0	95	3-4 days
Muskmelon	1.7-7.2	85-90	1.5-4.5 weeks

[Table Contd.

Contd. Table]

Fruit/vegetable	Temperature (^{0}C)	Relative humidity (%)	Approx. storage life
Okra	8-9	90	2 weeks
Onion	0	70-75	5-6 months
Pea	0	88-98	1-3 weeks
Potato	3-5	85	5- 10 months
Pumpkin	1.7-11.6	70-75	6-9 months
Radish	0	88-92	3-5 weeks
Spinach	0	95-100	10-14 days
Tomato	7.2-10	85-90	1-4.5 weeks
Turnip	0	90-95	8-16 weeks
Watermelon	7.2-15.6	80-90	2 weeks

General Recommended storage conditions for fruits

Fruit/vegetable	Temperature (^{0}C)	Relative humidity (%)	Approx. storage life
Apple	0-4	80-90	4-8 months
Apricot	0-2	80-85	1-2 weeks
Banana	12-13	80-90	1-2 weeks
Cherry	0-2	85-90	2 weeks
Dates	7-8	85-90	2 weeks
Fig	0-2	85-90	4 weeks
Grape	-1.1-2	80-95	3-8 weeks
Guava	7.2-10	85-90	2-3 weeks
Jackfruit	11-13	85-90	1.5 months
Kiwifruit	-0.5-0	90-95	3-5 months
Lemon	8-10	85-90	1-6 months
Litchi	0-2.1	85-95	3-10 weeks
Mandarin	5-8	85-90	10-14 weeks
Mango	8-12.8	85-90	2-7 weeks
Papaya	7.2-10	80-90	1-2 weeks
Passion fruit	7-8	80-85	4-5 weeks
Pear	0-1	85-90	3-6 months
Peach	0-3	85-90	2-4 weeks
Persimmon	0-2	85-90	7 weeks
Pineapple	8-13	85-90	3-6 months
Plum	0-2	85-90	4-8 weeks
Pomegranate	0-2	80-90	2-6 weeks
Sapota (Chiku)	3-4	85-90	6-8
Strawberry	0-2	80-95	5-7 days

6.4 ENERGY BALANCE AND TEMPERATURE FOR DRYING

If unsaturated air is passed over wet material, the air will take up water (moisture) from the material as described earlier. This water has to be evaporated and the heat to do this, comes from the air and the material. The air is thereby cooled. In particular, if a volume 'V' of air is cooled from T_1 to T_2 in the process of evaporating a mass m_w of water, then $m_w L = \rho C V (T_1 - T_2)$ where L is the latent heat of vaporization of water and ρ and C are the density and specific heat of air respectively at constant pressure at the mean temperature, for moderate temperature differences. The basic problem in designing a crop dryer is therefore to determine a suitable T_1 and V to remove a specified amount of water m_w. The temperature T_1 must not be too high, because this would make the grain crack and so bacteria and parasites would enter into the product causing adverse effects. Similarly, high humidity at elevated temperature for extended time will encourage the microbial growth in the grain.

6.5 QUALITY ATTRIBUTES OF DRIED PRODUCTS

Drying is one of the oldest food preservation techniques which is done by the heating effect through the natural environmental conditions, specially, with the help of the incident solar radiation. Product quality is the only attribute that assesses the acceptance of a product for safe consumption or use by human beings apart from marketable opportunities. This attribute is equally important in fresh, semi-dried and fully dried products both in edible and non-edible categories. Traditionally, most agriculture products are dried under direct sunlight to achieve the required degree of dryness. In general, the quality attributes of dried products can be classified into physical, chemical, biological and nutritional points of view depending on the type of product (food or non-food) as shown in Table 6.5

Table 6.5: Selected dried product quality attributes

Attributes	Parameters
Physical	Colour, texture, shrinkage, porosity, rehydration, breakage, split
Chemical	Flavour, odour, water activity, shelf life
Nutritional	Calorie, vitamins, minerals, fibres, lipids, proteins, carbohydrates, antioxidants
Biological	Mould, yeast, E. coli, Salmonella, mycotoxins, aflatoxins
Sensory	Appearance, odour, flavour, mouthfeel and texture

6.6 SOLAR DRYING OF AGRICULTURAL PRODUCES

Solar drying has been used since time immemorial to dry vegetables, plants, seeds, fruits, meat, fish, wood, and other agricultural and forest products. In order to get benefit from the free and renewable energy source provided by the sun, several attempts have been made in recent years to develop solar drying mainly for preserving agricultural and forest products. However, for large-scale use, the limitations of open-air drying are well known. Among these are high labour costs, large area requirement, lack of ability to control the drying process, possible degradation due to biochemical or microbiological reactions, insect infestation, and so on. The drying time required for a given commodity can be quite long and result in post-harvest losses (more than 30%). Solar drying of agricultural produces in enclosed structures by forced convection is an attractive way of reducing post-harvest losses as against the low quality of dried products associated with traditional open sun-drying methods. In many rural locations, specially, for the developing countries, grid-connected electricity and supplies of other non-renewable sources of energy are either unavailable, unreliable or, too expensive. In such conditions, solar dryers appear increasingly to be attractive as commercial propositions. The mode of drying in the presence of solar energy depends upon the method of solar energy collection and its conversion to useful thermal energy. In general, there are three modes of drying namely, open sun, direct and indirect modes.

6.6.1 Open Sun Drying

Drying/dehydration is a traditional method of preserving agri-produces and food materials. Drying in open sun may be termed as open sun drying or natural sun drying. In this technique, the product is spread in thin layers on a hard platform and the product is turned once or twice a day. This natural sun drying is simple and economical but suffers from many drawbacks such as (i) there is no control over the drying rate, the crop may be over dried resulting in discoloration, loss of germination power, nutritional changes, and sometimes complete damage (ii) there is no uniform drying, under dried condition favours the growth of harmful micro-organisms (iii) in case of slow drying, there can be deterioration of the materials due to the growth of fungi and bacteria (iv) the rain and dust storm may damage the crop, since in open drying there is no protection and (v) there will be considerable losses in the open sun due to birds, insects, rodents, etc. The moisture content and the temperature at which the food products need to be dried are always fixed, which is possible only in controlled drying not in open sun drying. There are several advantages of controlled drying of the products such as improvement in the product quality, storage capability and reduced wastage of materials, time, and space favouring better return from the dried products and improved transportability. Moreover, solar energy is more effective for food drying because of following reasons:

(i) Solar energy is diffuse in nature and provides low grade heat. This characteristics of solar energy is good for drying at low temperature, high flow rates with low temperature rise.

(ii) The intermittent nature of solar radiation will not hamper the drying performance at low temperature. Even the energy stored in the product itself will help in removing excess moisture during the period of no sun shine.

(iii) Solar energy is available at the site of use and saves transportation cost.

(iv) The high capital cost of solar dryers can be compensated if the dryer, developed can be used for drying of other products also.

The working principle of open sun drying is shown in Fig. 6.1a. The short wavelength solar radiation is incident on the uneven surface of crop/agri-produces. A part of this incident energy is reflected back to the atmosphere and the remaining part is absorbed by the crop surface. The absorbed radiation is converted into thermal energy and the temperature of the crop starts increasing. The attainment of higher temperature in the product, results in losses of energy due to emission of long wavelength radiation, convection, and mass transfer in the form of water vapour from the surface of the crop to the ambient through the surrounding air. The process is independent of any other source of energy except sunlight and hence the cheapest method, however has a number of limitations as discussed above. The convection heat loss depends on the speed of the blowing air over the crop surface. Evaporation of moisture takes place in the form of evaporative losses and so the crop gets dried. Further, a part of absorbed thermal energy in the crop is conducted into the interior of the product. This causes a rise in temperature and formation of water vapour inside the crop. The water vapour (moisture) diffuses towards the surface of the crop and is finally evaporated by absorbing the heat energy of the flowing air, thereby, mass is lost and transferred to the atmosphere in the form of water vapour. During initial stages, the moisture removal is rapid due to the availability of excess moisture on the surface of the product. Subsequently, drying depends upon the rate at which the moisture within the product moves to the surface by a diffusion process depending upon the type of the product. In general, open sun drying does not meet the required qualities of the products.

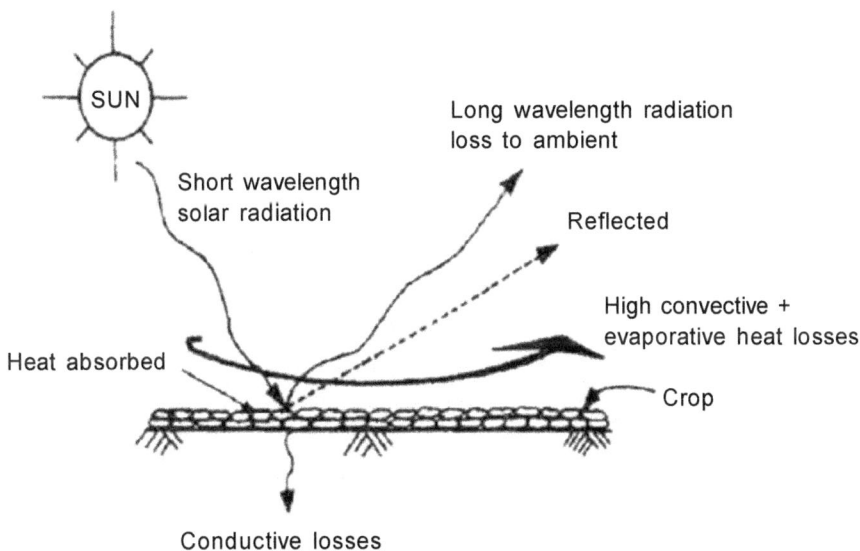

Fig. 6.1a. Working principle of open sun drying

6.6.2 Direct Solar Drying

Figure 6.1 b shows the working principle of direct solar crop drying. This system is also known as a solar cabinet dryer. Here the moisture is taken away by the air entering into the cabinet from the bottom part of the structure and escaping through the exit, located at the top as shown in the figure. In the cabinet dryer, of the total solar radiation impinging on the glass cover, a part is reflected back to atmosphere and the remaining is transmitted inside the cabinet. A part of the transmitted radiation is then reflected back from the crop surface and the rest is absorbed by the surface of the crop which causes its temperature to increase and thereby emit long wavelength radiations which are not allowed to escape to atmosphere due to the use of various glazing materials such as glass, fibre reinforced plastic, polythene sheet etc. as the covering component for the cabinet. The overall phenomena cause the rise in the temperature of the enclosure inside the cabinet. The glass cover in the cabinet dryer thus serves in reducing direct convective losses to the ambient which plays an important role in increasing the crop and cabinet temperature. The advantages of solar drying over open sun drying are that the method is simpler and cheaper to construct than the indirect-type for the same loading capacity and it offers protection of the materials from rains, dews, debris etc. The conductive heat losses to the ground is also reduced due to the bottom structure existing in the cabinet.

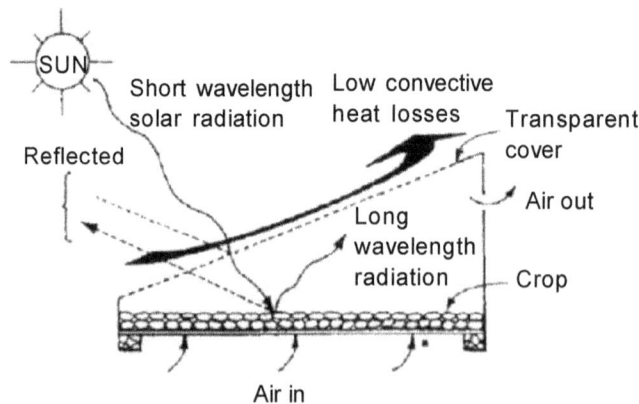

Fig. 6.1 b. Working principle of direct solar drying

6.6.3 Indirect Solar Drying

The working principle of indirect solar drying is shown in Fig. 6.1 c. It is generally known as conventional solar dryer. In indirect solar drying, the crop is not directly exposed to solar radiation, thus minimizing discolouration and cracking of the crop. The crops in these indirect solar dryers are placed in trays or shelves inside an opaque drying cabinet and a separate unit termed as solar collector is attached to it for heating of the entering air into the cabinet. The heated air is allowed to flow through wet crop. Here, the heat for moisture evaporation is provided by convective heat transfer between the hot air and the wet crop. The drying is basically achieved by the difference in the moisture concentration between the drying air and the air in the vicinity of crop surface. The advantages of indirect solar drying are that the method offers (i) a better control over drying and the product obtained is of better quality than sun drying, (ii) protection of crops due to the absence of the accumulation of localized heat and (iii) favourable conditions for deep bed drying and photo-sensitive crops. However, this method requires relatively elaborate structures involving more capital investment in equipment and incurs larger maintenance costs than the direct drying units.

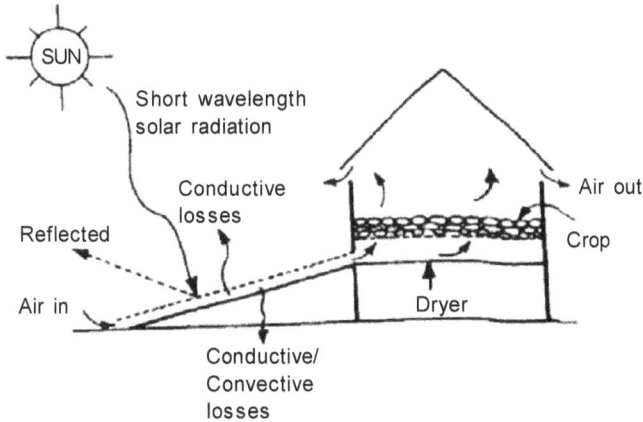

Fig. 6.1 c. Working principle of indirect solar drying

6.6.4 Hybrid Solar Drying

The hybrid solar drying combines the features of the direct and indirect type solar energy dryings as shown in figure (6.1 d) below. Here the combined action of incident direct solar radiation on the product to be dried and air pre-heated in a solar collector produce the necessary heat required for the drying process.

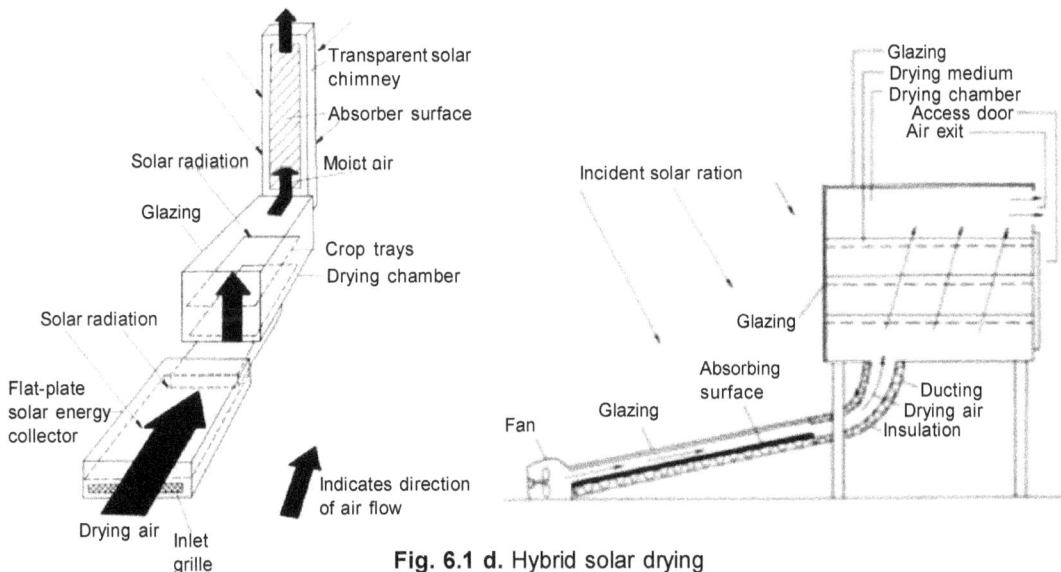

Fig. 6.1 d. Hybrid solar drying

6.6.5 Difference between open sun drying and solar drying

Open sun drying	Solar drying
Traditional method	More recent developments
Delayed drying	Fast drying
Problems of contamination by birds, insects etc.	No contamination
Less hygienic and less clean	Highly hygienic and very clean
Inferior quality products	Best quality products
Drying possible only on sunny days	Drying possible on all days including cloudy and rainy days with electrical backup
Poor sensory qualities to products	Highly acceptable sensory qualities to products
Poor appearance/colour and textures	Attractive appearance, color and texture
Uneven drying	Even/uniform drying
More nutrient loss	Better nutrient retention
Low profit margins	Best profit margins due to quality products
More space requirement	Less space requirement

6.7 CLASSIFICATION OF SOLAR DRYERS

On the basis of control over the drying of the agri-produces by using solar energy, the solar dryer can be broadly classified into open sun and controlled drying. The traditional way of drying agricultural products in developing countries is to spread the material in thin layer on a platform in open sun. This natural way of drying does not involve money except some labour, but results in poor quality of product due to no control over drying rate, unhygienic conditions, and spoilage. Solar energy can be effectively employed for controlled drying, resulting in good quality product. Actually, controlled drying means controlling the drying parameters like air temperature, humidity, drying rate, moisture content, and the air flow rate. Therefore, a solar dryer is to be designed carefully keeping all the above drying parameters in mind and the appropriateness in developing the dryer. Since, there are many options in the design of the solar dryers, hence there is large variety of solar dryers. These solar dryers, thus can be classified basically into three types i.e. the direct type or natural convection type dryers, mixed mode type dryers and indirect type dryer or forced circulation type dryers. The classification of controlled drying using solar energy is described as shown in the Fig. 6.2.

6.7.1 Natural Convection or Direct Type Solar Dryers

These dryers appear to be more attractive for use in developing countries since these do not use fan or blower to be operated by electrical energy. These dryers are also called as passive solar energy dryers. In this type of solar dryer, solar heated air is circulated through the crop either by buoyancy forces or as a result of wind pressure or in combination of both. Natural circulation solar energy dryers appear the most attractive option for use in remote rural locations. There are three types of passive solar energy dryers e.g. direct mode (integral type), indirect mode (distributed type) and the mixed mode type. The natural convection or direct type solar dryers are less costly and easy to operate. However, the problems with these dryers are of slow drying, not much control on temperature and humidity, small quantities can be dried and some products change colour and flavour due to direct exposure to sun. In its simplest form, they consist of some kind of enclosure and a transparent cover. The products get heated due to direct absorption of heat through solar radiation or due to high temperature of air in the enclosure and therefore, moisture from the product evaporates and goes out by the natural circulation of air. There are several designs of direct type dryers and these are developed keeping in mind either the availability of local materials required for its fabrication or for drying a particular product. Several dryers are fabricated, tested, and analysed in many countries. A simple cabinet dryer is a natural convection type. The solar cabinet dryer in its simple form consists of a wooden (or of any material) box of certain width and length (length is generally kept as three times its width), insulated at its base and also preferably at the sides and covered with a transparent roof. The details of the cabinet dryer are shown in Fig. 6.3.

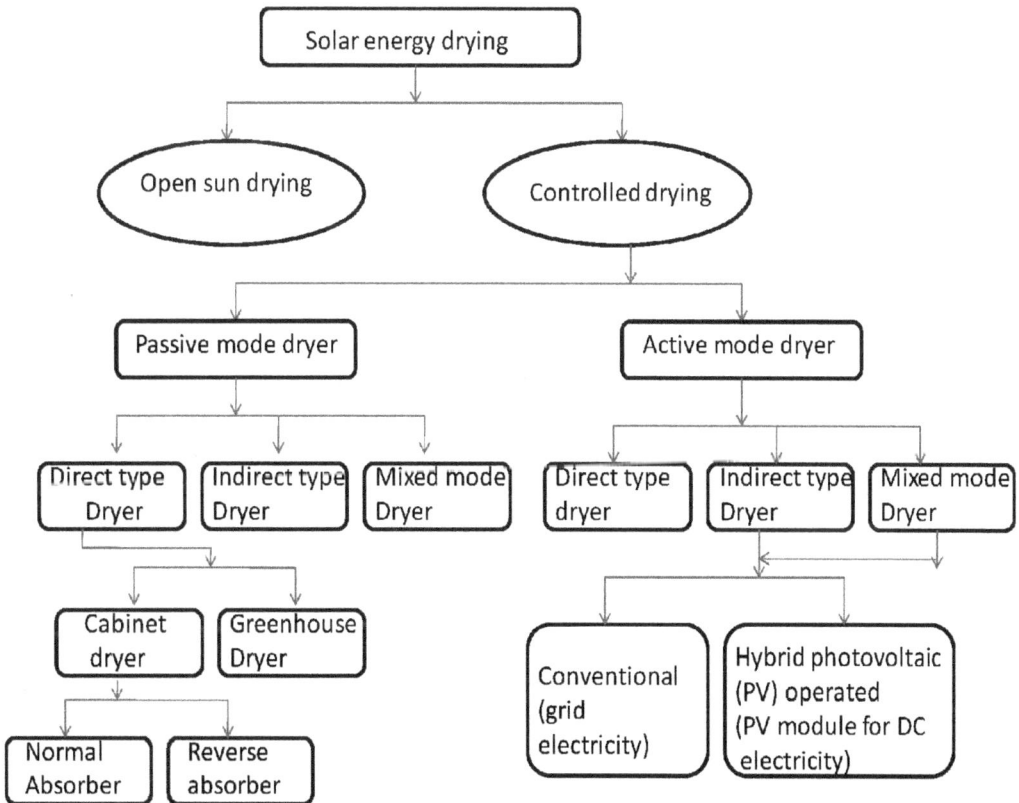

Fig. 6.2. Classifications of solar dryers

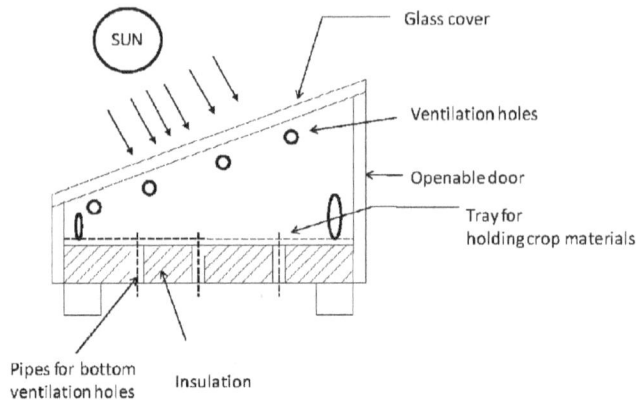

Fig. 6.3. Working principle of a cabinet type solar dryer

The inside surfaces of the box are coated with black paint and the product to be dried is kept in the trays made of wire mesh bottom. These loaded trays are kept through an openable door provided on the rear side of the dryer. Ventilation holes are made in the bottom through which fresh outside air is sucked automatically. Holes are also provided on the upper sides of the dryer through which moist warm air escapes. When the material is placed in the trays and exposed to solar radiation, the temperature of the product rises resulting in the evaporation of moisture. This warm moist air passes through upper ventilation holes by natural convection, creating a partial vacuum and drawing fresh air up through the holes provided in the base of the dryer. Temperatures as high as 70 ^{0}C have been reported in this dryer when it is empty. The drying time can be reduced from one half to one third of the time required during open sun drying.

6.7.2 Mixed Mode Type Solar Dryer

In the mixed-mode type of dryers, the solar air heater without any fan along with the drying bin is used. The flow of air is generally by natural convection. This type of dryer consists of a simple air heater, drying chamber, and a tall chimney used to increase the convection effect. The dryer as shown schematically in Fig. 6.4 consists of solar air heater made of a frame of bamboo poles and wire covered with 0.15 mm thick transparent PVC sheet. The base of the air heater is covered with burnt rice husk which absorbs the solar radiation and heats the air in contact. The hot air in this air heater rises to the drying chamber which either consists of transparent PVC sheets on bamboo frame absorbing directly the solar radiation or a bamboo frame covered from all the four sides with some opaque material. The drying material is kept on the tray in thin layer through which hot air heated from air heaters enters its bottom and goes up in to the chimney. The chimney is a long cylinder made of bamboo frame covered with black PVC to keep the inside air warm. There is a cap at the top of the chimney, leaving some space in between chimney top and cap to allow warm humid air to go out and protecting the product from rain and other foreign materials. The height of the chimney and the hot air inside it creates a pressure difference between its top and bottom thereby creating forced movement of air through the crop bed to the top of the chimney. This dryer is suitable for the materials requiring slow and low temperature drying. The drying rate will depend on the depth of the bed, initial moisture content of the material, solar insolation, ambient temperature, and the design of the dryer.

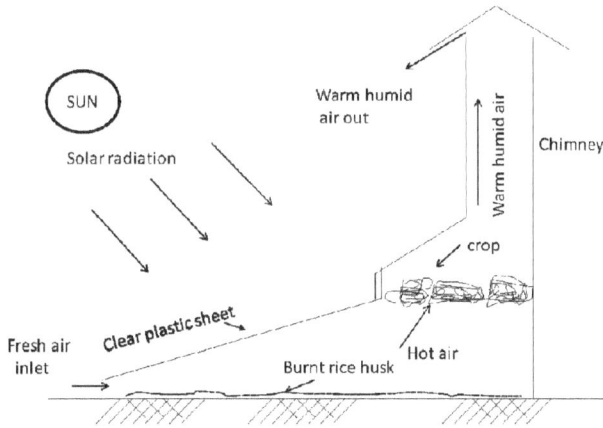

Fig. 6.4. Working principle of a mixed mode type solar dryer

6.7.3 Normal and Reverse Absorber Based Cabinet Solar Dryer

According to the arrangement of absorber plate and the incidence of solar radiation on the absorber plate, cabinet dryer can be classified as normal or reverse absorber type.

(i) Normal absorber cabinet Dryer: In this case, solar energy falls on the absorber from top (figure 6.3 above for cabinet dryer). A normal absorber cabinet dryer has the following limitations:

    a. Its use is limited to small scale applications due to its small capacity.

    b. Discolouration of crop due to direct exposure to solar radiation.

    c. Moisture condensation inside glass cover reducing its transmissivity.

    d. Sometimes the insufficient rise in crop temperature affects moisture removal.

    e. Limited use of selective coatings on the absorber plate.

(ii) Reverse Absorber Cabinet Dryer (RACD): In this case, the absorber plate is facing downward and the solar radiation is allowed to fall on it from the reflector placed underneath. The schematic view of RACD without and with glass has been shown in Figs. 6.5 (a) and (b). In this case, the crop is not directly exposed to solar radiation to minimize discolouration and cracking on the surface of the crop and convective heat loss from the absorber is suppressed due to facing of absorber in downward direction. The drying chamber is used for keeping the crop in wire mesh tray. A

downward facing absorber (reverse absorber) is fixed below the drying chamber at a sufficient distance (≈ 0.05 m) from the bottom of the drying chamber. A cylindrical reflector, placed under the absorber is fitted with the glass cover on its aperture to minimize convective heat losses from the absorber. The absorber can be selectively coated. The inclination of the glass cover is taken as 45^0 from the horizontal to receive maximum solar radiation. Solar radiation after passing through the glass cover is reflected by cylindrical reflector towards the absorber. After absorption, a part of this is lost to ambient through the glass cover and the remaining is transferred to the flowing air above it by convection. The flowing air is thus heated and passes through the crop placed in the drying chamber. The crop is heated and moisture is removed through a vent provided at the top of drying chamber.

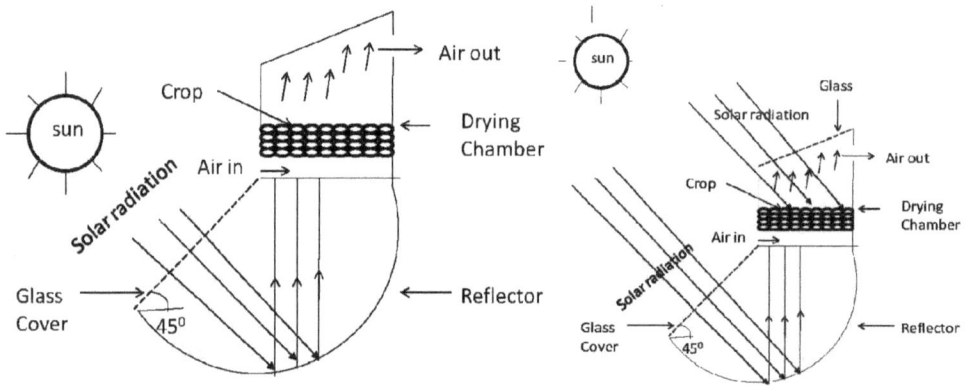

Fig. 6.5 a & b. Schematic view of RACD without and with glass

6.7.4 Forced circulation type dryer

All of the three principal solar dryer types (direct, indirect and mixed) can be further sub-divided depending on the method used to move air through the dryer, i.e., natural convection and forced convection.

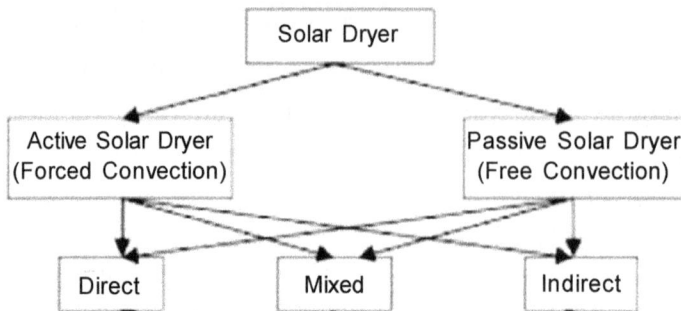

The forced convection or circulation type of dryer uses blower for the circulation of air which is mostly operated electrically. Such dryers are more efficient, faster, and can be used for drying large quantities of agricultural products. In a natural convection solar

dryer, the air moves through the system because of the difference in density between the ambient air and the air inside the dryer. The air inside the dryer is less dense than the ambient air because it has been heated. The less dense air rises up through the dryer and creates a small negative pressure that in turn induces fresh ambient air into the system. Because of the small pressure differences, the airflow rates are also small. Researchers have reported the air velocities in the range of 0.1-0.5 m s^{-1}. Since the air temperatures in the dryer (natural convection) are dependent on solar radiation, the airflow in a natural convection system is also variable. Airflow rates may be increased marginally through the addition of a solar chimney on the outlet side of the dryer to increase the density difference. However, the relatively low and variable airflow, inherent in a natural convection solar dryer is its main limitation. As a consequence, the performance is usually inferior to a forced convection system. On the other hand, a natural convection solar dryer has lower capital and running costs because there is no need of electric fan. This type of dryer is also sometimes the only choice in locations that do not have access to electricity.

However, in a forced convection system (fig. 6.6), a fan is used to move air through the dryer. Higher air velocities, up to 3 m s^{-1} through the crop usually improve the drying rates especially in the constant rate drying period. The airflow rate can also be controlled and varied depending on the stage of drying. In addition, the quantity of the crop in the dryer can be increased because the fan can overcome any additional resistance to airflow. The disadvantages of a forced convection system are the increased capital and running costs and the requirement for electricity. In commercial solar drying systems, however, the advantages mentioned usually far outweigh the disadvantages and an electric fan is used if possible.

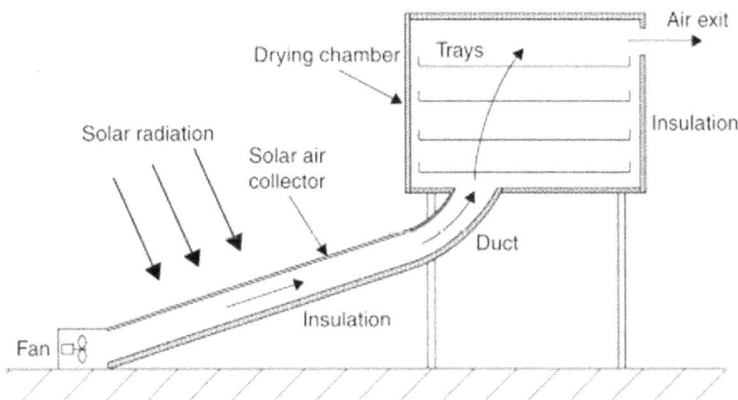

Fig. 6.6. Working principle of a forced circulation type solar dryer

6.7.5. Greenhouse dryers

The idea of greenhouse dryer is to replace the function of the solar collector by a greenhouse system. The air inside the greenhouse gets heated and its temperature rises due to the greenhouse effect. The glazing materials, such as glass and some plastics provided in the roofs and walls of the greenhouse, are transparent for the short wavelength radiations of sun light, but opaque to the infrared long wave radiations (thermal heat radiation). Placing glass or plastic over the agricultural commodity for drying will produce a greenhouse effect to trap solar radiation in two ways: (i) the glazing acts as on opaque cover to the thermal heat radiation radiated by the agricultural commodity, thus trapping its heat energy within the cover (though, some of the incident radiations are lost due to the reflection from the cover) and (ii) the glazing cover serves as an envelope to reduce the convective heat loss. The absorbed heat from the commodity is lost to the ground. However, the total resulting effect in the greenhouse is better than the open-sun drying. The working of greenhouse drying is shown in fig. 6.7

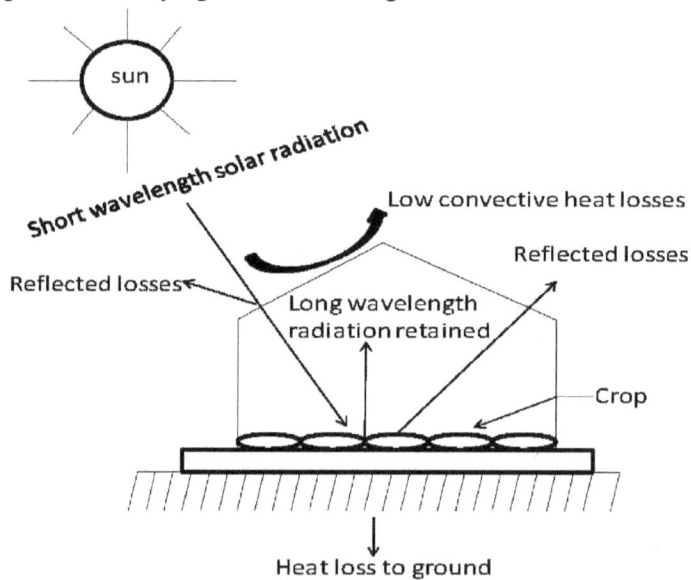

Fig. 6.7. Schematic diagram of a greenhouse dryer

The roofs and walls of this solar dryer can be made of transparent materials such as glass, fibre glass, UV stabilized plastic or polycarbonate sheets. The transparent materials are fixed on a steel frame support or pillars with bolts and nuts and rubber packing to prevent humid air or rain water leaking into the chamber other than those introduced from the inlet opening. To enhance solar radiation absorption, black surfaces should be provided within the structure.

Inlet and exhaust fans are placed at proper position within the structure to ensure even distribution of the drying air. Designed properly, the greenhouse dryers allow a greater degree of control over the drying process than the cabinet dryers and they are more appropriate for large scale drying. The schematic diagram of a greenhouse dryer is shown in Fig. 6.8.

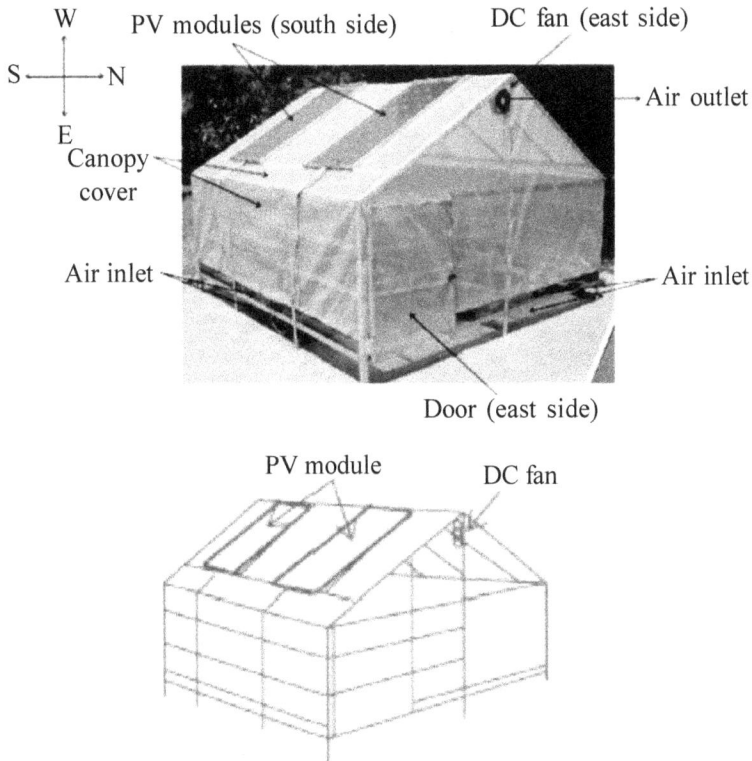

Fig. 6.8. Schematic diagram of a greenhouse dryer

6.8 SOME TERMS RELATING TO DRYING OF PRODUCTS

Drying of agri-produces/food materials involves transfer of water from the product to the air around it. Hence, it is required to determine the capacity of air to accept water vapour. The terms such as moisture content of the product, absolute humidity and relative humidity of the surrounding air, equilibrium moisture content of the product, water activity of a product, properties of surrounding air (temperature, water vapour content, air-vapour mixture, vapour pressure, thermodynamic relations etc.), drying rate, drying efficiency, etc. need to be discussed.

6.8.1 Moisture Content

Moisture content refers to the amount of water contained in a product, expressed as a percentage of the weight of its oven dried condition. No agricultural produce in its natural state is completely dry. Some water is always present. There are two ways to measure moisture content i.e., wet basis and dry basis. Although the most convenient way to express moisture for mathematical calculations is on dry basis but for agricultural products, moisture content normally is expressed in wet basis. However, use of the wet basis has one clear disadvantage i.e. the total mass changes as moisture is removed. Since the total mass is the reference base for the moisture content, the reference condition is changing as the moisture content changes. On the other hand, the amount of dry matter does not change. Thus, the reference condition for dry basis measurements does not change as moisture is removed. Hence, for engineering research purpose, the dry basis (moisture content) is preferred and used. The general equations for expressing moisture contents are as follows.

$$\text{Moisture content wet basis, } W_w = \frac{m_w}{m_w + m_d} = \frac{m_w}{m_t} \qquad \text{... (6.1)}$$

$$\text{Moisture content dry basis, } \quad W_d = \frac{m_w}{m_d} \qquad \text{... (6.2)}$$

where W_w = decimal moisture content wet basis (wb), W_d = decimal moisture content dry basis (db), m_d = mass of dry matter in the product, m_w = mass of water in the product, m_t = total mass of the product (water plus dry matter). The percent moisture content is found by multiplying the decimal moisture content by 100. In addition, relationships between wet and dry moisture content on a decimal basis can be derived from Equations 6.1 and 6.2. Those relationships are as follows,

$$\text{or,} \qquad W_d = \frac{W_w}{1 - W_w} \text{ or, } W_w = \frac{W_d}{1 + W_d} \qquad \text{... (6.3)}$$

For a given product, the moisture content dry basis is always higher than the wet basis moisture content. This is obvious from a comparison of Equations 6.1 and 6.2. The difference between the two bases is small at low moisture levels, but it increases rapidly at higher moisture levels. Figure 6.9 below shows the plot of relationship between the dry and wet basis moisture content.

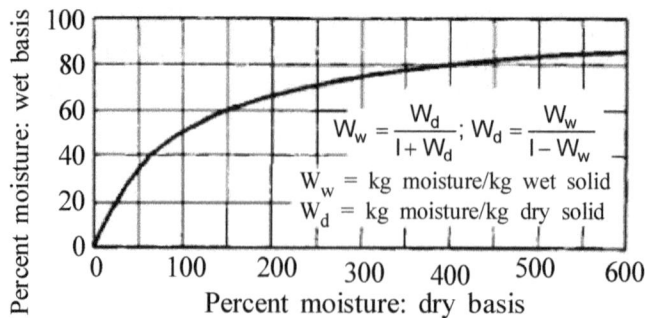

Fig. 6.9. Plot of relationship between the dry and wet basis moisture content

The usual method of measuring the moisture content of a product is by using an oven in which, the product is kept for 24 hours at a temperature of 105 ^{0}C. The weight of the sample is measured before and after oven drying and the moisture content on dry and wet basis is calculated using the equations 6.1 and 6.2.

The mass of water removed (M_w) from the wet product can be calculated as

$$M_w = \frac{m_0\left(m_i - m_f\right)}{100 - m_f} \qquad \dots (6.4)$$

where m_0 is the initial mass of the product, m_i is the initial m.c. (w.b.) on percent and m_f is the final m.c. (w.b.) on percent.

Example 6.1: 1000 kg of paddy with a moisture content of 24 % (d.b) is dried to a final moisture content of 12 % (d.b). Calculate the amount of moisture evaporated and final weight of the paddy after drying.

Solution: First of all, it is required to calculate the dry weight of the material. Moisture content (d.b) = (Initial weight-final weight)/final weight $\rightarrow$ 0.24 = (1000-final weight)/final weight $\rightarrow$ Final dry weight = 806 kg dry weight (which remains same throughout the calculation). For 12 % moisture content (d.b.), the dry weight will remain same but the final weight will change.

0.12 = (Initial weight-final weight)/final weight = (x-806)/806 $\rightarrow$ Initial weight = 903 kg. Hence the amount of moisture evaporated = 1000-903 = 97 kg.

Example 6.2: Calculate the weight of paddy, with 38 % moisture content (w.b.), required to produce 1500 kg of final product (paddy) with 14 % moisture content (w.b.).

Solution: For 14 % moisture content (w.b.) with 1500 kg final product, the dry weight can be calculated as 0.14 = (1500-dry weight)/1500 $\rightarrow$ dry weight of the product = 1290 kg.

For 38 % moisture content (w.b.), the weight of paddy can be calculated as 0.38 = (x-dry weight)/x $\rightarrow$ 0.38 = (x-1290)/x where x is the weight of the paddy for moisture content 38% (w.b.) and x = 2080.65 kg

Example 6.3: A bin holds 2000 kg of wet grain containing 500 kg of moisture. This grain is to be dried to a final moisture content of 14% (wb). (i) What are the initial and final moisture contents of the grain (wet basis, dry basis, decimal and percent)? (ii) How much moisture is removed during drying?

Solution: Using equations, 6.1 and 6.2, the initial moisture contents are

$W_w = \dfrac{500}{2000} = 0.25 = 25\%$. (wet basis) and $W_d = \dfrac{500}{2000 - 500} = 0.3333 = 33.33\%$. (dry basis). Final moisture content =14 % (wb). Using equation, 6.4, the quantity of water removed during drying, $M_w = \dfrac{m_0\left(m_i - m_f\right)}{100 - m_f} = \dfrac{2000(25 - 14)}{100 - 14} = 255.81 \approx 256\,\text{kg}.$

6.8.2 Equilibrium Moisture Content (EMC)

It refers to the moisture content at which the material neither gains nor loses moisture when surrounded by air at a given relative humidity and temperature. EMC is used to indicate the potential of the atmosphere to bring the material to a specific moisture content during a drying operation. Agricultural produces/food materials being hygroscopic, lose or gain moisture until equilibrium is reached with the surrounding air. A material held for a long time at a given temperature and relative humidity of the surrounding air, attains a moisture content that is in equilibrium with the neighbouring air. This does not mean that the material and the air have the same moisture content. It simply means that an equilibrium condition exists such that there is no net exchange of moisture between the material and the air. This EMC is a function of the temperature, the relative humidity and the product. EMCs for a range of grains are shown in Table 6.6.

Table 6.6: Grain Equilibrium Moisture contents (Fuller R.J., 1993)

Grain	Relative Humidity (%)							
	30	40	50	60	70	80	90	100
Equilibrium Moisture Content EMC (% *wb*) at 25 °C								
Barley	8.5	9.7	10.8	12.1	13.5	15.8	19.5	26.8
Shelled maize	8.3	9.8	11.2	12.9	14.0	15.6	19.5	26.8
Paddy	7.9	9.4	10.8	12.2	13.4	14.8	16.7	-
Milled rice	9.0	10.3	11.5	12.6	12.8	15.4	18.1	23.6
Sorghum	8.6	9.8	11.0	12.0	13.8	15.8	18.8	21.9
Wheat	8.6	9.7	10.9	11.9	13.6	15.7	19.7	25.6

Under no circumstances it is possible to dry to a moisture content lower than the EMC associated with the given temperature and humidity of the drying air; for example, the data in Table 6.6 above show that paddy can only dry to a moisture content of 16.7% when exposed to air at 25°C and 90% relative humidity. If paddy at moisture content less than 16.7% is required, then either the temperature of the drying air has to be increased or its humidity to reduce. After attaining EMC, further exposure of the material to this air (with given temperature and humidity of surrounding air) for indefinitely long periods does not bring about any additional loss of moisture. The moisture content in the material can be reduced further by exposing it to the air of lower relative humidity. At the equilibrium condition, the moisture desorption from the product is in dynamic equilibrium with the absorption of the moisture for the surrounding air. Relative humidity at this point is known as the *"Equilibrium Relative Humidity (ERH)"*. Figure 6.10 below presents the curve for variation of EMC and relative humidity (*RH*) for four temperatures of the surrounding air.

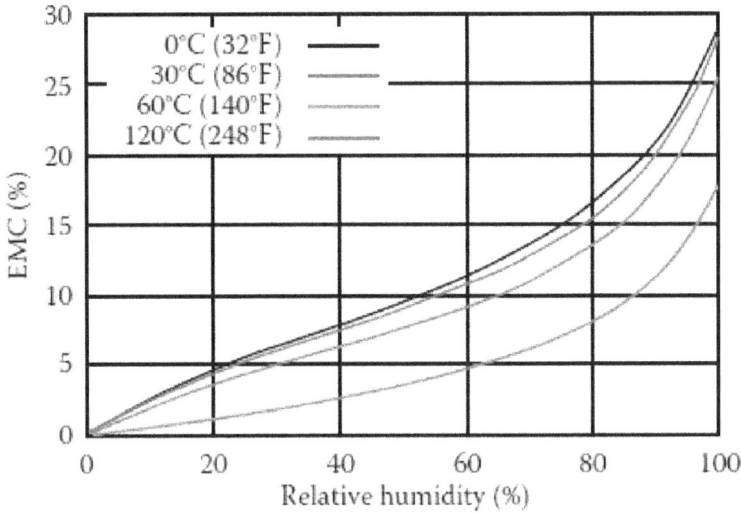

Fig. 6.10: Curves for EMC versus RH at four temperatures

Numerous equations have been proposed to represent the EMC curves for various products. No single equation is suitable for all products; however, most products can be represented by one of several equations available. The usual equations followed are Henderson's equation, modified Henderson's equation and Modified Halsey are mentioned below.

$$ERH = 1 - e^{KtM_e^N} \qquad \text{Henderson} \qquad \qquad \dots (6.5)$$

$$ERH = 1 - e^{-K(t+C)M_e^N} \qquad \text{Modified Henderson} \qquad \dots (6.6)$$

$$ERH = \exp\left\{ \frac{\exp(K+Ct)}{M_e^N} \right\} \qquad \text{Modified Halsey} \qquad \dots (6.7)$$

$$ERH = \exp\left\{ \frac{-K}{M_e^N} \right\} \qquad \text{Halsey} \qquad \dots (6.8)$$

where: ERH = Equilibrium relative humidity, decimal, M_e = equilibrium moisture content, percent, dry basis, t = temperature, °C and K, N, C are constants determined for each material (Table 6.7).

Example 6.4: Compute the equilibrium moisture content for popcorn at 20 ^{0}C and 50 % relative humidity?

Solution: We first rearrange Equation (6.6) to express equilibrium moisture content in terms of other parameters.

$$M_e = \left(\frac{\ln(1-ERH)}{-K(t+C)} \right)^{\frac{1}{N}}$$

Here, $ERH = 0.5$ and $t = 20$ ^{0}C. The values for N, K and C can be obtained from the table 6.7 above. N= 1.5978, K= -1.5593 $\times$ 10^{-4}, C = 60.754. Putting all those values, $M_e = 12.19$ % (db).

Table 6.7: Constants for Modified Henderson and Halsey Equations. (From ASAE, 2000.)

Grain	K	N	C	Equation
Beans	4.4181	1.7571	-0.011875	7.7
Beans	0.1633	1.567	87.46	7.6
Canola meal	0.000103	1.6129	89.99	7.6
Corn, shelled	6.6612×10^{-5}	1.9677	42.143	7.6
Popcorn	1.5593×10^{-4}	1.5978	60.754	7.6
Peanut, kernel	3.9916	2.2375	-0.017856	7.6
Pumpkin seed	3.3725×10^{-5}	3.4174	1728.729	7.6
Pumpkin seed	3.3045×10^{-5}	3.3645	1697.76	7.6
Rice	3.5502×10^{-5}	2.31	27.396	7.6
Soybean	2.87	1.38	-0.0054	7.7
Wheat	4.3295×10^{-5}	2.1119	41.565	7.6

Table 6.8: Selected EMC relationships for food products. Constants are valid only for the equation number listed.

Product	Temperature	RH range	Equation	K	N
Apple	30°C	0.10 – 0.75	7.5	0.1091	0.7535
Apple	19.5°C	0.10 – 0.70	7.8	4.4751	0.7131
Banana	25°C	0.10 – 0.80	7.5	0.1268	0.7032
Chives	25°C	0.10 – 0.80	7.8	11.8931	1.1146
Grapefruit	45°C	0.10 – 0.80	7.5	0.1519	0.6645
Mushrooms	20°C	0.07 – 0.75	7.8	7.5335	1.1639
Mushrooms	25°C	0.10 – 0.80	7.8	11.5342	1.1606
Peach	20 -30°C	0.10 – 0.80	7.5	0.0471	1.0096
Peach	40°C	0.10 – 0.80	7.5	0.0440	1.1909
Peach	50°C	0.10 – 0.80	7.5	0.0477	1.3371
Pear	25°C	0.10 – 0.80	7.5	0.0882	0.7654
Tomato	17°C	0.10 – 0.80	7.8	10.587	0.9704

6.8.3 Moisture Extraction Rate and Specific Moisture Extraction Rate During Drying

In order to determine the effectiveness of dryer systems, the specific moisture extraction rate (SMER) and moisture extraction rate (MER) are used. Specific moisture extraction rate (SMER) describes the effectiveness of the energy used in the drying process. SMER is defined as kilogram of moisture removed per kilowatt-hour of energy consumption and is related to the total power to the dryer including the fan power and the efficiencies of the electrical devices etc.

$$SMER = \left(\frac{Amount\ of\ water\ removed\ during\ drying}{Total\ energy\ supplied\ during\ drying\ process} \right) \frac{kg}{kWh}$$

Moisture extraction rate (MER) is defined as kilogram of moisture removed per hour and indicates the dryer capacity or throughput rate.

$$MER = \left(\frac{Amount\ of\ water\ removed\ during\ drying}{Drying\ time} \right) = \frac{kg}{h}$$

6.8.4 Relative Humidity (RH)

Humidity is the moisture content of air. It is important to differentiate between the absolute humidity and relative humidity of air. *Absolute humidity* is the moisture content of the air (mass of water vapour per unit mass of air, e.g. g/kg) whereas the *relative humidity (RH)* is the ratio, expressed as a percentage, of the moisture content of the air at a specified temperature to the moisture content of air if it were saturated at that temperature. Hence, 0 % *RH* refers to air completely dry and 100% for the air that is fully saturated with water vapour. Low *RH* (or dry) air must be blown over the products so that it has the capacity to pick up water vapour from them and remove it. If high *RH* (or wet) air is used, it quickly becomes saturated and cannot pick up further water vapour from the products to be dried. The temperature of the air affects the humidity (higher temperatures reduce the humidity and allow the air to carry more water vapour). Knowledge about relative humidity is required for assessing the drying rate of the products. At a given temperature, the air cannot absorb unlimited quantities of water vapor. The air is considered to be "saturated" when it is unable to absorb water vapor at a given temperature and has a relative humidity of 100%. Table 6.9 shows the maximal weights of water vapor contained in 1 kg of air, where *RH%* is the relative humidity of the air (in percent).

$$RH\left(\%\right) = \frac{Weight\ of\ water\ vapour\ in\ 1\ kg\ of\ air}{Weight\ of\ water\ vapour\ in\ 1\ kg\ of\ saturated\ air} \times 100,\ \text{at a given temperature}$$

Table 6.9: Maximal Weights of Water Vapour Contained in 1 kg of Air

Air temperature	0°C	10°C	20°C	30°C	40°C
Maximal water vapour weight (in grams)	3.9	7.9	15.2	28.1	50.6
Air temperature	50°C	60°C	70°C	80°C	90°C
Maximal water vapour weight (in grams)	89.5	158.5	289.7	580.0	1559

Air containing a given amount of water vapor tends to become saturated if its temperature is lowered. On the contrary, if the "drying power" of air is required to be increased (its capacity for absorbing more water vapor), it is necessary to heat it. For example, if air saturated at 15.2 g of water vapor per kilogram of air at a temperature of 20°C has a 100% relative humidity and when temperature is increased, a kilogram of air can hold more water vapor (28.1 g of water vapor is required to saturate a kilogram of air at 30°C). Therefore, *RH* is decreased to 15.2:28.1, or 54%.

6.8.5 Water Activity (a_w)

Dried food products are considered safe with respect to microbial hazards. Water plays an important role in the stability of fresh, frozen and dried foods. It acts as a solvent for chemical, microbiological and enzymatic reactions. The water activity (a_w) is a measure of the availability of water to participate in such reactions. There is a critical water activity below which no microorganism can grow. The water in an agricultural produce/food material exerts a vapour pressure. The extent of this pressure depends on the amount of water present, the temperature and the composition of the food. Different food components lower the water vapour pressure to different extents, with salts and sugars being more effective than larger molecules such as starches or proteins. Thus, two different foods with similar moisture contents may not necessarily have the same a_w. The amount of water in food and agricultural products affects the quality and perishability of these products. However, perishability is not directly related to moisture content. In fact, perishability varies greatly among products with the same moisture content. A much better indicator of perishability is the availability of water in the product to support degradation activities such as microbial action. The term '*water activity*' is widely used in the food industry as an indicator of water availability in a product. However, humidity and relative humidity refers to the water vapour present in the surrounding air. Hence, water activity a_w, is of great importance for food preservation as it is a measure and a criterion of microorganism growth and probably toxin release, of enzymatic and non-enzymatic browning development. For every food or agricultural product, there exists an activity limit below which microorganisms stop growing. The vast majority of bacteria grows at about a_w = 0.85, mould and yeast at about a_w = 0.61, fungi at a_w < 0.70, etc. In these cases, water activity is regulated in detail, after drying by the addition of some solutions of sugars, starch, etc.

Water activity can therefore be defined as the ratio of the vapour pressure exerted by the agricultural produces/food materials to the saturated vapour pressure of water at the same temperature.

$$a_w = \frac{vapour\ pressure\ of\ water\ exerted\ by\ agricultural\ produce/food\ materials}{saturated\ vapour\ pressure\ of\ water\ at\ the\ same\ temperature}$$

Values range from 0 for dried foods to 1.0 for foods such as milk, fresh fruit where the water is readily available. There is a relationship between a_w and *RH* of the form, $a_w = \frac{RH}{100}$. If we take a sample of a food product and place it in an enclosed container at a fixed temperature, the product will exchange moisture with the air surrounding it. After a period of time, as with the equilibrium moisture example noted in the previous section, an equilibrium condition will occur. The product no longer has any net change in moisture. The water activity is equal to the decimal relative humidity at that condition. Hence, the ratio of the equilibrium vapour pressure to the saturation vapour pressure is known as the equilibrium relative humidity (ERH), or water activity.

6.8.6 Properties of Surrounding Air

For effective drying, air should be hot, dry and moving. These factors are inter-related and it is important that each factor has its contributions for drying of a product as cold moving air or wet moving air are each unsatisfactory. The relationship between temperature, humidity, vapour pressure and other thermodynamic properties of air is referred through a psychometric chart.

(i) *Vapour Pressure and phase equilibrium: Vapour pressure:* The pressure in a gas container is due to the individual molecules striking the wall of the container and exerting a force on it. This force is proportional to the average velocity of the molecules and the number of molecules per unit volume of the container (i.e., molar density). Therefore, the pressure exerted by a gas is a strong function of the density and the temperature of the gas. For a gas mixture, the pressure measured by a sensor such as a transducer is the sum of the pressure exerted by the individual gas components, called the partial pressure. The partial pressure of a gas in a mixture is proportional to the number of moles (or the mole fraction) of that gas.

Atmospheric air can be viewed as a mixture of dry air (air with zero moisture content) and water vapour (*also referred to as moisture*), and the atmospheric pressure is the sum of the pressure of dry air P_a and the pressure of water vapour, called the vapour pressure P_v (Fig. 6.11).

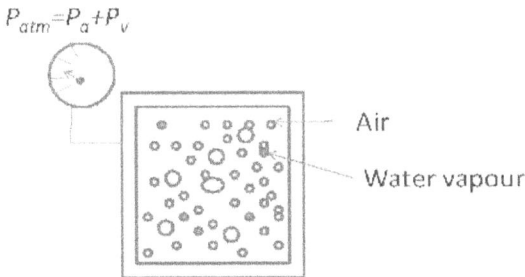

Fig. 6.11. Atmospheric pressure (Sum of dry air pressure and vapour pressure)

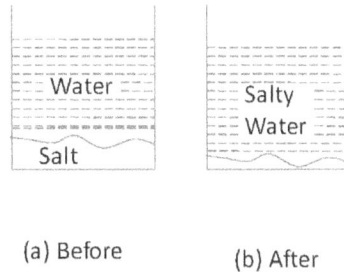

Fig. 6.12. Flow from high to low concentration (sum of dry air pressure and vapour pressure)

That is, $P_{atm} = P_a + P_v$. The vapour pressure constitutes a small fraction (usually under 3 percent) of the atmospheric pressure since air is mostly comprised of nitrogen and oxygen, and the water molecules constitute a small fraction (usually under 3 percent) of the total molecules in the air. However, the amount of water vapour in the air has a major impact on thermal comfort and many processes such as drying.

Air can hold a certain amount of moisture only, and the ratio of the actual amount of moisture in the air at given temperature to the maximum amount air can hold at that temperature is called the *relative humidity* (RH). The relative humidity ranges from 0 for dry air to 100 percent for saturated air. (air that cannot hold any more moisture). The vapour pressure of saturated air at a given temperature is equal to the saturation pressure of water at that temperature. For example, the vapour pressure of saturated air at 25^0C is 3.17 kPa. The amount of moisture in the air is completely specified by the temperature and the relative humidity, and the vapour pressure is related to relative humidity ϕ by $P_v = \phi P_{sat \, @ \, T}$ where $P_{sat \, @ \, T}$ is the saturation pressure of water at the specified temperature. For example, the vapour pressure of air at 25^0C and 60 percent relative humidity is $Pv = \phi P_{sat \, @ \, 25°C} = 0.6 \times (3.17 kPa) = 1.90 \; kPa$. The desirable range of relative humidity for thermal comfort is 40 to 60 percent.

Note that the amount of moisture air can hold is proportional to the saturation pressure, which increases with temperature. Therefore, air can hold more moisture at higher temperatures. Dropping the temperatures of moist air reduces its moisture capacity and may result in the condensation of some of the moisture in the air as suspended water droplets (fog) or as a liquid film on cold surfaces (dew). So it is no surprise that fog and dew are common occurrences at humid locations especially in the early morning hours when the temperature are the lowest. Both fog and dew disappear (evaporate) as the air temperatures rises shortly after sunrise.

Phase equilibrium: It is common observation that whenever there is an imbalance of a commodity in a medium, nature tends to redistribute it until a "balance" or "equality" is established. This tendency is often referred to as the driving force, which is the mechanism behind many naturally occurring transport phenomena such as heat transfer, fluid flow, electric current, and mass transfer. If we define the amount of a commodity per unit volume as the concentration of that commodity, we can say that the flow of a commodity is always in the direction of decreasing concentration, that is, from the region of high concentration to the region of low concentration (Fig. 6.12). The commodity simply creeps away during redistribution, and thus the flow is a diffusion process.

We know from experiences that a wet T-shirt hanging in an open area eventually dries, a small amount of water left in a glass evaporates, and the aftershave in an open bottle quickly disappears. These and many other similar examples indicate that there is a driving force between the two phases of a substance that forces the mass to transform from one phase to another. The magnitude of this forces depends on the relative concentrations of the two phases. A wet T-shirt dries much faster in dry air than it be in humid air. In fact, it does not dry at all if the relative humidity of the environment is 100 percent and thus the air is saturated. In this case, there is no transformation from the liquid phase to the vapour phase and the two phases are in *phase equilibrium*. For liquid water that is open to the atmosphere, the criterion for phase equilibrium can be expressed as follows,

The vapour pressure in the air must be equal to the saturation pressure of water at the water temperature. That is, (Fig. 6.13), phase equilibrium criterion for water exposed to air:

Water Vapour

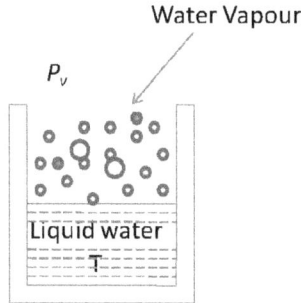

Fig. 6.13: When open to atmosphere, water remains phase equilibrium with vapour in air if vapour pressure equals to saturation temperature of water

Therefore, if the vapour pressure in the air is less than the saturation pressure of water at the water temperature, some liquid will evaporate. The larger the difference between the vapour and saturation pressures, the higher the rate of evaporation. The evaporation offers a cooling effect on water, and thus reduces its temperature. This, in turn, reduces the saturation pressure of water and thus the rate of evaporation until some kind of quasi-steady operation is reached. This explains why water is usually at a considerably lower temperature than the surroundings air, especially in dry climates. It also suggests that the rate of evaporation of water can be increased by increasing the water temperature and thus the saturation pressure of water.

Note that the air at the water surface will always be saturated because of the direct contact with water, and thus the vapour pressure. Therefore, the vapour pressure at the lake surface is simply the saturation pressure of water at the temperature of the water at the surface. If the air is not saturated, then the vapour pressure decreases to the value in the air at some distance from the water surface, and the difference between these two vapour pressure is the driving force for the evaporation of water.

Example 6.5: Temperature drop of a lake due to evaporation

On a summer day, the air temperature over a lake is measured to be 25^0C. Determine water temperature of the lake when phase equilibrium conditions are established between the water in the lake and vapour in the air for relative humidities of 10, 80 and 100 percent for the air (Fig. 6.14).

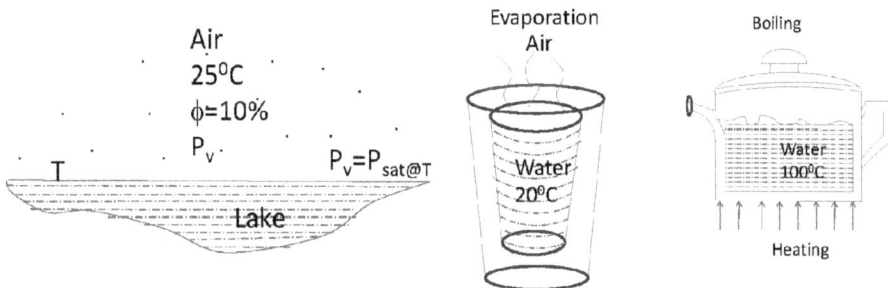

Fig. 6.14 Schematic diagram for the example   Fig. 6.15. Evaporation of water   Boiling of water

Solution: The data for saturation pressure of water 25°C is 3.17 kPa. Then the vapour pressures at relative humidities of 10, 80, and 100 percent are determined as per the above expression mentioned earlier.

Relative humidity = 10%: $P_{v1} = \phi_1 \, P_{sat \, @ \, 25°C} = 0.1 \times (3.17 \; kPa) = 0.317 \; kPa$

Relative humidity = 80%: $P_{v2} = \phi_2 \, P_{sat \, @ \, 25°C} = 0.8 \times (3.17 \; kPa) = 2.536 \; kPa$

Relative humidity = 100%: $P_{v3} = \phi_3 \, P_{sat \, @ \, 25°C} = 1.0 \times (3.17 \; kPa) = 3.17 \; kPa$

The saturation temperatures corresponding to these pressure are determined from Table below by interpolation to be

$T_1 = -8.0°C$ $\qquad\qquad T_2 = 21.2°C$ $\qquad$ and $\quad T_3 = 25°C$

Therefore, water will freeze in the first case even though the surrounding air is hot. In the last case the water temperature will be the same as the surrounding air temperature.

Table 6.10: Saturation pressure of water at various temperatures

Temperature (T°C)	Saturation pressure, P_{sat}, kPa
-10	0.26
-5	0.40
0	0.61
5	0.87
10	1.23
15	1.71
20	2.34
25	3.17
30	4.25
40	7.38
50	12.35
100	101.3 (1 atm)
150	475.8
200	1554
250	3973
300	8581

Discussion: It is surprising about the lake freezing when air is at 25°C, and it is right. The water temperature drops to -8°C in the limiting case of no heat transfer to the water surface. In practice, the water temperature drops below the air temperature, but it does not drop to − 8°C because (i) it is very unlikely for the air over the lake to be so dry (a relative humidity of just 10 percent) and (ii) as the water temperatures near the surface drops, heat transfer from the air and the lower parts of the water body tends to make up for this heat loss and prevents the water temperature from dropping too much. The water temperature stabilizes when the heat gain from the surrounding air and the water body equals the heat loss by evaporation, that is, when a dynamic balance is established between heat and mass transfer instead of phase equilibrium. This can be experimented using a shallow layer of water in a well-insulated pan, it is possible to freeze the water if the air is really dry and relatively cool.

Boiling and evaporation: These two processes are often used interchangeably to indicate phase change from liquid to vapour. Although they refer to the same physical process, they differ in some aspects. Evaporation occurs at the liquid-vapour interface when the vapour pressure of the surrounding air is less than the saturation pressure of the liquid at a given temperature. Water in a lake at 20 ^{0}C, for example, evaporates to air at 20 ^{0}C and 60 percent relative humidity since the saturation pressure of water at 20 ^{0}C is 2.34 kPa, and the vapour pressure of air at 20 ^{0}C and 60 percent relative humidity is 1.4 kPa. Other examples of evaporation are the drying of clothes, fruits, and vegetables; the evaporation of sweat to cool the human body: and the rejection of waste heat in wet cooling towers. However, evaporation does not involve any bubble formation or bubble motion (fig. 6.15).

Boiling, On the other hand, occurs at the solid-liquid interface when a liquid is brought into contact with a surface maintained at a temperature $T_{solid\ surface}$ sufficiently above the saturation temperature T_{sat} of the liquid. At 1 atm, for example, liquid water in contact with a solid surface at 110 ^{0}C will boil since the saturation temperature of water at 1atm is 100 ^{0}C. The boiling process is characterized by the rapid motion of vapour bubbles that are formed at the solid liquid interface, detach from the surface when they reach a certain size, and attempt to rise to the free surface of the liquid. When cooking, we do not say water is boiling unless we see the bubbles rising to the top.

(ii) Dry air: The pure dry air is mixture of a number of gases such as nitrogen, oxygen, carbon dioxide, hydrogen, argon, neon, helium etc. But the nitrogen and oxygen have the major portion of the combination. The dry air is considered to have the composition as given in the following table:

Table 6.11: Composition of dry air.

S.N.	Constituent	By volume	By mass	Molecular mass
1	Nitrogen (N_2)	78.03%	75.47%	28
2	Oxygen (O_2)	20.99%	23.19%	32
3	Argon (Ar)	0.94%	1.29%	40
4	Carbon-dioxide (CO_2)	0.03%	0.05%	44
5	Hydrogen (H_2)	0.01%	-	2

The molecular mass of dry air is taken as 28.966 and the gas constant of air (R_a) is equal to 0.287 kJ/kg K or 287 J/kg K. The molecular mass of water vapour is taken as 18.016 and the gas constant for water vapour (R_v) is equal to 0.461 kJ/kg K or 461 J/kg K. The pure air does not ordinarily, exist nature because it always contains some water vapour. The term air, where used usually means dry air containing moisture in the vapour form. Both dry air and water vapour can be considered as perfect gases because both exist in the atmosphere at low pressure. Thus all perfect gas terms can be applied to them individually. The density of dry air is taken as 1.293 kg/m^3 at pressure 1.0135 bar or 101.35 kN/m^2 and at temperature 0^0 C (273 K).

(iii) *Moist air.* It is mixture of dry air and water vapour. The amount of water vapour, present in the air, depends upon the absolute pressure and temperature of the mixture.

(iv) *Saturated air.* It is the mixture of dry air and water vapour, when water vapour has diffused maximum amount into it. The water vapours, usually, occur in the form of superheated steam as an invisible gas. However, when the saturated air is cooled, the water vapour in the air starts condensing, and the same may be visible in the form of moist, fog or condensation on cold surface.

(v) *Degree of saturation.* It is the ratio of actual mass of water vapour in a unit mass of dry air to the mass of water vapour in the same mass and pressure of dry air when it is saturated at the same temperature.

(vi) *Dry bulb temperature.* It is the temperature of air recorded by a thermometer, when its sensing element or bulb is uncovered.

(vii) *Wet bulb temperature.* It is the temperature of air recorded by a thermometer when its bulb is surrounded by a wet (water saturated) cloth exposed to the air.

(viii) *Wet bulb depression.* It is the difference between dry bulb temperature and wet bulb temperature at any point. *The wet bulb depression indicates relative humidity of the air.*

(ix) *Dew point temperature.* It is the temperature of air recorded by a thermometer, when the moisture (water vapour) present in its begins to condense. In other words, the dew point temperature is the saturation temperature (t_{sat}) corresponding to the partial pressure of water vapour (p_v). Since p_v is very small, therefore the saturation temperature of water vapour at p_v is also low (less than the atmospheric or dry bulb temperature). The dew point is therefore the temperature at which air becomes saturated with moisture (100% *RH*) and any further cooling from this point results in condensation of the water from the air. This is seen at night when air cools and water vapour forms as dew on the ground

For saturated air, the dry bulb temperature, wet bulb temperature and dew point temperature is same.

(x) *Dew point depression.* It is the difference between the dry bulb temperature and dew point temperature of air.

(xi) *Latent heat of vaporization.* The amount of energy absorbed or released during a phase- change process is called latent heat. The latent heat absorbed by the agricultural produce to vaporize the moisture from it is called latent heat of vaporization. This heat is absorbed by the produce from the surrounding air and is dependent on the type of crop, moisture content and temperature. The latent heat of vaporization is low for high moisture content and temperature of the product. At 1 atm. pressure, the latent heat of vaporization of water is 2257 kJ/kg.

(xii) *Drying rate:* It refers to the loss of the amount of moisture from the product per unit time. Drying rate can be calculated for each 10-minute period and as follows:

$$Drying\ rate = \frac{Initial\ weight - final\ weight}{Time\ interval\ (eg\ 10\ minutes)} [g/s], [kg/hr]$$

Drying efficiency: The efficiency of the drying operation is an important factor in the assessment and selection of the optimum dryer for a particular task. The drying efficiency, is defined as:

$$Drying\ efficiency = \frac{Heat\ utilized\ for\ moisture\ removal}{Heat\ available\ for\ moisture\ removal}$$

6.9 PSYCHROMETRY

It is a branch of science dealing with the study of moist air i.e. dry air mixed with water vapour or humidity. It also includes the study of the behavior of dry air and water vapour mixture under various sets of conditions (temperature, pressure, volume, density etc.).

For drying purpose, the moist air of the surrounding environment plays a key role. The post-harvest life of the perishable commodities is significantly affected by the environment to which they are exposed. The perishable commodities therefore require proper post-harvest attention to maintain the quality and reduce spoilage. Even though, the percentage of water vapour in the atmospheric air is very small compared to oxygen and nitrogen, it plays a significant role for air conditioning and drying of agricultural produces. In drying, there is a heat and mass transfer from wet solid or surfaces through moist air. Therefore, the properties (thermal, physical) of moist air changes and it becomes necessary to work out/evaluate the properties (thermal and physical) of moist air before and after drying process which decides the drying effectiveness. For effective drying, air should be hot, dry and moving. These factors are interrelated and it is important that each factor is correct as cold moving air or hot and wet moving air is unsatisfactory. Generally, hot and dry air is a major heat and mass transfer medium in drying. Therefore, it becomes necessary to ascertain the relationship between the air and amount of moisture it contains.

The physical and thermodynamic properties of moist air (psychrometric variables) are related by a number of physical laws. The psychrometric variables are dry bulb temperature, wet bulb temperature, dew point temperature, relative humidity, humidity ratio, enthalpy and specific volume. The knowledge of psychrometrics helps the researchers, entrepreneurs, crop growers, pack house operators and commercial cool chamber operators to improve post-harvest management and storage of agricultural produces or food materials.

Psychrometric relations: Some psychrometric terms have already been discussed above. These terms have some relations between one another. The following psychrometric relations are important from subject point of view.

(i) *Delton's Law of Partial Pressures:* It states, "The total pressure exerted by the mixture of air and water vapour is equal to the sum of the pressures, which each constituent would exert, if it occupies the same space by itself". Or, in other words, the total pressure exerted by air and water vapour mixture is equal to the barometric pressure. Mathematically, barometric pressure (P_b) of the mixture, $p_b = p_a + p_v$

where P_a = partial pressure of dry air, and p_v = partial pressure of water vapour.

(ii) *Specific humidity, humidity ratio or moisture content:* It is the mass of water vapour present in 1 kg of dry air (in the air-vapour mixture) and is generally expressed in g/kg of dry air. It may also be defined as the ratio of mass of water vapour to the mass of dry air in a given volume of the air-vapour mixture.

Let p_a, v_a, T_a, m_a and R_a be the pressure, volume, absolute temperature, mass and gas constant respectively for dry air, and p_v, v_v, T_v, m_v and R_v be the corresponding values for the water vapour. Assuming that the dry air and water vapour behave as perfect gases, we have for dry air, $p_a v_a = m_a R_a T_a$ and for water vapour, $p_v v_v = m_v R_v T_v$, also $v_a = v_v$ and $T_a = T_v = T_d$ (where is T_d dry bulb temperature). From above expressions, $\dfrac{P_v}{P_a} = \dfrac{m_v R_v}{m_a R_a}$. Hence, humidity ratio, $W = \dfrac{m_v}{m_a} = \dfrac{R_a P_v}{R_v P_a}$. Substituting $R_a =$ 0.287 kJ/kg K for dry air and $R_v =$ 0.461 kJ/kg K for water vapour in the above equation,

$$W = \frac{0.287 \times P_v}{0.461 \times P_a} = 0.622 \times \frac{P_v}{P_a} = 0.622 \times \frac{P_v}{P_b - P_v} (\because p_b = p_a + p_v)$$

Consider unsaturated air containing superheated vapour at dry bulb temperature t_d and partial pressure p_v as shown by point A on the T-S diagram in Fig. 6.16.

If water is added into this unsaturated air, the water will evaporate which will increase the moisture content (specific humidity) of the air and the partial pressure p_v will increase. This will continue until the water vapour becomes saturated at that temperature, as shown by point C in Fig. 6.16 and there will be more evaporation of water. The partial pressure p_v will increase to the saturation pressure p_s and it is maximum partial pressure of water vapour at temperature t_d. The air containing moisture in such a state (point C) is called saturated air. For saturated air (i.e. when the air is holding maximum amount of water vapour), the humidity ratio or maximum specific

humidity, $W_s = W_{max} = 0.622 \times \dfrac{P_s}{P_b - P_s}$

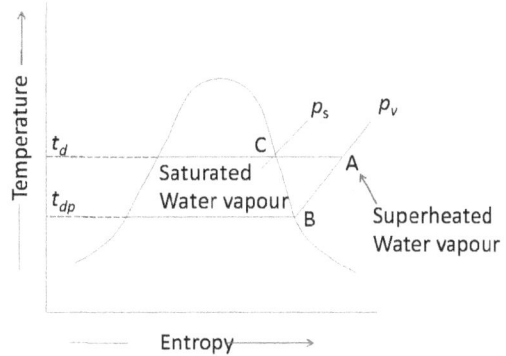

Fig. 6.16. T-S diagram

where P_s = partial pressure of air corresponding to saturation temperature (i.e. dry bulb temperature t_d).

(iii) Degree of saturation or percentage humidity: The degree of saturation is the ratio of actual mass of water vapour in a unit mass of dry air to the mass of water vapour in the same mass of dry air when it is saturated at the same temperature (dry bulb temperature). In other words, it may be defined as the ratio of actual specific humidity to the specific humidity of saturated air at the same dry bulb temperature. It is, usually, denoted by μ. Mathematically, degree of saturation.

$$\mu = \frac{W}{W_s} = \frac{\dfrac{0.622 p_v}{P_b - P_v}}{\dfrac{0.622 p_s}{P_b - P_s}} = \frac{p_v}{p_s}\left(\frac{P_b - P_s}{P_b - P_v}\right) = \frac{p_v}{p_s}\left[\frac{1 - \dfrac{P_s}{P_b}}{1 - \dfrac{P_v}{P_b}}\right].$$

The partial pressure of saturated air (p_s) is obtained from the steam tables corresponding to dry bulb temperature If the relative humidity, $\phi = p_v / p_s$ is equal to zero, then the humidity ratio, $W = 0$, i.e. for dry air, $\mu = 0$. Similarly, if the relative humidity, $\phi = p_v / p_s$ is equal to 1, then $W = W_s$ and $\mu = 1$. Thus μ varies between 0 and 1.

(iv) *Relative humidity:* Relative humidity is the ratio of actual mass of water vapour (m_v) in a given volume of moist air to the mass of water vapour (m_s) in the same volume of saturated air at the same temperature and pressure. Mathematically, relative

humidity, $\phi = \dfrac{m_v}{m_s}$. Let p_v, v_v, T_v, m_v and R_v be the the the pressure, volume, temperature,

mass and gas constant respectively for water vapour in actual conditions, and p_s, v_s, T_s, m_s and R_s be the corresponding values for water vapour in saturated air. For water vapour in actual conditions, $p_v v_v = m_v R_v T_v$, similarly, for water vapour in saturated air, $p_s v_s = m_s R_s T_s$. According to the definitions, $v_v = v_s$ and $T_v = T_s$.

Also $R_v = R_s = 0.461$ kJ/kg K. Hence, from the above expressions, $\phi = \dfrac{m_v}{m_s} = \dfrac{p_v}{p_s}$.

Thus, the relative humidity may also be defined as the ratio of actual partial pressure of water vapour in moist air at a given temperature (dry bulb temperature) to the saturation pressure of water vapour (or partial pressure of water vapour in saturated air) at the same temperature. For saturated air, the relative humidity is 100%.

(v) *Pressure of water vapour:* According to carrier's equation, the partial pressure of

water vapour, $p_v = p_w - \dfrac{(p_b - p_w)(t_d - t_w)}{1544 - 1.44 t_w}$ where p_w = saturation pressure

corresponding to wet bulb temperature (from steam table, p_b = barometric pressure, t_d = dry bulb temperature and t_w = wet bulb temperature).

(vi) *Vapour density or absolute humidity:* We have already discussed that the vapour density or absolute humidity is the mass of water vapour present in 1 m^3 of dry air. Let v_v = volume of water vapour in m^3 /kg of dry air at its partial pressure, v_a = volume of dry air in m^3 /kg of dry air at its partial pressure, ρ_v = density of water vapour in kg /m^3, corresponding to its partial pressure and dry bulb temperature t_d, and ρ_a = density of dry air in kg/m^3 of dry air. The mass of water vapour, $m_v = v_v$

ρ_v and mass of dry air, $m_a = v_a \rho_a$. Dividing the above expressions, $\dfrac{m_v}{m_a} = \dfrac{v_v \rho_v}{v_a \rho_a}$.

Since $v_a = v_v$, therefore humidity ratio, $W = \dfrac{m_v}{m_a} = \dfrac{\rho_v}{\rho_a}$ or $\rho_v = W \rho_a$. But $p_a v_a = m_a R_a$

T_d. Since $v_a = \dfrac{1}{\rho_a}$ and $m_a = 1$ kg, therefore substituting these values in the above

expression, $p_a \times \dfrac{1}{\rho_a} = R_a T_d$ or $\rho_a = \dfrac{p_a}{R_a T_d}$. Substituting the value of ρ_a in the

expression of ρ_v, $\rho_v = \dfrac{W p_a}{R_a T_d} = \dfrac{W(p_b - p_v)}{R_a T_d}$ $(\because p_b = p_a + p_v)$, where p_a = pressure of

air kN/m^2, R_a = gas constant for air = 0.287 kJ/kg K, and T_d = dry bulb temperature in K.

(vii) *Enthalpy (total heat) of moist air:* The enthalpy of moist air is numerically equal to the enthalpy of dry air plus the enthalpy of water vapour associated with dry air. Let us consider one kg of dry air. We know that enthalpy of 1 kg of dry air, $h_a = c_{pa} t_d$, where c_{pa} = Specific heat of dry air which is normally taken as 1.005 kJ/kg

K, and t_d = dry bulb temperature. Similarly, enthalpy of water vapour associated with 1 kg of dry air, $h_v = Wh_s$, where W = mass of water vapour in 1 kg of dry air (i.e. specific humidity), and h_s = enthalpy of water vapour per kg of dry air at dew point temperature (t_{dp}). If the moist air is superheated, then the enthalpy of water vapour = Wc_{ps} ($t_d - t_{dp}$) where c_{ps} = specific heat of superheated water vapour which is normally taken as 1.9 kJ/kg K, and ($t_d - t_{dp}$) = degree of superheat of the water vapour.

Hence, total enthalpy of superheated water vapour, $h = c_{pa} t_d + Wh_s + Wc_{ps} (t_d - t_{dp})$

$= c_{pa} t_d + W[h_{fdp} + h_{fgdp} + c_{ps} (t_d - t_{dp})]$ ... ($\because h_s = h_{fdp} + h_{fgdp}$)

$= c_{pa} t_d + W[4.2\, t_{dp} + h_{fgdp} + c_{ps} (t_d - t_{dp})]$ ... ($h_{fdp} = 4.2\, t_{dp}$)

$= c_{pa} t_d + 4.2\, W\, t_{dp} + W\, h_{fgdp} + W\, c_{ps}\, t_d - W\, c_{ps}\, t_{dp}$

$= (c_{pa} + Wc_{ps})\, t_d + W[h_{fgdp} + t_{dp} (4.2 - c_{ps})]$

$= (c_{pa} + Wc_{ps})\, t_d + Wh_{fgdp} + t_{dp} (4.2\text{-}1.9]$

$= (c_{pa} + Wc_{ps})\, t_d + W[h_{fgdp} + 2.3\, t_{dp}]$

The term ($c_{pa} + Wc_{ps}$) is called humid specific heat (c_{pm}). It is the specific heat or heat capacity of moist air, i.e. (1+W) kg/kg of dry air. At low temperature of air conditioning range, the value of W is very small. The general value of humid specific heat in air conditioning range is taken as 1.022 kJ/kg K. $\therefore h = 1.022\, t_d + w(h_{fgdp} + 2.3\, t_{dp})$ kJ, where h_{fgdp} = latent heat of vaporisation of water corresponding to dew point temperature (from steam tables).

(viii) *Specific volume:* It is defined as the total volume of dry air and water vapor mixture per kg of dry air and water vapor, (m^3/kg) The specific volume can be expressed as: $v = (V) / (m_a + m_w)$ where v = specific volume of moist air per mass unit of dry air and water vapor. Specific volume is a property of materials, defined as the number of cubic meters occupied by one kilogram of a particular substance

Example: 6.6: The readings from a sling psychrometer are as follows:

Dry bulb temperature = 30 ^{0}C; wet bulb temperature = 20 ^{0}C; Barometer reading = 740 mm of Hg. Using steam tables, determine: a. Dew point temperature; b. Relative humidity; c. Specific humidity; d. Degree of saturation; e. Vapour density; and f. Enthalpy of mixture per kg of dry air.

Solution: Given: t_d= 30 ^{0}C; t_w = 20 ^{0}C; p_b= 740 mm of Hg

a. *Dew point temperature:* First of all, let us find the partial pressure of vapour (p_v). From steam tables, we find that the saturation pressure corresponding to wet bulb temperature of 20 ^{0}C is p_w = 0.02337 bar

We know that barometric pressure, p_b= 740 mm of Hg = 740 × 133.3 = 98642 N/m^2 = 0.98640 bar (as 1 mm of Hg = 133.3 N/m^2 and 1 bar = 10^5 N/m^2).

$\therefore$ Partial pressure of vapour,

$$p_v = p_w - \frac{(p_b - p_w)(t_d - t_w)}{1547 - 1.44 t_w} = 0.02337 - \frac{(0.98642 - 0.02337)(30 - 20)}{1547 - 1.44 \times 20}$$

$= 0.02337 - 0.00634 = 0.01703$ bar. Since the dew point temperature is the saturation temperature corresponding to the partial pressure of water vapour (p_v), therefore from steam tables, we find that corresponding to a pressure of 0.01703 bar, the dew point temperature is $t_{dp} = 15°C$ Ans.

b. *Relative humidity:* From steam tables, we find that the saturation pressure of vapour corresponding to dry bulb temperature of 30 ^{0}C is $p_s = 0.04242$ bar. We know that

relative humidity, $\phi = \dfrac{p_v}{p_s} = \dfrac{0.01703}{0.04242} = 0.4015$ or 40.15%

c. *Specific humidity:* We know that specific humidity,

$$W = \frac{0.622 p_v}{p_b - p_v} = \frac{0.622 \times 0.01703}{0.98642 - 0.01703} = \frac{0.01059}{0.96939} = 0.010924 \; \frac{kg}{kg} \text{ of dry air} = 10.924 \text{ g /}$$

kg of dry air (Ans).

d. *Degree of saturation:* We know that specific humidity of saturated air,

$$W_s = \frac{0.622 P_s}{P_b - P_v} = \frac{0.622 \times 0.04242}{0.98642 - 0.04242} = \frac{0.02638}{0.944} = 0.027945 \frac{kg}{kg} \text{ of dry air.}$$

We know that degree of saturation, $\mu = \dfrac{W}{W_s} = \dfrac{0.010924}{0.027945} = 0.391$ or 39.1% Ans.

The degree of saturation (μ) may also be calculated from the following relation:

$$\mu = \frac{P_v}{P_s}\left(\frac{P_b - P_s}{P_b - P_v}\right) = \frac{0.01703}{0.04242}\left[\frac{0.98642 - 0.04242}{0.98642 - 0.01703}\right] = 0.391 \text{ or } 39.1\% \text{ Ans.}$$

e. *Vapour density:* We know that vapour density,

$$\rho_v = \frac{W(p_b - p_v)}{R_a T_d} = \frac{0.010924(0.98640 - 0.01703)10^5}{287(273 + 30)} = 0.01218 \text{ kg/m}^3 \text{ of dry air Ans.}$$

f. *Enthalpy of mixture per kg of dry air:* From steam tables, we find the latent heat of vaporisation of water at dew point temperature of 15 ^{0}C is $h_{fgdp} = 2466.1$ kJ/kg. $\therefore$ Enthalpy of mixture per kg of dry air, $h = 1.022\, t_d + W[h_{fgdp} + 2.3\, t_{dp}] = 1.022 \times 30 + 0.010924 \{2466.1 + 2.3 \times 15\} = 30.66 + 27.32 = 57.98$ kJ/kg of dry air sAns.

6.10 PSYCHROMETRIC CHART

It is a graphical representation of the various thermodynamic properties of moist air. The psychrometric chart is very useful for finding out the properties of air (which are required in the field of air-conditioning) and eliminate lot of calculations. Three is a slight variation in the charts prepared by different air-conditioning manufacturers but basically they are all alike. The psychometric chart is normally drawn for standard atmospheric pressure of 760 mm of Hg (or 1.01325 bar).

In a psychrometric chart, dry bulb temperature is taken as abscissa and specific humidity i.e. moisture contents as ordinate, as shown in Fig. 6.17.

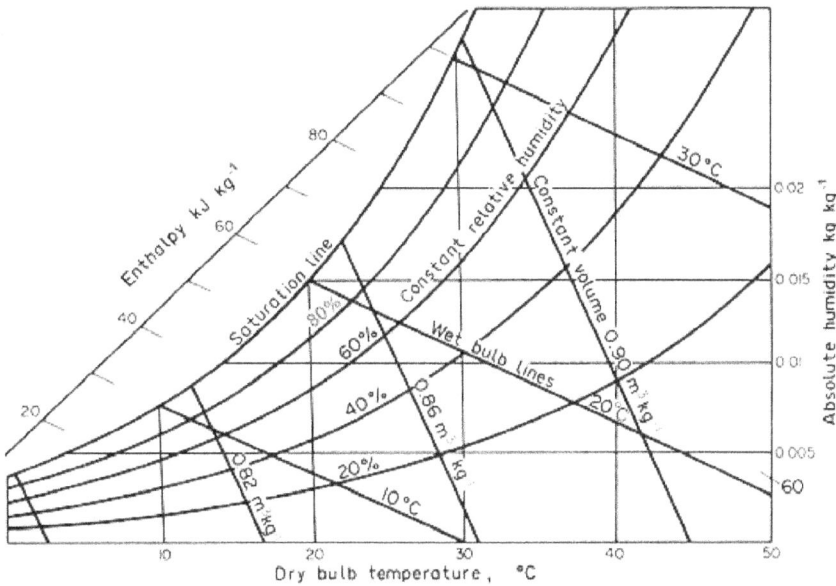

Fig. 6.17. Psychrometric chart

Now the saturation curve is drawn by plotting the various saturation points at corresponding dry bulb temperatures. The saturation curve represents 100% relative humidity at various dry bulb temperatures. It also represents the wet bulb and dew point temperatures.

Though the psychrometric chart has a number of lines, yet the following lines are important from the subject point of view:

i. *Dry bulb temperatures lines.* The dry bulb temperatures lines are vertical i.e. parallel to the ordinate and uniformly spaced as shown in Fig. 6.18.

Generally, the temperatures range of these lines on psychrometric chart is from -5^{0}C to 45^{0}C. The dry bulb temperature lines are drawn with difference of every 5 ^{0}C and up to the saturation curve as shown in the figure. The values of dry bulb temperatures are also shown on the saturation curve.

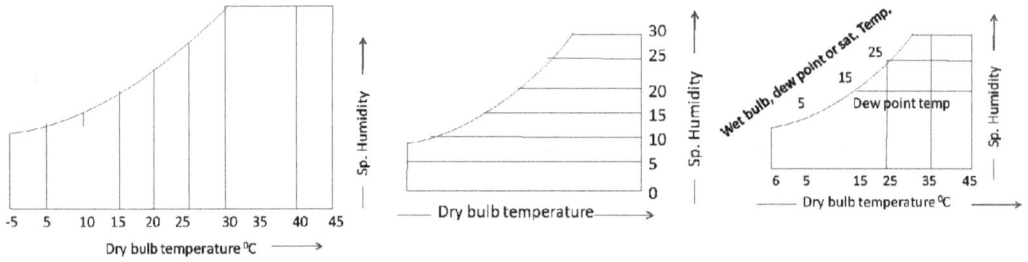

Fig. 6.18. Dry bulb temperature lines   Specific humidity line   Dew point temperature lines

ii. *Specific humidity or moisture content lines.* The specific humidity (moisture content) lines are horizontal i.e. parallel to the abscissa and are also uniformly spaced as shown in Fig. above. Generally, moisture content range of these lines on psychrometric chart is from 0 to 30 g/kg of dry air (or from 0 to 0.030 kg/kg dry air). The moisture content lines are drawn with a difference of every 1 g (or 0.001 kg) and up to the saturation curve as shown in the figure above.

iii. *Dew point temperature lines.* The dew point temperature lines are horizontal i.e. parallel to the abscissa and non-uniformly spaced as shown in the figure above. At any point on the saturation curve, the dry bulb and dew point temperatures are equal. The values of dew point temperatures are generally given along the saturation curve of the chart as shown in the figure above.

iv. *Wet bulb temperature lines.* The wet bulb temperature lines are inclined straight lines and non-uniformly spaced as shown in Fig. 6.19. At any point on the saturation curve, the dry bulb and wet bulb temperatures are equal. The values of wet bulb temperatures are generally given along the saturation curve of the chart as shown in the figure.

v. *Enthalpy (total heat) lines.* The enthalpy (or total heat) lines are inclined straight lines and uniformly spaced as shown in the figure above. These lines are parallel to the wet bulb temperature lines also. The values of total enthalpy are given on a scale above the saturation curve as shown in the figure above.

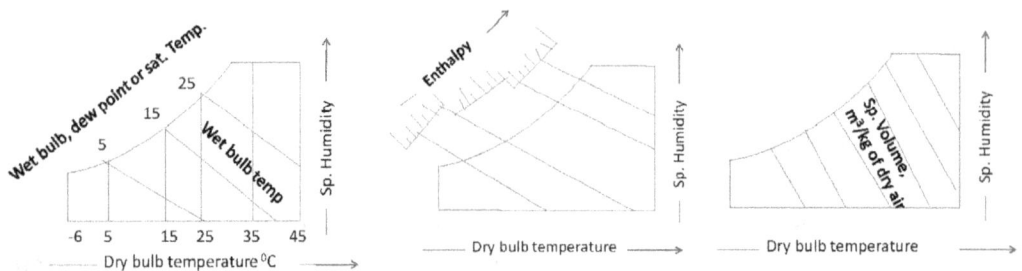

Fig. 6.19. Wet bulb temperature lines Enthalpy lines Specific volume lines

vi. Specific volume lines. The specific volume lines are obliquely inclined straight lines and uniformly spaced as shown in the figure above. These lines are drawn up to the saturation curve. The values of volume lines are generally given at the base of the chart.

vii. Vapour pressure lines. The vapour pressure lines are horizontal and uniformly spaced. Generally, the vapour pressure lines are not drawn in the main chart. But a scale showing vapour pressure in mm of Hg is given on the extreme left side of the chart as shown in Fig. 6.20.

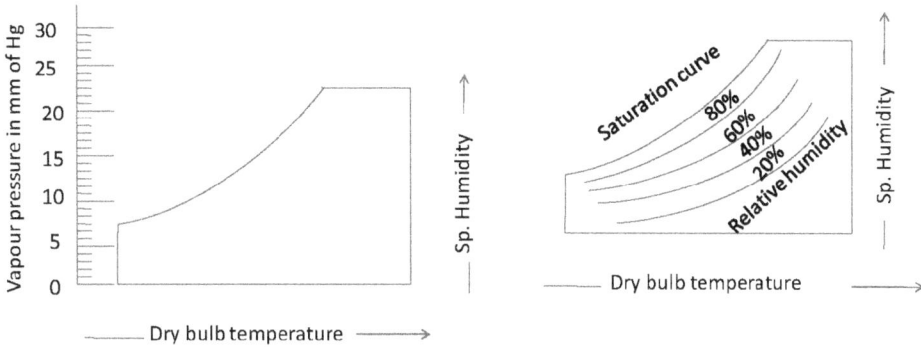

Fig. 6.20. Vapour pressure line Relative humidity line

vii. Relative humidity lines. The relative humidity lines are curved lines and follow the saturation curve. Generally, these lines are drawn with values 10%, 20%, 30% etc. and up to 100%. The saturation curve represents 100% relative humidity. The values of relative humidity lines are generally given along the lines themselves as shown in the figure above.

Example.6.7 For a sample of air having 22 ^{0}C DBT, relative humidity (ϕ) 30 per cent at barometric pressure of 760 mm of Hg, calculate: a. Vapour pressure, b. Humidity ratio, c. Vapour density, and d. Enthalpy. Verify the results by psychrometric chart.

Solution. Given : t_d = 22°C; ϕ = 30% = 0.3 ; p_h = 760 mm of H_g = 760 × 133.3 =101308 N/m^2=1.01308 bar

a. Vapour pressure: From steam tables, we find that the saturation pressure of vapour corresponding to dry bulb temperature of 22°C is p_s = 0.02642 bar.

We know that relative humidity (ϕ), $0.3 = \dfrac{p_v}{p_s} = \dfrac{p_v}{0.02642}$, hence, p_v = 0.3 × 0.02642

= 0.007926 bar (Ans)

b. Humidity ratio: We know that humidity ratio, $W = \dfrac{0.622\,p_v}{p_b - p_v} = \dfrac{0.622 \times 0.07926}{1.01308 - 0.007926} =$

0.0049 kg/m^3 of dry air

c. *Vapour density:* Vapour density $\rho_v = \dfrac{W(p_b - p_v)}{R_a T_d} = \dfrac{0.0049(1.01308 - 0.007926)10^5}{287(273+22)}$

= 0.00582 kg/m³ of dry air

d. *Enthalpy:* From steam tables, we find that saturation temperature or dew point temperature corresponding to a pressure of p_v = 0.007926 bar is t_{dp} = 3.8°C. and latent heat of vaporisation of water at dew point temperature of 3.8°C is h_{fgdp} = 2492.6kJ/kg. We know that enthalpy, $h = 1.022\, t_d + W(h_{fgdp} + 2.3\, t_{dp})$ = 1.022 × 22 × 0.0049 (2492.6 + 2.3 × 3.8)

= 22.484+12.256= 34.74 kJ/kg of dry air   Ans.

Verification from psychrometric chart: The initial condition of air i.e. 22 °C dry bulb temperature and 30% relative humidity is marked on the psychrometric chart at point A as shown in Fig. (6.21).

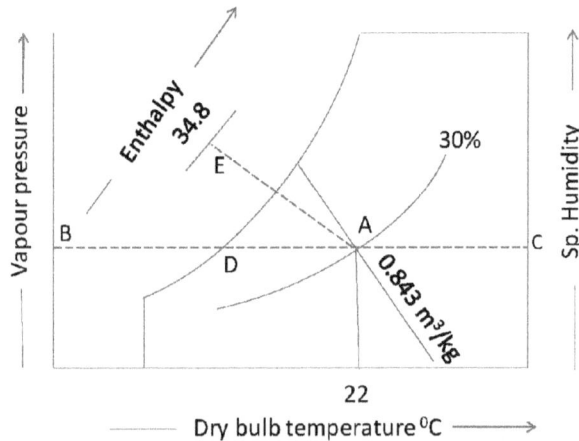

Fig. 6.21. Verification of calculated data on psychrometric chart

From point A, draw a horizontal line meeting the vapour pressure line at point B and humidity ratio line at C. From the psychrometric chart, we find that vapour pressure at point B, p_v= 5.94 mm of Hg = 5.94 × 133.3 = 791.8 N/m²= 0.007918 bar and humidity ratio at point C, W = 5g per kg of dry air = 0.005 kg/kg of dry air. We also find the psychrometric chat that the specific volume at point A is 0.843 m³/kg of dry air. Hence,

vapour density, $\rho_v = \dfrac{W}{\rho_a} = \dfrac{0.005}{0.843} = 0.0058 \dfrac{kg}{m^2}$ of dry air. Now from point A, draw a line parallel to the wet bulb temperature line, meeting the enthalpy line at point E. Now the enthalpy of air as read from the chart is 34.8 kJ/kg of dry air.

6.11 DRYING MECHANISM

Drying basically comprises of two fundamental and simultaneous processes: (i) heat is transferred to evaporate liquid, and (ii) mass is transferred as a liquid or vapour within the solid and as a vapour from the surface. The factors governing the rates of these processes determine the drying rate. The different dryers may utilize heat transfer by convection, conduction, radiation, or a combination of these. However, in almost all solar dryers and other conventional dryers, heat must flow to the outer surface first and then into the interior of the solid, with exception for dielectric and microwave drying.

The drying of product is a complex heat and mass transfer process which depends on external parameters such as temperature, humidity and velocity of the air stream; drying material properties like surface characteristics (rough or smooth surface), chemical composition (sugar, starches, etc.), physical structure (porosity, density, etc.) and size as well as shape of the product. The rate of moisture movement from the product inside to the air outside differs from one product to another and very much depends on whether the material is hygroscopic (agricultural produces/food materials) or non-hygroscopic (textiles in a laundry, sand, stone, dust or paper). The hygroscopic materials contain bound and unbound moisture while non-hygroscopic materials contain only unbound moisture. The unbound moisture is the 'free water', which is only loosely held in the cell pores and is therefore quickly lost when placed in the ambient condition. The remaining water (usually 30-40 %) is bound to the cell wall by hydrogen bonds and is therefore harder to remove. Free water/moisture is the water that can move through the product in an unrestricted way. Its movement is not dependent on the internal structure of the crop. Non-hygroscopic materials can be dried to zero moisture level while the hygroscopic materials, like most of the agricultural/food products, will always have a residual moisture content. The bound mixture exerts equilibrium vapour pressure less than that of pure water at the same temperature. The bound mixture can be considered as moisture trapped in closed capillaries, the water component of juices or water held by surface forces as well as unbound water held within the material by the surface tension of the water itself. The unbound mixture exerts equilibrium vapour pressure equal to that of pure water at the same temperature. The unbound moisture is the moisture in excess of bound moisture for hygroscopic material. When equilibrium vapour pressure of moisture becomes equal to that of partial water vapour pressure of environment, it is referred as equilibrium moisture. The moisture in excess of equilibrium moisture is referred as free moisture. The free moisture consists of unbound and some bound moisture. The equilibrium moisture content is, therefore, important in the drying since this is the minimum moisture to which the material can be dried under a given set of drying conditions. These moisture types are shown in Fig. 6.22.

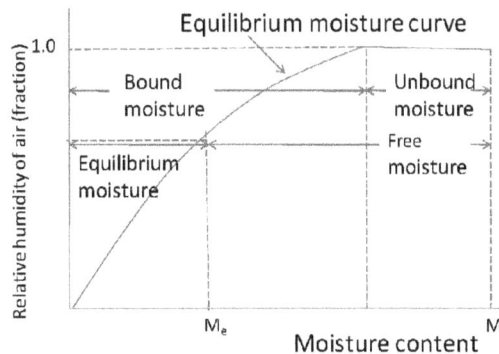

Fig. 6.22. Moisture in drying material

The water vapour pressure increases when a wet solid is heated. If the environment is at a lower vapour pressure, there is a moisture transfer from the wet solid to the environment. The rate of moisture transfer is almost proportional to its vapour pressure difference with the environment. Generally, moisture transfer takes place in two ways i.e. from inner portion of material to its surface through diffusion and from its surface to ambient or its surrounding environment through evaporation. As the moisture is evaporated, the moisture content of the product falls and the product temperature is close to the wet bulb temperature of the drying air. A series of drying characteristics curves can be plotted. The simplest one is, if the moisture content of the material is plotted versus time. Another curve can be plotted between drying rate versus time but more information can be obtained if a curve is plotted between drying rate versus moisture content. The drying characteristics curves are shown in the fig. 6.23 below.

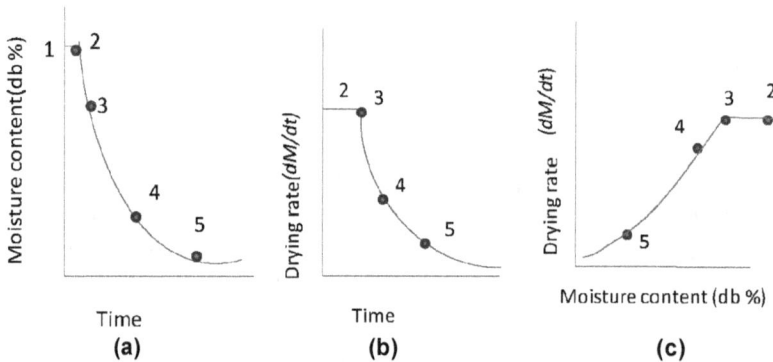

(a) Drying rate curve b) Drying rate variation with time (c) Drying rate variation with moisture content

Fig. 6.23   Illustration of drying rate periods: 1-2, the heating period (constant moisture content), 2-3: the constant rate drying period, 3, the critical moisture content, 3-4, the first falling rate period, 4-5, the second falling rate period (Ekechukwu, 1999).

6.11.1 Periods of Drying

Figure 6.24 shows the plotting of curves with respect to moisture content and drying rate versus time and drying rate versus moisture content, generally obtained by experimentally drying a solid. The curve represents a typical case when a wet solid loses moisture initially by evaporation from a saturated surface on a solid, followed by a period of evaporation from a saturated surface on a solid, followed by a period of evaporation from a saturated surface of gradually decreasing area and finally when the moisture is evaporated in the interior of the solid. Fig. 6.24 a indicates that the drying rate is subject to variation of moisture content with time or, further better illustrated by graphically or numerically differentiating the curve and plotting drying rate versus moisture content, as shown in Fig. 6.24 b, or as drying rate versus time as shown in Fig. 6.24 c. These rate curves illustrate that the drying process is not a smooth, continuous one in which a single mechanism controls throughout. Fig. 6.24 c has the advantage of showing how long each drying period lasts.

The section *AB* on each curve represents a warming-up period of the solids. Section *BC* on each curve represents the constant-rate period. Point *C,* where the constant rate ends and the drying rate begins falling, is termed as the critical-moisture content. The curved portion *CD* on Fig. 6.24 a is termed the falling-rate period and, as shown in Fig 6.24 *b* and *c,* is typified by a continuously changing rate throughout the remainder of the drying cycle. Point *E* (Fig 6.24 b) represents the point at which all the exposed surface becomes completely unsaturated and marks the start of that portion of the drying cycle during which the rate of internal moisture movement controls the drying rate. Portion *CE* in Fig 6.24 b is usually defined as the first falling-rate drying period; and portion *DE,* as the second falling-rate period (Perry 2007).

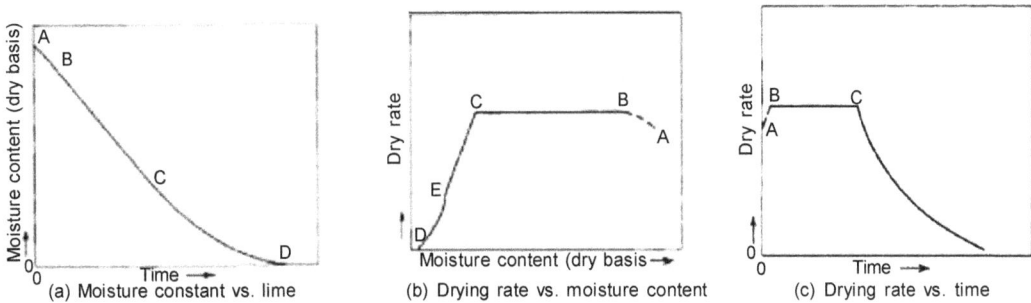

Fig. 6.24. Drying period curves (a) (b) (c)

The constant drying rate for both non-hygroscopic and hygroscopic materials is the same while the period of falling rate is little different (fig 6.25).

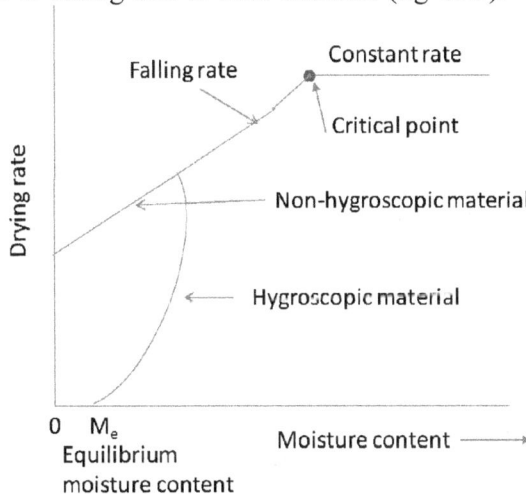

Fig. 6.25. Typical drying rate curve

For non-hygroscopic materials, in the period of falling rate, the drying rate goes on decreasing till the moisture content becomes zero. In the hygroscopic materials, the period of falling rate is similar until the unbounded moisture is completely removed, then it further decreases and some bound moisture is removed; this continues till the vapour pressure of material becomes equal to the vapour pressure of drying air. In the falling-rate period, the effect of external factors starts diminishing and the transport properties of the material becomes important.

6.11.2 Constant Drying Rate Period

For agricultural produces, in particular, there are two drying rate regimes: (i) the constant drying rate period and (ii) the falling drying rate period. During initial period of drying, the products having high moisture content follow the constant drying period and after that the drying is done under falling drying rate period.

During constant drying rate period, the evaporation of moisture takes place from the surface of the material, similar to that from the free-water surface and the rate of moisture transfer mainly depends on the environmental conditions e.g. vapour pressure difference between the drying air and the wet surface, the temperature and relative humidity of drying air, and the drying air velocity etc. The rate of drying depends on various factors e.g. the vapour pressure difference between the wet surface and drying air $(p_v\text{-}p_a)$, the surface area of the material (A_s) the mass transfer coefficient (h_D) and the drying air velocity etc. Now, from the concept of mass transfer, the mass flow rate per unit area (kg/m²s) can be expressed as $\dot{m} = h_D\left(\rho_w^0 - \rho_a^0\right)$, where $\dot{m}$ is the rate of mass flow per unit area, (kg/m²s), h_D, the mass transfer coefficient [(kg/m²s) / (kg/m³)] and ρ^0, the partial mass density of water vapour, (kg/m³). The subscripts 'w' and 'a' indicate the fluid (water) and the surrounding air, respectively.

Similarly, the drying rate per unit area can be expressed as

$$\frac{1}{A_s}\frac{dW}{dt} = \frac{h_D}{RT}(p_v - p_{va})$$

Where R is universal gas constant, p_v is water vapour pressure in the produce and p_{va} is the vapour pressure of drying air. W is the mass of the moisture removed.

From evaporative heat transfer balance, we have

$$\dot{Q}_e = \lambda\frac{dW}{dt} = h_c A_s \left(T_a - T_s\right)$$

Where λ is latent heat of vaporization, h_c convective heat transfer coefficient etc. Now, from the above two equations, we get the drying rate equation

$$\frac{dW}{dt} = \frac{A_s h_D}{RT}(p_v - p_{va}) = \frac{h_c A_s}{\lambda}(T_a - T_s)$$

Under constant drying period, the product surface becomes saturated with moisture at nearly constant temperature and almost equal to the wet bulb temperature i.e. T_s may be replaced by T_{wb}. Then from the above equation, $\dfrac{dW}{dt} = \dfrac{h_c A_s}{\lambda}(T_a - T_{wb})$

Thus, drying rate becomes a function of wet bulb depression $(T_a\text{-}T_{wb})$ and $(p_v\text{-}p_{va})$ becomes the difference in vapour pressure of unsaturated air at T_a and saturated vapour pressure of air at T_{wb}.

At particular moisture content, the constant drying rate period changes to falling drying rate period which is referred as critical moisture content. At this point, the rate of moisture transfer through diffusion from within the product to its surface becomes insufficient to replenish the moisture being evaporated from the surface.

6.11.3 Falling Rate Drying Period

During falling rate period, the moisture content is less than the critical moisture content of the product (Fig. 6.25) i.e. the rate of moisture transfer through diffusion from within the product to its surface is less than the moisture being evaporated from the surface. Generally, it is subdivided into two stages: (i) the first falling drying rate period which involves the unsaturated surface drying and (ii) the second falling drying rate period where the rate of moisture transfer through diffusion from within the product to its surface is slow and is the determining factor. The drying rate is governed by the rate of internal moisture movement and the influence of external variables diminishes.

For drying of agricultural produces, the initial moisture content and the safe storage moisture content are the determining factors for duration of these drying rate regimes. Generally, initial moisture content for grains is lower than its critical moisture content, therefore whole drying takes place under the falling drying rate regime. The drying of fruits, most vegetables and most tropical tuber crops take place under both the constant and falling drying rate periods because initial moisture content for these produces is usually above the critical moisture content. However, for overall drying rate, both external factors and internal mechanisms play a crucial role to determine the drying processes under constant and falling rate regimes.

6.12 COMMONLY ENCOUNTERED TERMS IN DRYING

S.N.	Term	Meaning
1	Bound moisture	is that in a solid liquid, which exerts a vapor pressure less than that of the pure liquid at the given temperature.
2	Capillary flow	is the flow of liquid through the interstices and over the surface of a solid, caused by liquid-solid molecular attraction.
3	Constant-rate period	is that drying period during which the rate of water removal per unit of drying surface is constant.
4	Critical moisture content	is the average moisture content when the constant-rate period ends.

[Table Contd.

Contd. Table]

S.N.	Term	Meaning
5	Dry-weight basis.	expresses the moisture content of wet solid as kilograms of water per kilogram of bone-dry solid
6	Equilibrium moisture content	Is the limiting moisture to which a given material can be dried under specific conditions of air temperature and humidity.
7	Falling-rate period	is a drying period during which the drying rate continually decreases.
8	Free-moisture content	is that liquid which is removable at a given temperature and humidity, may include bound/unbound moisture.
9	Hygroscopic material	is material that may contain bound moisture
10	Initial moisture distribution	refers to the moisture distribution throughout a solid at the start of drying
11	Internal diffusion	may be defined as the movement of liquid or vapor through asolid as the result of a concentration difference.
12	Moisture content	of a solid is usually expressed as moisture quantity per unit weight of the dry or wet solid.
13	Moisture gradient.	refers to the distribution of water in a solid at a given moment in the drying process
14	Non-hygroscopic material	is material that can contain no bound moisture
15	Unbound moisture	in a hygroscopic material, is that moisture in excess of the equilibrium moisture content corresponding to saturation humidity.
16	Wet-weight basis	expresses the moisture in a material as a percentage of the weight of the wet solid.

6.13 SIMULTANEOUS HEAT AND MASS TRANSFER

Many mass transfer processes encountered in practice occur isothermally and thus they do not undergo any heat transfer. But some engineering applications involve the vaporization of a liquid and diffusion of this vapour into the surrounding gas. Such processes require the transfer of the latent heat of vaporization h_{fg} to the liquid in order to vaporize it and thus such problems involve simultaneous heat and mass transfer.

To generalize, any mass transfer problem involving phase change (evaporation, sublimation, conduction, melting etc.) must also involve heat transfer and the solution of such problems needs to be analysed by considering simultaneous heat and mass transfer. Some examples of simultaneous heat and mass transfer are drying, evaporative cooling, transpiration (sweat) cooling etc.

To understand the mechanism of simultaneous heat and mass transfer, evaporation of water from a swimming pool into air can be considered. Let us assume that the water and air are initially at the same temperature. If the air is saturated (= 100 percent), there will be no heat or mass transfer as long as isothermal conditions remain. But if air is not saturated (< 100 percent), there will be a difference between the concentration of water vapour at the water-air interface (which is always saturated) and some distance above the interface. Thus concentration difference will drive the water vapour into air. But the water must vaporize first, and it must absorb the latent heat of vaporization in order to be vaporized. Initially, the entire heat of vaporization will come from the water near the interface since there is no temperature difference between the water and the surroundings and thus there cannot be any heat transfer. The temperature of water near the surface must drop as a result of sensible heat loss which also drops the saturation pressure and thus the vapour concentration at the interface.

This temperature drop creates temperature difference within the water at the top as well as between the water and surrounding air. These temperature differences drive heat transfer towards the water surface from both the air and deeper parts of the water as shown in fig 6.26 below

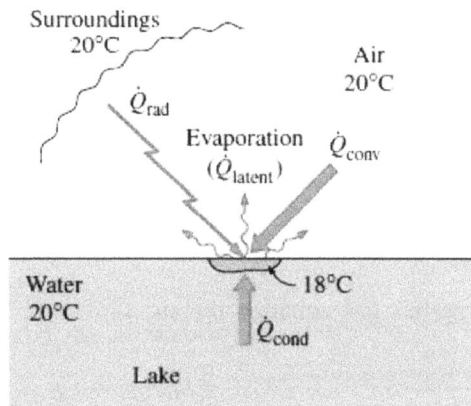

Fig. 6.26. Various mechanisms of heat transfer involved during the evaporation of water from free water surface (lake)

Evaporative heat transfer coefficient

Both evaporation and condensation are simultaneous heat and mass transfer processes. The heat transfer is based on Newton's law of cooling, which is $\dot{q} = h(Tw - T_a)$ where, h = convective heat transfer coeffcinet, T_w and T_a are the fluid (water) and the surrounding air temperature respectively. If heat transfer occurs from water to air, then the above equation becomes $\dot{q} = h_{ew}(T_w - T_a)$ where h_{ew} be the convective heat transfer coefficient from water surface to the surrounding air. If mass transfer is taken into account, then $\dot{m} = h_{mass}(\rho_w - \rho_a)$ where $\dot{m}$ is rate of mass flow per unit area, h_{mass} = convective mass transfer coefficient of water vapour ρ_w = partial mass density of water vapour $\left(kg/m^3\right)$, ρ_a = partial mass density of air at far distance from the liquid-air interface.

But from Lewis relation, it is clear that $\frac{h_{cw}}{h_{mass}} = \rho C_p$ (Y.A. Cengel)

Where ρ = mass density of air and C_p = specific heat of air = 1.005 KJ/kg^{0}C

Hence $\dot{m} = \frac{h_{cw}}{\rho C_p}(\rho_w - \rho_a)$. Using perfect gas equation, $PV = RT$ (1 mole of gas) and

R = universal constant, $V = \frac{M}{\rho}$, so, $PM = \rho RT$. If we will assume $T_w \simeq T_a \simeq T$ then PM_w

$= \rho RTw$ and $PM_a = \rho RT_a$, then $M_a = M_w$. So, for water vapour $\rho_w = \frac{P_w M_w}{RT}$.. Substituting

this expression in the equation of mass flow rate $(\dot{m})$ then

$$\dot{m} = \frac{h_{cw} M_w}{\rho.C_p RT}(P_w - P_a)$$

The rate of heat transfer on account of mass transfer of water vapour is $\dot{q}_{ew} = \dot{m}L$

Or, $\dot{q}_{ew} = \frac{h_{cw}L}{\rho C_p}\frac{M_w}{RT}(P_w - P_a)$ Or $\dot{q}_{ew} = h_e(P_w - P_a)$ where $h_e = \frac{h_{cw}L}{\rho C_p}\frac{M_w}{RT}$ and

$\frac{h_e}{h_{cw}} = \frac{L}{\rho C_p}\frac{M_w}{RT}$. Using perfect gas equation for air, $\rho_a = \frac{P_a M_a}{RT}$ (for 1 mole of air) and

substituting ρ_a in the above equation, $\frac{h_e}{h_{cw}} = \frac{L}{C_p \frac{P_a M_a}{RT}}\frac{M_w}{RT} = \frac{LRT}{C_a P_a M_a}.\frac{M_w}{R_T} =$

$\frac{L}{C_p}\frac{M_w}{M_a}.\frac{1}{P_a} = \frac{L}{C_p}\frac{M_w}{M_a}.\frac{1}{P_T}$ (as for small value of $P_w, P_T \simeq P_a$, So, $\frac{h_e}{h_{cw}} = \frac{L}{C_p}\frac{M_w}{M_a}.\frac{1}{P_T}$

where L = latent heat of vaporisation = 2200 × 10^3 J/kg = 2200 kJ/kg, C_p = specific heat of air = 1.005 KJ/kg °C, M_w = 18 Kg/mol (molar mass of water), M_a = 29 Kg/mol (molar mass of air), P_T= Total pressure of air-vapour mixture = 1 atm (1 atm = 101325 N/m^2). Substituting all those values, h_e/h_{cw} =0.013, or h_e = 0.013 h_{cw}. But the best representation of the mass-heat transfer phenomenon is obtained if the value of h_e/h_{cw} = 16.273 × 10^{-3} (Cooper). Hence $\dot{q}_e$ = 16.273 × 10^{-3} h_{cw} ($P_w - P_a$). If the surface is exposed to atmosphere, then the above equation is turned to $\dot{q}_e$ = 16.273 × 10^{-3} h_{cw} ($P_w - \gamma P_a$), where γ is the relative humidity of atmospheric air in decimal. Also $\dot{q}_{e\,w}$ = h_{ew} ($T_w - T_a$) where h_{ew} is the evaporative heat transfer coefficient. Therefore, h_{ew} = 16.273 × 10^{-3} h_{cw} ($P_w - \gamma P_a$)/($T_w - T_a$). The values of P_w and P_a (for the range of temperature 10°C

- 90°C) can be obtained from the expression $P(T) = \exp\left[24.317 - \dfrac{5144}{T+273}\right]$ (Ref. Fernandez and Chargoy 1990).

6.14 CONVECTIVE HEAT TRANSFER COEFFICIENT FROM CROP TO AIR

Bansal et al (1993) have proposed the expression for convective heat transfer coefficient from crop to the surrounding air as

$h_c = 2.38\ (T_p - T_r)^{0.25}$ for V, the wind speed less than 1.0 m/s

$h_c = 12.1\sqrt{V}$ for $1.0 < V < 2.6\ m/s$ where T_p crop temperature and T_r surrounding air temperature.

Example 6.8: Estimate the convective heat transfer coefficient from agricultural produce to the surrounding for the air velocity of 0.75 m/s with the temperatures of 30 ^{0}C and 25 ^{0}C for the produce and air respectively.

Solution: The related equation, $h_c = 2.38\ (T_p - T_r)^{0.25}$ for V less than 1.0 m/s. where h_c, is the convective heat transfer coefficient and V, is the wind velocity. Putting the values, $h_c = 3.55$ W/m^2^{0}C.

Example 6.9: Calculate the rate of evaporation for exposed wetted surface (agricultural produce) having a temperature of 35 ^{0}C and relative humidity of 50 %. Ambient air temperature is 15 ^{0}C. Also calculate the evaporative heat transfer coefficient.

Solution: Temperature of wetted surface, $T_w = 35^0$C and temperature of surrounding air, $T_a = 15\ ^0$C. The partial vapour pressure, $P(T_w) = \exp\left(25.317 - \dfrac{5144}{273+35}\right) = 5517.6$ N/m^2, similarly, partial pressure of ambient air $P(T_a) = \exp\left(25.317 - \dfrac{5144}{273+35}\right) = 1730$ N/m^2.

Using $h_c = 3.55$ W/m^2^{0}C as per example 6.8, the rate of evaporation, $\dot{q}_{e\,w} = 16.237 \times 10^{-3} \times 3.55\ (5517.6 - 0.5 \times 1730) = 268.07$ W/m^2. The evaporative heat transfer coefficient can be calculated as $h_{ew} = \dfrac{\dot{q}_{ew}}{(T_w - T_a)} = \dfrac{268.07}{(35-15)} = 13.40$ W/m^2°C.

6.15 HYBRID SOLAR PHOTOVOLTAIC/THERMAL DRYING SYSTEM AND ITS APPLICATIONS

A Hybrid Solar Photovoltaic/Thermal (HPV/T) system is one which can supply simultaneously both electrical and thermal energy. It is well established that the dryer in forced convection mode is better than that in natural convection mode. For forced convection mode, a fan or blower is required to be operated for removal of moisture in the form of vapour, evaporated from the crop surface. The power required to operate the fan or blower may be supplied either from the centralized grid electricity or from the solar photovoltaic (PV) panel. The PV operated solar dryer is useful for the drying applications in rural areas where grid electricity is not readily available. The utilization of solar energy through photovoltaic conversion system is gaining momentum in the present context of reducing the reliance on fossil fuels. Accelerated research and development is going on worldwide, to develop more efficient solar energy devices to harness and utilize solar energy in a wider scale. More attention has also been given on enhancing the efficiency of solar photovoltaic panel. One controllable approach for increasing the efficiency of solar panel is to decrease the operating temperatures of the panel to a permissible level, mostly nearer to the temperature, specified as per the standard test condition. This can be achieved by incorporating various cooling devices. The use of cooling devices reduces the operating temperatures of the panel by extracting the absorbed heat energy at its back surface. The extracted heat energy by the way, may be utilized for low energy consuming applications such as greenhouse heating, water heating, crop drying and thermo-electric power generation etc. Such type of system is therefore called as a *hybrid system* due to the dual advantages of improving electrical efficiency of the panel and utilizing unused thermal energy in a productive and sustainable manner. There should therefore be an integrated approach to focus on the enhancement of the efficiency of solar panel by cooling its back surface either with the flow of air or water through forced circulation mode with the help of the DC electricity, generated from the same panel. The drying activities of the agricultural produces may be accomplished with the entry of the extracted heat energy from the panel to the greenhouse, integrated with the developed hybrid system. The developed system may therefore be called as Hybrid Photovoltaic/Thermal (HPV/T) solar greenhouse system. The combined greenhouse set up with the hybrid system may be used for effective and rapid drying of high value horticultural produces mostly, the fresh flower. The trading and commercial value of dry flower is increasing day by day both in the international and national markets. The performance of HPV/T solar greenhouse dryer (fig. 6.27) can also be evaluated and compared with greenhouse system alone and open sun drying method. This improved method of drying i.e. hybrid photovoltaic thermal (HPV/T) greenhouse drying may even be followed in a small scale by the resource poor farmers for earning livelihood because of efficient drying and improvement in the quality of the product. The HPV/T solar greenhouse system consists of a solar photovoltaic panel, cooling arrangement and a greenhouse. The cooling agents generally taken are either water or air, which needs to be circulated around the PV panels for cooling the solar cells, such that the warm water or air leaving the panels may be used for low thermal applications.

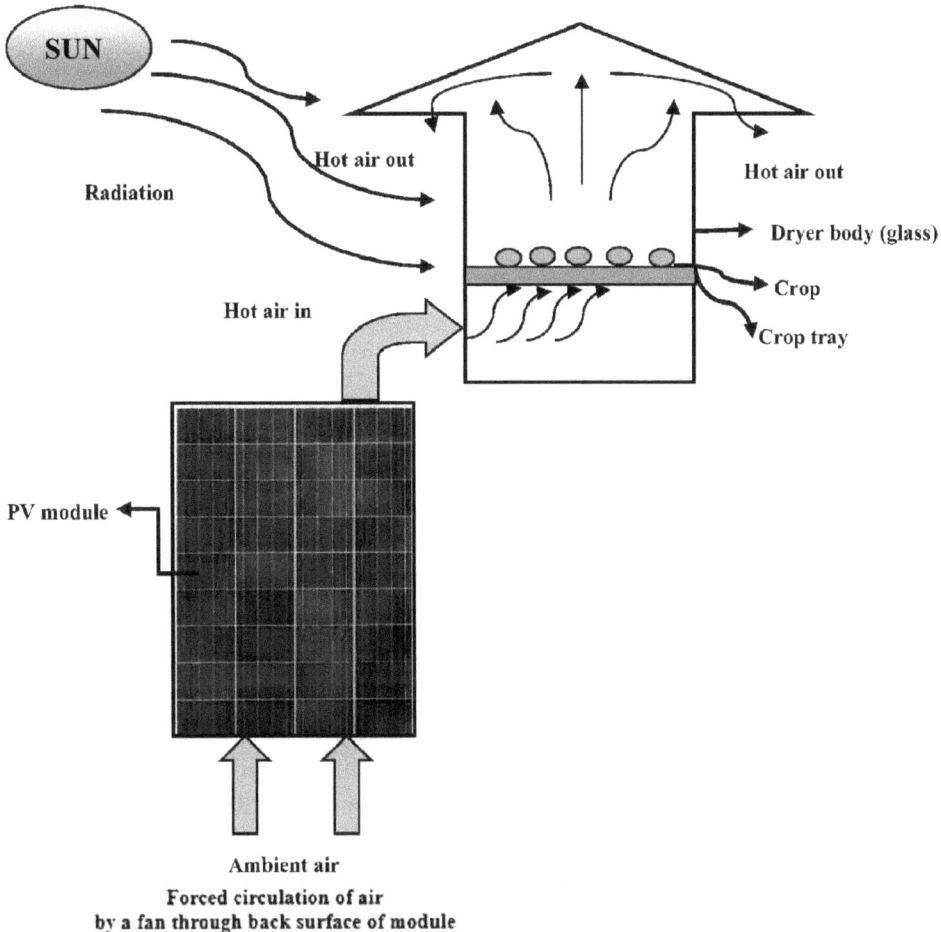

Fig. 6.27. Working principle of Hybrid Photovoltaic/Thermal (HPV/T) greenhouse drying system

6.15.1 Importance of Hybrid Photovoltaic/Thermal (HPV/T) System

About 15 to 20% of incident solar radiation is converted into electricity while rest of the radiation gets absorbed and converted into heat energy resulting in increase of the temperature of the solar module. In general, the wavelengths of solar radiation from 400-1200 nm are strongly absorbed by PV cell and converted to electricity. The other parts of the radiation are transformed into heat. Solar PV cell is a solid state device, strong, durable, require little maintenance and generate electricity without going through any thermal process. However, the major disadvantages of PV cell are its lower conversion efficiency and further decrease in performance with increasing operating temperatures due to the ambient air and higher intensity of incident solar radiation. The PV cells are sensitive to temperature variations. When the ambient temperature and intensity of solar radiation incident on the PV cells increases, the operating temperatures of PV cell also increase linearly. This increase in operating temperatures of the PV cell leads to reduction in open circuit voltage, fill factor and power output of PV cells. The net results lead to the loss of conversion efficiency, power output (fig.6.28). and subsequently irreversible damage to the PV cell materials.

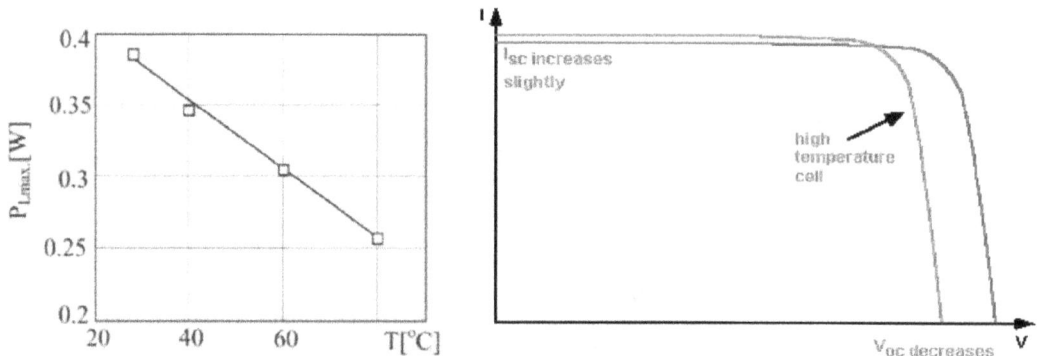

Fig. 6.28. Temperature dependence of maximum power output

Higher cell temperature reduces the voltage by 0.04 to 0.1 volts for every one-degree Celsius rise in temperature

To overcome the temperature effects of the solar cell on its performance and to maintain the operating temperatures of the PV cells nearer to the manufacturer's specified value, it is necessary to remove heat from PV cells during operating period and practically from the panel by adopting proper cooling methods. Solar module output is specified as rated output under standard test conditions (solar intensity =1000 W/m^2, temperature= 25°C). These conditions practically never occur during normal outdoor operations as they do not take into consideration the actual geographical and metrological conditions at the installation site. This undesirable effect can be partially avoided by applying a heat recovery unit with the circulation of air or water. A hybrid photovoltaic-thermal (PV/T) solar system is an easy and simple way for cooling the PV cell. An effective way of improving the efficiency of a PV module is by reducing the operating temperature of its front/back surface. This can be achieved by cooling (water/air flowing) the module/panel and reducing the heat stored inside the PV cells during operation. Hybrid photovoltaic/thermal solar system therefore consists of panels coupled to the heat extraction devices, in which air or water of lower temperature than that of PV modules is heated while at the same time the PV module temperature is reduced.

6.16 HYBRID SOLAR PHOTOVOLTAIC THERMAL (HPV/T) GREENHOUSE SYSTEM FOR FLOWER DRYING (A CASE STUDY OF SMALL SCALE APPLICATIONS)

The work related to the development of HPV/T solar greenhouse system and drying of rose flower has been undertaken to evaluate its quality with respect to the appearance and decorative value and economic viability in a small scale adoption for supporting livelihood to the resource poor flower growers and to take up this venture on an entrepreneurship basis for employment generation (Kamal and Ghosal 2020). Attempt has been made for studying the feasibility of drying and packaging of rose flower because of its increased demand at present in national and international markets. A study was conducted to evaluate the performance of Hybrid Photovoltaic/Thermal (HPV/T) solar system combined with a greenhouse dryer and air cooling arrangements. The experiments were carried out in warm and humid climatic condition of Odisha, India for the winter and summer months during the year 2020. The enhancement in the electrical efficiency of the solar panel was studied by lowering its temperature during the operating period with the help of extracting absorbed heat energy from its back surface through an air cooling arrangement. The extracted heat energy in this study, was utilized in a low thermal energy consuming application such as greenhouse heating for drying of rose flower.

6.16.1 Importance of Flower Drying

In floriculture trade, fresh flowers constitute a major part but due to their reduced shelf-life, flowers remain fresh only for a short duration. Therefore, to overcome this problem, techniques of drying play a vital role. The dried flowers can be stored in moisture-free atmosphere for longer periods without losing their appearance and decorative value. Thus, the availability of flowers may not be dependent on season only. The demand for dry flowers and attractive plant parts (branches, twigs, foliage etc.) for floral arrangements and floral crafts has increased manifold during the last decade. Flower drying is an important post-harvest technique for enhancing the keeping-quality of fresh flowers. The flower-drying technique involves reducing moisture content of flowers to a point at which biochemical changes are minimized while maintaining cell structure, pigment level and shape. The demand for dry flowers is increasing at an impressive rate, thus offering good opportunity for the entrepreneurs to enter the global floriculture trade. There is large potential to develop the dry flower industry in India and to provide employment to housewives, unemployed youth and rural women. Simplified indigenous techniques have been developed by which flowers retain their freshness for several months or even years. The original shape, colour and size before dehydration are retained, thus, making them highly suitable raw materials for interior decoration. Dried flowers and foliage can be used for designing artistic decorative items, e.g. greeting cards, wall plates, calendars, landscapes, etc. Drying of flowers and foliage by various methods like air drying, sun drying, oven and microwave oven drying and embedded drying can be used for making decorative floral crafts items like cards, floral segments, wall hangings, landscapes, calendars, potpourris etc. and for other various purposes. Dried flowers offer good standby for the florists, since decorative floral activities can be taken up during the slack periods and arrangement can be displayed where fresh flowers are unavailable or unsuitable from the grower's point of view and the price is less than for equivalent fresh flowers. The Indian export basket comprise of 71% of dry flowers which are exported to USA, Europe, Japan, Australia, far East and Russia. Dry flowers constitute more than two-thirds of the total floriculture exports. Only a few research and development projects have been undertaken in the flower drying industry across the globe, in contrast to other areas of floriculture, due to the unavailability of improved dryers.

6.16.2 Development of Dryer

The work was conducted and the data were recorded in an interval of one hour from 9 am to 4 pm from December 2019 to April 2020 by installing the set up at the roof top of College of Agricultural Engineering and Technology, Bhubaneswar, Odisha (latitude 20.50°N and longitude 85.81°E). The place is coming under warm and humid climatic condition where the annual average rainfall is 1450 mm and average daily solar insolation is 4.8 kWh/m^2. During the course of investigation, important parameters recorded were solar radiation, ambient air temperature, wind velocity, relative humidity, voltage, current and power output from the solar panel. These parameters were recorded at an interval of 1 hour. Solar radiation was measured using solar irradiance meter. Two solar panels, each of 100 W$_p$ were used for the study. Of the two, one panel was used without incorporating any air cooling arrangement to evaluate its electrical efficiency and to compare with the other panel with air cooling device. The air cooling arrangement consisted of four number of DC fans and an air duct system of aluminium tubes along with aluminium fins spaced in between the rows of the tubes. The power used for operating the fans was only from the same experimental panel. The cooling arrangement was fitted in the back side of the panel in order to extract the absorbed thermal energy from it and to maintain its temperatures nearing to its normal operating temperatures. The extracted thermal energy was ultimately allowed to enter into the greenhouse in which fresh rose flower was kept for drying. One fan was also fitted in the greenhouse to remove moisture from the materials kept for drying. Hence, in this study, the reduction in the temperatures of the panel, improvement in its electrical efficiency along with drying characteristics of rose flower have been studied. The performance of HPV/T solar greenhouse dryer was evaluated and compared with greenhouse system alone and open sun drying method. Forced air was used to extract the heat from the back surface of the solar panel. Open sun drying, in which the product is spread on the ground in open condition, is the simplest and cheapest method of drying. But there are considerable losses associated with it. So, an advanced method of drying i.e. hybrid photovoltaic thermal (HPV/T) greenhouse drying has been followed for efficient drying and improving the quality of the dried product. The details of each of the component in the experimental set-up are shown in the fig. 6.29 below.

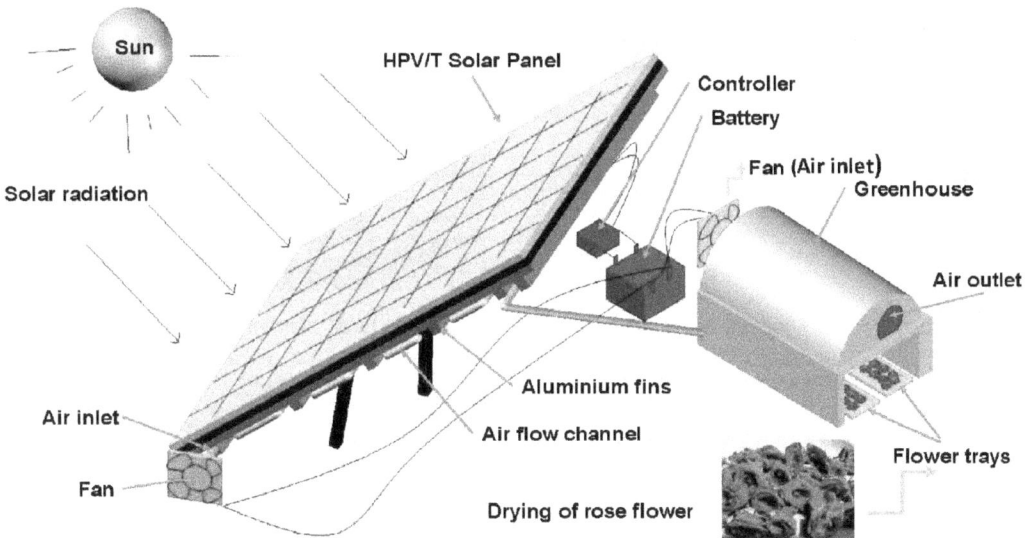

Fig. 6.29. Schematic diagram of the experimental set up

Two poly crystalline solar panels were used for the study with the following specifications:

No of solar panel	2
Type of solar panel	Poly crystalline
Panel connection	series
Panel dimensions	100cm×166cm×4cm
Maximum power	100 Wp
Total power of panel	1×100 Wp = 100 Wp
Open circuit voltage	37.69V
Short circuit current	8.89A
Voltage at maximum power	30.33V
Current at maximum power	8.41A

The solar panels were installed at a height of 0.91 m from ground level and were oriented in the south direction. Two panel were installed at 10 m distance apart. Arrangement was made to vary the inclination of solar panel with respect to the horizontal plane. In order to measure different electrical parameters for the panel, a solar testing kit was used. It can wirelessly capture and record real-time solar irradiance, ambient air temperature and PV panel temperature. The drying behaviour of rose flower was recorded and compared with open sun drying, greenhouse drying and HPV/T greenhouse drying. The moisture content was calculated every day during the experiment.

Hybrid Photovoltaic/Thermal solar greenhouse system

Air duct made of aluminium, fixed at back side of panel

Air duct with aluminium fins at back side of panel

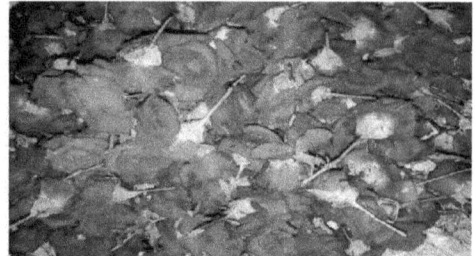

Open sun drying of rose flower

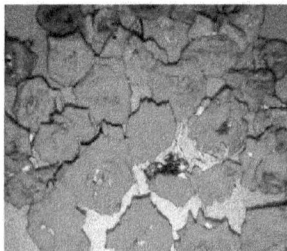

Greenhouse drying

HPV/T greenhouse drying

HPV/T greenhouse drying

6.16.3 Performance Evaluation

The performance of the developed dryer was evaluated by following the fixing of tilt angle, measurement of all electrical parameters and the drying behaviour of rose flower. In the first part, the tilt angle of the solar module was decided to fix for the study on the basis of the maximum availability of incident solar radiation on it in the experimental site from January to April (Fig 6.30). In the second part, all the electrical parameters such as short circuit current, maximum current, open circuit voltage, maximum voltage, fill factor of the used solar module were measured on hourly basis from 9 am to 4 pm along with the incident solar radiation on the surface of the module with the help of solar PV testing kit (Fig 6.31). In the third part, the drying behaviours of the rose flower were studied. The data for temperatures of panel were recorded by following air cooling method and compared these without any cooling arrangement (Fig 6.32). The efficiency of the solar module is calculated from the measured data and compared with those in case of without using any cooling method. Similarly, the drying behaviours of the rose flower were compared with the open sun, greenhouse alone and hybrid solar photovoltaic thermal drying (Fig 6.33 and Table-6.12). The method of sensory analysis was used to evaluate the quality of the dried flowers.

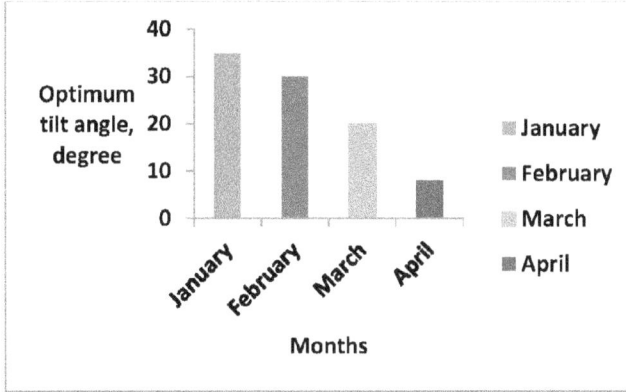

Fig. 6.30. Fixation of Tilt Angle of Panel during January-April 2020

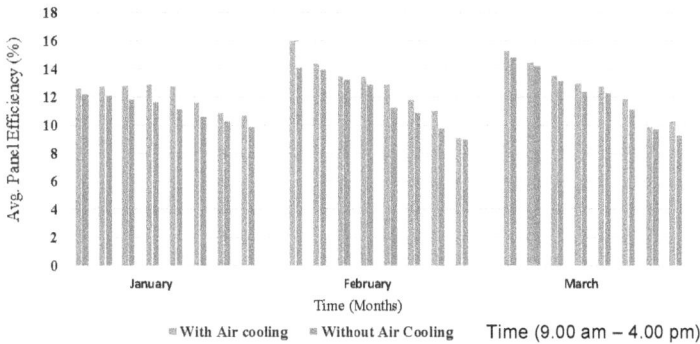

Fig. 6.31. Average efficiency of solar panel in different months with and without air cooling

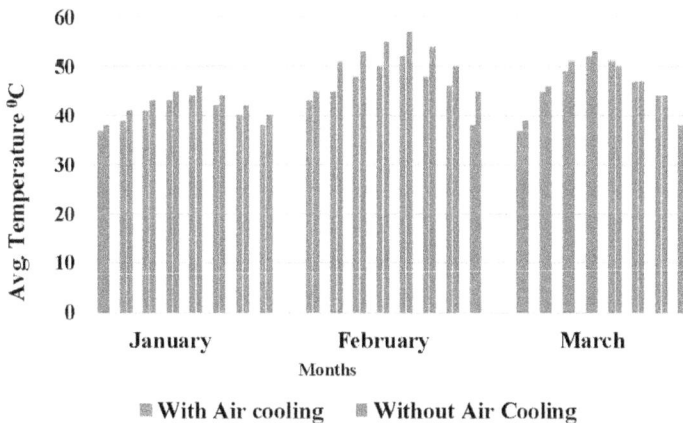

Fig. 6.32. Average temperature of solar panel in different months with and without air cooling

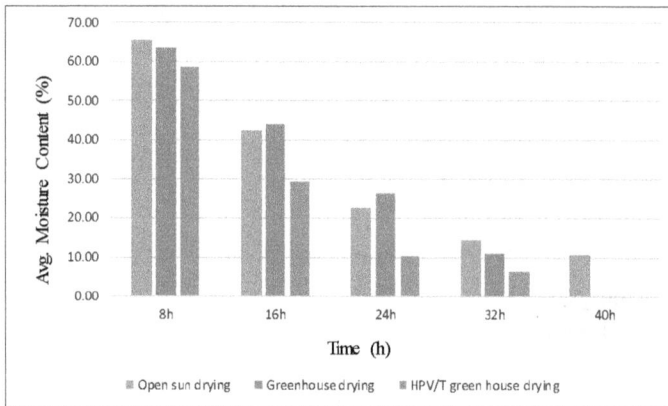

Fig. 6.33. Variations of moisture content of rose flower for different drying systems

Table 6.12: Mean values of different parameter of Open sun, Greenhouse alone and HPV/T greenhouse dried sample for rose

Parameter	Open sun drying	Greenhouse drying	HPV/T Greenhouse drying
Total drying time (h)	41	32	24
Reduction in diameter (%)	18.85	16.86	7.45
Shape	1	1.98	2.80
Colour	1.2	1.88	3.84
Intactness	81.22	89.87	94.56
Brittleness	3.97	3.28	1.07
Texture	1.7	2	3.86

Sensory analysis is the method which employs human senses (sight, smell, touch, taste and hearing) for evaluation of quality of dried flowers. Different properties like shape, colour, texture, intactness and brittleness were evaluated by a panel of five members. Each treatment was evaluated by the panel after drying of the produces and the respective scores were recorded. The scale used for scoring was similar to that followed by Ugaale et al 2016. (Colour: 1- Not acceptable, 2- less acceptable, 3- acceptable, 4- highly acceptable. Shape: 1- Distorted, 2- maintained, 3- very well maintained. Intactness: 100% - No damage, 90%, -80% partially damage, 80%-10% damage. Texture: 1- Not acceptable, 2 - less acceptable, 3 - acceptable, 4 - highly acceptable. Brittleness: 1 – Very low, 2 – low, 3- moderate, 4 – high). As the temperature inside the HPV/T greenhouse dryer is fairly constant throughout the drying process and shields the flower from direct solar radiation, the drying rate for the flower was found to be higher than that in open sun and greenhouse dryer. Thus the drying time was reduced by 65%, when the flower was dried using the HPV/T greenhouse dryer. As the solar radiation varies throughout the day, with solar intensity reaching maximum at noon, the quality parameters like shape, texture, brittleness and colour were adversely affected and high in open sun drying and moderate in greenhouse drying.

6.16.4 Dryer Performance

Dryer performance can be expressed in terms of efficiency factors. The daily drying efficiency of the hybrid PV/T greenhouse dryer can be expressed as:

$$\eta_{dryer} = \frac{M_{ev} \times \lambda}{I(t) \times A} \times 100$$

where M_{ev} is the amount of moisture evaporated (kg)λ, is the latent heat of evaporation (J/kg), $\overline{I(t)}$ is the average solar radiation incident on the dryer (W/m^2), A is the area of the greenhouse structure (m^2), and N is the number of sunshine hours per day (h).

Drying time 24 hours, Drying duration = 2 days. λ = 226 × 10^4 J/kg, A = 1.0 m × 0.5 m = 0.5 m^2. This area of greenhouse can easily accommodate 1.5-2 kg. of flowers for drying.

$\overline{I(t)}$ = 480 W/m^2, N= 6h/day. No. of clear days in a year = 300. Initial moisture content, m_i = 85 % (w.b.). Final moisture content, m_f = 10 % (w.b.). Initial weight of product, m_o = 1.5 kg. As per Eq. (6.4), the mass of water removed (M_{ev}) from the wet

product can be calculated as $M_{ev} = \dfrac{m_o \left(m_i - m_f \right)}{100 - m_f}$ where m_o is the initial mass of the

product, m_i is the initial m.c. (w.b.) on percent and m_f is the final m.c. (w.b.) on percent. M_{ev} = 1.25 kg. So, 1.25 kg moisture will be removed in 2 days during drying of 1.5 kg of rose flower. Therefore, average daily thermal output of the dryer can be calculated as follows,

Daily thermal output of the dryer $\left(\dot{Q}_u \right)_{daily}$ = moisture evaporated (kg) × latent heat

of vaporization (J/kg) $= \dfrac{\left(1.25 \times 226 \times 10^4 \right)}{3.6 \times 10^6}$ kWh = 0.784 kWh/day.

Therefore, annual thermal output of the dryer, E_{out} = Daily thermal output of the

dryer (kWh) $\times \dfrac{300}{2}$ = 0.784 × 150 = 117.6 kWh.

Input energy = {Average solar radiation (W/m^2) × No. of sun shine hours/day drying time (day) × area × 10^{-3}} (kWh) = 480 × 6 × 2 × 0.5 × 10^{-3} = 2.88 kWh.

$$\eta_{dryer} = \frac{Output\ energy}{Input\ energy} = \frac{0.784}{2.88} \times 100 = 27.22\%$$

Therefore, the daily drying efficiency for the said dryer of floor area 0.5m^2 with an average solar intensity 480W/m^2 over a period 2 days (12 sunshine hours) was calculated as 27.22 %.

6.16.5 Cost economics

For the success and commercialization of any new technology, it is essential to know whether the technology is economically viable or not. Therefore, an attempt has been made to assess the economics of the hybrid photovoltaic/thermal solar system with conventional greenhouse dryer. The economic evaluation was done for the drying of rose flower in the hybrid PV/T greenhouse dryer.

The costs and the key economic parameters based on the prevailing price in Bhubaneswar, Odisha, India are presented in the Table 6.13. The moisture content of rose flower has been decreased to about 10 % (w.b.) for its safe storage. The capacity of greenhouse for drying of flower is around 1 kg. With this capacity, the flowers can be spread uniformly inside the greenhouse.

Table 6.13: Cost of unit electricity generation from HPV/T with greenhouse dryer with air cooling arrangement

Sl. No.	Items	Cost (Rs.)
1	One number of PV module of 100 Wp @ Rs. 40 per watt	4,000
2	2 nos. of 12 V and 3 Amp DC fan @ Rs.750 per unit	1500
3	1 no. of 75 Ah battery	5000
3	Angle iron frame for erecting solar photovoltaic module	2500
4	Back side wall aluminum pipes, sheet and fins arrangement	2000
5	Greenhouse structure	1000
	Total	**16,000**

$P_{net} = P_i + P_M - P_S$, where P_{net} = Net present worth, P_i = Initial investment = Rs. 16,000, P_S = Present worth of salvage value. Salvage value of the system = S = 5% of initial investment = Rs. 800. Life span of HPV/T module and solar DC fans = 15 years, Interest rate = i = 10%

Annual maintenance cost = 1 % of total cost of the system = M= Rs. 160.

Present worth of HPV/T module with cooling arrangement (life of 15 years) = Rs. 16,000

Present value of annual maintenance cost $= = P_m = M\left[\dfrac{(1+i)^{15}-1}{i(1+i)^{15}}\right] = \text{Rs. }1220$

Where $\left[\dfrac{(1+i)^{n}-1}{i(1+i)^{n}}\right]$ is the uniform series present worth factor

Present worth of Salvage value $= \dfrac{S}{(1+i)^{15}} = \text{Rs. }190$

$P_{Net} = P_i + P_M - \dfrac{S}{(1+i)^{25}} = \text{Rs.}16{,}000 + 1220 - 190 = \text{Rs. }17{,}030$

$A_A = $ Annualized cost of the system $= P_{Net}\left[\dfrac{i(1+i)^{15}}{(1+i)^{15}-1}\right] = \text{Rs. }2240$

where $\left[\dfrac{i(1+i)^{n}}{(1+i)^{n}-1}\right]$ is the uniform series capital – recovery factor

With cooling arrangement, the power output of one module of 100 $W_p = 100 \times\dfrac{11}{16}$ = 70 Watt (on site module efficiency with cooling arrangement = 11 % and 16 % efficiency at standard test condition). Without cooling arrangement, the power output of one module of 100 $W_p = 100 \times\dfrac{9}{16} = 60$ Watt (on site module efficiency without cooling arrangement = 9 % and 16 % efficiency at standard test condition).

Energy output from the system with cooling arrangement = Net output (70 watt) x 6 hours/day × 300 sunny days/year = 126 kWh/year

Energy output from the system without any cooling arrangement = Net output (60 watt) x 6 hours/day x 300 sunny days/year = 108 kWh/year

Cost of unit electricity generation from HPV/T module with cooling arrangement = 2240 (annualized cost)/126 = Rs. 17/kWh

Cost of unit electricity generation from HPV/T module without cooling arrangement = 1540 (annualized cost)/108 = Rs. 20/kWh

Percentage decrease in cost of unit electricity from HPV/T module with cooling arrangement against without cooling arrangement = {(20-17)/20} × 100 = 15 %

6.16.5.1 Cost Economics of Rose Flower Drying in HPV/T Greenhouse Dryer

Cost of fresh rose flower of 1kg. = Rs.300

Cost of dried rose flower with packing of 100 g = Rs.240

The weight of dry rose flower (10 % moisture content, wet basis) from 1 kg fresh flower = 250 g

Therefore, cost of 250 g of dried rose flower = Rs. 600

Profit for 1 kg rose flower after drying = 600-300 = Rs. 300

Taking 10 batches of drying in a month and 8 months in a year, the total income per year = Rs 24,000

Simple payback period = 16000/24,000 = 8 months

6.16.5.2 Comparative Cost of Drying in Various Drying Systems Under Study

Comparative economics studies for drying of rose flower in open sun, greenhouse and hybrid photovoltaic thermal greenhouse drying have been evaluated on the basis of quality of flowers, with respect to shape, colour, texture, brittleness and intactness to fetch its market value.

(i) *Cost economics of rose flower drying in greenhouse dryer*

Cost of fresh Rose flower of 1kg. = Rs.300

Cost of dried rose flower with packing of 100 g = Rs.180 (due to poor quality)

The weight of dry rose flower (10 % moisture content, wet basis) from 1 kg fresh flower = 250 g

Therefore, cost of 250 g of dried rose flower = Rs. 450

Profit for 1 kg rose flower after drying = 450-300 = Rs. 150

Taking 10 batches of drying in a month and 8 months in a year, the total income per year = Rs 12,000

Cost of greenhouse = Rs. 1000

Simple payback period = 1000/12,000 ≈ 1 month

(ii) *Cost economics of rose flower drying in open sun drying*

Cost of fresh Rose flower of 1kg. = Rs.300

Cost of dried rose flower with packing of 100 g = Rs.150 (due to very poor quality)

The weight of dry rose flower (10 % moisture content, wet basis) from 1 kg fresh flower = 250 g

Therefore, cost of 250 g of dried rose flower = Rs. 375

Profit for 1 kg rose flower after drying = 375-300 = Rs. 75

Taking 10 batches of drying in a month and 8 months in a year, the total income per year = Rs 6,000. The simple payback period is very nominal.

It was observed that open sun drying of products takes long time for drying to the required moisture level. It can even take two to three days or even more. Moreover, the qualities of dried products do not meet as per the requirement of the markets. Therefore, need is felt to introduce advanced solar drying techniques. Rose flowers dried under hybrid solar photovoltaic thermal (HPV/T) greenhouse dryer were observed to be of better quality for trading commercially compared to greenhouse drying alone and open sun drying. The monthly income of Rs. 2,000, Rs.1000 and Rs. 500 may be earned from drying of 1 kg of rose flower in one batch of 3 days' duration (10 batches /month and 8 months/year) in HPV/T greenhouse dryer, greenhouse dryer and open sun drying respectively.

6.17 ENTREPRENEURSHIP SCOPE FOR HPV/T GREENHOUSE DRYING SYSTEM

Hybrid solar photovoltaic/Thermal (HPV/T) greenhouse system is at present gaining more importance among the researchers, users and entrepreneurs in order to enhance the electrical efficiency of the solar panel by lowering its temperature during the operating period with the help of extracting absorbed heat energy at its back surface through various cooling devices and to utilize the extracted thermal energy into the greenhouse for drying agricultural and horticultural produces. This type of effort is also under the control of the user. Crop drying has got more relevance through the above practice especially by the use of various air cooling techniques compared to water cooling arrangement which needs more complex structure and incurs more expenditure. There is also the scope for drying of other products such as leaves, roots and twigs of medicinal plants, high value flowers, most perishable vegetables, fish etc. in HPV/T greenhouse system for value addition and commercialization to raise income and to achieve livelihood security in a sustainable manner. The following findings for the drying of the rose flower in HPV/T greenhouse system provide the references to the entrepreneurs and users for adopting the technology either at individual level or mass scale basis.

(i) The temperatures of solar panel may be decreased in the range of 9-12 °C in forced air cooling compared to without cooling in a clear day.

(ii) The efficiencies of solar panel may be increased in the range of 8-14 % in air cooling compared to without cooling in a clear day.

(iii) The reduction in moisture content of the rose flower compared to fresh flower may be about 67-73 % in HPV/T greenhouse drying, 54-61% in greenhouse drying alone and 32-41% in open sun drying in 3 days' duration of drying. The colour, appearance and texture of the flower can also be maintained in HPV/T greenhouse drying for its marketability in national/international market.

(iv) The dryer performance in terms of its efficiency was about 30 %.

The HPV/T greenhouse drying system may be encouraged among the growers, entrepreneurs, traders and users with respect to reduction of the drying time and maintaining the quality of the products. This type of venture is economically viable, technologically feasible, environmentally sustainable and income generating scope with less efforts and cost involvement. Conventional energy may be conserved and saved by using solar energy which is available plentifully in each and every corner of a tropical climate. The device is simple to operate, affordable, durable, applicable to various agricultural and horticultural produces and requiring negligible maintenance.

REFERENCES

Cooper P.I. 1969. The absorption of solar radiation in solar stills. Solar Energy. 12(3), pp. 333-346.

Ekechukwu, O.V., and Norton B. 1999. Review of solar energy drying systems II: an overview of solar drying technology. Energy Conversion and Management, 40(6), pp. 615-655.

Fernandez J. and Chargoy N. 1990. Multistage, indirectly heated solar still. J. Solar Energy, 44(4), pp. 215-223.

Fuller R.J. 1993. Solar drying of horticultural produce: present practice and future prospects, Post-harvest News and Information, 4(3).

Ghosal, M.K. and Das Kamal Kumar. 2020. Performance Evaluation of Hybrid Photovoltaic/Thermal Solar System Combined with Greenhouse for Drying of Rose Flower. Accepted for publication in International Journal of Chemical Studies.

Perry, J.H., Chemical Engineering Handbook, 2007, 8th ed., McGraw-Hill, New York, pp. 28.

Ugale H, Alka S, Timur A, Palagani NJ. 2016. Ornament. Hortic., 2016, 19(1&2), 34–38.

Y.A. Cengel. 1998. Heat Transfer. A practical Approach. McGraw-Hill Publication. Chapter 11, pp. 666-669.

CHAPTER - 7

MICROWAVE ASSISTED PYROLYSIS OF AGRICULTURAL RESIDUES FOR BIOFUELS PRODUCTION

7.1 SOURCES OF AGRICULTURAL RESIDUES

After crop harvesting, the left over plant material including leaves, stalk and roots is known as crop residue or agricultural residue. Agricultural residues are the biomass left in the field after harvesting of the economic components i.e., grain. The term agricultural residue is used to describe all the organic materials which are produced as the by-products from harvesting and processing of crops. These residues can further be categorized into primary residues and secondary residues. Agricultural residues, which are generated in the field at the time of harvest are defined as primary or field based residues (e.g. rice straw, sugar cane tops), whereas those co-produced during processing are called secondary or processing based residues (e.g. rice husk and bagasse) (Table 7.1). Crop residue can contain both the field residues that are left in an agricultural field or orchard after the crop has been harvested and the process residues that are left after the crop is processed into a usable resource. Stalks and stubble (stems), leaves, and seed pods are some common examples for field residues. Sugarcane bagasse and molasses are some good examples for process residues. Residues generated by different crops can also be grouped in four categories based on the type of crop, namely cereals (rice, wheat, maize, jowar, bajra, ragi and small millets), oilseeds (groundnut and rapeseed mustard), fibers (jute, mesta and cotton) and sugarcane. The amount of crop residue generated is estimated as the product of crop production, residue to crop ratio and dry matter fraction in the crop biomass. The residue to grain ratio varied 1.5–1.7 for cereal crops, 2.15–3.0 for fiber crops, 2.0–3.0 for oilseed crops and 0.4 for sugarcane (Table 7.2). There is a large variation in crop residues generation across the different places depending on the crops grown, their cropping intensity and productivity.

Table 7.1: Crop residues produced by major crops

Source	Composition
Rice	Straw, husk, bran
Wheat	Bran, straw
Maize	Stover, husk,
Millet	Stover
Sugarcane	Sugarcane tops, bagasse, molasses

Table 7.2: Crop wise production, residue generated and dry matter fraction in India

Crop	Annual production	Dry residue generated	Residue to crop ratio	Dry mater fraction
	Mt/yr (million tons/year)			
Rice paddy	153.35	192.82	1.50	0.86
Wheat	80.68	120.70	1.70	0.88
Maize	19.73	26.75	1.50	0.88
Jute	18.32	31.51	2.15	0.80
Cotton	37.86	9.86	3.00	0.80
Groundnut	7.17	11.44	2.00	0.80
Sugarcane	285.03	107.50	0.40	0.80
Rapeseed & Mustard	7.20	17.28	3.00	0.88
millets	18.62	21.57	1.50	0.80
Total	627.96	620.43		

7.2 AGRO RESIDUE GENERATION IN INDIA

India is an agrarian country and generates a large quantity of agricultural residues. This amount will increase in future and as with growing population there is a need to increase the productivity also. Specifically, India is the second largest producer of rice and wheat in the world, two crops that usually produce large volume of residue. Large quantities of crop residues are generated every year, in the form of cereal straws, woody stalks, and sugarcane leaves/tops during harvest periods. Processing of farm produce through milling also produces large amount of residues. Traditionally crop residues have numerous competing uses such as animal feed, fodder, fuel, roof thatching for rural homes, residential cooking fuel, industrial fuel, packaging and composting. However, a large portion of the crop residues is not utilized and left in the fields. The disposal of such a large amount of crop residues is a major challenge. To clear the field rapidly and inexpensively and allow tillage practices to proceed unimpeded by residual crop material, the crop residues are burned *in situ*. Farmers opt for burning because it is a quick and easy way to manage the large quantities of crop residues and prepare the field for the next crop well in time.

India being an agriculture-dominant country produces more than 500 million tons of crop residues annually (GOI, 2016) with wide regional variability. A large portion of unused crop residues are burnt in the fields primarily to clear the left-over straw and stubbles after the harvest. The uneven distribution and use of crop residue depends on the crops grown, cropping intensity and productivity across the nation. The highest crop residue estimate was recorded for Uttar Pradesh (109 Mt). Other high crop residue producing regions were Maharashtra (52 Mt), Madhya Pradesh (45 Mt) and Punjab (37 Mt) (Table 7.3). Cereals, sugarcane, oilseeds, pulses and fibre contributed the majority of crop residue with the estimated quantity of 305 Mt, 141Mt, 29 Mt, 27 Mt and 15 Mt, respectively (Table 7.3). Among cereal crops, rice, wheat, maize and millets together contributed about 60% of crop residue followed by sugarcane (27%). However, there is still a surplus of 140 Mt out of which 92 Mt is burned each year (Table 7.4). Table 7.5 compares the agricultural residues generated by selected Asian countries in Mt/year. It is also interesting to note that the portion burnt as agricultural residue in India, the volume is much larger than the entire production of agricultural residue in other countries in the region.

Non-availability of labour, high cost of residue removal from the field and increasing use of combines in harvesting the crops are main reasons behind burning of crop residues in the fields. Burning of crop residues causes environmental pollution, is hazardous to human health, produces greenhouse gases causing global warming and results in loss of plant nutrients like N, P, K and S. (Table 7.6). Burning of crop residue not only leads to pollution but also results in loss of nutrients present in the residues. The entire amount of C, approximately 80–90% N, 25% of P, 20% of K and 50% of S present in crop residues are lost in the form of various gaseous and particulate matters, resulting in atmospheric pollution. Therefore, appropriate management of crop residues assumes a great significance.

7.3 ADVERSE EFFECTS OF CROP RESIDUE BURNING

Agricultural crop residue burning contributes towards the emission of greenhouse gases (CO_2, N_2O, CH_4), air pollutants (CO, NH_3, NO_x, SO_2, NMHC, volatile organic compounds), particulates matter and smoke thereby posing threat to human health. Non-methane hydrocarbons (NMHCs) are important reactive gases in the atmosphere since they provide a sink for hydroxyl radicals and play key roles in the production and destruction of ozone in the troposphere. NMHCs generally refer to the C_2–C_4 series, notably ethane, ethene, acetylene, propane, propene, and n-butane. Crop residue burning is a cheap and easy method of crop residue management used by the farmers. One tone straw on burning releases 3 kg particulate matter, 60 kg CO, 1460 kg CO_2, 199 kg ash and 2 kg SO_2. These gases and aerosols consisting of carbonaceous matter have an important role to play in the atmospheric chemistry and can affect regional environment, which also has linkages with global climate change (Gupta et al., 2004). Crop burning increases the PM

Table 7.3: Total crop residue generation (million tonnes) in different states of India during 2014-15

State/UT	Rice	Wheat	Coarse cereal	Pulse	Oilseed	Sugarcane	Cotton	Jute & Mesta	Total
Andhra Pradesh+ Telangana	13.5	0.0	8.1	1.8	1.3	5.2	2.2	0.02	32.1
Assam	5.7	0.1	0.1	0.2	0.2	0.4	0.0	0.27	6.9
Bihar	7.5	5.0	3.8	0.7	0.1	5.5	0.0	0.50	23.1
Chhattisgarh	7.0	0.2	0.5	1.0	0.2	0.0	0.0	0.00	8.9
Gujarat	1.9	4.0	2.6	0.9	4.3	5.5	3.6	0.00	22.9
Haryana	4.7	14.7	1.4	0.1	0.8	3.0	0.8	0.00	25.4
Himachal Pradesh	0.1	0.9	1.4	0.1	0.0	0.0	0.0	0.00	2.5
Jammu & Kashmir	0.5	0.4	0.8	0.0	0.0	0.0	0.0	0.00	1.8
Jharkhand	3.9	0.4	0.8	0.9	0.2	0.2	0.0	0.00	6.4
Karnataka	4.3	0.3	11.4	2.3	1.1	16.4	0.7	0.00	36.6
Kerala	0.7	0.0	0.0	0.0	0.0	0.1	0.0	0.00	0.7
Madhya Pradesh	4.2	17.6	5.1	7.3	8.4	1.8	0.6	0.00	45.0
Maharashtra	3.4	1.5	7.6	2.7	3.1	32.1	2.3	0.00	52.7
Orissa	9.7	0.0	0.4	0.7	0.2	0.3	0.1	0.02	11.4
Punjab	13.0	19.6	0.9	0.1	0.1	2.8	0.5	0.00	36.9
Rajasthan	0.4	12.2	12.9	3.0	5.8	0.2	0.5	0.00	35.1
Tamilnadu	6.8	0.0	5.1	1.0	1.0	9.6	0.3	0.00	23.8
Uttar Pradesh	14.3	31.3	6.1	2.2	0.9	54.4	0.0	0.00	109.2
Uttarakhand	0.7	0.8	0.5	0.1	0.0	2.4	0.0	0.00	4.6
West Bengal	17.2	1.2	1.1	0.3	1.0	0.8	0.0	3.00	24.6
Others	3.0	0.1	0.8	1.3	0.2	0.4	0.0	0.03	5.8
All-India	122.6	110.3	71.3	26.7	28.9	141.1	11.6	3.85	516.3
Dry Matter Fraction	0.86	0.88	0.88	0.80	0.80	0.88	0.80	0.80	

Table 7.4: State-wise generation and remaining surplus of crop residues in India (Pathak et al 2010)

States	Crop residue generation (MNRE, 2009)	Crop Residue surplus (MNRE, 2009)	Crop residues burnt (Pathak et al., 2010)
Andhra Pradesh	43.89	6.96	2.73
Arunachal Pradesh	0.40	0.07	0.04
Assam	11.43	2.34	0.73
Bihar	25.29	5.04	3.19
Chhattisgarh	11.25	5.12	0.83
Goa	0.57	0.14	0.04
Gujarat	28.73	8.9	3.81
Haryana	27.83	11.22	9.06
Himachal Pradesh	2.85	1.03	0.41
Jammu and Kashmir	1.59	0.28	0.89
Jharkhand	3.61	0.89	1.10
Karnataka	33.94	8.98	5.66
Kerala	9.74	5.07	0.22
Madhya Pradesh	33.18	10.22	1.91
Maharashtra	46.45	14.67	7.41
Manipur	0.90	0.11	0.07
Meghalaya	0.51	0.09	0.05
Mizoram	0.06	0.01	0.01
Nagaland	0.49	0.09	0.08
Odisha	20.07	3.68	1.34
Punjab	50.75	24.83	19.62
Rajasthan	29.32	8.52	1.78
Sikkim	0.15	0.02	0.01
Tamil Nadu	19.93	7.05	4 08
Tripura	0.04	0.02	0.11
Uttarakhand	2.86	0.63	0.78
Uttar Pradesh	59.97	13.53	21.92
West Bengal	35.93	4.29	4.96
India	501.76	140.84	92.81

Table 7.5: Crop residues generation in India compared to other select nations

Country	Crop residues generated (million tons/year)
India	500
Bangladesh	72
Indonesia	55
Myanmar	19

Table 7.6: Loss of nutrients due to burning of crop residues per year in India (Jain et al 2014)

Crop residues	N loss	P loss	K loss	Total
		Million ton (Mt)/Yr		
Rice	0.236	0.009	0.200	0.45
Wheat	0.079	0.004	0.061	0.14
Sugarcane	0.079	0.001	0.033	0.84
Total	**0.394**	**0.014**	**0.295**	**1.43**

Niveta Jain, Arti Bhatia, Himanshu Pathak. Emission of Air Pollutants from Crop Residue Burning in India. Aerosol and Air Quality Research, 14: 422–430, 2014.

in the atmosphere and contributes significantly to climate change. One contributor to global climate change is the release of fine black and also brown carbon that contributes to the change in light absorption. Usually PM in the air is categorized as $PM_{2.5}$ and PM_{10} based on the aerodynamic diameter and chemical composition ($PM_{2.5}$ or fine, particulate matter with aerodynamic diameter < 2.5 micro meter and PM_{10} or coarse, particulate matter with aerodynamic diameter <10 micro meter). Lightweight particulate matter can stay suspended in the air for a longer time and can travel a longer distance with the wind. The effect of particulate matter gets worsened by the weather conditions, as the particles are lightweight, stay in air for a longer time and causes smog which forms a thick haze.

Loss of nutrients: It is estimated according to NPMCR, 2019 that burning of one tonne of rice straw accounts for loss of 5.5 kg nitrogen, 2.3 kg phosphorus, 25 kg potassium and 1.2 kg sulphur besides, organic carbon. Generally, crop residues of different crops contain 80% of Nitrogen (N), 25% of Phosphorus (P), 50% of Sulphur (S) and 20% of Potassium(K). If the crop residue is incorporated or retained in the soil itself, it gets enriched, particularly with organic C and N.

Impact on soil properties: Heat from burning residues elevates soil temperature causing death of beneficial soil organisms. Frequent residue burning leads to complete loss of microbial population and reduces level of N and C in the top 0-15 cm soil profile. The N and C components in the residue are important for crop root development.

Emission of greenhouse and other gases: Crop residues burning is a potential source of Green House Gases (GHGs) and other chemically and radiative important trace gases and aerosols such as CH_4, CO, N_2O, NO_X and other hydrocarbons. It is estimated that upon burning, Carbon (C) present in rice straw is emitted as CO_2 (70% of carbon present), CO (7%) and CH_4(0.66%) while 2.09% of Nitrogen (N) in straw is emitted as N_2O. Emission of nitrous oxide (NO_2) contributes to a very powerful GHG effect (300 times that of CO_2), even relatively small emissions can have a significant impact on the overall GHG balance. Besides, burning of crop residue also emits large amount of particulates that are composed of wide variety of organic and inorganic species. Many of the pollutants found in large quantities in biomass smoke are known as suspended carcinogens which lead to various air borne/lung diseases.

Pathak, H., Singh, R., Bhatia, A. and Jain, N. (2006). Recycling of Rice Straw to Improve Wheat Yield and Soil Fertility and Reduce Atmospheric Pollution. Paddy Water Environ. 4.

NPMCR. National policy for management of crop residues Available online: http://agricoop.nic.in/sites/default/files/NPMCR 1.pdf.

7.4 BIOFUELS FROM CROP-RESIDUES

Plant biomass is mainly comprised of cellulose, hemicellulose and lignin with smaller amounts of pectin, protein extractives, sugars, and nitrogenous material, chlorophyll and inorganic waste. Compared to cellulose and hemicellulose, lignin provides the structural support and it is almost impermeable. Lignin resist fermentation as it is very resistant to chemical and biological degradation. The non-food-based portion of crops such as the stalks, straw and husk are categorized under ligno-cellulosic biomass. The major agricultural crops grown in the world are rice, wheat, maize and sugarcane which account for most of the ligno-cellulosic biomass. Ligno-cellulosic biomass, composed of cellulose, hemicellulose, and lignin, are increasingly recognized as a valuable commodity, due to its abundant availability as a raw material for the production of biofuels.

Production of Biochar

As a measure for controlling GHG emissions, the agricultural research community is constantly looking for ways to effectively enhance natural rates of carbon sequestration in the soil. This has made an increased interest in applying charcoal, black carbon and biochar as soil amendment to stabilize soil organic content. These techniques are viewed as a viable option to mitigate the GHG emissions while considerably reducing the volume of agricultural waste. The process of carbon sequestration essentially requires increased residence time and resistance to chemical oxidation of biomass to CO_2 or reduction to methane, which leads to reduction of CO_2 or methane release to the atmosphere. The partially burnt products are biochar/carbon black and becomes a long-term carbon sink with a very slow chemical transformation, ideal for soil amendment.

Biochar is a fine-grained carbon rich porous product obtained from the thermo-chemical conversion called the pyrolysis at low temperatures in an oxygen free environment. It is a mix of carbon (C), hydrogen (H), oxygen (O), nitrogen (N), sulphur (S) and ash in different proportions.

When amended to soil, highly porous nature of the biochar helps in improved water retention and increased soil surface area. It mainly interacts with the soil matrix, soil microbes, and plant roots, helps in nutrient retention and sets off a wide range of bio-geochemical processes. Many researchers have reported an increase in p^H, increase in earthworm population and decreased fertilizer usage. Specifically, biochar is used in various applications such as the water treatment, construction industry, food industry, cosmetic industry, metallurgy, treatment of waste water and many other chemical applications. In India currently, the biochar application is limited and mainly seen in the villages and small towns. Based on its wide applicability, it could be more valuable to promote biochar in India.

Production of bio-oil

Bio-oil can be produced from crop residues by the process of fast pyrolysis, which requires temperature of biomass to be raised to 400-500 ^{0}C within a few seconds, resulting in a remarkable change in the thermal disintegration process. About 75% of dry weight of biomass is converted into condensable vapours. If the condensate is cooled quickly within a couple of seconds, it yields a dark brown viscous liquid commonly called bio-oil. The calorific value of bio-oil is 16-20 MJ/kg. The bio-oil has also wide applicability in agricultural, industrial and pharmaceutical sectors.

Generation of producer gas

The producer gas obtained from the gasification of biomass can be used as an alternative to the diesel to generate motive power to run engine on dual fuel mode. The diesel can be replaced by the producer gas to some extent for use in the engine. It can also be used to produce heat for thermal applications in industrial sector. The various applications of biofuels are shown in Fig. 7.1.

7.5 CONVERSION ROUTES OF BIOMASS INTO BIOFUELS

There are many different ways of extracting energy from biomass. Biomass is burned by direct combustion to produce steam, the steam turns a turbine and the turbine drives a generator, producing electricity. Because of potential ash build-up (which fouls boilers, reduces efficiency and increases costs), only certain types of biomass materials are used for direct combustion. Heat is used to convert biomass chemically into a pyrolysis oil. The oil, which is easier to store and transport than solid biomass material, is then burned like petroleum to generate electricity. Pyrolysis can also convert biomass into phenol oil, a chemical used to make wood adhesives, molded plastics and foam insulation. Wood adhesives are used to glue together plywood and other composite wood products. Gasification is a form of pyrolysis, which is performed at high temperatures in order to optimize gas production. The resulting gas, known as producer gas, is a mixture of carbon monoxide, hydrogen and methane, together with carbon dioxide and nitrogen. The gas is more versatile than the original solid biomass (usually wood or charcoal). It can be burnt to produce process heat and steam, or used in gas turbines to produce electricity. Gasifiers are used to convert biomass into a combustible gas (producer gas). The producer gas is then used to drive a high efficiency, combined cycle gas turbine. These energy-conversion technologies may be grouped into four basic types: (i) physical method, (ii) incineration (direct combustion), (iii) thermochemical method, and (iv) biochemical method. The general outlines of these technologies are briefly described here.

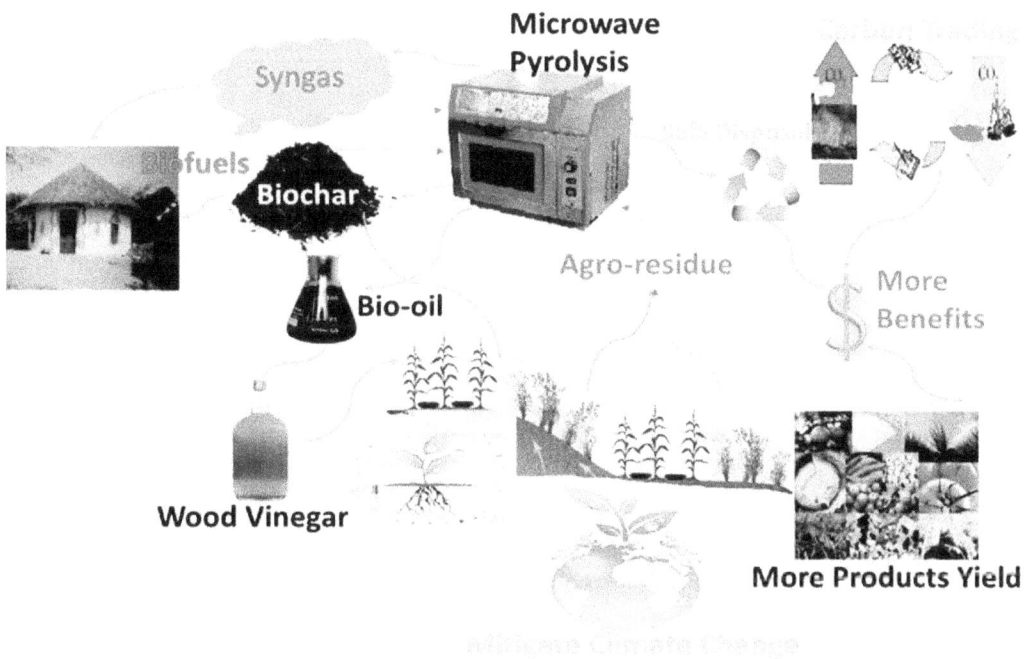

Fig. 7.1. Applications of biofuels derived from pyrolysis of agricultural residues

Physical method

The simplest form of physical conversion of biomass is through compression of combustible material. Its density is increased by reducing the volume by compression through the processes called briquetting and pelletization. Fuel oils can be extracted from plant products by expelling oil from them. Also, light hydrocarbons may be obtained from certain plants in the same way as the production of rubber.

(i) Pelletization: Pelletization is a process in which unused biomass is pulverized, dried and forced under pressure through an extrusion device. The extracted mass is in the form of pellets (rod, 5 to 10 mm dia. and 12 mm long), facilitating its use in steam power plants and gasification system. Pelletization reduces the moisture to about 7 to 10 per cent and increases the heat value of the biomass.

(ii) Briquetting: Biomass briquettes, made from woody matter (e.g., agricultural residues and saw dust), are a replacement for fossil fuels such as oil or coal and can be used to heat boilers in manufacturing plants. Burning a biomass briquette is far more efficient than burning the raw firewood. The moisture content of a briquette can be as low as 4%, whereas for freshly harvested biomass, it may be as high as 65%.

Briquetting is brought about by compression and squeezing out moisture and breaking down the elasticity of the wood and bark. If elasticity is not sufficiently removed, the compressed wood will regain its pre-compression volume. Densification is carried out by compression under a die at high temperature and pressure. It is a process similar to forming a wood pellet but on a large scale. Thera are no binders involved in this process. The natural lignin in the wood binds the particles of wood together to form a solid piece.

(iii) Agro-chemical method: Seed crops, which contain a high proportion of oil, can be crushed and the oils are extracted and used directly to replace diesel (called biodiesel) or as a heating oil. The energy content of vegetable oils is 39.3–40.6 MJ/kg. There are a wide range of crops that can be used for biodiesel production, but the most common crop is rapeseed. Other raw materials used are palm oil, sunflower oil, soya bean oil and recycled frying oils. The cost of the raw material is the most important factor affecting the overall cost of production. Concentrated vegetables oils may be obtained from certain agro products and may be used as fuel in diesel engines. However, difficulties arise with direct use of plant oil due to high viscosity and combustion deposits. Therefore, these oils are upgraded by a chemical method known as transesterification to overcome these difficulties. Categories of certain materials with examples are as follows:

- Seeds of sunflower, rapeseed, soya beans
- Nut oil palm, coconut copra
- Fruits Olive
- Leaves Eucalyptus

(iv) Fuel extraction: Occasionally, milky latex is obtained from the freshly cut plants. The material is called exudates and it is obtained by cutting (tapping) the stems or trunks of living plants (a technique similar to that used in rubber production). Some plants are not amenable to tapping and in such cases, the whole plant (usually a shrub) is crushed to obtain the product. For example, the Eupborbia latbyris plant is crushed to extract hydrocarbons of less molecular weight than rubber, which may be used as a petroleum substitute.

Incineration

Incineration means direct combustion of biomass for immediate useful heat. The heat and then steam produced are either used to generate electricity or provide the heat for industrial process, space heating, cooking or district heating. Furnaces and boilers have been developed for large-scale burning of various types of biomass such as wood, waste wood, black liquid from pulp industry, food industry waste, and MSW. The moisture content in the biomass and wide range of composition tends to decrease the efficiency of conversion. However, the economic advantages of cogeneration make it attractive for adoption.

Thermo-chemical processes: The basic thermochemical process to convert biomass into a more valuable and/ or convenient product is known as pyrolysis. Biomass is heated either in absence of oxygen or by partial combustion of some of the biomass in restricted air or oxygen supply. Pyrolysis can process all forms of organic materials including rubber and plastics, which cannot be handled by other methods. The products usually consist of three types of fuels i.e. a gas mixture (H_2, CO, CO_2, CH_4 and N_2), an oil-like liquid (a water-soluble phase including acetic acid, acetone, methanol and non-aqueous phase including oil and tar) and nearly pure carbon char. The distribution of these products depends upon the type of feedstock, the temperature and pressure during the process and its duration of reaction and the heating rate. It is a process to decompose biomass with various combinations of temperature and pressure with no or limited availability of oxygen.

Pyrolysis

Pyrolysis is the thermochemical process that converts biomass into liquid (bio-oil or bio-crude), charcoal and non-condensable gases, acetic acid, acetone and methanol by heating the biomass to about 480 ^0C in the absence of air. The process can be adjusted to favor charcoal, pyrolytic oil, gas or methanol production. Pyrolysis can be used for the production of bio-oil if flash pyrolysis processes are used. Pyrolysis produces energy fuels with high fuel-to-feed ratios, making it the most efficient process for biomass conversion and the method is most capable of competing with and eventually replacing non-renewable fossil fuel resources. The crude oil produced during pyrolysis can be used in engines and turbines. Its use as feedstock for refineries is also being considered. Some problems in the conversion process and use of the oil need to be overcome. These include poor thermal stability and corrosivity of the oil. Upgrading by lowering the oxygen content and removing alkalis by means of hydrogenation and catalytic cracking of the oil may be required for certain applications.

Pyrolysis is therefore the basic thermochemical process for converting biomass to a more useful fuel. Biomass is heated in the absence of oxygen, or partially combusted in a limited oxygen supply, to produce a hydrocarbon rich gas mixture, an oil-like liquid and a carbon rich solid residue. Traditionally in developing countries, the solid residue produced is charcoal, which has a higher energy density than the original fuel and is smokeless and, thus, ideal for domestic use. The traditional kilns are simply mounds of wood covered with earth, or pits in the ground. However, the process of carbonization is very slow and inefficient in these kilns and more sophisticated kilns are replacing the traditional ones. The thermal degradation properties of ligno-cellulosic biomass i.e. hemicelluloses, celluloses and lignin can be summarized in the order of thermal degradation of hemicelluloses > of cellulose >>> of lignin. If the purpose is to maximize the yield of liquid products resulting from biomass pyrolysis, a low temperature, high heating rate, short gas residence time process would be required. For high char production, a low temperature, low heating rate process would be chosen. If the purpose is to maximize the yield of fuel gas resulting from pyrolysis, a high temperature, low heating rate, long gas residence time process would be preferred.

Gasification

The word gasification (or thermal gasification) implies converting solid fuel into a gaseous fuel by thermochemical method without leaving any solid carbonaceous residue. Gasification involves partial combustion (oxidation restricted quantity of air/ oxidant) and reduction operations of biomass. In a typical combustion process, generally the oxygen is surplus, while in a gasification process, the fuel is surplus. The combustion products, mainly carbon dioxide, water vapour, nitrogen, carbon monoxide and hydrogen pass through the flowing layer of charcoal for the reduction process to occur. During this stage, both carbon dioxide and water vapour oxidize the char to form CO, H_2 and CH_4. The following are the typical reactions, which occur during gasification:

$$C + O_2 \rightarrow CO_2 \text{ (Combustion, 240 kJ/mole)}$$

The moisture available in the biomass is converted to steam and generally no extra moisture is required. Thus the product of combustion of pyrolysis gases results in CO_2 and H_2O (steam), which further react with char:

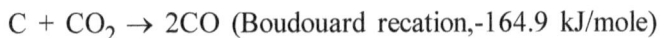

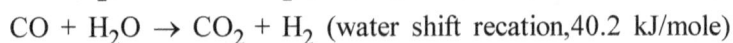

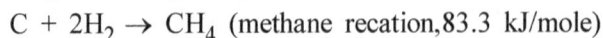

$$C + CO_2 \rightarrow 2CO \text{ (Boudouard recation, -164.9 kJ/mole)}$$
$$C + H_2O \rightarrow CO + H_2 \text{ (water gas recation, -122.6 kJ/mole)}$$
$$CO + H_2O \rightarrow CO_2 + H_2 \text{ (water shift recation, 40.2 kJ/mole)}$$
$$C + 2H_2 \rightarrow CH_4 \text{ (methane recation, 83.3 kJ/mole)}$$

The comparison between combustion, pyrolysis and gasification is shown in fig. 7.2

Biochemical processes: In biochemical conversion, microbial digestion of biomass takes place for its decomposition to derive biofuels (biogas and ethanol). In biochemical conversion, there are two principal conversion processes i.e. (i) Anaerobic digestion and (ii) Alcoholic fermentation

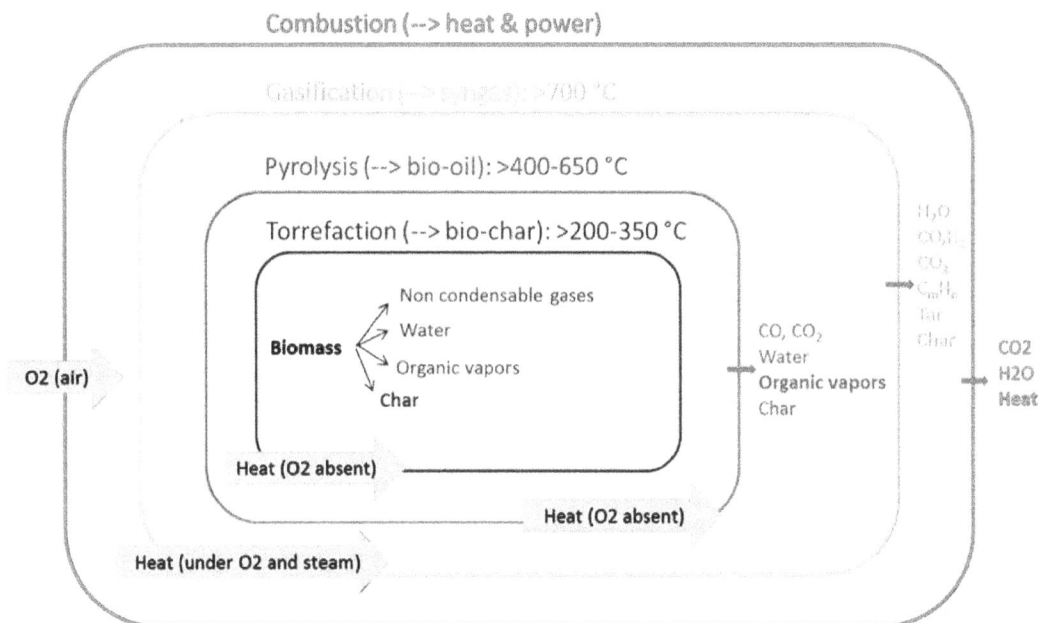

Fig. 7.2. Comparison between combustion, gasification, and pyrolysis, with major products, (Matsakas et al., 2017).

(i) Anaerobic digestion: Anaerobic digestion is the decomposition of biomass through bacterial action in the absence of oxygen. It is essentially a fermentation process and produces a mixed gas output of methane and carbon dioxide. The product generated by the decay of sewage or animal waste in the absence of air is known as biogas. The anaerobic digestion of MSW (municipal solid waste) buried in landfill sites produces a gas known as landfill gas. Biogas is most commonly produced by using animal manure mixed with water which is stirred and kept inside an airtight container, known as a digester. The biogas produced can be burnt directly for cooking and space heating, or used as fuel in internal combustion engines to generate electricity. The methane gas produced in landfill sites eventually escapes into the atmosphere. However, the landfill gas can be extracted from existing landfill sites by inserting perforated pipes into the landfill. In this way, the gas will travel through the pipes under natural pressure for use as an energy source, rather than simply escaping into the atmosphere to contribute to greenhouse gas emissions.

(ii) Alcoholic fermentation: It is the process of decomposing biomass by micro-organism especially bacteria and yeasts. Ethanol can be produced from certain biomass materials which contain sugars, starch or cellulose. The best known source of ethanol is sugar cane, but other materials can be used, including wheat and other cereals, sugar beet etc. The choice of biomass is important as feedstock costs typically make up 55–80 % of the final alcohol selling price. Starch based biomass is usually cheaper than sugar based materials but requires additional processing. Similarly, cellulose materials,

such as wood and straw, are readily available but require expensive preparation. Ethanol is produced by a process known as fermentation. Typically, sugar is extracted from the biomass crop by crushing, mixed with water and yeast and kept warm in large tanks called fermenters. The yeast breaks down the sugar and converts it to ethanol. A distillation process is required to remove the water and other impurities in the diluted alcohol product (l0–15% ethanol). The concentrated ethanol (95% by volume with a single step distillation process) is drawn off and condensed to a liquid form. Ethanol can be used as a supplement or substitute for petrol fuel.

The most useful biomass materials appear to be animal, algae, kelp, hyacinth, plant residues and other organic waste materials with high moisture content. The energy available from various biomass resources is listed in Table 7.7.

Table 7.7: Energy available from various biomass resources

SN	Biomass source	Biofuel produced	Conversion technology	Available energy (MJ/kg)
1	Wood chips, saw mill dust, forest residues	(direct heat)	Incineration	16-20
2	Wood chips, saw mill dust, forest residues	Gas	Pyrolysis	40 (nitrogen removed)
		Oil		40
		Char		20
3	Crop residues	Straw	Incineration	14-16
4	Sugar-cane residue	Bagasse	Incineration	5-8 (fresh cane)
5	Urban refuse	(Direct heat)	Incineration	5-16 (dry input)
6	Sugar-cane juice	Ethanol	Fermentation	3-6 (fresh cane)
7	Animal waste	Biogas	Anaerobic digestion	(4-8 (dry input)
8	Municipal sewage	Biogas	Anaerobic digestion	2-4 (dry input)

The various routes for conversion of crop residue are shown in the fig. 7.3.

7.6 MICROWAVE ASSISTED PYROLYSIS (MAP)

7.6.1 Pyrolysis principles and mechanisms

Pyrolysis refers to the decomposition or cracking of organic materials into a range of useful products in absence of oxygen by the application of thermal energy and in a temperature range of 300-650 °C (Basu, 2018; Andrade et al., 2018). It thermally reprocesses the lignocellulosic biomass i.e., crop residues into more usable forms of energy and value-added products from the environmental and economic standpoints which makes them useful in many applications (fig. 7.4). When exposed to heating process, the large molecules of biomass undergo primary decomposition. The product of cracking reaction comprises of condensable vapors (bio-oil) and solid biochar. The condensable

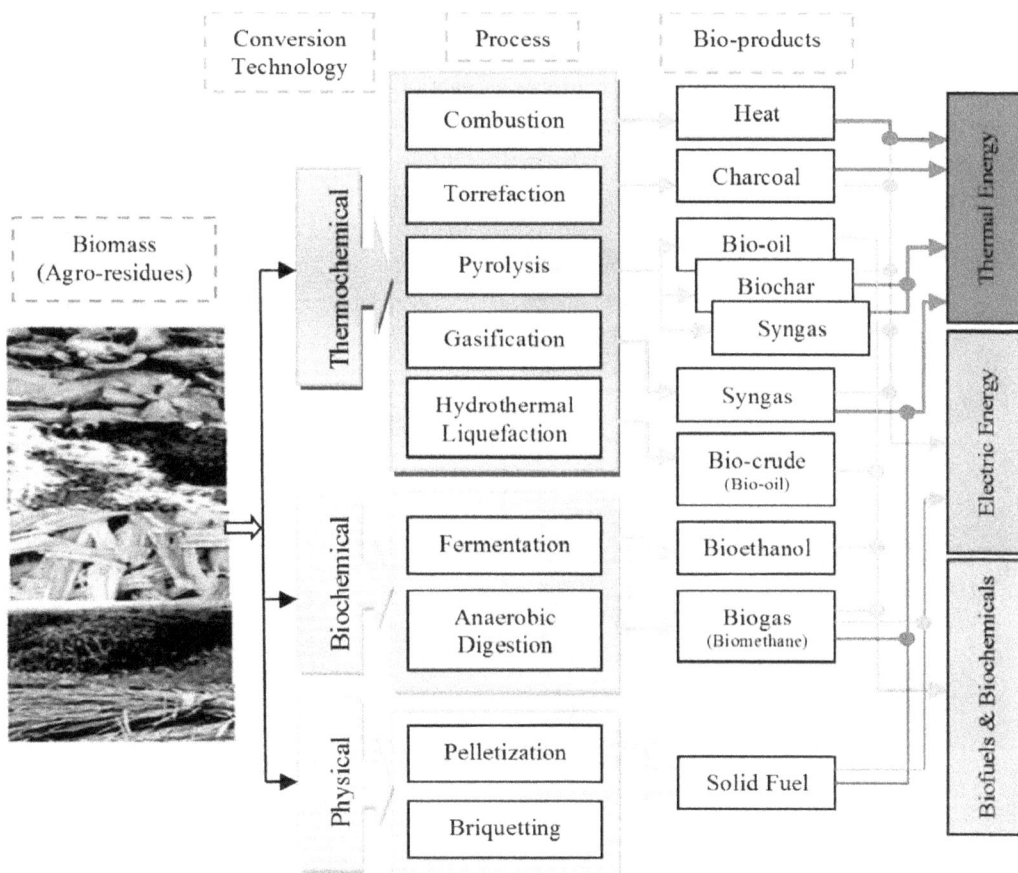

Fig. 7.3. Conversion processes of biomass to various bio-products

vapors may undergo secondary cracking breaking down further into non-condensable gases (CO, CO_2, H_2, and CH_4), bio-oil, and biochar. This decomposition occurs partly through gas-phase reactions and partly through gas-solid-phase thermal reactions. In gas-phase reactions, the condensable vapors are cracked into smaller molecules of non-condensable permanent gases such as CO and CO_2, but these are of less interest in pyrolysis. The pyrolysis process may be represented by the following reaction (Basu, 2018).

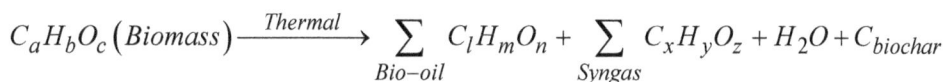

$$C_aH_bO_c\left(Biomass\right)\xrightarrow{\ Thermal\ }\underbrace{\sum C_lH_mO_n}_{Bio-oil}+\underbrace{\sum C_xH_yO_z}_{Syngas}+H_2O+C_{biochar}$$

The thermochemical reaction of biomass (i.e., pyrolysis process) is a complex mechanism. For more specific, the pyrolysis of biomass passes through primary reaction mechanisms (three stages) in addition to secondary reaction mechanisms. As can be seen from Fig. 7.5, the primary mechanism stages cover the devolatilization of the main constitutes in the biomass feedstock and it involves the reactions of char formation, depolymerization, and fragmentation. Char formation occurs at a reaction temperature below 500 °C under a low heating rate. In this stage, the reactions are rearranged resulting in the formation of a solid char product which is a thermally stable product and has a structure of polycyclic aromatic. The dehydration reaction for water forming and the formation of

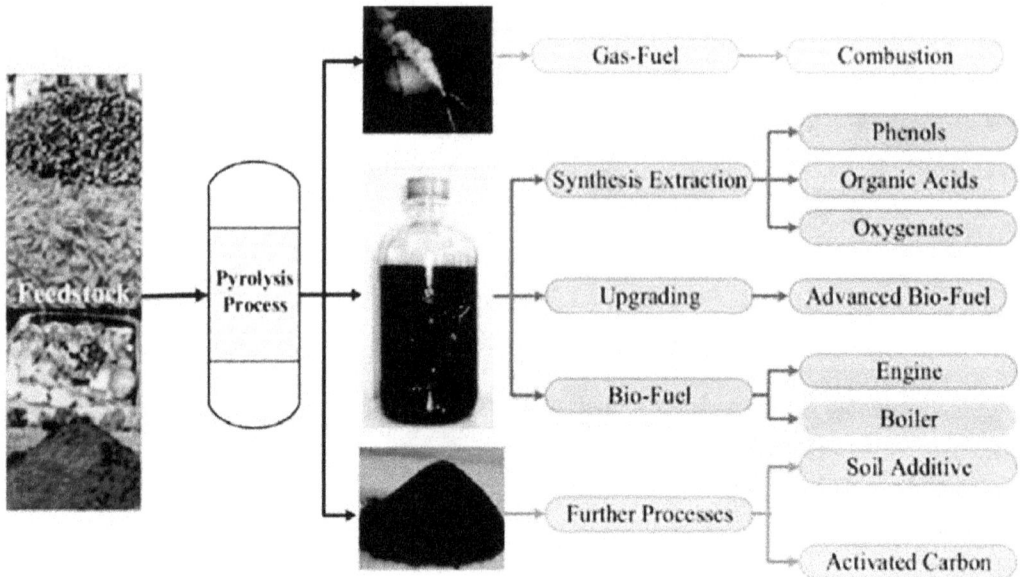

Fig. 7.4. Pyrolytic products from pyrolysis of biomass and their applications

The depolymerization reaction includes the breakage of the bonds between the monomer units resulting in the formation of smaller chains with continuous reaction until the produced molecules become volatiles. The depolymerization reactions vary between the cellulose, hemicellulose, and lignin constitutes. Fragmentation reactions occur at a reaction temperature above 600 °C resulting in the formation of incondensable gases. Whilst the secondary reaction in pyrolysis includes, the catalytic thermal cracking of heavy compounds or solid char to produce gases phase (CO, CO_2, CH_4, and H_2). The gases produced in this pathway could be similar to the gases formed under the fragmentation reaction pathway. The secondary pyrolysis reaction occurs when the volatile products are in an unstable stage during the pyrolysis reactions.

7.6.2 Pyrolytic Products and Biomass Constituent's Mechanism

In respect of products distribution of pyrolysis and their quality, it has been believed that the yield and quality of each of pyrolytic products are functions of the type of the biomass materials, used, and the pyrolysis conditions which include temperature, heating rate, and reaction time, etc. Therefore, according to these factors, the pyrolysis process is often broadly classified as slow, fast, and flash pyrolysis as shows in Table 7.8. The biochar and bio-oil produced from the pyrolysis process have versatility in their applications at both users' and commercial levels compared to the gaseous product because of their easy collection, storability and transportation, and higher commercial value. Hence, the bio-oil product is usually the target product from the pyrolysis process followed by the biochar product instead of the gaseous product.

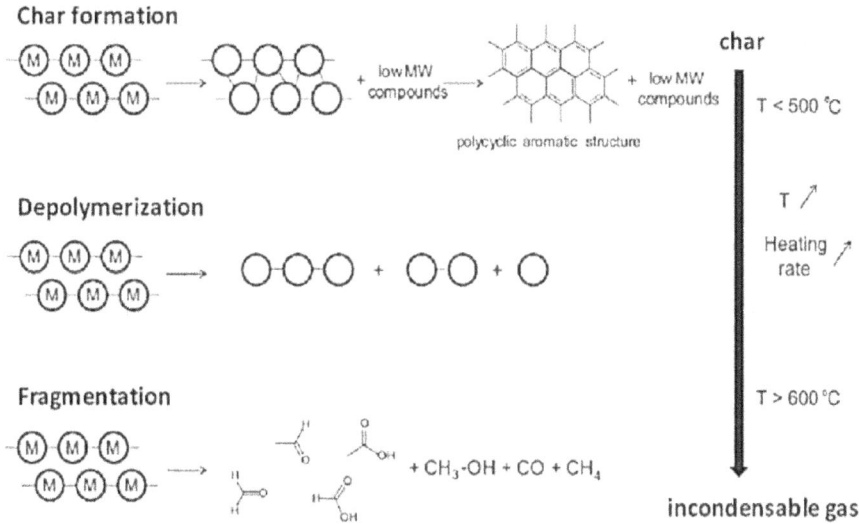

Fig. 7.5. Three stages primary mechanisms in biomass pyrolysis, redrawn from (Collard and Blin, 2014; Uddin et al., 2018).

Table 7.8: Conditions for different types of pyrolysis process.

Pyrolysis type	Temperature (°C)	Heating rate (°C/s)	Residence time (s)	Particle size (mm)	Major product
Slow	300-700	0.1-1.0	450-550	5-50	Biochar
Fast	550-1000	10-200	0.5-10	<1	Bio-oil
Flash	800-1100	<1000	< 0.5	< 0.5	Bio-oil and syngas

On the other hand, for biomass materials, crop residues (rice straw, corn stover, bagasse, etc.,), as lignocellulosic materials can be used successfully as a feedstock for the pyrolysis process (Parvez et al., 2019). These lignocellulosic biomasses are made up of three main constituents that are cellulose, hemicellulose, and lignin. Therefore, each component in the biomass (hemicellulose, cellulose, and lignin) has its preferred temperature range to decompose and undergoes differently during the pyrolysis process making varying contributions to pyrolytic yields and their properties (Basu, 2018). In general, both cellulose and hemicellulose are carbohydrate polymers, while lignin is a complex and highly aromatic non-carbohydrate polymer consisting of three primary monomers. It is also known that hemicellulose is more reactive and less thermally stable, which makes it more prone to condensation and repolymerization, while cellulose is more stable and lignin is considered the most difficult component to decompose (Zhang et al., 2017c). Hemicellulose is a primary source of non-condensable gases and decomposing in a temperature range of 150-350 °C. Cellulose, on the other hand, yields more condensable vapor and decomposes at a reaction temperature of 300-400 °C. The lignin degrades slowly (350-500 °C), due to its aromatic content, making its major contribution to the biochar yield (Basu, 2018).

7.6.3 Microwave-assisted Pyrolysis of Biomass

The pyrolysis process, usually carried out conventionally, is called conventional pyrolysis (CP), with the help of an electrical heating mechanism. This heating mechanism is usually inefficient and energy-intensive. The microwave-assisted pyrolysis (MAP) of biomass has several advantages over CP such as energy and time saving, enhancing heating efficiency, more control during the process, and increased yield of the pyrolytic products (Parvez et al., 2019a). Also, the temperature is maintained uniformly in MAP resulting in the avoidance of the undesirable secondary reactions, and the production of desired pyrolytic products compared to CP. Therefore, MAP has been introduced as an effective technique for selectivity in the yield of the pyrolytic products and improves their quality as compared with CP for different biomass feedstocks such as wood sawdust, corn stover, corn cob, rice straw, gumwood, rapeseed shell, rice husk, sugarcane bagasse, sugarcane peel, coconut shell, as well as palm kernel shell, wood chips, and sago wastes etc. (Ravikumar et al., 2017).

7.6.3.1 Basic Information of Microwave Heating

Microwaves are normally defined as the electromagnetic waves. The electromagnetic waves as microwaves consist of two main components electric and magnetic fields which are perpendicular to each other. Hence it is categorized as electromagnetic radiation. The position of microwaves within the electromagnetic spectrum is characterized by a frequency ranged from 300 MHz to 300 GHz and the wavelengths ranged from 0.001 m to 1 m. The microwave region of the electromagnetic spectrum lies between infrared and radio frequencies. Fig. 7.6 represents the position of microwaves in electromagnetic spectrum (top) with the propagation of electromagnetic field, electric and magnetic fields, (bottom). Most microwave reactors for chemical synthesis and all domestic microwave ovens operate at 12.25 cm wavelength with corresponding frequency of 2.45 GHz (Fuad et al., 2019).

7.6.3.2 Microwave Heating Mechanism

The most important parameter in microwave heating is the penetration of microwave through the materials or penetration depth property. It is reported that microwaves can penetrate the materials up to a depth range of 1-2 cm of it. However, the depth of penetration in the microwaves varies according to a material type and its microstructure properties in addition to the reaction temperature. Water can absorb microwave upto 1.4 cm and 5.7 cm depth at 25 °C and at 90 °C respectively with 2.45 GHz frequency

(Mutyala et al., 2010). For biomass, the microwave penetration depth is according to its density and water content (Mushtaq et al., 2014). The interaction between the microwave and the material can be categorized into three groups. First type is insulator, when the microwave penetrates through the material without energy losses. Second type is the conductor when the microwave reflects back and cannot pass through the material. The third type is the dielectrics or absorptive. When the microwave is absorbed by the material while passing through it, the material is usually called dielectrics (Yin, 2012; Komarov, 2012). Therefore, the material can be heated by absorbing microwave energy which is called as dielectric heating. The dielectric heating is mainly due to the interaction of the electric field component of electrons and charged particles of some materials (Mushtaq et al., 2014). In addition, the ability of the material to absorb microwave is determined by its dielectric properties i.e., dielectric constant (ε') as well as the dielectric loss (ε''). The capacity of the material molecules to be polarized when it is under the electric field is described by the dielectric constant. The efficiency of converting the electromagnetic radiation into heat through the material is also measured by its dielectric constant property. While the quantity of microwave energy that is dissipated through heat is described by dielectric loss. The ratio between dielectric loss and dielectric constant ($\tan \delta = \varepsilon'' / \varepsilon'$) called loss factor or dissipation factor (Yin, 2012). Thus, one can say, the dissipation factor is used to assess the ability of a material to heat in an electric field. Therefore, if two materials have the same dielectric loss, then the material with a lower dielectric constant would heat better where it will have a higher dissipation factor (Menéndez et al., 2007).

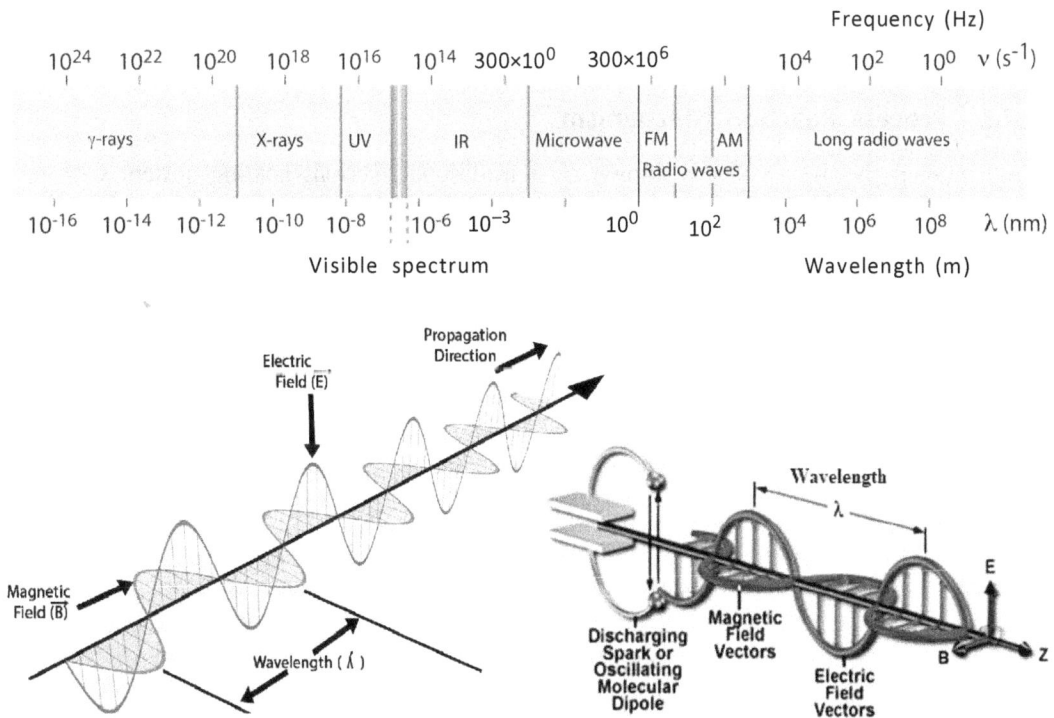

Fig. 7.6. The electromagnetic spectrum includes microwave region (top) and the electromagnetic field propagation (bottom).

In the case of biomass materials, it is reported that several factors affect the biomass dielectric properties. These factors are frequency, temperature; density, and moisture content. But, in general, biomass materials are weak in absorbing the microwave due to their poor dielectric property. Dielectric properties are decreased slightly and remain constant during the pyrolysis process with increasing the temperature from 200 °C to 450 °C. Whilst, the penetration depth of the microwave increases significantly from 0.4 m to 3.8 m when the temperature increased from 200 °C to 400 °C. Therefore, the biomass has poor absorbability of microwave in the pyrolysis process due to poor dielectric property i.e., high dielectric constant (ε') and lower dielectric loss (ε'') resulting in low dissipation factor (tan δ). Due to the poor dielectric property of some biomass feedstock, it is necessary to explore the possible alternative mechanism for maximizing heating efficiency and improving the yield and quality of the products during the MAP process. A possible mechanism for improving the microwave heating of biomass, is that the biomass feedstock needs to be mixed with a microwave absorber. Carbon-based materials are good microwave absorbers with a high capacity to absorb and convert microwave energy into heat energy. The microwave is first absorbed by the microwave absorber and is added to the biomass particles as the temperature of the microwave absorber becomes much higher than the biomass particles. Finally, the absorbed energy is transferred from the microwave receptor to biomass by conduction method of heat transfer (Menéndez et al., 2010).

7.7. MICROWAVE-ASSISTED PYROLYSIS VERSES CONVENTIONAL PYROLYSIS OF BIOMASS

7.7.1 Process Reaction Mechanism

Conventional pyrolysis (CP) is known as a traditional thermal heating system with an external heat source (electrical heating). In this heating mechanism, the heat is transferred from the external heat source to the surface of the material then transferred to the center of the material by conduction, convection, and radiation. This makes the heating system slow, inefficient, energy-intensive as well as highly depends on convection currents and on the thermal conductivity of the material. On the contrary, microwave heating (dielectric heating) transfers electromagnetic energy to thermal energy which is the form of energy conversion rather than heating. In this heating mechanism, the microwave penetrates into the materials directly and is accumulated as the energy, then the energy is converted to the heat inside the core of the materials rather than the external heating. Therefore, losses of heat through the heat transfer process from the external heating source to the material is avoided in the microwave heating mechanism because of volumetric heating. Fig. 7.6 explains the difference in the mechanism of heating for conventional and microwave systems. As seen from Fig. 7.7, the material subjected to microwave heating has a higher temperature at the center as opposed to the outer surface while in conventional heating it is in the opposite trend. Thus, the material under the microwave is at a higher temperature than its surroundings, unlike conventional heating.

In the MAP, when the microwave is penetrating, the microwave energy is transformed into thermal energy inside the biomass particle and the gradient of the temperature is from inside to outside of the biomass particle. The released volatile materials during the pyrolysis are diffused from the inside core of the biomass feedstock to its outside surface through a lower temperature region.

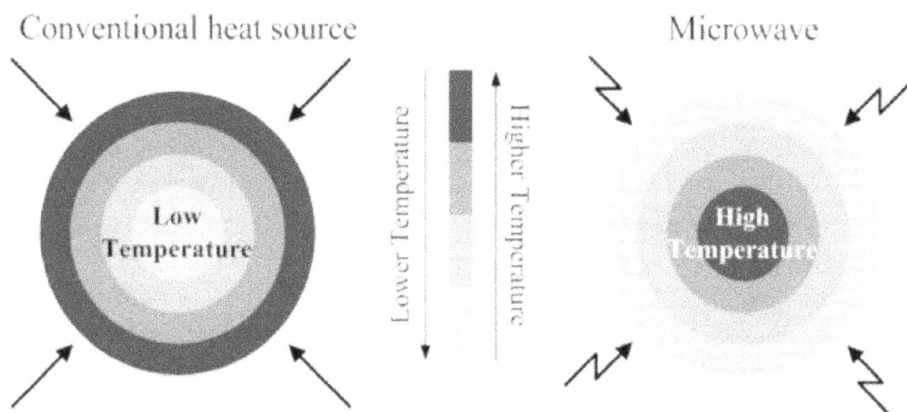

Fig. 7.7. Temperature distribution during conventional and microwave heating (Bundhoo, 2018).

In CP, the gradient of the temperature is from the outer furnace (heating source) to the inner biomass feedstock particle. The heat is transferred from the heating source to the outer surface of the biomass particle after that to the inside core of it. In this case, the released volatile materials during pyrolysis are diffused from the inside core of the biomass to its outside surface through a higher temperature region. Therefore, the heating and reaction mechanisms of MAP offer several advantages over CP due to improved heating mechanisms as explained. The advantages are such as higher heating rate, non-contact heating, uniform heating, energy transfer rather than heat transfer, lower thermal inertia, higher heating efficiency, less feedstock pre-treatment, faster response rate, as well as greater control and safety, etc. This uniform and better control of temperature in the reactor during the MAP process lead to avoidance of the undesirable secondary reactions and the production of desired pyrolytic products compared to CP. The energy-saving is also one of the advantages in MAP because it accelerates thermo-chemical reactions and reduces the process reaction time.

7.7.2 Effects on Pyrolytic Products

Comparing the improvements of the products through MAP, Parvez et al. (2019a) studied the comparison between CP and MAP of gumwood. The conventional pyrolysis (CP) performed in a vertical tube furnace reactor, while MAP performed in multimode microwave cavity (2.45 GHz) with maximum power of 3 kW. Both of MAP and CP were adjusted to achieve 600, 700 and 800 °C temperature with heating rate 40 °C/min. Approximately 20 g of gumwood was loaded in both of MAP and CP under N_2 atmosphere for 45 min. The results showed that at each temperature, the corresponding energy and exergy rates of gas under microwave heating were found to be 23% and 26%, respectively, higher than those of conventional one. The results also exhibited that the biochar yield in MAP is higher than CP by approximately 42, 44, and 46 % at 600, 700, and 800 °C respectively. Also the syngas yield in MAP showed higher than CP by approximately 5, 6, and 5% at 600, 700, and 800 °C respectively, while the bio-oil yield in MAP showed lower than CP by approximately 53, 62, and 68 % at 600, 700, and 800 °C respectively. The qualities of the biochar produced from algae biomass were found to higher in the case of MAP compared to CP (Parvez et al., 2019). It was also found that MAP contributed to high reactivity of biochar through the formation of hot spot in addition to the high specific surface area and pore volume compared to the biochar produced through CP. Omar and Robinson (2014) investigated the comparison between CP and MAP of rapeseed oil waste in the batch reactors system. The effects of different reaction temperatures of 500, 550, and 600 °C under 300 W to 500 W of microwave power as well as adding 10% HSZM-5 catalyst were examined. The results indicated that better products were produced in the case of MAP within a short reaction time of 15 min, while CP required a long reaction time to achieve the products. Wu et al. (2014) also studied the comparison between CP and MAP of wood biomass under different operating conditions. The results indicated that the heating in the MAP is much faster compared to CP. The CP produced higher gas and biochar yields than MAP, whilst the bio-oil yield was approximately the same in both heating systems. Both bio-oil and biochar produced through MAP was found to be of much higher qualities than those produced from CP. Table 7.9 also shows the main chemical properties of bio-oil from MAP as compared with CP. As can be seen from the Table 7.9, the bio-oil produced from MAP is found to be of much higher carbon and hydrogen contents with low oxygen content.

Table 7.9: Elemental analysis and HHV of bio-oil from MAP as compared to CP (Zhang et al., 2017a).

Property	Bio oil	
	Microwave	**Conventional**
C	60.66	54-58
H	7.70	5.5-7.0
N	2.02	0-0.2
S	0.15	-
O	29.4	35-40
HHV (MJ/kg)	17.51	14-19

It is reported that during MAP, the microwave radiation could break the heavier hydrocarbon component of the biomass into lighter hydrocarbon component, resulting in generation of light hydrocarbon fuel products of the bio-oil product. As a result, the MAP is considered as an effective conversion method for converting the biomass into high quality bio-oil and biochar products in a short time with maximum energy saving (Omoriyekomwan et al., 2016).

7.7.3 The Additives/Catalysts in Microwave Pyrolysis

As mentioned earlier, the dielectric properties of some crop residues are poor and they have low capacity in absorbing the microwave. Therefore, it is necessary to mix the biomass feedstock with a microwave absorbent during the MAP process. The selection of the additive/catalyst should be based on achieving one or more of the objectives of improving the microwave absorption for increasing the heating performance, maximizing the yields of desirable pyrolysis products, and improving the quality of these products. Thus, a good additive/catalyst is required to help catalysing the reaction in a short time and with minimum microwave power consumption. Also, if the bio-oil and biochar are the desirable pyrolytic products as a biofuel source or to be processed for producing chemicals, it requires an improvement through selecting appropriate additive/catalyst with the necessity of appropriate blending ratio.

The main purpose of additive/catalyst in MAP is to remove the oxygenated compounds from the products and cracking high molecular weight components to lower chain compounds. In MAP, with an additive/catalyst, several thermo-chemical reactions such as cracking, hydrocracking, polymerization, hydrogen transfer, cleavage of C–C bonds, scission of aromatic side-chain, dehydration, decarboxylation, decarbonylation, condensation, alkylation, and aromatization, etc. occur. Due to decarbonylation (CO), decarboxylation (CO_2), and dehydration (H_2O), the bio-oil produced by MAP with a catalyst/additive has low oxygen content in addition to low molecular weight. The most desirable reaction is decarboxylation because two moles of O_2 are removed in exchange for one C mole and H_2 is retained.

There are two methods for using additive or catalyst in the MAP process i.e., catalytic bed and catalyst mixing. In the catalyst mixing (in-bed) method, the biomass and catalyst are mixed physically or mechanically before being inserted in the pyrolysis reactor. Under this method, both the biomass and the additive/catalyst absorb the microwave energy, then the temperature increases quickly in the reactor. This makes the heating of the biomass to occur more quickly and efficiently. While, on the catalytic bed method, also called ex-bed, the pyrolytic vapors are allowed to pass from the first reactor through a catalytic reactor then the vapors pass to the condensation system resulting in bio-oil and gaseous products. This method requires more microwave power to attain the suitable pyrolysis temperature of biomass, due to the poor absorbability of the biomass for microwave.

It is reported that the in-bed and one step method is an effective and promising method for using additive or catalyst in MAP. Also, the maximum yield and the best quality of bio-oil are achieved using the in-bed method. However, the two-step method could promote bio-oil quality more but required high microwave power. Furthermore, in-situ (in-bed) catalytic upgrading is more commonly used than ex-situ (ex-bed). This is because of the fact that the catalyst directly reacts with the pyrolyzed vapors resulting in the improvement of the product in the in-situ upgrading method.

A large class of additives used in in-situ upgrading consists of zeolites, metal oxides, and carbon-based materials which are either used in their existing form or modified with metals. Two of the most commonly used carbon-based additives, activated carbon (AC) and silicon carbide (SiC) are usually used in most of the research activities. The SiC improves the heating by absorbing more microwave resulting in attaining the required pyrolysis temperature in short time with minimum microwave power consumption. But this leads to the occurrence of secondary cracking reaction resulting in more gas and less biochar yields. On the other hand, AC is considered to be an effective additive to improve the heating during MAP (Lam et al., 2019). It is reported that the utilization of AC through in-bed mixing improves the microwave absorption, helping in rapid heating and uniformity in heat distribution, and selectivity of producing the desirable pyrolysis products. AC provides extensive pyrolysis cracking in MAP with efficient heat transfer. Further, reduction reaction occurs during MAP preventing the oxidation and avoids formation of oxygenated compounds in the bio-products causing the beneficial effects when using AC in-bed reaction (Lam et al., 2019). AC can also help to increase the bio-oil yield up to 41wt.% with high-quality (Bu et al., 2014).

The metal base oxides are also used as an additive/catalyst in MAP. The results indicate that metal oxides are suitable for ketonization and aldol condensation of carboxylic acid and carbonyl compounds. Transitional metal oxides used as catalysts may decrease the bio-oil and biochar yields with the increase of gas, water, and solid yields. The high-water content in the bio-oil and the low calorific value of the biochar are reported when using most of the metal oxides additives in MAP (Huang et al., 2013). Although, the metal oxides additive decreases the bio-oil yield, some of them (CuO and Al_2O_3) could improve the bio-oil quality (Chen et al., 2008). Zeolites (HZSM-5) can be used as a catalyst but with less yield of bio-oil and high yield of gas (Fan et al., 2017).

Sodium carbonate (Na_2CO_3) catalyst shows a positive effect on the enhancement of the bio-oil and biochar yields compared to other catalysts (NaOH, HZSM-5, Na_2SiO_3) and non-catalytic conditions. Also, good quality bio-products (bio-oil and biochar) are achieved. In addition, the effect of Na_2CO_3 as a catalyst on the MAP of wheat straw has been studied by Zhao et al. (2014). The results showed that using Na_2CO_3 catalyst (3 wt.% of feedstock) helped in decreasing the gas yield by 50%, while the bio-oil yield was increased by 84%. The aromatics compounds in the bio-oil are increased by adding

Na_2CO_3 catalyst during pyrolysis of microalgae, while the acidity of the bio-oil is decreased significantly compared to non-catalytic samples. Compared with non-catalytic conditions, the yield is increased by 20 wt.%, oxygen content is decreased by 47.1%, calorific value is increased by 19.4%, and the acidity is lowered by 83.2% in the bio-oil samples when Na_2CO_3 is used (Tirapanampai et al., 2019).

Summarizing the studies on the use of additives/catalysts in microwave pyrolysis, it can be concluded that the selection of the additive/catalyst in MAP is the key in improving the quality of bio-oil and biochar for several applications. Therefore, the selection should be based on achieving one or more of the aims such as improving the microwave absorption for increasing the heating performance, maximizing the yields of desirable pyrolysis products, and improving the quality of these products. Thus, a good additive/catalyst is required to help catalysing the reaction in a short time and with minimum consumption of microwave power as well as improving the quality of bio-oil and biochar. Carbon-based additives such as SiC and AC and catalyst-based Na_2CO_3 show the most effective additives on the absorption of the microwave as well as increasing the yields of desirable products and enhance their quality. Therefore, the selection of additives along with their blending ratios are very important to achieve the better performance from the pyrolysis, desirable product distribution and the good quality of the pyrolytic products.

7.7.4 Characteristics of Pyrolysis Products

In the MAP process, the reaction involves a breakdown of large complex molecules into several small molecules to produce a liquid fraction (bio-oil), a solid fraction (biochar), and pyrolytic gas. The yields of these products depend on several factors including the reaction temperature, heating rate, reaction time, type of inert gas, as well as feedstock composition, etc. (Basu, 2018). The bio-oil and biochar products have versatility in their applications both at users' and commercial levels compared to the gaseous product because of their easy collection, storability and transportation, and higher commercial value (Zhang et al., 2020). Hence, the bio-oil and biochar products are usually the target products from the pyrolysis process instead of the pyrolytic gas (Kumar ct al., 2019).

7.7.5 Bio-oil Characteristics

The bio-oil or bio-crude is a dark brown tarry organic and viscous liquid containing more than 300 chemical compounds, which make it a complex liquid. The bio-oil from MAP of biomass differs in the composition and properties as compared with the diesel and other petroleum oils. It is a mixture of complex hydrocarbons with large amounts of oxygen and water. The compounds detected in the bio-oil fall under the following chemical categories i.e., hydroxy aldehydes, carboxylic acids, sugars and dehydrosugars, phenols, and hydroxy-ketones compounds (Basu, 2018).

The bio-oil produced from lignocellulosic materials, such as corn stover, has higher water, oxygen, and ash contents as well as higher viscosity than heavy petroleum oil. The p^H is reported to be also very low (2-2.6) with acetic acid as the major acid product which results from the deacetylation of hemicellulose. In addition, the carbon content in the bio-oil produced from biomass, corn stalk briquettes, was found to be in the ranges of 7.85-9.11%, with high oxygen content ranging from 81 to 83% (Salema et al., 2017). The higher oxygen content with less carbon content in the bio-oil indicates that the bio-oil is of poor quality and resulting in a low calorific value of it (1.7-2.5 MJ/kg). The same calorific value (2.65 MJ/kg) was also reported by Ravikumar et al. (2017) for the bio-oil from corn stover. This is due to higher water and oxygen contents. Furthermore, the calorific value of the bio-oil produced from the wood pellets is also reported to be low, 6-8 MJ/kg, (Undri et al., 2015). The poor qualities in terms of low p^H, high water content, low calorific value, and high viscosity lead to the unsuitability of using the bio-oil directly for its application in engine combustion or as a biofuel. So, reducing the acidity of the bio-oil is considered the main task, in fuel applications, because the bio-oil with high acidity leads to several problems associated with the bio-oil transfer in pipes and use in engines. Thus, without upgrading, the bio-oil product is unsuitable for direct use as an alternative fuel. Therefore, the challenge remains on the upgrading of bio-oil for use as liquid biofuels applications. The upgrading of the bio-oil to reduce its water content can be achieved by a high heating rate through MAP combined with proper additive/catalyst. Higher heating rates make the water inherent in the biomass to evaporate quickly making the minimum possibility for water to participate in the decomposition reactions. This may help to improve the quality of bio-oil and increase its calorific value. Eventually, improving the bio-oil quality to make it suitable for biofuel applications is much required and can be achieved through upgrading the bio-oil by using proper additive/catalyst during the MAP process.

7.7.6 Biochar Characteristics

Biochar is the solid product obtained during the thermochemical decomposition of biomass and it offers various applications in day-to-day life because of its environmental friendliness. The yields of biochar from MAP vary widely (3 wt.% to 84 wt.%) due to the difference in the chemical composition of the feedstock i.e., cellulose, hemicellulose, and lignin which are subjected to different mechanisms during the pyrolysis process. The main element in the biochar is carbon (38%-85%). It also contains hydrogen, oxygen, ash in addition to a trace amount of nitrogen and sulfur. The biochar free from sulphur with low nitrogen and ash contents is considered as an environmental friendly solid biofuel compared to the conventional biochar and coal and emitting no or low harmful pollutants i.e., SO_x and NO_x during the combustion (Ge et al., 2020).

One of the most important properties of the biochar is its heating value, which is found to be in the range of 25-32 MJ/kg, substantially higher than that of the parent biomass or its liquid product. The calorific value of biochar from corn stover is found in the range of 24.12-26.32 MJ/kg with the corresponding energy yield of 20.16-33.17% (Huang et al., 2013). These wide ranges are due to the use of different catalysts in the pyrolysis process. The calorific value of biochar briquettes from corn stalk is reported to be about 31 MJ/kg. This value is found to be higher than that reported by Sahoo and Remya, 2020 for biochar produced from rice husk (25.46 MJ/kg). Specific surface area and pores volume are also important properties of the biochar making it useful in various applications i.e., wastewater treatment and soil improvements. The specific surface of biochar from corn stover was found to be low as 45 m^2/g with average pores volume of 0.021 cm^2/g (Zhu et al., 2015). These results were lower than the surface area of wood pellets biochar (180.12 m^2/g) and average pores volume of 0.0951 cm^2/g (Nhuchhen et al., 2018). While the surface area of the biochar produced from rice husk was found to be 190 m^2/g (Sahoo and Remya, 2020). However, the specific surface areas of the biochar from different biomass feedstock show the lower values than those found in activated carbon (600-1200 m^2/g) (Zhu et al., 2015).

Summarizing the characteristics of pyrolytic products (bio-oil and biochar), it can be concluded that several researchers have studied the microwave pyrolysis of various biomasses, however additional works are still needed to improve the qualities of both bio-oil and biochar products (Lam et al., 2019). The use of a proper additive/catalyst in the MAP could be a better choice to improve the quality of bio-oil by reducing the water and oxygen content resulting into improving the bio-oil calorific value as well as reducing the bio-oil acidity and viscosity (Ge et al., 2020). The quality of biochar could also be improved by selecting a proper additive/catalyst in the MAP process. The improvements in reducing sulfur, nitrogen, and ash contents would make the biochar environmental friendly biofuel with high calorific value (Lam et al., 2019; Zhang et al., 2020). Also, specific surface area and pores volume of the biochar may also be improved resulting into its uses in wastewater treatment and soil improvements applications.

7.8 A CASE STUDY OF SOLAR PV POWERED MICROWAVE ASSISTED PYROLYSIS FOR BIOCHAR AND BIO-OIL PRODUCTION (FODAH AND GHOSAL 2020)

7.8.1 Perspectives of Valorisation of Rice Straw

Valorisation of underutilized agricultural residues represents a significant challenge in the context of their safe disposal and potential recovery of energy in an environmentally sustainable manner. Several strategies for their transformations to high value products have been proposed or a major focus of research in the past few years. Rice straw is one such crop residue available abundantly and cheaply every year in an agricultural based country like India where rice being a major food grain crop is cultivated in about 43.95 million hectares with the total production of about 106.54 million tonnes and approximately the annual availability of 160 million tonnes of straw in a ratio of 1:1.5 for grain to straw. Currently, a considerable portion of this voluminous residue is used for fodder, organic manure, thatching of rural houses, domestic cooking, mushroom cultivation etc. and around 40 % of the gross residue generated still remains surplus. This surplus residue, considered to be a sustainable bioenergy resource because of the availability from the most important crop, is either simply discarded or burnt in the field itself, causing excessive particulate matter emissions and air pollution. There is therefore a need to explore and evaluate the viable options for its effective utilization in recovering both energy and chemical values out of it.

The non-food portion of crop such as straw, stalk, husk etc. are categorized under lingo-cellulosic biomass which is mainly composed of cellulose, hemicellulose and lignin. These basic chemical components in cell wall of biomass are increasingly recognized as valuable commodity, due to its plentiful availability as a raw material for the production of biofuel. Both cellulose and hemicellulose are polysaccharides. Lignin is an amorphous polymer with no exact structure and consists of an irregular array of various bonded phenyl-propane units. Generally, the percentages of cellulose, hemicellulose, and lignin content are 35-50, 25-30, and 15-30 wt.%, respectively. Therefore, lignocellulosic biomass consists of approximately 60-80 wt.% polysaccharides and can thus be a potential feedstock for fermentation in producing large amounts of free sugars through biochemical processes. However, it is difficult to process the biomass biologically because of the natural resistance of plant cell walls to microbial and enzymatic degradation, collectively known as biomass recalcitrance. In addition, it is difficult to ferment lignin because of its high resistance to chemical and biological degradation by the microorganisms causing a real obstacle to degrade cellulose and hemicellulose materials to their corresponding monomers and sugars for effective conversion into high value fuels, gas and chemicals. Moreover, the biochemical conversion processes are complex, time consuming with low conversion efficiency and requiring tedious pre-treatments for transforming biomass into energy and value-added products. Hence, thermochemical conversion route has been proved convenient and may be a better option for recovering both energy and chemical values of lignocellulosic biomass particularly in case of valorisation of agro-residues. Though, the cost involvement through biological conversion pathway may be less than the thermochemical pathway, however, thermochemical pathway does have higher conversion efficiency than biological pathway. Therefore, thermochemical methods, such as combustion, torrefaction, pyrolysis and gasification would be more favorable to convert lignocellulosic biomass into bioenergy quickly and completely. Both combustion and incineration routes of conversion are used to produce heat and power but pyrolysis and gasification conversion routes are followed to produce liquid and gaseous fuel respectively. However, direct combustion and incineration recover only the heating value of the biomass materials and not its chemical value. These methods are thus becoming increasingly impracticable due to the concerns of release of greenhouse gases and environmental pollution associated with toxic emissions. Furthermore, the cleaning of flue gases produced is complex and expensive due to strict regulations on atmospheric emissions. Pyrolysis is therefore a better solution from an environmental and economic standpoint by thermally reprocessing the lignocellulosic biomass into more useful forms of energy. The main advantage of pyrolysis is that it has the potential to recover both the energy and chemical values of not only agricultural residues but also other types of wastes such as municipal solid waste, plastic wastes and sludges etc.

The usual practice of carrying out pyrolysis by heating the biomass in an oxygen-starved environment uses traditional external heating sources such as heat delivered by electrical energy, flue gas from burning of wood and agro-residues or through partial combustion. These methods of heating for the pyrolysis process to carry out, have been existing since time immemorial and still prevailing in the rural areas. Traditional method of carrying out torrefaction/pyrolysis is commonly practised by the use of metal drums or in-ground fire pits covered in clay mud with bamboo made chimney. However, disadvantages of this pyrolysis technology include slow process, low energy yield and excessive air pollution (Fig. 7.8).

Fig. 7.8. Traditional method of biomass torrefaction

To overcome the problem, nowadays an improvement in the process is through the use of electrical energy and incorporation of a fixed bed stainless steel reactor along with condensing units to get liquid products, gas collection unit and nitrogen purging cylinder to create oxygen-free environment inside the reactor in order to make a complete pyrolysis set up, commonly known as conventional pyrolysis. The pyrolysis of biomass is a very versatile process since conditions such as temperature, heating rate, residence time, etc. can be optimised to maximize the yield of char, oil or gas depending on which product is required. On the basis of the process conditions, the pyrolysis method has been categorized as slow, fast and flash pyrolysis. Slow pyrolysis refers to the practice of low temperature, low heating rate and long residence time favouring the yield of char. Fast pyrolysis occurs in moderate temperature, high heating rate and short vapour residence time maximizing the yield of liquid product. On the contrary, flash pyrolysis occurs at high temperature, low heating rate and long gas residence time conducing the yield of fuel gas. The most important and the studied process out of the three is the fast pyrolysis due to the yield of higher amount of bio-oil which can be easily and economically collected, stored and transported with considerable potential of producing a number of valuable chemicals that offer the attraction of much higher added value than fuels. Lesser amounts of char (solid) and syngas (gas) are also formed as the intermediate products during the fast pyrolysis process. Out of these, char too has the advantages in terms of collection, storage and transportation and can be used as energy source for combustion due to its higher fixed carbon and calorific value compared to raw biomass feedstock. Due to the ability to produce potentially the valuable products, efforts have therefore been made by the researchers to perfect the pyrolysis process maintaining suitable operating conditions for the recovery of both energy and chemical values of the lignocellulosic biomass.

In recent years, a new approach for carrying out biomass pyrolysis is the integration of microwave radiation as a heating source and considered as an emerging pyrolysis technology for saving of time, energy, achieving higher heating performance, precise control over the process and higher yield of the desired product compared to conventional pyrolysis (CP). The microwaves are the electromagnetic radiations whose frequencies range from 300 MHZ (wavelength 1m) to 300 GHZ (wavelength 1 mm). Heating mechanism for the feedstock in the reactor of Microwave Assisted Pyrolysis (MAP) is different from that of conventional pyrolysis process. In CP, the heat from the outside source of electrical energy is transferred to the feedstock in the reactor through convection, conduction and then by radiation causing the loss of heat and maintaining a temperature gradient from outside to the inside due to poor thermal conductivity of the biomass material. On the contrary, in MAP, microwave penetrates directly into the feedstock through the transparent reactor without any loss and the radiation is then transformed into thermal energy in the biomass particle after absorption. As the surface of the particle, then loses heat, a temperature gradient is formed from inside to outside, just opposite to CP, resulting into the uniform and volumetric heating of the released volatiles with higher heating efficiency. The heating mechanism in MAP is therefore a type of energy conversion rather than transfer of heat. Due to uniform and better control of temperature in the reactor during the process, the undesirable secondary reactions can be controlled or avoided and desired thermochemical products can be obtained. Higher heating rates can also be accomplished due to the transfer of energy at molecular and atomic level within the material.

One difficulty commonly encountered during MAP is the low capacity in absorption of microwave by the biomass feedstock as it has poor dielectric properties causing the necessity of addition of microwave receptor for maximizing heating efficiency and better yield of the products. One such easily and conveniently available receptor is the bio-char which may be obtained from the same pyrolyzer without any additional cost involvement. Moreover, it has been established that carbon-based additives such as activated carbon and silicon carbide have the good capacity in absorptance of microwave. However, water is also a good microwave absorber due to its dielectric property and polar behaviour. Hence, moisture present in the biomass feedstock assists to some extent in absorption of microwave initially and its subsequent availability due to release of water vapour through various chemical reactions during the pyrolysis process. It is therefore necessary to add exogenous microwave absorbents for better outputs particularly in case of fast pyrolysis.

Selection of material for the reactor in MAP is another important consideration because of its capacity in transparency for microwave. The material should be selected such that it should allow the waves to pass through and capability to withstand high temperature, pressure and having low coefficient of thermal expansion. Quartz is such a material suitable for the reactor material because of fulfilling the above characteristics. It can also resist temperature in the range of 1000-1500 ^{0}C. However, the only drawback is its easy fragility and cracking.

MAP is generally carried out with the help of electrical energy. For promotion and sustainability of this improved practice, availability of reliable and secured source of power is also an important factor especially in off-grid or remote areas. Solar photovoltaic electricity may be a promising and viable alternative for the regions located mostly in tropical and sub-tropical zones in the globe. This may ultimately enhance the easy acceptability and sustainability in rural areas compared to CP by eliminating or reducing the amount of conventional power input. Hence, inclusion of solar PV system with MAP set up may be a better and viable option from economical and environmental points of view for the users in rural areas.

The bio-char and bio-oil produced from the pyrolysis process have versatility in their applications both at users' and commercial level compared to the gaseous product (syngas) because of their easy storability and transportability. These two products can directly be utilized by the users or among the rice growers in various ways for their energy needs and livelihood earning at the individual level. The simplest way of using bio-char, a solid carbon rich product, is as a cooking fuel in domestic sector after making even the hand press briquettes out of it. It can also be used as a solid fuel in boiler, in production of activated carbon, as a microwave absorber, soil conditioner, in making nanotubes etc. Likewise, bio-oil, a liquid product from the biomass pyrolysis, also known as pyrolysis oil or bio-crude oil is considered to be a substitute for combustion fuel in stationary applications for heat and electricity generation especially in furnaces, boilers, turbines etc. However, it needs to be pre-heated to the temperatures in the range of 70-100 ^{0}C before its use for reducing viscosity lower than 10 cSt. Likewise, the bio-oil is required to be upgraded to meet the necessary qualities for a transportation fuel. The upgradation is generally accomplished through two methods i.e. hydro-deoxygenation with suitable hydro-treating catalyst and catalytic vapour cracking. Without upgradation, it cannot be used directly as an alternative for transportation fuel due to its unfavourable qualities such as higher viscosity, lower heating value, higher acidity, higher oxygen content, lower thermal/chemical stability compared to conventional fuels i.e. gasoline and diesel. The above upgradation technologies are however complicated, expensive and economically not viable for small scale transaction or at individual level. Nevertheless, it has the potential to extract high-value chemicals in a convenient way for their wide applications in timber, food, pharmaceutical, fertilizer industries etc. One of the simple, easy and inexpensive ways is to produce wood vinegar (pyroligneous acid) from the bio-oil through sedimentation process.

The crude bio-oil is kept in a closed container for about 2-3 months at ambient temperature and decanted from sedimentation tar. The separated liquid, called as wood vinegar, a natural product, has its high commercial value in industrial sector. Wood vinegar offers its applications as herbicides in agriculture to combat against diseases and pests, as additives and flavouring agents for food materials due to anti-fungal and anti-bacterial properties, wood preservatives for anti-termite properties, detoxification pad in pharmaceutical and medical applications etc. By adopting microwave pyrolysis for various agro-residues available abundantly in rural areas, an user can easily produce a better quality and higher fraction of bio-oil and wood vinegar and tar out of it in a small scale at own level and earn the economic benefits through marketing in nearby places.

From the above discussions, an attempt has therefore been made in this study to valorise the most underutilized agro-residue (rice straw) at the user's level in an agrarian society from economical, environmental and societal stand points. Efforts have also been made to experimentally investigate the potential of extracting more useful and value added products from rice straw in a convenient manner by following microwave pyrolysis assisted with the reliable solar PV electricity with an aim of disposing them safely, mitigating greenhouse gas emissions by preventing their improper uses and creating avenues for earning livelihood particularly among the resource poor rice growers. Improvements in higher yield of two chosen products i.e. biochar and bio-oil in this study (Fig. 7.9) have too been assessed with respect to variations in microwave powers and addition of required amount of biochar as an additive.

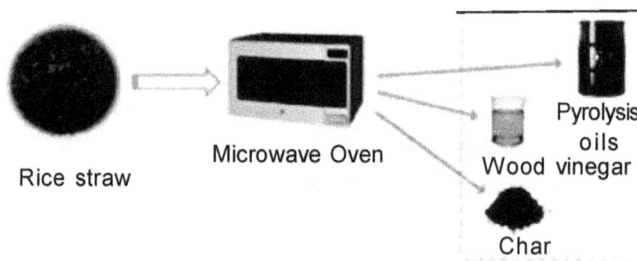

Fig. 7.9. Pyrolysis products desired under study

Percentage of energy recovery of the pyrolysis products from the raw residues and exergy analysis of the method, followed, have as well been studied for sustainability of the practice as against the present concern of rampant on-farm burning of rice straw.

7.8.2 Experimental Details

Biomass Feedstock

Biomass feedstock such as rice straw (RS) used for the study was collected locally. The variety of rice from which straw was chosen for the experiment was Lalata (semi-dwarf,

120-125 days' duration crop) as it is widely cultivated in the state of Odisha, India. The collected feedstock was naturally air-dried, shredded and sieved progressively through 3/ 8 mesh screen to get the particle size of about 10 mm. The straw was used without any pre-treatment. The processed feedstock was tested in hot air oven at 105 ±2 °C for 24 h as per ASTM E871 for determining its moisture content before using it in the microwave reactor for pyrolysis.

Experimental Set-up

The experimental set-up developed for the study consists of a modified domestic microwave oven (LG, 28L convection oven, model, India) operating at a frequency of 2.45 GHz having maximum incident power of 1000W, a quartz reactor (14 cm dia × 19.5 cm height) with a three-neck lid, a K-type thermocouple sensor, inserted inside the reactor and regulated to submerge in the sample for avoiding any reaction with the microwave radiation and saving it from sparks, a temperature controller (PID) with a digital display (0.01 accuracy) connected to the thermocouple for recording temperature in every one minute, a nitrogen cylinder with 99.99 purity, used for the flow rate of 100 ml/min before commencing the experiment for about 2 min to create an anoxic condition inside the reactor and with subsequent decrease of flow rate to 50 mL/min during the experiment, a two-stage condensation system quenched by cold water from the water tank (20-25°C) for liquefying the condensable vapours, tedlar bag only for collecting the exiting non-condensable gas, not for analysis and use, solar PV modules, battery and an inverter. The experiments for the study were undertaken in the Department of Farm Machinery and Power, Odisha University of Agriculture and Technology, Bhubaneswar, Odisha, India during 2019-2020. The schematic diagram of solar powered microwave assisted pyrolysis system for the study is shown in Fig. 7.10

Apparatus used for sample analysis

The proximate analysis of raw biomass and its bio-char was carried out according to the standard test method E1131-08 of the American Society for Testing and Materials (ASTM). The analysis determines the components such as fixed carbon, volatile matter, moisture content and ash content present in solid biomass and bio-char for assessing the combustible and non-combustible constituents present among them. Different ASTM standards were followed separately to find out the fraction of individual components such as ASTM E871 for moisture content, ASTM E872 for volatile matter, ASTM E-1755-01 for ash content and the rest, the fixed carbon through mass difference. The fixed carbon value deviates from the elemental analysis of carbon content.

Ultimate analysis was done to determine the elemental composition such as C, H, N, S of biomass or its pyrolysis products. Oxygen is determined by mass balance. This analysis is done to know the molecular formula of biomass and the requirement of theoretical air for oxidation. It indicates the higher heating value of the biomass and pyrolysis fuels. The elemental analyses were conducted using Elementar (Germany), model UNICUBE CHNS/O analyser as per standards ASTM D 5291 and ASTM D1552.

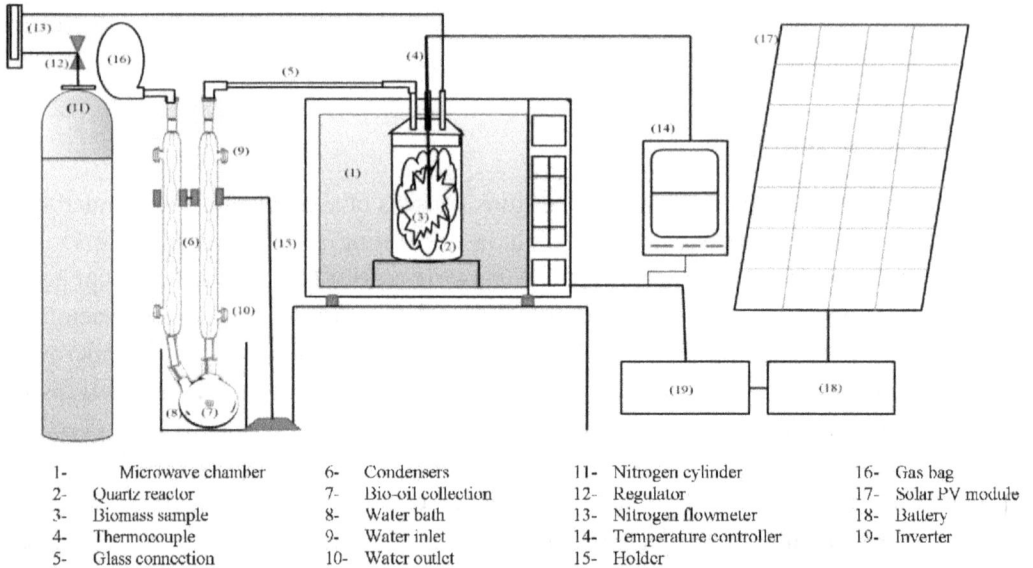

1-	Microwave chamber	6-	Condensers	11-	Nitrogen cylinder	16-	Gas bag
2-	Quartz reactor	7-	Bio-oil collection	12-	Regulator	17-	Solar PV module
3-	Biomass sample	8-	Water bath	13-	Nitrogen flowmeter	18-	Battery
4-	Thermocouple	9-	Water inlet	14-	Temperature controller	19-	Inverter
5-	Glass connection	10-	Water outlet	15-	Holder		

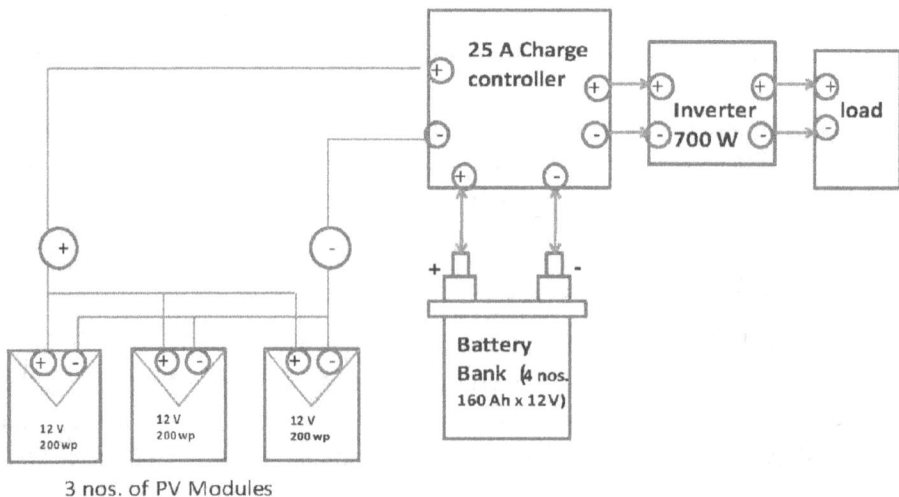

Fig 7.10. Schematic diagram of the experimental set-up with PV system

Higher heating value

The higher heating value (HHV) for raw rice straw and its biochar were measured by bomb calorimeter, while for bio-oil, it was calculated with the help of standard equation by using its contents such as carbon, hydrogen, and oxygen The procedure of using the bomb calorimeter is as per IP-12, IS 1448 (P-6) and ASTMD-4809.

Bio-oil p^H

Bio-oil acidity was measured by a standard p^H meter, which was calibrated earlier by different standard buffer solutions.

Bio-oil dynamic viscosity

To measure the bio-oil dynamic viscosity, the Ostwald viscometer, Cannon-Fenske (SO), No. 150-0.035 026110-0008 was used with the standard method i.e. ASTM D 2515 at 25°C. The method of determining viscosity with this instrument consists of measuring the time for a known volume of the liquid to flow through the capillary tube under the influence of gravity.

Bio-oil water content

The water content in the bio-oil was measured using VEEGO/ MATIC-D Volumetric Karl Fischer Titration according to ASTM E203.

Bulk density

For biomass and its biochar, the density was measured as the ratio of weight of the sample to their respective volume. While, the density of bio-oil was determined by a Pycnometer at 25 °C as per standard ASTM D369. The Pycnometer volume was previously calibrated with distilled water.

Experimental procedure followed

For each batch of experiment, 450 g of shredded and sieved rice straw was kept in the quartz reactor. A comparative study was undertaken with respect to the yields of pyrolytic products with and without the addition of microwave absorption enhancer i.e. biochar (5 % of the weight of input feedstock) in the reactor. The quartz reactor placed in the pathway of microwave was tightly sealed with the help of an O-ring and a stainless sealer. A thermocouple sensor was positioned in contact with the feedstock to record the temperatures during the process of pyrolysis. Before starting the experiment, a constant flow (100 mL/min) of nitrogen gas was purged into the reactor for maintaining an inert atmosphere. When the purging was sufficient to create an anoxic environment inside reactor, the nitrogen flow rate was decreased to 50 ml/min and the power supply switch was turned on along with the fixing of the adjusting knob for the input power from the microwave oven and reaction time at their designated and chosen levels. The power level and reaction time were decided on the basis of preliminary studies with an aim of higher yield of bio-oil followed by biochar because of their direct utilization by the users. Power supply to the microwave oven was provided from solar module through battery and inverter. When reaction time reached to its pre-set value and no appreciable volatiles were released along with no further rise of temperature inside the reactor, power supply switch was tuned off, but the flow of nitrogen was continued until the pyrolysis products were allowed to cool down to ambient temperature. Meanwhile, the condensable volatiles, passing through two condensers were collected after cooling with the circulating water around the condensing flasks from the normal water supply unit. Similarly, the non-condensable gases were only collected in the gas bag to prevent the experimental site from the emitted pollutants and not for any of its analysis. After allowing self-cooling to the room temperature, biochar was removed from the reactor, weighed and stored for analysis. Bio-oil was also collected, weighed and stored for experimental investigations. The yields of solid residue and liquid fraction were calculated from the weight of each fraction, but the yield of gas was calculated on the basis of mass balance. All the experiments were carried out at least three times to get an average value for the experimental results. Energy recovery and exergy analysis of pyrolysis products along with the techno-economic evaluation of the practice followed were studied for the sustainability of the above method/technology among the rice growers.

Preliminary studies undertaken

The quality and quantity of pyrolysis products depend significantly on the various operating parameters such as size of biomass feedstock, effects of nitrogen gas flow rate, types and amount of microwave absorbers, input power, heating rate and residence time. The size of the feedstock and nitrogen gas flow rate were decided as per the previous studies of the researchers and were kept fixed for the present study. The size of rice straw at about 10 mm and nitrogen gas flow rate at 50 ml/min were kept constant all through the experiments. Biochar (5 % of the weight of the feedstock) was considered as an additive due to its low cost and easy availability among the users. Heating rate generally varies linearly with the level of input power. Hence the levels of input power from the microwave oven and the residence time for the pyrolysis process were considered the two important factors for the yield of the products during microwave pyrolysis. Preliminary studies were therefore carried out to optimize the input power levels and reaction times for better yield of liquid product before undertaking the final set of experiments for the study. The levels of input power as 300 W, 500 W, 700 W and 900 W and reaction times as 15, 20 and 25 minutes were considered and varied during the preliminary studies from which one level of input power and one reaction time were chosen during further experimental investigations.

Product Yield

The weight of biochar and bio-oil was calculated in percentage, based on the ratio of product collected to the amount of original biomass loaded and as follows:

$$Biochar\ yield\ (\%) = \frac{Mass\ of\ biochar}{Mass\ of\ raw\ biomass\ sample} \times 100$$

$$Bio-oil\ yield\ (\%) = \frac{Mass\ of\ bio-oil}{Mass\ of\ raw\ biomass\ sample} \times 100$$

$$Gas\ yield\ \% = 100 - \{Biochar\ yield\ (\%) + Bio\text{-}oil\ yield\ (\%)\}$$

Energy and Exergy analysis of pyrolysis products

Energy analysis of a system provides the information about its quantity but not quality. Quality of the available energy actually signifies the useful energy that can be obtained or utilized for practical purposes. The quality of energy or utilizable fraction of the available energy i.e. exergy, therefore, represents quality besides quantity which identifies the factor of inefficiencies due to unavoidable irreversibilities and explores the scopes for improvement in a process. The exergy is thus defined as the maximum amount of useful work that can be obtained from an energy system when it is brought to the thermodynamic equilibrium with its reference to the surrounding environment. During the past decades, exergy analysis has been proved to be an essential/useful tool while assessing and improving the performance of various thermochemical conversion paths of biomass along with the sustainability of the technology. Both energy and exergy analyses have hence been evaluated in case of biochar and bio-oil, the desirable products in this study, by the use of additives for studying the improvement in the efficiency of the practice/technology with respect to the yield and change in chemical compositions of the pyrolysis products along with the reduction in the loss of heat energy from system to the surroundings. The flow of input energies into the system and exit of the pyrolysis products have been shown in the figure 7.11 below.

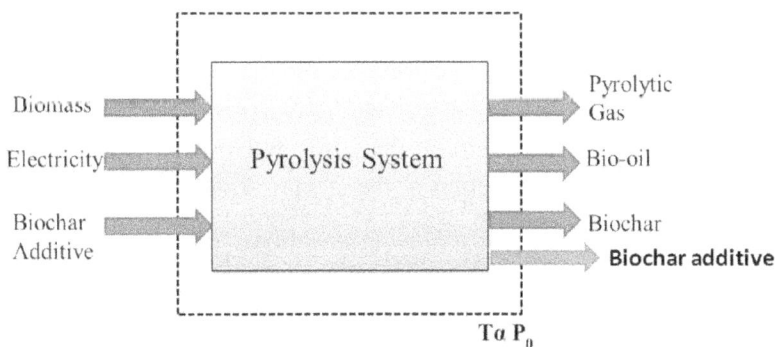

Fig. 7.11. Energy and Exergy flows for the pyrolysis system

The system boundary is defined by the ambient conditions i.e. T_a and P_0 which represent the ambient temperature and atmospheric pressure respectively and eliminates the heat transfer as an energy contributor. The energy contained in ash, energy released due to condensation by the cooling of water and energy required by the flow of N_2 gas for maintaining inert atmosphere in the reactor, have been neglected due to their small quantities.

Energy analysis

Energy balance for the above mentioned pyrolysis process (Fig. 7.11) can be written as

$$\dot{E}n_{biomass} + \dot{E}n_{electrical} + \dot{E}n_{biochar, additive} = \dot{E}n_{biochar} + \dot{E}n_{bio-oil} + \dot{E}n_{gas}$$
$$+\dot{E}n_{biochar, additive} + \dot{E}n_{loss} \qquad \ldots (7.1)$$

where $\ddot{E}n_{biomass}$, $\ddot{E}n_{electrical}$, and $\ddot{E}n_{biochar, additive}$ denote the energy rates for the inputs such as biomass, electrical energy for microwave reactor and biochar additive respectively. Likewise, $\ddot{E}n_{biochar}$, $\ddot{E}n_{bio-oil}$, $\ddot{E}n_{gas}$, represent the energy rates for pyrolysis products i.e. biochar, bio-oil, and gas respectively and $\ddot{E}n_{loss}$, the energy loss rate of the pyrolysis process.

While considering the total energy of a stream, the kinetic energy and potential energy may be neglected and it can be calculated as per Eq. (7.2).

$$E_n = E_n^{phy} + E_n^{che} \qquad \ldots (7.2)$$

Where E_n^{phy} is the physical energy and E_n^{che} is the chemical energy. However, the physical energy, which represents the energy available if the system is not in thermal equilibrium with its environment, is also assumed zero since all inputs and outputs considered are at ambient temperature (25°C) and pressure (1 atm). The chemical energy of the components in the pyrolysis process can be written as per Eq. (7.3)

$$\dot{E}_n^{che} = \dot{m}.HHV \qquad \ldots (7.3)$$

Where $\dot{m}$ and *HHV* are mass flow rate and higher heating value for the stream respectively. To comprehensively evaluate the performance of the pyrolysis system, energy efficiencies for biochar ($\eta_{biochar}$), bio-oil ($\eta_{bio-oil}$), and pyrolysis system (η_{system}) are calculated based on energy rate of output of the streams divided by the sum of the energy rates of each input ($\dot{E}n_{input}$) by the following equations:

$$\dot{E}n_{input} = \dot{E}n_{biomass} + \dot{E}n_{electrical} \qquad \ldots (7.4)$$

$$\eta_{biochar} = \frac{\dot{E}n_{biochar}}{\dot{E}n_{inputs}} \times 100 \qquad \ldots (7.5)$$

$$\eta_{bio-oil} = \frac{\dot{E}n_{bio-oil}}{\dot{E}n_{inputs}} \times 100 \qquad \ldots (7.6)$$

$$\eta_{system} = \frac{\dot{En}_{outputs}}{\dot{En}_{inputs}} \times 100 \qquad \dots (7.7)$$

The energy and exergy efficiencies of the gaseous product from the pyrolysis system have been ignored as biochar and bio-oil are the two desirable products considered in the present study.

Exergy Analysis

The exergy balance for the same pyrolysis system as shown in the above figure 7.11 can be written as

$$\dot{Ex}_{biomass} + \dot{Ex}_{electrical} + \dot{Ex}_{biochar,\ additive} = \dot{Ex}_{biochar} + \dot{Ex}_{bio-oil} + \dot{Ex}_{gas} +$$
$$\dot{Ex}_{biochar,\ additive} + \dot{Ex}_{loss} \qquad \dots (7.8)$$

Where $\dot{Ex}_{biomass}$, $\dot{Ex}_{electrical}$ and $\dot{Ex}_{biochar,\ additive}$ denote the exergy rates of inputs such as biomass, electrical energy to the microwave reactor and additive respectively. The energy and exergy value of the electrical component are same. Likewise, $\dot{Ex}_{biochar}$, $\dot{Ex}_{bio-oil}$, and $\dot{Ex}_{gas}$ represent the rates of output products of pyrolysis process for biochar, bio-oil and gas respectively and $\dot{Ex}_{loss}$ is the rate of exergy loss. The exergy rate for biomass, biochar, and bio-oil can be calculated by equation (7.9):

$$Ex = (h - h_0) - T_0 (s - s_0) + \frac{v^2}{2} + gz + Ex_{ch} \qquad \dots (7.9)$$

Where the last term in the above equation is the chemical exergy, while the first four terms are the thermomechanical exergy. In the pyrolysis system as shown in the figure 7.11, all entering and exiting streams as well as heat transfer for the boundary to be at reference conditions. Hence the first four terms of Equation (7.9) are assumed to be zero and therefore the chemical exergy is the only contribution to the exergy balance of the streams. Like energy, the exergy contained in ash, exergy released due to condensation by the cooling of water and exergy required by the flow of N_2 gas for maintaining inert atmosphere in the reactor, have also been neglected due to their small quantities. Chemical exergy is found by formulating the reactions of a given chemical with the elements in the environment and finding the maximum theoretical work that could be obtained from this reaction. The chemical exergy of the biomass, biochar, and bio-oil can be calculated by the equation (7.10):

$$Ex_{ch} = \beta. \dot{m} LHV \qquad (7.10)$$

Where β is the ratio between chemical exergy and LHV of the organic fraction of biomass, and $\dot{m}$ and LHV are mass flow rate and lower heating value of the streams respectively. The LHV is calculated from the equation (7.11):

$$HHV = LHV + 21.978\ H \qquad \text{... (7.11)}$$

Where H, is the weight fraction of element H in the ultimate analysis. The value of β can be determined by the equations (7.12, 7.13, 7.14) for the biomass, biochar, and bio-oil by correlating the mass fractions of Carbon (C), Hydrogen (H), Nitrogen (N) and Oxygen (O) of the streams:

$$\beta_{biomass} = \frac{1.0412 + 0.2160\dfrac{H}{C} - 0.2499\dfrac{O}{C}\left[1 + 0.7884\dfrac{H}{C}\right] + 0.0450\dfrac{N}{C}}{1 - 0.3035\dfrac{O}{C}} \qquad \text{... (7.12)}$$

$$\beta_{biochar} = 1.0437 + 0.1896\frac{H}{C} + 0.0617\frac{O}{C} + 0.0428\frac{N}{C} \qquad \text{... (7.13)}$$

$$\beta_{bio-oil} = 1.0401 + 0.1728\frac{H}{C} + 0.0432\frac{O}{C} + 0.2169\frac{S}{C}\left(1 - 2.0628\frac{H}{C}\right) \quad \text{... (7.14)}$$

Like energy analysis, exergy efficiencies for biochar ($\Psi_{biochar}$), bio-oil ($\Psi_{biochar}$), and pyrolysis system (Ψ_{system}) are calculated based on exergy output of the streams divided by the total exergy inputs (Ψ_{input}) by equations (7.15, 7.16, 7.17):

$$\Psi_{biochar} = \frac{\dot{Ex}_{biochar}}{\dot{Ex}_{inputs}} \qquad \text{... (7.15)}$$

$$\Psi_{bio-oil} = \frac{\dot{Ex}_{bio-oil}}{\dot{Ex}_{inputs}} \qquad \text{... (7.16)}$$

$$\Psi_{system} = \frac{\dot{Ex}_{outputs}}{\dot{Ex}_{inputs}} \qquad \text{... (7.17)}$$

7.8.3 Experimental Findings

Small-scale batch reactor using solar powered microwave pyrolysis was used in this study with the key focus on obtaining higher yield of bio-oil followed by the biochar for their easy collection, storage and direct marketing to earn income for the users. The preliminary study was therefore undertaken to choose the desired microwave power and reaction time with a view to obtain higher yield of bio-oil. Biochar was also collected ignoring syngas as the gaseous product is difficult to store and marketing. Before starting the experiment, the proximate and ultimate analysis of raw rice straw were done to assess its fuel and combustion characteristics compared to the pyrolysis products i.e. biochar and bio-oil.

Products characteristics

The yield of bio-oil from the microwave assisted pyrolysis (MAP) is affected by the type biomass feedstock. This is due to the differences in the chemical composition and proximate fractions of the feedstock. More the oxygenated compounds, less is the calorific value of the products. On the other hand, feedstock with high ash content produces less bio-oil, because of the non-conversion of ash into bio-oil. The ultimate, proximate analysis and higher heating value of a fuel provide broad understanding of its combustion characteristics. The elements such as nitrogen and sulphur produces greenhouse gases such as NO_x and SO_x by the burning of fuel. The components such as fixed carbon, volatile matter, ash concentration, moisture etc. provide a rough estimate of how the fuel will behave when it is burned. Among the proximate components, the fixed carbon and volatile matters represent the combustible components of the fuel. Similarly, in the elemental analysis, the components such as carbon and hydrogen contribute to the heating value of the fuel but oxygen reduces the heating value. The fuel with higher volatile matter favours the easy ignition and fast burning compared to the fixed carbon which is hard to ignite and burns slowly. The proximate and ultimate analysis and higher heating value of raw rice straw and biochar have been presented in Tables 7.10 and 7.11 respectively and ultimate analysis, heating value and physical characteristics of bio-oil in Table 7.12. The decrease in volatile component and increase in fixed carbon for biochar compared to initial biomass is mainly due to the increase of reaction temperature causing more devolatilization of the organic material. The lower atomic ratio of H/C and oxygen content in the ultimate analysis of biochar with respect to raw biomass are due to the breaking of more functional groups (deoxygenation) with the increase of temperature during pyrolysis resulting into the higher heat value of biochar. However, bio-oil with higher atomic ratio of O/C makes it unsuitable for engine fuel and therefore, the better option is for its use as chemical feedstock. But the use of additive (5 % biochar) in the pyrolysis process further lowered the H/C ratio in biochar and O/C ratio in bio-oil compared to without additive causing the improvement in their heating value. This may be due to further increase of process temperature by more absorption of microwave radiation leading to enhanced devolatilization and deoxygenation. Lowering in the percentage of water content fraction also favoured the increase of heating value in bio-oil with additive. Hence, the use of additive in the pyrolysis process decreased the volatile matters and increased the fixed carbon in biochar. Likewise, use of additive increases elemental carbon and decreases oxygen content in bio-oil.

Microwave power and reaction time during pyrolysis process

Preliminary study was carried out to decide the optimal microwave power and reaction time for obtaining higher yield of bio-oil. Twenty-four batches of experiments were conducted initially by varying 4 levels of microwave power (300 W, 500 W, 700 W and 900 W) and

Table 7.10. Proximate, ultimate analyses, bulk density, and higher heating value for raw rice straw biomass.

Proximate analysis, (%, db)			Ultimate analysis (%, db)					Bulk density (kg/m³)	HHV, (MJ/kg)
VM	ASH	FCᵃ	C	N	H	S	Oᵇ		
68.2	12.5	19.3	41.25	0.87	5.58	0.038	52.26	56.78	12.28

Note: VM = Volatile Matter; ASH= Ash content; FC= Fixed Carbon; C= Carbon; N= Nitrogen; H= Hydrogen; S= Sulfide; O= Oxygen.
ᵃ Calculated by difference (dry basis), FC= 100 - (VM + ASH).
ᵇ Calculated by difference (dry and ash-free basis), O= 100 - (C+H+N+S).

Table 7.11. Proximate analysis, ultimate analysis, bulk density and higher heating value of rice straw biochar at chosen microwave power level 700W with and without additive.

Sample name	MWP (W)	RT (min)	Additive	Proximate analysis, (%, db)			Ultimate analysis (%, db)					Bulk density (kg/m³)	HHV, (MJ/kg)
				VM	ASH	FCᵃ	C	N	H	S	Oᵇ		
RS 700W	700	20	without	26.78	44.67	28.55	61.15	1.06	3.18	0.731	33.88	30.12	19.28
RS 700W (with 5% biochar)			With 5% biochar	20.56	44.18	35.26	66.27	1.17	3.10	0.713	28.75	27.31	21.18

Note: MWP = Microwave power, RT = Reaction time, VM= Volatile Matter; ASH= Ash content; FC= Fixed Carbon; C= Carbon; N= Nitrogen; H= Hydrogen; S= Sulfide; O= Oxygen.
ᵃ Calculated by difference (dry basis), FC= 100 - (VM + ASH).
ᵇ Calculated by difference (dry and ash-free basis), O= 100 - (C+H+N+S).

Table 7.12: The pH value, density, viscosity, moisture content, ultimate analysis and higher heating value (HHV) of rice straw bio-oil at chosen microwave power level 700W with and without additive.

Sample name	MWP (W)	RT (min)	Additive	pH value	Density (g/ml)	Dynamic viscosity (cp)[a]	Water content (%)	Ultimate analysis (%)					HHV, (MJ/kg)
								C	N	H	S	O[b]	
RS 700W	700	20	Without	3.68	0.976	3.17	79.21	11.51	1.08	9.27	0.347	77.79	3.24
RS 700W (5% biochar)			5% biochar	3.84	0.963	2.98	71.54	13.21	1.07	9.64	0.003	76.08	4.66

[a] Viscosity measured at 25 °C.

[b] Calculated by difference (ash-free), O= 100 - (C+H+N+S).

[c] Calculated by the equation Higher Heating Value (HHV) MJ/kg= 0.3382 C% + 1.4428 (H% - 0.125 O%).

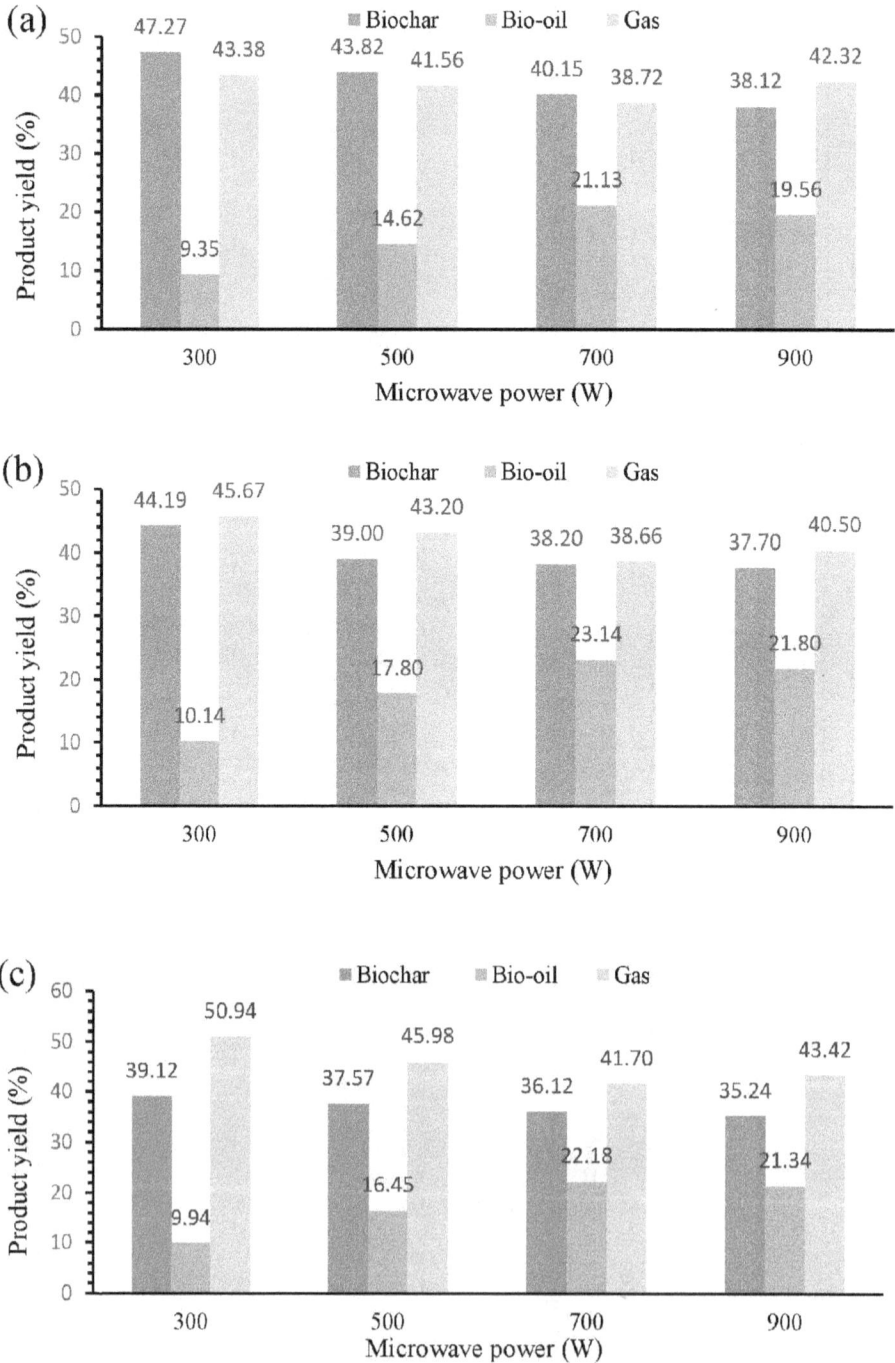

Fig. 7.12. Percentage yield of pyrolysis products under different microwave power levels and reaction times without additive (a) 15 min reaction tim, b) 20 min, c) 25 min.)

3 levels of reaction time (15, 20 and 25 minutes) using additives and without additives taking one power level and one reaction time once. The percentage yields of the pyrolysis products under the above levels of microwave powers and reaction times are summarised in Figs. 7.12 and 7.13, keeping other operating parameters (450 g of feedstock in each batch, 10 mm particle size and 50 mL/min of N_2 flow rate) same. The fractional yields of bio-oil, the most desirable product in the study were observed to be maximum at the microwave power of 700 W and reaction time of 20 minutes both with and without additive, compared to the other levels. The variations of temperature in general were observed to be linear and positively correlated with the change of microwave power. Attainment of moderate temperature with the microwave power of 700 W along with reaction time of 20 minutes favour the pyrolysis process for higher yield of bio-oil compared to the other products and chosen the optimum level for further studies of experimental investigations with respect to time, energy and cost saving. Lower power with lower heating rate favoured the higher yield of biochar (biomass torrefaction, a mild pyrolysis), moderate power with moderate heating rate led to formation of more bio-oil while higher power and higher heating rate promoted secondary cracking reaction causing the decrease of oil as well as biochar and increase of gas. These results are also similar to those reported by the other researchers.

Product fractional yield and net energy recovery

The three-phase product distributions under different levels of microwave power and reaction times as shown in figures 7.12 and 7.13 indicated that the yields of bio-oil were increasing from 300 to 700 W of microwave power and then decreasing for further rise of power to 900 W in all reaction times such as 15, 20 and 25 minutes both for additives and without additives. The reverse trend of decrease in yield of biochar was observed with the increase of power and reaction time levels. The maximum yields of bio-oil i.e. 23.14 and 25.18 wt. % were observed at 700 W microwave power and 20 minutes reaction time for without and with additives respectively and their corresponding yields of biochar be 38.20, 36.00 and gas as 38.66, 38.82 wt %. The reason for this is due to favouring fast pyrolysis causing higher yield of bio-oil.

Energy consumption and net energy recovery in the pyrolysis process are the important factors for the commercial application of the products. They determine not only the energy efficiency of the system but also the financial cost as well. In the microwave pyrolysis, the energy consumption is mostly from the usage of grid electricity. The data relating to the net energy recovery from the raw rice straw in microwave pyrolysis with the use of grid and PV electricity have been presented in Table 7.13. The energy recovery in the microwave pyrolysis process has been estimated ignoring energy supplement from the gaseous product in this study. Also no attempt has been made to recover energy during the cooling of the products in the condensation system. It was observed that the amount of input energy (electricity) consumed in the microwave pyrolysis was equal to

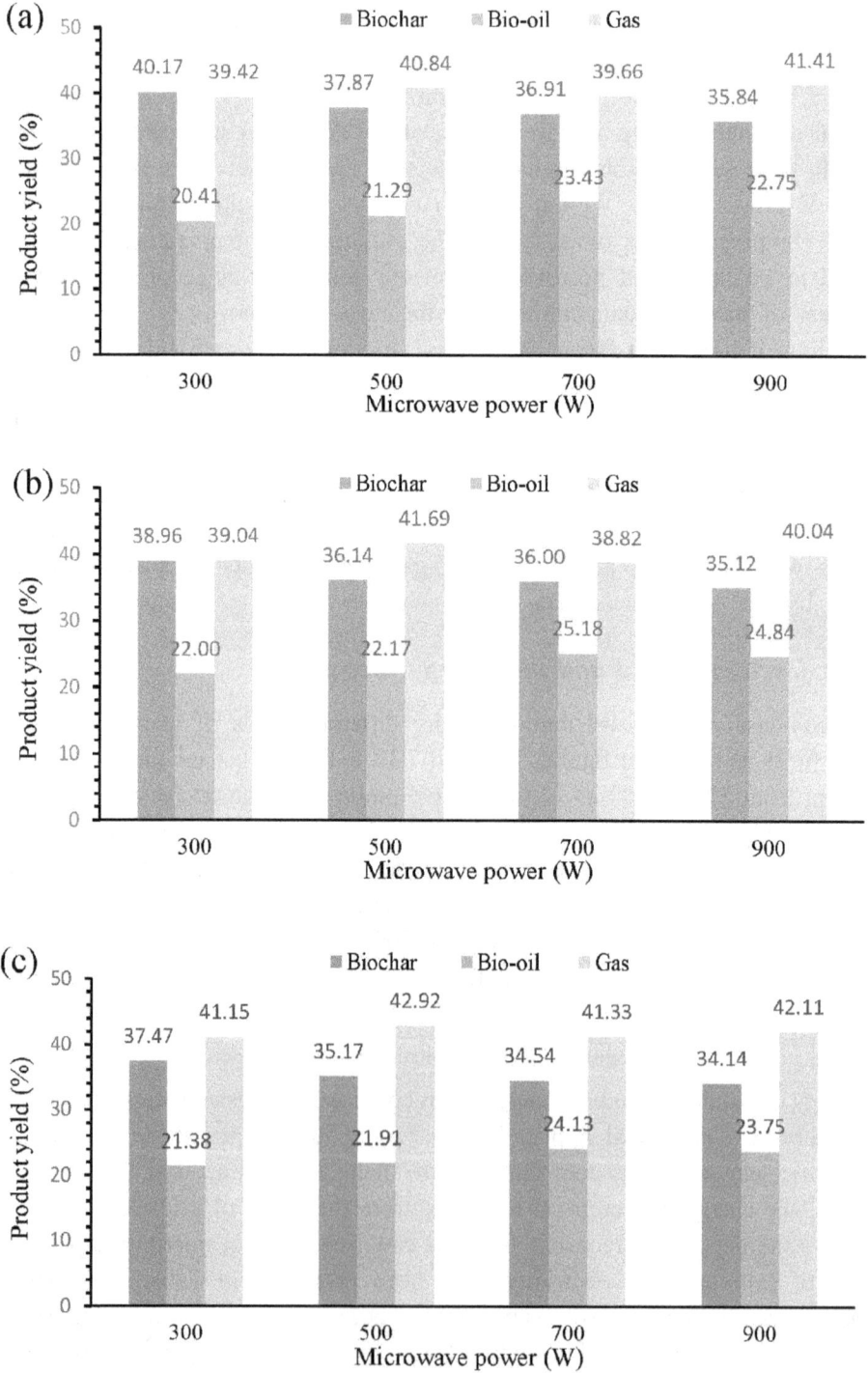

Fig. 7.13. Percentage yield of pyrolysis products under different microwave power levels and reaction times with additive (a) 15 min reaction tim, b) 20 min, c) 25 min.)

Table 7.13: Energy recovery from rice straw (RS) through microwave pyrolysis using solar PV electricity

Source of electricity	RS feeding rate (kg/h) (a)	E_{RS} (kJ/h) (b)	E_{PP} (kJ/h) (c)	$E_{recovery}$ (%) (c/b) x 100 (d)	$E_{pyrolysis}$ (kJ/h) (e)	$E_{pyrolysis}/E_{RS}$ (%) (e/b) (f)	$E_{balance}$ (kJ/h) (c-e) (g)	Net ($E_{recovery}$) (%) = (g/b) x 100 (h)
Conventional grid electricity	1.35	16578	11857	71.52	2520	15.20	9337	56.32
Solar PV electricity	1.35	16578	11857	71.52	0	0	11857	71.52

(a) Rice straw feed rate = 450 g for 20 min = 1.35 kg/h

(b) Estimated calorific value of raw rice straw based on the CV and its feeding rate i.e. CV of RS $\times$ feed rate.

(c) Estimated calorific value of the pyrolysis products (PP) ignoring syngas i.e. only biochar and bio-oil based on their CVs and wt.% yields i.e. CV of pyrolysis products$\times$ wt.% yield of pyrolysis oil RS feed rate/100. (taking feed rate 1.35 kg/h)

(d) Energy recovery (%) in the pyrolysis products was calculated based on the energy recovered from rice straw, i.e. E_{PP}/E_{RS} x 100.

(e) Electrical energy consumed during the pyrolysis process (700 W for 60 minutes) 700 J/s x 3600 s = 2520 kJ/h

(f) Amount of energy (from E_{RS}) consumed by $E_{pyrolysis}$.

(g) Energy balance, defined as the energy content of the pyrolysis products minus the electrical energy input needed to operate the system, i.e. $E_{PP} - E_{pyrolysis}$.

(h) Net energy recovery (%) from rice straw is the ratio of net energy obtained from the pyrolysis products to the energy content of raw rice straw.

15 % of the energy content of raw rice straw. That much input energy was saved in this study by using solar PV electricity which can be generated freely and environmental friendly from the sun. Hence, the net energy recovery percentage with respect to the energy content of raw rice straw was calculated to be 71. 52 and 56.32 in case of using solar PV and grid electricity respectively ignoring the energy from the gaseous product. The net energy recovery percentage may also be enhanced by considering the calorific value of gas and use of suitable (based on quality and quantity) additives for promoting higher yield of the energy rich product.

Energy and Exergy Analysis of Pyrolysis Products

As per the theoretical considerations, discussed earlier, the energy and exergy efficiencies of the microwave pyrolysis system have been calculated with respect to using additive and without additive and shown in figure 7.14.

The energy and exergy efficiencies of pyrolysis system under study were estimated to be about 58 and 55 % respectively. The less in exergy efficiency may be due to the practical utilization of products or available useful energy based on their lower heat values. The results also showed that both the efficiencies in case of using additives were higher than without additives. This improved performance was mostly due to the higher yields of oil and biochar than without additives and less loss of heat because of more heat accumulation in the feedstock at molecular level.

7.8.4 Techno-economic Environmental Assessment

Techno-economic environmental assessment for the present study comprises of sizing of solar photovoltaic (PV) system, cost estimation of the experimental set-up, life cycle cost analysis, mitigation of CO_2 emissions and carbon credit potential of the technology. The microwave oven was used 5 times in a day for the pyrolysis of 5 batches of rice straw, each of 450 g feedstock and 20 minutes duration. But, the total duration recorded for completion of one batch of experiment taking into account the time of preparation for pre and post phases of the trial was one and half hours. In the first attempt, experiments were conducted in the laboratory for 25 days in a month during June 2019 for rabi season crop and then December 2019 for kharif season crop in order to assess the techno-economic feasibility of adopting solar powered microwave pyrolysis of rice straw among the growers. However, an user may take up the technology in a small scale throughout the year for effective utilization of huge quantity of available and underutilized rice straw at individual level. Techno-economic analysis has therefore been assessed throughout the year and one may continue up to the end of the useful life of the components included in the technology. The findings of the assessment would ultimately provide an insight to the rice growers, entrepreneur, planners and policy makers regarding the merits and demerits of the technology, either to adopt or neglect in the efforts of strategic approach for the concerns of on-farm crop residue burning.

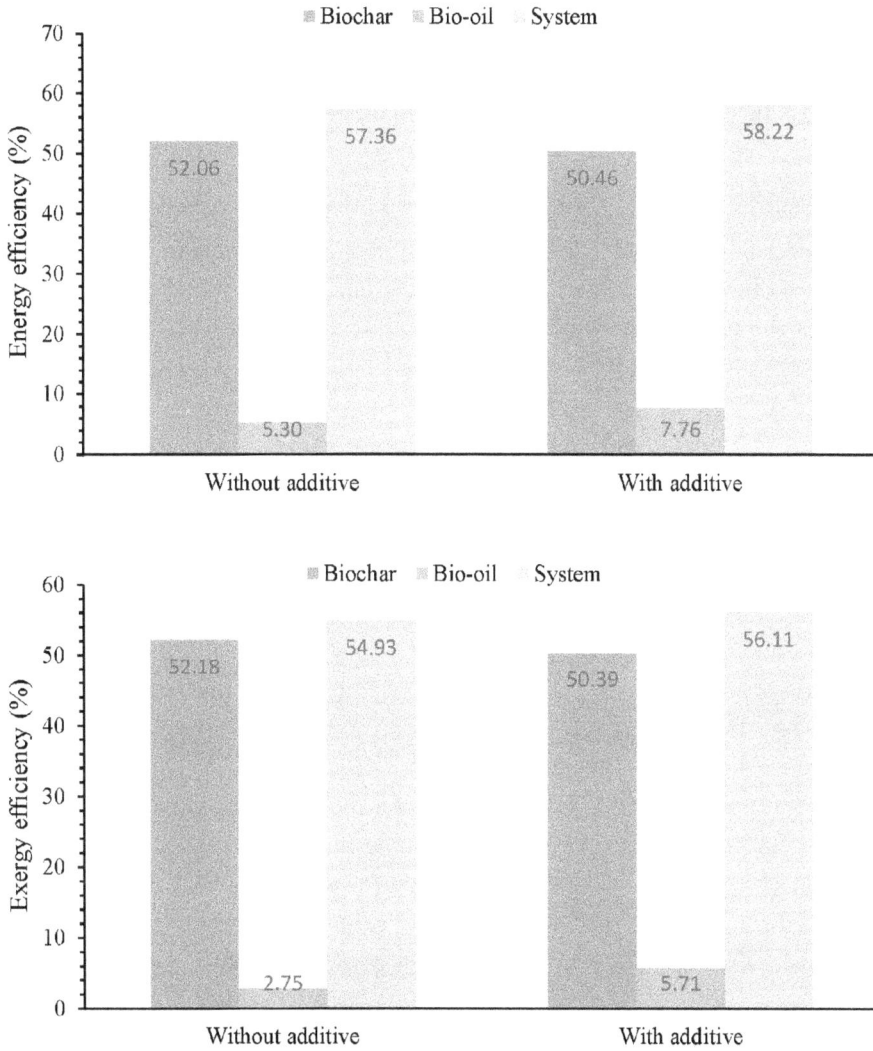

Fig. 7.14. Energy and exergy efficiency of biochar, bio-oil, and the system with and without additive.

Sizing of Solar PV system

Sizing of the solar PV system used in the study has been explained in this section. The electrical load for the experimental set up of solar powered microwave pyrolysis is mentioned in the table 7.14.

Table 7.14. Electrical load on Solar PV system

Name of component	Number	W	Hours/day	Wh/day
Microwave oven	1	1000	1.66	1660
Water pump	1	18	1.66	29.88
Total				1689.88 ≈ 1690

Determination of designed electrical load

In order to meet the losses occurred in the components of the PV system, the designed load should be higher than the actual load. The designed load was determined as 2.53 kWh/day based on the factor of safety to be 1.5.

$$\text{Design electrical load} = 1.5 \times \text{electrical load} \qquad \ldots (7.18)$$

The factor 1.5 accounts for losses in the PV system components and the uncertainty to anticipate daily electrical load of appliances and also accounts for all the system efficiencies, including wiring and interconnection losses as well as the efficiency of the battery charging and discharging cycles.

Determination of PV array size (W_p)

$$PV\ array\ size\ required\ (W_p) = \frac{Designed\ load\ \left(\dfrac{Wh}{day}\right)}{Peak\ sunshine\ hours\ per\ day} \qquad \ldots (7.19)$$

$$Number\ of\ PV\ modules\ (N) = \frac{PV\ array\ size\ \left(W_p\right)}{Peak\ watt\ of\ PV\ module\ (W_p)} \qquad \ldots (7.20)$$

The average peak sun shine hours per day on the optimum tilted surface in India is considered as 5.0 h/day. Considering the available peak watt of one module as 200W_p in the market, the number of module for the purpose was 3, each of 200 W_p.

Determination of inverter size (W) and charge controller capacity (A)

The input rating of the inverter should never be lower than the PV array rating. The maximum continuous input rating of the inverter (W) should be about 10% higher than the PV array size to allow safe and efficient operation of PV power system. Hence, the inverter size can be determined using Eq. (7.21). The inverter converts the DC power to AC power for operating AC electrical load. The charge controller saves the battery from over charging and over discharging and feeds the right configuration of current and voltage to the load. The charge controller capacity can be determined based on the PV array size and the operating voltage (such as 24 V) designed for battery bank using Eq. (7.22).

$$Inverter\ capacity\ (W) = 1.1 \times PV\ array\ size\ (W_p) \qquad \ldots (7.21)$$

$$Charge\ controller\ capacity\ (A) = \frac{PV\ array\ size\ \left(W_p\right)}{Peak\ bank\ design\ voltage\ (V)} \qquad \ldots (7.22)$$

Determination of battery bank size (Ah)

In actual practice, the battery bank size (*Ah*) can be determined by using the Eq. (7.23)

$$\text{Battery bank size }(Ah) = \text{Autonomy days} \times \frac{\text{Electrical load}\left(\dfrac{Wh}{day}\right)}{\text{Battery voltage }(V)} \qquad \text{... (7.23)}$$

The number of autonomy days for the PV system is taken to be 3 to account for three consecutive cloudy days and to improve battery life with shallow depth of discharge.

$$\text{Number of batteries } = \frac{\text{Battery bank size }(Ah)}{\text{Ampere hour capacity of each battery }(Ah)} \qquad \text{... (7.24)}$$

As per Eq. (7.23), the battery bank size was determined to be 316.25 *Ah* and hence, the next higher size was considered as 320 *Ah*. However, for obtaining 24 Volt and 320 *Ah*, from battery bank, 4 number of 160 *Ah*, 12 *V* batteries were required. Two strings of series connections were formed, each consisting of two number of (160 *Ah* x 12 V) batteries and then two parallel strings were formed by connecting the above two series connected strings for obtaining 24 *V* and 320 *Ah* from the total of four batteries. This battery bank has the energy storage capacity of 7,680 *Wh* (320 *Ah* x 24*V*) and be capable of fulfilling the operating load of 1.69 *kWh/day* for even four consecutive cloudy days. The sizing of the PV system for the experimental set-up is as Table 7.15.

Table 7.15. Sizing and cost of solar PV components

Name of PV component	Number	Sizing	Cost per unit	Cost (INR) (Indian Rupees)
Solar module	3	200 W_p	Rs. 35/W_p	21,000
Charge controller	1	25 ampere x 24 Volt	Rs. 60/Amp	1,500
Batteries	4	160 *Ah* x 12 *V*	Rs. 43.75/*Ah*	28,000
Inverter	1	700 *W*	Rs. 10/*W*	7, 000
Wiring and connections				2,000
Total				**59,500 ($774)**

Experimental observations related to solar PV system components

The observations of various electrical parameters, measured during the operation of solar PV system were utilized to evaluate its performance. The electrical parameters such as open circuit voltage (V_{OC}), short circuit current (I_{SC}), fill factor (*FF*), solar radiation (I_T) (W/m^2) were measured throughout the year 2019 by using the instrument namely solar PV analyser (SEAWARD PV 200). The output and input power of the inverter were measured by AC/DC multi meter. The hourly experimental observations were measured for typical clear days for a month and continuing throughout the year 2019 in the warm and humid climatic condition as per the following equations:

$$\text{Conversion efficiency of module } = \eta_{Module} = \left[\frac{FF \times (V_{OC} \times I_{SC})}{(I_T \times A)} \right] \times 100(\%) \quad \dots (7.25)$$

Due to the increased cell temperatures of the module compared to standard test condition in open field situation and other environmental factors such as dust deposition and reduced solar intensity, the module efficiency is assumed to be reduced by 0.2 % annually. Therefore average value for the efficiencies during 1st and last year has been considered i.e. $\eta_{module(average)}$. A is the area of module (m^2), η is efficiency

$$\text{Efficiency of DC/AC inverter: } = \eta_{Inverter} = \left[\frac{(Out\ power)}{(Input\ power)} \right] \times 100(\%) \quad \dots (7.26)$$

Overall conversion efficiency of solar PV system:
$$\eta_{system} = [\eta_{module\ (average)} \times \eta_{battery} \times \eta_{inverter}] \times 100\ (\%) \quad \dots (7.27)$$

Energy output of PV system

Energy output from the PV system is usually cite-specific and variable. It depends on the solar radiation, ambient air temperatures and efficiency of the module in on-field condition. The output of the PV system has been calculated with the following equations.

Annual average insolation (kWh/m^2/year) = hourly average solar radiation (kW/m^2)/ day × peak sunshine hours per day (h) × number of clear sunny days in a year ... (7.28)

Annual electrical energy output, E$_{out}$ = Insolation (kWh/m^2/year) × system efficiency ... (7.29)

Alternatively, E_{out} can also be calculated as per Eq. (7.30)

E$_{out}$ (kWh/year) = Daily electrical load (kWh/day) × Number of days of operation/year ... (7.30)

E$_{out}$ (kWh/year) as per Eq. (7.30) = 1.69 × 300 = 507 = Rs. 2535/year @ 5.00 per unit of electricity.

Life cycle cost assessment

Life cycle cost assessment was carried out for the solar powered microwave pyrolysis set up of rice straw taking into account all the relevant present and expected future costs as well as revenues involved with it. The useful lives of PV system, microwave oven, battery bank and inverter have been assumed to be respectively 30, 10, 10, and 10 years respectively. Breakup costs of each component is mentioned in the Table 7.16.

Table 7.16: Capital cost estimation of experimental set-up

Sl. No.	Item	Cost (Rs/-) (Indian rupees)
1	Microwave oven 28L (LG convection type)	18, 000.00
2	Quartz reactor	5, 000.00
3	Nitrogen cylinder (47L), 99.99 purity	10, 000.00
4	Nitrogen flowmeter	2, 000.00
5	Glassware connection and two glass condensers	2, 000.00
6	pipes	500.00
7	Water pump (18 W)	200
8	Water storage tank 200 litre capacity	1, 500
9	Shredder units and sieves	1,300
10	Solar PV system	59,500
	Total	**1,00,000 ($ 1300)**

The line diagram of various cash flows at different intervals of time during the useful life of the experimental set-up is shown in figure 7.15. The life cycle cost assessment of the set up includes the following items.

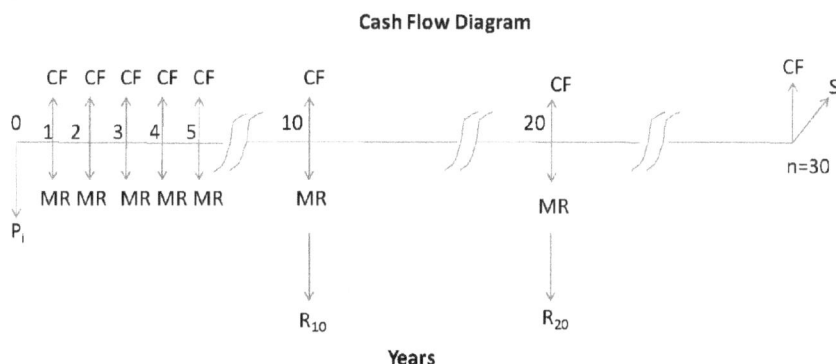

Fig. 7.15. Cash flow diagram for life cycle costs and revenues of the experimental set-up

Capital cost (P_i)

Capital cost is the total cost involved for all the components in the set-up. The capital cot is mentioned in the Table 7.16.

Maintenance and repair cost (P_{mr})

It is the cost, incurred during the operation of a technology/practice on annual basis. It includes the costs for repair, maintenance, consumables, nitrogen gas refilling and quartz reactor replacement as and when required. It is assumed to be 2.5 % of the total capital cost (P_i) in 1st year. The inflation in the cost of maintenance and repair (MR) has been considered at the rate of 1 % per annum. The net maintenance and repair cost in terms of present value (P_{mr}) is calculated by using the formula for present worth of geometric gradient series as per Eq. (7.31).

$$(P_{mr}) = \frac{A_1}{(i-g)}\left[1-\left(\frac{1+g}{1+i}\right)^n\right] \qquad \text{... (7.31)}$$

Where A_1 is the 1st year maintenance and repair cost (2.5 % of P_i), i = interest rate (4 %), g = inflation rate (constant percentage increase rate in each year) and n is the useful life of the set-up (30 years). The inflation rate (g) in annual maintenance and repair cost is assumed to be 1 %. The annual interest rate has been considered as 4 %, a subsidized interest rate offered by Government of India to promote the use of renewable energy applications in India.

P_{mr} = Rs. 48, 702

Replacement cost (P$_r$)

It is the cost incurred during replacement of microwave oven, battery bank and inverter at every 10 years interval. If R_{10} and R_{20} are the replacement costs after 10 and 20 years respectively, hence the net replacement cost in terms of present value is

$$(P_r) = R_{10} \times \left[\frac{1}{(1+i)^{10}}\right] + R_{20} \times \left[\frac{1}{(1+i)^{20}}\right] \qquad \text{... (7.32)}$$

Where (P_p) is the replacement cost in terms of present value. However, inflation rate per annum for the equipment, need to be replaced is considered as 2 % of their current price as per the present market survey. Hence, $R_{10} = R \times (1+0.02)^{10}$ and $R_{20} = R \times (1+0.02)^{20}$ where R is the capital cost included for microwave oven, battery bank and inverter. P_r = Rs.79,613.

Salvage value (P$_s$)

It is the cost incurred in demolition and disposal of the system. If S is the salvage value at the end of the useful life of the set up, then the salvage value in terms of present value is

$$(P_s) = S \times \left[\frac{1}{(1+i)^n}\right] \qquad \text{... (7.33)}$$

Salvage value is considered to be 20 % of capital cost (P_i), the depreciation of balance of system was considered equivalent to the rate of escalation in the price of the structural components, mainly steel and aluminium per kg. $P_s = 6167$

Hence the overall life cycle cost of the existing set-up in terms of present value is expressed as

$$P_{net} = P_i + P_{mr} + P_r - P_s = 1,00,000 + 48702 + 79613 - 6167 = \text{Rs. } 2,22,148 \qquad \text{... (7.34)}$$

Annualized uniform cost (C_A)

Annualized uniform cost is defined as the product of net present value of the system and Capital Recovery Factor (CRF) and can be written as

$$C_A = P_{net} \times \left[\frac{\left(i \times (1+i)^n \right)}{\left((1+i)^n - 1 \right)} \right] = \text{Rs. } 12,776 \qquad \text{... (7.35)}$$

Cost per unit electricity (C_u)

Cost per unit electricity (C_u) = (Annualized uniform cost) / (Annual energy output) =

$$\frac{C_A}{E_{out}} (Rs/kWh) = 12776/507 = \text{Rs. } 25 \qquad \text{... (7.36)}$$

Payback Period (n_{pp})

Payback period is the time (generally in years) required to recover the investment costs. Let us assume that net cash flow (CF) is same for each year, then the net present value (P_{net}) in terms of payback period can be expressed as

$$P_{net} = CF \times \left[\frac{(1+i)^{n_{pp}} - 1}{i(1+i)^{n_{pp}}} \right] \rightarrow n_{pp} = \frac{\ln\left[\dfrac{CF}{CF - i P_{net}} \right]}{\left[\ln(1+i) \right]} \qquad \text{... (7.37)}$$

The net cash flow per year (CF) includes the annual income from the pyrolysis products such as biochar, wood vinegar and tar produced in the present study.

In one batch of microwave pyrolysis for 450 g of rice straw, 90 g wood vinegar, 25 g tar and 165 g biochar were produced. The experiments were conducted 5 times in a day completing 300 days/year during 2019-20. The current prices per kg of wood vinegar, tar and biochar are respectively Rs.140, Rs.50 and Rs. 50.

The net cash inflow per year (CF) = Rs. 18900 +1875 + 12375 = Rs 33150/-

Payback period = 6.5 years

Carbon Dioxide (CO_2) emission, mitigation and carbon credit

CO_2 emission

On-farm burning of rice straw is now-a-days the most prevailing practice followed throughout India causing the emission of CO_2 to the environment. It has been reported that by burning 1 kg of rice straw, 1.5 kg CO_2 is released to the environment. Hence, through the practice of microwave pyrolysis of rice straw in a small scale, energy can be produced and emission of CO_2 from burning can also be prevented.

CO_2 mitigation

CO_2 mitigation can be accomplished by the use of solar PV system and preventing on-farm burning of rice straw. The electricity is mostly generated in India from the coal based thermal power plant. It has been reported that a coal based thermal power plant produces 0.98 kg of CO_2/kWh. In Indian conditions, considering 40% and 20 % losses of energy in transmission and distribution systems respectively, the amount of emission is 1.568 kg of CO_2/kWh of electricity generated from coal thermal power plant. Hence,

CO_2 emission mitigated (kg/life) = 1.57 (kg/kWh) $\times$ E_{out} (kWh/year) $\times$ life of system + 1.5 kg CO_2 emitted /kg of rice straw burning $\times$ kg of rice straw pyrolysed/year $\times$ life of system ... (7.38)

Carbon credit earned

One carbon credit represents the mitigation of 1 ton of CO_2 emission. The emissions of CO_2 due to various energy utilizations and generations pose a major contributing factor to the greenhouse effects. International treaties for mitigating emissions of greenhouse gases across the globe have therefore introduced carbon credit trading which is an administrative approach for controlling pollution by providing economic incentives for the efforts to reduce the emissions of pollutants. Carbon credits are therefore a tradable permit scheme prevailing among the countries included in the strategic planning of controlling the emissions of greenhouse gases. The amount of carbon credit earned in the present study due to the use of PV system and preventing rice straw burning can be calculated by the following equation:

Carbon credit earned = € 27/ton CO_2 emission mitigated by the set up (tons/life) ... (7.39)

1€ = Rs 83. The amount € 27 ($ 30) /ton represents the monetary value of mitigating one ton of CO_2 emission. The price of carbon credit is as per European Climate Exchange (www.ecx.eu). The summary of techno-economic environmental parameters of the existing set-up is presented in Table 7.17.

Table 7.17: Estimated techno-economic environmental parameters of the study

S.N.	Parameters	Unit	Calculated values
1	Capital cost of set-up	Rs.	1,00,000
2	Life of set-up	year	30
3	Net present value (considering the costs for initial investment, repair and maintenance, replacement of equipment and revenues from salvage value during life of set-up)	Rs.	2,22,148
4	Annualized uniform cost	Rs.	12,776
5	Cash inflow/year (from wood vinegar, tar and biochar)	Rs.	33,150
6	Electricity generated from PV system/year	kWh	507
7	Cost of unit electricity from PV system	Rs.	25
8	Payback period	year	6.5
9	Payback period (without PV system but with grid electricity)	year	3.12
10	Total reduction of CO_2 emission from the present study during life of the set-up	ton	55
11	Total reduction of CO_2 emission from the present study during life of the set-up (without PV system but with grid electricity)	ton	31
12	Carbon credit earned from the present practice during life of the set-up	Rs.	1,23,255 ($1600)
13	Carbon credit earned from the present practice during life of the set-up (without PV system)	Rs.	68,059 ($885)
14	Monthly income of the user, expected after payback period	Rs.	2500 ($33)

7.9 INFORMATION FOR THE USERS

This study focusses the valorisation of one of the most underutilized lignocelulosic biomasses namely rice straw, available abundantly and cheaply in the rice growing areas. Its utilization has been assessed with an integrated approach of improved method of thermochemical conversion via microwave pyrolysis powered through a reliable and sustainable source of energy by solar PV electricity. The current escalating concern of on-farm burning of rice straw has diverted the attention of planners, policy makers and environmentalists in search of suitable technology for proper disposal of this agro-residue with an objective of energy recovery out of it in an environmentally sustainable manner. Acute shortage of manpower, untimely rainfall, cyclone and other natural disasters during the harvesting of crop have forced the growers to use combine harvester for completing the work in short period leaving the straw as such in the field without any interest to collect and utilize it. They find no other way but to burn it in the field, an easiest and convenient approach for making the field ready for the next crop. However, massive burning of rice straw has now become a major source of greenhouse gas emission in the country like India. Efforts have therefore been made in the present study for effective utilization of this sustainable resource, from technical feasibility, economic viability, environmental sustainability and income generating points of view mostly for the rice growing areas. Emphasis has also been given to derive only the energy-rich solid biofuel (biochar) and bio-oil ignoring gaseous product during microwave pyrolysis due to their easy storage, transport and high commercial value and a good scope for direct marketing to earn income at the individual level. Microwave assisted pyrolysis, an efficient and controllable method of thermochemical conversion path has been followed in a small scale with the source of microwave power from solar PV electricity to evaluate the techno economic viability of the proposition at users' level. Pyrolysis being a very versatile process needs to optimize the various operating parameters such as microwave power, heating rate, temperature, reaction times etc. to maximize the yields of char, oil and gas depending on which one is required. Based on the preliminary study, the microwave power of 700 W and reaction time of 20 minutes have been chosen for higher yield of bio-oil through solar powered microwave pyrolysis of rice straw in laboratory scale experiments and the findings of the study showed that the practice is quite environment-friendly, energy-saving, affordable and income generating even in this small scale set-up, suitable for resource poor users too. The net energy recovery percentage with respect to the energy content of raw rice straw was calculated to be around 71 and 56 in case of using solar PV and grid electricity respectively. The payback period of the set-up is 6.5 and 3.12 years if solar PV system (involving more initial investment and operating cost compared to grid electricity) and grid electricity are respectively used. Expected monthly income of Rs. 2500/- may be achieved from the pyrolysis products of biochar, wood vinegar and tar by pyrolysing 675 kg of rice straw annually (2.25 kg per day @ 450 g per batch with 5 batches per day and 300 days per year). The potential of mitigating CO_2 emission during the total life time of 30 years from the existing set-up is 55 tons and the earning of carbon credit is Rs. 1,23,255/- considering the current price of carbon credit as € 27/ton. The added advantages of this technology are that it does not require any skilled manpower to operate. The set-up may also be used for pyrolysis of other agro-residues, the surplus electricity from the set-up may also be utilized in operating post-harvest machinery and its promotion would favour more trading of solar PV modules encouraging power generation through solar energy in a wider scale. This approach would thus make a strong foundation and become a sustainable solution for reducing the major concern of on-farm burning of rice straw and encouraging massive scale-up of the technology for inclusion of its findings in policy making and planning for waste to energy.

REFERENCES

Basu P. 2018. Pyrolysis. in Biomass Gasification, Pyrolysis and Torrefaction Practical Design and Theory Handbook. *Academic Press, London,* pp. 155-187.

Bu Q, Lei H, Wang L, Wei Y, Zhu L, Zhang X, Tang J. 2014. Bio-based phenols and fuel production from catalytic microwave pyrolysis of lignin by activated carbons. *Bioresource technology,* **162**: 142-147.

Bundhoo ZM. 2018. Microwave-assisted conversion of biomass and waste materials to biofuels. *Renewable and Sustainable Energy Reviews,* **82**: 1149-77.

Collard FX, Blin J. 2014. A review on pyrolysis of biomass constituents: Mechanisms and composition of the products obtained from the conversion of cellulose, hemicelluloses and lignin. *Renewable and sustainable energy reviews,* **38**: 594-608.

Fuad MA, Hasan MF, Ani FN. 2019. Microwave torrefaction for viable fuel production: A review on theory, affecting factors, potential and challenges. *Fuel,* **253**: 512-26.

Chen MQ, Wang J, Zhang MX, Chen MG, Zhu XF, Min FF, Tan ZC. 2008. Catalytic effects of eight inorganic additives on pyrolysis of pine wood sawdust by microwave heating. *Journal of analytical and applied pyrolysis,* **82**(1), 145-150.

Fodah AE, Ghosal MK, Behera D. 2020. Studies on microwave-assisted pyrolysis of rice straw using solar photovoltaic power. *BioEnergy Research,* 1-9.

Fan L, Chen P, Zhang Y, Liu S, Liu Y, Wang Y, Ruan R. 2017. Fast microwave-assisted catalytic co-pyrolysis of lignin and low-density polyethylene with HZSM-5 and MgO for improved bio-oil yield and quality. *Bioresource technology,* **225**: 199-205.

Ge S, Foong SY, Ma NL, Liew RK, Mahari WA, Xia C, Yek PN, Peng W, Nam WL, Lim XY, Liew CM. 2020. Vacuum pyrolysis incorporating microwave heating and base mixture modification: An integrated approach to transform biowaste into eco-friendly bioenergy products. *Renewable and sustainable energy reviews,* **127**: 109871.

Huang HJ, Yang T, Lai FY, Wu GQ. 2017 Komarov VV. 2012. Handbook of dielectric and thermal properties of materials at microwave frequencies. *Artech house.*

Huang HJ, Yang T, Lai FY, Wu GQ. 2017. Co-pyrolysis of sewage sludge and sawdust/rice straw for the production of biochar. *Journal of analytical and applied pyrolysis,* **125**: 61-68.

Komarov VV. 2012. Handbook of dielectric and thermal properties of materials at microwave frequencies. *Artech house.*

Kumar PS, Varjani SJ, Saravanan A. 2019. Advances in production and application of biochar from lignocellulosic feedstocks for remediation of environmental pollutants. *Bioresource technology,* **292**: 122030.

Lam SS, Mahari WA, Ok YS, Peng W, Chong CT, Ma NL, Chase HA, Liew Z, Yusup S, Kwon EE, Tsang DC. 2019. Microwave vacuum pyrolysis of waste plastic and used cooking oil for simultaneous waste reduction and sustainable energy conversion: Recovery of cleaner liquid fuel and techno-economic analysis. *Renewable and sustainable energy reviews,* **115**: 109359.

Matsakas L, Gao Q, Jansson S, Rova U, Christakopoulos P. 2017. Green conversion of municipal solid wastes into fuels and chemicals. *Electronic journal of biotechnology,* **26**: 69-83.

Menéndez JA, Arenillas A, Fidalgo B, Fernández Y, Zubizarreta L, Calvo EG, Bermúdez JM. 2010. Microwave heating processes involving carbon materials. *Fuel processing technology,* **91**(1):1-8.

Menéndez JA, Domínguez A, Fernández Y, Pis JJ. 2007. Evidence of self-gasification during the microwave-induced pyrolysis of coffee hulls. *Energy & Fuels,* **21**(1): 373-378.

Mushtaq F, Mat R, Ani FN. 2014. A review on microwave assisted pyrolysis of coal and biomass for fuel production. *Renewable and sustainable energy reviews,* **39**: 555-74.

Mutyala S, Fairbridge C, Paré JJ, Bélanger JM, Ng S, Hawkins R. 2010. Microwave applications to oil sands and petroleum: A review. *Fuel processing technology,* **91**(2): 127-35.

Nhuchhen DR, Afzal MT, Dreise T, Salema AA. 2018. Characteristics of biochar and bio-oil produced from wood pellets pyrolysis using a bench scale fixed bed, microwave reactor. *Biomass and bioenergy,* **119**: 293-303.

Omar R, Robinson JP. 2014. Conventional and microwave-assisted pyrolysis of rapeseed oil for bio-fuel production. *Journal of analytical and applied pyrolysis,***105**: 131-42.

Omoriyekomwan JE, Tahmasebi A, Yu J. 2016. Production of phenol-rich bio-oil during catalytic fixed-bed and microwave pyrolysis of palm kernel shell. *Bioresource technology,* **207**: 188-196.

Parvez AM, Wu T, Afzal MT, Mareta S, He T, Zhai M. 2019a. Conventional and microwave-assisted pyrolysis of gumwood: A comparison study using thermodynamic evaluation and hydrogen production. *Fuel processing technology,* **184**: 1-11.

Ravikumar C, Kumar PS, Subhashni SK, Tejaswini PV, Varshini V. 2017. Microwave assisted fast pyrolysis of corn cob, corn stover, saw dust and rice straw: Experimental investigation on bio-oil yield and high heating values. *Sustainable materials and technologies,* **11**: 19-27.

Sahoo D, Remya N. 2020. Influence of operating parameters on the microwave pyrolysis of rice husk: biochar yield, energy yield, and property of biochar. *Biomass Conversion and Biorefinery.*

Salema AA, Afzal MT, Bennamoun L. 2017. Pyrolysis of corn stalk biomass briquettes in a scaled-up microwave technology. *Bioresource technology,* **233**: 353-62.

Tirapanampai C, Phetwarotai W, Phusunti N. 2019. Effect of temperature and the content of Na2CO3 as a catalyst on the characteristics of bio-oil obtained from the pyrolysis of microalgae. *Journal of analytical and applied pyrolysis,* **142**: 104644.

Uddin MN, Techato K, Taweekun J, Rahman MM, Rasul MG, Mahlia TM, Ashrafur SM. 2018. An overview of recent developments in biomass pyrolysis technologies. *Energies,* **11**(11): 3115.

Wu C, Budarin VL, Gronnow MJ, De Bruyn M, Onwudili JA, Clark JH, Williams PT. 2014. Conventional and microwave-assisted pyrolysis of biomass under different heating rates. *Journal of analytical and applied pyrolysis,* **107**: 276-283.

Yin C. 2012. Microwave-assisted pyrolysis of biomass for liquid biofuels production. *Bioresource technology,* **120**: 273-284.

Zhao X, Wang W, Liu H, Ma C, Song Z. 2014. Microwave pyrolysis of wheat straw: product distribution and generation mechanism. *Bioresource technology,* **158**: 278-85.

Zhang Y, Chen P, Liu S, Fan L, Zhou N, Min M, Cheng Y, Peng P, Anderson E, Wang Y, Wan Y. 2017a. Microwave-assisted pyrolysis of biomass for bio-oil production. In Pyrolysis, Samer M. *IntechOpen, London, UK,* pp. 129-166.

Zhang H, Ma Y, Shao S, Xiao R. 2017b. The effects of potassium on distributions of bio-oils obtained from fast pyrolysis of agricultural and forest biomass in a fluidized bed. *Applied energy,* **208**: 867-877.

Zhang Y, Cui Y, Liu S, Fan L, Zhou N, Peng P, Wang Y, Guo F, Min M, Cheng Y, Liu Y. 2020. Fast microwave-assisted pyrolysis of wastes for biofuels production-A review. *Bioresource technology,* **297**: 122480.

Zhou H, Long Y, Meng A, Chen S, Li Q, Zhang Y. 2015. A novel method for kinetics analysis of pyrolysis of hemicellulose, cellulose, and lignin in TGA and macro-TGA. *RSC Advances,* **5**(34): 26509-16.

SOLAR PHOTOVOLTAIC WATER PUMPING SYSTEM

8.1 PERSPECTIVES OF SOLAR PHOTOVOLTAIC (SPV) PUMP

Renewable energy sources in general and solar energy in particular has the potential to provide energy services with zero or almost zero emission. The solar energy is abundant and available freely in nature. The solar photovoltaic (SPV) powered water pumping system can be used anywhere but it is more appropriate and applicable in rural areas where the availability of grid-electricity is quite erratic and even, many places are non-electrified. Due to the geographical position in the globe, India is a tropical country and experiences ample sunshine hours throughout the year which makes it ideal location for utilization of solar energy. Small farms, villages and animal herds in the developing countries require hydraulic output power of less than a kilowatt. Many of the potential users are even too far from an electrical grid to economically tap that source of power. The engine-driven pumping system tends to be very expensive as well as unreliable due to the rising price of fuels and insufficient maintenance as well as repair facilities. A SPV water pump is therefore the right option for its use in rural areas particularly for irrigating land in crop cultivation.

8.2 SOLAR PHOTOVOLTAIC (SPV) WATER PUMP

A SPV water pump is a normal pump with an electric motor. Electricity for the motor is generated onsite through a solar panel which converts solar energy to direct-current (DC) electricity. Because the nature of the electrical output from a solar panel is DC, a SPV pump requires a DC motor if it is to operate without additional electrical components. If a pump has an alternating-current (AC) motor, an inverter would be required to convert the DC electricity produced by the solar panels to AC electricity. Due to the increased complexity and cost, and the reduced efficiency of an AC system, most SPV pumps have DC motors. DC motor is used to drive the water pump. The system consists of frame of solar water pump, DC motor, pump, solar panel, suction pipe, delivery pipe, On/Off control switch and water tanks. SPV pump generally works on sunny days. For the operation of SPV pump during off-sunshine hours and cloudy days, there is the requirement of storage device (batteries) which is able to operate the pump to fulfil the water requirements for about three or four days. The components comprising a SPV pump therefore depend on whether the pumping system is a direct-drive system or a battery-operated system (fig. 8.1). In a direct drive system, solar energy incident on the solar panel converts the radiant energy into direct current and drives the DC motor pump through which water is transferred from water source to water storage device. In case of battery operated system, solar energy incident on the solar panel converts sunlight into direct current which is stored in the battery and the stored electricity is supplied from battery to DC motor to drive pump through which water is transferred from water source to water storage device. Both systems incorporate a water-storage facility, but the water-storage component of the battery-powered system can be reduced in size relative to the direct-drive system due to the fact that the batteries effectively provide some storage of electricity. For the direct-drive systems, it is important to match the power output of the solar array with the power requirements of the pump to maximize efficiency. For the battery-powered systems, it is important to use good quality deep-cycle batteries and to incorporate electrical controls such as blocking diodes and charge regulators to protect the batteries. Some solar panels incorporate the electrical control elements. SPV pumping systems can be configured to meet a wide variety of demands. The amount of water a SPV pump can deliver is a function of how far the water has to be lifted, the distance it has to travel through a delivery pipe (and the size of pipe), the efficiency of the pump being used and how much power is available to the system. Power can be increased by sizing the solar panels. One of the main advantages of a SPV pumping system is its simplicity and durability. The purpose of using solar water pump is to (i) replace the non-renewable energy resources to renewable energy resources (ii) utilize solar energy and (iii) produce electricity in remote area.

Advantages

(i)   This (SPV) water pumping system has the added advantage of storing water for use when the sun is not shining, eliminating the need for battery, simplicity and reducing overall system costs.

(ii) The source of energy is an abundant renewable resource.

(iii) It is a non-polluting and low carbon technology, which means that it does not release greenhouse gases.

(iv) It is a noiseless technology as there are few moving parts involved in energy generation.

(v)   This technology requires low maintenance because of lack of moving parts.

(vi) It can be installed on modular basis and expanded over a period of time.

(vii) Most viable alternative for providing electricity in remote and rural areas as it can be installed where the energy demand is high and can be expanded on modular basis.

Block diagram of a direct coupled PV DC water pumping system.

Block diagram of a PV AC water pumping system.

Block diagram of a PV water pumping system with battery storage.

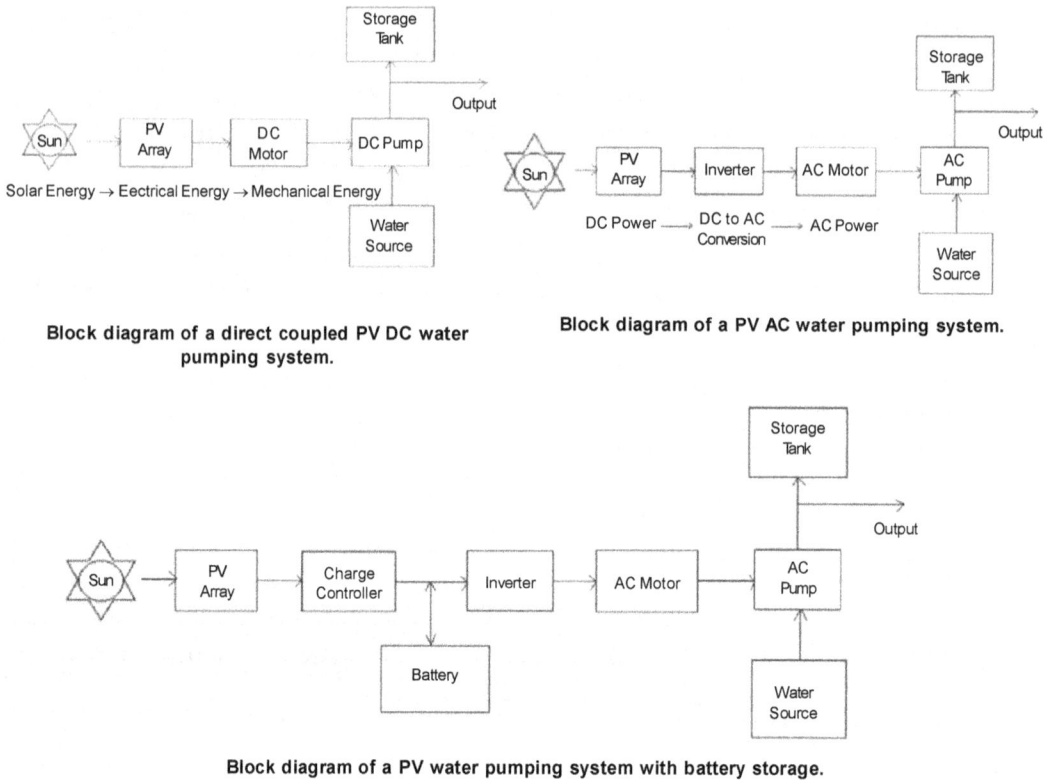

Fig. 8.1. Various arrangements of SPV water pumping system

Limitations

(i)   The efficiency levels of conversion from sunlight to electricity is in the range of 10 to 17%, depending on the technology used.

(ii) The initial investment cost of this technology is high, though it is decreasing day by day.

(iii) Solar energy is available only during daytime. Most load profiles indicate the peak load in the evening/night time. This necessitates expensive storage devices like battery, which need to be replaced in every 3 to 5 years. Generally, the cost of the batteries is about 30 to 40% of the system cost.

(iv) As the efficiency levels are low, the space required is relatively high. For instance, with the existing levels of technologies, the land required for installing a 1 MW solar PV power plant is about 4-6 acres. However, research is going on to increase the efficiency levels of the cell.

(v)   It is heavily dependent on atmospheric conditions.

Applications

A SPV pump can be used in various fields such as (i) for supply of drinking water (ii) village water supply (iii) livestock watering (iv) irrigation (v) process industry etc.

8.3 DESIGN OF SOLAR PV WATER PUMPING SYSTEM

Water and energy are the key drivers of the agricultural sector which is at present facing severe energy and water crisis. Increasing crop production per unit area for the increasing population of the world is the dire need of time. Thus, sustainable approaches are required to ensure food security and energy security. One of the sustainable approaches is the SPV water pumping system. It is a promising alternative to the conventional pumping systems and a cost-effective application especially in remote off-grid areas of developing countries. In recent years, through continuous improvement, the SPV pumping system has been widely used in agricultural, industrial, and domestic sectors. The use of solar water pump for water pumping is one of the most attractive applications particularly in the agricultural sector. In general, a solar water pumping system consists of PV panel, motor, pump and storage tank. The block diagram of a solar PV water pumping system is presented in Fig. 8.2. The storage tank can be thought of as an energy storage media like batteries. Therefore, the use of batteries is not required for water pumping application. Also a DC motor can directly be coupled with a solar panel, avoiding the use of any inverter. An AC motor can also be used with an inverter which converts DC power of a PV panel into AC power. Additionally, a PV water pumping system can also have a Maximum Power Point Tracking (MPPT) (explained in next section) device to match the PV panel output impedance with that of motor to extract maximum of the available power throughout the day. Similar to solar home lighting system, a PV pumping system can be designed for sizes ranging from a very small water pumping requirement for drinking water to large volume water requirement for irrigation purpose.

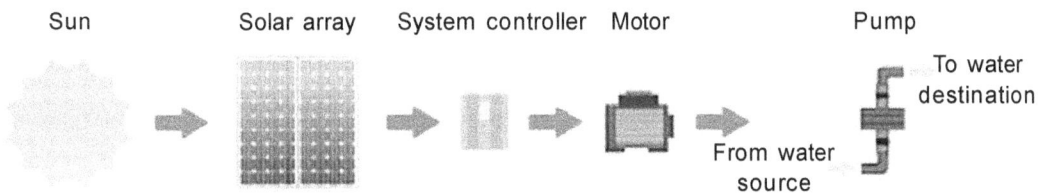

Fig. 8.2. Block diagram of a solar PV water pumping system

Before looking at the design aspects of water pumping system, some of the basic terms need to be discussed.

Daily water requirement (in l/day or m³/day): It is an important parameter. The size and cost of the water pumping system depends on the amount of water required per day. It should be kept in mind that the water requirement may vary daily, over the months or seasons. If the amount of water used every day varies, then weekly average or monthly average can be considered for calculation. For a reliable sizing of SPV water pumping system, in worst case, maximum water requirement needs to be considered.

Total dynamic head (TDH) (meter): TDH primarily consists of two parameters, total vertical lift and total frictional losses (fig.8.3). The total vertical lift is the sum of elevation, standing water level and drawdown. The elevation is the difference between the ground and the height at which the water is discharged. Standing water level is the difference between the ground surface and the water level in the well, when the well is fully charged condition. Drawdown is the height by which the standing water level drops due to pumping.

Frictional loss (equivalent meter): Frictional loss is the pressure required to overcome friction in the pipes from water pump outlet to the point of water discharge. It is given in equivalent meters and added to the total vertical lift for TDH

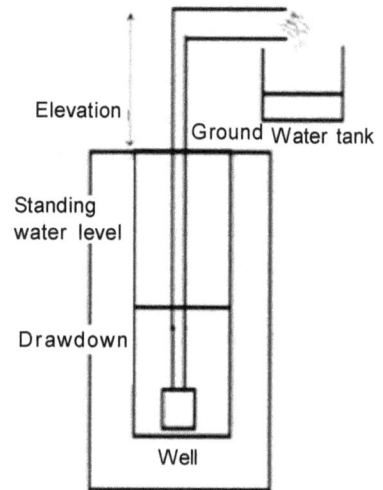

Fig. 8.3. A typical solar PV water pumping system

calculation. The friction loss depends on many factors such as size of the pipe, flow rate, type of fittings, number of bends etc. Usually relevant tables are used to calculate the frictional loss. But if the water discharge point or tank is close to the well, then approximation is used. If the tank is within 10 m of well, then frictional loss is taken as 5 % of total vertical lift.

All the above terms are useful while doing calculations for the design of the water pumping system.

The overall sizing of SPV water pumping system can be divided into the following 5 steps.

(i)   Determine the amount of water required per day as per the applications

(ii) Determine the total dynamic head (TDH) for water pumping

(iii) Determine the hydraulic energy required per day (watt-hour/day) to pump the required amount of water

(iv) Determine the solar radiation available at given location [in terms of equivalent of peak sunshine radiation ($1000 \ W/m^2$) hours for which the solar panel is characterized, typically this number is 5 to 8 hours varying from season to season and location to location

(v)   Determine the size and number of solar PV panel required, size of motor, motor efficiency and other losses.

8.3.1 An Example of Sizing Electric Motor for SPV Water Pumping System

Design of a solar PV system for pumping 25000 litres of water every day from a depth of about 10 metre is considered.

The data required for calculation and the steps of calculation are as follow:

Amount of water to be pumped per day = 25000 litre=25 m^3

Total vertical lift= 12 metres (5m-elevation,5 m-standing water level, 2m-drawdown)

Water density= 1000 kg/m^3, Acceleration due to gravity, g= 9.8 m/s^2, Solar PV module used = 75 W$_p$. Operating factor= 0.75 (PV panel mostly does not operate at peak rated power)

Pump efficiency = 30% or 0.30 (it is typically between 0.25 to 0.40)

Mismatch factor = 0.85 (PV panel usually don't operate at maximum power point, should be 1 if MPPT circuit is used)

Calculations for SPV water pumping system

Step 1: Determination total daily water requirement

Daily water requirement = 25 m^3/day

Step 2: Determination total dynamic head

Total vertical lift= 12 m

Frictional losses= 5% of the total vertical lift = 12 × 0.05 = 0.6 metre

Total dynamic head (TDH) = 12 + 0.6 =12.6 m

Step 3: Determine the hydraulic energy required per day

Hydraulic energy required to raise water level = Mass × g × TDH = density × volume × g × TDH = (1000 kg/m^3) × (25 m^3/day) × (9.8 m/s^2) × 12.6 m , (multiply by 1/3600 to convert second in hours) = 857.5 watt-hour/day.

Step 4: Determine the solar radiation data

The solar radiation may be taken as 6 h/day (peak of 1000 W/m^2 equivalent), actual day length is longer.

Step 5: Determine the number of PV panels and pump size

$$\text{Total wattage of PV panel} = \frac{Total\ hydraulic\ energy}{No.\ of\ hours\ of\ peak\ sunshine\,/\,day} = 857.5/6 = 142.9 \text{ watt}$$

Considering system losses = (Total PV panel wattage)/(Pump efficiency) × Mismatch factor = 142.9/0.3 0.85 = 560 Watt

Considering operating factor for PV panel = Total PV panel wattage after losses/ operating factor = 560/0.75 = 747.3 watt

Number of 75 W$_p$ solar PV panels required = 747.3/75 = 9.96 =10 (round figure)

Power rating of the motor = 747.3/746 ≈ 1 hp motor.

In this way, a solar PV water pumping system can be designed. The above design has been done assuming the use of a DC motor. A system, can also be designed for an AC motor but one must consider inverter and its efficiency in the calculations. Also, the cost of the solar PV irrigation system can be estimated by considering the individual component cost, e.g. cost of solar panel, cost of motor and cost of pump and wiring cost.

8.3.2 Techno-economic Analysis of SPV Water Pumping System: An Example

The techno-economic analysis of solar pump (1hp) has been done for cultivating okra in 1 acre (0.4 ha) of land. The hourly cost of operating 1 hp solar pump (DC) and its pay-back period has been calculated. Analysis has been done on the basis of present worth of the setup, its present value of annual maintenance cost, annualized cost and hourly cost of operation following the cash flow diagrams and considering the inflation in the price of the components during the life of the set up. Submersible screw type pump; Head: 70-125 metre; Pumping water capacity: 18,000-22,000 litres per day.

Initial investment of SPV water pumping for drip irrigation system (1 hp DC solar pump)

S.N.	Item	Cost (Rs.)
1	PV module of 300 W_p @ Rs. 60 per watt, and total 3 nos. of modules	54,000
2	Boring of deep well (200 feet)	60,000
3	Drip system for 1 acre of land	1,00,000
4	1 hp solar pump and controller	75,000
5	Transportation and installation cost	11,000
	Total	**3,00,000**

$P_{net} = P_i + P_{drip \ (replacement \ cost)} + P_M - P_S$

P_{net} = Net present worth

P_i = Initial investment = Rs. 3, 00,000

Initial cost of drip irrigation system for 1-acre land = Rs. 1, 00,000

$P_{drip \ (replacement \ cost)}$ = Present worth of replacement of drip system after each 10 years

P_M = Present value of annual maintenance cost

The cost of the drip system is assumed to be increasing @ 2% of the present cost of drip per annum as per market survey

P_{dripN1}=New set up of drip system after 10 years = $1,00,000 \times (1+0.02)^{10}$ = Rs. 1,21,899

P_{dripN2} = New set up of drip after 20 years = $1,00,000 \times (1+0.02)^{20}$= Rs. 1,48,594

P_S = Present worth of salvage value

Salvage value of the system = S = 10% of initial investment

Life span of solar pump = 25 years

Life of drip system = 10 years

Interest rate = i = 10%

Annual maintenance cost = 1 % of total cost of the system = M= Rs. 3000

Present worth of drip (with replacement of 10 years) $= \dfrac{121899}{(1+i)^{10}} + \dfrac{148594}{(1+i)^{20}}$ = Rs. 69084

Present value of annual maintenance cost $= P_M = M\left[\dfrac{(1+i)^{25} - 1}{i(1+i)^{25}}\right]$ = Rs. 27230

Present Salvage value $= \dfrac{S}{(1+i)^{25}}$ = Rs. 2770

P_{drip} (replacement cost) $= \dfrac{P_{drip\,N1}}{(1+i)^{10}} + \dfrac{P_{dripN2}}{(1+i)^{20}}$

$P_{Net} = P_i + P_M + \dfrac{P_{dripN1}}{(1+i)^{10}} + \dfrac{P_{dripN2}}{(1+i)^{20}} - \dfrac{S}{(1+i)^{25}}$ = Rs. 393544

Energy output from the system = Net output (746 watt) × 6 hours/day × 300 sunny days/ year = 1342 kWh/year

Energy cost = Rs. 32.29 /kWh

Cost per one hour of operation = Rs.32.29 × 0.746 = Rs. 24.00/hour

8.4 MAXIMUM POWER POINT TRACKING

An MPPT, or maximum power point tracker is an electronic DC to DC converter that is used to extract maximum of the available power from the PV panel and to supply to the battery or the load at any time during its operation. With the help of MPPT, about 20 to 30 % more power can be obtained. Maximum power varies with solar radiation, ambient temperature and solar cell temperature. Maximum power point tracking regulates the voltage and current going from the solar PV system to the load. It enables to supply full rated power from the panels to the load, rather than wasting a portion of it as heat. Compared to a standard (non MPPT) charge controller, it provides 20-30% more power.

8.4.1 An Analogous Explanation for Working of MPPT

An MPPT tracker is analogous to a thumb placed over a garden hose. If you put your thumb over part of the opening of the hose (adding resistance to the circuit), the pressure (voltage) goes up and the stream of water flows faster, but less water (current) is getting through. If you completely cover the opening, nothing gets through. If you remove your thumb entirely, the maximum flow rate gets through, but the stream falls slowly at your feet. That's the basic mechanism of the MPPT tracker which performs the work of varying the resistance in the circuit to modify current and voltage. It may now be imagined that there are hundreds of pumps (solar panels) upstream of the hose and they are delivering water (energy). Further, the problem may arise when some of these pumps go offline at certain parts of the day (partial shading of the array). So the force behind the delivery of water will be constantly varying. But the purpose is to wash the car, which is situated about 15 feet away from the water source. You need to keep moving your thumb as the upstream pump force varies in order to avoid undershooting or overshooting the car. The "car" in this case is the Maximum Power Point. For any array of solar panels, there is a configuration of current and voltage that aligns with maximum power availability for the load:

8.5 REVIEW OF SOME DEVELOPMENTS IN SOLAR PHOTOVOLTAIC WATER PUMPING SYSTEM

The approaches for energy security through solar photovoltaic power system and improved water use efficiency measures through micro-sprinkler irrigation system compared to traditional flooded method of irrigation, have thus been thought up now-a-days among the researchers, scientists and agriculturists not only for achieving assured water availability to the crops but also protecting the environment against the release of greenhouse gases and noise pollution by the use of rising diesel pump sets in the state. Attempts have already been made by some researchers in assessing the viability of solar PV water pumping system for domestic drinking water and irrigation purposes. Some developments in this area have been discussed in this section.

Hamidat et.al 2003 studied on small-scale irrigation with photovoltaic water pumping system in Sahara regions. The authors have developed a mathematical program to test the performance of photovoltaic arrays under Saharan climatic condition. Their work showed that it is possible to use a photovoltaic water pumping system for low heads for small scale-irrigation of crops in Algerian Sahara regions. Thus, the solar photovoltaic (SPV) water pumping system could easily cover the daily water need rates for small-scale irrigation with an area smaller than 2 ha. They also concluded that the SPVPS (solar photovoltaic pumping system) could improve the living condition of the farmer with the development of local farming and thus the migration of rural work force would be brought to an end.

Kala Meah et. al. 2008 studied on solar photo voltaic water pumping for remote locations in rural western US. They realized that solar photo voltaic water pumping system (SPVWPS) is a cost effective and environmental friendly way to pump water in remote locations where 24 hours electrical service is not necessary and maintenance is an issue. From their survey, it was indicated that a total of 88 number of solar photo voltaic water pumping systems are being installed in all 23 countries of the United States, of which 75 systems are in operation till 2005. They have observed that drought affected areas like Wyoming, Montana, Idaho, Washington, Oregon and part of Texas could use solar photo voltaic water pumping systems to improve the water supply to livestock in remote locations. They have been convinced that successful demonstration of these systems is encouraging other ranchers to try this relatively new technology as another viable water supply option. They concluded that SPVWPS had excellent performance in terms of productivity, reliability and cost effectiveness and the system could reduce the CO_2 emission considerably over its 25-year life span.

Kala Meah et.al 2008 studied on solar photovoltaic water pumping opportunities and challenges in United States. According to their views, they stated that some improvements could be done to lower the capital investment cost and to reduce the cost of operation and maintenance services using local level operation and maintenance. The authors have demonstrated that by using local resources such as skills, materials and finances, the solar photovoltaic water pumping system (SPVWPS) could be economically viable in developing countries and competitive with the conventional diesel generator water pumping systems. They concluded that the SPVWPS should be compatible with the local culture and practices to satisfy local wishes and needs, which also could be achieved by using local resources.

Leah C. Kelley et.al 2010 studied on the feasibility of solar powered irrigation in the United States. They developed a method for determining the technical and economic feasibility of photovoltaic power irrigation systems applicable to any geographic location and crop type in USA and applied it to several example cases. According to the opinion of authors, the results of technical feasibility analysis agreed with the results obtained from past studies and also showed that there is no technological barrier to implementation of SPV irrigation if land is available for installation of solar panels. The results of economic feasibility study suggested that the price of diesel has increased sufficiently within the last ten years to make SPV irrigation economically feasible, despite the high capital costs of photovoltaic systems. The authors concluded that as the price of the solar panels is decreasing, the capital costs would decrease making SPV systems even more economically attractive.

Gopal, C. et al. 2013 reviewed the research developments on renewable energy source water pumping systems referring 168 research papers across the globe. They concluded that solar photovoltaic water pumping systems are identified as an alternative source for replacing conventional pumping methods. The integration of renewable energy sources with water pumping systems plays a major role in reducing the consumption of conventional energy sources and their environmental impacts, particularly for irrigation applications. The solar photo voltaic water pumping systems are the most widely used for irrigation and domestic applications, followed by wind energy water pumping systems. The solar thermal and biomass water pumping systems are less popular due to their low thermal energy conversion efficiencies.

Narela et al. 2013 studied the feasibility of solar photovoltaic water pumping system for irrigating banana plants. They presented the design and economic analysis of efficient solar PV water pumping system for irrigation of banana. The system was designed and installed in solar farm of Jain Irrigation System Limited (JISL), at Jalgaon (Maharashtra). The study area falls at 21° 05' N – latitude, 75° 40'E–longitude and at an altitude of 209 m above mean sea level. The PV system sizing was made in such a way that it was capable of irrigating 0.165 ha of banana plot with a daily water requirement of 9.72 m^3/day and total head of 26 m. Also, the life cycle cost (LCC) analysis was conducted to assess the economic viability of the system. The results of the study encouraged the use of the PV systems for water pumping application to irrigate orchards. The installed system of solar PV water pumping system was capable of irrigating 0.165 ha area of banana crop within 6.02 hrs with a daily water requirement of 9.72 m^3/day.

8.6 A CASE STUDY OF SPV WATER PUMPING BASED MICRO-IRRIGATION IN VEGETABLE CULTIVATION (GHOSAL et al., 2020)

8.6.1 Need of SPV Water Pumping System in Vegetable Cultivation

The growing demands of energy and water particularly in the present agricultural sector have necessitated the adoption of reliable, environment-friendly and water saving technologies so as to combat against the energy crisis and water stress in near future. It has been established that conventional sources of energy like oil, gas, coal etc. will not be able to provide the desired levels of energy security to mankind in foreseeable future. Hence, there is a global consensus for exploitation and utilization of different renewable energy resources. The search for new options should be eco-friendly as well as abundant in nature. Among the different available renewable energy resources, solar energy seems to be more promising and sustainable. Solar powered agricultural irrigation may be an attractive application for renewable energy in replacing fossil fuel powered irrigation devices to achieve energy security. The use of solar photovoltaic systems may provide good solution not only for all energy related problems of the present society but also perform excellently in terms of productivity, reliability, sustainability and environmental protection ability. Solar photovoltaic water pumping systems can provide water for irrigation without the need for any kind of fuel or the extensive maintenance as required by diesel and electric pump sets. Therefore, an attempt was made in this study to develop an affordable solar PV water pumping system for irrigating vegetable crops in the state of Odisha, India. Micro-irrigation method through sprinkler system was integrated with the solar PV device to achieve judicious utilization of water. Monthly income of Rs. 15,000/- throughout the year was possible by adopting remunerative tomato cultivation in 1 acre of land both during rabi and summer seasons in the year 2017 in coastal region of Odisha. Pay- back period of the developed set up was calculated to be only a half year, due to which, it may be easily accepted by the small and marginal farmers of the state inspite of its high initial cost. The popularization of this technology would not only achieve assured water availability to the crops with improved water use efficiency measures by micro-sprinkler irrigation system compared to traditional flooded method of irrigation but also protect the environment against release of greenhouse gases and noise pollution by the use of rising diesel and electric pump sets in the state.

8.6.2 Use of SPV Water Pumping System in Agriculture: The Need of Hour

Energy demand is at present growing exponentially in each segment of the national developments due to the continuous growth and expansion in different sectors like industry, agriculture, irrigation, transportation, communication, housing, health, education, city modernization, entertainment etc. To meet the increasing demands of energy, the share of coal based power plants for power generation in India is also rising day by day causing severe environmental hazards and thus global warming by releasing a considerable amount of greenhouse gases to the atmosphere. The only alternative in this context is to supplement to the existing power sector with non-conventional energy sources. Among the non-conventional energy sources, the solar energy appears to be an attractive and viable proposition because of the abundant and free availability of sun shine in the tropical areas. Moreover, electricity from solar photovoltaic system is now gaining more importance because of the rapid decline in the cost of solar PV modules through advances in research and development in this area. The attention of planners, policy makers and researchers is also now diverted to the applications of solar photovoltaic system for water pumping in irrigation sector due to recent increased water demands in agricultural sector and availability of water has become more crucial than ever before. In India, electrical and diesel powered water pumping systems are most widely used for irrigation applications. A source of energy to pump water is also a big problem in developing countries like India. Developing a grid system is often too expensive because rural villages are frequently located too far away from existing grid lines. Even if fuel is available within the country, transporting that fuel to remote and rural villages can be difficult. There are no roads or supporting infrastructure in many remote villages. The use of renewable energy is therefore of utmost importance for water pumping applications in remote areas of many developing countries. Transportation of renewable energy systems, such as photovoltaic (PV) pumps, is much easier than the other types because they can be transported in pieces and reassembled on site (Khatib, 2010). Photovoltaic (PV) energy production is recognized as an important part of the future energy generation. Because it is non-polluting, free in its availability, and is of high reliability. These facts make the PV energy resource more attractive for many applications, especially in rural and remote areas of the developing countries like India. SPV water pumping has been recognized as suitable for grid-isolated rural locations in places where there are high levels of solar radiation. The state Odisha in India also receives a good amount of solar radiation for about 4-5 hours in a day over a period of nearly 300 days in a year (Solar Policy 2013, Govt. of Odisha). Solar photovoltaic water pumping systems can provide water for irrigation without the need for any kind of fuel or the extensive maintenance as required by diesel and electric pump sets. They are easy to install and operate, highly reliable, durable and modular, which enable future expansion. They can be installed at the site of use, avoiding the spread of long pipelines and infrastructures (Andrada and Castro, 2008).

Odisha is blessed with highly fertile soil due to flowing of many rivers through it namely the Mahanadi, the Baitarani, the Bramhani, the Subarnarekha, the Budhabalanga, the Bansadhara etc. (Economic Survey of Odisha, 2013). As Odisha receives an average annual rainfall of 1500 mm, there is no dearth of water resources. Farmers of the state grow different vegetable crops round the year using hand pump, electric and diesel pumps for lifting of irrigation water. Lifting of water by hand pump is a most tedious and labour consuming operation. Similarly, non-availability and erratic supply of grid connected electric supply in the remote areas and the rising cost of diesel day by day necessitate the search of a sustainable source of power for assured irrigation particularly for vegetable cultivation which is now-a-days more remunerative and profitable. Cost of lifting water in the above pumping systems is many folds compared to lifting water by solar photo voltaic water pumping system (Leah C. Kelley et al., 2010). Development of an affordable, durable and with a very little repair and maintenance of the device, would be preferred by the small and marginal farmers of Odisha, India. Installation of electric pump sets is not at all possible at most of the locations as the agricultural fields are far away from the electric grid station. In addition, the electric tariff is increasing in every year and thus increasing the cost of water pumping operation. Further, the repair and maintenance cost of electric motor operated pump sets is generally more than that of solar photo voltaic water pumping system (Sako et al. 2011). When not much research work was conducted on solar photo voltaic water pumping system, then, diesel pumping system was very popular among the farming community due to its low cost and portability. During this time, the diesel cost was also cheaper. But it causes a lot of environmental pollution and global warming by the emission of substantial quantity of CO_2 into the atmosphere. The repair and maintenance cost of diesel pump set is also more than that of solar photo voltaic water pumping system.

Hence, solar photo voltaic water pumping system is today a viable option left for the farming community as its pumping cost is cheaper compared to electric and diesel pump sets. Moreover, the risk of environmental pollution is less and its repair and maintenance cost is very low. It can be installed at any location as per the desire of the farmers as solar energy is available profusely and free of cost. Hence, the use of solar powered micro-irrigation system is also the need of the hour looking into the present day's concerns of energy crisis and water scarcity particularly in agricultural sector. Therefore, an attempt has been made to develop an affordable solar water pumping system along with micro-irrigation device for irrigating vegetable crops in the coastal regions of Odisha and to study its feasibility among the farmers of the state for growing vegetable crops in their fields for strengthening their livelihoods and socio-economic status. The specific objectives of popularizing the SPV water pumping system are to (i) develop an affordable solar photo voltaic powered water pumping device integrated with sprinkler system for irrigating vegetable crops (ii) study the feasibility and performance study of solar photo voltaic powered sprinkler irrigation system in off-grid remote areas and (iii) the techonomic analysis of the device

8.6.3 Experimental Location

Development of solar photovoltaic (SPV) micro-sprinkler irrigation system has been made for cultivating tomato in 1 acre (0.4 ha) of land to achieve secured irrigation and to improve water use efficiency mostly in vegetable cultivation. The details of the design and developments are mentioned below. The experiments were carried out during the year 2017 in the Central Farm of Orissa University of Agriculture and Technology (OUAT), Bhubaneswar, Odisha which lies at the latitude of $20\ ^0\ 15'$ N and longitude of $85\ ^0\ 52'$ E and coming under warm and humid climatic condition. Tomato was cultivated both in rabi and summer season.

8.6.3.1 Design of Solar Photovoltaic Powered Sprinkler Irrigation System

(A) Water Requirement for Vegetable Crop

$W_r = (\text{Crop area} \times PE \times P_c \times K_c \times w_a)/E_u$

W_r = Peak water requirement (m^3/day); crop area (m^2); PE = Pan Evaporation rate (mm/day) converted to m/day; P_c = Pan Coefficient (0.7 to 0.9); K_c = Crop Coefficient (0.8 to 1); w_a = wetted area (%) (90 % for sprinkler irrigation); E_u = Emission uniformity of sprinkler irrigation (Approx. 0.8)

Putting the values of crop area = 4000 m^2; PE = 8 mm/day; P_c = 0.85; K_c = 0.9; w_a = 0.9 and E_u = 0.8

W_r = 27.54 m^3/day (27540 lit/day)

Taking irrigation interval to be 2 days, W_r = 27540/2 = 13770 lit/day = 13.77 m^3/day ≃ 14 m^3/day

(B) Sizing of SPV Module for above water requirement

$E = (\rho\, g\, H\, V)/(3.6 \times 10^6)$

E = Hydraulic energy required (kWh/day); ρ = density of water (1000 kg/m^3); g = gravitational acceleration (9.81 m/s^2); H = Total hydraulic head (m) (15 m in this case); V = volume of water required (14 m^3/day in this case)

Putting all values, E = 0.572 kWh/day = 572 Wh/day

Assuming actual sun shine hours in a day = 6 hours, the total wattage of PV module = 572/6 = 95.33 watt

Assumptions:

i. Operating factor = 0.75-0.85 (PV panel mostly does not operate at peak rated power)

ii. Pump efficiency = 70-80 % (can be taken 75 %)

iii. Motor efficiency = 75-85 % (can be taken 80 %)

iv. Mismatch factor = 0.75-0.85 (PV panel does not operate at maximum power point)

Considering system losses, wattage requirement = (Total PV panel wattage) / (pump efficiency x mismatch factor) = (95.33)/ (0.75 x 0.8) = 158.88 watt

Considering operating factor for PV panel, the total wattage requirement for the PV panel = (Total PV panel wattage after losses) / (operating factor motor efficiency) = (158.88)/(0.8 0.8) = 248.25 watt

Number of 75 W_p solar PV panel required = 248.25/75 = 3.31 $\simeq$ 4 modules

Power rating of motor = 248.25/746 = 0.33 hp

For sprinkler irrigation system, the minimum rating of pump may be taken to be 1.5 hp. Hence, accordingly, the size of PV system needs to be decided.

(C) Sizing of PV system for 1.5 hp rating motor

(i) Battery Sizing: 1.5 hp = 1119 watt

Daily water requirement for 1-acre land = 14 m^3/day

Hour of operation of motor per day for discharge of 14 m^3 water is around 1 hour

Daily energy that needs to be supplied by battery is 1119 × 1 = 1119 watt-hr

System voltage be 24 volt

In solar PV system, depth of discharge of battery may be from 70-80 % (80 % may be taken)

Hence, required charge capacity of batteries = 1119/24 = 46.6 Ah

Total Ah capacity of battery = (Energy input to motor × No. of days of autonomy)/ (depth of discharge x system voltage) = (1119 3)/(0.8 × 24) = 174.89 Ah

We need to find out how many batteries of 12 V, 100 Ah should be used for supply required energy to the motor.

Total number of batteries = (Total Ah capacity required)/ (Ah capacity of one battery) = 174.89/100 = 1.74 $\simeq$ 2 batteries

From two batteries, 100 Ah + 100 Ah = 200 Ah is available instead of 174. 89 Ah (1119 × 3)/(0.8 × 24). These two batteries need to be connected parallel to get 200 Ah.

We have battery of 100 Ah with 12 V. Hence to get 24 V system voltage, two batteries need to be connected in series. Hence, in total 4 batteries of 100 Ah 12 V are required, two of them connected in series and two such series connected batteries to be connected in parallel.

(ii) PV Sizing to meet the required daily energy requirement: Normally battery efficiency varies from 80 - 90 % (85 % may be taken)

Charge controller efficiency may be taken 90 %

The energy to be supplied by the PV system to the input of battery terminal be (1119)/ (0.85 × 0.9) = 1462 Wh

Taking daily sun shine hours to be 6 hours, the wattage requirement for the solar panel is 1462/6= 243.66 W. The number of panels required considering a 75 W_p solar panel =243.66/75 = 3.24 ≃ 4

Components required for PV system

i.   Four batteries of 100 Ah 12 V

ii. Four modules of 75 W_p

iii. One charge controller

(iii) Cost of Experimental Solar Photovoltaic Powered Sprinkler Irrigation System

i.   Solar PV Module (4 × 75 w_p) = 300 watt @ Rs. 60 per W_p =Rs. 18,000

ii. Batteries (4 x 100 Ah, 12 V) @ Rs. 5000 per battery = Rs. 20,000

iii. Charge controller   = Rs. 3,000

iv. 1.5 hp DC motor with pump set   = Rs. 15,000

v.   Sprinkler set up for 1-acre land   = Rs. 32,000

vi. Pipes, fittings, wiring etc.   = Rs. 2,000

   Total   **= Rs. 90,000**

The figure for the experimental set-up is shown below.

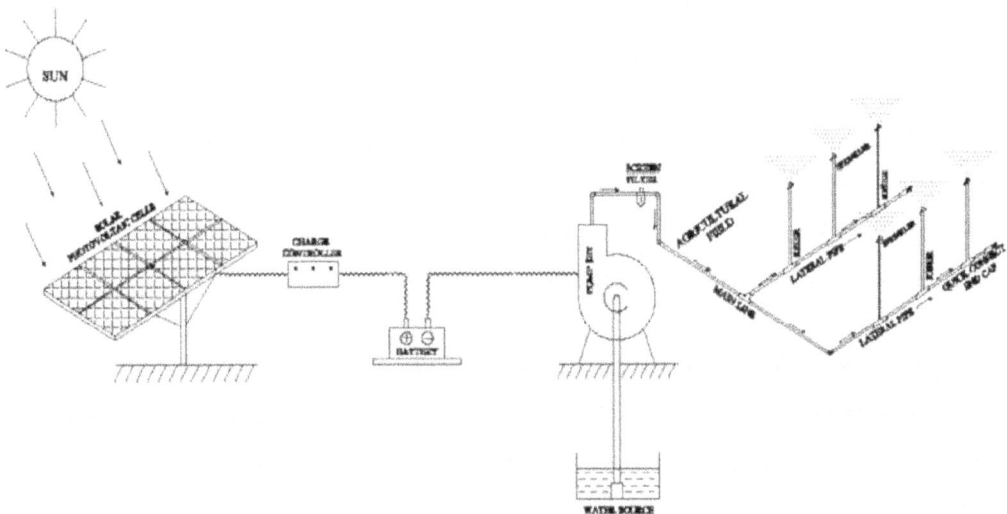

Solar Photovoltatic Integrated Sprinkler System

(D) Hourly Cost of Operation of Various Water Pumping Devices

Information for cost analysis

i.   Cost of 1.5 hp electric pump set = Rs. 10,000

ii.   Cost of 1.5 hp diesel pump set = Rs. 13,000

iii. Cost of 1.5 hp PV powered pump set = Rs. 90,000

iv.   Prevailing interest rate may be taken as 10 %

v.   Efficiencies of motor varies from 70-80 % (70 % taken)

vi.   Efficiencies of pump varies from 70-80 % (70 % taken)

vii. Efficiencies of diesel engine varies from 30-40 % (40 % taken)

viii. Useful life of PV panel varies from 20-25 years (can be taken 22 years)

ix.   Useful life of diesel engine pump set = 8 years

x.   Useful life of electric pump set = 8 years

xi.   Maintenance cost of PV system with sprinkler as 0.5 per cent of total capital cost per year

xii. Maintenance cost of diesel engine pump set as 10 per cent of total capital cost per year

xiii. Maintenance cost of electric pump set as 10 per cent of total capital cost per year

xiv. Annual working hours of diesel, electric pump sets and PV system be 500 hours

xv.   One hp engine consumes about 250ml. diesel per hour (present cost of diesel Rs. 80/lit)

xvi. One unit of electric energy (1 kWh) = Rs. 5.00

xvii. Salvage value of diesel pump set be taken as 20 % of capital cost

xviii. Salvage value of electric pump set be taken as 20 % of capital cost

xix. Salvage value of PV powered pump set be taken as 5 % of capital cost

xx.   Operator's time spent in the proposed system be 1 hr/day (labour charge Rs. 400/day)

xxi. Energy consumption (kWh) of electric pump set = (BHP) / (motor efficiency × pump efficiency) 0.746 × 1 hour

xxii. Cost per hour of operation of diesel pump set = (BHP) / (motor efficiency × pump efficiency) × fuel consumed in litres/hour/BHP × cost of fuel/lit

(E) Hourly operating cost of PV powered water pumping device with sprinkler system

Fixed Cost

(i) Depreciation

D = (C-S)/ (L x H) where C= capital cost; S = Salvage Value; L = Useful life of device; H= Annual working hour

Putting the values of all necessary data, D= Rs. 7.77/hour

(ii) Interest (I) = (C + S)/(2) x (Interest rate/100) x (1/H) = Rs. 9.45/hour

(iii) Insurance and taxes and housing are not applicable

Total fixed cost = 7.77 + 9.45 = Rs. 17.22/hour

Variable Cost

(i) Fuel cost = Nil

(ii) Lubricants = Nil

(iii) Repair and maintenance = (C) x (0.5/100) x (1/H) = Rs. 0.9/hour

(iv) Operator's wages Rs. 400/8 = Rs. 50/hour

Total variable cost = 0.9 + 50 = Rs. 50.9/hour

Total operation cost per hour = Total fixed cost/hour + Total variable cost/hour = **Rs. 68/hour**

(F) Hourly operating cost of diesel pump set

Fixed Cost

(i) Depreciation

D = (C-S)/ (L x H) where C= capital cost; S = Salvage Value; L = Useful life of device; H= Annual working hour

Putting the values of all necessary data, D= Rs. 2.6/hour

(ii) Interest (I) = (C + S)/(2) x (Interest rate/100) x (1/H) = Rs. 1.56/hour

(iii) Insurance and taxes and housing are not applicable

Total fixed cost = 2.6 + 1.56 = Rs. 4.16/hour

Variable Cost

(i) Fuel cost = $(1.5)/(0.4 \times 0.7) \times 0.25 \times 80$ = Rs. 107/hour

(ii) Lubricants = 20 % of cost of fuel = Rs. 21.4/hour

(iii) Repair and maintenance = $(C) \times (10/100) \times (1/H)$ = Rs. 2.6/hour

(iv) Operator's wages Rs. 400/8 = Rs. 50/hour

Total variable cost = 107 + 21.4 + 2.6 + 50 = Rs. 175.6/hour

Total operation cost per hour = Total fixed cost/hour + Total variable cost/hour = **Rs. 180/hour**

(G) Hourly operating cost of electric pump set

Fixed Cost

(i) Depreciation

D = (C-S)/ (L x H) where C= capital cost; S = Salvage Value; L = Useful life of device; H= Annual working hour

Putting the values of all necessary data, D= Rs. 2/hour

(ii) Interest (I) = (C + S)/(2) x (Interest rate/100) x (1/H) = Rs. 1.2/hour

(ii) Insurance and taxes and housing are not applicable

Total fixed cost = 2.0 + 1.2 = Rs. 3.2/hour

Variable Cost

(i) Energy consumption (kWh) = (1.5)/(0.7 x 0.7) x 0.746 = 2.28 kWh

(iii) Electric energy cost = 2.28 x 5 = Rs. 11.40/hour

(ii) Lubricants = 20 % of cost of fuel = Rs. 2.28/hour

(iii) Repair and maintenance = (C) x (10/100) x (1/H) = Rs. 2/hour

(iv) Operator's wages Rs. 400/8 = Rs. 50/hour

Total variable cost = 11.40 + 2.28 + 2 + 50 = Rs. 65.68/hour

Total operation cost per hour = Total fixed cost/hour + Total variable cost/hour = **Rs. 68/hour**

8.6.4 Experimental Results

The results of the experiment conducted during the course of the study are presented in this section. Tomato is one of the most important and remunerative crops in Odisha and is grown in an area of 97,018 ha (Agricultural Statistics, 2016, Govt. of Odisha) covering 11.02 % area of the total tomato cultivation in all India level. It ranks second in the state in vegetable production. Odisha also ranks fourth among the tomato producing states in India. It is considered as one of the most important supplementary sources of minerals and vitamins in human diet. However, targeted production and productivity is not achieved so far at par with the national level due to lack of assured irrigation facilitates both in rabi and summer season. The most prevailing variety of tomato i.e. Utkal Kumari (BT-10) has been cultivated for the present study in order to evaluate the effectiveness of the developed solar PV sprinkler irrigation device with respect to production and productivity, without depending upon conventional source of energy and flooded system of watering practice. The cost of cultivating tomato in 1 acre of land has been calculated in order to know the annual profits out of it and its expected pay-back period. Similarly, the mitigation of greenhouse gases with the use of the developed set-up has been estimated compared with traditional diesel and electric pump sets for its contribution in combating global warming and climate change and thus achieving sustainable agriculture.

8.6.5 Cost-Benefit calculation of tomato cultivation in 1Acre (0.4 ha) Land

(A) Cost of cultivation of tomato in 1.0 Acre land

S.N.	Name of operation	Implements used	No. of operation	Man-hr /Ac	Operation cost (Rs.)	Input (kg)	Cost of input (Rs.)	Total cost (Rs.)
1	Tillage	Tractor drawn rotavator	11	21	1200 600	-	-	1800
2	Planking	Wooden planker (manual)	1	2	31.25/hour	-	-	62.50
3	Seed (Hybrid)	-	-	-	-	-	-	500
4	Planting (manual)		1	16	31.25/hour	-	-	500
5	Fertilizer	FYM	Once			1 tractor load	4000	4000
		Gromer	Twice			100 kg	2500	2500
		Potash	Twice			100 kg	2500	2500
6	Interculture	Manual	Thrice	40	31.25/hour	-	-	3750
7	Plant protection	Knapsack sprayer	Thrice	2	31.25/hour	Pesticides	4000	4187
8	Irrigation	Solar PV powered sprinkler system	45 (2 days interval)	1	Rs. 49/hour	-	-	2205
9	Harvesting	manual	Twice/ week	120/ month	-	-	-	3750
10	Miscellaneous	-	-	-	-	-	-	4000
	TOTAL COST							**29,754**
								≈ 30,000

(B) Benefit

Without assured irrigation, production of tomatoes = 40 quintals/acre @ Rs. 20/kg = Rs. 80,000

With assured irrigation, production of tomatoes with 15 % increase in yield = 46 quintals/acre @ Rs. 25/kg = Rs. 1, 15, 000

Net gain = Rs. 1, 15,000 –Rs. 30,000 = Rs. 85,000 (in Rabi season)

Net gain = Rs. 1, 15,000 –Rs. 30,000 = Rs. 85,000 (in Summer season)

Considering tomato cultivation in both the seasons in a year with assured irrigation, total gain = Rs. 1, 70,000/annum

Monthly income from tomato cultivation with assured irrigation = Rs. 14,167 ≃ Rs. 15,000 per month

Simple payback period = (Initial investment cost) / (Net annual gain) = 90,000/1, 70,000 = 0.5 years or 6 months

(C) Mitigation of CO_2 emission by use of SPV Water Pumping System

Diesel and electricity are the two mostly used fuels to operate diesel and electric pump sets for water pumping in irrigating cultivable lands in our state of Odisha. Burning of diesel in the internal combustion engines and generation of electricity in power plants contribute a lot in the emissions of greenhouse gases to the atmosphere causing more to the present concerns of global warming and climate change. This may be due to the strong initiatives being taken by the Government to achieve more areas under assured irrigation. The replacement of diesel and electric pump sets with a reliable solar photovoltaic powered water pumping system particularly in the irrigation sector would definitely reduce to a greater extent in the emissions of greenhouse gases to the atmosphere. The existing diesel and electric pump sets in our state is 2.47 lakhs and 1.38 lakhs respectively in the power rating range of 1-5 hp. Taking the average power rating of both diesel and electric pump sets as 3 hp, the amount of emissions of CO_2 are as follows;

i.   One hp engine consumes about 250 ml of diesel per hour

ii. Burning of 1 litre of diesel releases 3 kg of CO_2 to the atmosphere (Manfredi et al. 2009)

iii. The average carbon dioxide emission for electricity generation from coal based thermal power plant is approximately 1.58 kg of CO_2 per kWh at the source.

iv. Annual working hours of diesel and electric pump sets can be taken 500 hours

v.   Annual CO_2 emissions from 2.47 lakhs diesel pump sets to be 30 crore kg in our state

vi. Annual CO_2 emissions from 1.38 lakhs electric pump sets to be 25 crore kg in our state

vii. Total annual CO_2 emissions can be mitigated by 55 crore kg with the replacement of existing diesel and electric pump sets in our state by the adoption of solar photo voltaic powered system in irrigation sector.

viii. Total annual electrical energy consumption from 1.38 lakhs electric pump sets can be saved in the tune of 15×10^7 kWh (saving around 15 crore units of electricity costing about Rs. 75 crores/annum)

ix. Total annual diesel consumption from 2.47 lakhs diesel pump sets can be saved in the tune of 10×10^7 litres of diesel (saving around Rs. 800 crores/annum)

8.6.6 Information for the users

Sustainable energy source along with the adoption of possible water management practices may be achieved with the help of solar photovoltaic micro-irrigation system in order to solve the problem of inadequate availability of two critical inputs such as energy and water for assured irrigation in agricultural sector. Micro-irrigation method through sprinkler system may also be an added advantage if integrated with the solar PV device to achieve judicious utilization of water. Hence, use of solar PV system may be a sustainable proposition of energy source for water pumping to achieve assured irrigation in the state. The findings of the present study would definitely give an insight to the farming community of the state to go for adopting the technology to strengthen their agricultural production system with secured availability of energy and water. The summary of the study are as follows;

i.   Wide popularization of solar photo voltaic powered water pumping system for achieving assured irrigation through sustainable energy source.

ii. Monthly income of Rs. 15,000/- throughout the year may be possible by adopting remunerative vegetable cultivation in 1 acre of land during rabi and summer seasons only.

iii. The small and marginal farmers of the state may be attracted to adopt solar photo voltaic powered water pumping system as the hourly operating cost is lowest i.e. Rs. 68/hour and same of Rs. 68/hour for electric pump set and Rs. 180/hour for diesel pump set.

iv. The existing area under vegetable cultivation in the state may be enhanced by adopting vegetable cultivation in the unutilized land mostly during summer season due to the assured irrigation facility through solar photo voltaic system.

v.   The developed set up may also be utilized for irrigating land in rainy season in case of irregular rainfall

vi. Pay- back period of the developed set up is only ½ year, due to which, it may be easily accepted by the small and marginal farmers of the state in spite of its high initial cost.

vii. Total annual CO_2 emissions can be mitigated by 55 crore kg with the replacement of existing diesel and electric pump sets in our state by the adoption of a reliable solar photo voltaic powered system in irrigation sector.

viii. Total annual electrical energy consumption from 1.38 lakhs electric pump sets can be saved in the tune of 15×10^7 kWh (saving around 15 crore units of electricity costing about Rs. 75 crores/annum)

ix. Total annual diesel consumption from 2.47 lakhs diesel pump sets can be saved in the tune of 10×10^7 litres of diesel (saving around Rs. 800 crores/annum)

8.7 A CASE STUDY ON OFF-GRID SPV BASED MICRO-IRRIGATION SYSTEM IN AEROBIC RICE CULTIVATION (GHOSAL ET AL 2020)

8.7.1 Perspectives of SPV Water Pumping in Aerobic Rice Cultivation

Aerobic rice cultivation is now-a-days gaining importance due to the constraints in the availability of required amount of water for traditional rice growing system. An attempt was therefore made to develop a portable solar photovoltaic powered (off-grid) drip irrigation system for aerobic rice cultivation, which is a water saving and less water consuming rice production system without any compromise with decline in yield. It is suitable mostly in the water-deficient, non-irrigated and off-grid areas. There may be the saving of 40-45 % of water for irrigation purpose compared to the conventional method, mitigation of 0.55 million tons of CO_2 with the replacement of existing diesel and electric pump sets and 0.2 million tons of CH_4 from 4.0 million hectares of rice fields in the state of Odisha, India, through the system, developed by adopting aerobic rice cultivation. The pay- back period of the set-up was estimated to be 4 years and total annual saving of Rs. 675 crores due to reduction in the use of electrical energy and petroleum fuels through the existing pump sets in Odisha. Monthly income of Rs. 4000/- throughout the year was achieved by adopting aerobic rice cultivation in 1 acre (0.4 ha) of land.

8.7.2 Need of SPV Pumping Based Micro-irrigation in Aerobic Rice Cultivation

Sustainability of rice production in the present context of fast growing population is a big challenge before the farming community with respect to achieving food security and controlling over the increased concerns of water scarcity, energy crisis and global warming due to the emission of greenhouse gases through anthropogenic activities particularly in the agricultural sector (Hossain et al 2015 and Nazmul 2016). In order to attain self-sufficiency in food grain, the production and productivity of rice crop alone play a crucial role not only for our state Odisha but also for India, Asia and the world as well, as rice is the principal food of more than 60 % of the world's population and around 90 % of the rice area worldwide is in Asia and low land rice fields produce about 75 % of the world's rice supply (Singh et al. 2010). Agriculture too in Odisha to a considerable extent means growing rice. It is the staple food for almost the entire population of Odisha and therefore, the state economy is directly linked with the improvements in production and productivity of rice (Anon, 2017). Rice production practices mostly followed in the state are through wet and low land cultivation, covering about 80 % of the total rice area and the methods of cultivation are usually transplanting of seedlings and broadcasting of sprouted seeds in the puddled soil. In the above method of cultivation, the field remains either in fully or partially flooded condition most of the time of the growing season, creating favourable environment for emission of methane, which is one of the principal and potent greenhouse gases relative to global warming potential of 25 times higher than that of CO_2 and accounts for one-third of the current global warming phenomenon (Gag et al, 2011). Rice cultivation is a major source of atmospheric CH_4 and contributes about 10-20 % of total methane emission to the atmosphere (Pathak and Agarwal, 2012). The environmental experts have now recognized and expressed in intergovernmental panel on climate change (IPCC, 2010) that the submerged rice fields are the most significant contributors of atmospheric methane (Jain et al. 2004). According to the current estimate, the production of rice needs to be enhanced by around 70 % to meet the demands for ever-increasing population by 2030 and thus making rice cultivation, a potential major cause of growing atmospheric methane (Singh 2010).

Aerobic rice cultivation where fields remain unsaturated or nearer to saturation throughout the season like an upland crop offers an opportunity to produce rice with less water (Patel 2010). Aerobic and upland rice are both grown under aerobic condition, however, the former is under controlled water management system but latter is not (Parthasarathi et al 2012). Aerobic rice is a new cropping system in which specially developed varieties are directly seeded in well-drained, un-puddled and unsaturated soil conditions for most of the crop growing period. The rice is grown like an upland crop with adequate inputs and supplementary irrigation when rainfall is insufficient. Aerobic rice varieties are bred by combining the draught resistance of upland varieties with high yielding characteristics of low land cultivars. Some of the suitable varieties like CRDhan-200 (pyari), Naveen and Annada are also available in our state Odisha. Aerobic rice is therefore grown with the use of external inputs such as supplementary irrigation and fertilizers with an aim to maintain the productivity at par with the traditional production system without compromise with yield decline. The micro-irrigation in general and drip irrigation in particular has received considerable attention from researchers, policy makers, economists etc. for its perceived ability to contribute significantly to ground water resources development, agricultural productivity, economic growth and environmental sustainability (Singh et al 2010, Kashiv et al 2016, Hoque et al 2016, Campana et al 2015 and Shinde VB 2015). Currently, our state has around 1.38 lakhs (35 %) grid based (electric) and 2.47 lakhs (65 %) diesel irrigation pump sets. However, erratic grid supply of electricity and high cost of diesel pumping continue to remain as a big problem for the farmers. The rising cost of using diesel for powering irrigation pump sets are often beyond the means of small and marginal farmers. Consequently, the lack of required amount of water often leads to poor growth of plants, thereby, reducing yields and income. Hence, use of conventional diesel/gasoline powered pumping systems poses an economic risk to the farmers. Scientific studies reveal that timely and required amount of water availability for the crop favours the increase in the yield by 10-15 % (Narale et al 2013). The burning of petroleum fuels also creates threats by polluting environment and causing global warming by releasing a considerable amount of CO_2 into the atmosphere. The continuous exhaustion of limited stocks of conventional energy sources and their environmental impacts have, therefore, forced researchers, planners, and policy makers to search for the reliable, environment-friendly and cost effective energy resources to power water pumping system in a sustainable manner (Pradhan and Ali 2016). Hence to keep pace with the growing demands of energy and acute shortage of water, solar photovoltaic water pumping device may be integrated with drip irrigation system for rice cultivation in addressing the issues of food security under climate change scenario (Gopal et al. 2013). Thus an integrated approach of water saving technology and reliable source of energy along with the reduced greenhouse gases emissions from the conventional method of rice cultivation has been thought up in the study to achieve food security of the nation. An attempt was therefore made to study the feasibility and performance of solar photovoltaic powered drip irrigation system in aerobic rice cultivation

8.7.3 Experimental Location

Development of solar photovoltaic (SPV) drip irrigation system (Fig. 8.4) has been made for cultivating paddy in 1 acre (0.4 ha) of land to achieve secured irrigation and to improve water use efficiency mostly in aerobic method of cultivation. The details of the design and developments are mentioned below. The experiments were carried out during the year 2017-18 in OUAT farm, Bhubaneswar, Odisha, which lies at the latitude of $20\,^0\,15'$ N and longitude of $85\,^0\,52'$ E and coming under warm and humid climatic condition. Paddy was cultivated in rabi and summer seasons. The soil type of the experimental site is sandy loam and the climate of the study area is humid and sub-tropical in nature.

8.7.4 The Cost Estimate for Solar Photovoltaic Powered Drip Irrigation System

i.   Solar PV Module of 1000 watt @ Rs. 50 per w_p = Rs. 50,000

ii. 1 hp DC motor with pump set   = Rs. 80,000

iii. Mounting structure   = Rs. 15,000

iv. Civil works/Balance of system   = Rs. 20,000

v.   Drip set up for 1-acre land   = Rs. 35,000

     Total   = **Rs. 2, 00,000**

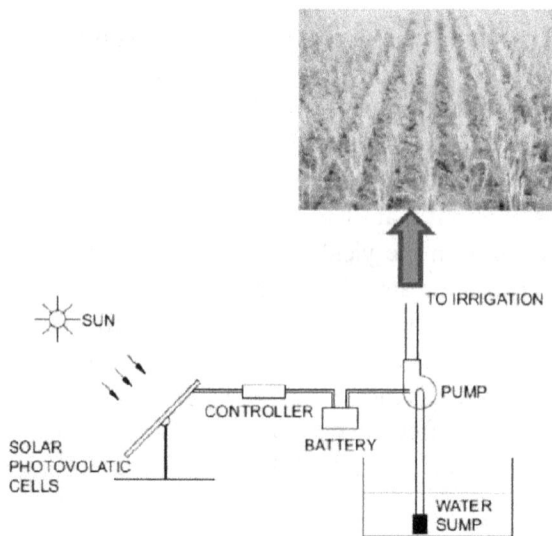

Fig. 8.4. Experimental Set-up for Solar Water Pumping Based Drip Irrigation in Aerobic Rice

8.7.5 Economics of Using Various Water Pumping Devices

Information for cost analysis

i.   Cost of 1 hp electric pump set = Rs. 7,000

ii. Cost of 1 hp diesel pump set = Rs. 10,000

iii. Cost of 1 hp PV powered pump set with drip irrigation system = Rs. 2,00,000

iv. Prevailing interest rate may be taken as 10 %

v.   Efficiencies of motor varies from 70-80 % (70 % taken)

vi. Efficiencies of pump varies from 70-80 % (70 % taken)

vii. Efficiencies of diesel engine varies from 30-40 % (40 % taken)

viii. Useful life of PV panel varies from 20-25 years (can be taken 22 years)

ix. Useful life of diesel engine pump set = 8 years

x.   Useful life of electric pump set = 8 years

xi. Maintenance cost of PV system with drip as 0.5 per cent of total capital cost per year

xii. Maintenance cost of diesel engine pump set as 10 per cent of total capital cost per year

xiii. Maintenance cost of electric pump set as 10 per cent of total capital cost per year

xiv. Annual working hours of diesel, electric pump sets and PV system be 500 hours

xv. One hp engine consumes about 250ml. diesel per hour (present cost of diesel Rs. 80/lit)

xvi. One unit of electric energy (1 kWh) = Rs. 5.00

xvii. Salvage value of diesel pump set be taken as 20 % of capital cost

xviii. Salvage value of electric pump set be taken as 20 % of capital cost

xix. Salvage value of PV powered pump set be taken as 5 % of capital cost

xx. Operator's time spent in the proposed system be 1 hr/day (labour charge Rs. 400/day)

xxi. Energy consumption (kWh) of electric pump set = (BHP) / (motor efficiency × pump efficiency) × 0.746 × 1 hour

xxii. Cost per hour of operation of diesel pump set = (BHP) / (motor efficiency × pump efficiency) × fuel consumed in litres/hour/BHP × cost of fuel/lit

(A) Hourly operating cost of PV powered water pumping device with drip system

Fixed Cost

(i) Depreciation

D = (C-S)/ (L x H) where C= capital cost (Rs. 2,00,000); S = Salvage Value (5% of C); L = Useful life of device (22 years); H= Annual working hour (500 hours)

Putting the values of all necessary data, D= Rs. 17.27/hour

(ii) Interest (I) = (C + S)/(2) x (Interest rate/100) x (1/H) = Rs. 21/hour

(ii) Insurance and taxes and housing are not applicable

Total fixed cost = 17.27 + 21 = Rs. 38.27/hour

Variable Cost

(i) Fuel cost = Nil

(ii) Lubricants = Nil

(iii) Repair and maintenance = (C) x (0.5/100) x (1/H) = Rs. 2/hour

(iv) Operator's wages Rs. 400/8 = Rs. 50/hour

Total variable cost = 2 + 50 = Rs. 52/hour

Total operation cost per hour = Total fixed cost/hour + Total variable cost/hour = **Rs. 90/hour**

(B) Hourly operating cost of diesel pump set

Fixed Cost

(i) Depreciation

D = (C-S)/ (L x H) where C= capital cost (Rs. 10,000); S = Salvage Value (20 % of C); L = Useful life of device (8 years); H= Annual working hour (500 hours)

Putting the values of all necessary data, D= Rs. 2.0/hour

(ii) Interest (I) = (C + S)/(2) x (Interest rate/100) x (1/H) = Rs. 1.2/hour

(iii) Insurance and taxes and housing are not applicable

Total fixed cost = 2.0 + 1.2 = Rs. 3.20/hour

Variable Cost

i. Fuel cost = (1)/(0.4 x 0.7) x 0.25 x 80 = Rs. 71.42/hour

ii. Lubricants = 20 % of cost of fuel = Rs. 14.28/hour

iii. Repair and maintenance = (C) x (10/100) x (1/H) = Rs. 2.0/hour

iv. Operator's wages Rs. 400/8 = Rs. 50/hour

Total variable cost = 71.42 + 14.28 + 2.0 + 50 = Rs. 97.53/hour

Total operation cost per hour = Total fixed cost/hour + Total variable cost/hour = **Rs. 137/hour**

(C) Hourly operating cost of electric pump set

Fixed Cost

(i) Depreciation

D = (C-S)/ (L × H) where C= capital cost (Rs. 7000); S = Salvage Value (20% of C); L = Useful life of device (8 years); H = Annual working hour (500 hours)

Putting the values of all necessary data, D = Rs. 1.4/hour

(ii) Interest (I) = (C + S)/(2) × (Interest rate/100) × (1/H) = Rs. 0.84/hour

(ii) Insurance and taxes and housing are not applicable

Total fixed cost = 1.4 + 0.84 = Rs. 2.24/hour

Variable Cost

i.   Energy consumption (kWh) = (1) / (0.7 × 0.7) × 0.746 = 1.52 kWh

ii.   Electric energy cost = 1.52 × 5 = Rs. 7.6/hour

iii. Lubricants = 20 % of cost of fuel = Rs. 1.52/hour

iv.   Repair and maintenance = (C) × (10/100) × (1/H) = Rs. 1.4/hour

v.   Operator's wages Rs. 250/8 = Rs. 50/hour

Total variable cost = 7.6 + 1.52 + 1.4 + 50 = Rs. 60.52/hour

Total operation cost per hour = Total fixed cost/hour + Total variable cost/hour = **Rs. 60/hour**

8.7.6 Experimental Results

The results of the experiment conducted during the course of the study are presented in this section. Rice is one of the most important and major crops in Odisha and is grown in an area of 40 lakh ha in kharif season and only 2.5 lakh ha in rabi season, (Agricultural Statistics, 2017, Govt. of Odisha). Kharif rice is entirely monsoon-fed. The area for rice cultivation during rabi season has not been covered widely due to lack of assured irrigation causing most of the cultivable land to remain unutilized. The yield of rice from the areas during rabi season is not satisfactory due to erratic supply of grid electricity for operating irrigation pumps. Farmers are also not interested to use diesel pump sets due to frequent rise in the cost of diesel fuel. Hence captive power and water source along with water saving technology for rice cultivation are the only alternative for the resource poor farmers of the state. The harnessing of solar energy for electricity generation through PV system is a viable option in the state due to abundant availability of solar radiation in about 300 days in a year. The variety chosen for the study was CR Dhan-200 (Pyari). This variety of rice was cultivated for the present study in order to evaluate the effectiveness of the developed solar PV drip irrigation device with respect to production and productivity, without depending upon conventional source of energy and flooded system of watering practice. The cost of cultivating rice in 1 acre of land has been mentioned (in table below) in order to know the annual profits out of it and its pay-back period. Similarly, the mitigation of greenhouse gases with the use of the developed set-up has been estimated compared with traditional diesel and electric pump sets for its contribution in combating global warming and climate change and thus achieving sustainable agriculture.

8.7.7 Cost-benefit of the Study

(i) Cost of Cultivation in 1.0 Acre Land

Cost-Benefit calculation of aerobic rice cultivation in 1Acre (0.4 ha) Land

S.N.	Name of operation	Implements used	No. of operation	Man-hr /Ac	Operation cost (Rs.)	Input (kg)	Cost of input (Rs.)	Total cost (Rs.)
1	Tillage	Tractor drawn rotavator	1	2	1200	-	-	1200
2	Seed		-	-	-	20 kg	-	500
3	Direct line sowing (by rope method)	-	1	8	31.25/hour	-	-	250
4	Manures and	FYM	Once	-	-	1 tractor load	1500	1500
	Fertilizer	Gromer	Twice			100 kg	1000	1000
		Potash	Twice			100 kg	1000	1000
5	Interculture (cono weeder)	Manual	Thrice	16	31.25/hour	-	-	1500
6	Plant protection	Knapsack sprayer	Thrice	2	31.25/hour	Pesticides	2000	2062
7	Irrigation	Solar PV powered drip system	45 (2 days interval)	1	Rs. 51/hour	-	-	2295
8	Harvesting	Hired Reaper	Once	1				750
9	Threshing	Pedal thresher	Once	50-60 kg/hr	Six man days	-	-	1500
	TOTAL COST							**13,557** ≈ **13,500**

(ii) Benefit

Yield of paddy in aerobic rice practice = 2.5 tones/acre; By- product yield = 1.5 tones/acre; Returns from paddy @ Rs. 1350/quintals = 33,750; Returns from by-product @ Rs. 250/ quintal = 3,750; Total returns = 33,750 + 3,750= Rs. 37,500; Net gain = Rs. 37,500– Rs. 13,500 = Rs. 24,000 (kharif); Net gain = Rs. 37,500– Rs. 13,500 = Rs. 24,000 (rabi); Total gain from 1 acre of aerobic rice cultivation in a year = Rs. 48,000; Monthly income from aerobic rice cultivation with assured water supply = Rs. 48,000/12 = Rs. 4000 per month

Simple payback period = (Initial investment cost) / (Net annual gain) = 2, 00,000/ 48,000 = 4.1 years ≃ 4 years

(iii) Estimation for mitigation of CO_2 emission by use of SPV Water Pumping System

The existing diesel and electric pump sets in our state is 2.47 lakhs ad 1.38 lakhs respectively in the power rating range of 1-5 hp. Taking the average power rating of both diesel and electric pump sets as 3 hp, the amount of emissions of CO_2 is calculated and as follows;

1. One hp engine consumes about 250 ml of diesel per hour
2. Burning of 1 litre of diesel releases 3 kg of CO_2 to the atmosphere
3. The average carbon dioxide emission for electricity generation from coal based thermal power plant is approximately 1.58 kg of CO_2 per kWh at the source.
4. Annual working hours of diesel and electric pump sets can be taken 500 hours
5. Annual CO_2 emissions from 2.47 lakhs diesel pump sets to be 30 crore kg in our state
6. Annual CO_2 emissions from 1.38 lakhs electric pump sets to be 25 crore kg in our state
7. Total annual electrical energy consumption from 1.38 lakhs electric pump sets can be saved in the tune of 15×10^7 kWh (saving around 15 crore units of electricity costing about Rs. 75 crores/annum)
8. Total annual diesel consumption from 2.47 lakhs diesel pump sets can be saved in the tune of 10×10^7 litres of diesel (saving around Rs. 800 crores/annum)

(iv) Estimation for mitigation of CH_4 emission by shifting from anaerobic to aerobic rice cultivation

Rice fields have been identified as a major source of atmospheric methane. The global methane emission rate from rice fields was recently estimated to be 40 Tg/year (1 Tg = 10^{12} g) which accounts for about 8 % of the total methane emission. The reduction in the amount of methane emission from the conventional method of rice cultivation is estimated as follows;

1. On an average, 50 kg methane is emitted from 1 hectare of transplanted rice crops in one crop growing season.
2. Annual Area under rice cultivation in the state Odisha is 4.0 million hectares (both kharif and rabi).
3. Annual Area under rice cultivation in India is 43.0 million hectares (both kharif and rabi).
4. Mitigation of CH_4 emissions through aerobic rice cultivation is 0.2 million tones and 2.15 million tones per annum in Odisha and India respectively.

8.7.8 Information to users

A solar photovoltaic powered drip-irrigation system for aerobic rice cultivation in warm and humid climate of Odisha appears to be a viable proposition looking into the present day's concerns of water scarcity and energy crisis in agricultural sector. The following conclusions may be drawn from the study.

i.   Monthly income of Rs. 4000/- throughout the year may be possible by adopting aerobic rice cultivation in 1 acre of land both during rabi and summer seasons.

ii. The small and marginal farmers of the state may be attracted to adopt off-grid solar photo voltaic powered water pumping system as the hourly operating cost is Rs. 90/hour and Rs. 60/hour for electric pump set and Rs. 137/hour for diesel pump set.

iii. Pay- back period of the proposed set up is 4 years, due to which, it may be easily accepted by the small and marginal farmers of the state inspite of its high initial cost.

iv. Total annual CO_2 emissions can be mitigated by 0.55 million tones with the replacement of existing diesel and electric pump sets in our state by the adoption of a reliable off-grid solar photo voltaic powered system in irrigation sector.

v.   Total annual CH_4 emissions can be mitigated by 0.2 million tones from 4.0 million hectares rice fields in Odisha

vi. Total annual electrical energy consumption from 1.38 lakhs electric pump sets can be saved in the tune of 15×10^7 kWh (saving around 15 crore units of electricity costing about Rs. 75 crores/annum)

vii. Total annual diesel consumption from 2.47 lakhs diesel pump sets can be saved in the tune of 10×10^7 litres of diesel (saving around Rs. 800 crores/annum)

REFERENCES

A. Hamidat, B. Benyoucef and T. Hartani. 2003. Renewable Energy, Vol. 28, pp. 1081-1096.

Amit Garg, Bhushan Kankal and P.R. Shukla. 2011. Methane emissions in India: Sub-regional and sectoral trends. Atmospheric Environment. 45 (2011): 4922-4929.

Andrada, P. and Castro, J. 2008. Solar photovoltaic water pumping system using a new linear actuator. Grup d'Accionaments Electrics amb Commutació Electrònica (GAECE), England.

Annonymous. 2017. Agricultural Statistics 2017, Govt. of Odisha.

Anonymous. 2013. Economic Survey of Odisha, 2013, Govt. of Odisha

Anonymous. 2013. Solar Policy, Govt. of Odisha

Anonymous. 2015. Agricultural Statistics, 2015, Govt. of Odisha.

Campana PE, Li H, Zhang J, Yan J (2015), Economic optimization of photovoltaic water pumping systems for irrigation, *Energy conservation and management.*

D.P. Patel, Anup Das, G.C. Munda, P.K. Ghosh, Juri Sandhya Bordoloi and Manoj Kumar. 2010. Evaluation of yield and physiological attributes of high-yielding rice varieties under aerobic and flood-irrigated management practices in mid-hills ecosystem. Agricultural Water Management 97 (2010) 1269–1276.

Ghosal M.K., N. Sahoo and Sonali Goel. 2020. Studies on Off-Grid Solar Photovoltaic-Powered Micro-Irrigation System in Aerobic Rice Cultivation for Sustainable Agriculture and Mitigating Greenhouse Gas Emission. Book Chapter. Innovation in Electrical Power Engineering, Communication, and Computing Technology. © Springer Nature Singapore Pte Ltd. 2020, https://doi.org/10.1007/978-981-15-2305-2_11. Page 135-145.

Ghosal MK, Soni Badra and N Sahoo. 2020. Studies on solar water pumping based micro-irrigation for sustainable vegetable cultivation. International Journal of Chemical Studies 2020; 8(6): 1202-1208.

Gopal, C., Mohanraj, M. P., Chandra Mohan and Chandrasekhar, P.. 2013. Renewable energy source water pumping systems-A literature reviews. Renewable and sustainable energy reviews, 25 (2013), 351-370.

H. Pathak and P.K.Aggarwal. 2012. Low carbon technologies for agriculture: A study on rice and wheat systems in Indo-Gangetic plains. Division of Environmental Science, IARI, New Delhi.

Hoque N, Roy A, Beg RAB, Das BK (2016), Techno-economic evaluation of solar irrigation plants installed in Bangladesh, *Int. Journal of Renewable Energy Development*, 5(1); 73-78.

Hossain MA, Hassan MS, Mottalib MA, Ahmmed S (2015), Techno-economic feasibility of solar pump irrigations for eco-friendly environment, *Procedia Engineering*, 105; 670-678.

Jay Shankar Singh, 2010. Capping methane emission. Science reporter, September, 2010, page 29-30.

Kala Meah, Sadrul Ula and Steven Barret. 2008. Solar photovoltaic water pumping-Opportunities and challenges. Renewable and Sustainable Energy reviews, vol. 12, pp. 1162-1175.

Kala Meah, Steven Fletcher and Sadrul Ula. 2008. Solar photovoltaic water pumping for remote locations. Renewable and Sustainable Energy reviews. 12 (2008), 472-487.

Kashiv A, Bilala A, Shirazi N, Dwivedi A, Joshi R (2016), Solar drip irrigation system, *Intt. Journal of Emerging Technology and Advanced Engineering*, Volume 6, Issue 4.

Khatib, T. 2010. Design of photovoltaic water pumping system at minimum cost for Palestine: a review. Journal of applied sciences, **10**(22):2773-2784.

Khattab NM, Badr MA, Shenaway ET, Sharawy HH, Shalaby MS (2016), Feasibility of hybrid renewable energy water pumping system for a small farm in Egpyt, I*ntt. Journal of Applied Engineering Research*, 11;0973-4562.

Leah C.Kelley, Eric Gilbertson, Anwar Sheikh, Steven D. Eppinger and Steven Dubowsky 2010. On the feasibility of solar-powered irrigation, Renewable and Sustainable Energy reviews, vol. 14, pp. 2669 – 2682.

Manfredi, Simone, Tonini, Davide and Christensen, T. 2009. Land filling of waste: accounting of greenhouse gases and global warming contributors. Waste Management and Research, 27: 825-836.

N. Jain, H, Pathak, S. Mitra and A. Bhatia. 2004. Emission of methane from rice fields-A review. Journal of scientific and Industrial research. Vol. 63, February 2004, pp. 101-115.

Narale P.D; Rathore N.S. and Kothari S. 2013. Study of Solar PV Water Pumping System for Irrigation of Horticulture Crops. International Journal of Engineering Science Invention Volume 2 Issue 12 December. 2013 pp.54-60.

Nazmul K M (2016), Municipal soild waste to energy generation in Bangladesh: possible scenarios to generate to generate renewable electricity in Dhaka and Chittagong city, *Journal of Renewable Energy*, vol. 2016, ID 1712370.

P.D.Narale, N.S.Rathore and S.Kothari. 2013. Study of Solar PV Water Pumping System for Irrigation of Horticulture Crops. International Journal of Engineering Science Invention. Volume 2 Issue 12, December. 2013, PP.54-60.

Pradhan A, Ali SM (2016), Analysis of solar performance with change in temperature, *International Journal of Applied Engineering Research*, 11; 5225-5227.

Ravender Singh, D.K. Kundu and K.K. Bandyopadhya. 2010. Enhancing Agricultural Productivity through Enhanced Water Use Efficiency. Journal of Agricultural Physics. Vol. 10, pp. 1-15.

Sako, K.M., N'guessan, Y., Diango, A.K., and Sangare, K.M. 2011. Comparative Economic Analysis of Photovoltaic, Diesel Generator and Grid Extension in Cote D'ivoire. Asian Journal of Applied Sciences, 4: 787-793.

Shinde VB, Wandre SS (2015), Solar photovoltaic water pumping system for irrigation: A review, *African Journal of Agricultural Research* ,10(2);2267-2273.

T. Parthasarathi, K. Vanitha, P. Lakshamanakumar and D. Kalaiyarasi. 2012. Aerobic rice-mitigating water stress for the future climate change. International Journal of Agronomy and plant Production. 3(7): 241-254.

C H A P T E R - 9

FINANCIAL EVALUATION OF ESTABLISHING AN ENTERPRISE

INTRODUCTION

Investment for a business venture necessitates and demands appropriate financial evaluation for the profitability, survivability and sustainability of today's increasing growth of business enterprises, in order to secure more benefit in long run. Whether they may be the industrial firm, agricultural farm or a day-to-day event in somebody's personal life, there is the requirement of a large number of financial decisions for smooth and efficient functioning of any system in order to provide goods/services at a lower price fulfilling the consumers' needs and desires. In the process of managing a system, the entrepreneur should at various levels take necessary financial decisions which would ultimately help in minimizing investment, operating and maintenance expenditures besides enhancing the revenues, savings and other related gains of the enterprise. These can be achieved through the application of sound knowledge of economics, dealing mainly with various methods and tools that enable one to make financial decisions towards minimizing costs and maximizing benefits to any business enterprise. Hence, the necessity of systematic evaluation of financial aspect is to optimize the allocation of funds by focusing on those parameters which would ultimately lead the enterprise economically more rewarding besides being socially acceptable and environmentally sustainable. The economic analysis or financial evaluation is therefore a consideration of primary importance from the increasingly competitive business point of view. In any project, the organizer should think for the alternatives. In mutually exclusive projects, the value of each alternative is closely observed before concluding with a final decision. In today's competitive world of business, it has become essential to decide the right alternative which is based upon financial evaluation and economic feasibility of one proposed venture.

This chapter covers the elementary economic analysis, interest formulae, bases for comparing alternative ventures (e.g. present worth method, future worth method, annual equivalent method, rate of return method etc.) and cost-benefit analysis. Inflation and depreciation have also been considered in the financial analysis.

9.1 BASIC TERMS AND DEFINITIONS

Cost: The word cost refers to all expenses incurred in setting up or running an enterprise. *First cost:* It is the initial cost of setting up an enterprise including transportation, installation and other related preliminary expenditures.

Life cycle cost: The Life Cycle Cost (LCC) for a project or system or structure is its total cost of purchase and operation over its entire service life. This cost includes the expenses for acquisition, operation, maintenance and disposal.

Capital investment: This refers to the amount of money, spent at the beginning of the project for creation of capacity to produce goods or services and also called initial investment.

Gestation period: It is the time period between the investment and the time when project starts yielding returns.

Cash flow: It refers to frequent receipts and disbursement taking place over a particular interval of time for an enterprise. Cash flow diagram is a convenient way to display the revenues (savings, cash receipt, income) and costs (expenses, disbursement) associated with an investment. Cash flow is of two types, i.e. cash inflow and cash outflow. The difference between the two is the net cash flow i.e. net cash flow = receipts − disbursements. Cash inflow occurs because of the sale of product, sale of assets and borrowing. Cash outflow occurs in the business because of purchase of raw material, purchase of assets and repayment of loans. One has to ensure that the net cash flows are positive in the business. A cash flow diagram is therefore simply a graphical representation of cash flows drawn on a time scale as shown in the figure (9.1) below. Cash receipts or revenues or incomes are represented by cash inflows, similarly disbursements or expenses are represented by cash outflows. Hence receipts and disbursements are the cash flows with a plus sign representing cash inflows and a minus sign representing cash outflows. The diagram convention is as follows;

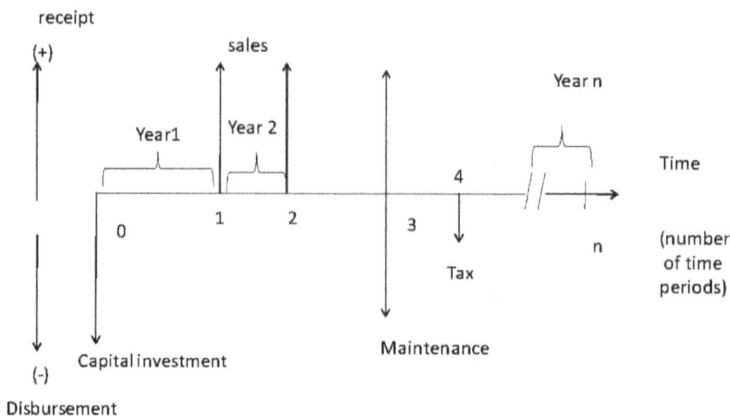

Fig. 9.1. Cash flow diagram

Horizontal Axis: It is marked off in equal increments, one per period, up to the duration of the project.

Revenues: They are represented by upward pointing arrows.

Disbursements: They are represented by downward pointing arrows.

All disbursements and receipts (i.e. cash flows) are assumed to take place at the end of the year in which they occur.

Life of the project: Every project has a defined life period based on the life of the most crucial assets on which the investment is made. The life of a project is important to determine the benefits from or costs of a project.

Present value: It gives the present value of all future costs or benefits that appear as discrete values with a given discount rate of '*i*'.

Future value: It gives the amount that will be accumulated after *n* years for the amount invested at the beginning at an interest rate of *i* percent.

Compounding: Two alternative procedures are used to find the value of cash flows; compounding and discounting. In compounding, future values of cash flows at a given interest rate at the end of a given period of time are found. In the compounding process, the interest is added to the principal at the end of each time period which, in turn earns interest. The future value of present investment is calculated by using the well-known formula of compound interest.

$$F = P(1 + i)^n \qquad \qquad \text{... (1)}$$

Discounting: Discounting is the inverse procedure of compounding. A present sum is compounded to know the future value and the future sum is discounted to know the present value of the future amount.

Annuity: Annuity refers to a stream of payments or returns over time. It is a fixed payment (or receipt) each year for a specified number of years.

Sinking fund: It is a fund, which is created out of fixed payments in each period to accumulate a future sum after a specified period. For example, companies generally create sinking funds to return bonds on maturity. It gives the uniform amount that must be invested at '*i*' percent at the end of each of *n* years to accumulate a given future sum. The factor used to calculate the annuity for a given future sum is called the sinking fund factor.

Capital recovery: It is the annuity of an investment made today, for a specified period of time at a given rate of interest. If we make an investment today for a given period of time at a specified rate of interest, we may like to know the annual income. It shows how much can be withdrawn in equal amounts at the end of each of *n* years if some amount is deposited initially at '*i*' percent interest rate.

Capital: The financial resources involved in establishing and sustaining an enterprise or a project.

Interest factors: In financial evaluation, cash flows are based on either the present worth or its future worth. In the present worth, we usually convert a single future sum of a series of future value to an equivalent amount at an earlier date. Similarly, in case of future worth, we convert the values occurring at any time to an equivalent amount at a future date. Equivalent value can be determined by calculating the compound amount of each sum for each period. There are seven interest factors for discrete compounding. The notations which are used in various interest formulae are as P = Principal amount, n = No. of interest period, i = Interest rate, F = Future amount at the end of year n, A = Equal amount of payment or receipt at the end of every interest period, G = Uniform amount which will be added / subtracted in period/ after period to / from the amount of deposit $A1$ at the end of period 1.

Salvage/Junk value: The value of the unspent assets remaining after the life of a project is referred to as salvage/Junk value. Though the most crucial assets have spent their life span, some amount may be recovered from the sale of the assets after the expiry of their useful life. This amount also helps in determining whether the project can be continued with minimum additional investment.

Time value of money

The change in the amount of money over a given time period is called time value of money. If an amount of Rs. 1000/- now is put into an investment (bank, bond, stock or other) and would produce a value greater than Rs. 1000/- at some time in the future, this refers to time value of money. The amount of the future value depends on the interest rate or rate of return earned on the investment. Because money can earn at a certain interest rate through its investment for a period of time, rupee received at some future date is not worth as much as rupee in hand at present. A rupee today is worth more than a rupee that will be received tomorrow. The value of money is more today because it gives an opportunity to invest and earn returns (interest).

Thus, as money has earning and purchasing power, it has time value. Most financial evaluation studies on any project proposal are likely to involve cash flows occurring at different periods of time in future. For a proper evaluation of such cash flows, it may be necessary to explicitly consider the time value of money and consequently develop methods to deal with the same.

Interest: Interest may be defined as money paid for the use of borrowed money and may be regarded alternatively as the return obtained by the productive investment of capital.

Interest rate (i): The interest rate for a sum of money (principal) is usually expressed as the sum that is to be paid (as a percentage) for the use of the principal for an interest period of one year. It is also quoted for interest periods other than one year. However, unless otherwise stated, interest rates are customarily understood to be specified on annual (per annum) basis. The interest rate is therefore basically the rental value of money. It represents the growth of capital per unit period. The period may be a month, a quarter, semi-annual or a year. It is the manifestation of the time value of money. Thus, an interest rate of 10 per cent would mean that the investor regards a sum of money worth 10 per cent less next year than it is worth this year.

Number of time periods(n): This is the number of time intervals over which the amount of money is being invested or borrowed. Although '*n*' is usually the number of years, it may represent other time periods such as the number of quarters or months.

9.2 TECHNIQUES FOR ADJUSTING TIME VALUE OF MONEY OR INTEREST FORMULAS

In time value analysis, more importance is given to interest as this refers to earning power of money. While making investment decisions, it is extremely important to know how to use interest formulas more effectively. Two approaches are used in calculating the interest i.e. simple interest and compound interest.

(a) **Simple Interest**: In simple interest, the interest accrued is directly proportional to the capital involved, the number of interest period and the interest rate. When money is borrowed for a certain period of time, the borrower pays interest to the lender on the principal amount. In simple interest, a fixed percentage is charged on the borrowed amount which remains unchanged. It is called as simple because it ignores the effects of compounding. If P represents the principal, '*n*' the number of interest periods and '*i*' the interest rate per period (expressed as fraction), the total interest accrued, I, may be computed as

$$I = P \times n \times i \qquad \qquad \dots (2)$$

The total future amount F, may thus be calculated as

$$F = P + I = P + Pni = P\,(1 + ni) \qquad \dots (3)$$

(b) Compound Interest: This approach assumes that the interest earned is not withdrawn at the end of an interest period and is automatically re-deposited with the original principal in the next interest period and so on. The interest thus accumulated is called compound interest. The compound interest is always more than the simple interest (for a given set of values of principal, interest rate and number of interest periods), the difference being the result of compounding which essentially is the calculation of interest on previously earned interest. Now-a-days, compound interest approach is used more frequently than that of simple interest.

The formula to find the future value of investment P in case of interest rate compounded annually is $F = P(1+i)^n$ with usual rotations as mentioned earlier $\qquad \dots (4)$

Nominal and Effective Interest Rates

Cash flows in a project proposal are considered to occur once in a year. In practice, cash flows could occur more than once a year. For example, banks may pay interest on savings account quarterly. On bonds, debentures and public deposits, financial institutions may pay interest semi-annually. Hence when compounding / discounting has to be done at intervals less than a year, it is necessary to introduce the terms nominal interest rate and effective interest rate to describe the nature of compounding more precisely and the effective distinction should be made between i) Nominal interest rate ii) Effective interest rate. The interest rate is usually specified on an annual basis in a loan agreement and is known as the nominal interest rate. If compounding is done more than once a year, the actual annualized rate of interest would be higher than the nominal interest rate and it is called the effective interest rate.

Example 9.1: A person invested Rs. 100 now in a bank, interest rate being 10 percent a year and that bank would compound the interest semi-annually (i.e., twice a year). How much amount will he get after a year?

The bank will calculate interest on his deposit of Rs. 100 for first six months at 10 per cent and add this interest to his principal. On this total amount accumulated at the end of first six months, he will again receive interest for next six months at 10 percent. Thus the amount of interest for first six months would be the interest = 10 % × ½ = Rs. 5 % and the outstanding amount at the beginning of the second six-month period will be: Rs. 100 + Rs. 5 = Rs. 105. Now he will earn interest on Rs. 105. The interest on Rs. 105 for next six months will be equal to Rs. 105 × 10% × ½ = Rs. 5.25. Thus the person will get Rs. 100 + Rs. 5 + Rs. 5.25 = Rs. 110.25 at the end of a year. If the interest were compounded annually, he could have received: Rs. 100 + 10% of Rs. 100 = Rs. 110. But he is getting more under semi-annual compounding because the person earned interest on interest earned during the first six months. He would get still higher amount if the compounding is done quarterly, monthly or daily.

What effective annual interest did he earn on his deposit of Rs. 100?

On an annual basis, he earned Rs. 10.25 on his deposit of Rs. 100, so the effective interest rate (*EIR*) is $EIR = 5 + 5.25 = 10.25\%$. This implies that Rs. 100 compounded annually at 10.25 percent or Rs. 100 compounded semi-annually at 10 per cent would accumulate the same amount. Mathematically EIR for the above example can be expressed as

$$EIR = (1 + \frac{i}{2})^{1 \times 2} - 1 = (1 + \frac{0.1}{2})^2 - 1 = 0.1025 = 10.25\%$$ and if the interest is

compounded once in a year $EIR = (1 + i)^1 - 1 = (1 + 0.1)^1 - 1 = 0.1 = 10\%$, same as the annual interest rate specified. Here it is noticed that the annual interest rate 'i' has been divided by 2 to find the semi-annual interest, since the interest has been compounded twice in a year and since there are two compounding periods in one year, the term $\left(1+\dfrac{i}{2}\right)$ has been squared. If, compounding is done quarterly, the annual interest rate 'i' will be divided by four and there will be four compounding periods is one year. If 'i' represents the nominal interest rate per year and 'm' be the number of interest (compounding) periods per year, the effective annual interest rate 'i_{eff}' may be expressed

as $i_{eff} = \left(1+\dfrac{i}{m}\right)^{m} - 1$ and for one interest period per year which is the case of annual

compounding (i.e. $m=1$), $i_{eff} = (1+i)^{1} - 1 = i$. Thus with annual compounding, the nominal and effective interest rates are same.

For continuous compounding (m = ∞), as a limiting case, the interest may be compounded an infinite number of times per year, the effective annual interest is

$i_{eff(C)} = e^{i} - 1$.

Nominal interest Rate: Nominal interest rate is the annual interest rate obtained simply by converting the interest rate specified for one period (where the period is less than one year) to a year disregarding the effect of compounding after each period. For example, an interest rate of 1.5% compound each month for a year is typically quoted as 18 % (1.5 % × 12) compounded monthly. When the interest is expressed in this manner, the 18 percent interest rate is called a nominal interest rate or annual percentage rate.

Effective interest rate: It represents the actual or exact rate of interest upon the principal during a specified period. The effective interest rate based on a year is referred as the effective annual interest rate.

Now for 18% nominal interest rate, the effective interest rates are as follow;

For quarterly compounding (*m*=4),　$i_{eff} = (1+\dfrac{0.18}{4})^{4} - 1$　or 19.25 % per annum

For monthly compounding (*m* = 12),　$i_{eff} = (1+\dfrac{0.18}{12})^{12} - 1$　or 19.56 % per annum

For daily compounding (*m* = 365),　$i_{eff} = (1+\dfrac{0.18}{365})^{365} - 1$　or 19.71 % per annum

For continuous compounding (*m*→∞),　$i_{eff} = e^{0.18} - 1$.　or 19.72 % per annum

It is seen that with the increase in the number of compounding in a year, the effective interest rate also increases.

If F is the future value, P, the cash flow to-day, 'i', the annual rate of interest, 'n' is the number of years of availing financial transaction; and 'm' is the number of compounding per year, the general formula for the above computation is $F_n = P[1 + \dfrac{i}{m}]^{m \times n}$

and for continuous compounding $F_n = P(e^{i \times n})$. Continuous compounding is seldom used in engineering economic studies.

Example 9.2. What amount of money must be invested to realize Rs. 100 after 10 years when interest at 3.5 percent per annum is compounded continuously?

Solution: Here, F_{10} = Rs.100; n = 10 yeras i = 3.5% = 0.035, then, the amount to be invested is $P = F_n e^{-in} = 100 \times e^{-0.35}$ $\therefore \log_{10} P = \log_{10} 100 - 0.35 \log_{10} e = 2 - 0.35(0.4343)$ =1.848 or P= anti log (1.848) = Rs.70.47

Example 9.3. What is the amount due after 5 years at the interest 3.5% per annum on the principal Rs. 100 in continuous compounding?

Solution: P = Rs. 100, n = 5 years, i = 3.5% = 0.035

$\therefore$ $F_5 = Pe^{in} = 100 \ e^{(0.035 \times 5)} = 100 \ e^{0.175}$ = Rs.119.10

The most obvious computational effect of using continuous interest is that, it produces a larger future amount than does the same rate compounded discretely.

9.3 CLASSIFICATION OF CASH FLOW TRANSACTIONS

In any financial analysis of an investment, there are usually cash receipts (income) and cash disbursements (costs) which occur over a particular interval of time. These receipts and disbursements in a given time period are referred to as cash flows. The cash flow analysis always holds an important position in the evaluation of financial and economic aspects of each individual project. In practice, cash flow transactions are basically classified into five categories.

(a) Single payment cash flow (b) Uniform payment series cash flow (c) Linear gradient series cash flow (d) Geometric gradient series cash flow and (e) Irregular payment services cash flow

Single payment (or receipt) cash flow: It involves a single present or future cash flow. The cash flow diagram of this cash flow pattern is shown in fig. 9.2 below.

(i) *Single–payment compound amount factor or future-value factor:* When an amount P is invested for n equal interest periods with i as the interest rate, the future value of the investment obtained is

$$F = P(1 + i)^n \qquad \qquad \text{... (5)}$$

As shown in the cash-flow diagram, (fig. 9.2), the present value P is an investment (expense) and it is represented by a downward arrow at time zero '0' while its future value F is an inflow represented by an upward arrow. The factor $(1+i)^n$, which is used to multiply P in order to derive F, is called 'single payment compound-amount factor or future-value factor'. It is usually designated as $(F/P, i,n)$. Therefore, the above relationship can be written as

$$F = P(F/P, i, n) \qquad \qquad \text{... (6)}$$

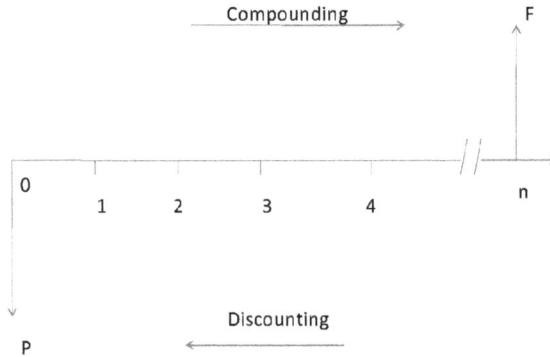

Fig. 9.2. Single payment and future value

Example 9.4. What will be the amount if Rs. 3000 invested at 12% interest is compounded annually at the end of 5 years?

Solution: Given, P = Rs. 3000 i = 0.12, n= 5, $F = P (1 + i)^n = 3000 (1 + 0.12)^5 = 3000 (1.12)^5$ = Rs. 5287

Example 9.5. Find the number of years in which an investment of Rs. 10 lakhs will be doubled in value if the interest rate is 12.25%.

Solution: Given, P = 10 lakhs, F = 20 lakhs, i = 0.1225. From the single payment compound amount formula, $20 = 10(1.1225)^n = 2 \Rightarrow (1.1225)^n$

$$\log 2 = n \log (1.1225) \Rightarrow n = \frac{\log 2}{\log (1.1225)} = 6 \text{ year}$$

(ii) Single-payment discount amount factor or present-worth factor: If the future value of an investment (or payment) is known, we can derive its present value with a given interest rate and the number of discounting periods by using Eq. 5, as

$$P = F \left[\frac{1}{(1+i)^n} \right] \qquad \qquad \text{... (6.1)}$$

The expression in brackets, which is used to multiply F to obtain P is called 'single-payment present-worth factor' and is designated as $(P/F, i, n)$. Therefore, Eq. 6.1 can be written as

$$P = F(P/F, i, n) \qquad \qquad \ldots (7)$$

The process used in converting the future value to its present value is called discounting and has wide application in engineering economy. In the above expression, the present value is calculated by discounting the future value of a given amount. The term 'i' is known as the discount rate.

Example 9.6. Mr. X wishes to have a future amount of Rs. 5, 00,000 for establishing of dairy farm after 6 years now. What is the exact amount that he should deposit now to get the desired amount after 6 years? The bank gives 15% interest rate compounded annually.

Solution: Given $F = 5, 00,000$, $i = 15\% = 0.15$, $n = 6$ year

$$P = F\left[\frac{1}{(1+i)^n}\right] = 5,00,000\left[\frac{1}{(1.15)^6}\right] = 2,16,163.$$

The initial investment of Rs. 2, 16,163 on behalf of Mr. X will entitle to get Rs. 5, 00,000 for setting up a dairy farm after 6 years at 15% interest rates compounded annually.

Uniform payment (or receipt) series cash flow: It involves a series of flows of equal amounts at regular intervals. The relevant cash flow diagrams are shown in Fig. 9.3 below.

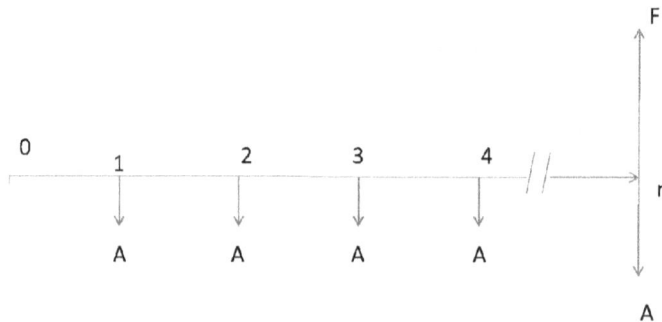

Fig. 9.3. Cash flow diagram of equal payment series compound amount

(i) *Uniform-series compound amount factor:* In this type of investment mode, the objective is to find the future value of 'n' equal payments which are made at the end of every interest period till the end of the nth interest period at an interest rate of 'i' compounded at the end of each interest period. Business firm or individual investors, in order to meet their future capital investment requirement 'F', make a certain amount of savings over the time period resulting into accumulation of an amount that, hopefully will meet their expected needs. In this case, investment process involves investing an amount 'A' each year. At the end of the nth year, the principal plus the interest earned compounded annually to become F, the future value of investment as shown in the fig. 9.3. The accumulated future amount F after 'n' years may be

calculated as $F = A\left[\dfrac{(1+i)^n - 1}{i}\right]$ where A is the equal amount deposited at the end

of each interest period and compounded to accumulate F. The expression in the bracket is known as 'uniform series compound – amount factor' and is designated as. (F/A, i, n), Hence, $F = A(F/A, i, n)$.

Example 9.7. What will be the future amount of a uniform payment of Rs. 1000 deposited at the end of each year with an interest rate of 12 percent at the end of 5 years?

Solution: - Given A = Rs. 1000, i = 12% = 0.12, n = 5

$$F = A\left[\frac{(1+i)^n - 1}{i}\right] = 1000\left[\frac{(1+0.12)^5 - 1}{0.12}\right] = \text{Rs. } 6353$$

(ii) Uniform series sinking fund factor: In this method, we have to find out the equal amount that should be deposited at equal intervals to get a future sum (F) after n^{th} interest period at an interest rate of 'i' applicable. The corresponding cash flow diagram is also same as Fig. 9.3 above. When the future capital (F) expenditure for a certain activity or project is known, the required amount 'A' to be invested annually at a given interest rate can be determined from the following equation.

$$A = F\left[\frac{i}{(1+i)^n - 1}\right]$$

The expression in the bracket is known as 'Uniform series sinking fund factor' and is designated as $(A/F, i, n)$, hence $A = F(A/F, i, n)$.

Example 9.8. The replacement of an inverter of a solar PV system at the end of 3 years is likely to cost Rs. 10,000. What amount should a user deposit every year to accumulate the desired amount if he earns 10 percent interest on his deposit.

Solution: - Given F = Rs. 10,000, i = 0.1, n = 3

From the formula of equal payment series sinking fund amount

$$A = 10,000\left[\frac{0.1}{(1.1)^3 - 1}\right] = 10,000\left[\frac{0.1}{0.331}\right] = \text{Rs. } 3,021$$

(iii) Uniform-series present worth factor: The objective of this mode of investment is to find the present worth of an equal payment made at the end of every interest period for n interest periods at an interest rate of 'i' compounded at the end of every interest period. The corresponding cash flow diagram is shown below (fig. 9.4).

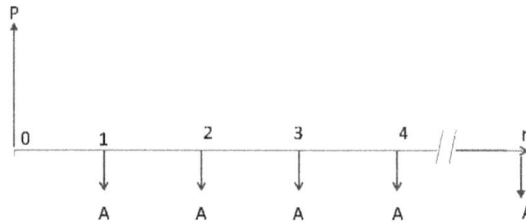

Fig. 9.4. Cash flow diagram of equal payment series present worth amount

The formula to compute P is

$$P = A\left[\frac{(1+i)^n - 1}{i(1+i)^n}\right] = A(P/A,i,n) \qquad \dots (8)$$

The expression in the bracket is known as 'Uniform-series present worth factor' and is designated as $(P/A, i, n)$

Example 9.9. If a machine undergoes a major overhaul now, its output can be increased by 10%, which translates into additional cash flow of Rs. 5,000 at the end of each year for six years due to reduction in fuel consumption and increase of its performance. If the interest rate is 12 percent, how much expenditure may be allowed for investment to overhaul this machine.

Given $A = 5,000$, $n = 6$, $i = 0.12$. From the formula of equal payment series present worth amount,

$$P = A\left[\frac{(1+i)^n - 1}{i(1+i)^n}\right] = 5000\left[\frac{(1.12)^6 - 1}{0.12(1.12)^6}\right] = \text{Rs. } 20,562$$

Therefore, at the most, the above amount of Rs. 20,562 may be spent on the overhauling of the machine at present.

Example 9.10. A solar cooker is expected to save Rs. 3000 worth of LPG fuel each year during its useful life of 15 years. Assuming a discount rate of 12 percent, calculate the present worth of saving of LPG fuel through the use of solar cooker.

Given, $A = 3000$, $n = 15$, $i = 0.12$. From the formula of equal payment series present worth amount,

$$P = A\left[\frac{(1+i)^n - 1}{i(1+i)^n}\right] = 3000\left[\frac{(1.12)^{15} - 1}{0.12(1.12)^{15}}\right] = \text{Rs. } 20{,}433$$

Thus the present worth of the monetary value of the life cycle full savings of solar cooker is Rs. 20,433.

Example 9.11. A 100 litre-per-day domestic solar water heater saves consumption of electricity in an electric geyser on 100 days of the year by heating 100 litres of water from 15^0 C to 60 ^{0}C. The useful life of the solar heater is estimated as 12 years, determine the present worth of saving through the use of the solar water heater, if the efficiency of the electric geyser is 90% and the cost of electricity is Rs 6 per kWh. Assume interest rate as 12%.

Solution: Given, *n=12, i=0.12*

Electricity saved by the solar water heater in one day:

$$= \left(\frac{1}{0.9}\right) \times 100kg \times 4.2\frac{kJ}{kg°C} \times (60-15)°C \times \left(\frac{1KWh}{3{,}600\,kJ}\right) = 5.83 \text{ kWh/day}$$

Monetary worth of electricity saved by solar water heater during one year = (5.83 kWh/day) (100 days / year) ×(Rs. 6/kWh) = Rs 3498. Therefore, annual saving A = Rs. 3498

Present worth of overall saving during its lifetime of 12 years may be calculated from

the Eq. (8) as $P = A\left[\dfrac{(1+i)^n - 1}{i(1+i)^n}\right] \rightarrow P = 3498 \times \left[\dfrac{(1+0.12)^{12} - 1)}{0.12(1+0.12)^{12}}\right] = 21668$

(iv) Uniform series capital recovery factor: The objective of this mode of investment is to find the annual equivalent amount 'A' which is to be recovered at the end of every interest period for 'n' interest periods for a loan 'P' which is sanctioned now at an interest rate of 'i' compounded at the end of every interest period. In other way, it can also be expressed in terms of the present value of an investment if made in order to generate an equal annual receipt in the future at a given rate of interest. The corresponding cash flow diagrams for both the cases are shown in Fig. 9.5.

The formula for this case can be calculated by rearranging the formula for equal

payment uniform series present worth amount $A = P\left(\dfrac{i(1+i)^n}{i(1+i)^n - 1}\right)$

The expression in the bracket is known as uniform-series capital recovery factor and is designated as (*A/P, i, n*). Hence the above equation can also be written as $A = P\ (A/P,\ i,\ n)$

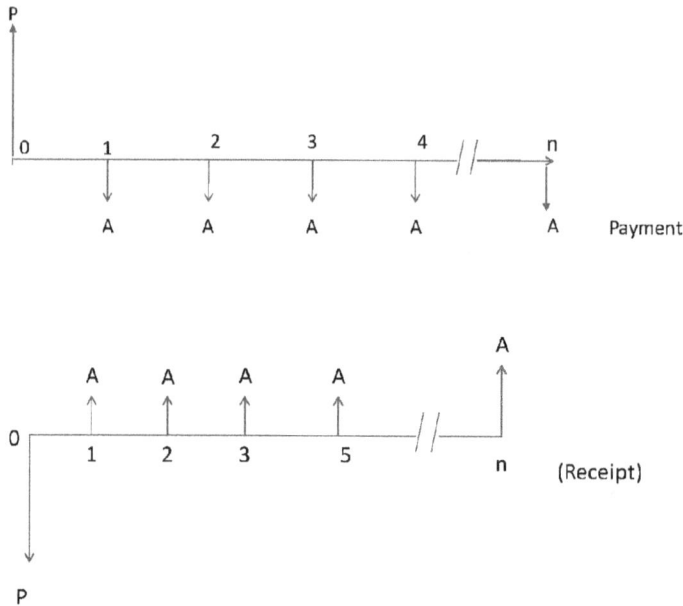

Fig. 9.5. Cash flow diagram of equal payment series capital recovery amount

Example 9.12. A bank gives a loan to a person to purchase a 100-liter domestic solar water heater costing to Rs. 20,000 at an interest rate of 12% compounded annually. This amount should be repaid in 10 yearly equal instalments. Find the instalment amount that the person has to pay to the bank.

Solution: - Given, $P = 20,000$, $n = 10$, $i = 0.12$, we are to calculate A

Using the corresponding formula, $A = P\left[\dfrac{i(1+i)^n}{i(1+i)^n - 1}\right] = 20,000\left[\dfrac{0.12(1.2)^{10}}{(1.12)^{10} - 1}\right] = \text{Rs. } 3,540$

The annual equivalent instalment to be paid by the person to the bank is Rs. 3,540.

Example 9.13. A 1 kW$_p$, PV rooftop plant costs Rs. 1, 00, 00 and has a useful life of 25 years. The annual average maintenance cost is 1% of the initial cost and the interest rate is 12%. Calculate the unit cost of a solar PV generated electricity if it supplies a load of 40% of its power rating for 10 hours daily.

Solution: - Given, $P = 1, 00,000$, $n = 25$, $i = 0.12$. The annualized capital cost is

$$A = P\left[\dfrac{i(1+i)^n}{(1+i)^n - 1}\right] = 1,00,000\left[\dfrac{0.12(1.12)^{25}}{(1.12)^{25} - 1}\right] = \text{Rs. } 4,765$$

Annual maintenance cost = $0.01 \times 1,00,000$ = Rs. 1000. Thus, total annual cost of a solar PV system = $4,765 + 1000$ = Rs. 5765

Annual electricity generation = $1 \times 0.4 \times 10 \times 365$ = 1460 kWh. Therefore, unit cost of electricity generation $= \dfrac{5765}{1462} = $ Rs. 3.94

Example 9.14. A 2m^3 biogas plant costs Rs. 25,000 and has a useful life of 20 years. Calculate the unit cost (Rs. /m^3) of biogas produced if the annual average biogas production efficiency is 80 percent. The average annual maintenance cost is 5 percent of the initial cost and the discount rate is 12 percent.

Solution: Given, $P = 25,000$, $n = 20$ years, $i = 0.12$

The annualized capital cost of biogas plant using capital recovery factor is

$$A = 25,000 \left[\frac{0.12(1.12)^{20}}{(1.12)^{20} - 1} \right] = \text{Rs. } 3347$$

Annual maintenance cost = $0.05 \times 25,000$ = Rs. 1250. Thus the total annual cost of biogas plant to the user = Rs. 3347 + Rs. 1250 = Rs. 4597

Annual biogas generation = $(2)(365)(0.8)$ = 584 m^3

$$\text{Unit cost of biogas produced} = \frac{Total\ annual\ cost}{Annual\ biogas\ generation} = \frac{5497}{584} = \text{Rs.} 7.87 / \text{m}^3$$

Uniform gradient series of payments (or receipt)

Sometimes, an investment generates a uniform series of payments that either increase or decrease by a constant amount as shown in fig. 9.6. Such a series is known as uniform (or arithmetic) gradient series of payments.

The objective of this mode of investment is to find the annual equivalent amount of a series with an amount A1 at the end of the first year and with an equal increment or decrement (G) at the end of each of the following (n-1) years with an interest rate 'i' compounded annually. In a variety of situations in financial evaluation, disbursements or receipts are increased or decreased by a uniform amount each period, thus constituting an arithmetic sequence of cash flows. For example, the repair and maintenance expenses of a system may increase by a relatively constant amount each period as it grows older. Cost inflation can also be treated by increasing the cost by a constant amount in each period. In a similar manner, a constant periodic increase in rents can be handled by adding a fixed increment each time, the rent is paid. Such situations can be formulated in the best and most convenient manner by suing a uniform or arithmetic gradient.

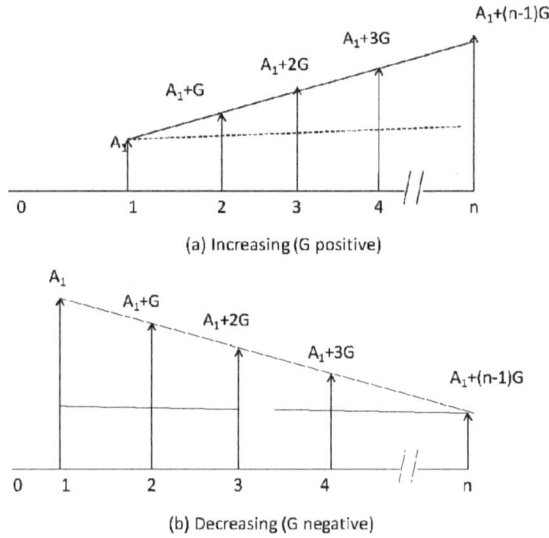

(a) Increasing (G positive)

(b) Decreasing (G negative)

Fig. 9.6. Uniform gradient series

To evaluate such a series, an equivalent uniform series of payment/receipts with constant cash flows A_g is found and then present value (P_g) and future value (F_g) are computed for this equivalent series by using appropriate formulae as mentioned earlier.

The equivalent cash flows of uniform series of payments may be obtained from $A_g = A_1 + G$ $(A/G, i, n)$ where G is the constant amount of change or gradient per period (which may be positive or negative), A_1 is the payment at the end of the first year and n is the number of periods. $(A/G, i, n)$ is known as 'Uniform gradient annual series factor' or Gradient to uniform series conversion factor and is given by $(A/G, i, n)$

$$= \left[\frac{1}{i} - \frac{n}{\left(1+i\right)^n - 1} \right].$$ It may be noted that the first payment starts from the year 1 and

continues up to the year n and grading starts from the year 2.

Example 9.15. A business firm plans to start manufacturing solar cooker and expects the sales revenues of Rs. 5 lakhs at the end of first year. In the subsequent 8 years, sales are expected to increase by 0.5 lakh annually. The firm has applied for a loan of Rs. 22 lakhs from a bank to finance the project. As per the Bank's policy, the loan must not exceed 80% of the expected revenues. If the discount rate is 12%, will the firm get the loan?

Solution: Given, A_1 = 5 lakhs, G = 0.5 lakhs, i = 0.12, n = 8

The equivalent cash flows of uniform series,

$$Ag = A_1 + G\left[\frac{1}{i} - \frac{n}{(1+i)^n - 1}\right] = 5 + 0.5\left[\frac{1}{0.12} - \frac{8}{(1.12)^8 - 1}\right] = Rs.6.46 \text{ lakhs}$$

The present worth of the series of the cash flow may be

$$P_g = A_g\left[\frac{(1+i)^n - 1}{i(1+i)^n}\right] = 6.46\left[\frac{(1.12)^8 - 1}{0.12(1.12)^8}\right] = 32.10 \, lakhs$$

80% of the expected revenues = 32.10 × 0.8 = 25.68 lakhs. This amount is more than the amount of loan (i.e., 22 lakhs). Therefore, the firm will get loan from the bank.

Example 9.16. A person is planning to start a solar lantern manufacturing business after 5 years from now. He is capable to invest Rs. 1 lakh at the end of 1st year and thereafter he wishes to deposit the amount with an annual increase of 20,000 for the next 4 years with an interest of 12%. Find the total amount at the end of 5th year of the above cash flows. The expected amount of Rs. 10 lakhs is required to start the above proposal after 5 years.

Solution: Given $A1$ = Rs. 1, 00,000, G = Rs. 20,000, i = 12%, n = 5 years

$$A_g = A1 + G\left(\frac{1}{i} - \frac{n}{(1+i)^n - 1}\right) = 1,00,000 + 20,000\left(\frac{1}{0.12} - \frac{5}{(1.12)^6 - 1}\right) = Rs. \ 1,35,446$$

This is equivalent to paying an equivalent amount of Rs. 1,35,446 at the end of every year for the next 5 years. The future worth sum of this revised series at the end of the 5th year will be as follows: Using equal payment series compound amount factor,

$$F = A(F/A, i, n) = 1,35,446\left[\frac{(1+i)^n - 1}{i}\right] = 1,35,446\left[\frac{(1.12)^5 - 1}{0.12}\right] = Rs. \ 8,60,468$$

As the future amount is less than the expected project proposal of Rs. 10 lakhs, the person should not go for this project proposal.

Example 9.17. A person is planning to start manufacturing of the tubular battery of solar PV system after 4 years from now. He is capable of investing Rs. 1 lakh at the end of 1st year and thereafter he wishes to deposit the amount with an annual decrease of Rs. 10,000 for the next 3 years with an interest rate of 12%. Find the total amount at the end of 4th year of the above cash flows. The expected amount of Rs. 4 lakhs is required to start the above project proposal after 4 years.

Solution: - Given, $A1$ = Rs. 1, 00,000, G= Rs. 10,000, i = 12%, n = 4 years.

$$A_g = A1 - G\left(\frac{1}{i} - \frac{n}{(1+i)^n - 1}\right) = 1,00,000 - 10,000\left(\frac{1}{0.12} - \frac{4}{(1.12)^4 - 1}\right) = \text{Rs. } 86,446$$

Using equal payment series compound factor, $F = 86,466\left[\dfrac{(1+i)^n - 1}{i}\right] = \text{Rs. } 4,13,250$

Hence, the above project proposal can be taken up.

Geometric gradient series of payments (or receipt): It is a series of cash flows increasing or decreasing over time, not by a constant amount (gradient) but by a constant percentage rate i.e. a geometric growth through time as shown in Fig. 9.7. Escalation in the prices of different fuels at a constant rate each year (known as inflation) is a typical situation that can be modelled with geometric sequence of cash flow.

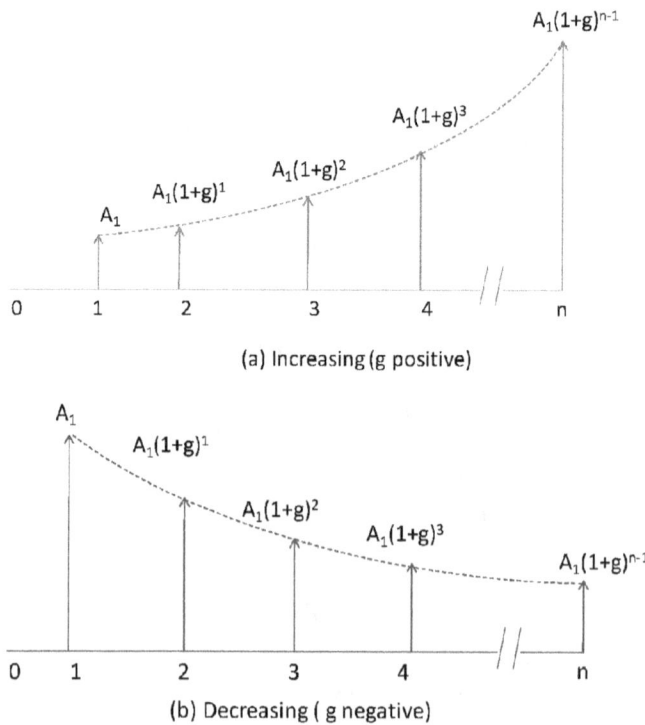

(a) Increasing (g positive)

(b) Decreasing (g negative)

Fig. 9.7. Geometric gradient series

If 'g' is used to designate the percentage change in the magnitude of the payment from one year to the next (as a fraction of the first period), the magnitude of jth payment is related to the payment A1 in 1st year as

$$A_j = A1 (1 + g)^{j-1}, j = 1, 2, \ldots \ldots \ldots \ldots \ldots \ldots n$$

When 'g' is positive, the series will increase and when 'g' is negative, the series will decrease. To derive an expression, for the present amount P_{gg} of the geometric gradient series at interest rate 'i' per period, the relationship between $A1$ and A_j given by the above equation can be used, together with the single payment present worth factor.

$$P_{gg} = A1\left[\frac{(1+g)^0}{(1+i)^0}\right] + A1\left[\frac{(1+g)^1}{(1+i)^1}\right] + A1\left[\frac{(1+g)^2}{(1+i)^2}\right] \ldots\ldots\ldots + A1\left[\frac{(1+g)^{n-1}}{(1+i)^{n-1}}\right]$$

Multiplying each term by $(1+g)/(1+g)$ and simplifying,

$$P_{gg} = \frac{A1}{1+g}\left[\frac{(1+g)^1}{(1+i)^1} + \frac{(1+g)^2}{(1+i)^2} + \ldots \frac{(1+g)^n}{(1+i)^n}\right]$$

Let $\frac{1}{(1+g')} = \frac{1+g}{1+i}$ and $P_{gg} = \frac{A1}{1+g}\left[\frac{1}{1+g'} + \frac{1}{(1+g')^2} \ldots\ldots\ldots + \frac{1}{(1+g')^n}\right]$

The term within the bracket constitute the equal payment series present worth factor for n year. $\quad P_{gg} = \frac{A1}{1+g}\left[\frac{(1+g')^n - 1}{g'(1+g')^n}\right]$

The factor within the bracket is called the geometric gradient series factor. From the expression, $\frac{1}{(1+g')} = \frac{1+g}{1+i}, g' = \frac{1+i}{1+g} - 1$

Putting the value of g' in the above equation,

$$P_{gg} = \frac{A1}{(i-g)}\left[1 - \left(\frac{1+g}{1-i}\right)^n\right] \text{ when } i \neq g \qquad \ldots (9)$$

and $\qquad P_{gg} = \frac{nA1}{(i+g)} \text{ when } i=g \qquad \ldots (10)$

Example 9.18. The use of a box type solar cooker saves cooking gas equivalent to 4 LPG cylinders every year during its useful life of 15 years. If the current price of an LPG cylinder is Rs. 430 and likely to increase by 15% annually, what is the present monetary worth of the life cycle saving of a solar cooker. Assume a discount rate of 12%.

Solution: A1 = Cost of saving of LPG cylinder per annum = 430×4 =Rs. 1720, g = 0.15, i = 0.12, n = 15.

Using the formula for present worth of geometric gradient series (Eq. 9),

$$P_{gg} = \frac{1720}{0.12 - 0.15}\left[1 - \frac{(1.15)^{15}}{(1.12)^{15}}\right] = \text{Rs. } 27,907$$

9.4 ANNUITY (UNACOST) 'A'

An annuity is a series of equal payments made at fixed interval for a specified number of periods. It is a cash flow which is either incoming or outgoing, involves the same sum in each period. For example, Rs. 1000 at the end of each of the next five years is a five-year annuity. Hence annuity is characterized by (i) equal payment (A) (ii) equal periods between n payments and (iii) the first payment occurring at the end of the first period. When the payment occurs at the end of each period, the annuity is called an ordinary annuity. Payments on mortgages, car loans, educational loans etc. are few examples of ordinary annuity.

When the payment occurs at the beginning of each period, the annuity is called *annuity due*. For example, rent for the house, premiums for insurance policies etc. are few examples of annuity due.

The annuity (A) is also called as the term unacost (the uniform end of year annual amount). The future value of an annuity is calculated on the basis of 'Uniform series compound amount factor' i.e. $F = A\left[\dfrac{(1+i)^{n} - 1}{i}\right]$

The present value of an annuity is calculated on the basis of 'Uniform-series present worth factor' i.e., $P = A\left[\dfrac{(1+i)^{n} - 1}{i(1+i)}\right]$

Annuity Due

A series of payments made at the beginning instead at the end of each period is referred to as 'annuity due'. In this case, calculation will be slightly different from general annuity (or ordinary annuity). It will differ in the following ways.

i)   The series should be divided into two parts.

ii) The first payment should be treated separately

iii) The remaining payments should follow the rule of general (or ordinary) annuity calculation.

The present worth is the sum of first payment plus the product of annuity and the series of present worth factor. The number of payments in the present worth factor will be n-1 where n is the total number of periods. The present worth of annuity due

$$P_{(annuity\ due)} = A + A(P/A,\ i,\ (n\text{-}1)) \qquad \text{... (11)}$$

Concepts of annuity due: The concepts of compound value and present value of annuity (uniform end of year annual amount) discussed earlier are based on the assumption that series of cash flows occur at the end of the period. In practice, cash flows in some situations take place at the beginning of the period. When any home appliance is purchased on an installment basis, the dealer requires to make the first payment immediately (i.e., in the beginning of the first period) and subsequent installments in the beginning of each period. It is also common in lease or hire purchase contracts that the lease or hire purchase payments are required to be made in the beginning of each period. Lease is a contract to pay lease rentals (payments) for the use of an asset. Hire purchase contract involves regular payments (installments) for acquiring (owning) an asset. 'Annuity due' is therefore a series of fixed receipts or payments starting at the beginning of each period for a specified number of periods. It is seen that the compound value of an annuity due is more than of an annuity because it earns extra interest for one year.

$$\text{Future value of annuity due} = \text{Future value of an annuity} \times (1+i) = A\left[\frac{(1+i)^n - 1}{i}\right][1+i]$$

$$\text{Present value of annuity due} = \text{Present value of an annuity} \times (1+i) = A\left[\frac{1}{i} - \frac{1}{i(1+i)^n}\right](1+i)$$

The present value of annuity due is also more than of an annuity by the factor of $(1 + i)$.

Example 9.19. What is the present worth of a series of 10 years – end payments of Rs. 1000 each when the first payment is due today and interest rate is 12 percent.

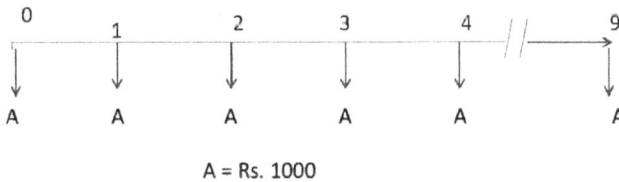

A = Rs. 1000

Solution: Given, A = Rs. 1000, $P = A + A\ (P/A,\ 12\%,\ 9) = 1000 + 1000\left[\frac{(1+i)^n - 1}{i(1+i)^n}\right] =$

$$1000 + 1000\left[\frac{(1.12)^9 - 1}{0.12(1.12)^n}\right] = 1000 + 1000\ [5.329] = \text{Rs. } 6329$$

Alternatively, the present value can be calculated by putting the formula directly,

$$P = A\left[\frac{(1+i)^n - 1}{i(1+i)^n}\right][1+i], \quad P = 1000\left[\frac{(1.12)^{10} - 1}{0.12(1.12)^{10}}\right][1.12],$$

$$P = 1000\left[\frac{2.1058}{0.372}\right][1.12] = \text{Rs. } 6329$$

9.5 SUMMARY OF INTEREST FORMULAE DERIVED SO FAR

The interest formulae derived so far have been summarized in the following Table. The nature of payments has been considered discrete and compound interest has been used.

Table 9.1 Different Interest formulae

Types of transaction	Factor	To find	Given	Formulae
Single payment	Compound amount or future value	F	P	$F = P(1+i)^n = P(F/P, i, n)$
	Present worth	P	F	$P = \dfrac{F}{(1+i)^n} = F(P/F, i, n)$
Equal payment series	Compound amount	F	A	$F = A\left[\dfrac{(1+i)^n - 1}{i}\right] = A(F/A, i, n)$
	Sinking fund	A	F	$A = F\left[\dfrac{i}{(1+i)^n - 1}\right] = F(A/F, i, n)$
	Present worth	P	A	$P = A\left[\dfrac{(1+i)^n - 1}{i(1+i)^n}\right] = A(P/A, i, n)$
	Capital Recovery	A	P	$A = P\left[\dfrac{i(1+i)^n}{(1+i)^n - 1}\right] = P(A/P, i, n)$
Gradient Series	Uniform or arithmetic gradient series	A_g	g	$A_g = A1 + G\left[\dfrac{1}{i} - \dfrac{n}{(1+i)^n - 1}\right]$
	Geometric gradient series	P_{gg}	g	$P_{gg} = \dfrac{A_1}{(i-g)}\left[1 - \left(\dfrac{1+g}{1+i}\right)^n\right]$ When $i \ne g$

9.6 EFFECT ON INFLATION ON CASH FLOWS

The section discusses the understanding and calculating the effects of inflation in the computation of time value of money. Inflation refers to increase in price and fall in purchasing power of money. It causes increased prices of goods and services in future. Inflation is a reality that we deal with nearly every day in our professional and personal lives. We are all very well aware that Rs. 100 now does not purchase the same amount as Rs. 100 did in previous years. This can be explained with the help of inflation and purchasing power of money.

Inflation is an increase in the amount of money necessary to obtain the same amount of goods or services before the inflated price was present. It is a global phenomenon. It is the process in which there is a continuous increase in general price level of goods and services and the money is continuously losing its value. Purchasing power or buying power measures the value of a currency in terms of the quantity and quality of goods or services that one unit of money will purchase. Inflation decreases the purchasing ability of money in which less goods or services can be purchased for the same one unit of money. Due to decrease in value of money, more amount of money is required for the same amount of goods or services. On the contrary, deflation reflects an upward change in the purchase power of the currency.

A general inflationary trend in the cost of goods is common everywhere due to various reasons. If the rate of inflation is very high, it produces extremely serious consequences for both individuals and institutions.

Inflation is the rate of increase in the prices of goods per period. So, it has a compounding effect. Thus, prices that are inflated at a rate of 5% per year will increase 5% in the first year and for the next year; the expected increase will be 5% of these new prices. The same is true for succeeding years and hence the rate of inflation is compounded in the same manner that an interest rate is compounded. If the average inflation over five years is 5%, then the prices at the beginning of the sixth year would be 127% [100 $(1+0.05)^5$] that of first year by assuming 100% for the prices at the beginning of the first year of five-years period.

If economic decisions are taken without considering the effect of inflation into account, most of them would become meaningless and, as a result, the organization would end up with unpredictable return. But there is always difficulty in determining the rate of inflation. For practical decision making, an average estimate may be assumed depending on the period of the proposals under consideration. Hence, we need a procedure which combines the effects of inflation rate and interest rate to take a realistic economic decision.

The impact of inflation is that the "basket of goods" a consumer buys today with Rs. 100 contains more than the "basket" the consumer could buy one year from today. The decrease in purchasing power is the result of inflation.

9.7 NOMINAL VERSUS REAL RATES OF RETURN:

Suppose a person deposits Rs. 100 in the State Bank of India for one year at 10 percent rate of interest. This means that the bank agrees to return Rs. 110 to him after a year, irrespective of how much goods or services, this money can buy for him. The sum of Rs. 110 is stated in nominal terms due to nominal interest rate. Thus 10 per cent is a nominal rate of return on his investment. Let us assume that the rate of inflation is expected to be 7 percent next year. Consideration of inflation indicates that the prices prevailing today will rise by 7 percent next year. In other words, a 7 percent rate of inflation implies that what can be bought for Rs. 1 now can be bought for Rs. 1.07 next year. We can thus say that the purchasing power of Rs. 1.07 next year is the same as that of Rs. 1.00 today. Likewise, the purchasing power of Rs. 110 received next year can buy goods worth Rs. 102.80 (110/1.07 = 102.80). Hence, Rs. 110 next year and Rs. 102.80 today are equivalent in terms of the purchasing power if the rate of inflation is 7 percent. Thus, Rs. 110 is expressed in nominal terms since they have not been adjusted for the effect of inflation. The person thus earns 10 per cent nominal rate of return, but only 2.8 [{(102.8-100)/ 100} × 100] percent real rate of return. It should be noted that the rate of inflation is an expected rate, therefore the real rate of return is also expected.

10 % interest rate might seem to be an adequate return for an investor, but when 7 % inflation rate is considered and compounded annually and deducted from the 10 % return, the real return on investment is obtained.

Hence, the real interest rate 'i_r' is a function of the nominal interest rate 'i_n' and the rate of inflation 'i_i' and is mathematically expressed as $i_r = \dfrac{(i_n - i_i)}{(1 + i_i)} = \dfrac{(0.1 - 0.07)}{(1 + 0.07)} = 0.028 = 2.8\%$

Let us consider for Rs. 1.00 with 10 % interest rate and 7 % inflation rate.

Table 9.2: Computation of real rate of return

Period	Value at 10 % interest (A)	Annual inflation value (7 %) (B)	Actual return (C) = {(A-B)/B}+ 1.000	Real rate of return (D) = {(C_n-C_{n-1})/A} × 100
Today	1.000	1.000	1.000	------
1styr	1.100	1.070	1.028 = $(1.028)^1$	2.8 %
2ndyr	1.210	1.145	1.057 = $(1.028)^2$	2.8 %
3rdyr	1.331	1.225	1.086 = $(1.028)^3$	2.8 %
4thyr	1.464	1.310	1.117 = $(1.028)^4$	2.8 %
5thyr	1.610	1.403	1.147 = $(1.028)^5$	2.8 %

Due to inflation, the real rate of return is 2.8 % instead of 10 %.

9.8 EXPRESSION FOR AVERAGE RATE OF INFLATION

The average rate of inflation is usually expressed as an annual percentage rate that represents the annual increase in the sample prices of some stipulated market basket of commodities (usually both goods and services) over one-year period. The inflation rate has a compounding effect because its value for each year is based on the previous year's price. If C_0 is the present (initial) cost of a commodity and $j_1, j_2, \ldots\ldots\ldots j_n$ is the rate of inflation in years 1, 2, ... n, then its cost at the end of corresponding years will be

$$C_1 = C_0\left(1+j_1\right), C_2 = C_0\left(1+j_1\right)\left(1+j_2\right)\ldots\ldots C_n = C_0\left(1+j_1\right)\left(1+j_2\right)\ldots\ldots\ldots\left(1+j_n\right)$$

If j is the average rate of inflation during the year 0 to year n, the cost C_0 changes to C_n as

$$C_n = C_0 \ (1 + j)^n$$

Conversely, if C_n is the cost of a commodity in current currency, that has inflated by an average annual rate j, to obtain its cost in currency with worth that prevailed at the year 0 (real value), the current value is deflated i.e. divided by a deflator; $C_0 = \dfrac{C_n}{(1+j)^n}$

Example 9.20. A person wants to start a solar lantern manufacturing company after 5 years. The capital investment for the above business is Rs. 10 lakhs in terms of today's rupee value. He is to save the required amount each year with 12% rate of interest; compounded annually. What equal amount should he save each year, so that he will be able to start the business after 5 years from now taking the inflation rate into account. Assuming the annual average rate of inflation is 6% for next 5 years.

Solution: The inflated amount after 5 years

$F = P(1 + j)n = 10$ lakh $(1+ 0.06)^5 = 13.38$ lakhs

Present worth after considering inflation is

10 lakhs × inflation factor $\times\left(\dfrac{P}{F}, 12\%, 5\right) = 5\times\left(F/P, 6\%, 5\right)\times\left(P/F, 12\%, 5\right) = 7.6\,\text{lakhs}$

Total present worth of the proposed project is 10 lakhs + 7.6 lakhs = 17.6 lakhs. Putting this amount into equal payment series capital recovery factor.

$$A = P\left[\dfrac{i(1+i)^n}{(1+i)^n}\right] = P\left(A/P, i, n\right); \ A = 17.6\left[\dfrac{0.12(1.12)^5}{(1.12)^5 - 1}\right] = 4.88\,\text{lakhs}$$

Hence the person is to save Rs. 4.88 lakhs annually for 5 years in order to enable him to start the business after 5 years.

Example 9.21. A restaurant plans to install 400 liters capacity solar water heating system. It received prices from two firms from its recent advertisement for the purchase and installation of solar water heating system. The data are as per the estimate in today's rupee value. Assuming an average annual inflation of 5% for the next five years, choose the firm from which the solar water heating system should be purchased on present worth method. Interest rate is 15% compounded annually.

Information of solar water heater	Firm 1	Firm 2
Purchase price (Rs.)	1,00,000	1,40,000
System life (years)	8	8
Salvage value at the end of the system (Rs)	20,000	40,000
Annual operating & maintenance cost (Rs.)	9,000	5,000

Solution: Average annual inflation rate = 5%, Interest rate = 15% compounded annually.

Firm – 1, Purchase price Rs. 1, 00,000; System life = 8 years, salvage value at the end of system's life = Rs. 20,000; Annual operating & maintenance cash = Rs. 9,000.

Computation of the present worth of the annual operating and maintenance cost of System for firm – 1.

End of year	Annual operating & maintenance cost (Rs.)	Inflation factor (F/P, 5%, n)	Inflated annual operating & maintenance cost (Rs.)	Present worth factor (P/F, 15%, n)	Present worth of inflated annual operation & maintenance cost
A	B	C	D	E	F (Rs.)
			B x C		D x E
1	9,000	1.050	9450	0.8696	8217
2.	9,000	1.102	9918	0.7561	7499
3.	9,000	1.158	10,422	0.6575	6852
4.	9,000	1.216	10944	0.5718	6258
5.	9,000	1.276	11484	0.4972	5710
6.	9,000	1.340	12060	0.4323	5214
7.	9,000	1.407	12662	0.3759	4760
8.	9,000	1.477	13293	0.3269	4345
				Total	Rs. 48855

The equation for the present worth of system for firm 1

$PW_{(Firm\ 1)}(15\%)$= Purchase price + Present worth of inflated annual operating and maintenance cost – present worth of salvage value.

$$= 1,00,000 + 48855 - 20,000 \times \text{(inflation factor)} \times \text{(P/F, 15\%, 8)}$$

$$= 1,00,000 + 48855 - 20,000 \,(1.477)\,(0.327)$$

$$= 1,00,000 + 48855 - 9660 = \text{Rs. } 1,39,195$$

For Firm 2

A	B	C	D	E	F
			(B x C)		(D x E) Rs.
1.	5000	1.050	5250	0.8696	4565
2.	5000	1.102	5510	0.7561	4166
3.	5000	1.158	5790	0.6575	3807
4.	5000	1.216	6080	0.5718	3477
5.	5000	1.276	6380	0.4972	3172
6.	5000	1.340	6700	0.4323	2896
7.	5000	1.407	7035	0.3759	2644
8.	5000	1.477	7385	0.3269	2414
				Total	27,141

The equation for the present worth of system for firm – 2

$$PW_{(\text{Firm 2})}\,(10\%) = 1,40,000 + 27141 - 40,000\,(1.477)\,(0.327)$$

$$= 1,40 + 2741 - 19319 = 1,47,822$$

Since the present worth cost of firm 1 is less than of firm 2, we are to choose firm1.

Inflation rate (j): The inflation rate, 'j' is a measure of the decline in the value of money over time. Thus if inflation rate is 5%, the value of money will decrease by 5% in a year and as compared to today, a product would cost 5% more in the next year. If the cost of a product today is C_0, then its cost, a year later, will be $C_0\,(1+j)$. Two years later, the cost of the same product would be $C_0\,(1+j)\,(1+j)$. Here 'j' is inserted in decimal and not in percentage. Extending this logic to 'n' years, the future cost of the product in n years, $C(n)$, due to inflation rate, 'j' can be given as $C(n) = C_0\,(1 + j)^n$

Discount rate (d): The value of money over time increases due to the interest it can earn. It gives the rate at which value of saved money increases. If we have 'X' amount of money today, it will have some purchasing power. The purchasing power of this 'X' amount will be higher next year due to the interest that it will earn. Thus, money has increased its value with time. In order to represent the true increase in the value of money over time, discount rate, 'd' (instead of inflation rate) is used. The discount rate gives the increase in value of money with respect to future value of money, while interest rate gives the increase in the value of money with respect to current value of money as given below:

$$Discount\ rate\,(d) = \frac{Future\ value - present\ value}{Future\ value}$$

$$Interest\ rate\,(i) = \frac{Future\ value - present\ value}{Present\ value}$$

As mentioned earlier, the process used in converting the future value to its present value is called discounting i.e. $P = F\,(P/F,\ i,\ n)$ and it is through discount rate. Similarly, the process of converting present value to its future value is called compounding and it is through the use of interest rate. The inflation rate has a compounding effect because its value for each year is based on the previous year's price. If j is the average rate of inflation during year 0 to year n, the cost C_0 changes to C_n as $C_n = C_0\,(1 + j)^n$.

9.9 COMPARATIVE ECONOMIC EVALUATION OF INVESTMENT PROJECTS

A comparison of the costs, benefits and timings of a project is often required to differentiate its feasibility or profitability of the investment compared to the other. When investment projects are considered individually, any of the cash flow criteria – net present worth, net future worth or internal rate of return may be followed for obtaining a correct 'accept' or 'reject' signals. Because of economic dependency, a need arises for comparing projects in order to accept one and reject others. This is known as the evaluation of project alternatives. The technique of evaluation of projects, be it in industry or agriculture, throws light on the capacity of different projects to offer returns on investments. In order to evaluate the projects, one should have sufficient information about the various concepts and techniques of capital investment among alternative prospects. These aspects have been discussed in this section.

A project is basically an investment activity where we spend capital resources to create productive assets for realizing benefits over time. It is meant for increasing the output from the given resources. Evaluation of projects needs projecting the future trend of output, sales, expected costs, returns, flow of funds etc. It also refers to specific activity, with specific starting point and specific end point to achieve a specific objective. It should be measureable in cost and returns.

Any investment activity has six important aspects to be looked into. They are technical, administrative, organizational, commercial, financial and economic aspects.

(i) *Technological aspect:* All the technological aspects of the project must be thoroughly studied under technicality points of view. Goods and services required for the execution of project, need a detailed assessment. The awareness of the supporting agency regarding the technology to be used, i.e. capital intensive technology, labour intensive technology, latest technology or the existing technology is to be assessed.

Another important factor is the technical feasibility which determines the size of the project based on capital requirement, future and present demand for product, cost – benefit aspects etc. The selected area for the project must be adequate in the resource endowment base and infrastructural facilities.

(ii) *Administrative aspect*: In this aspect, managerial features, project staff, extension personnel, credit agencies and beneficiaries are assessed.

(iii) Organizational aspects: In this aspect, the relationship of project administration and the Government training arrangements, disbursement of wages etc. are dealt.

(iv) Commercial aspects: In this aspect, arrangements for the supply of input materials, services needed for the project, marketing of output etc. are to be assessed.

(v) *Financial aspects:* The items which fall under this category are sources of funds, cost of funds, repayment etc. The estimated costs based on technical aspects, estimated sales based on commercial analysis and probable profits from the operation of the project are to be properly evaluated. Financial gains accrued as well as incentives offered to the participants in the project should also be viewed. The financial analysis compares benefits and costs to the enterprise only, while the economic analysis compares the benefits and costs to the whole economy of the society where the project is proposed to be executed. Financial analysis is largely confined to individual organization or its unit. It involves a fairly quantitative, fund-based approach that directly compares the expenses and revenues from a venture to determine profitability and hence sustainability. Such evaluation may often employ the financial statement of an enterprise i.e. the balance sheet, the income statement and the cash flow statement etc. However, economic analysis measures the project's positive and negative impacts to the society and its economic development as a whole.

(vi) *Economic analysis:* The economic analysis concentrates in determining project's contribution to the development of the economy as a whole and justifying the use of scare resources. Economic analysis, on the other hand, takes a much wider view and entails the impact of a project on society as a whole. It considers the viewpoints of all stakeholders and how the results of a project align with the broader economic and social policies as well as the International scenario. The costs in an economic analysis are a measure of the resources that a society collectively invests for the fulfilment of the project. The benefits, however, need not be just monetary and often include intangible benefits. Proper identification of costs and benefits is an important aspect in the economic analysis of projects.

For economic evaluation of projects among the alternatives, a wide variety of appraisal criteria have been suggested. However, no best appraisal is found for all types of conditions. Each method has its own merits and limitations. The methods discussed here merely aid in the decision making process, which may include several non-quantitative and non-economic criteria as well. In this section, the concept of present worth, future worth, annual worth and internal rate of return methods are used for assessing the feasibility of the project as well as the selection of the best alternatives from a given set of mutually exclusive projects.

A present worth comparison converts all the cash flows to a present worth or value. A future worth comparison converts all the cash flows to a future value at a common future time (usually at the end of the study period). An annual worth analysis converts all the cash flows to a uniform annual series over the study period. The decision rules for present worth, future worth and annual worth are the same. A positive value of the present worth, future worth and annual worth indicates that the project is economically feasible and cost effective for a given minimum annual worth indicates that the project is economically feasible and cost effective for a given minimum annual rate of return. It is important to note that the three measures are economically equivalent.

The basic difference between financial and economic analysis is that the financial analysis compares benefits and costs to the enterprise, while the economic analysis compares the benefits and costs to the whole economy. **Economic analysis** *is concerned with the **true value** a project holds for the society as a whole. It considers all sections of society and measures the project's positive and negative impacts.*

Net present Value (NPV): NPV is one of the economic methods of evaluating the investment proposals. It is a discounted cash flow (DCF) technique that explicitly recognizes the time value of money. It correctly postulates that cash flows arising at different time periods differ in value and are comparable only when their equivalents i.e. present values are found out. It is calculated by subtracting present value of cash outflows from present value of cash inflows. The investment proposal may be accepted if NPV is positive (i.e. NPV>0). Wealth is also sometimes called as net present value. Hence NPV of a financial decision is the difference between the present value of cash inflows and the present value of cash outflows.

Non-discounted Cash flow: The non-discounted cash flow method is a simple method of choosing among the alternative projects. It ignores the time value of money and is therefore unaffected by discount rates. It is easy to understand, compute and inexpensive to use. Simple payback period and accounting rate of return method are included under this, for appraisal of the project.

Discounted Cash flow: Discounted cash flow (DCF) method is called time adjusted method in which timing of the cash (not the profit) is important in capital investment appraisal. It therefore, requires estimates of cash flows which is a tedious task. The methods like discounted payback period, internal rate of return (IRR), benefit cost ratio and sensitivity analysis are included under discounted cash flow method. It involves the application of discounting arithmetic to the estimated future cash flows from a capital investment project, in order to decide that the project is expected to earn a satisfactory rate of return.

9.10 BASES FOR COMPARISON OF ALTERNATIVES

In most of the practical decision environments, the organizer is forced to select the best alternative from a set of competing alternatives. It is not advisable to invest the required amount in every project. A project which is giving highest return would be selected for the investment proposal. The feasibility of the project can be analysed with various methods. Broadly, feasibility methods can be divided into two i.e. discounted and non-discounted method. The details of the methods of analysis have been shown in Fig. 9.8

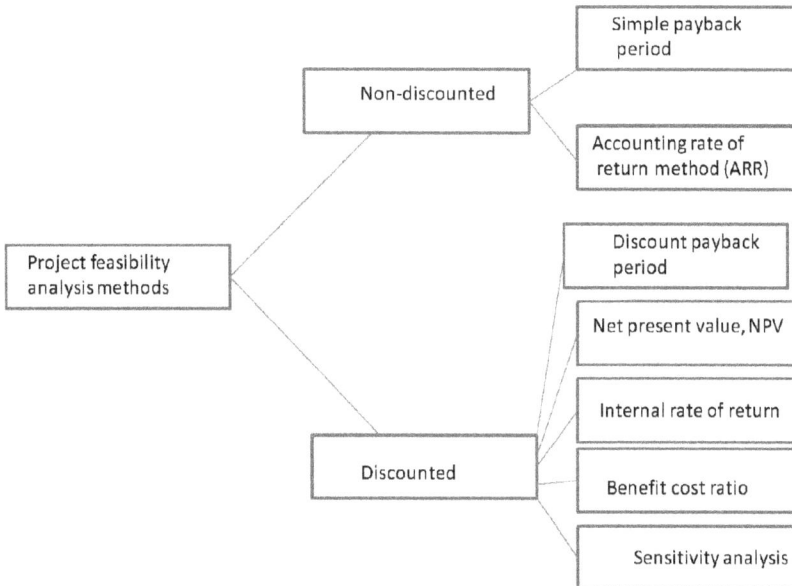

Fig. 9.8. Project feasibility analysis methods

Payback period: - The payback period is one of the most popular and widely used traditional methods of evaluating investment proposals. Sometimes business firms are concerned with the number of years required to recover the initial outlay of an investment. Payback period is therefore defined as the number of years required to recover the original cash outlay invested in a project. The payback period is calculated in two ways such as simple or conventional and discounted payback period.

Simple (conventional) payback period (n_{sp}): In case of simple payback period, the interest rate is as summed to be equal to zero and therefore time value of money is ignored. It is computed by counting the number of years for which the cumulative cash receipts (benefits B_n) be equal to the cash expenses (Costs, C_n).

$$\sum_{n=0}^{n_{sp}} (B_n - C_n) = 0$$

If payback period is less than the useful life time of the investment project, then the proposed investment is accepted. If the project generates constant annual cash inflows, the payback period can be computed by dividing cash outlay by the annual cash inflow. In case of unequal cash inflows, the payback period can be found out by adding up the cash inflows until the total is equal to the initial cash outlay.

Example 9.22. A project which requires an initial investment of Rs. 50,000 would give a return of Rs. 10,000 cash for next 12 years. Find the payback period of this project.

Given, Initial investment C_0 = Rs. 50,000

Annual Cash inflow = Rs. 10,000

$$Payback\ period = \frac{Initial\ investment}{Annual\ cash\ inflow} = \frac{50,000}{10,000} = 5\ years$$

which means the initial investment can be recovered in 5 years for this project.

Example 9.23. An investment of Rs. 20,000 in a project will generate income of Rs. 10,000, 8,000, 6,000, 4,000 and 2,000 in next 5 years. Find the payback period for the above investment.

Year	Cash in flows	Cumulative cash inflows
1.	10,000	10,000
2.	8,000	18,000
3.	6,000	24,000
4.	4,000	28,000
5.	2,000	30,000

From the above table, we can find that, at the end of 2nd year, the project recovers Rs. 18,000 and at the end of 3rd year Rs. 24,000. But we will have to find out at what time period, the initial investment of Rs. 20,000 can be recovered. The cash inflow in 3rd year is Rs. 6,000. For one month, the recovered amount will be $\frac{6000}{12} = $ Rs. 500. But we need Rs. 2000 more to recover (after 2nd year), the project's initial investment. Thus time required to recover Rs. 2000 is $\frac{2000}{500} = 4$ months.

Hence payback period = 2 years + 4 months.

Note: - While comparing the projects (mutually exclusive) on the basis of payback period, the project with shortest payback period will be chosen.

Example 9.24. A proposed solar PV street light will cost Rs. 12, 000 and last for 25 years. The saving of electricity bill is estimated as Rs. 1000 per year. What is the simple payback period of the proposal?

Solution: Given, Initial investment, C_0 = 12,000, Net annual savings = B = 1000 per year

Now, Simple payback period, $n_{sp} = \dfrac{12000}{1000} = 12$ years

This period is much less than the expected useful life of the street light. Hence the proposal may be accepted.

Merits of simple payback period:

- It is simple both in concept and application.
- It is a ready method for dealing with risk.
- It favors projects which generate substantial receipts in earlier years and discriminates against projects which bring substantial receipts in later years but not in earlier years.

Its Limitations:

- It ignores cash flows beyond the pay-back period.
- It fails to consider time value of money. Receipts, in calculation of pay-back period are simple added without suitable discounting.
- It is a measure of the project's capital recovery not profitability.

Discounted Payback period (n_{dp}): As a modification of the simple payback period (which fails to consider the time value of money), the costs and benefits may be adjusted to account for the changing value of money over time. This leads to the estimation of a discounted payback period which is the length of time required for the project's equivalent receipts to be equal to the project's equivalent capital outlays. Therefore, in discounted payback period, n_{dp} is the smallest 'n' that satisfies

$$\sum_{n=0}^{n_{dp}} (B_n - C_n)\left(\frac{1}{(1+i)^n}\right) = 0$$

where 'i' is the interest rate or the minimum rate of return.

Example 9.25. The cash flows of three investment proposals are given in the following table. The life time of the project proposals is 6 years. The rate of interest is considered to be 15%. Calculate both simple and discounted payback period for each of them and comment on the results.

Cash flow (Rs.)

Year	Project I	Project II	Project III
0	-2500	-2500	-2500
1	500	700	400
2	500	700	800
3	500	700	1000
4	500	700	1100
5	500	700	1300
6	500	700	1400

Calculation for simple payback period

Cumulative Cash flows (Rs.)

Year	Project I		Project –II		Project – III	
	Current	Cumulative	Current	Cumulative	Current	Cumulative
0	-2500	-2500	-2500	-2500	-2500	-2500
1	500	-2000	700	-1800	400	-2100
2	500	-1500	700	-1100	800	-1300
3	500	-1000	700	-400	1000	-300
4	500	-500	700	300	1100	800
5	500	0	700	1000	1300	2100
6	500	500	700	1700	1400	3500

For Project – I, Payback period is 5 years, For Project – II, Payback period is $3 + \dfrac{400}{700} = 3.57$ years. For Project – III, Pay back is $3 + \dfrac{300}{1100} = 3.27$ years. As per simple payback period, Project III is accepted due to shortest period for recovery of investment.

Computation for discounted payback period (Rate of interest = 15%)

Year	Project I		Project –II		Project – III	
	Discounted	Cumulative	Discounted	Cumulative	Discounted	Cumulative
0	-2500	-2500	-2500	-2500	-2500	-2500
1	435 = {500 × (1/1.15)}	-2065	609 = {700 × (1/1.15)}	-1891	348 = {400 × (1/1.15)}	-2152
2	378 = {500 × (1/1.15)^2}	-1687	529 = {700 × (1/1.15)^2}	-1362	604 = {800 ×(1/1.15)^2}	-1548
3	329	-1358	460	-902	658	-890
4	286	-1072	362	-540	629	-261
5	249	-823	348	-192	646	385
6	216	-607	303	113	605	990

For project I, the investment is not recovered till the useful life of the project (6 years). Therefore, this is a net loss to the project and should be rejected. For project II, discounted payback period $= 5 + \dfrac{192}{303} = 5.63$ years. For project III, discounted payback period $= 4 + \dfrac{261}{646} = 4.40$ years. From the discounted payback point of view, the project (III) to consider compared to project II as the investment is recovered earlier.

Though in simple as well as in discounted payback period computation, Project III is the best, but the payback period is shorter in simple payback period than the discounted payback period.

Hence discounted payback period is more effective than simple payback period.

Accounting Rate of Return (ARR): The accounting rate of return method is based on the accounting approach rather than cash flow approach. Under this method, various projects are ranked in order of rate of return. A project which is giving highest rate of return is selected. It can be calculated as follows;

$$Accounting\ rate\ of\ return = \left(\frac{Total\ profits\ (after\ depreciation\ and\ taxes)}{Net\ Investment\ in\ the\ project} \right) \times (100)$$

or, Accounting rate of return $= \dfrac{Average\ Annual\ profit}{Net\ investment} \times 100$

This is a non-discounted method of analysing the project's feasibility. Procedure of calculating accounting rate of return is as follows.

i) Calculate average profit for the whole period for which the profit is earned.

ii) The net investment in the project can be calculated after deduction of scrap value (if any)

iii) The ARR can be calculated by dividing the average annual profit by net investment in the project with a multiplication of 100 to know the percentage of return.

Example 9.26. A project costs Rs. 4, 00,000 and has a scrap value of Rs. 40,000 after five years. It is expected to yield a profit after depreciation and taxes during the five years as Rs. 20,000, Rs. 50,000, Rs. 60,000, Rs. 40,000 and Rs. 70,000 for the 1st, 2nd, 3rd, 4th and 5th year respectively. Calculate the accounting rate of return.

Solution: Total profit for the five years $= 20,000 + 50,000 + 60,000 + 40,000 + 70,000$

$= 2, 40,000$. Average profit $= \dfrac{2,40,000}{5} = 48,000$

Net investment in the project = Rs. 4,00,000 – Rs. 40,000 = Rs. 3,60,000

$$\text{Accounting rate of return} = \frac{Average\ annual\ profit}{Net\ investment} = \frac{48,000}{3,60,000} \times 100 = 13.33\%$$

Hence, the accounting rate of return of the project is 13.33%.

Net Present Value (NPV): The net present value (NPV) method is also one of the economic methods of evaluating the investment proposals. It is a discounted cash flow technique that clearly focuses the time value of money. It correctly considers the cash flows arising at different time periods and is comparable only when their equivalents i.e. present values are found out.

The net present value is therefore defined as the difference between the present value of cash inflows or benefits and the present value of cash outflows or costs.

Assuming that project 'X' costs Rs. 2,500 now and is expected to generate year-end cash inflows of Rs. 900, Rs. 800, Rs. 700, Rs. 600 and Rs. 500 in years 1 through 5. The interest rate may be assumed to be 10 percent.

The net present value for project 'X' can be calculated as follows.

$$NPV = \left[\frac{900}{(1+0.1)^1} + \frac{800}{(1+0.1)^2} + \frac{100}{(1+0.1)^3} + \frac{600}{(1+0.1)^4} + \frac{500}{(1+0.1)^5} \right] = 2500$$

$$= 2,725 - 2,500 = + Rs.\ 225$$

Project 'X''s present value of cash inflows (Rs. 2,725) is greater than the of cash outflows (Rs. 2500). Thus it generates a positive net present value. (NPV = +Rs. 225). Project 'X' adds to the wealth of the owners, therefore, it should be accepted. The general formula for the net present values can be written as

$$NPV = \left[\frac{C_1}{(1+i)^1} + \frac{C_2}{(1+i)^2} + \frac{C_3}{(1+i)^3} + \ldots + \frac{C_n}{(1+i)^n} \right] - C_0 = \sum_{t=1}^{n} \frac{C_t}{(1+i)^t} - C_0$$

Where C_1, C_2, ... represent net cash inflows in year 1, 2, ... 'i' is the rate of interest. C_0 is the initial cost of the investment and 'n' is the expected useful life of the investment. 't' represents the end of the period and varies from 1 to n.

If C_0 is the initial capital investment in the project and 'A' is the uniform annual cash flow of the project, then NPV would become as follows;

$$NPV = C_0 - A \left[\frac{(1+i)^n - 1}{i(1+i)^n} \right]$$

It should be clear that the acceptance rule using the NPV method is to accept the investment project if its net present value is positive (NPV > 0) and to reject it if the net present value is negative (NPV < 0). Positive NPV contributes to net wealth of the owner. The positive net present value will result only if the project generates cash inflows at a rate higher than the interest rate. A zero NPV implies that project generates cash flows at a rate just equal to the interest rate. Hence the NPV acceptance rules are

- Accept the project when NPV is positive. NPV > 0.
- Reject the project when NPV is negative. NPV < 0.
- May accept the project when NPV is zero. NPV = 0.

The NPV method can be used to select between mutually exclusive projects; the one with the higher NPV should be ranked in order of net present values, i.e. the first rank will be given to the project with highest positive net present value and so on. The interest rate used in the calculation of NPV is referred to as the discount rate.

Example 9.27. For the cash flows of three projects, I, II & III mentioned in the previous example-9.25, calculate the net present value for each of them, considering the rate of interest as 15%. Comment on the result.

Solution: The net present value of each project for the lifetime of the investment (6 years) is calculated. This is in fact, the net cumulative discounted cash flows. Therefore,

NPV of project I = -607

NPV of project II = 113

NPV of project III = 990

Comments: Project I has a negative NPV and therefore, it should be rejected. As seen earlier, the project III is better than the project II from the point of view of discounted payback period. In the net present value of the project, Project III is more favourable. This is due to higher cash flows is later years.

Example 9.28. A 100 litre per-day domestic solar water heater costs Rs. 20,000 to purchase. It is expected to save Rs. 4000 annually by reducing the electricity consumption through electric geyser. The annual maintenance cost is estimated to be Rs. 500. Calculate the NPV of the investment in the domestic solar water heater if the useful life is 15 years and interest rate is 15 percent.

Solution: Net annual benefits of using domestic solar water of 100 litre per day capacity = 4000 − 500 = Rs. 3500. Since the amount of net annual benefits is constant over the useful life of the solar water heater, then

$$NPV = -C_0 + A\left[\frac{(1+i)^n - 1}{i(1+i)^n}\right] = -20,000 + 3500\left[\frac{(1.15)^{15} - 1}{0.15(1.15)^{15}}\right] = \text{Rs. } 475$$

Therefore, the investment in the domestic solar water heater is a financially viable investment for the owner.

Merits and demerits of NPV

Merits	Demerits
Considers all cash flows	Requires estimates of cash flows which is a tedious task
True measure of profitability	Caution needs to be given when alternative (mutually exclusive) projects with unequal lives.
Based on the concept of time	Sensitive to discount rates value of money
Satisfies the value – additivity principle (i.e., NPV's of two or more projects can be added)	Ranking of investment projects as per the NPV rule is not independent of the discount rates.

Example of last point Demerits: Let two projects A & B costing Rs. 50 each. Project A returns Rs. 100 after one year and Rs. 25 after two years. Project B returns Rs. 30 after one year and Rs. 100 after two years. At discount rates of 5 percent and 10 percent, the NPV of projects and their rankings are changed.

Projects	NPV at 5%	Rank	NPV at 10%	Rank
A	67.92	II	61.57	I
B	69.27	I	59.91	II

It is seen than the project ranking is reversed when discount rate is changed from 5 percent to 10 percent.

Internal Rate of Return (IRR)

The internal rate of return method is another discounted cash flow technique, which considers the magnitude and timing of cash flow. The internal rate of return (IRR) is the discount rate that makes the present worth of the costs of project equal to the present worth of the benefits of the project. Another way of stating this concept is to define the IRR as the interest or discount rate that reduces the present worth of a series of receipts and disbursement to zero. It is the rate of return that makes the NPV of an investment equal to zero. Thus, the internal rate of return for an investment proposal is the interest

rate (r) which satisfies the equation, $NPV(r) = \sum_{t=0}^{n} \frac{(B_t - C_t)}{(1+r)^t} = 0$ where 't' represents

the end of the period and varies from 0 to n. 'n' is the expected useful life of the investment. B and C represent the benefit and cost of the cash flow.

In other words, IRR is the interest rate 'r' that causes the discounted present value of the benefits in a cash flow to be equal to the present value of the costs i.e.,

$$\sum_{t=0}^{n} \frac{B_t}{(1+r)^t} = \sum_{t=0}^{n} \frac{C_t}{(1+r)^t}$$

Concept of IRR: The concept of IRR is quite simple to understand in the case of a one-period project. Assume that one person has deposited Rs. 10,000 in a bank and would get back Rs. 10,800 after one year. The true rate of return on his investment would be

$$\text{Rate of return} = \frac{10,800 - 10,000}{10,000} = 1.08 - 1 = 0.08 \quad \text{or 8\%. The amount that he would}$$

obtain in the future (Rs. 10,800) would consist of his investment (Rs. 10,000) plus return on his investment (0.08 x Rs. 10,000) = 10,000 + 0.08 (10,000) = 10,000 (1.08) = 10,800

and $10,000 = \dfrac{10,800}{(1.08)}$. It is observed that the rate of return of his investment (8 percent)

makes the discounted value of his cash inflow (Rs. 10,800) equal to his investment (Rs. 10,000). The formula can now be developed for the rate of return (r) on an investment (C_0) that generates a single cash flow after one period (C_1) as follows;

$$r = \frac{C_1 - C_0}{C_0} \Rightarrow r = \frac{C_1}{C_0} - 1$$

The above equation can be rewritten as follows; $\dfrac{C_1}{C_0} = 1 + r$ or, $C_0 = \dfrac{C_1}{(1+r)}$

It is noticed from the above equation that the rate of return 'r' depends on the project's cash flows, rather than any outside factor. Therefore, it is referred to as the internal rate of return. The internal rate of return approach is designed to calculate a rate of return that is "internal" to the project. This also implies that the rate of return is the discount rate which makes NPV to be zero.

It can be noticed that the IRR equation is the same as the one used for the NPV method. In the NPV method, the required rate of return is known and the net present value is found, while in the IRR method, the value of 'r' has to be determined at which the net present value becomes zero.

The accept-or-reject rule using the IRR method, is to accept the project if its internal rate of return is higher than the interest rate of the capital investment. The project would be rejected if its internal rate of return is lower than the interest rate of the capital investment. The decision maker may remain indifferent if the IRR is equal to the interest rate of the investment of the project.

Interpretation of Rate of return and Concept of IRR

From the perspective of someone who has borrowed money, the interest rate is applied to the unpaid balance so that the total loan amount and interest are paid in full exactly with the last loan payment. From the perspective of a lender of money, there is an unrecovered balance at each time period. The interest rate is the return on this unrecovered balance so that the total amount lent and the interest is recovered exactly with the last receipt. Rate of return describes both of these perspectives.

Hence rate of return is the rate paid on the unpaid balance of borrowed money, or the rate earned on the unrecovered balance of an investment, so that the final payment or receipt brings the balance to exactly zero with interest considered. The rate of return is expressed as percent per period. For example, $r = 10\%$ per year. The concept of present value can be used to find out the rate of return or yield. Let us discuss with some examples.

A bank offers someone to deposit Rs. 100 and promises to pay Rs. 112 after one year. What rate of interest would he earn? The answer is 12 percent. $100\ (1+r)=112 \rightarrow r=0.12 \rightarrow 12\%$. Similarly, what rate of interest would he earn if he deposits Rs. 1000 today and receives Rs. 1762 at the end of five years? The problem can be discussed as follows; Rs. 1000 is the present value of Rs. 1762 to be received at the end of the fifth year. Thus,

$$1000 = \frac{1762}{(1+r)^5} = 1762 \times (P/F,\ 5\ \%,\ r) = 1762\ \times present\ value\ factor$$

$$\Rightarrow present\ value\ factor = \frac{1000}{1762} = 0.576$$

From the interest rate table of present and future value factor, the rate of interest would become 12 percent. Hence Rs. 1000 can earn at the rate of 12 percent interest to generate Rs. 1,762 after 5 years. (Rs. 1000 × $(1.12)^5$ = Rs. 1,762).

Likewise, another example of annuity cash flow can also be taken. Assume that someone has borrowed Rs. 70,000 from a finance institution to buy a solar PV system. He is required to pay Rs. 11, 396.93 annually for a period of 15 years. What interest rate would he to pay. It is clear that Rs. 70,000 is the present value of a fifteen-year annuity of Rs. 11,396.93 × Present value annuity factor

$$\text{Present value annuity factor } = \frac{70,000}{11,396.93} = 6.142$$

From the interest table, it can be seen that the interest rate would become 14 percent. Thus the finance institution is charging 14 percent interest from him.

Finding the rate of return for an uneven series of cash flow is a bit difficult. By practice and using trial and error method, one can find it. A following example may be taken to illustrate the calculation of rate of return for an uneven series of cash flows.

Example 9.29. Calculation of rate of return for uneven series of cash flow

A person wants to borrow from his friend Rs. 1,600 today and would return him Rs. 700, Rs. 600 and Rs. 500 in year 1 through year 3 as principal plus interest. What rate of return would his friend earn?

It is to be computed that the rate of return at which the present value of Rs. 700, Rs. 600 and Rs. 500 received respectively, after one, two and three is Rs. 1600. Suppose (arbitrarily), this rate is 8 percent. While calculating the present value of cash flows at 8 percent, the following amount may be obtained.

Cash flow		Present value (PV) of cash flow	
Year	(Rs.)	Present value factor (8%)	(Rs.)
1	700	0.926	648.20
2	600	0.857	514.20
3	500	0.794	397.00
			1559.40

Since the present value at 8 percent is less than Rs. 1600, it implies that his friend is allowing him a lower rate of return, so for example a rate of return of 6 percent may be tried; and the following amount may be obtained.

Cash flow		Present value (PV) of cash flow	
Year	(Rs.)	Present value factor (6%)	(Rs.)
1	700	0.943	660.00
2	600	0.890	534.00
3	500	0.840	420.00
			1,614.00

The present value at 6 percent is slightly more than Rs. 1600, it means that his friend is offering him approximately 6 percent interest. In fact, the actual rate would be a little higher than 6 per cent. At 7 percent, the present value of cash flows is Rs. 1, 586. Hence the interpolation method can be tried to calculate the actual rate.

$$= 6\% + \left(7\% - 6\%\right) \times \frac{(1614 - 1600)}{1614 - 1586} = 6.5\%$$

Check: At 6.5% rate of return, present value of Rs. 700, Rs. 600 and Rs. 500 occurring respectively in year one through three is equal to Rs. 1,600.

$$= \frac{Rs.\,700}{(1.065)} + \frac{Rs.\,600}{(1.065)^2} + \frac{Rs.\,500}{(1.065)^3} = Rs.\,1600$$

The rate of return of an investment is therefore called as internal rate of return (IRR) or yield, since it depends exclusively on the cash flows of the investment.

Graphical representation of NPV (net present value) with interest rate

The variation of NPV with interest rate can be obtained by taking a generalized cash flow diagram to demonstrate the rate of return method for comparison.

Generalized cash flow diagram (Fig. 9.9)

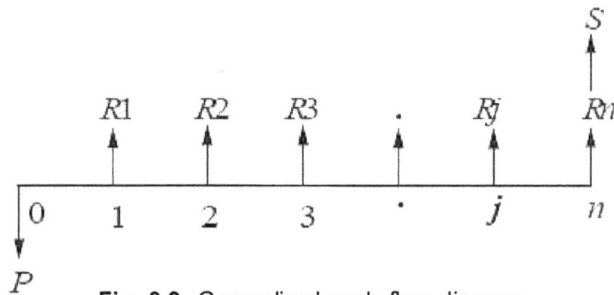

Fig. 9.9. Generalized cash flow diagram

In the above cash flow diagram, P represents an initial investment, R_t, the net revenue at the end of tth year and S, the salvage value at the end of the nth year. The first step is to find the net present value of the cash flow diagram using the following expression at a given interest rate, i.

$$NPV(i) = -P + R_1/(1+i)^1 + R_2/(1+i)^2 + ...R_t/(t+i)^t + ...+$$
$$+Rn/(1+i)^n + S/(1+i)^n$$

Now, the above function is to be evaluated for different values of 'i' until the net present value reduces to zero, as shown in the figure below (Fig. 9.10).

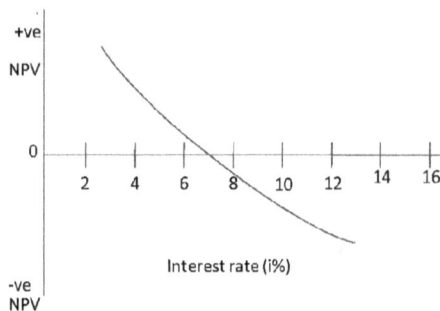

Fig. 9.10. Graphical interpretation of NPV

In the figure above, the NPV goes on decreasing when the interest rate is increased. The value of 'i' at which the NPV curve cuts the X-axis is the rate of return of the given proposal/project. It will be very difficult to find the exact value of 'i' at which the NPV reduces to zero. So, one has to start with an intuitive (trial and error) value of 'i' and check whether, the present worth function is positive. If so, increase the value of 'i' until NPV(i) becomes negative. Then the rate of return is determined by interpolation method in the range of values of 'i' for which the sign of the NPV changes from positive to negative.

Computation of IRR: - The following step-by-step procedure is adopted for computation of IRR by the iterative approach.

i) Make an initial guess for internal rate of return (IRR), 'r'.

ii) Calculate the value of NPV (r) at this value of 'r'.

iii) If the calculated value of NPV (r) is positive, it indicates that the actual value of 'r' would be more than the trial value. Increment of the value 'r' by a small step for the next trial. If the value of NPV (r) is negative, it indicates that the actual value of 'r' would be less than the trial rate. Decrement of the value 'r' by a small step for the next trial. Go to the step 2 for the net trial.

iv) Repeat steps 2 and 3 until a situation is reached where a value 'r', gives NPV (r_1) as positive and the next higher value r_2 gives NPV (r_2) as negative

v) Find the value of IRR by interpolation as follows;

$$IRR = r_1 \left\{ \frac{(r_2 - r_1)}{NPV(r_1) - NPV(r_2)} \right\} NPV(r_1)$$

When the IRR of an investment is greater than interest rate of the capital amount of the project, the investment is accepted. The cost of the capital is the minimum acceptable rate of return or required rate of return of the investment.

Example 9.30. A 100 liter per day capacity solar water heating system requires initial costs of Rs. 20,000. It is expected to save Rs. 4000 per year by reducing electrical energy consumption through geyser. Calculate the IRR on this investment if the life of the solar water heater is 15 years and salvage value after its useful life is negligible.

Solution: C_0 = Rs. 20,000, Annual benefit = B = 4000, n = 15

$$NPV(r) = \sum_{t=0}^{n} \frac{B_t - C_t}{(1+r)^t} = \sum_{t=0}^{n} \frac{B_t}{(1+r)^t} - C_0 = 0 = 4000 \left[\frac{(1+r)^{15}}{r(1+r)^{15}} \right] - 20,000$$

With 0.15 as the initial guess for 'r', the value of NPV (r) is calculated and tabulated as follows;

Number of interaction	r	NPV (r)
1	0.15	3389
2	0.17	1313
3	0.19	-497.28

A better estimate of the true IRR may be obtained by using smaller incremental changes in the interest rate.

$$IRR = 0.17 + \frac{(0.19 - 0.171313)}{1313 + 497} = 0.184 = 18.4\%$$

Use of Linear Interpolation to find the approximation of IRR

The use of linear interpolation to find the approximation of IRR with little error can be discussed through graphical representation of an illustration.

Example 9.31. A capital investment of Rs. 10,000 can be made in a project that will produce a uniform annual revenue of Rs. 5,310 for five years and then have a salvage value of Rs. 2000. Annual expenses will be Rs. 3000. The organizer is willing to accept any project that will earn at least 10% per year on all invested capital. Determine whether it is acceptable by using IRR method.

Solution: NPV $= 0 = -10,000 + (5310 - 3000)(P/A, r\%, 5) + 2000(P/F, r\%, 5)$

We do not know the rate of return. We would probably try a relatively low 'r' such as 5% and a relatively high 'r' such as 15%. Linear interpolation will be used to solve for 'r' and the procedure illustrated in the figure below should not exceed a range of 10%.

At 'r' = 5%; NPV = -10,000 + 2310 (4.3295) + 2000 (0.7835) = +1568

At 'r' = 15%; NPV = -1262

Because we have both a positive and a negative NPV, the answer is in between the two rates of return. The dashed curve in the figure below is what we are linearly approximating. The answer, r % can be determined by using the similar triangles dashed in the figure 9.11.

$$\frac{Line\ BA}{Line\ BC} = \frac{line\ dA}{line\ de}$$

BA in this line segment is; B – A = 15% - 5%. Thus

$$\frac{15\% - 5\%}{Rs.1568 - (-Rs.1262)} = \frac{r\% - 5\%}{Rs.1568 - 0}$$

$$or\ r\% = 5\% + \frac{1568}{(1568 + 1262)}(15\% - 5\%) = 5 + 5.5 = 10.5\%$$

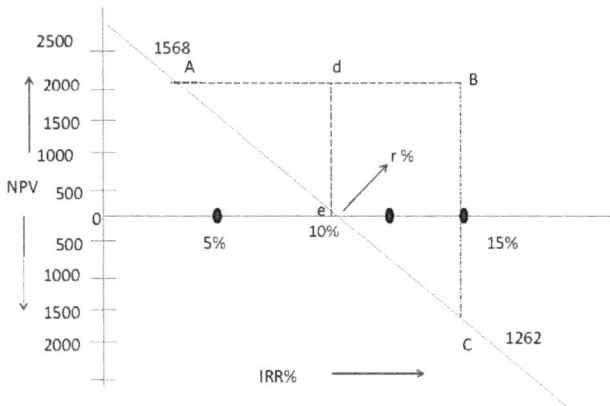

Fig. 9.11. Linear interpolation for IRR

This approximate solution illustrates the trial – and – error process, together with linear interpolation. The error in this answer is due to nonlinearity of NPV and would be less if the range of interest rates used in the interpolation is smaller. The minimum attractive rate of return (MARR) in this example is 10% per year. As IRR > MARR, hence, the project is acceptable.

Merits and Demerits of IRR

Merits	Demerits
Considers all cash flows	Requires estimates of cash flows which is a tedious task
True measure of profitability	At times, fails to indicate correct choice between mutually exclusive projects
Based on the concept of the time value of money	At times yields multiple rates
Generally consistent with wealth maximization principle	Relatively difficult to compute

Benefit to Cost Ratio

Benefit to Cost ratio is a popular method for deciding upon the economic justification of a public project. In evaluating alternatives of private organizations, the criterion is to select the alternative with the maximum profit. The profit maximization is the main goal of private organization. But the same criterion cannot be used while evaluating public alternatives. The main objective of any public alternative is to provide goods/services to the public at the minimum cost. In this process, one should see whether the benefits of the public activity are at least equal to its costs. If yes, then the public activity can be undertaken for implementation. Otherwise, it can be rejected.

Private project is owned and regulated by private individuals. It is always profit oriented and motivated. Whereas public project looks for social welfare not for profit motive. Private projects are evaluated by Net present value (NPV) and Internal rate of return (IRR), but public project is analysed by cost benefit analysis.

The benefits may occur at different time periods of the public activity. For the purpose of comparison, these are to be converted into a common time base (present worth or future worth or annual equivalent). Similarly, the costs consist of initial investment and yearly operation and maintenance cost. These are to be converted to a common time base as done in the equivalent benefits. Now the ratio between the equivalent benefits and equivalent costs is known as the "Benefit-Cost ratio." If this ratio is at least one, the public activity is justified, otherwise, it is not justified.

$$BC\ ratio = \frac{Equivalent\ benefits}{Equivalent\ costs}$$

Let B and C be the present values of the net cash inflows (benefits) and outflow (costs). Then

$$B = \sum_{t=0}^{n} \frac{B_t}{(1+i)^t} \text{ and } C = \sum_{t=0}^{n} \frac{C_t}{(1+i)^t}$$

The ratio B/C is known as benefit – Cost ratio. The cost-benefit of a project can be evaluated by different methods like present worth, future worth and equivalent annual worth methods.

Let B_P = Present worth of the total benefits, B_F = Future worth of the total benefits, B_A = Annual equivalent of the total benefits, P = Initial investment, P_F = Future worth of the initial investment, P_A = Annual equivalent of the initial investment, C_A = Annual cost of operation and maintenance, C_P = Present worth of yearly cost of operation and maintenance, C_F = Future worth of yearly cost of operation and maintenance.

Present worth method: $B/C\ ratio = \dfrac{B_p}{P + C_p}$

Future worth method: $B/C\ ratio = \dfrac{B_F}{P_F + C_F}$

Equivalent annual worth method: $B/C\ ratio = \dfrac{B_A}{P_A + C_A}$

If B/C > 1 ⇒ project is accepted, B/C = 1 ⇒ project is rejected

Example 9.32. Out of three given mutually exclusive project, suggest which one is feasible. Their respective costs and benefits are included in the table below. Each of the project has a useful life of 20 years and nominal rate of interest is 12% per annum.

Item	A (Rs.)	B (Rs.)	C (Rs.)
Capital Investment	7,50,000	9,00,000	11,00,000
Annual maintenance cost	65,000	60,000	50,000
Annual benefits	2,00,000	3,00,000	4,00,000
Salvage value	1,00,000	2,00,000	3,00,000

Solution: This problem can be solved by taking into consideration, the present worth of benefits and present worth of costs.

$$PV \text{ (Cost A)} = 7,50,000 + 65,000 \text{ (P/A, 12\%, 20)} - 1,00,000 \text{ (P/F, 12\%, 20)}$$
$$= 7,50,000 + 65,000 \text{ (7.4694)} - 1,00,000 \text{ (0.1037)} = \text{Rs. } 12,25,141$$

$$PV \text{ (Cost B)} = 9,00,000 + 60,000 \text{ (P/A, 12\%, 20)} - 2,00,000 \text{ (P/F, 12\%, 20)}$$
$$= 9,00,000 + 60,000 \text{ (7.4694)} - 2,00,000 \text{ (0.1037)} = \text{Rs. } 13,27,424$$

$$PV \text{ (Cost C)} = 11,00,000 + 50,000 \text{ (P/A, 12\%, 20)} - 3,00,000 \text{ (P/F, 12\%, 20)}$$
$$= 11,00,000 + 50,000 \text{ (7.4694)} - 3,00,000 \text{ (0.1037)} = \text{Rs. } 15,42,360$$

$$PV \text{ (Benefit A)} = 2,00,000 \text{ (P/A, 12\%, 20)} = 2,00,000 \text{ (7.4694)} = \text{Rs. } 14,93,880$$
$$PV \text{ (Benefit B)} = 3,00,000 \text{ (P/A, 12\%, 20)} = \text{Rs. } 22,40,820$$
$$PV \text{ (benefit C)} = 4,00,000 \text{ (P/A, 12\%, 20)} = \text{Rs. } 29,87,760$$

Hence benefit cost ratio of Project $A = \dfrac{14,93,880}{12,25,141} = 1.2194 > 1$

Benefit cost ratio of Project $B = \dfrac{22,40,820}{13,27,424} = 1.688 > 1$

Benefit cost ratio of Project $C = \dfrac{29,87,760}{15,42,360} = 1.937 > 1$

The benefit cost ratio of Project C is highest among the projects A and B. Hence out of these three projects, project proposal of C can be chosen.

Example 9.33. For the cash flows of three project proposals given in the earlier example-9.25, calculate the benefit to cost ratio for each of them, considering the rate of interest as 15%.

Solution: The discounted benefits and costs are calculated and tabulated in the following table.

Cash flows (Rs.)

	Project-I		Project-II		Project-III	
Year	Discounted benefits	Discounted Cost	Discounted benefits	Discounted costs	Discounted benefits	Discounted costs
0	0	2500	0	2500	0	2500
1	435	0	609	0	348	0
2	378	0	529	0	604	0
3	329	0	460	0	658	0
4	286	0	362	0	629	0
5	249	0	348	0	646	0
6	216	0	303	0	605	0
Total	1893	2500	2611	2500	3490	2500

$$\text{The B} - \text{C ratio of Project I} = \frac{1893}{2500} = 0.75$$

$$\text{The B} - \text{C ratio of Project II} = \frac{2611}{2500} = 1.04$$

$$\text{The B} - \text{C ratio of Project III} = \frac{3490}{2500} = 1.40$$

As the B – C ratio of Project III is highest, it may be chosen compared to the other two project proposals, I & II.

Decision criteria for benefit cost ratio (BCR)

$$BCR \geq 1$$

Relationship between NPV and BCR

BCR = 1, NPV = 0

BCR < 1, NPV = –ve

BCR > 1, NPV = +ve

Projects shall be arranged in the descending orders of BCR for prioritization of investment

Importance of benefit – cost analysis (BCA)

Followings are the various merits of BCA

a. It is a quantitative technique used to evaluate the economic costs and social benefits associated with a particular course of action.

b.   Compare costs and benefits using equal terms.

c.   Provide a clear indication of net costs or benefits.

d.   Simplifies complex concepts and processes

e.   It is mostly applicable for decision on public projects.

f.   Easily accepted by the society than any other economic methods.

g.   Can be carried out at many levels (i.e, local, regional, national and international)

Limitations

a.   In case of those projects whose benefits are diverse and whose beneficiaries are widely dispersed, it may be difficult even to enumerate all the benefits. In such cases, omission or double counting of benefits may also be difficult to avoid.

b.   A number of problem may arise with regard to the evaluation of benefits and costs. Some of the externalities of a project may be important yet difficult to measure.

c.   It neglects the problem of joint benefits and costs arising from a project.

d.   There are also difficulties involved in finding an appropriate rate of discount for making proper allowance for uncertainty. In view of the uncertainty generated by political instability, natural disasters etc, the costs, benefits and even the rate of discount might be difficult to estimate and evaluate.

e.   Cost estimates are made on the basis of the choice of techniques, locations and prices of services used. Market prices of factors of production are used for this purpose provided they reflect opportunity cost. But in underdeveloped countries, market prices usually don't reflect the opportunity cost.

f.   The correct estimation from a project becomes difficult due to uncertainty regarding the future demand and supply of the products from a new project and their prices.

Annual Equivalent Method of evaluating alternatives

Annual equivalent worth is another important method for comparing the project alternatives. In this method, all the receipts and disbursements occurring over a period are converted to an equivalent annual amount. Then the alternative with the maximum annual equivalent revenue in the case of revenue-based comparison or with the minimum annual equivalent cost in case of cost-based comparison will be selected as the favourable alternative. In the annual equivalent method of comparison, either annual equivalent cost or revenue of each alternative is computed. The cash flow diagrams for revenue – dominated and cost-dominated are as follows.

Revenue – dominated cash flow diagram

A generalized revenue-dominated cash flow diagram to demonstrate the annual equivalent method of comparison is presented below (fig. 9.12).

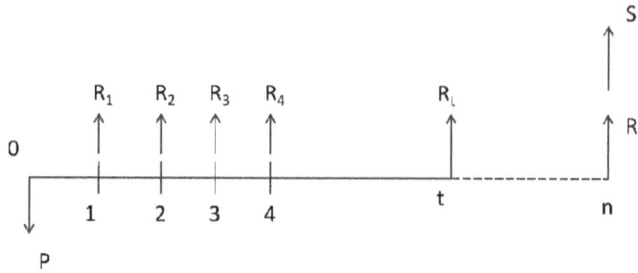

Fig. 9.12. Revenue dominated cash flow diagram

P represents an initial investment. R_t, the net revenue at the end of tth year and S, the salvage value at the end of the nth year. The first step is to find the net present value of the cash flow diagram using the following expression for a given interest rate, 'i'.

$$NPV = -P + \frac{R_1}{(1+i)^1} + \frac{R_2}{(1+i)^2} + ... + Rt\big/(1+i)^t + ... + Rn\big/(1+i)^n + S\big/(1+i)^n$$

In the above formula, the expenditure is assigned with a negative sign and the revenues are assigned with a positive sign.

In the second step, the annual equivalent revenue is computed using the following formula

$$AE = NPV(i) \left[\frac{i(1+i)^n}{(1+i)^n - 1} \right] = NPV(i)(A/P, i, n) \text{ where } (A/P, i, n) \text{ is called uniform}$$

series capital recovery factor.

If we have some more alternatives, which are to be compared with this alternative, then the corresponding annual equivalent revenues are to be computed and compared. Finally, the alternative with the maximum annual equivalent revenue should be selected as the favourable alternative. The cash flow diagram finally converted to annual equivalent revenue based cash flow is as follows;

Cost-dominated Cash flow diagram

A generalized cost-dominated cash flow diagram to demonstrate the annual equivalent method of comparison is illustrated below (fig. 9.13).

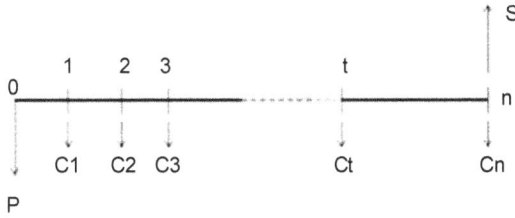

Fig. 9.13. Cost dominated cash flow diagram

P represents an initial investment, C_t the net cost of operation and maintenance at the end of tth year and S, the salvage value at the end of the nth year.

The first step is find the net present cost of the cash flow diagram using the following relation for a given interest rate, 'i'.

$$C_{net} = P + C1/(1+i)^1 + C_2/(1+i)^2 + ... + Ct/(1+i)^t + ... + C_n/(1+i)^n - S/(1+i)^n$$

In the above formula, each expenditure is assigned with positive sign and the salvage value with negative sign. Then, in the second step, the annual equivalent cost is computed using the following equation.

$$C_A = C_{net}\left[\frac{i(1+i)^n}{(1+i)^n - 1}\right]$$

As in the previous case, if we have some more alternatives which are to be compared with this alternative, then the corresponding annual equivalent costs are to be computed and compared. Finally, the alternative with the minimum annual equivalent cost should be selected in the favourable alternative. The cash flow diagram finally converted to annual equivalent cost based cash flow is as follows;

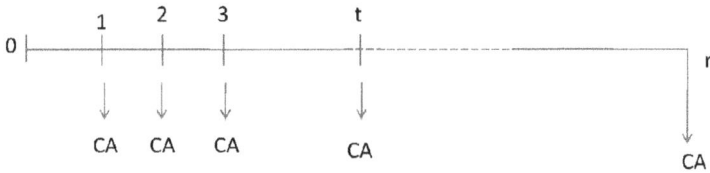

If C_o is the initial cost and C_A is the annual maintenance cost, the $C_{net} = C_0 - S/$

$(1+i)^n + C_A\left[\dfrac{(1+i)^n - 1}{i(1+i)^n}\right]$ where the term $\left[\dfrac{(1+i)^n - 1}{i(1+i)^n}\right]$ is called uniform series present

worth factor.

Example 9.34. A business organizer invests in one of the two mutually exclusive alternatives. The life of both alternatives is estimated to be 5 years with the following investments, annual return and salvage values.

	Alternative	
	A	**B**
Investment (Rs.)	- 2,00,000	-2,15,000
Annual equal return (Rs.)	+ 50,000	+ 60,000
Salvage value (Rs.)	+ 20,000	+ 30,000

Determine the best alternative based on the annual equivalent method by assuming $i = 15\%$.

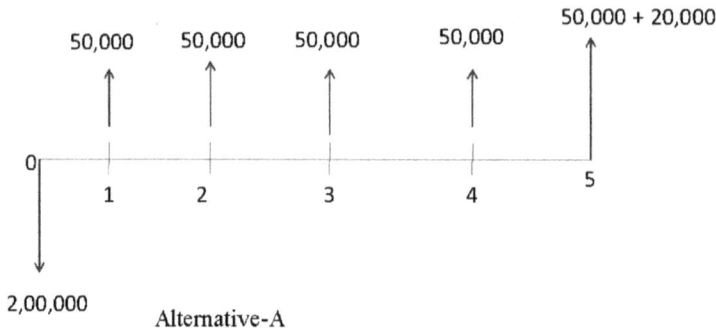

Alternative-A

Solution: Alternative A

The annual equivalent revenue expression of the above cash flow diagram is as follows;

$$AE_A \ (15\%) = -2,00,000(A/P, \ 15\%, \ 5) + 50,000 + 20,000 \ (A/F, \ 15\%, \ 5)$$

$$= -2,00,000\left\{\frac{(0.15)(1.15)^5}{(1+i)^n - 1}\right\} + 50,000 + 20,000\left\{\frac{i}{\left[(1+i)^n - 1\right]}\right\}$$

$$= -2,00,000\left\{\frac{(0.15)(1.15)^5}{(1.15)^5 - 1}\right\} + 50,000 + 20,000\left\{\frac{0.15}{\left[(1.15)^5 - 1\right]}\right\}$$

$$= -2,00,000 \ (0.2984) + 50,000 + 20,000 \ (0.1483)$$

$$= -59,680 + 50,000 + 2966 = -6,714$$

$$AE_B(15\%) = -2,15,000(A/P, \ 15\%, \ 5) + 60,000 + 30,000 \ (A/F, \ 15\%, \ 5)$$

$$= -2,15,000 \ (0.2954) + 60,000 + 30,000 \ (0.1483)$$

$$= -64,156 + 60,000 + 4449 = Rs. \ 293$$

By comparing annual equivalent net returns, Project 'B is +ve and more than Project A, which is –ve, and hence the firm should select Project B.

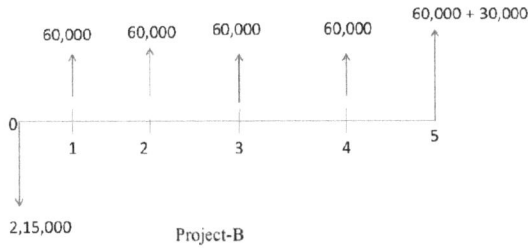

2,15,000   Project-B

The term $(A/F, i, n)$ is called uniform series sinking fund factor.

Example 9.35. A certain firm desires an economic analysis to determine which of the two energy devices is attractive in a given interval of time. The minimum attractive rate of return for the firm is 15%. The following data are to be used in the analysis.

Item	Device X	Device Y
First cost	Rs. 2,00,000	Rs. 2,50,000
Estimated life	15 years	15 years
Salvage value	0	Rs. 10,000
Annual maintenance cost	0	Rs. 4,00

Which machine should be chosen based on annual equivalent cost

The annual equivalent cost (AEC) for the device X is

$$AEC(X)\ (15\%) = 2,00,000(A/P,\ 15\%,\ 15)$$

$$= 2,00,000\left\{\frac{i(1+i)^n}{(1+i)^n -1}\right\} = 2,00,000\left\{\frac{(0.15)(1.15)^{15}}{(1.15)^{15}-1}\right\} = 2,00,000(0.171) = Rs.\ 34,204$$

$$AEC(Y)\ (15\%) = 2,50,000\ (A/P,\ 15\%,\ 15) + 4,000 - 10,000\ (A/F,\ 15\%,\ 15)$$

$$= 2,50,000\ (0.171) + 4000 - 10,000\left\{\frac{i}{(1+i)^n -1}\right\}$$

$$= 2,50,000\ (0.171) + 4000 - 10,000\left\{\frac{0.15}{(1.15)^{15}-1}\right\}$$

$$= 2,50,000\ (0.171) + 4000 - 10,000(0.021) = 42,750 + 4000 - 210 = Rs.\ 46,\ 540$$

The annual equivalent cost of device X is less than that of device Y, So, device 'X' is more cost effective and to be selected.

Example 9.36. A firm is to decide whether to buy the device A or device B.

Item	Device A	Device B
Initial cost (Rs.)	3,00,000	6,00,000
Useful Life (years)	4	4
Salvage value (Rs.)	2,00,000	3,00,000
Annual maintenance (Rs.)	30,000	0

At 15% interest rate, which device should be purchased?

Cash flow diagram for device (A)

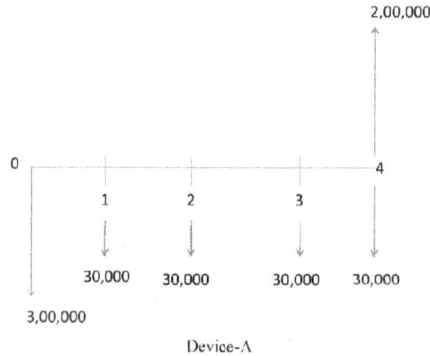

Device-A

$AEC(A)(15\%) = 3,00,000\ (A/P,\ 15\%,\ 4) + 30,000 - 2,00,000\ (A/F,\ 15\%,\ 4)$

$= 3,00,000\ (0.3503) + 3000 - 2,00,000\ (0.2003) = $ Rs. 95,030

Cash flow diagram for device B

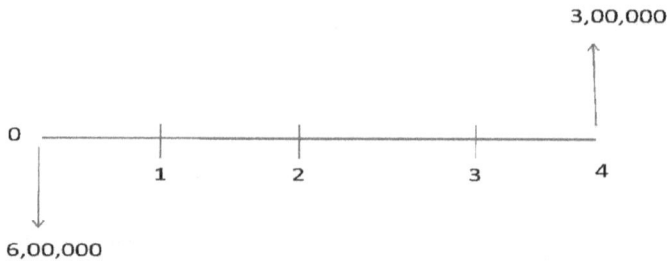

The annual equivalent cost expression for device B is

$AEC(B)\ (15\%) = 6,00,000\ (A/P,\ 15\%,\ 4) - 3,00,000\ (A/F,\ 15\%,\ 4)$

$= 6,00,000\ (0.3503) - 3,00,000\ (0.2003) = $ Rs. 1,50,090

Since the annual equivalent cost of device 'A' is less than that of device 'B', it is advisable to buy device 'A'.

Example 9.37. The cost of a box type solar cooker is 6,000. During its useful life of 15 years, its routine annual maintenance will cost Rs. 200. The replacement of reflecting mirror in 10ᵗʰ year will cost Rs. 2000. If its salvage value is Rs. 1000. Determine the annual equivalent cost for an interest rate of 15%.

Solution: Given C_0 = Rs. 6,000, C_a = Rs. 200, C_{10} = Rs. 2000, S = Rs. 1000, n = 15, i = 0.15

The net present cost of the solar cooker =

$$C_{net} = C_0 - S/(1+i)^n + C_a(P/A, \ i, \ n) + 2000(\frac{1}{(1+i)^{10}})$$

$$= C_0 - S(P/F, i, n) + C_a(P/A, i, n) + 2000(0.2472)$$

$$= 6000 - 1000\left(\frac{1}{(1+0.15)^{15}}\right) + 200\left\{\frac{(1.15)^{15} - 1}{0.15(1.15)^{15}}\right\} + 495$$

$$= 6000 - 1000(0.1228) + 200\left(\frac{7.127}{1.220}\right) + 495$$

$$= 6000 - 122.8 + 1170 + 495 = Rs.\,7,542$$

$$\text{Annual equivalent cost (AEC)} = C_{net}\left[\frac{i(1+i)^n}{(1+i)^n - 1}\right] = 7542\left\{\frac{(0.15)(8.137)}{7.137}\right\} = Rs.\,1290.$$

Hence, the annual equivalent cost of the solar cooker is Rs. 1290.

9.11 COMPARISON OF ALTERNATIVES HAVING UNEQUAL DURATION OR LIFE

Often it is necessary to compare alternative investment options with different economic lives. In using the NPV method to compare alternative investments, it is important to evaluate costs and benefits of each alternative over an equal number of years. This is usually done by least common multiple method.

In least common multiple method, the costs and benefits can be measured over a time period that is a common multiple of the economic lives of different alternatives. Benefits and costs are, however, assumed to be the same during the successive lives. For example, if it is desired to compare alternatives which have lives of 3 years and 5 years respectively, alternatives must be compared over a period of 5 × 3 = 15 years, with reinvestment assumed at the end of each life cycle. The terminal salvage value of the investments must also be included as an income in the corresponding year (s).

Example 9.38. Two devices have the following cost comparison. State which device is preferred when they are compared having unequal lives with rate of interest as 15%.

Items	Device (A)	Device (B)
Initial cost (Rs.)	40,000	50,000
Annual maintenance Cost (Rs.)	6, 000	5,000
Salvage value (Rs.)	2, 000	5,000
Useful life (years)	2	3

Solution: As the lives of two devices are unequal, the LCM of lives 2, 3 is 6. Hence, NPVs of two devices are to be compared for taking lives to be 6 years each.

Cash flow diagram for device (A)

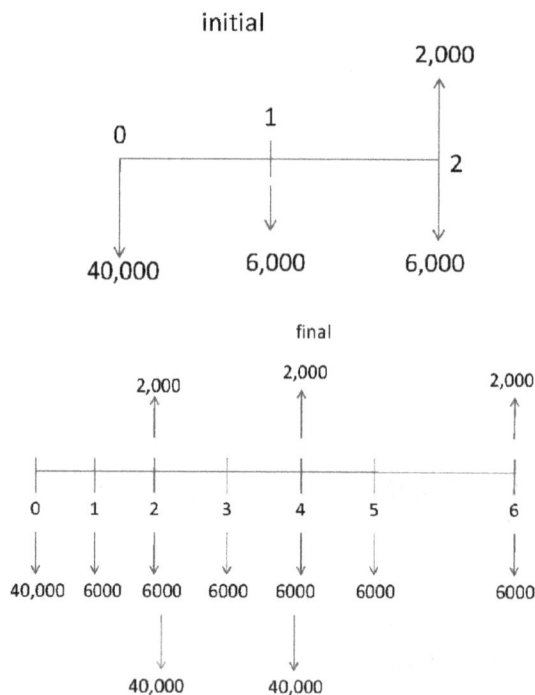

$NPV_A = -40,000 - 40,000 \,(P/F, 15\%, 2) + 2000 \,(P/F, 15\%, 2) - 40,000 \,(P/F, 15\%, 4)$
$+2000 \,(P/F, 15\%, 4) + 2000 \,(P/F, 15\%, 6) - 6000 \,(P/A, 15\%, 6)$

$$NPV_A = -40,000 - 40,000 \left(\frac{1}{(1.15)^2}\right) + 2000 \left(\frac{1}{(1.15)^2}\right) - 40,000 \left(\frac{1}{(1.15)^4}\right)$$

$$+ \; 2000 \left(\frac{1}{(1.15)^4}\right) + 2000 \left(\frac{1}{(1.15)^6}\right)$$

$$-6000 \left\{ \frac{(1+0.15)^6 - 1}{0.15(1+0.15)^6} \right\} = -40,000 - 40,000(0.756) + 2000(0.756) - 40,000(0.571)$$

$$+2000(0.571) + 2000(0.432) - 6000 \left\{ \frac{(1.15)^6 - 1}{0.15(1.15)^6} \right\}$$

$$= -40,000 - 40,000(0.756) + 2000(0.756) - 40,000(0.571)$$

$$+2000(0.571) + 2000(0.432) - 6000(3.79)$$

$$= -40,000 - 30,240 + 1512 - 22840 + 1142 + 864 - 22740$$

$$= -1,12,302$$

Hence net present cost of the device 'A' is + 1, 12,302

Cash flow diagram for device (B)

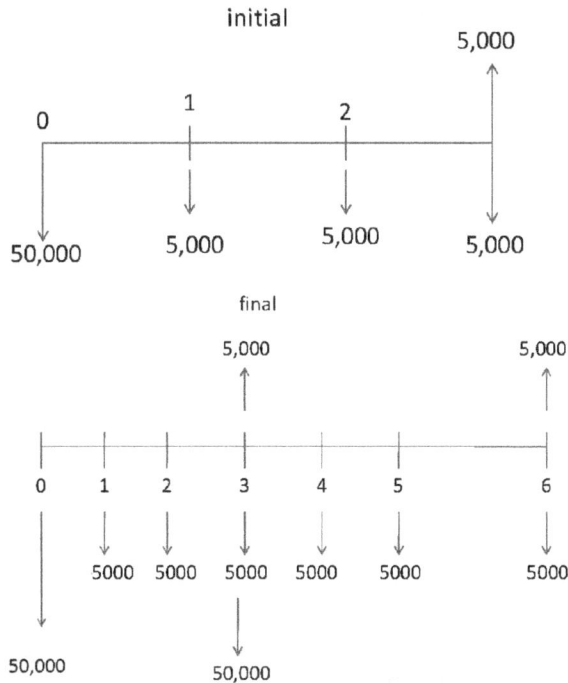

NPV_B = –50,000 –50,000 (P/F, 15%, 3) + 5000 (P/F, 15%, 3) + 5,000 (P/F, 15%, 6) –5000 (P/A, 15%, 6)

$$= -50,000 - 50,000 (0.657) + 5000 (0.657) + 5000 (0.432) - 5000 \left\{ \frac{(1.15)^6 - 1}{0.15(1.15)^6} \right\}$$

$$= -50,000 - 32,850 + 3285 + 2160 - 18900 = - 96,305$$

Hence net present cost of device 'B' is + 96,305. Since $NPV_B > NPV_A$ for 6 years of service life, hence the device 'B' should be preferred.

Also Net Present cost of device 'B' is less than device 'A' and hence device 'B' is preferred.

Alternative methods for cost comparison with unequal duration

Besides, using single present method as explained above for the comparison of two systems with unequal durations through converting them into equal durations, the other methods may also be followed as solved below.

Example 9.39 Two devices have the following cost comparison. Which system is more economical if the money is worth 10 percent per year?

Cost components	System (A)	System (B)
First cost (Rs)	20,000	30,000
Uniform end-of-year maintenance (Rs)	4,000	3,000
Salvage value (Rs)	500	1,500
Service life (n)	2	3

Solution:

(i) The cash-flow diagram for both the systems are first reduced to single present value of cost.

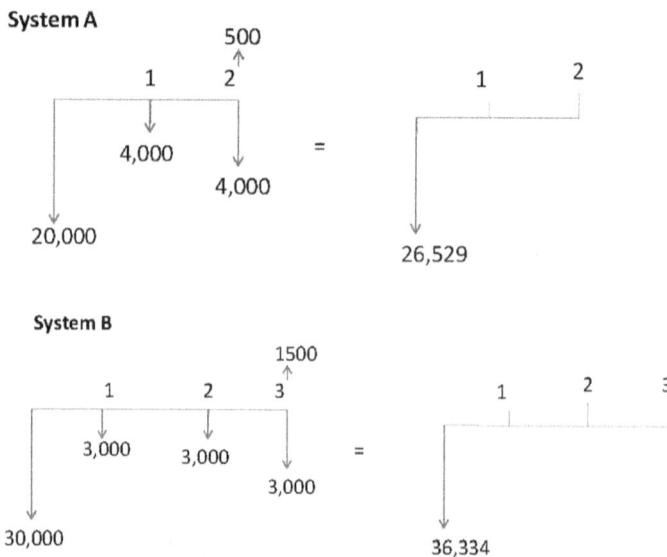

The simplified diagrams are now repeated to obtain six-years duration

For the present value of system A (6-years duration)

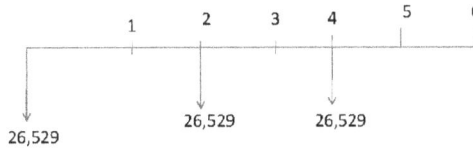

For the present value of system B (6-years duration)

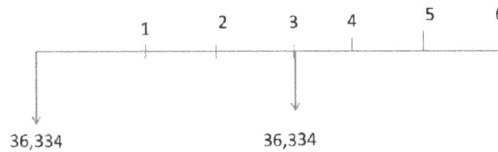

The present value of each of the preceding diagrams at 10 percent interest is

P_{A6} = 26,529 + 26,529 (P/F, 10%, 2) + 26,529 (P/F, 10%, 4)

$\quad$ = 26,529 + 26,529 $(1.10)^{-2}$ + 26,529$(1.10)^{-4}$

$\quad$ = 26,529 + 21924.79 + 18,119.66 = Rs. 66,573.43

Similarly, P_{B6} = Rs 63,632.27

The ratio of cost is $\dfrac{P_{A6}}{P_{B6}} = \dfrac{66,573.45}{63,632.22} = 1.0462$

Thus, system 'B' is more economical than system A.

(i) *Cost comparison by annual cost method (second method)*

In this case, uniform end-of –year annual amount, unacost (A) will be calculated by

using uniform series capital recovery factor, i.e. $A = P\left[\dfrac{i(1+i)^n}{(1+i)^n - 1}\right]$ and $A = P(A/P,$

$i,n)$ for P_{A2} = Rs 26,529 and P_{B2} = 36,334 calculated above.

The unacost for two systems are A_A = P_{A2} (A/P, 10%, 2) = 26,529 × (0.57619)

= Rs 15, 285.74

And A_B = 36, 334 (A/P, 10%, 3)= 36,334 × (0.40211)

= Rs 14, 610.26

The ratio is $\dfrac{A_A}{A_B} = \dfrac{15,285.74}{14,610.26} = 1.0462$

The system 'B' is more economical than system 'A' as concluded earlier.

(ii) Third method

It is possible to convert a present value P_A of n_1 years duration for system A to an equivalent present value of P_B of n_1 years duration for system B and then to compare.

As calculated above, P_{A2} = Rs. 26,529

$$P_{B3} = \text{Rs. } 36,334$$

The present value of system B to an equivalent value for 2 years-duration can be calculated as follows,

(a) Converting 3 years-duration equivalent present value of system B into unacost value (uniform equivalent annual value) i.e

$$P_{B3} \times (A/P_{B3} \ 10\%, \ 3) = 36334 \ [\frac{i((1+i)^3}{(1+i)^3 - 1}] = 14610.26$$

(b) Converting uniform equivalent annual value (unacost) to present equivalent value for 2 years-duration. i.e.

$$A_B \times (P_{B2}/A_B, \ 10\%, 2) = 14610 \times \frac{(1+i)^2 - 1}{i(1+i)^2} = 25356.21 \rightarrow P_{B2} = 25356.21$$

The ratio of cost $\dfrac{P_{A2}}{P_{B2}} = \dfrac{26,529}{25,356} = 1.0462$

The result of cost ratio is the same at that obtained by the above methods.

9.12 EFFECT OF DEPRECIATION AND INCOME TAX ON CASH FLOWS

Up to this point, income tax has not been considered in financial evaluation of a project. We have compared different alternatives related to the project proposals by Net present value (NPV), Internal rate of return (IRR), Payback period (PBP) without taking into the consideration of income tax. The alternative that often appears superior prior to income tax considerations, may become interior after income tax effects. Thus, results of financial evaluations are more realistic when they internalize relevant income tax considerations. Income tax is always a cost and often a substantial part of the cost of private investment as evaluation of public projects does not take taxes into consideration. Income tax considerations can sometimes make the difference between acceptance and rejection of a project. Depreciation and taxes have a significant effect on the financial evaluation of a project and should be considered for a business enterprise that pays taxes.

Depreciation is not a cash flow, but it is considered a business expense by the government and therefore, lowers the taxable income. It is treated as non-cash expenses and provides a tax shield, equal to depreciation times the marginal tax rates and thus in effect increases the cash flow of an investment. The effect of depreciation causes the reduction in the payment of taxes and taxes are the cash flow. Income taxes paid are just one type of expense but income taxes saved through depreciation are identical to other kinds of reduced expenses. Hence, Taxable income = Gross income – all expenses except capital investments – depreciation

Net income after taxes = Taxable income – income tax.

The computation of cash flows for income tax calculation must consider the tax rate structure, the depreciation methods used and other relevant regulations. Depreciation refers to decrease in value of assets with time. There are number of methods for calculating the depreciation. However, the method, a business firm may use to depreciate its capital assets depends upon on the available options allowed under current tax laws. The simplest method is the straight line depreciation. The annual amount of depreciation is computed by dividing the initial cost C_0 of the asset by the number of years 'N' of its estimated life. If a salvage value 'S' is expected, then the annual amount of depreciation is computed as $D= (C_0\text{-}S)/N$. The salvage value may be positive, zero or even negative. A negative salvage value is estimated if dismantling or carry-away costs are anticipated. The detailed analysis of various methods of computing depreciation have been done in the subsequent section.

Some terminologies on Income tax calculation

Personal income: It is the sum of wages, salaries, fees etc. paid to an individual.

Business income: It is the total revenue received from a business or profession minus the total cost of conducting it. It is income on which income tax is levied.

Income tax: It is the amount of payment (taxes) (out of total income in a specified period) on income or profit that is to be paid to the Government as per the prevailing laws and regulations.

Gross income: The total of all income from revenue producing sources (including all items listed in the revenue section of an income statement) is defined as gross income.

Taxable income: It is the amount upon which taxes are to be paid and is computed as Taxable income = Gross income – expenses – depreciation

Expenses: All costs incurred while transacting business are considered expenses.

Effective income tax rates: An effective income tax rate is a single rate, which on multiplying by the taxable income of a venture under consideration provides the income tax attributable to the venture. Effective income tax rates are essentially average rates that are applicable over increments of income.

Income tax computation relation: Taxes = (taxable income) (effective income tax rate) = (gross income – expenses – depreciation) (effective income tax rate)

In many circumstances, income taxes are actually computed using a graduated tax-rate schedule with different tax rates for various slabs of taxable income.

Income tax benefits on the interest paid on borrowed funds: In addition to effects of depreciation, one can also earn income tax benefits on the interest paid on his borrowed funds. The interest paid is often deductible from the income as expenses of carrying out an activity. Since the interest paid is deductible from the gross income as an expense, the amount of the borrowed funds may have a marked effect upon the amount of income tax that must be paid. A simple example is shown below in which two cases are identical in all respects except the fact that in case 'B', borrowed funds are used on which interest is paid.

Effect of interest paid on borrowed funds on the income tax liability

Income and expenses statement	Case A	Case B
Gross annual income (Rs.)	30,00,000	30,00,000
Borrowed fund (Rs.)	—	55,00,000
Annual interest rate (%)	—	12
Annual interest paid (expenses) (Rs.)	—-	6,60,000
Taxable income (Rs.)	30,00,000	23,40,000
Effective income tax rate (%)	30	30
Annual income tax to be paid (Rs.)	9,00,000	7,02,000
Tax benefits (Rs.)		1,98,000

It is seen that even with a flat value of the effective income tax rate, there is a considerable saving in the amount of taxes to be paid for case 'B' with borrowed funds.

Income tax benefits on depreciation

Since the amount of taxes to be paid during any one year is dependent upon deductions made for depreciation, a proper handling of depreciation for tax purposes is quite important.

The cash flows that represent the actual receipts and disbursements associated with an investment alternative can be either before or after-tax cash flows. Since taxes constitute a substantial portion of the disbursements (payments) that are related to an alternative, it is vital to correctly tabulate cash flows after taxes so that the computation made for estimation of the Net Present Value (NPV), equivalent uniform annual cost or internal rate of return or any other figure of merit reflects the correct after-tax situation. Using the appropriate formulae, these figures of merit can be determined by considering after-tax cash flows.

Depreciation of an asset plays an important role in computing after tax cash flows. Depreciation being a non-cash expense provides a *tax shield, equal to depreciation times the effective tax rate resulting ultimately into the increase of cash inflow for an investment.*

If E is the annual earning, T is the rate of tax, D is the depreciation, then for the purpose of calculation of tax, depreciation is deducted from the earning. Tax is calculated on this earning, known as pre-tax earnings. After subtracting the tax amount, the depreciation amount D is again added to find the net cash flows. 'D' is added to compute net cash flow, because 'D' is a non-cash expenses and the depreciation amount actually lies with the accounts of the investment without being spent.

Thus, pre-tax earning $= E - D.$, Tax $= T(E - D)$, Earning after tax $= (E - D) - T$ $(E - D)$ and cash flow $= (E - D) - T(E - D) + D = E(1 - T) + DT$

The amount DT is known as tax shield to protect, save and reduce the expenses for income tax payment. This is due to the effect of depreciation. The business firms generally prefer accelerated methods of depreciation as it increases the cash flow.

Example 9.40. An investment for a producer gas plant is of Rs. 1,60,000 to manufacture and install and has negligible salvage value after a useful life of 8 years. Use straight line depreciation and 40% tax on earnings. The discount rate is 12%. The year wise earnings are given below. Calculate the net present value (NPV) of the plant.

Solution:

Input data

In rupees								
Year	1	2	3	4	5	6	7	8
Earnings (E)	30,000	35,000	38,000	40,000	45,000	48,000	50,000	52,000

Given $C_0 = 1, 60,000$, Useful life $= 8$ years, Tax rate $T = 40\%$. Discount rate $i = 12\%$ $= 0.12$

Using straight line method, depreciation 'D' can be calculated as $1,60,000/8 = 20,000$. The steps of computation and intermediate values are tabulated in the following table.

Steps of computation and intermediate values.

Year	1	2	3	4	5	6	7	8
Earning, E (Rs)	30,000	35,000	38,000	40,000	45,000	48,000	50,000	52,000
Depreciation, D (Rs)	20,000	20,000	20,000	20,000	20,000	20,000	20,000	20,000
Pretax Earnings (E-D) (Rs)	10,000	15,000	18,000	20,000	25,000	28,000	30,000	32,000
Tax, {(E-D) × tax rate T (40 %)}	4,000	6,000	7,200	8,000	10,000	11,200	12,000	12,800
Earnings after tax, {(E-D) - T (E-D)} (Rs.)	6,000	9,000	10,800	12,000	15,000	16,800	18,000	20,000

Cash flow (Rs) earnings after tax + D	26,000	29,000	30,800	32,000	35,000	36,800	38,000	40,000
Discounted cash flows (Rs) converting to present value	23,214	23,119	21,923	20,336	19,860	18,644	17,189	16,155
Net Discounted Income (Rs.)				1,60,440				
Initial Cost (Rs)				1,60,000				
Net Present value (NPV) (Rs)				440				

Single payment present worth factor, $P = F/(1 + i)^n$. As the NPV is +ve, hence the project may be accepted.

Example 9.41. The Characteristics of the proposed purchase plan of a new energy device are as following,

Initial cost = Rs. 5, 00,000, Useful life = 5 years, Salvage value = Rs. 50,000, Expected annual income = Rs. 2, 00,000, Expected annual disbursements = Rs. 60,000

Tabulate the cash flows after tax if the effective tax rate is 40 percent and straight line method of depreciation is used.

Solution: Cash flow after tax

Year	Income (Rs)(a)	Disbursements (Rs)(b)	Cash flows before taxes (Rs)(c) = (a) – (b)	Deprecia-tion* (Rs)(d)	Taxable income (e) = (c-d)	Tax (Rs) (f) (f) = 0.4 (e)	Cash flow after tax (Rs)(g) = (c-f)
0	-	5,00,000	-5,00,000	-	-	-	-5,00,000
1	2,00,000	60,000	1,40,000	90,000	50,000	20,000	1,20,000
2	2,00,000	60,000	1,40,000	90,000	50,000	20,000	1,20,000
3	2,00,000	60,000	1,40,000	90,000	50,000	20,000	1,20,000
4	2,00,000	60,000	1,40,000	90,000	50,000	20,000	1,20,000
5	2,00,000	60,000	1,40,000	90,000	50,000	20,000	1,20,000

* Straight line depreciation = (5, 00,000 – 50,000)/5 = 90,000

Cash flows can be verified as earning after tax + depreciation.

Earning after tax = Earning before tax – Depreciation – Tax

In row 3 of the above Table, Earning after tax = 1, 40,000 – 90,000 – 20,000 = 30,000

Hence cash flow = Earnings after tax + depreciation = 30,000 + 90,000 = 1, 20,000.

Comparative effects of straight line and declining balance methods of depreciation in Income tax calculation

The following example illustrates the comparative effect of the straight line and the fixed percentage on the declining balance method of depreciation on the income tax calculation.

Example 9.42. A business enterprise has purchased a device whose first cost is Rs. 1,00,000 with an estimated life of 8 years. The estimated salvage value of the asset at the end of lifetime is Rs. 20,000. The device is estimated to provide a constant operating income of Rs. 20,000 per year before depreciation and income taxes. The business owner estimates the applicable effective income tax rate during the service life of the device to be 30 percent. Considering the worth of money to be 15%, compare the income tax liabilities using straight line depreciation and declining balance method of depreciation.

Solution: Let the first method be represented by Alternative 'A' following straight line method of depreciation and the second method by alternative 'B' following declining balance method of depreciation. These two alternatives are summarized in the following tables.

Income tax calculations for the case of straight line method of depreciation (Alternative 'A')

Year end (a)	Capital cost (Rs) (b)	Income before depreciation and IT (Rs) (c)	Annual deprecia- tion (Rs) (d)	Cumulative deprecia- tion (Rs) (e)	Taxable income (Rs) (c - d) (f)	Income Tax rate (fraction) (g)	Income tax = f × g (Rs) (h)
0	1,00,000	-	-	-	-	-	-
1	-	20,000	10,000	10,000	10,000	0.3	3000
2	-	20,000	10,000	20,000	10,000	0.3	3000
3	-	20,000	10,000	30,000	10,000	0.3	3000
4	-	20,000	10,000	40,000	10,000	0.3	3000
5	-	20,000	10,000	50,000	10,000	0.3	3000
6	-	20,000	10,000	60,000	10,000	0.3	3000
7	-	20,000	10,000	70,000	10,000	0.3	3000
8	-	20,000	10,000	80,000	10,000	0.3	3000

Thus the cumulative present worth of annual income tax payments in alternative 'A' can be calculated as = 3000 (P/A, 15%, 8) = 3000 × (Uniform series present worth factor)

$$= 3000\left[\frac{(1+i)^n - 1}{i(1+i)^n}\right] = 3000\left[\frac{(1.15)^8 - 1}{0.15(1.15)^8}\right] = 3000\left(\frac{3.06 - 1}{0.459}\right) = Rs.13,464$$

Income Tax calculation in case of declining balance method of depreciation (20%) (Alternative 'B')

Year end (a)	Capital cost (Rs) (b)	Income before depreciation and Income tax (Rs) (c)	Annual deprecia- tion (Rs) (d)	Cumulative deprecia- tion Σd (Rs) (e)	Taxable income (c–d) (Rs) (f)	Income tax rate (fraction) (g)	Income Tax = f x g (h)
1	1,00,000	-	-	-	-	-	-
2	-	20,000	20,000	20,000	0	0.3	0
3	-	20,000	16,000	36,000	4000	0.3	1200
4	-	20,000	12,800	48,800	7200	0.3	2160
5	-	20,000	10,240	59,040	9760	0.3	2928
6	-	20,000	8,192	67,232	11,808	0.3	3542
7	-	20,000	6,553	73,785	13,447	0.3	4034
8	-	20,000	5,243	79,028	14,757	0.3	4427
9	-	20,000	4194	83,222	15,806	0.3	4742
						Total	23,033

Thus the cumulative present worth of the annual income tax payment made in alternative 'B'

$$= \sum_{t=1}^{8} \text{column } h_t \times (P/F, 15\%, t) = 23,033 \frac{1}{(1+i)^t}$$

$$= 23,033 \times \text{(single payment present worth factor)} = 23,033 \frac{1}{(1.15)^8} = \text{Rs.}7,527$$

The above two examples demonstrate the important economic advantage of charging depreciation or other expense items as early as possible over the life of an asset. Hence by examining the book value of an asset over its life for the various methods of depreciation, one can easily see which methods produce larger depreciation deductions in the early portion of the asset's life. Since the book value of an asset is the first cost of the asset less accumulated depreciation charges, the method with the lowest book value in early life is most favorable to the taxpayer.

9.13 CONCEPT OF DEPRECIATION

The concept of depreciation is explained as follows.

Any equipment/device/asset which is purchased today will not work for ever. This may be due to wear and tear of the equipment or obsolescence of technology. Hence, it is to be replaced at the proper time for continuance of any business. The replacement of the device at the end of its life involves money. This must be internally generated from the earnings of the equipment/device/asset. The recovery of money from the earnings of an equipment for its replacement purpose is called depreciation fund since we make an assumption that the value of device decreases with the passage of time. Thus, the word 'depreciation' means decrease in value of any physical asset with the passage of time.

The term depreciation is derived from the Latin word 'Depletion' which means declining worth. It refers to loss or diminution in the value of an asset consequent upon wear and tear, obsolescence, passage of time or fall in the market value.

9.13.1 Characteristics of Depreciation

The depreciation charged on fixed assets has the following features;

i) It is always charged on the fixed assets excepts land.

ii) It causes gradual and continuous fall in the value of an asset.

iii) As it is a continuous process, it does not matter whether the asset was put to use during the period or not.

iv) It is the fall in book value of the asset and not in the market value of the asset.

v) It is the charge against the revenue of an accounting period.

vi) It is the result of the use of the assets, passage of time and obsolescence.

vii) Total depreciation of an asset cannot exceed its depreciable value (cost less salvage value).

viii) It is a non-cash expense.

Causes of depreciation: Depreciation basically occurs due to two types of causes. They are (i) Internal (ii) External

(i) *Internal:* When depreciation occurs for certain inherent normal causes, it is known as internal depreciation. The cause of internal depreciation is

   (a) Wear and tear. It is otherwise known as the physical deterioration of the asset due to continued use i.e. building, machinery, device etc.

ii) *External:* When depreciation occurs for certain external reasons, it is called external depreciation. The causes of external depreciation are as follows;

   (a) *Obsolescence:* Due to advancement in technology, old technology becomes outdated and loses its value overtime. Obsolesce of an asset also results from the invention of another asset that is sufficient superior and which makes it uneconomical to continue using the former.

   b) *Passage of time:* There are some fixed assets which lose their value because of passage of time. They are lease, licenses, patent, copyright etc. Even if an asset is not used at all, it loses its working capacity over a period of time.

c) *Accidents:* Sometimes due to natural calamities such as fire, earthquake, flood etc., the asset may lose its value which are abnormal in nature. Assets may also lose its value due to accident or sudden failure of the technological characteristics inherent on them.

d) *Fall in the market value:* The assets lose their market value due to innovations of the modern assets.

e) *Deferred maintenance:* Sometimes the loss of value of an asset begins very quickly due to deferred (delayed) maintenance. If proper materials are not used or instructions to operate the machine are not properly followed, the loss of value starts early.

9.13.2 Need for Providing Depreciation

The need for depreciation arises because of the following reasons.

- *To know the true profits:* As depreciation is an expense, it becomes an important element of the cost of production. Though it is not visible like other expenses and never paid to the outside party, yet it is desirable to charge depreciation on fixed assets as these are used for earning purposes. So their depreciation must be deducted out of the income earned from their use in order to calculate true net the income earned from their use in order to calculate true net profit or loss.

- *To show true financial position:* Financial position can be studied from the balance sheet and for the preparation of the balance sheet, fixed assets are required to be shown at their true value. If assets are shown in the balance sheet without any charge made for their use or depreciation, then their value must have been overstated in the balance sheet and will not reflect the true financial position of the business. So, for the purpose of reflecting true financial position, it is necessary that depreciation must be deducted from the assets and then at such reduced value, these may be shown in the balance sheet.

- *To make provision for replacement of assets:* The amount charged as depreciation on fixed assets is a non-cash expenses. This amount is not paid to outsider and kept in the business with an intention of replace old assets. The accumulated funds during the working life of the assets helping the business to purchase a new asset without facing any difficulty. In the absence of depreciation fund, it would become difficult to replace an old asset when that becomes obsolete.

9.13.3 Depreciable Property

Before discussing the different methods involved in the calculation of depreciation, we should have sufficient knowledge about depreciable property. It is that property which can amortize or depreciated. Amortization is the writing off intangible assets such as goodwill, patents, copyright, leasehold etc. Periodic write off is called amortization. Depreciable property may be tangible or intangible. Tangible property is any property that can be seen or touched. Intangible property is that which does not exist as a physical thing but still has values. Examples are the copyrights and patent rights. Depreciable tangible property is of two types; (i) Real and (ii) Personal. Personal property is not real estate and includes machinery and equipment. Real property is land and anything that is built on it. Land is never depreciable. A property is depreciable in the following cases.

(a) The property must be used in business or to produce income.

(b) It must be something that wears out, decays, deteriorates, becomes obsolete and loses value from natural causes.

(c) The property must have determinable life and that life must be longer than one year.

In general, if a property does not fulfil the above conditions, cannot be regarded as depreciable property.

9.13.4 Factors for Estimation of Depreciation

The following factors are taken into consideration while calculating the depreciation expenses.

- *Cost of asset:* It includes all expenses involved in the asset in workable condition. It is the total cost of asset including all freight, insurance and installation charges.
- The salvage value at the end of its life
- Estimated number of years of its usefulness

9.13.5 Some Terminologies for Depreciation Calculations

Depreciation: It is defined as the decrease in the value of a physical asset such as equipment, device and machinery with the passage of time and use. With the possible exception of land this phenomenon is a normal characteristic of the physical assets. Therefore, it is common to allocate the cost of such assets to different time periods over which such use occurs rather than at a single point in time, such as when the asset is purchased.

Depreciation rate or recovery rate: It is the fraction of the first cost removed by depreciation in each year to compute annual depreciation deduction. This rate may be same in each year or different for each year of the recovery period.

Recovery period: It is the depreciable life of the asset in years over which the first cost of the asset is recovered through the accounting process. This period is also called the useful life of the asset.

First cost: It is the initial price of the asset including purchase price, delivery charges, installation fees and other direct costs incurred to make the asset ready for use.

Book value: It represents the remaining, undepreciated capital investment after the total amount of depreciation charges to date has been subtracted from the first cost. In other words, it refers to the current worth of an asset as shown by the books of account. Since depreciation is normally charged once a year, the book value is computed at the end of each year. The book value of an asset decreases every year by the amount of depreciation till it reaches the salvage value at the end of the lifetime of depreciating asset.

Market value: The market value of an asset refers to price at which an asset could actually be sold in the free market. In some cases, the market value bears very little relation to the book value. For example, commercial buildings tend to increase in market value, whereas their book value decreases as a result of depreciation charges. However, a computer workstation may have a market value much lower than its book value due to rapidly changing technology. In the calculation of financial evaluation for an enterprise, it is the market value which must be taken into consideration.

Useful life: The expected (estimated) period in year that an asset will be used in a business to produce income prior to its disposal and/or replacement. It is not how long the asset will last but how long the owner expects to use it productively. It is also called the recovery period or depreciable life, which may be somewhat different for depreciation and tax purposes than the actual expected productive life span. This difference may come about because of rulings, management policy, anticipated product changes etc.

Physical and economic life of an asset or machinery: The physical life of an asset/ machinery means the period for which the asset/machinery can provide service to the company whereas the economic life of an asset implies the period for which the asset/ machinery can provide profitable service to the enterprise. Hence as far as the enterprise or a business venture is concerned, the economic life of an asset is far more important than physical life because companies want to make profits and economic life is relevant when it comes to earning profits. An example of physical life will be human beings, suppose, human being has age of 80 years then his or her physical life will be 80 years, whereas in case of enterprises where retirement age is 60 years the actual economic life of human being is 60 years and not 80 years as they can work only until 60 years for the enterprise. Similarly, in the case of machines if the machine is producing units at profitable margins for the enterprise then it is economical for the company and if it stops producing units at profitable margins, then its physical life has no importance for the company. Hence, economic life is defined as the length of time from the purchase of a machine to that point where it is more economical to replace it with a second machine than to continue with the first. At this time, a machine may still have considerable service life but be uneconomic because of high rate-of-repair costs, technological obsolescence, or a change in the farm enterprise.

Salvage value: It is the expected value of an asset at the end of its useful life. It is the expected selling price of the asset when it can no longer be used productively by the owner. It can be positive, zero or negative. A negative salvage value is estimated if dismantling or carry-away costs are anticipated. Sometimes, it is expressed as a percentage of the first cost.

Depreciation Methods: There are four commonly used depreciation methods – Straight line depreciation, declining balance depreciation, sum of the years–digits method of depreciation, sinking fund method of depreciation. A brief description of each of these methods is given below.

9.13.6 Depreciation Calculation using Various Methods

(a) Straight Line method of Depreciation

Straight line method of depreciation is the simplest depreciation method. It assumes that a constant amount is depreciated each year over the depreciable (useful) life of the asset. The meaning of straight line depreciation is from the fact that the book value decreases linearly with time. The depreciation rate is the same ($1/n$) each year of the recovery period 'n'. In the straight line depreciation method, the full service life of an asset is estimated along with the prospective salvage value at the end of the service life. Yearly depreciation is then determined by dividing the first cost (C_0) minus its salvage value (S) by the service life (n) of the asset. Thus if an asset has a first cost of Rs. 5000 and an estimated salvage value of Rs. 1000, the total depreciation over its life will be 5000-1000 = Rs. 4000. If the estimated life is 5 years, the depreciation per year will be Rs. 4000/ 5 = Rs. 800. This is equivalent to a depreciation rate of 1/5 or 20% per year. The rate of depreciation, 'd' is the fraction by which the depreciable amount (C_0-S) is decreased each year. For the straight line depreciation method, this rate is same for each year i.e. $d=1/n$. For this example, the annual depreciation and book value for each year are given in the followings Table.

The straight line method

End of year t	Depreciation charge during year 't'	Book value at the end of year 't' (Rs.)
0	-	5000
1	800	4200
2	800	3400
3	800	2600
4	800	1800
5	800	1000

Mathematically, for the tth year, $D_t = \dfrac{C_0 - S}{n}$ where t = 1,2,3,4............, n. D_t is the depreciation charge, same for all the year during useful life of the asset. The book value,

$$BV_t = C_0 - \frac{t(C_0 - S)}{n}$$

The following table shows the depreciation charge (D_t) and book-value (BV) expression for each year.

General expression for the straight-line method

End of year 't'	Depreciation charge (D_t) during year 't'	Book value (BV) at end of year 't'
0	–	C_0
1	$\dfrac{C_0 - S}{n}$	$C_0 - \left(\dfrac{C_0 - S}{n}\right)$
2	$\dfrac{C_0 - S}{n}$	$C_0 - 2\left(\dfrac{C_0 - S}{n}\right)$
3	$\dfrac{C_0 - S}{n}$	$C_0 - 3\left(\dfrac{C_0 - S}{n}\right)$
t	$\dfrac{C_0 - S}{n}$	$C_0 - t\left(\dfrac{C_0 - S}{n}\right)$
n	$\dfrac{C_0 - S}{n}$	$C_0 - n\left(\dfrac{C_0 - S}{n}\right)$

If we plot the book value of an asset with each depreciable year, the graph will be straight line.

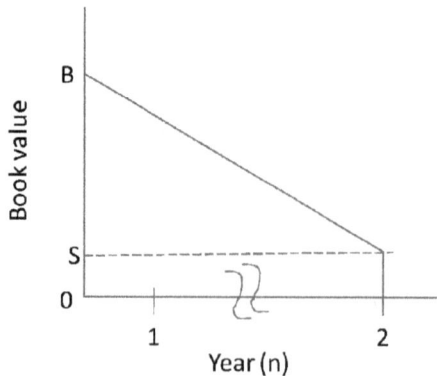

In straight line method of depreciation, we make an important assumption that inflation is absent.

Example 9.43. A heat exchanger costs Rs. 2,00,000 and has an estimated salvage value of Rs.15000. Compute the annual straight line depreciation and book value using a lifetime of 10 years.

Solution:

$$D_t = \frac{2,00,000 - 15,000}{10} = Rs.18,500$$

Thus the depreciation each year is Rs. 18,500 and the book value decreases each year by that amount.

$BV_1 = 2,00,000 - 18500 = 1,81,500$

$BV_2 = 1,81,500 - 18500 = 1,63,000$

$BV_{10} = 33,500 - 18500 = 15000$.

At the end of the depreciation, the book value equals to the salvage value that was estimated in the beginning.

Example 9.44: A business enterprise has purchased a device whose first cost is Rs. 1,00,000 with an estimated life of eight years. The estimated salvage value of the device at the end of its lifetime is Rs. 20,000. Determine the depreciation charge and book value at the end of various years using the straight line method of depreciation.

Solution: $C_0 = Rs.1,00,000$

$S = Rs.\ 20,000,\ n=8$ years

$D_t = (C_0\text{-}S)/n = Rs.\ 10,000$

The value of D_t, depreciation charge is the same for all the years. The calculation pertaining to BV_t for different values of t are summarized in the Table below.

D_t and BV_t values under straight line method of depreciation.

End of year (t)	Depreciation (D_t) (Rs.)	Book value (Rs.)
0	-	1,00,000
1	10,000	90,000
2	10,000	80.000
3	10,000	70,000
4	10,000	60,000
5	10,000	50,000
6	10,000	40,000
7	10,000	30,000
8	10,000	20,000

Example 9.45: Consider the above example (9.44) and compute the depreciation and the book value for the period 5.

Solution: The general expression for Book value

$$BV_t = C_0 - \frac{t(C_0 - S)}{n}$$

Hence $t = 5$, $n = 8$, $C_0 = 1,00,000$, $S = 20,000$

$$BV_t = 1,00,000 - 5\left(\frac{1,00,000 - 20,000}{8}\right) = \text{Rs. } 50,000$$

(b) Declining Balance Method

The contribution of assets to income is often much greater in the early years of life than in the later years. It is therefore, reasonable to write off the cost of assets more rapidly in the early years of life than in the later years. Declining balance method offers a way of making this rapid write-off in the early years.

The declining balance method of depreciation assumes that an asset decreases in value at a faster rate in the early portion of its service life than in the later portion of its life. By this method, depreciation is calculated at a fixed percentage each year on the decreasing book value of the asset. This method ignores the salvage value of the asset while calculating the depreciation of the asset. The book value of the asset gradually reduces on account of changing depreciation. As the depreciation is charged on the reducing balance of asset, it is therefore called the declining – balance method of depreciation.

For an asset, with Rs. 5,000 as first cost and Rs. 1000, as estimated salvage value with an estimated life of 5 years and a depreciation rate of 30% per year, the depreciation charge per year and book value at each year are shown in the following table.

If this fixed fraction is denoted as f where $0 < f < 1$, then the depreciation during year 't' is given by

$D_t = f\, B_{t-1}$ Where B_{t-1} is the book value of the asset at the beginning of the year t (end of year t-1).

The declining balance method

End of year 't'	Depreciation charge during year 't' (Rs.)	Book value at end of year 't' (Rs.)
0	–	5000
1	0.30 (5000) = 1500	3500
2	0.30 (3500) = 1050	2450
3	0.30 (2450) = 735	1715
4	0.30 (1715) = 515	1200
5	0.30 (1200) = 360	840

We can also determine the general expressions for the depreciation charge and the book value for any point in time as shown in the followings table.

General expression for the declining balance method of depreciation

End of year 't'	Depreciation charge during year 't'	Book value (B) at end of year 't'
0	-	-
1	$D_1 = fC_0$	$B_1 = C_0 - fC_0 = C_0(1-f)$
2	$D_2 = fB_1 = f(1-f)C_0$	$B_2 = C_0(1-f) - C_0 f(1-f) = C_0(1-f)^2$
3	$D_3 = fB_2 = (1-f)^2 C_0$	$B_3 = C_0(1-f)^2 - C_0 f(1-f)^2$ $= C_0(1-f)^3$
4	$D_t = f(1-f)^{(t-1)} C_0$	$B_t = C_0(1-f)^t$
5	$D_n = f(1-f)^{(n-1)} C_0$	$B_n = C_0(1-f)^n$

If the declining–balance method of depreciation is used for income tax purposes, the maximum rate that may be used is double the straight line rate that would be allowed for a particular asset or group of assets being depreciated. Thus, for an asset with an estimated life of n years, the maximum rate that may be used with the method is $2(1/n)$. Such a method of depreciation is commonly referred to as the double declining balance method of depreciation.

Many firms and individuals choose to depreciate their assets using double-declining-balance method. Since the salvage value is not used directly in declining balance methods, an implied salvage value after n years may be computed as

$$S = B_n = C_0(t-f)^n$$

Example 9.46. Consider the example (9.44) as solved for straight line method and demonstrate the calculations of the declining balance method of depreciation by assuming 0.2 for 'f'.

Solution: C_0 = Rs. 1, 00,000, S = Rs. 20,000, n = 8 years, f = 0.2

D_t and B_t for declining balance method of depreciation

End of year	Depreciation (D_t)	Book value (B_t)
0	-	1,00,000
1	20,000	80,000
2	16,000	64,000
3	12,800	51,200
4	10,240	40,960
5	8,192	32,768
6	6553.60	26,214.40
7	5242.88	20,971.52
8	4194.30	16,777.22

If we are interested is computing D_t and B_t for a specific period 't', the general expressions for them can also be used.

Example 9.47. Considering the above example, calculate the depreciation and the book value for period 5 using the declining balance method of depreciation by assuming 0.2 for 'f'.

$$D_t = f(1-f)^{t-1} C_0 \Rightarrow D_5 = 0.2(1-0.2)^4 \times 1,00,000 = Rs. 8192$$

$$B_t = (1-f)^t C_0 = (1-0.2)^5 \times 1,00,000 = Rs. 32,768$$

The implied salvage value at $n = 8$ is $S = (1 - 0.2)^8 \times 1,00,000 = $ Rs. 16,777. Since the salvage value is anticipated to be Rs. 20,000, the lower limit on book value is Rs. 20,000. It may be noted that in declining balance and double balance methods, the salvage value is not subtracted from the first cost when the depreciation is calculated. However, when the book value reaches the expected salvage value, no additional depreciation may be taken.

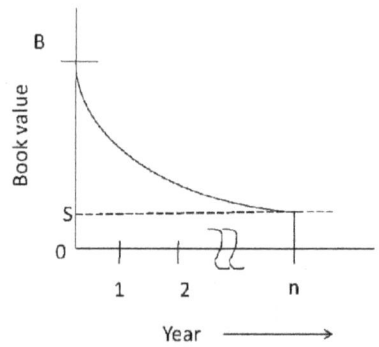

If we plot the book value with respect to each depreciable year, the graph would become a curve line instead of a straight line in straight line depreciation methods.

Distinction between straight line and declining balance methods of depreciation

S.N.	Straight line method	Declining balance method
1.	The depreciation rate percent is calculated on the cost of the asset every year	The depreciation rate percent is calculated on the book value of asset every year
2.	The value of asset is reduced to zero at the end of its life	The value of asset is never reduced to zero at the end of its life.
3.	The depreciation amount remains unchanged throughout the useful life	The depreciation amount always reduces annually proportionately
4.	The amount of depreciation remains same, while cost of its repair increases. It reduces the annual profit	The amount of depreciation decreases gradually, while the cost of repair increases. It causes no change in profit
5.	It is very much simple as compared to declining balance method	It is not as simple as compared to straight line method.

(c) Sum of the years – digits method of depreciation

It is a depreciation method that results in a more accelerated write-off than straight line, but less than declining – balance method. This method assumes that the value of an asset decreases at a decreasing rate. It is an accelerated or rapid write-off technique by which much of the value of the asset is written off in the first one third of its life. Thus, the depreciation charges are very high in the first few years but decrease rapidly in later years of the asset's life. In this case also, book value follows a convex curve.

To compute the depreciation by this method, the digits corresponding to the number of each permissible year of life are first listed in reverse order. The sum of these digits is then determined. The depreciation factor for any year is the number from the reverse-ordered listing for that year divided by the sum of the digits. For example, an asset having a depreciable (useful) life of 8 years, the sum of the years-digits method depreciation factors are as follows.

Year	Number of the year in reverse order (digits)	Depreciation factor
1	8	8/36
2	7	7/36
3	6	6/36
4	5	5/36
5	4	4/36
6	3	3/36
7	2	2/36
8	1	1/36

Sum of the digits = 36

The depreciation for any year is the product of depreciation factor for that year and the difference between first cost and the estimated salvage value.

D_t = depreciation rate $\times (C_0 - S)$, and $B_t = B_{t-1} - D_t$

In alternative way, depreciation amount and book value of the asset can be expressed as

$$D_t = \frac{Depreciable\ years\ remaining}{Sum\ of\ years\ digits}(first\ cost - salvage\ value)$$

$$= \left[\frac{n-(t-1)}{\frac{n(n+1)}{2}} \right](C_0 - S)$$

It may be noted that the remaining depreciable years must include the year for which the depreciation charge is being estimated.

The book value for any given year can be calculated without making the year – by – year depreciation determinations through the use of the following equation.

$$B_t = C_0 - \left[\frac{t(n - t/2 + 0.5)}{\frac{n(n+1)}{2}} \right](C_0 - S)$$

The general expression of book value can be known very well by computing it in the value of each year.

When t = 1 (1st year) $D_1 = \left\{ \frac{n}{n(n+1)/2} \right\}(C_0 - S)$

When t = 2 (2nd year) $D_2 \left\{ \frac{n-1}{n(n+1)/2} \right\}(C_0 - S)$

When t = 3 (3rd year) $D_3 \left\{ \frac{n-2}{n(n+1)/2} \right\}(C_0 - S)$

When $t = n$ (final depreciable year), $D_n = \frac{1}{n(n+1)/2}(C_0 - S)$

B_3 (Book value of 3rd year) =

$$C_0 \left[\frac{n}{n(n+1)/2}(C_0 - S) + \frac{n-1}{n(n+1)/2}(C_0 - S) + \frac{n-2}{n(n+1)/2}(C_0 - S) \right]$$

$$B_3 = C_0 - \frac{(C_0 - S)}{n(n+1)/2}\{n + n - 1 + n - 2\}$$

$$B_3 = C_0 - \frac{(C_0 - S)}{n(n+1)/2}(3n - 3) = C_0 - \frac{(C_0 - S)}{n(n+1)/2} \times 3(n - 1)$$

$$B_3 = C_0 - \left[\frac{3(n - 3/2 + 0.5)}{n(n+1)/2} \right](C_0 - S)$$

As a general expression

$$B_t = C_0 - \left\{ \frac{t(n - t/2 + 0.5)}{n(n+1)/2} \right\}(C_0 - S)$$

Example 9.48. Considering the example (9.44) as solved in straight line method of depreciation, calculate the same in the sum of the year's digits method of depreciation.

Solution:
$C_0 = 1,00,000$; $S = 20,000$, $n = 8$ years.
Sum of the years digits $= n(n + 1)/2 = 36$
D_t = Depreciation rate $\times (C_0 - S)$
D_t and B_t under sum – of – the years digits method of depreciation

End of year (n)	Depreciation (D_t) (Rs.)	Book value (B_t) (Rs.)
0	-	1,00,000
1	17,777.77	82,222.23
2	15,555.55	66,666.68
3	13,333.33	53,333.35
4	11,111.11	42,222.24
5	8,888.88	33,333.36
6	6,666.66	26,666.70
7	4,444.44	22,222.26
8	2,222.22	20,000.04

Example 9.49. Considering the above example (9.48), find the depreciation and book value for the 5th year using the sum-of-the-years digits method of depreciation.

Solution: $C_0 = 1,00,000$; $S = 20,000$, $n = 8$ years, $t = 5$ years.

$$D_t = \left[\frac{n-(t-1)}{n(n+1)/2}\right](C_0 - S)$$

$$D_5 = \left[\frac{8-(5-1)}{8(8+1)/2}\right](1,00,000 - 20,000) = Rs.8888.88$$

$$B_5 = C_0 - \left[\frac{5(8-5/2+0.5)}{8(8+1)/2}\right](1,00,000 - 20,000) = 33,333.36$$

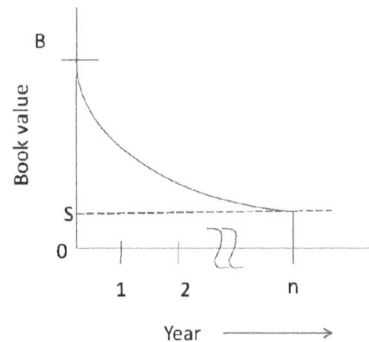

(d) Sinking fund method of depreciation: In this method, the amount of depreciation is ascertained in such a way that if it is invested every year with compound interest, it will yield an amount equal to the cost of the asset. By creating a sinking fund, money is accumulated to replace the existing asset at the proper time. At the end of the useful life of the asset, the total amount in depreciation plus compound interest should become equal to the original cost of the fixed asset. The amount in the form of depreciation charge is invested every year outside the business along with the amount of interest received at the beginning of the year. When the asset is to be replaced, the investments are exhausted to purchase the new one.

Let C_0 = first cost of the asset, S = salvage value of the asset, n = useful life of the asset

i = rate of return compounded annually, A = Annual equivalent amount, D_t = Depreciation amount at the end of the period 't', B_t = Book value of the asset at the end of the period 't'.

The loss in value of the asset $(C_0 - S)$ is made available in the form of cumulative depreciation amount at the end of the life of the asset by setting up an equal depreciation amount (A) at the end of each period during the lifetime of the asset.

$$A = (C_0 - S) \times [A/F, i, n]$$

The fixed sum depreciated at the end of every time period earns an interest at the rate of i % compounded annually and hence the actual depreciation amount will be in the increasing manner with respect to the time period.

A generalized formula for D_t is $D_t = (C_0 - S) \times (A/F, i, n) \times (F/P, i, t - 1)$

The formula to calculate the book value at the end of period 't' is

$$B_t = C_0 - (C_0 - S) \times (A/F, i, n) \times (F/P, i, t)$$

The above two formulae are very useful if we have to calculate D_t and B_t for any specific period.

Example 9.50. Considering the example (9.44) as mentioned in the straight line method of depreciation, calculate the sinking fund method of depreciation with an interest rate of 12%, compounded annually.

Solution: C_0 = Rs. 1, 00,000; S = Rs. 20,000, n = 8 years, i = 12%

$$A = (C_0 - S)[A/F, 12\%, 8] = (C_0 - S)\left[\frac{i}{(1+i)^n - 1}\right]$$

$$= (1,00,000 - 20,000)\left[\frac{0.12}{(1.12)^8 - 1}\right] = Rs.\ 6,504$$

The expression $\left[\dfrac{i}{(1+i)^n - 1}\right]$ is known as 'Uniform series sinking fund factor'.

In this method of depreciation, a fixed amount of Rs. 6,504 will be depreciated at the end of every year from the earning of the asset. The depreciated amount will earn interest for the remaining period of life of the asset at an interest rate of 12%, compounded annually. For example, the calculations of effective depreciation for some periods are as follows.

Depreciation at the end of year $1(D_1)$ = Rs. 6,504

Depreciation at the end of year 2 (D_2) = 6,504 + 6504 × 0.12 = 6,504(1+i) = 7284.48

Depreciation at the end of year 3 $(D_3) = 6,504 \ (1.12)^2 = 8158.62$

Or, $6,504 + (6,504 + 7284.48) \times 0.12$

Depreciation at the end of year 4 $(D_4) = 6,504(1.12)^3 = 9137.65$

Or, $6,504 + (6504 + 7254.48 + 8158.62) \times 0.12$

These calculations along with book values are summarized in the following Table.

D_t and B_t according to sinking fund method of depreciation

End of year 't'	Fixed depreciation (Rs.)	Effective depreciation D_t (Rs.)	Book value (B_t) (Rs)
0	-	-	1,00,000
1	6504	6504	93,496.00
2	6504	7284.48	86,211.52
3	6504	8158.62	78,052.90
4	6504	9137.65	68,915.25
5	6504	10,234.17	58,681.08
6	6504	11,462.27	47,218.81
7	6504	12,837.74	34,381.07
8	6504	14,378.27	20,002.80

$B_t = B_{t-1} - D_t$

Example 9.51. Considering the previous example (9.44), compute D_5 and D_7 using the sinking fund method of depreciation with an interest rate of 12% compounded annually.

Solution: $C_0 = 1,00,000$, $S = 20,000$, $n = 8$ years, $i = 12\%$

$$D_t = (C_0 - S)(A/F, i, n)(F/P, i, t-1) = (C_0 - S)\left[\frac{i}{(1+i)^n - 1}\right]\left[(1+i)^{t-1}\right]$$

$= (C_0\text{-}S) \times (\text{Uniform series sinking fund factor}) \times (\text{Future value factor})$

$$D_5 = (1,00,000 - 20,000)\left[\frac{0.12}{(1.12)^8 - 1}\right]\left[(1.12)^4\right]$$

$= 80,000 \ (0.0813) \ (1.574) = \text{Rs. } 10,237.30$

This is almost the same as the corresponding value given in the table above. The minor difference is due to truncation error.

$$B_t = C_0 - (C_0 - S)(A/F, i, n)(F/A, i, t)$$

$$B_7 = 1,00,000 - (1,00,000 - 20,000)\left[\frac{i}{(1+i)^n - 1}\right]\left[\frac{(1+i)^t - 1}{i}\right]$$

$$B_7 = 1,00,000 - (80,000)(0.0813)\left\{\frac{(1.12)^7 - 1}{0.12}\right\}$$

$$B_7 = 1,00,000 - (80,000)(0.0813)(10.089) = \text{Rs. } 34,381$$

Graph of sinking fund depreciation

The graph from the above example is as follows.

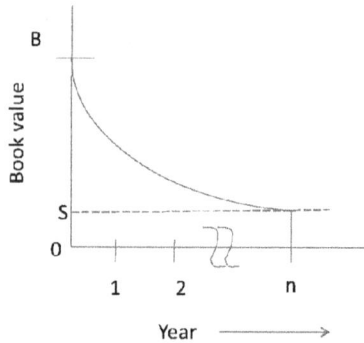

9.13.7 Effect of Inflation on Depreciation

If there will be constant rate of inflation, the future price of the equipment in n^{th} year

$$F = P(1 + j)^n$$

If the inflation rate changes in each year, the future price of the device would be

$$F = P(1 + j_1)(1 + j_2)(1 + j_3)(1 + j_4) \dots\dots\dots(1 + j_n)$$

The depreciation of the device under constant inflationary condition using straight line method can be written as $CD_n = \frac{n}{L}[P(1 + j)^n - S]$ where CD_n, the cumulative depreciation charges up to n^{th} year (Rs.), $n=$ number of year elapsed after the purchase of the device. L, the life of the device, P, the purchase price and S, the salvage value of the device.

Annual depreciation charge (D_n) in the nth year will be $D_n = CD_n - CD_{n-1}$

Future value of the device at the end of nth year (V_n) =future price at the end of nth year - CD_n

$$= P(1 + j)^n - \frac{n}{L}[P(1 + j)^n - S]$$

$$CD_{n-1} = \frac{n-1}{L}[P(1 + j)^{n-1} - S]$$

Variation in future price, total depreciation cost and remaining value of the device as a function of purchase price 'P', at constant 10 % annual inflation rate under straight line method for total life of machine of 15 years with salvage value as 10% of the inflated price of device are given in the following table.

$$CD_n = \frac{n}{15}[P(1+j)^n - 0.1P(1+j)^n] = \frac{n}{15}P(1+j)^n \ (0.9) = 0.06nP(1+j)^n$$

Future   price $F = P(1+j)^n$

Comparison of future price, total depreciation cost and remaining value of a device

End of year	Future price	Total depreciation	Remaining value (col.2-col. 3)
1	1.100P	0.0660P	1.0340P
2	1.210P	0.1452P	1.0648P
3	1.331P	0.2396P	1.0914P
4	1.464P	0.3514P	1.1126P
5	1.611P	0.4833P	1.1277P
10	2.594P	1.5564P	1.0376P

Example 9.52: A device costing Rs. 3,00,000 has a total life of 15 years. If the rate of inflation is taken as 10 % constant throughout its life period, find the future value of the device, total depreciation, remaining value and annual depreciation charge at the end of 12th year. Take salvage value as 10% of inflated price.

Solution:

$$F = P(1+j)^n = 3,00,000(1+0.1)^{12} = 941528.51$$

$$CD_n = \frac{12}{15}[P(1+j)^n - 0.1P(1+j)^n] = \frac{12}{15}P(1+j)^n(0.9)$$
$$= 0.8 \times 0.9 \times 3,00,000(1.1)^{12} = 677900.53$$

Remaining value of the device at the end of 12th year will be 941528.51-677900.53 = 263627.98

Annual depreciation charge in the 12th year = CD_n-CD_{n-1}=

$$0.06P[n(1+j)^n - (n-1)(1+j)^{n-1}] = 113040.00$$

Thus it can be seen that at the higher inflation rate, the financial burden on the enterprise will be very high. In straight line method, depreciation charge being constant = $(C-S)/L$ = (300,000-30,000)/15 = 18,000 per year.

9.14 LIFE CYCLE COSTING (LCC)

Life Cycle Costing (LCC) is also known as cradle-to-grave costing. The purpose of this type of accounting is to provide a complete record of all the costs associated with the product or service. It is a process to determine the sum of all the costs associated with an asset or part thereof, including acquisition, installation, operation, maintenance and disposal costs.

Life-cycle cost is therefore defined as all costs, both non-recurring and recurring, that occur over the life cycle of a system. Many systems and products are planned, designed, produced and operated with very little concern for their life-cycle cost. But life-cycle costing helps determining the true cost of expenditure over time and a way for making valid economic comparisons between the alternatives. Many business and organizations still use economic analysis methods that do not include all costs and do not use the time value of money. The Simple Payback Period (SPP), method is commonly used by business and other organizations, but it is not a life cycle cost analysis approach. Some organization still make purchase decisions based on lowest initial costs and do not consider the operating and maintenance costs at all. Costs that occur after the devices or equipment are purchased and installed are ignored. This omission generally creates a very inaccurate view of the economic viability of a project in many cases. Use of LCC can lead to more rational purchase decisions and can often lead the business to higher profits.

Life cycle begins with the identification of the economic need or want (the requirement) and ends with phase out and disposal activities. This life cycle is basically divided into two general time periods, the acquisition phase and the operation phase. The acquisition phase begins with an analysis of the economic need or want i.e. the analysis necessary to make explicit requirement for the product, structure, system or service. The next group of activities is the acquisition phase that involves the detailed design and planning for production or construction. This step is followed by the activities necessary to prepare, acquire and make ready for the operation of the facilities and other resources needed for the production, delivery or construction of the product, structure, system or services. Again the engineering economy studies are the essential part of the design process to analyse and compare alternatives and to determine the final detailed design.

In the operation phase, the production, delivery or construction of the end item (s) or service (s) and their operation or customer use occur. This phase ends with retirement from active operation or use and often, disposal of the physical assets involved. The priorities for the business economy studies during the operation phase are (i) achieving efficient and effective support to operations (ii) determining whether (and when) replacement of assets should occur and (iii) projecting the timing of retirement and disposal activities.

In brief, the phases of the life cycle and their relative costs can be summarized as follows;

Acquisition phase

i)   Needs assessment and requirement

ii) Preliminary design, advanced development, prototype testing.

iii) Detailed design, production or construction planning, facility and resource acquisition.

Operation phase

i)   Production or construction

ii) Operation or customer use, maintenance and support.

iii) Phase out and disposal phase covering all activities to transition to a new system, removal / recycling / disposal of old system.

9.14 (i) LCC Calculation

This concept of life cycle costing was promoted to the concept of 'terotechnology', which is a combination of management, engineering, financial and other practices applied to the physical assets in pursuit of economic life cycle costs. Terotechnology includes the (i) acquisition costs, (ii) the product distribution costs (iii) the maintenance cost (iv) the inventory costs (v) the phase out costs and (vi) the disposal costs

The life cycle cost of an asset/device/project can be calculated using the formula,

$$LCC = C_0 + M_{pw} + E_{pw} + R_{pw} - S_{pw}$$

Where 'C_0' is the initial capital expense for procurement and installation costs. It is always considered as a single payment occurring in the initial year of the project, regardless of how the project is financed. 'M' tends for all yearly scheduled operations and maintenance (O & M) costs. It basically includes items such as operator's salary, inspection, insurance, property tax and all scheduled maintenance. 'E' tends for the energy cost of the system and is the sum of the yearly fuel cost. 'R' tends for replacement cost which is the sum of all repair and equipment replacement cost anticipated over the life of the system. 'S' tends for the salvage value of a system which is its net worth in the final year of the life-cycle period. The subscript 'PW' signifies the present worth of all costs.

The life cycle costing favours decision in choosing between more than one alternative. Therefore, a simple methodology for making valid economic comparisons between alternatives is necessary. The present worth method is mostly followed for this comparison. Several other methodologies exist, including future worth, annual worth, internal rate of return, cost / benefit ratio etc. Future worth and annual worth are economically equivalent to present worth and can be interchanged if desired. Internal rate of return and cost / benefit ratio methods require an "incremental" approach to make valid comparisons between the projects. In addition, multiple solutions for the value of IRR can occur depending on the form of the cash flow. Since present worth is considered the standard against which other methods are judged, the present worth method is therefore recommended. In this method, all the cash flows in LCC analyses are brought to a single, present-day base line. It allows the analyst to compare the projects with different cash flows occurring at different times. The present worth of the total costs for each alternative is calculated and the alternative with the lowest LCC is selected. If multiple projects need to be selected, then the projects can be ranked according to increasing LCC.

Example 9.53. An energy efficient air compressor is proposed by a vendor. The compressor will cost Rs. 30,000 and will require Rs. 1000 worth of maintenance each year for its life of 10 years. Energy cost will be Rs. 6000 per year. Another standard air compressor will cost Rs. 25,000 and will require Rs. 500 worth of maintenance each year. Its energy cost will be Rs. 10,000 per year. If the discount rate permitted would be 10%, would it be advisable to invest in the energy efficient compressor.

Solution: Alternative 1: Energy efficient air compressor

Alternative 2: Standard air compressor

Cash flows:

End of Year	Alternative 1 (Rs)	Alternative 2 (Rs)
0	30,000	25,000
1	1000 + 6000 = 7000	500 + 10,000 = 10,500
2	7000	10,500
3	7000	10,500
4	7000	10,500
5	7000	10,500
6	7000	10,500
7	7000	10,500
8	7000	10,500
9	7000	10,500
10	7000	10,500

Note: In a life cycle cost analysis, all cash flows are costs and the signs are ignored.

LCC (alternative-1) = Rs. 30,000 + Rs. 7000 (P/A, i, n) = Rs. 30,000+ 7000 $\left\{\dfrac{(1+i)^n-1}{i(1+i)}\right\}$

$= Rs.\,30,000 + 7000\left(\dfrac{(1.1)^{10}-1}{0.1(1.1)^{10}}\right) = Rs.\,30,000 + 7000\left(\dfrac{1.6}{0.26}\right) = Rs.\,73,076$

LCC (alternative -2) = Rs. 25,000 + Rs. 10,500 (P/A, i, n) = Rs. 25,000 + 10,500 (6.15) = Rs. 89,615

The decision rule for LCC analysis is to choose the alternative with the lowest LCC. Since alternative – 1 has the lowest LCC, it should be chosen.

Example 9.54. Compare the use of fluorescent lamp with incandescent lamp on the basis of LCC and with the information in the following table. The time span of study is 5 years. Roughly, the cost of fluorescent lamps is about 10 to 15 times higher for same wattage rating, but they are about 4 to 5 times more efficient and last about 4 to 5 times longer than incandescent lamp.

Comparison of the LCC of Incandescent Lamp and Fluorescent Lamp for the operation of 5 years. All costs are in (Rs.). The discount rate is 10%

| Lamp cost and replacement | Incandescent lamp | | | Fluorescent Lamp | | |
	Present cost	Present worth factor	Present worth	Present cost	Present worth factor	Present worth
1st year	20	1	20	200	1	200
2nd year	20	0.909	18.18	-	-	0
3rd year	20	0.826	16.52	-	-	0
4th year	20	0.751	15.02	-	-	0
5th year	20	0.683	13.66	-	-	0
Annual electricity cost	540	$\left\{ \dfrac{(1+i)^5 - 1}{i(1+i)^5} \right\}$ $= 3.82$	2,063	140	3.82	535
LCC			2146			735

It can be seen from the table that the LCC of the fluorescent lamp for 5 years of operation is much lower than that of the incandescent lamp, therefore fluorescent lamp should be the preferred choice.

9.14 (ii) Annualized LCC (ALCC)

It is sometimes useful to compare the LCC of a system on an annual basis. In this case, dividing the LCC by the expected life of the system will not give us the annualized LCC or ALCC of the system. Because, it is assumed than the annual cost of operation of the system will not be the same throughout its lifetime, therefore, in order to find out the annual cost in terms of current value of money, it is necessary to divide the LCC by the present worth factor. In the above example, the ALCC for both the incandescent Lamp ($ALCC_{IL}$) or fluorescent lamp ($ALCC_{FL}$) can be given as

$$ALCC_{(IL)} = \frac{LCC}{Present\ worth\ factor} = \frac{2146}{3.82} = Rs.\ 562$$

$$ALCC_{(FL)} = \frac{LCC}{Present\ worth\ factor} = \frac{735}{3.82} = Rs.\ 192$$

Thus the annualized cost of operation in the above example is much higher in case of the use of incandescent lamp compared to fluorescent camp.

9.15 SENSITIVITY ANALYSIS

Sensitivity analysis measures the impact on the investment decision by changing one or more key input values which are not known with certainty. If an input parameter can be varied over a wide range without affecting the investment decision, the decision under consideration is said to be insensitive to the particular parameter. On the other hand, if a small change in the relative magnitude of a parameter can reverse an investment decision, the decision is highly sensitive to that factor. Investment decision can be made on a more rational basis by the knowledge of the more uncertain factors in a given investment situation and an understanding of how sensitive the alternative is to those factors.

In the evaluation of an investment project, we work with the forecasts of cash flows. Forecasted cash flows depend on the expected revenue and costs. Further, the expected revenue is a function of sales volume and unit selling price. Similarly, sales volume depends on the market size and the firm's market share. Costs include variable costs (sales volume and unit variable cost) and fixed costs. The net present value (NPV) or the internal rate of return (IRR) of a project is determined by analysing cash flows arrived at by combining forecasts of various variables. It is difficult to arrive at an accurate and unbiased forecast of each variable. We cannot be certain about the outcome of any of these variables. The reliability of the NPV or IRR of the project depends on the reliability of the forecasts of variables underlying the estimates of net cash flow. To determine the reliability of the project's NPV or IRR, we can work out how much difference it makes if any of these forecasts goes wrong. We can change each of the forecast, one at a time, to at least three values; pessimistic (unfavourable direction), expected and optimistic (favourable direction). The NPV or IRR of the project is recalculated under these different assumptions. This method of recalculating NPV or IRR by changing each forecast is called sensitivity analysis. While assuming the pessimistic condition for an input factor, there is 95 percent chance of the input parameter being exceeded by the actual outcome. Similarly, for the optimistic condition, there is 5 per cent chance of being exceeded by the actual outcome. Besides these two extreme possibilities, the set of most likely values of the input parameters is also considered in the evaluation.

Hence, sensitivity analysis is a way of analysing change is the project's NPV (or IRR) for a given change in one of the variables. It indicates how sensitive a project's NPV (or IRR) is to change in a particular variable. The more sensitive the NPV, the more critical is the variable. Sensitivity analysis identifies parameters that have the most impact on an economic decision.

The decision maker, while performing sensitivity analysis, computes the project's NPV (or IRR) for each forecast under three assumptions i.e. pessimistic, expected and optimistic. It allows him to ask 'what if' questions. For example, what (is the NPV) if volume increases or decreases? What (is the NPV) if variable cost or fixed cost increases or decreases? What (is the NPV) if the selling price increases or decreases? What (is NPV) if the project is delayed or outlay escalates or the project's life is more or less than anticipated? A whole range of questions can be answered with the help of sensitivity analysis.

Example 9.55. Make a sensitivity analysis of the optimal decision resulting from an investment proposal with the following expected values of variables.

Initial investment	-	Rs. 1,00,000
Annual revenues	-	Rs. 40,000
Life of the project	-	5 years
Discount rate	-	12%

The sensitivity analysis should explore the sensitivity of net present worth (NPV) to changes in annual revenue, initial cost and MARR (Minimum attractive rate of return) over the range -10% to +10%.

$$NPV = -1,000,000 + 40,000\left(P/A, 12\%, 5\right) = -1,00,000 + 40,000\left\{\frac{(1+i)^n - 1}{i(1+i)^n}\right\}$$

The expression in the bracket is known as 'uniform series present worth factor'.

$$= -1,00,000 + 40,000\left\{\frac{(1.12)^5 - 1}{0.12(1.12)^5}\right\}$$

$$=-1,00,000 + 40,000\ (3.62) = -1,00,000 + 1,44,800 = 44,800$$

If annual revenue decreases by 10%, it becomes, 40,000-4,000=36,000 and NPV becomes

$$NPV = -1,00,000 + 36,000\ (P/A,\ 12\%,\ 5)$$

$$= -1,00,000 + 36,000(3.62) = 30,320$$

If annual revenue increases by 10%, it becomes $40,000 + 4,000 = 44,000$

NPV=$-1,00,000+44,000$ (P/A, 12%, 5) $= -1.00,000 + 44,000 (3.62) = 59.280$

The sensitivity of NPV to changes in annual revenue over the range $-$ 10% to +10% is Rs. 28,960 from Rs. 30,320 to Rs. 59,280.

Change of initial cost

NPV at the expected input variables is Rs. 44,800

If initial cost decreases by 10%, it becomes $1,00,000 - 10,000 = 90,000$ and NPV becomes

NPV $= -90,000 + 40,000$ (P/A, 12%, 5)

$= -90,000 + 40,000 (3.62) = -90,000 + 1,44,800 = $ Rs. 54, 800

If initial cost increases by 10%, it becomes $1,00,000 + 10,000 = 1,10,000$

NPV $= -1,10,000 + 40,000$ (P/A, 12%, 5)

$= -1,10,000 + 40,000 (3.62) = -1,10,000 + 1,44,800 = $ Rs. 34,800

The sensitivity of NPV to changes in initial cost over the range -10% to +10% is 20,000 from Rs. 54,800 to Rs. 34,800.

Change of MARR

NPV at expected input variables is Rs. 44,800

If MARR decreases 10%, it becomes 12% - 0.10 × 12% =10.8% and NPV becomes

NPV $= -1,00,000 + 40,000$ (P/A, 10.8%, 5)

$$= -1,00,000 + 40,000\left\{\frac{(1+i)^{n}-1}{i(1+i)^{n}}\right\} = -1,00,000 + 40,000\left\{\frac{(1.1)^{5}-1}{0.1(1.1)^{5}}\right\}$$

$= -1, 00, 000 + 40, 000(3.78) = -1,00,000 + 1, 51, 200 = $ Rs. 51, 200

If MARR increase 10%, it becomes 12% + 0.1 x 12% = 13.2% and NPV becomes

$$NPV = -1,00,000 + 40,000 \left(P/A, 13.2\%, 5\right) = -1,00,000 + 40,000\left\{\frac{(1.13)^{5}-1}{0.132(1.13)^{5}}\right\}$$

$= -1,00,000+40,000(3.46) = -1,00,000 + 1, 38, 400 = $ Rs. 38, 400

The sensitivity of NPV to changes in MARR over the range $-$ 10% to 10% is 12,800 from Rs. 38,400 to Rs. 51,200.

Sensitivity analysis data table for NPV

Parameter/Percent change	-10%	Base	+10%
Initial cost	54,800	44,800	34,800
Annual revenue	30,320	44,800	59,280
MARR	51,200	44,800	38,400

A review of the table reveals that the investment proposal remains attractive (NPV > 0) within the range of 10% changes in annual revenue initial cost and MARR. An appealing way to summarize single factor sensitivity data is using a 'spider' graph.

A spider graph plots the NPVs determined in the examples and connects them with lines, one line for each factor evaluated for NPV. The figure below illustrates the spider graph for the data mentioned in above table. On this graph, lines with large positive or negative slopes (angle relative to horizontal regardless of whether it is increasing or decreasing) indicate factors to which the NPV is sensitive. The figure shows that NPV is least sensitive to changes in MARR (the MARR line is the most clearly horizontal) and most sensitive to changes in annual revenue (the annual revenue line has the steepest slope).

Sensitivity Analysis "Spider graph"

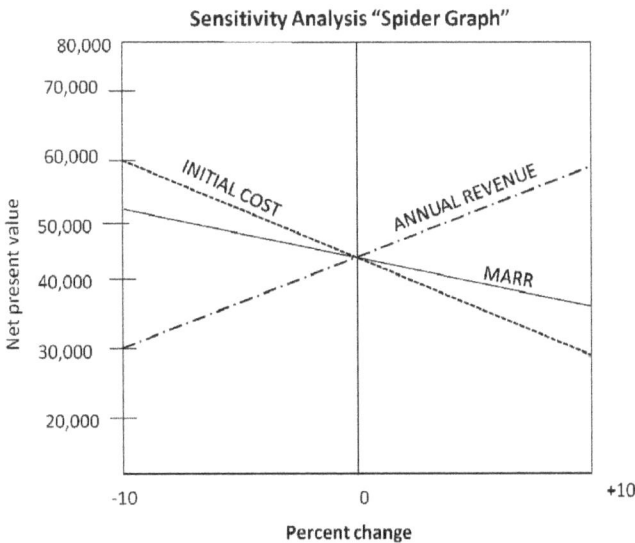

Sensitivity Analysis "Spider Graph"

.... Initial cost,   — . — . — Annual revenue,   _____ MARR

Example 9.56. Consider a proposed solar hybrid dryer for which the optimistic, pessimistic and most likely (or best) estimates are given in the following table. Process these estimates to facilitate decision-making.

Optimistic, pessimistic and most likely estimates for a solar hybrid dryer.

Parameters	Cost and performance estimates		
	Optimistic	Pessimistic	Most likely
Capital Cost (Rs.)	2,00,000	2,00,000	2,00,000
Useful life (years)	25	15	20
Salvage value (Rs.)	30,000	0	20,000
Annual revenue (Rs)	40,000	20,000	32,000
Annual expense (Rs.)	5,000	8,000	4,000
Minimum attractive rate of return (MARR) (%)	12	12	12

(i) Optimistic estimate

Annual cost contribution of capital cost $= P\left[\dfrac{i(1+i)^n}{(1+i)^n-1}\right] = P[A/P, i, n]$

The expression in the bracket is called uniform series capital recovery factor.

$= 2,00,000\left[\dfrac{0.12(1.12)^{25}}{(1.12)^{25}-1}\right] = 2,00,000\left(\dfrac{2.04}{16}\right) = \text{Rs. } 25,500$

Annual worth of salvage value $= F\left[\dfrac{i}{(1+i)^n-1}\right] = F[A/F, i, n]$

The expression in the bracket is called uniform series sinking fund factor.

$= 30,000\left[\dfrac{0.12}{(1.12)^{25}-1}\right] = \dfrac{0.12}{16} \times \dfrac{0.12}{16} \times 30,000 = \text{Rs. } 225$

Equivalent uniform annual worth of the optimistic scenario
$= 40,000 + 225 - 25,500 - 5000 = \text{Rs } 9,725$

(ii) Pessimistic estimate

Annual cost contribution of capital $= 2,00,000\ [\dfrac{0.12(1.12)^{15}}{(1.12)^{15}-1}] = 29,322$

Salvage value is zero.

Equivalent uniform annual worth of the pessimistic estimate = 20,000 – 29,322 – 8,000 = - Rs. 17,322

Most likely estimate: Annual cost contribution of capital cost

$$= 20,000 \left[\frac{0.12 (1.12)^{20}}{(1.12)^{20} - 1} \right] = \text{Rs. } 26,765$$

Annual worth of salvage value $= 20,000 \left[\frac{0.12}{(1.12)^{20} - 1} \right] = \text{Rs. } 277$

Equivalent annual worth of most likely estimate = Rs. 32,000 + 277 – 26,765 – 4000 = Rs. 958

Comments: In the above example, the AW (annual worth) for the optimistic estimate is very favourable (Rs. 9,725), while the AW for the pessimistic estimate is quite unfavourable (-Rs. 17,322). If both extreme AW values were positive, we would make a 'go' decision with respect to the device without further analysis because no combination of factor values based on the estimates would result in AW < 0. By similar reasoning, if both AW values were negative, a 'no go' decision would be made regarding the proposed device. In the above situation, a sensitivity graph (spider plot) is needed to show explicitly the sensitivity of the AW to three factors of concern: useful life, annual revenue and annual expense. The most sensitive factor is to be considered in all three estimates (optimistic, pessimistic and most likely), if AW would become positive and then the proposed device can be accepted; otherwise not.

9.15 (i) Sensitivity involving Multiple Alternatives

Very often, multiple alternatives of a proposed project come before the business organizer. He/she is to make the complex sensitivity analysis before judging the best alternative from an economic desirability point of view. It is required to study the variation of NPV of each alternative with respect to various input parameters and to compare before accepting one alternative, for which sensitivity involving multiple alternatives is to be done. Let us consider the following cash flows for three alternatives for making sensitivity analysis.

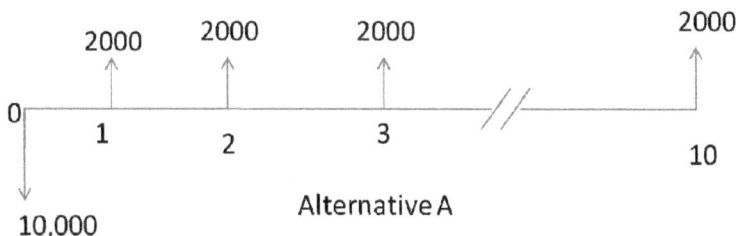

Alternative A

$NPV_A = -10,000 + 2000 \ (P/A, 10\%, 10)$

$$= -10,000 + 2000 \left\{ \frac{(1+i)^n - 1}{i(1+i)^n} \right\} = -10,000 + 2000 \left\{ \frac{(1.1)^{10} - 1}{0.1(1.1)^{10}} \right\}$$

$= -10,000 + 2000 \ (6.15) = +2300$

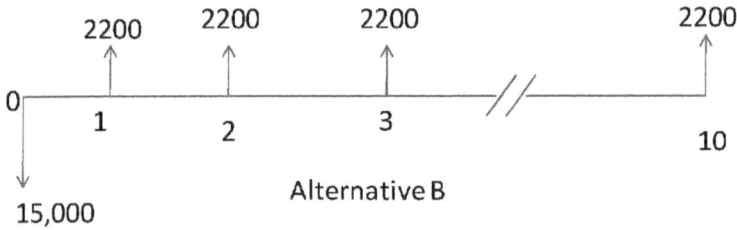

Alternative B

$NPV_B = -15,000 + 2200 \ (6.15) = -1470$

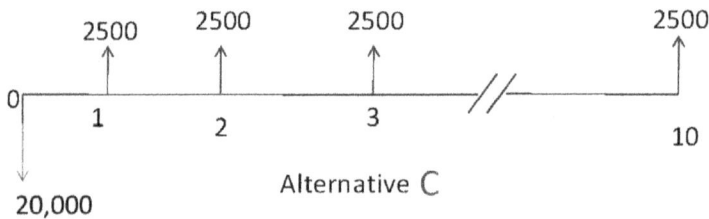

Alternative C

$NPV_C = -20,000 + 2500(6.15) = -4625$

The sensitivity analysis of three alternatives with regard to annual revenue, interest rate and life of the project may be done. It is expected that the lives for each of the three alternatives be 10 years and the appropriate interest rate is 10%.

Sensitivity of NPV to annual revenue, given i (10%) and n (10 years)

Annual revenue	NPV_A	NPV_B	NPV_C
1500	-775	-5775	-10,775
1800	1070	-3930	-8,930
2000	+2308	-2700	-7700
2200	3530	-1470	-6470
2500	5375	375	-4625
3000	8450	3450	-1550

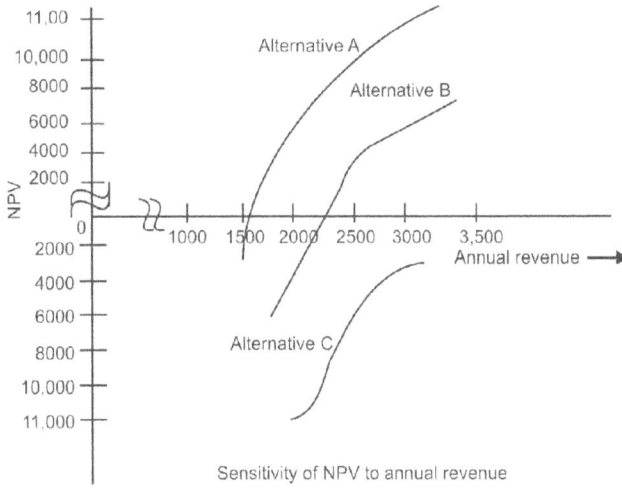

Sensitivity of NPV to annual revenue

Sensitivity of NPV to interest rate (i), given annual revenue (Rs. 2200) and n (10 years)

$i\%$	NPV$_A$	NPV$_B$	NPV$_C$
5	7056	2336	-2664
6	6243	1236	-3764
7	5416	400	-4600
8	4812	-282	-5282
9	4120	-898	-5898
10	3539	-1470	-6470
11	2974	-2020	-7020
12	2420	-2570	-7570
13	1972	-3010	-8010
14	1423	-3560	-8560
15	1147	-3868	-8868

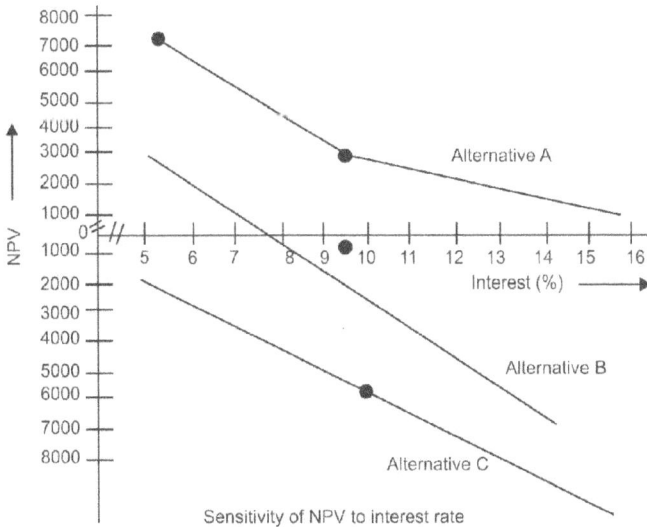

Sensitivity of NPV to interest rate

Sensitivity of NPV to the life of the project given annual revenue (2200) and interest rate (8%)

Life (n)	NPV$_A$	NPV$_B$	NPV$_C$
5	-4170	-9170	-14170
6	120	-4880	-9880
7	1440	-3560	-8560
8	2540	-2460	-7460
9	3750	-1250	-6250
10	4828	-172	-5172
11	5730	730	-4270
12	6610	1610	-3390
13	7424	2424	-2576
14	8216	3216	-1784
15	8876	3876	-1124

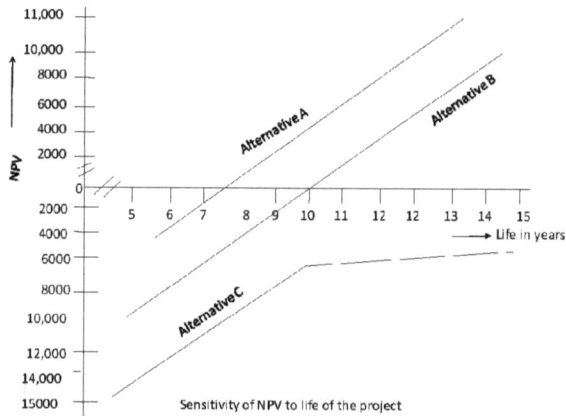

Sensitivity of NPV to life of the project

Preferred alternative for annual revenue

Preferred Alternative	Range of annual revenue
A	Rs. 1800 - Rs. 3000
B	Rs. 2500 - Rs. 3000
C	> Rs. 3000

Preferred alternative for interest rate

Preferred Alternative	Range of annual revenue
A	5 - 15%
B	5 - 7%
C	> 15%

Preferred alternative for life of the project	
Preferred Alternative	**Range of life span**
A	6 - 15
B	11 - 15
C	> 15

If it is anticipated that the interest rate, life span and annual revenue should be in the vicinity of 10%, 10 years and Rs. 2000/-, then alternative 'A' is the favoured choice. This type of information can provide the decision maker with a better understanding of the effects of estimates for the future outcomes.

Pros and Cons of Sensitivity Analysis

There are several advantages of using sensitivity analysis in proper selection of an alternative for a business venture. It shows how significant is any given input variable in determining a project's economic worth. It does this by displaying the range of possible project outcomes for a range of input values, which show to the decision maker, the input values that would make the project a looser or a winner. Hence the sensitivity analysis has the following advantages.

● It compels the decision maker to identify the variables which affect the cash flow forecasts. This helps him in understanding the investment project in totality.

● It indicates the critical variables for which additional information may be obtained. The decision maker can consider actions, which may help in strengthening the 'weak spots' in the project.

● It helps to expose inappropriate forecasts and thus guides the decision maker to concentrate on relevant variables.

Therefore, sensitivity analysis is not a remedy for a project's uncertainties. It helps a decision maker to understand the project better. It has also the limitation.

● It does not provide the clear-cut results. The terms 'optimistic' and 'pessimistic' could mean different things to different persons in an organization. Thus the range of values suggested may be inconsistent.

9.16 REPLACEMENT ANALYSIS OF MACHINE / EQUIPMENT

The business organizations providing goods and services use various machines and equipment which need to work efficiently to produce goods of desired quality and quantity in order to compete with the present market with respect to cost, quality and services. The operation of the machines needs to be monitored continuously for their efficient functioning. Otherwise, the quality of goods and services would be poor and the organization would not cope up with the increasingly competitive market resulting into deriving less benefit. In addition to this, the cost of their operation and maintenance would increase with the passage of time. Hence, it becomes an absolute necessity to go for replacement of inefficient equipment as and when required for which the replacement analysis needs to be done.

In replacement decision, one is concerned with the machinery that deteriorate with the passage of time. The efficiency of a machine goes on decreasing with the elapse of time of use and the machine gets older. It is then decided to replace it when the maintenance and capital costs of the existing machine is more than the average capital and operating costs of the asset proposed for replacement. Many people feel that a machine should not be replaced until it is physically worn out. But it is not correct, preferably machines must be constantly renewed and updated, otherwise, there is the increasing risk that they will become obsolete. If an organization wants to be in the same business to face the existing competitions, it has to take decision on whether to replace the old machine or to retain it by taking the cost of maintenance and operation into account.

The replacement analysis is used to make decisions related to keeping and maintaining the existing equipment or replacing the aging equipment with the new one. If any equipment or machine is used for a long period of time, the item tends to worsen due to wear and tear. The remedial action is to bring the equipment to the original level by replacing the same with an alternative. The equipment replacement planning is a methodology; it is a means to an end, a process that focuses on the equipment mostly in need of replacement. It is the last attempt which has ideally not having any second alternative i.e., condition beyond repair. Repair is not a replacement but an attempt to restore the lost values of an asset.

Repair work can be performed in case of broken, damaged or failed equipment. The repair can make the asset acceptable, operating or usable condition.

In the world of competition and availability of the better quality in goods and services, the type of decision of replacement is occurring more frequently. An organization not operating efficiently will be out of the business sooner or later. In the fast-growing emerging markets, the challenge is to meet the huge demand and gain market share while ensuring efficiency and the quality of end-user services. The business enterprises who have the plan and keep their set up updated are most likely to streamline the operations. The question of replacement of an equipment arises when the same cannot meet the production requirements. Every organization is looking for ways to reduce the manufacturing costs and at the same time increasing the production. When the existing machine cannot meet production requirements in terms of either throughput or manufacturing cost, it should be replaced.

9.16.1 Reasons for Replacement

A machine is generally considered for replacement for the following reasons;

i) *Deterioration:* It is the decline in the performance of a machine as compared to a new one. Deterioration may occur due to wear and tear as well as misalignments of the machine. Deterioration raises maintenance costs, reduces the quality of product and work, decreases production, causes loss is operating time, increases labour costs and reduces the efficiency of the machine. So the machine needs to be replaced.

ii) *Obsolescence:* Obsolescence is the process of an existing machine to become obsolete or out of date due to the improvement in technology. If the performance of a machine is better than the existing one, then the same unit has to be replaced to withstand market competition.

iii) *Inadequacy:* When current operating condition of a set-up changes, the older machine occasionally lacks the capacity to meet the new requirements. A machine can be considered to be in inadequacy level due to less productivity of machine and growth and changes in size of the firm.

iv) *Working Condition:* It may be thought of replacing old machines which create unpleasant and hazardous working conditions causing unsafe to worker and leading to accidents.

Replacement Terminologies: The following terminologies are generally used in the study of replacement analysis.

i) *Defender:* It is the existing machine (old one)

ii) *Challenger:* It is the best available replacement machine (new one)

iii) *Current market value:* It is the value to use in preparing a defender's economic analysis.

iv) *Sunk costs:* It is the past costs that cannot be changed by any future investment decision and should not be considered in defender's economic analysis.

v) *Opportunity cost:* It is the cost associated with the decision to retain the defender in its current market value. This cost must be allocated over the defender's remaining life.

vi) *Operating cost:* It is the sum of the various costs related to the operation of the machine. It includes repair and maintenance costs, wages for the operators, energy consumption costs and costs of materials.

9.16.2 Different Types of Life of an Asset

The period over which a machine / equipment / asset is expected to be usable with normal repair and maintenance is known as the life of the asset. There are various types of life for an asset such as economic life, ownership life, physical life and useful life.

Economic life: It is the period of time (years) that results in minimum annual equivalent cost of owning and operating an asset.

Ownership life: It is the period between the date of acquisition and the date of disposal by a specific owner.

Physical life: It is the period between the date of acquisition and final disposal of an asset over its succession of owners. For example, a car may have several owners over its existence.

Useful life: It is the time period (years) that an asset is kept in productive service. It is an estimate of how long an asset is expected to be used in a business to produce income.

Factors to be considered in replacement studies: There are several factors which are taken into consideration in replacement studies. The factors to be required for replacing machine are classified into two categories i.e. (i) technical factors such as technical soundness of the machine, capacity of the machine etc. (ii) financial factors such as initial costs of the new equipment, direct and indirect material cost, labour cost, interest on capital, cost of replacing the parts etc. Once the decision has been taken to replace the equipment, careful attention has to be taken into consideration. The factors to be considered are as follows;

i) Recognition and acceptance of past errors, ii) Sunk costs, iii) Existing asset value, iv) Economic life of the proposed replacement asset v) Remaining life of old asset vi) Income tax consideration

9.16.3 Determination of Economic Life of an Asset

Economic life of an asset refers to the period during which economic value can be obtained out of it. Any asset will have the following cost components;

- Capital recovery cost (average first cost), computed from the first cost (purchase price) of the machine.
- Average operating and maintenance cost
- Total cost which is the sum of capital recovery cost (average first cost) and average maintenance cost.

A typical shape of each of the above costs with respect to life of the machine is shown below.

With the machine run, the value of asset goes on decreasing due to excessive or rigorous use throughout its life time. Here the first or the initial cost (purchase cost), goes on decreasing with each period of machine run. The cost of maintenance goes high with each period of machine run because more and more, the machine or asset gets old, it needs some extra care or additional maintenance. From the beginning, the total cost continues to decrease up to a particular period and then it starts increasing. The point, where, the total cost is minimum is called the economic life of the machine. If the interest rate is more than zero percent; then we use interest formulas to determine the economic life. The replacement alternative can be evaluated based on present worth criterion and annual equivalent criterion.

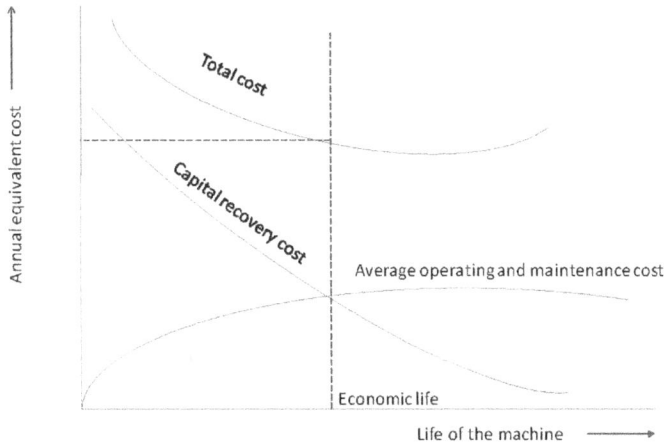

Graph showing economic life a machine

9.16.4 Replacement of existing Asset with a New Asset

In the replacement analysis, the decision is taken not only to replace the old machine but also to do the same at which period. If an existing asset is considered for replacement with a new one, then the existing asset is known as the defender and the new asset is known as challenger, which is considered to be the best available replacement asset. The defender-challenger comparison decides exactly which asset is the best challenger.

Generally, replacement alternatives are described by cash flows. Once the proper cash flows are identified for a replacement alternative, the methods for selecting the most economical alternative become easy. Basically annual equivalent cost method and present worth method are used for comparison.

There are two important principles that guide the identification of the appropriate cash flows for replacement studies. These are;

i)   The value of the defender that should be used a study of replacement is not what it costs when originally purchased but what it is worth in the present time. Thus, the defender's sunk cost, which has no effect on future action, is irrelevant to any decision about the defender.

ii) When an explicit decision is made to retain rather than dispose of an asset already owned, an opportunity to receive compensation has been foregone. This economic opportunity foregone must also be associated with the alternative that causes it to be lost.

Assume that a machine has been purchased about four years back for Rs. 6,00,000 and it is considered for replacement with a new machine. The supplier of the new machine will take the old one for some money, say Rs. 4,00,000. This should be treated as the present value of the existing machine and it should be considered for all further economic analysis. The purchase value of the existing machine before four years is now known as sunk cost and it should not be considered for further analysis.

Example 9.57. A business organization is considering replacement of a machine; whose first cost is Rs. 5000 and the salvage value is negligible at the end of any year. Based on experience, it was found that the maintenance cost is zero and Rs. 300 during the first and second year respectively and it increases by Rs. 200 every year thereafter.

(a) When should the machine be replaced if i = 0%

(b) When should the machine be replaced if i = 12%

(a) Determination of economic life of an asset (First cost = Rs. 5,000, Interest = 0%)

End of year (n)	Maintenance cost at end of year	Summation of maintenance costs	Average cost of maintenance through year given	Average first cost if replaced at year end given	Average total cost through year given
		ΣB	C/A	5000/A	D+E
A	B (Rs.)	C (Rs.)	D (Rs.)	E (Rs.)	F (Rs.)
1	0	0	0	5000	5000
2	300	300	150	2500	2650
3	500	800	267	1667	1934
4	700	1500	375	1250	1625
5	900	2400	480	1000	1480
6	1100	3500	584	834	1418
7	1300	4800	686	714	**1400**
8	1500	6300	788	625	1413

The average total cost decreases till the end of year 7 and then it increases. Therefore, the optimal replacement period is 7 years i.e. the economic life of the asset is 7 years.

(b)Determination of Economic Life (First cost = Rs. 5000, Interest 12%)

End of year (n)	Maintenance cost at end of year	(P/F, 12%, n)	Present worth as the beginning of year 1 of maintenance costs	Summation of present worth of maintenance costs through year given	Present worth of cumulative maintenance cost & first cost	(A/P, 12%, n)	Annual equivalent total costs through year given
			(B x C)	ΣD	E+Rs. 5000		F x G
	B (Rs)	C	D (Rs)	E (Rs)	F (Rs)	G	H (Rs)
1	0	0.8929	0	0	5000	1.1200	5600
2	300	0.7972	239.16	239.16	5239.16	0.5917	3100
3	500	0.7118	356	595.16	5595.16	0.4163	2329.26
4	700	0.6355	445	1040.16	6040.16	0.3292	1988.42
5	900	0.5674	511	1,551.16	6551.16	0.2774	1817.29
6	1100	0.5066	557.26	2,108.42	7108.42	0.2432	1728.76
7	1300	0.4524	588.12	2696.54	7696.54	0.2191	1686.31
8	1500	0.4039	605.85	3302.39	8302.39	0.2013	1671.27
9	1700	0.3606	613.02	3,915.41	8915.41	0.1877	1673.42
10	1900	0.3220	611.8	4527.21	9527.21	0.1770	1686.31

Hence the economic life of the asset is 8 years.

The steps followed for calculation of economic life are summarized below.

i) Discount the maintenance costs to the beginning of year 1

$$\text{Column D} = \text{Column } B \times \frac{1}{(1+i)^2} = \text{column } B \times (P/F, i, n) = \text{column } B \times \text{column } C$$

ii) Find the summation of present worth of maintenance costs through the year given (column $E = \Sigma$ column D).

iii) Find column F by adding the first cost of Rs. 5,000 to Column E

iv) Find the annual equivalent total cost through the year given

$$\text{Column H} = \text{column } F \times \frac{i(1+i)^n}{(1+i)^n - 1}$$

$= \text{column } F \times (A/P, 12\%, n) = \text{column } F \times \text{column } G$

v) Identify the end of year, for which the annual equivalent total cost is minimum.

In this example, the annual equivalent total cost is minimum at the end of year 8. Therefore, the economic life of the machine is eight years.

Example 9.58. Four years ago, a machine was purchased at a cost of Rs. 4,00,000 to be useful for ten years. Its salvage value at the end of its life is Rs. 50,000. The annual maintenance cost is Rs. 40,000. The market value of the present machine is Rs. 2,50,000. Now a new machine to cater to the need of the present machine is available at Rs. 3,00,000 to be useful for six years. Its annual maintenance cost is Rs. 25,000. The salvage value of the new machine is Rs. 30,000. Using an interest rate of 12%, find whether it is worth replacing the present machine with the proposed new one.

Solution: Alternative I (Present machine)

Purchase price = Rs. 4,00,000; Present value (*P*) = Rs. 2,50,000, Salvage value (*S*) = Rs. 50,000; Annual maintenance cost (*A*) = Rs. 40000, Remaining life = 6 years, Interest rate = 12%

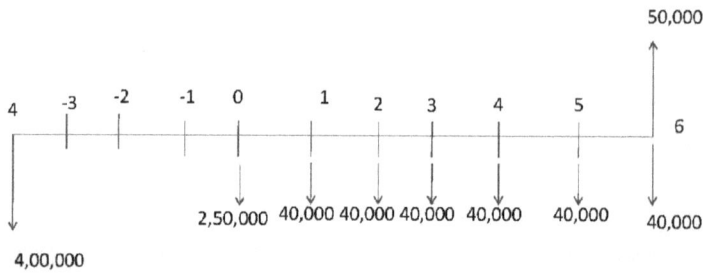

The annual equivalent cost is the computed as
$$AE\ (12\%) = P(A/P,\ 12\%,\ 6) + 40{,}000 - S(A/F,\ 12\%,\ 6)$$

$$= 2{,}50{,}000\left\{\frac{i(1+i)^n}{(1+i)^n-1}\right\} + 40{,}000 - 50{,}000\left\{\frac{i}{(1+i)^n-1}\right\}$$

$$= 2{,}50{,}000\left\{\frac{0.12(1.12)^6}{(1.12)^6-1}\right\} + 40{,}000 - 50{,}000\left\{\frac{0.12}{(1.12)^6-1}\right\}$$

$=2{,}50{,}000(0.243) +40{,}000-50{,}000(0.123) = 60{,}750+40{,}000-6150=$Rs. 94,600

Alternative II (New proposed machine)

Purchase price (*P*) = Rs. 3,00,000; Salvage value (*S*) = Rs. 30,000, Annual maintenance cost = Rs. 25,000; Life = 6 years, Interest rate = 12%

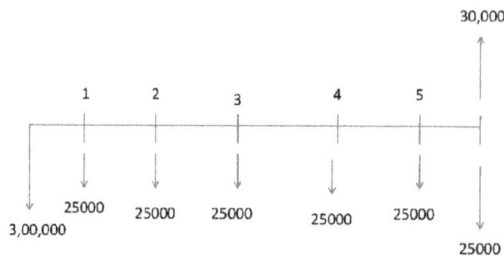

Cash flow diagram for alternative -II

Cash flow diagram for alternative-II

$AE(12\%) = P(A/P, 12\%, 6) + 25000 - S(A/F, 12\%, 6)$

$$= 3,00,000 \left\{ \frac{i(1+i)^n}{(1+i)^n - 1} \right\} + 25,000 - 30,000 \left\{ \frac{i}{(1+i)^n - 1} \right\}$$

$= 3,00,000 \,(0.243) + 25,000 - 30,000 \,(0.123)$

$= 72,900 + 25,000 - 3690 = \text{Rs. } 94,210$

Since the annual equivalent cost of the new machine is less than that of the present machine, it is suggested that the present machine may be replaced with the new machine.

Present Worth Method: Present worth method may also be followed to decide for the replacement of old machine with the new one. In order to compare the result, the future costs are translated into today's value of money.

The above example may be solved for the present worth method.

Alternative I (Present machine)

Present worth, $P.W.(12\%) = P +$ annual maintenance cost $(P/A, 12\%, 6) - S(P/F,$

$$12\%, \ 6) = 2,50,000 + 40,000 \left\{ \frac{(1+i)^n - 1}{i(1+i)^n} \right\} - 50,000 \left(P/F, 12\%, 6 \right)$$

$= 2,50,000 + 40,000 \,(4.11) - 50,000 \quad (0.506) = Rs.3,89,100$

Alternative II (New machine)

Present worth, P.W. (12%)

$= P +$ annual maintenance cost $(P/A, 12\%, 6) - S(P/F, 12\%, 6)$

$= 3,00,000 + 25,000 \,(4.11) - 30,000 \,(0.506) = \text{Rs. } 3,87,570$

The present worth of new machine is less and should be selected.

Example 9.59. Three years ago, a water supplying organization purchased a 10 hp motor for pumping drinking water. Its useful life was estimated to be 10 years. Due to increase in population of the locality, the organization is unable to meet the current demand for water with the existing motor. The organization can cope with the situation either by augmenting an additional 5 hp motor or replacing the existing 10 hp motor with a new 15 hp motor. The details of these motors are given below.

	Old 10 hp motor	New 5 hp motor	New 15 hp motor
Purchase cost (*P*) Rs.	25,000	10,000	35.000
Life in years	10	7	7
Salvage value at the end of machine life (Rs.)	1,500	800	4,000
Annual operating & maintenance cost (Rs.)	1,600	1,000	500

The current market value of the existing 10 hp motor is Rs. 10,000. Using an interest rate of 15%; find the best alternative.

Solution: There are two alternatives to cope with the situation.

i) Augmenting the present 10 hp motor with an additional 5 hp motor.

ii) Replacing the present 10 hp motor with a new 15 hp motor

Alternative I: Augmenting the present 10 hp motor with an additional 5 hp motor.

Total annual equivalent cost = Annual equivalent cost of 10 hp motor + Annual equivalent cost of 5 hp motor.

Calculation of annual equivalent cost of 10 hp motor:

Present market value of the 10 hp motor (P) = Rs. 10,000

Remaining life (n) = 7 years

Salvage value at the end of motor life (S) = Rs. 1,500

Annual operation and maintenance cost = Rs. 1,600

Interest rate, i = 15%

The cash flow diagram of this alternative is shown below

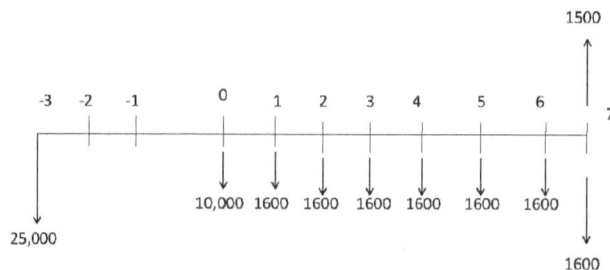

Cash flow diagram of 10 hp motor

The annual equivalent cost of the 10 hp motor is calculated as

$AE(15\%) = 10,000(A/P, 15\%, 7) + 1600 - 1500 (A/F, 15\%, 7)$

$$= 10,000 \left\{ \frac{i(1+i)^n}{(1+i)^n - 1} \right\} + 1600 - 1500 \left\{ \frac{i}{(1+i)^n - 1} \right\}$$

$= 10,000(0.24) + 1600 - 1500(0.09) = 2400 + 1600 - 135 = $ Rs. 3865.

Calculation of annual equivalent cost of 5 hp motor

Purchase value of 5 hp motor (P) = Rs. 10,000; Life (n) = 7 years

Salvage value at the end of motor life (S) = Rs. 800

Annual operation and maintenance cost = Rs. 1000

Interest rate, i = 15%.

The cash flow diagram of 5 hp motor is shown below.

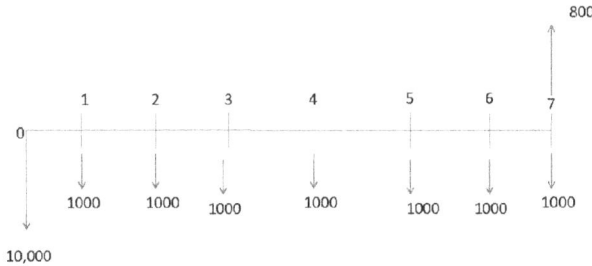

Cash flow diagram of 5 hp motor

The annual equivalent cost of the 5 hp motor is calculated as

$AE(15\%) = 10,000(A/P, 15\%, 7) + 1000 - 800(A/F, 15\%, 7) = $ Rs. 3328

The annual equivalent cost of the alternative I = Rs. 3856 + Rs. 3328 = Rs. 7193

The annual equivalent cost of alternative-II is calculated as

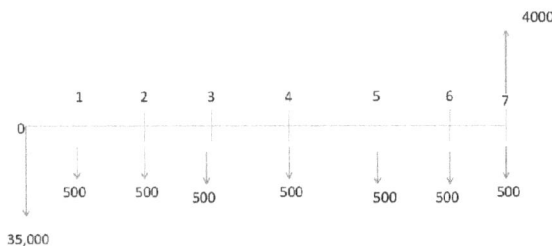

Cash flow diagram of 15 hp motor

$AE(15\%) = 35000(A/P, 15\%, 7) + 500 - 4000\ (A/F, 15\%, 7)$

$= 35000(0.24) + 500 - 4000(0.09) = 8400 + 500 - 360 = $ Rs. 8, 540

The total annual equivalent cost of alternative I is less than that of alternative II. Therefore, it is suggested that 10 hp motor be augmented with a new 5 hp motor.

Example 9.60. A machine was purchased two years ago for Rs. 10,000. Its annual maintenance cost is Rs. 750. Its life is six years and its salvage value at the end of its life is Rs. 1000. Now, an organization is offering a new machine at a cost of Rs. 10,000. Its life is four years and its salvage value at the end of its life is Rs. 4000. The annual maintenance cost of the new machine is Rs. 500. The organization which is supplying the new machine is willing to take the old machine for Rs. 8000 if it is replaced by the new machine. Assume the interest rate of 12%, compounded annually,

(a) Find the comparative use value of the old machine?

(b) Is it advisable to replace the old machine?

Solution: Old machine: - Let the comparative use value of old machine now be X.

Remaining life (n) = 4 years, Salvage value of the old machine (S) = Rs. 1000

Annual maintenance cost = Rs. 750

Interest rate, i = 12%.

The cash flow diagram of the old machine is shown below;

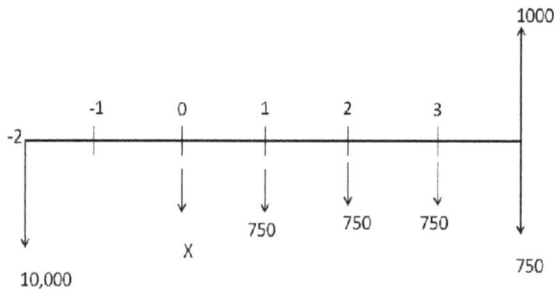

Cash flow diagram for old machine

The annual equivalent cost of the old machine is calculated as

$AE(12\%) = X(A/P, 12\%, 4) + 750 - 1000 \ (A/F, 12\%, 4)$

$$= X\left\{\frac{i(1+i)^n}{(1+i)^n - 1}\right\} + 750 - 1000\left\{\frac{i}{(1+i)^n - 1}\right\}$$

$$= X\left\{\frac{0.12(1.12)^4}{(1.12)^4 - 1}\right\} + 750 - 1000\left\{\frac{0.12}{(1.12)^4 - 1}\right\}$$

$X(0.33) + 750 - 1000 \ (0.209) = X(0.33) + 750 - 209 = X(0.33) + 541$

New machine

Cost of the new machine = Rs. 10,000

Life (n) = 4 years

Salvage value of new machine (S) = Rs. 4000

Annual maintenance cost = Rs. 500

Interest rate, i = 12%.

The cash flow diagram of the new machine is shown below.

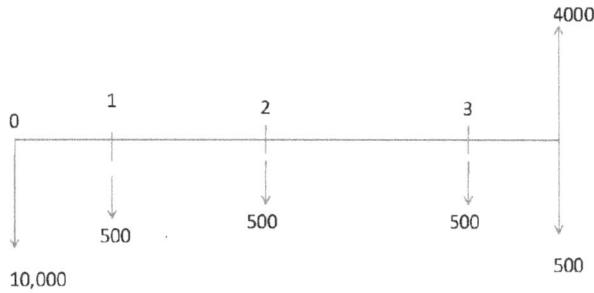

Cash flow diagram for new machine

The annual equivalent cost of the new machine is calculated as

$AE(12\%) = 10000(A/P, 12\%, 4) + 500 - 4000(A/F, 12\%, 4)$

$=10,000 (0.33) +500-4000 (0.209) = 3300 + 500-836 = 2,964$

Now, equating the annual equivalent costs of the two alternatives and solving for X,

$X(0.33) + 541 = 2964 \Rightarrow X = Rs.7342$

The comparative use value of the old machine is Rs. 7342, which is less than the price (Rs. 8000) offered by the selling organization which is supplying the new machine in the event of replacing the old machine by the new machine. Therefore, it is advisable to replace the old machine with the new one.

9.17 BREAK-EVEN ANALYSIS

Break-even analysis is an important analytical technique used to study the relationship between the total cost, total revenue and total profits as well as revenues over the whole range of stipulated output of a business organization. The analysis is particularly useful when selection from different alternatives, highly sensitive to an uncertain parameter has to be made. Most of the input data employed in financial evaluation exercises are uncertain as they represent the estimates of the future. The break-even values represent those values of uncertain parameters in which the decision-maker is to produce a basis or benchmark for comparing alternatives. This analysis is also concerned with finding the stage at which revenues and costs exactly agree i.e. they are equal. It therefore indicates the volume of output at which neither a profit is made nor a loss is incurred. The volume of output needs to be always that point in order to make the business profitable, as profit is the ultimate objective of any business organization. For a business to become sustainable, there should be the consistent profit making despite the various risks it has to face. Since profits are vital, they cannot be left to be earned by chance or luck. Proper planning and control of profits is therefore, of utmost importance for a business enterprise. Unless a business organization is prepared to face the uncertainties during its activity period, its profits would be left to chance. Hence, there should be the proper planning for profits. In this regard, a thorough understanding of the relationship between cost, volume and price becomes extremely important to the organizer. Break-even analysis is therefore most important to the organizer. Break-even analysis is therefore most important method for determining cost-volume-profit relationship.

Break-even point in the break-even analysis is an indicator to indicate the economic viability of a business organization. *It is defined as that point of activity (production/ sales volume) at which the total revenues be equal to the total costs and the net income is zero.* It is the point of zero profit. This is also known as no profit no loss point. Break-even point is, therefore, a point where losses cease to occur while profits have not yet begun. In case, the organization produces and sells less than what is suggested by the break-even point, it would make losses. If it produces and sells more than what is suggested by the break-even point, it would make profits. It can be determined either graphically or algebraically.

When the selection between two alternative projects is dependent on a single factor, we can solve for the value of that factor at which the decision-maker is indifferent between them. That value is known as the break-even point, that is, the value at which we are indifferent between the two alternatives.

In mathematical terms, we have $EW_A = f_1(y)$ and $EW_B = f_2(y)$

Where

$EW_A=$ an equivalent worth calculation for the net cash flow of alternative A.

$EW_B=$ the same equivalent worth calculation for the net cash flow of alternative B.

y = a common factor of interest affecting the equivalent worth values of alternative A and alternative B.

Therefore, the break-even point between alternative A and alternative B is the value of factor 'y' for which the two equivalent worth values are equal. That is,

$EW_A = EW_B$ or $f_1(y) = f_2(y)$, which may be solved for y.

Similarly, when the economic acceptability of one project is considered and which depends upon the value of the single factor, say Z, mathematically we can set an equivalent worth of the project's net cash flow for the analysis period equal to zero $EW_p = f(Z) = 0$ and solve for the break-even value of Z. That is, the value of Z at which we would be indifferent (economically) between implementing and rejecting the project.

The following are the examples of common factor for which break-even analysis might provide useful insights into decision-making problem.

a. *Annual revenue and annual cost:* Solve for the annual revenue required to meet annual costs.

b. *Rate of return:* Solve for the rate of return at which two given alternatives are equally desirable.

c. *Useful life of the equipment/system:* Solve for the useful life required for an alternative to be justified.

d. *Capacity utilization:* Solve for the annual hours of operation at which one project is justified or at which two alternatives are equally desirable.

e. *Salvage value:* Solve for future resale value that would make the decision maker indifferent to choose between the two alternatives.

Hence the usual break-even problem involving two alternatives can be most easily approached mathematically by equating an equivalent worth of the two alternatives expressed as a function of factor of interest. Using the same approach for the economic acceptability of a project, we can mathematically equate an equivalent worth of the project to zero as a function of the factor of concern.

The following examples illustrate the break-even analysis for one project or among two alternative projects.

Example 9.61. The cost and performance details of two alternative electric motors of 25 kW rating are given in the following table. Annual taxes and insurance costs on either motor are 3 percent of the investment. How many hours per year would the motors have to be operated at full load for annual costs to be equal if the minimum attractive rate of return is 15 percent. Assume that the unit cost of electricity is Rs. 4 per kWh and the salvage values for both the motors are negligible.

Cost and performance details of two electric motors

S.N.	Motor	Capital cost (Rs)	Efficiency (%)	Useful Life (Years)	Annual maintenance Cost (Rs.)
1	A	50,000	80	10	17,000
2	B	70,000	90	10	13,000

Solution: The electricity input required to each electric motor can be determined as the ratio of the rated output capacity and the efficiency of the respective motors. Let 'h' be the break-even number of hours of operation at which the annual costs of both the motors be equal. The annual cost of motor 'A' may then be calculated as

$50,000(A/P, 15, 10) + 25 (h/0.8)(4) + 17,000 + 0.03 (50,000)$

$$= 50,000 \left\{ \frac{i(1+i)^n}{(1+i)^n - 1} \right\} + \frac{25}{0.8}(h)(4) + 17,000 + 0.03(50,000)$$

$$= 50,000 \left\{ \frac{(0.15)(1.15)^{10}}{(1.15)^{10} - 1} \right\} + 125h + 17,000 + 1500$$

$= 50,000(0.2) + 125h + 17,000 + 15000 = 10,000 + 125h + 17000 + 15000 = 28,500 + 125h$

Similarly, the annual cost of motor 'B' can be determined as

$$70,000\,(A/P,15\%,10) + 25(h/0.9)(4) + 13,000 + 0.03(70,000)$$

$$= 70,000(0.2) + 111.11h + 13,000 + 2100 = 14000 + 111.11h + 13000 + 2100$$

$$= 29,100 + 111.11h$$

At break-even point, both the annual costs should be equal. Therefore,

$28,500 + 125h = 29,100 + 111.11h$ or $13.89h = 600 \Rightarrow h = 43$ hours/ year.

Example 9.62. An investment of Rs. 2,00,000 on an earth air heat exchange for a greenhouse system promises to earn an annual return of Rs. 20,000 in terms of fuel savings to meet both for heating and cooling loads. The prospective life of the earth air heat exchanger system is uncertain. How many years would the returns need to continue for the investment to be attractive, assuming a discount rate of 15 percent?

Solution: For the investment to be attractive, the present value of net benefits would be greater than Zero. If the break-even value of the useful life of the earth air heat exchanger system is *n* years, then $-2,00,000 + 20,000\,(P/A, i, n) > 0$

$$\text{or, } -2,00,000 + 20,000 \left\{ \frac{(1+i)^n - 1}{i(1+i)^n} \right\} > 0 \text{ ,or, } -2,00,000 + 20,000 \left\{ \frac{(1.15)^n - 1}{0.15(1.15)^n} \right\} > 0$$

$$\text{or, } -20 + 2 \left\{ \frac{(1.15)^n - 1}{0.15(1.15)^n} \right\} > 0 \text{ or, } -20 + 13.33 \left\{ \frac{(1.15)^n - 1}{(1.15)^n} \right\} > 1.5$$

$$\text{or, } 13.33 \left\{ \frac{(1.15)^n - 1}{(1.15)^n} \right\} > 20 \text{ or, } \frac{(1.15)^n - 1}{(1.15)^n} > \frac{20}{13.33} \Rightarrow \frac{(1.15)^n - 1}{(1.15)^n} > 1.5$$

$$\text{or, } 1 - \frac{1}{(1.15)^n} > 1.5 \text{ or, } -\frac{1}{(1.15)^n} > 0.5 \Rightarrow \frac{1}{(1.15)^n} < 0.5 \Rightarrow \frac{1}{(1.15)^n} < \frac{1}{2}$$

$$\text{or, } (1.15)^n > 2 \Rightarrow n\log(1.15) > \log 2$$

$$\text{or, } n > \frac{\log 2}{\log(1.15)} \text{ or } n > 5 \text{ years.}$$

9.17.1 Determination of Break-even Point in Terms of Physical Units (No. of Products) or in Money (Sales Volume)

Any business proposals cannot meet human needs if they remain in the form of plans and specifications. They must be converted to products through construction or production operations. In the production process, the relationship among the cost of inputs, the price obtained from the output and the levels of output, it is possible to determine the optimum operating conditions. Any production activity whether it represents a single machine or a complete manufacturing facility, can be estimated in terms of its cost and revenue generating characteristics. For every production activity, there is a potential in maximum level of output that defines its capacity and its efficiency, then derives the output as a percentage of the input. Break-even analysis specially involves the determination of volume of production at which, there is equal amount of cost and revenue. It basically captures the relation of fixed cost, variable cost, the value of the output, sales volume etc. to the profitability of a business organization. Hence profit is the ultimate objective of the any business venture along with providing quality products and services to the society. Hence, the break-even point can be determined either in terms of physical units (products) to be produced or in terms of money (sales value in rupees). Fixed cost and variable costs are the primary considerations in production operations and the ability to distinguish these cost types is essential in economy studies of the production systems. For the analysis of breakeven point, some terminologies are required to know. They are as all follows.

9.17.2 Terminologies

Fixed Cost (FC): In break-even analysis, fixed costs and variable costs are the major categories of costs. Fixed costs are those costs which do not change but remain constant within a given period of time in spite of fluctuations in production. In other words, fixed costs remain constant at different levels of output, even if output is zero. The sources of fixed costs are interest on investment, insurance, depreciation, property taxes, rent, advertising budget, technical services etc.

Variable Costs (VC): Variable costs are those which vary in direct proportion to any change in the volume of output. When there is no output, variable costs are zero. An increase in output will lead to an increase in total variable cost and decrease in output will lead to reduction in total variable cost. Some sources of variable costs are direct material cost, direct wages, power, maintenance, repairs, spoilage, royalties etc.

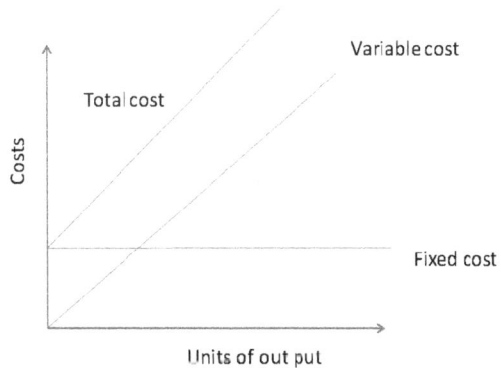

Total Cost: The total cost of providing a particular level of output is the sum of the fixed costs and the variable costs incurred for that output level.

Profit: It is the difference between revenue (income from sales of output) and total costs. It is the cause of effect of competitiveness and a handy yardstick of success. There are, basically three ways to increase profit;

(i) Increase the selling price

(ii) Increase the value to increase sales

(iii) Decrease the selling price to increase sales

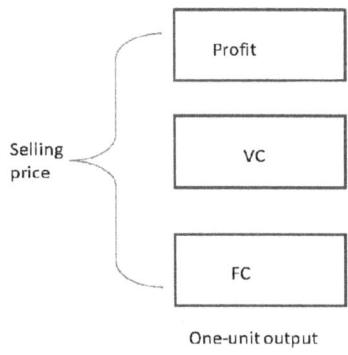

The profit expansion descriptions are oriented to consumers' interest.

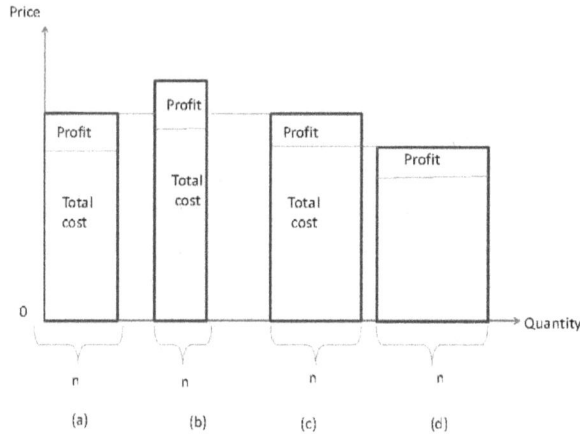

(a) Original Price structure (b) Increase in price (c) Maintain price and increase value (d) Decrease price, decrease total cost and increase value and more profit

One way to increase profit without changing the selling price is to sell more units by increasing the value. The greater value perceived by the consumer can result from better quality, more quantity and more effective advertising. All these measures increase the total cost of the product. Higher total costs lead to a lower margin of profit per unit sold. A straight forward means to increase profit while holding prices constant is to reduce total cost. Similarly, the decrease in price leads to the sale of more units and getting more profits.

Contribution: Contribution is the difference between the sales and the variable costs. It can be represented as Contribution = Sales – Variable cost

Contribution / Unit = Selling price / unit – variable cost / unit

Contribution / unit = Fixed cost / unit + variable cost / unit + profit / unit – variable cost / unit = Fixed cost / unit + profit/unit

The concept of contribution is a valuable aid to management for taking managerial decisions. It helps the management in the fixation of selling and in determining the break-even point. Contribution will first go to meet fixed expenses and then to earn profit.

Margin of Safety: The margin of safety refers to the difference between the actual sales and break-even sales. At break-even sales, there is zero profit. The excess of sales over the break-even sales is known as the margin of safety. It indicates the strength or weakness of business. A high margin of safety indicates the soundness of a business, because even with substantial fall in sale, some profit will be made. On the other hand, small margin of safety is an indicator of the weak position of the business, because even a small reduction in sale will adversely affect the profit position of the business. Margin of safety (M.S.) can be represented as

$$M.S. = \frac{Profit}{contribution} \times sales$$

M.S as a percent of sales $= \frac{M.S.}{Sales} \times 100$

Let's adopt the following symbols for deriving the formula of M.S.

Q = Volume of production, p = Selling price per unit, v = variable cost per unit, TR = Total sale revenue = $p \times Q$, FC = fixed cost per period, Total variable cost = $v \times Q$, TC = total cost = Total variable cost + Fixed cost = $v \times Q + FC$

Profit = Sales – (Fixed cost + Variable cost) = $p \times Q - (FC + v \times Q)$.

Break-even quantity $= \dfrac{Fixed\ cost}{Selling\ price/unit - variable\ cost/unit} = \dfrac{FC}{p-v}(in\ units)$

Break-even sales

$$= \frac{Fixed\ cost}{selling\ price\ per\ unit - variable\ cost\ per\ unit} \times selling\ price\ per\ unit$$

$$= \frac{FC}{p-v} \times p \ (Rs.)$$

Contribution = Sales – Variable cost = $pQ - vQ$

M.S. = Actual sales – break-even sales

$$= p \times Q - \frac{FC}{p-v} \times p = p\left\{Q - \left(\frac{FC}{p-v}\right)\right\} = p\left\{\frac{Q(p-v) - FC}{(p-v)}\right\} = p\left\{\frac{pQ - (FC + vQ)}{(p-v)}\right\}$$

$$= p\left(\frac{profit}{p-v}\right) = pQ\left\{\frac{profit}{(p-v)Q}\right\} = \frac{profit}{contribution} \times sales$$

Example 9.63. A business company has the following details;

Fixed cost = Rs. 15,00,000; Variable cost per unit (v) = Rs. 130, selling price per unit (p) = Rs. 200

Find the (a) Break-even sales quantity (b) Break-even sales (Rs.) (c) If the actual production quantity is 50,000, also find (i) contribution and (ii) margin of safety by the methods mentioned above

Solution: Fixed cost (*FC*) = Rs. 15,00,000, Variable cost per unit (*v*) = Rs. 130 and Selling price per unit (*p*) = Rs. 200

(a) Break-even quantity $= \dfrac{FC}{p-v} = \dfrac{15,00,000}{200-130} = 21429 \; units$

(b) Break-even sales $= \dfrac{FC}{p-v} \times p = 21429 \times 200 = Rs \; 42,85,800$

(c) (i) Contribution = Sales – Variable cost $= p \times Q - v \times Q = 200 \times 50,000 - 130 \times 50,000 = 10,000,000 - 6,500,000 = $ Rs. 3,500,000

(ii) Margin of safety (M.S.)

Method I: M.S. = Sales – Break-even sales= $50,000 \times 200 - 42,85,800 = 10,000,000 - 42,85,800 = 5,714,200$

Method – II

M.S. $= \dfrac{Profit}{Contribution} \times Sales$

Profit=Sales- $(FC + v \times Q) = 50,000 \times 200 - \{(15,00,000 + (130 \times 50,000)\}$

$= 10,000,000\text{-}(15,00,000+6,500,000) = 10,000,000\text{-}8,000,000$

= Rs. 2,000,0000

$M.S. = \dfrac{2,000,000}{3,500,000} \times 10,000,000 = 5,714,200$

M.S. as a percent of sales $= \dfrac{5,714,200}{10,000,000} \times 100 = 57.14\%$

9.17.3 Profit – Volume Ratio (P/V ration)

The profit-volume ratio (P/V ration) is a valid ratio which is also useful for studying the profitability of operations of a business at a glance. This ratio establishes the relationship between contribution and sales. Higher the (P/V) ratio, more will be the profit and lower the (P/V) ratio, lower will be the profit. That is why, every management aims at increasing the (P/V ratio). This analysis is also the result of the attempts to apply the break-even analysis to the situations of a multi-product business organization, where break-

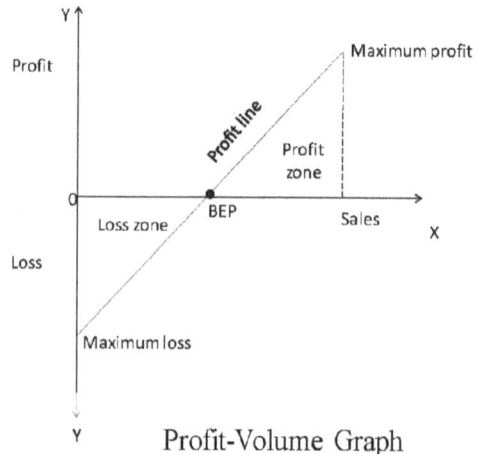

Profit-Volume Graph

even analysis is done separately for different divisions or products of the organization. The profit-volume graph will show the relationship of the profit of an organization to its volume. The total profit or loss is measured on the Y-axis. The profit zone is above the X-axis and the loss zone is below the X-axis as shown below. The volume is measured on the X-axis which is drawn at the point of zero profit.' The maximum loss which occurs at zero sales volume is equal to the fixed cost and is shown below the X-axis, on Y-axis. The maximum profit is earned when the firm works at full capacity and the maximum profit is shown on Y-axis above the X-axis. The line joining both the points is called the profit-volume line or P/V line.

P/V ratio can be expressed in the following ways.

$$\text{P/V ratio} = \frac{Contribution}{Sales} = \frac{Sales - variable\ costs}{Sales}$$

The relationship between Break-even Point (BEP) and P/V ratio is as follows

$$BEP = \frac{Fixed\ cost}{P/V\ ratio}$$

$$\text{Similarly, } M.S. = \frac{Profit}{P/V\ ratio}$$

Example 9.64. Consider the following data of a business organization for the year 2019. Sales = Rs. 1,00,000; Fixed cost = Rs. 20,000; Variable cost = Rs. 40,000. Find the following (a) Contribution (b) Profit (c) BEP (d) M.S.

Solution:

(a) Contribution = Sales − Variable costs = Rs. 1,00,000 − Rs. 40,000 = Rs. 60,000.

(b) Profit = Contribution − fixed cost = Rs. 60,000 − Rs. 20,000 = Rs. 40,000

(c) To calculate BEP

$$\frac{P}{V}\ ratio - \frac{Contribution}{Sales} = \frac{60,000}{1,00,000} \times 100 = 60\%$$

$$BEP = \frac{Fixed\ cost}{P/V\ ratio} = \frac{20,000}{60.00} \times 100 = Rs.\ 33,333$$

(d) $$M.S. = \frac{Profit}{P/V\ ratio} = \frac{40,000}{60.00} \times 100 = Rs.\ 66,667$$

9.17.4 Graphical Method of Break-even Analysis

The graphical approach to break-even analysis is generally in terms of break-even charts. The break-even chart depicts fixed cost, variable costs, total costs, break-even point, profit or loss, margin of safety and the angle of incidence.

Break-even point is calculated graphically on the basis of the behaviour of revenue and cost curves. In this respect, break-even point is shown by using two methods;

(i) Linear break-even analysis

(ii) Non-linear break-even analysis.

When the revenue and variable costs are directly proportional to output, the analysis is known as linear break-even analysis. This linear analysis assumes the constancy of fixed costs, per unit variable costs and per unit sales price over time and over output. This analysis rests on the assumption that all units of output produced must be sold out. But it is very often observed that cost and sales or revenue functions do not always assume the direct and proportional relationship to output. In this case, cost and revenue functions become non-linear; and the corresponding break-even analysis is called non-linear break-even analysis.

(a) Procedure of drawing a break-even chart

The procedure of drawing break-even chart is as follows;

i) Draw co-ordinate axes on a graph sheet. X-axis will measure sales volume in monetary value or output in units. Y-axis will measure costs and revenues.

ii) Select suitable scales for both

iii) Plot the sales volume or units against sales revenue or output. This will be straight line from origin to right. If the scales on both the axes are the same, the sale line will be at 45^0 angle to the base.

iv) Draw the fixed cost line parallel to X-axis. This lines indicates that fixed expenses remain constant at various levels of output.

v) Plot the total cost line for a given sales volume or units.

vi) The point at which the total cost line intersects the sales line is the break-even point. A line drawn perpendicular to the X-axis from the break-even point represents loss (below the point) and above the point, represents profit.

vii) The angle formed at the point of intersection of the total cost and sales line is called the angle of incidence.

BEP is that level of sales or production at which the sales revenue is exactly equal to total cost(both variable and fixed cost). BEP is that level of activity at which the firm neither earns any profit nor suffers any loss. It is that point at which the contribution by a product just covers fixed cost. The vertical distance between the revenue line and the total cost line indicates a profit to the right of BEP and a loss to the left. The angle of incidence is the angle between sales and total cost line. This angle formed at the point of intersection of the sales and total cost lines, indicates the profit earning capacity and as such, the wider the angle, the greater is the profit and vice-versa. Actual sales minus break-even sales is known as "Margin of safety."

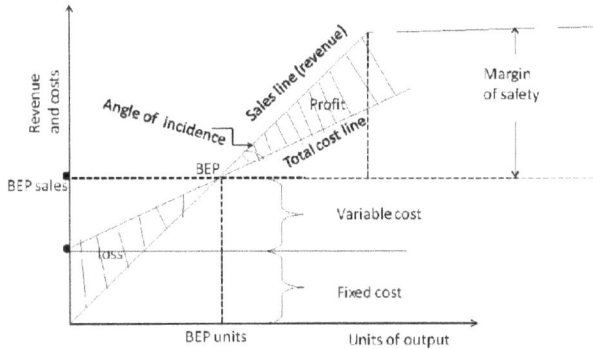

(b) Effects of change in costs and selling price for BEP

Any change in costs and selling price affects the BEP. This can be verified with the following example.

Let variable cost per unit = Rs. 7, Price per unit = Rs. 12, Units of products (Q) = 100, Break-even units = 80, Fixed cost = Rs. 400

If fixed cost is reduced

Let fixed cost be reduced to half i.e. Rs. $\dfrac{400}{2} = Rs.\ 200 = FC_1$

$$\text{New break-even point (BEP)}_1 = \dfrac{Changed\ fixed\ cost}{Price\ per\ unit - Variable\ cost\ per\ unit}$$

$$= \dfrac{200}{12-7} = \dfrac{200}{5} = 40\ units$$

Changed variable cost per unit, $(VC)_1$ is as follows

$$(BEP)_1 = \dfrac{(FC)}{Price\ per\ unit - (VC)_1} \Rightarrow price\ per\ unit = \dfrac{(FC)}{(BEP)_1} + (VC)_1 \Rightarrow$$

$$(VC)_1 = price\ per\ unit - \dfrac{(FC)}{(BEP)_1} = 12 - \dfrac{400}{40} = Rs\ 2$$

So variable cost per unit decreases to Rs. 2 when fixed cost is reduced to half.

Changed selling price per unit, p_1 is as follows;

$$\text{Break-even quantity} = \frac{FC}{Contribution / Unit}$$

$$\text{Contribution / Unit} = \frac{FC}{Break - even\ quantity} = \frac{400}{40}\ Rs.\ 10\ per\ unit$$

Contribution / unit = Selling price / unit – Variable cost / unit

Selling price / unit = Contribution / unit + Variable cost / unit

Changed selling price / unit = Contribution / unit + Variable cost / unit

$p_1 = 10+7 = Rs.\ 17.$ Hence selling price / unit increases with the change of fixed cost.

(c) Non-linear Break-even Analysis

The break-even analysis becomes non-linear when the revenue or sale and cost function is non-linear. In actual practice, the selling prices do not remain constant for all levels of output due to competition and changes in the general price level. Further, it may not be possible to increase the sales volume without offering concessions in price to the consumers. In the same manner, variable cost per unit may also increase with the increase in the level of production due to operating inefficiencies and the law of diminishing returns operates. Thus profit can be increased only up to a certain point and then it will decrease until it is converted into a loss. The break-even chart will then become curvilinear or non-linear instead of linear. Two break-even points arise. One is before the profit starts and the 2nd is after profit or profit ends. In other words, one at a lower level of output and another at a higher level of output. In such a case, increasing output sales volume beyond the first break-even point will increase profit but increase in volume beyond the second break-even point will result in a loss.

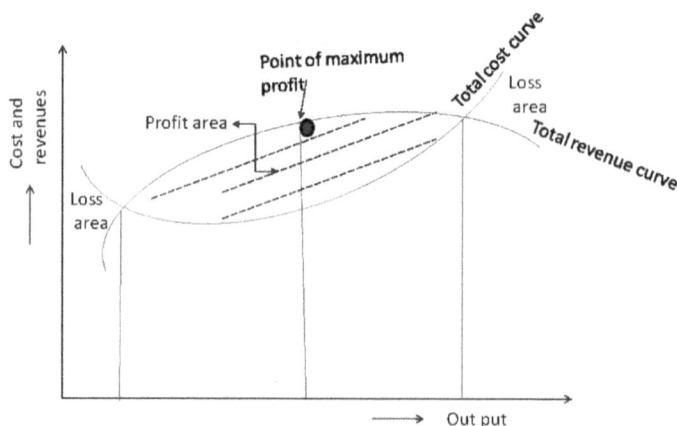

Non-Linear Break-even Graph

The optimum level of output shall be reached at the point where the difference between the total revenue and the total cost is highest. The point of maximum profit is especially important when two break-even points are present. The maximum profit point can be obtained by the classical technique of optimization i.e. by differentiating the profit function and setting the same equal to zero. The non-linear break-even chart is shown above. Between two break-even pints, the firm operates at profitable situation. At maximum profit point, the vertical distance between the total cost curve and the total revenue curve is greatest.

Example 9.65. If a revenue and cost functions of a business organization are $R = 100q - 4q^2$ and $C = q^2 - 4q + 100$ respectively, find (a) Break-even quantities (b) Break-even sales (c) maximum profit and (d) for what selling price, profit is maximized.

Solution: The revenue or sale function is $R = 100q - 4q^2$ and cost function is $C = q^2 - 4q + 100$, where 'q' is the unit of output of the firm.

(a) To determine break-even quantities, we must equate revenue and cost. i.e. $R = C$

$$\Rightarrow 100q - 4q^2 = q^2 - 4q + 100 = 5q^2 - 104q + 100 = 0$$

Or, $q = \dfrac{104 \pm \{(104)^2 - (4)(5)(100)\}^{\frac{1}{2}}}{10} = \dfrac{104 \pm 94}{10} = 19.8 \text{ or } 1$

Thus, 'q' value is either 19.8 or 1.

It means, two break-even quantities exist i.e. $q = 1$ unit and $q = 19.8$ units.

(b) The break-even sales for $q = 1$, is

$R\ (q=1) = 100-4 = \text{Rs. } 96$

The break-even sales for $q = 19.8$ units is $R(q=19.8) = \text{Rs. } 411.85$

(c) In order to find maximum profit, we have to define the profit function (P_r).

i.e. $P_r = 100q - 4q^2 - q^2 + 4q - 100 = 104q - 5q^2 - 100$

The marginal profit function is $\dfrac{d(P_r)}{dp} = \dfrac{d}{dq}\left(104q - 5q^2 - 100\right) = 104 - 10q$

And $\dfrac{d^2(P_r)}{dp} = \dfrac{d}{dq}\left\{\dfrac{d(P_r)}{dq}\right\} = \dfrac{d}{dq} = (104 - 10q) = -10$

The first order condition of profit maximization is $\dfrac{d(P_r)}{dp} = 0$

i.e. $104 - 10q = 0$ or $q = 10.4$ units

The second order condition is $\dfrac{(P_r)}{dq^2} < 0$; here $\dfrac{d^2(P_r)}{dq^2} = -10 < 0$

Therefore, for the business organization, profit maximizing output level is 10.4 units. The maximum profit is max $(P_r = \{104(10.4) - 5 (10.4)^2 - 100\} = $ Rs. 440.80

d) The price function for the organization is $p = \dfrac{R}{q} \Rightarrow p = 100 - 4q$

Thus the profit maximized selling price is $p = $ Rs$\{100 - 4(10.4)\} = $ Rs. 58.40

9.17.5 Effect of Price Reduction on P/V Ratio, Break-even Point and Margin of Safety

The P/V ratio gives the relationship between contribution and sales and indicates the change in percentage of contribution in relation to change in the volume of sales. When there is a reduction in the selling price, it will result in a reduction of contribution (variable cost per unit being constant) and a reduction in the sales volume, thus reducing the P/V ratio.

Break-even point is that point where total cost is equal to the sales (Rs.) and there is no profit or no loss. When selling price falls, the sales volume will also be reduced, but total cost remains the same. To meet the total cost equal to sales, the sales volume has to be increased. Thus, the break-even point will rise i.e., it will be achieved at a higher sales volume. Margin of safety is the excess of actual sales over the break-even sales volume. With the reduction of price, the sales volume will be reduced, so margin of safety will also be shortened.

Example 9.66. If the selling price of a product was Rs. 200/- per unit as against its variable cost of Rs. 100 per unit. The total fixed cost was Rs. 2,00,000. Calculate the effect of a reduction in price by Rs. 40, on the P/V ratio, break-even point and margin of safety if 4000 units were produced and sold.

Solution:

Item	Before price reduction	After price reduction
Selling price per unit	Rs. 200	Rs. 160 (Rs. 200 – Rs. 40)
Variable cost per unit	Rs. 100	Rs. 100
Contribution per unit	Rs. 100	Rs. 60
Total fixed cost	Rs. 2,00,000	Rs. 2,00,000
$P/V\ ratio = \left[\dfrac{Contribution}{Sales}\right] \times 100$	$\dfrac{Rs.\ 100}{Rs.\ 200} \times 100 = 50\%$	$\dfrac{60}{100} \times 100 = 37.5\%$
Break-even point		
$\left[\dfrac{F.C.\ (Fixed\ cost)}{P/V\ ratio}\right]$	$\dfrac{2,00,000}{50\%} = 4,00,000$	$\dfrac{2,00,000}{37.5\%} = 5,33,333$

[Table Contd.

Contd. Table]

Item	Before price reduction	After price reduction
Total sales	4000×200=Rs 800,000	4000×160 = Rs 6,40,000
Margin of safety = [Total sales – Break-even point]	= Rs. 8,00,000-4,00,000 = Rs. 4,00,000	= Rs. 6,40,000-5,33,333 = 1,06,667

From the above, it is clear that effect of price reduction is always to reduce ratio, to raise break-even point and to shorten the margin of safety.

9.17.6 Importance of Break-even Analysis

Break-even analysis is needed due to the following reasons

- Being future-oriented, it is useful for the organization in planning and forecasting the costs and profits for future.
- It is very much useful to the management for taking managerial decision, because it studies the relationship between cost, volume and profit at various levels of output.
- Information provided by the break-even chart is simple and easily understandable by a common man.
- Besides determining the break-even point, profit and losses at various levels of output can also be determined with the help of break-even charts.
- It is helpful in the determination of the sale price which would give a desired profit or break-even point.
- It is helpful in knowing the effect of increase or decrease in selling price.
- It is a tool for exercising cost control because it shows the relative importance of the fixed cost and the variable cost.

9.17.7 Limitations of Break-even Analysis

The important limitations of break-even analysis are as follows;

- Break-even analysis considers only cost, volume and profit relationship. It ignores other important factors like amount of capital investment, demand for the product and government restrictions etc.
- In reality, fixed costs do not remain constant at all levels of output and variable costs do not vary in direct proportion to the changes in the volume output.
- It is based on number of assumption which may not be possible under all circumstances.
- It does not provide any suitable remedial measure to the management for taking decisions.
- Output is not the only factor that influences cost.

9.18 GLOSSARY OF TERMS RELATING TO FINANCIAL EVALUATION

Accounting period: The time for which profits are being calculated, normally months, quarters or years.

Accounting life and economic life of an asset.

Accounting life is the reduction phase of assets through depreciation charges, whereas economic life measures the benefit derived from the asset during its total life time.

Annuity: This is a type of insurance policy. Upon retirement a lump sum is paid into it and the insurance company then provides a regular income.

Annuity due: An annuity due is an annuity whose payment is to be made immediately at the beginning of the period, rather than at the end of the period. Rent is an example of annuity due

Balance sheet: A balance sheet is a summary of an organisation's financial position. It lists the values, in the books of account on a particular date, of all the organisation's assets and liabilities. The assets and liabilities are grouped in categories, to show the clear picture of the organisation's strengths and weaknesses.

Break-Even Point: The point at which total revenues equals total costs.

Capitalization cost: Capitalized costs are those expenses that are incurred financing a fixed asset. For example, capitalized costs include the labour expenses incurred in developing a fixed asset or interest expenses incurred as a result of financing for the construction of a fixed asset.

Cash flow: The flow of money into and out of a company, project, or activity. Revenues are cash inflows and carry a positive (+) sign whereas expenses are cash outflows and carry a negative (-) sign. If only costs are involved, the- sign may be omitted.

Cost of capital: The interest rate incurred to obtain capital investment funds. COC is usually a weighted average that involves the cost of debt capital (loans, bonds, and mortgages) and equity capital (stocks and retained earnings).

Copyright: Copyright, a term that is more associated with writers and skilled personnel, is a form of protection for published and unpublished literary, scientific and artistic works that a business may create to associate with or promote their product or brand. The exclusive legal right, owned by the individual or group who created a work, or by an individual or group assigned by the originator, to use certain material and to allow others the right to use the material.

Compound interest: Compound interest is interest on the money lent, plus interest on any interest already added to the loan.

Corporation: A corporation is a legal entity that is separate and distinct from its owners. Corporations enjoy most of the rights and responsibilities that an individual possesses. That is, a corporation has the right to enter into contracts, loan and borrow money, hire employees, own assets and pay taxes.

Capital: Money invested into a company or project by its owners.

Cash flow: The movement of cash into and out of a business

Capital recovery (CR): It is the equivalent annual amount; an asset or system must earn to recover the initial investment plus a stated rate of return. Numerically, it is the annual worth value of the initial investments at a stated rate of return. The salvage value is considered in CR calculations.

Due Diligence: The evaluation of an opportunity, which is often performed by investors in order to allow them to make a more confident and prudent investment decision by evaluating risk. Entrepreneurs should also perform due diligence to verify that they are working with a legitimate investor. Both parties should always know whom they are dealing with. It is an investigation of a business or person prior to signing a contract, or an act with a certain standard of care. Due diligence is done to collect other information which may influence the outcome of a transaction.

Debtor: A debtor is a person or an organization that agrees to receive money from another party in exchange for a liability to pay back the obtained money in due course of time. A debtor is somebody who has taken out a loan at a bank for certain purpose.

Defender and challenger: These are the terms used for two mutually exclusive alternatives. The defender is the currently installed asset and the challenger is the potential replacement

Dividend: Money paid regularly by a company to its shareholders.

Discount Rate: Rate of return used to value future cash flows. It essentially measures the worth of tomorrow's money in today's terms. For example, if $1 one year from now is worth $0.83333 today, then the discount rate is 20%: $1/ (1 + 0.2) = $0.83333. Of course, knowing the discount rate and future cash flows, we can also find their today's value. For example, with a discount rate of 10% per year, $100 one year from now plus $150 two years from now will be worth, $100/ (1 + 0.1) + $150 / (1 + 0.1)^2 = $214.88 today.

Discounting: The process of finding the present value of a series of cash flows. It is the reverse process of compounding. For example, if you are scheduled to receive $2,000 one year from now and $3,500 two years from now, and the interest rate is 10% p.a., then we discount the two payments to get the present value: 2000 / (1+0.1) + 3500 / (1+0.1)^2 = $4,710.74.

Entrepreneur: A person who organizes, operates and assumes the risk for a business venture. An entrepreneur is someone who exercises initiative by organizing a venture to take benefit of an opportunity and, as the decision maker, decides what, how, and how much of a good or service will be produced.

Economic growth: This is the term used to describe an increase in the amount of goods and services produced by the county, known as gross domestic product (GDP).

Economic Equivalence: A combination of time value of money and interest rate that makes different sums of money at different times have equal economic value.

Economic service life: The ESL is the number of years 'n' at which the total annual worth of costs, including salvage and AOC (annual operating cost), is at its minimum, considering all the years, the asset may provide service.

End-of-period convention: To simplify calculations, cash flows (revenues and costs) are assumed to occur at the end of a time period. An interest period or fiscal period is commonly 1 year. A half-year convection is often used in depreciation calculations.

Fiscal year: Also known as a financial year. The fiscal year is a set period used to calculate financial statements. The period used differs between countries and between businesses, although in the UK, the year between 6th April and 5th April is most often used for personal taxation. The 'official' period for corporation tax runs from 1st April to 31st March, however companies can adopt any yearly period for corporation tax.

First cost: The initial investments in a project including transportation, installation, preparation for service, and other related initial expenditures.

Flat interest rate: Flat interest rate is the interest charged on the full amount of a loan throughout its entire term and commonly known as 'pre-determined' credit charge

Gross: The total amount of money, one has earned in a period of time before deductions such as taxes.

Gross domestic product (GDP): GDP is the sum of all goods and services produced in the country's economy.

Gross national product (GNP): GNP is another way to measure the economy, but also the welfare of a nation. This is GDP plus the profits, interest and dividends received from the residents of a nation abroad and minus those profits, interest and dividends paid from the concerned nation to overseas residents. GDP measures the value of goods and services produced within a country's borders. GNP measures the value of goods and services produced by only a country's citizens but both domestically and abroad.

Hyperinflation: This is the inflation that is rapid or out of control. It usually only occurs during wars or during severe political instability.

Insolvency: When a company becomes unable to pay off to its creditors, or its liabilities exceed its assets. Insolvency is a state of financial distress in which someone is unable to pay off their liabilities. It can lead to insolvency proceedings, in which legal action will be taken against the insolvent entity, and assets may be liquidated to pay off outstanding debts.

Insolvent: If debts cannot be paid when they are due for payment, the person or organisation owing the money is insolvent.

Intellectual property: Any work or invention that is original, innovative and of creative design. The individual or company responsible for the designs will be entitled to apply for a copyright or trademark on the designs.

Investor: A person that puts money into a project or small business, with the hope of eventually receiving profit. Example, when the entrepreneurs have created their eco-friendly cleaning products business, they are fortunate to secure several investors to support their growth.

Incubator: An organization or space dedicated to supporting new business ventures, An incubator is a way for the community to help entrepreneurs who have good ideas but do not have the resources to start their activities independently. A business incubator is a company that helps new and start-up companies to develop by providing services such as management training or office space.

Independent project: A project whose acceptance or rejection is independent of the acceptance or rejection of other projects is known as independent project.

Inflation: Inflation is an increase in the general level of prices of a basket of goods and service in an economy over a period of time and is measured by the help of consumer price index. It is expressed as a percentage per time (% per year), it is an increase in the amount of money required to purchase the same amount of goods or services over time. Inflation occurs when the value of a currency decreases. Economic evaluation is performed using either a market (inflation-adjusted) interest rate or an inflation-free rate (constant-value terms).

Intangible fixed assets: Intangible fixed assets have no physical presence. Examples include patents, goodwill, trademarks, and brand names

Joint Venture: A joint venture is a business arrangement in which two or more parties agree to pool their resources for the purpose of accomplishing a specific task. This task can be a new project or any other business activity. It is important to note that a joint venture differs from a strategic alliance because it is a legal entity created by two or more businesses joining together to conduct a specific business enterprise with both parties sharing profits and losses. It differs from a strategic alliance in that there is a specific legal entity created.

Leverage: A way of magnifying returns. To borrow money (use debt capital rather than equity capital) to fund a business or project, with the goal of increasing return on equity. Leverage is an investment strategy of using borrowed money—specifically, the use of various financial instruments or borrowed capital—to increase the potential return of an investment.

Liquid asset: Any asset which can be easily converted into cash.

Liquidity: The ease with which a company's assets can be converted into cash. Ease with which a security can be converted into cash. Stocks and bonds can be sold quickly for cash, so they are liquid assets; but antiques and real estates cannot be sold quickly for cash and they are illiquid assets.

Life cycle costing: Life-cycle costing is method of calculating the total cost of a product (covering both goods and services) induced throughout its life cycle. The life cycle consists of number of stages, which depend upon the goals and scope of the problem and the method of investigation chosen.

Mortgage: A mortgage is using a property as security for a debt.

Mutually exclusive projects: Projects are said to be mutually exclusive where in, the taking up of one project prevents the taking up of the other project. For example, if there are two projects i.e., A and B, one can execute project A or B, but not both.

Minimum attractive rate of return (MARR): A reasonable rate of return established for the evaluation of an economic alternative. Also called the hurdle rate, MARR is based on cost of capital, market trend, risk, etc.

Net value: The amount of profit remaining after deductions such as taxes have been made.

Net asset value: A way of measuring investment trusts. Take the total number of its assets minus its liabilities.

Nominal interest rate: An interest rate that isn't adjusted for inflation.

Nominal values: These values do not take inflation into account.

NPV and IRR: NPV is used in capital budgeting to analyse the profitability of an investment or project whereas IRR is the return that a company would earn if it invests by itself, rather than investing that money elsewhere.

Nominal and effective interest rate (r and i): A nominal interest rate does not include any compounding; for example, 1% per month is the same as nominal 12% per year. Effective interest rate is the actual rate over a period of time because compounding is imputed; for example, 1% per month, compounded monthly, is an effective 12.683% per year. Inflation or deflation is not considered.

Opportunity Cost: The value of the next-best alternative that must be given up or sacrificed to obtain something. A forgone opportunity caused by the inability to pursue a project. Numerically, it is the largest rate of return of all the projects, not funded due to the lack of capital funds.

Outcome: A result or effect that is caused by or attributable to the project, program or policy. Outcome is often used to refer to more immediate and intended effects. Related terms are also result or effect.

Outcome Evaluation: This form of evaluation assesses the extent to which a program achieves its outcome oriented objectives. It focuses on outputs and outcomes (including unintended effects) to judge program effectiveness but may also assess program processes to understand how outcomes are produced.

Outputs: The products, goods, and services which result from an intervention. An amount of something produced, especially during a given period of time.

Overheads: Costs that do not vary regardless of the level of production and are not usually directly involved with the cost of production, such as rent.

Outstanding: Not settled or paid.

Outsourcing: Obtain goods or resources from a source outside your company, for example, we do all of our creative work in-house, but we outsource the production of our actual products. It is the act of outsourcing in the acquisition of standard operational services from another business. Outsourced services typically include accounting, payroll, telemarketing, IT support, advertising and more.

Payback period: The payback period in capital budgeting refers to the length of time it will take to recover the principal amount of an investment. It can be calculated by using the following formula:

Payback period indicates the time '*n*', mostly in years, before recovery of the initial capital investment is expected.

Simple payback period and discounted payback period: Payback period is used to evaluate a capital investment project. It is the number of years needed to recover the initial investment. Suppose that a three-year project requires an initial investment of $1.5 million, and it will generate the following cash flows for the three years: $0.6 million, $0.8 million, and $2.0 million. As can be seen, after the first two years, we recover only $1.4 million of the investment; we still need to recover $0.1 million, which will take only a fraction of the third year: ($0.1 million / $2.0 million) (1 year) = 0.05 years. Therefore, the total payback period for this project is 2 + 0.05 = 2.05 years. This is called simple payback period.

To reflect the time value of money, sometimes people calculate the so-called "discounted payback period." All you have to do is to discount the future cash flows first to get their present values, and then go through the same procedure as before. To continue the above example, suppose the discount rate is 12% p.a., then the present values of the three future cash flows are, $0.6/ (1+0.12) = $0.5357 million, $0.8 / (1+0.12)^2 = $0.6378 million, and $2.0 / (1+0.12)^3 = $1.4236 million. Then the discounted payback period is 2.229 years. The calculation is as follows: 2 + (1.5 − 0.5357 − 0.6378)/1.4236 = 2.229.

Patent: A patent is an exclusive right granted for an invention, such as a product or a process. This is a property right granted to an inventor to exclude others from making, using, offering for sale, or selling the invention for a limited time in exchange for public disclosure of the invention when the patent is granted. An official legal document confirming that an individual or company has the sole right to make, use or sell a particular invention.

Profit: Financial gain, for example, entrepreneurs take financial risks with the hopes of profiting from their business venture.

Project: A discrete activity (or 'development intervention') implemented by a defined set of implementers and designed to achieve specific objectives within specified resources and implementation schedules. A set of projects make up the portfolio of a program. Related terms, activity or intervention.

Project Appraisal: A comprehensive and systematic review of all aspects of the project i.e. technical, financial, economic, social, institutional, environmental etc. to determine whether an investment should go ahead.

Project Evaluation: An evaluation of a discrete activity designed to achieve specific objectives within specified resources and implementation schedules, often within the framework of a broader program.

Placement of present worth (PW): In applying the (*P/A, i%, n*) factor, *PW* is always located on interest period (year) prior to the first amount. The *AW* is a series of equal, end-of-period cash flows for '*n*' consecutive periods, expressed as money per time (say, $/year; €/year).

Placement of future worth (FW): In applying the (*F/A, i%, n*) factor, *FW* is always located at the end of the last interest period (year) of the '*A*' annual payment series.

Project evaluation: For a specified MARR, determine a measure of worth for net cash flow series over the life or study period. Guidelines for a single project to be economically justified at the MARR (or discount rate) follow.

Present worth : if $PW \geq 0$ annual worth : if $AW \geq 0$

Future worth: if $FW \geq 0$ Benefit/cost: if $B/C \geq 1.0$

Start-Up: A newly established business.

Seed financing: Financing to fund an early-stage company, generally provided by either angel investors and/or a venture capital firm, to fund the early stages of a company's business plan.   Seed financing typically occurs before a company has commercially released its product or service, and therefore before the business is generating revenue.

Stakeholders: Any individual or party that has an interest in or may be affected by a business and/or its activities. This can include anyone, from shareholders to residents of the local community.

Sensitive analysis: Sensitivity analysis is the study of how the uncertainty in the output of a model can be allocated to different sources of uncertainty in the model input. It indicates how a measure of worth is affected by change in estimated values of a parameter over a stated range. Parameters may be any cost factor, revenue, life, salvage value, inflation rate, etc.

Sinking fund: Generally, the term refers to a situation whereby money is being periodically put into a fund which can be used to make capital purchases, retire debts or facilitate any other causes.

Sunk cost: Costs that have already been incurred and thus cannot be recovered are known as sunk cost, for example, cost of plant, machinery, furniture etc. which has been incurred. Sunk cost capital (money) that is lost and cannot be recovered. Sunk costs are not included when making decisions about the future. They should be handled using tax laws and write-off allowances, not the economic study.

Tangible fixed assets: These are fixed assets that have a physical presence and include things like land, buildings, machinery, equipment, computers and so on.

Time Value of money: It is a fact that money makes money. This concept explains the change in the amount of money over time for both owned and borrowed funds.

Trade: Buying or selling goods and services among companies, states, or countries, called also commerce.

Trademark: Distinctive name, symbol, motto, or emblem that identifies a product, service, or firm. A trademark is another form of legal protection for intellectual properties. They are more about words, names, symbols, sounds or colours that distinguish brand, goods and services. Unlike patents they have a limited time span and trademarks can be renewed forever as long as they are being used.

Trader: Anyone who buys and sells goods or services for profit; a dealer or merchant.

Trade-Offs: The sacrifice of some or all of an economic goal, good, or service to achieve a different goal, good, or service. (See also "Opportunity Cost," above.)

Turnover: The total sales of a business or company during a specified period.

Trade discount, quantity discount and cash discount.

Trade discounts are allowed by one businessman to another businessman to sell the product at the printed price. Quantity discounts gets available on bulk purchase. Cash discounts are allowed by retailers to the customers for prompt payments.

Value engineering: Value engineering is the systematic review of a project, product, or process to improve the performance without sacrificing the core manufacturing issues like safety, quality, operations, maintenance, and environment.

Venture: A risky project or business idea

Venture Capital: Financial capital made available by an investor or group of investors. The venture capitalists are often willing to assume a higher level risk than banks or lending institutions in return for a greater profit.

Winding up: Winding up a company is done by paying the company's creditors, and then distributing any money left (if any) among the members.

GLOSSARY

Absorber plate: A component of the solar flat plate collector that absorbs solar radiation and converts it into heat.

Absorptance: It is the ratio between the radition absorbed by a surface (absorber) and the total amount of solar radiation striking the surface.

Actinometer: This is the instrument used to measure direct radiation from the sun. Also known as pyrheliometer

Active solar heating: This refers to heating by solar energy using an additional energy source (usually electricity) for pumping water or blowing air

Air heating system: This refers to air heating by solar energy.

Air mass: It is atmospheric attenuation of solar radiation.

Albedo: The ratio of the amount of light reflected by a surface to the light falling onto it.

Alternating current(AC): An electric current that alternates directions between positive and negative cycles, usually 50 or 60 times per second. Alternating current is the current typically available from power outlets in a household.

Altitude: The sun's angle above the horizon, as measured in a vertical plane.

Amorphous silicon: A thin-film PV silicon cell having no crystalline structure. It is manufactured by depositing layers of doped silicon on a substrate.

Ampere: Electrical current; a measure of flowing electrons.

Ampere-hour(Amp-hr): Measure of flowing electron for a period of time.

Anaerobic digestion: It is a process by which organic matter is decomposed by bacteria in absence of oxygen to produce methane and other by product

Anemometer: This is the instrument used for measuring the wind speed.

Annual mean daily insolation: The average solar energy per m^2 available per day over the whole year.

Annual solar savings: The annual energy saving due to solar energy device.

Antifreeze: This refers to the substance added to water to lower its freezing point. Solar water heaters usually use a mixture of water and propylene glycol instead of water to prevent freezing.

Antireflection coating: A thin coating of a material, which reduces the light reflection and increases light transmission, applied to a photovoltaic cell surface.

Array: Any number of photovoltaic modules/panels connected together electrically to provide a single electrical output. An array is a mechanically integrated assembly of modules or panels together with support structure (including foundation and other components, as required) to form a free-standing field installed unit that produces DC power.

Audit: An energy audit seeks energy inefficiencies and prescribes improvements.

Automatic Tracking: A device that permits solar collectors to "track" or to follow the sun during the day without manual adjustment. Usually for concentrating collectors.

Azimuth: The horizontal angle between the sun and due south in the northern hemisphere, or between the sun and due north in the southern hemisphere.

Balance of system(BOS): Term used in photovoltaics, which represents all components and costs other than the PV modules.

Band Gap: It is the difference in energy between the state of the highest valence band and conduction band.

Battery: A collection of cells that store electrical energy; each cell converts chemical energy into electricity or vice versa, and is interconnected with other cells to form a battery for storing useful quantities of electricity.

Battery capacity: The maximum total electrical charge, expressed in ampere-hours (Ah), that a battery can deliver to a load under a specific set of condition.

Battery cell: The simplest operating unit in a storage battery. It consists of one or more positive electrodes or plates, an electrolyte that permits ionic conduction, one or more negative electrodes or plates, separators between plates of opposite polarity, and a container for all the above.

Battery available capacity: The total maximum charge, expressed in ampere-hours, that can be withdrawn from a fully-charged cell or battery. The energy capacity of a given cell varies with temperature, rate, age, and cut off voltage.

Battery energy capacity: The total energy available, expressed in watt-hours (kilowatt-hours), that can be withdrawn from a fully-charged cell or battery. The energy capacity of a given cell varies with temperature, rate, age, and cutoff voltage

Battery cycle life: The number of cycles, to a specified depth of discharge, that a cell or battery can undergo before failing to meet its specified capacity of efficient performance criteria.

Beam radiation: It is the radiation propagating along the line joining the receiving surface and sun. it is also known as direct radiation

Biofuel: It is a product derived from biomass.

Biogas: It is a mixture of methane (CH_4), carbon dioxide (CO_2) and some trace gases.

Biomass: Organic material of a non-fossil origin (living or recently dead and animal tissue) including aquatic, herbaceous and woody plants, animal wastes, and portions of municipal wastes.

Biopower: It is electricity produced from biomass fuel.

Black body: A perfect absorber and emitter of radiation. A cavity is a perfect black body.Lampblack is close to black body, while aluminium (polished) is a poor absorber and emitter of radiation

Brightness: The subjective human perception of luminance.

BTU: British thermal unit, the amount of heat required to raise the temperature of one pound of water one degree Fahrenheit; 3411 BTUs equals one kilowatt-hour.

Cadmium Telluride (CdTe): A polycrystalline thin-film photovoltaic material.

Calorie: The amount of the heat required to raise the temperature of one gram of water by one degree Celsius.

Calorific value: The energy content per unit mass (or volume) of a fuel, which will be released in combustion (kWh/kg, MJ/kg, kWh/m^3, MJ/m^3)

Candela (cd): An SI unit of luminous intensity. An ordinary candle has a luminous intensity of one candela.

Carbon dioxide (CO_2): The colourless and odorless gas that is formed during normal human breathing. It is also emitted by combustion activities used to produce electricity. CO_2 is a major cause of the greenhouse effect that traps radiant energy near the earth's surface.

Carbonization: It is a process whereby wood is heated with restricted air flow to form high carbon product by removing volatile materials from it.

Cell:A device that generates electricity, traditionally consisting of two plates or conducting surface placed in an electrolytic fluid.

Celsius: The international temperature scale in which water freezes at 0 [degree] and boils at 100 [degree] and named after Anders Celsius

Central power Tower: A configuration of independently tracking solar collectors focusing all the reflected solar radiation onto a receiver placed on a top of tower.

Charge rate: The current applied to a cell or battery to restore its available capacity. This rate is commonly normalized by a charge control device with respect to the rated capacity of the cell or battery.

Charge controller: A component of photovoltaic system that controls the flow of current to and from the battery to protect the batteries from over-charge and over-discharge. The charge controller may also indicate the system operational status.

Central receiver power plants: This refers to solar thermal power plants using a large array of tracking mirrors (heliostats) to reflect the sun's rays to a receiver on top of a tower to produce steam used to generate electricity.

Circuit: A system of conductors (i.e. wires and appliances) capable of providing a closed path for electric current.

Clear sky: A sky condition with few or no clouds, usually taken as 0-2 tenths covered with clouds. Clear skies have high luminance and high radiation, and create strong shadows relative to more cloudy conditions. The sky is brightest nearest the sun, whereas away from the sun, it is about three times brighter at the horizon than at the zenith.

Closed cycle: In this case, the working fluid is returned to the initial stage at end of cycle and is recirculated.

Clerestory: A wall with windows that is between two different (roof) levels. The windows are used to provide natural light into a building

Coefficient of performance (COP): An efficiency term to compare the performance of refrigerators and heat pumps.

Cogeneration: Joint production of heat and work, most often electricity and heat.

Collector: The name given to the device, which converts the incoming solar radiation to heat.

Collector efficiency: This is obtained as the ratio of the useful (heat) energy converted by the solar collector to the solar radiation incident on the device.

Collector plate: A component of the solar flat plate collector that absorbs solar radiation and converts it into heat.

Collector tilt angle: The angel between the horizontal plane and the surface of a solar collector.

Compact fluorescent light (CFL): A modern light bulb with an integral ballast using a fraction of the electricity used by a regular incandescent light bulb.

Concentrating collector: A solar collector which reflects the solar radiation (direct radiation) to an absorber plate to produce high temperature.

Condensation: The process of vapour changing into the liquid state. Heat is released in the process.

Conductance(C): A measure of the ease with which heat flows through a specified thickness of a material by conduction. Units are $W/m^2\,^0C$.

Conduction: The process by which heat energy is transferred through materials (solids, liquids or gases) by molecular excitation of adjacent molecules.

Conduction band: This is the energy band in which the electrons move freely.

Conductivity: The quantity of heat that will flow through one square metre of material, one-meter-thick, in one second, when there is a temperature difference of $1\,^0C$ between its surface.

Conductor: A substance or body capable of transmitting electricity, heat or sound.

Convection: The transfer of heat between a moving fluid medium (liquid or gas) and a surface, or the transfer of heat within a fluid by movements within the fluid.

Conservation of energy: The total amount of energy in any closed system remains constant.

Core: The central region of the earth, having a radius of about 3,470 kilometres, the radius of earth is 6,370 km. outside of which lies the mantle and crust.

Crystalline silicon: A type of PV cell made from a single crystal or polycrystalline slices of silicon.

Current: The flow of electrons through a conductor.

Declination: The angle of the sun north or south of the equatorial plane.

Deep discharge battery: A type of battery that is not damaged when a large portion of its energy capacity is repeatedly removed (i.e. motive batteries)

Depth discharge (DOD): The ampere-hours removed from a fully charged cell or battery, expressed as a percentage of rated capacity. For example, the removal of 25 ampere-hours from a fully charged 100 ampere-hours rated cell results in a 25% depth of discharge.

Design heat load: The total heat loss from a building during the most severe winter condition the building is likely to experience.

Design month: The month has the lowest mean daily insolation value, around which many standalone systems are planned.

Diffuse radiation: The solar radiation reaching the surface due to reflection and scattering effect.

Diffusion length: The man distance through which a free electron or hole moves before recombining with another hole or electron.

Direct combustion: Burning of biomass in the presence of oxygen.

Direct current (DC): The complement of AC or alternating current, presents one unvarying voltage to a load. This is used in automobiles.

Direct radiation: Radiation coming in a beam from the sun which can be focused.

Direct solar gain: The thermal energy gain in building through glazed window.

Discharge: The removal of electric energy from battery.

Diurnal: Recurring every day or having a daily cycle.

Dopants: A chemical impurity added usually in minute amounts to a pure semiconductor in order to alter its electrical properties.

Doping: The addition of dopants to a semiconductor.

Dry-bulb temperature: The temperature of a gas or mixture of gases indicated by an accurate thermometer after correction for radiation.

Duct: A pipe, tube or channel that conveys a substance usually warm or cold air.

Efficacy: Special term which refers to the efficiency by which lamps convert electricity to visible radiation, measured in lumens per watt (lux).

Efficiency: The ratio of output power (or energy) to input power (or energy) expressed as a percentage.

Electromagnetic spectrum: The entire range of wavelengths or frequencies of electromagnetic radiation extending from gamma rays to the longest radio waves including visible light.

Electronic ballasts: An improvement over core/coil ballasts used to drive fluorescent lamps.

Embodied energy: Literally the amount of energy required to produce an object in its present form; an inflated balloon's embodied energy includes the energy required to manufacture it and inflate it.

Emissivity: The ratio of the radiant energy emitted by a body to that emitted by a perfect black body.

Emittance: A measure of the ability of a material to give off heat as radiant energy.

Energy density: Energy per unit area.

Energy intensity: Energy intensity is measured by the quantity of energy required per unit output or activity, so that using less energy to produce a product reduces the intensity. The ratio of energy use in a sector to activity in that sector, for example, the ratio of energy use to constant-dollar production in manufacturing.

Energy: The ability to do work

Equinox: The times of the year when the sun passes over the celestial equator and when the length of the day and night are almost equal. It happens twice a year.

Evacuated tube collector: A solar collector that uses a vacuum between the absorber to glass to reduce the top loss coefficient (insulate the absorber plate).

EVA: Ethylene-Vinile-Acetate Foil, it will be used by module production for covering the cells.

Fill Factor (FF): For an I-V curve, the ratio of the maximum power to the product of the open-circuit voltage and the short-circuit current. Fill factor is a measure of the "squareness" of the I-V curve.

Flat plate collector: A solar collection device for gathering the sun's heat, consisting of a shallow metal container covered with one or more layers of transparent glass or plastic; either air or a liquid is circulated through the cavity of the container, whose interior is painted "black" and exterior is well insulated.

Focusing collector: See concentrating collector.

Fresnel collector: A type of concentrating solar collector consisting of a concentric series of rings with reflecting surfaces.

Fuel cell: A device combining a fuel with oxygen in an electrochemical reaction to generate electricity directly without combustion

Gallium Arsenide (GaAs): A crystalline, high-efficiency semiconductor/photovoltaic material.

Generator: A device that produces electricity

Geothermal energy: It is the energy contained in the earth's interior.

Glare: The perception caused by a very bright light or a high contrast of light, making it uncomfortable or difficult to see.

Glazing: Transparent or translucent materials, usually glass or plastic, used to cover an opening without obstructing (relative to opaque materials) the admission of solar radiation and light.

Global radiation: The sum of direct, diffuse and reflected radiation.

Greenhouse effect: The global warming resulting from the absorption of infrared solar radiation by carbon dioxide and other trace gases present in the atmosphere.

Greenhouse gases: Gases which contribute to the greenhouse effect by absorbing infrared radiation in the atmosphere. These gases include carbon dioxide, nitrous oxide, methane, water vapour, and a variety of chlorofluorocarbons (CFCs)

Grid: A utility term for the network of wires that distribute electricity from a variety of sources across a large area.

Heat capacity: The quantity of heat required to raise one kilogram of a substance by one degree centigrade.

Heat Exchanger: Device that passes heat from one substance to another; in a solar hot water heater, for example, the heat exchanger takes heat harvested by a fluid circulating through the solar panel and transfers it do domestic hot water.

Heat loss: It is a thermal energy loss to atmosphere due to temperature difference.

Hydropower: Energy in a falling water is converted into mechanical energy and then electrical energy.

I-V curve: The plot of current versus voltage characteristics of a solar cell, module or array. I-V curves are used to compare various solar cell modules, and to determine their performance at various levels of insolation and temperatures.

Illuminance: It is defined as the amount of lumens per unit area.

Incandescent bulb: A light source that produces light by heating a filament until it emits photons.

Incident radiation: The quantity of radiant energy striking a surface per unit time and unit area.

Infrared radiation: The part of the electromagnetic radiation (waves) whose wavelength range lies between 0.75 to 1000 micrometers.

Insolation (or incident solar-radiation): The amount of sunlight failing on a place (energy of sunlight per unit area).

Insulation: A material that keeps energy from crossing from one place to another: on electrical wire, it is the plastic or rubber that covers the conductor; in a building, insulation makes the walls, floor, and roof more resistant to the outside (ambient) temperature.

Insulator: A material that is a poor conductor of electricity or heat.

Inverter: The electrical device that changes direct current (DC) into alternating current (AC)

Irradiance: The solar radiation incident on a surface per unit time(W/m^2)

Joule: The unit of energy of work. One joule is equal to one watt second.

Kilowatt (kW): 1000 watts, energy consumption at a rate of 1,000 joules per second.

Kilowatt/hour (kWh): One kilowatt of power used for one hour.

Latitude: The angular position of a location north or south of the equator.

Life-cycle costing: A method for estimating the comparative costs of alternative energy or other systems. Life- cycle costing takes into consideration such long-term costs as energy consumption, maintenance, and repair.

Life-cycle costs: The entire cost of an energy device, including the capital cost in present dollars, and the costs and benefits, discounted to the present.

Light emitting diode (light emitting diode); An efficient source of electrical lighting, typically lasting 50,000 to 100,000 hours.

Load: The set of equipment or appliances that use the electrical power from the generating source, battery or PV module.

Longitude: The angular position east or west of Greenwich.

Low-E windows: Windows that reflect the infrared (IR) back into the room, instead of absorbing and transmitting the IR heat to the outside.

Maximum power point (MPP): The voltage at which a PV array is producing maximum power

Maximum power point tracker (MPPT): A power conditioning unit that increases the power of a PV system by ensuring operation of the PV generator at its Maximum Power Point (MPP). The ability to do so can depend on climate and the battery's state of charge.

Medium temperature Solar Collector: A solar thermal collector designed to operate in the temperature range of 80-100 ^{0}C.

Megawatt (MW): 1,000,000 watts.

Module: The smallest self-contained environmentally protected structure housing interconnected photovoltaic cells and providing a single (DC) electrical output.

Monthly mean daily insolation: The average solar energy per square meter available per day of a given month.

N-type semiconductor: A semiconductor produced by doping an intrinsic semiconductor with an electron donor impurity (e.g. phosphorous in silicon)

Natural convection: The natural convection of heat through the fluid in a body that occurs when warm, less dense fluid rises and cold, dense fluid sinks under the influence of gravity.

Newton: The Newton is the basic unit of force.

Night Sky radiation: A method of cooling through radiant energy exchange. Relatively warm surface are exposed directly to the colder night sky to which they radiate the heat they collected during the day

NOCT: Nominal operating cell temperature; the estimated temperature of a PV module when operating under 800 W/m^2 irradiance, 20 ^{0}C ambient temperature and wind speed of 1 metre per second. NOCT is used to estimate the nominal operating temperature of a module in its working environment.

Non renewable Energy Sources: It is the energy derived from finite and static stocks of energy.

One-axis tracking: A solar system capable of rotating about one axis and track the sun east to west.

Open circuit voltage (V_{OC}): The maximum possible voltage across a solar module or array. Open circuit voltage occurs in sunlight when no current is flowing.

Open cycle: In this case, the working fluid is renewed at the end of each cycle.

Orientation: The arrangement of solar device along a given axis to face in a direction best suitable to absorb solar radiation.

OTEC:(Ocean thermal energy conversion): A process which exploits the natural temperature gradient between shallow and deep ocean waters as the driving potential for a simple thermodynamic cycle that can extract work out of the temperature gradient.

Off-the-grid: Not connected to the power grid.

Overcharging: Leaving batteries on charge after they have reached their full state of charge (100 %).

Parabola: The geometrically curved shape to focus sunlight on a single point.

Parabolic concentrating cooker: A solar cooker that uses a parabolic disk to focus sunlight.

Parabolic mirror: A device with a large, shiny, curved surface that focuses solar radiation on a specific point.

Passive solar design: A building design that makes use of structural elements, using no moving parts to heat or cool the space in the building.

Passive solar heater: A solar water or space heating system that moves heated water or air without using fan/motor/pump.

Passive solar water heaters: Solar water heating system with natural/thermosiphon circulation is known as passive solar water heater.

Peak Sun shine Hours: The number of hours per day during which solar radiation averages 1000 W/m^2

Peak watt(W_p): Power output of PV module under standard test condition i.e. 1000 W/m^2 and 25 ^0C.

Pelletization: It is a process in which wood is compressed and extracted in the form of rods and cubes.

Periodic motion: Any motion that repeats itself in equal interval of times.

Photon: The elementary particle of electromagnetic energy; light. (Greek photos, light).

Photovoltaic array: A number of PV modules that are electrically connected in a series or parallel to provide the required rated power.

Photovoltaic conversion efficiency: The ratio of the electric power produced by a photovoltaic device to the power of the sunlight incident on the device.

Photovoltaic device: A device that converts light directly into DC directly.

Photovoltaic module: The basic building block of a photovoltaic device, which consists of a number of interconnected solar cells.

Photovoltaic device(PV): A device that converts light directly into DC electricity.

Polycrystalline silicon: A material used to make PV cells which consist of many crystals as contrasted with single crystal silicon.

Potential energy: It is the energy that an object possesses as a result of its elevation in gravitational field.

Power: The rate at which energy is consumed or produced. The unit is Watt.

Power density: It is power per unit area. (W/m^2)

Power factor: It is the cosine of the phase angle between the voltage and current of a circuit.

ppb: It is parts per billion.

ppm: It is parts per million.

Producer gas: It is a mixture of combustible and non-combustible gas

PV: See photovoltaic.

Pyranometer: A device that measures total (global) radiation.

Pyrheliometer: A device to measure direct solar radiation.

Pyrolysis: It is carbonization at high process temperature.

R-value: It is thermal resistance of material which is inverse of heat transfer coefficient.

Radiant Energy: Energy in the form of electromagnetic waves that travels outwards.

Radiation: Electromagnetic waves that directly transport energy through space. Sunlight is a form of radiation.

Reflectivity: The ratio of radiant energy reflected by a body to that falling upon it.

Reflector: A device that can be used to reflect solar radiation.

Renewable energy source: An energy source that renews itself without effort, fossil fuels, once consumed, are gone forever, while solar energy is renewable in that the solar energy we harvest today has no effect on the solar energy we can harvest tomorrow.

Resistor: Any electronic component that restricts the flow of electrical current in circuits.

Selective surface: A specially surface which has high absorptions and low emissive power.

Semiconductor: A material such as silicon, which has a crystalline structure that will allow current to flow under certain conditions. Semiconductors are usually less conductive than metals but not an insulator like rubber.

Short circuit current (I_{sc}): Current across the terminal when a solar cell or module in strong sunlight is not connected to a load (measured with ammeter).

Silicon: A semiconductor material commonly used to make PV cells.

Single-Crystal Structure: A material having a crystalline structure such as that a repeatable or periodic molecular pattern exists in all three dimensions.

Solar altitude: The sun's angle above the horizon, as measured in vertical plane.

Solar azimuth: The horizontal angle between the sun and due south in the north hemisphere, or between the sun and due north in the southern hemisphere.

Solar cabinet dryer: A device that uses solar radiation for crop drying.

Solar cell: It is a device that converts light energy or solar radiation (photons) directly into DC electricity.

Solar cell module: Groups of encapsulated solar cells framed in glass or plastic units, usually the smallest unit of solar electric equipment available to the consumer.

Solar air collector: A device that gathers and accumulates solar radiation to produce heat.

Solar concentrator: A device which uses reflective surface in a planar, parabolic trough, or parabolic bowl configuration to concentrate solar radiation onto a smaller surface.

Solar constant: An amount referring to radiation arriving from the sun at the edge of the earth's atmosphere. The accepted value is about 1367 watts per square meter.

Solar cooker: A device that uses solar radiation for food cooking.

Solar declination: The angle of the sun north or south of the equatorial plane.

Solar distillation: A process in which solar energy is trapped and used for to evaporate impure or salty water.

Solar electricity: Electricity that is obtained by using solar energy.

Solar energy: The electromagnetic radiation generated by the sun.

Solar incident angle: The angle at which the incoming solar beam strikes a surface.

Solar pond: A shallow body of salt water with a black or dark bottom.

Solar radiation: The radiant energy received from the sun, from both direct and diffuse or reflected sunlight.

Solar spectrum: The total distribution of electromagnetic radiation emitted from the sun.

Solar still: A device consisting of one or several stages in which brackish water is converted to potable water by successive evaporation and condensation with the aid of solar heat.

Solar water heater: A water heater that depends on solar radiation at its source of power.

Standard test condition: It is condition having 1000 W/m^2 solar radiation and 25 ^0C ambient air temperature with air mass of 1.5.

Sustainable: Material or energy source that, if managed carefully, will provide at current levels indefinitely.

Temperature: Degree of hotness or coldness measured on one of several arbitrary scales based on some observable phenomenon (such as the expansion).

Thermal conductivity: The ability of a material to conduct heat.

Thermal mass: A material used to store heat, thereby slowing the temperature variation within a space.

Thermal storage wall: A south-facing glazed wall generally known as Trombe wall.

Thermosyphon: A close loop system in which water automatically circulates between a solar collector and a water storage tank above it due to the natural difference in density between the warmer and cooler portions of a liquid.

Thermal storage: Any of several techniques to store heat energy by utilizing either the heat capacity of materials, the latent heat of phase change, or the heat of chemical dissociation.

Thin-Film Silicon: Most often this is amorphous (non-crystalline) material used to make photo-voltaic (PV) cells.

Tidal power: Energy obtained by using the motion of the tides to run water turbines that drives electricity generation.

Tilt angle: The angle at which a solar collector is tilted upward from the horizontal surface to receive the maximum solar radiation.

Tracking: The adjustments made to a solar collector concentrating collectors to track or follow the sun's path across the sky.

Transfer medium: A substance (air, water, or antifreeze solutions) that carries heat from a solar collector to a storage area or from a storage area to in a collector.

Transmission: Transporting bulk power over long distance.

Transmittance: The ratio of the solar radiation transmitted through a glazing to the total radiant energy falling on its surface.

Trombe wall: A solar radiation facing glazed thermal storage wall.

Trough: This is a type of concentrating collector with one axis-tracking.

Turbine generator: A device that uses steam, heated gases, water flow or wind to cause spinning motion that activates electromagnetic forces and generate electricity.

U-value: The amount of heat that flows in or out of a system at steady state, in one hour, when there is a one degree difference in temperature between fluid inside and outside.

Ultra-violet radiation: A portion of the electromagnetic radiation in the wavelength range of 4 to 400 nanometres.

Valence band: The highest energy band in a semiconductor that can be filled with electron

Ventilation: The exchange of room air between rooms to outside ambient air as per requirement. Generally, it is measured in the terms of number of air change per hour.

Volt: It is the measure of potential difference between two electrodes.

Wafer: Raw material for a solar cell, a thin sheet of crystalline semiconductor material is made by mechanically sawing it from a single-crystal boule or by casting it.

Water heating: The process of generating domestic hot water by employing a flat plate collector and utilizing solar radiation.

Watt: Measure of power (or work) equivalent to 1/746 of a horsepower.

Watt hour (Wh): A common energy measure arrived at by multiplying the power times the hours of use grid power is ordinarily sold and measured in kilowatt hours.

Wavelength: The distance between two similar points of a given wave.

Wind turbines: It is a device to convert kinetic energy associated with wind into mechanical energy.

Zenith: The top of the sky dome. It is a point directly overhead, 90^0 in altitude angle above the horizon.

APPENDIX

APPENDIX-1

The seven SI Base Units

Quantity	Name of Unit	Symbol
Length	Meter	M
Mass	Kilogram	Kg
Time	Second	S
Electric current	Ampere	A
Thermodynamics temperature	Kelvin	K
Luminous intensity	Candela	Cd
Amount of a substance	Mole	mol

Units and symbols of some physical quantity

Quantity	Name of Unit	Symbol
Acceleration	Meters per second squared	m/s^2
Area	Square meters	m2
Density	Kilogram per cubic meter	kg/m^3
Dynamic viscosity	Newton-second per square meter	$N\ s/m^2$
Force	Newton (=1 kg m/s^2)	N
Frequency	Hertz	Hz
Kinematic viscosity	Square meter per second	m^2/s
Plane angle	Radian	rad
Potential difference	Volt	V
Power	Watt(=1 J/s)	W
Pressure	Pascal (1 N/m^2)	Pa
Radiant intensity	Watts per steradian	W/sr
Solid angle	Steradian	sr
Specific heat	Joules per kilogram–Kelvin	J/kg K
Thermal conductivity	Watts per meter–Kelvin	W/m K
Velocity	Meters per second	m/s
Volume	Cubicmeter	m^3
Work,energy,heat	Joule(=1N/m)	J

English prefixes

Multiplier	Symbol	Prefix	Multiplier	Multiplier symbol
10^{12}	T	Tera	10^3	M(thousand)
10^9	G	Giga	10^6	MM (million)
10^6	m	Mega		
10^3	k	Kilo		
10^2	h	Hecto		
10^1	da	Deka		
10^{-1}	d	Deci		
10^{-2}	c	Centi		
10^{-3}	m	Milli		
10^{-6}	m	Micro		
10^{-9}	n	Nano		
10^{-12}	p	Pico		
10^{-15}	f	Femto		
10^{-18}	a	Atto		

Value of physical constants in SI Units

Quantity	Symbol	Value
Avogadro constant	N	6.022169×10^{26} kmol^{-1}
Boltzmann constant	k	1.380622×10^{-23} J/K
First radiation constant	$C_1 = 2\pi hC^2$	3.741844×10^{-16} W m^2
Gas constant	R	8.31434×10^3 J/kmol K
Planck constant	h	6.626196×10^{-34} Js
Second radiation constant	$C_2 = hc/k$	1.438833×10^{-2} m K
Speed of light in a vacuum	C	2.997925×10^8 m/s
Stefan–Boltzmann constant	σ	5.66961×10^{-8} W/m^2 K^4

APPENDIX-2

Conversion of units

(i) **Length, m**

1 yard = 3 ft = 36 inches = 0.9144 m

1 m =39.3701 inch = 3.280839 ft = 1.093613 yd = 1650763.73 wavelength

1 ft = 12 in = 0.3048

1 in = 2.54 cm = 25.4 mm

1 mil = 2.54 × 10^{-3} cm

1 micro meter = 10^{-6} m

1 nm = 10^{-9} m=10^{-3}mm

(ii) **Area, m^2**

1 ft^2 =0.0929 m^2,

1 in^2 = 6.452 cm^2 = 0.00064516m^2

1 cm^2 = 10^{-4} m^2 = 10.764× 10^{-4} ft^2 = 0.1550 in^2

1 ha = 10,000 m^2

(iii) **Volume, m^3**

1 ft^3 = 0.02832 m^3 = 28.31681 (litre)

1 in ^{3} = 16.39 cm^3 = 1.639 × 10^{-2} 1

1 yd^3 = 0.764555 m^3 =7.646 × 10^2 1

1 UK gallon = 4.54609 1

1 US gallon = 3.7851 = 0.1337 ft^3

1 m^3 = 1.000 × 10^6 cm^3 = 2.642 × $10^{1\ 2}$ US gallons = 109 litre

1l=10^{-3} m^3

1 fluid ounce = 28.41 cm^3

(iv) **Mass. Kg**

1 kg = 2.20462 1b = 0.068522 slug

1 ton (short) = 2000 1b (pounds) = 907. 184 kg

1 ton (long) = 1016.05 kg

1 1b = 16 0z (ounces) = 0.4536 kg

1 oz = 28.3495 g

1 quintal = 100 kg

1 kg=1000g =10 000 mg

$1\ \mu g = 10^{-6}\ g$

$1\ ng = 10^{-9}\ g$

(v)   Density and specific volumes, kg/m^3, m^3/kg

$1\ 1b/ft^3 = 16.0185\ kg/m^3 = 5.787 \times 10^{-4}\ 1b/in^3$

$1\ g/cm^3 = 10^3\ kg/m^3 = 62.43\ 1b/ft^3$

$1 lb/ft^3 = 0.016\ g/cm^3 = 16\ kg/m^3$

$1\ ft^3\ (air) = 0.08009\ lb = 36.5\ g\ at\ N.T.P.$

$1\ gallon/1b = 0.010 cm^3/kg$

$1\ \mu g/m^3 = 10^{-6}\ g/m^3$

(vi) Pressure, Pa (pascal)

$1\ lb/ft^2 = 4.88\ kg/m^2 = 47.88\ Pa$

$1\ lb/in^2 = 702.7\ kg/m^2 = 51.71\ mm\ Hg = 6.894757 \times 10^3\ Pa = 6.894757 \times 10^3\ N/m^2$

$1\ atm = 1.013 \times 10^5\ N/m^2 = 760\ mmHg = 101.325\ kPa$

$1\ in\ H_2\ O = 2.491 \times 10^2\ N/m^2 = 248.8\ Pa = 0.036\ 1b/in^2$

$1\ bar = 0.987\ atm = 1.000 \times 10^6\ dynes/cm^2 = 1.020\ kgf/cm^2 = 14.50\ 1bf/in^2 = 10^5\ N/m^2 = 100\ kPa$

$1\ torr\ (mm\ Hg\ 0^0\ C) = 133\ Pa$

$1 pascal\ (Pa) = 1\ N/m^2 = 1.89476\ kg$

$1\ inch\ of\ Hg = 3.377\ kPa = 0.489\ 1b/in^2$

(vii)   Velocity, m/s

$1\ ft/s = 0.3041\ m/s$

$1\ mile/h = 0.447\ m/s = 1.4667\ ft/s = 0.8690\ knots$

$1\ km/h = 0.2778\ m/s$

$1\ ft/min = 0.00508\ m/s$

(viii) Force, N

$1\ N\ (Newton) = 10^5\ dynes = 0.22481\ 1b\ wt = 0.224\ 1b\ f$

$1\ pdl\ (poundal) = 0.138255\ n\ (Newton) = 13.83\ dynes = 14.10\ g\ f$

$1\ lb\ f\ (i.e.\ wt\ of\ 1\ lb\ mass) = 4.448222\ N = 444.8222\ dynes$

$1\ ton = 9.964 \times 10^3\ N$

$1\ bar = 10^5\ Pa\ (Pascal)$

$1\ ft\ of\ H_2O = 2.950 \times 10^{-2}\ atm = 9.807 \times 10^3\ N/m^2$

$1\ in\ H_2O = 249.089\ Pa$

$1\ mm\ H_2O = 9.80665\ Pa$

1 dyne = 1.020×10^{-6}kgf = 2.2481×10^{-6}lbf = 7.2330×10^{-5}pdl = 10^{-5} N

1 mm of Hg = 133.3 Pa

1 atm = 1 kg f/cm^2 = 98.0665 kPa

1 Pa (pascal) = 1 N/m^2

(ix) **Mass flow rate and discharge, kg/s, m^3/s**

1 lb/s = 0.5436 kg/s

1 ft^3/min = 0.4720 l/s = 4.179×10^{-4} m^3 /s

1 m^3/s = 3.6×10^6 l/h

1 g/cm^3 = 10^3 kg/m^3

1 lb/h ft^2 = 0.001356 kg/s m^2

1 lb/ft^3 = 16.2 kg/m^3

1 litre/s (l/s) = 10^{-3} m^3/s

(x) **Energy, J**

1 cal = 4.187 J (joules)

1 kcal = 3.97 Btu = 12×10^{-4}kWh = 4.187×10^3 J

1 watt = 1.0J/s

1 Btu = 0.252 kcal = 2.93×10^{-4} kWh = 1.022×10^3 J

1 hp = 632.34 kcal = 0.736 kWh

1 kWh = 3.6×10^6 J = 1 unit

1 J = 2.390×10^{-4} kcal = 2.778×10^{-4}Wh

1 kWh = 860 kcal = 3413 Btu

1 erg = 1.0×10^{-7} J = 1.0×10^{-7} Nm = 1.0 dyne cm

1 J = 1 Ws = 1 Nm

1 eV = 1.602×10^{-19} J

1 GJ = 10^9 J

1 MJ = 10^6 J

1 TJ(TeraJoules) = 10^{12} J

1 EJ (Exajoules) = 10^{18} J

(xi) **Power, watt (J/s)**

1 Btu/h = 0.293071 W = 0.252 kcal / h

1 Btu/h = 1.163 W = 3.97 Btu/h

1 W = 1.0 J/s = 1.341×10^{-3}hp = 0.0569 Btu/min = 0.01433 kcal/min

1 hp (F.P.S.) = 550 ft lb f/s = 746 W = 596 kcal/h = 1.015 hp (M.K.S.)

1 hp (M.K.S.) = 75 mm kg f/s = 0.17569 kcal/s = 735.3 W

1 W/ft^2 = 10.76 W/m^2

1 ton (Rifrigeration)= 3.5 k W

1 kW = 1000 W

1 GW = 10^{9}W

1 W/m^2 = 100 lux

(xii) Specific heat, J/kg °C

1 Btu/lb °F= 1.0 kcal/kg °C = 4.187 × 10^3 J/kg °C

1 Btu/lb = 2.326 kJ/kg

(xiii) Temperature, °C and K used in SI

T(Celcius, °C) = (5/9) [T(Fahrenheit, °F)+40]-40

T(°F) =(9/5)[T(°C)+40]-40

T(Rankine, °R)=460 +T (°F)

T(Kelvin, K) = (5/9) T (°R)

T(Kelvin, K) = 273.15 + T(°C)

T(°C) = T(°F)/1.8 = (5/9) T (°F)

(xiv) Rate of heat flow per unit area of heat flux, W/m^2

1 Btu/ft^2 h = 2.713 kcal /m^2 h= 3.1552 W/m^2

1 kcal/m^2 h = 0.3690 Btu/ft^2 h = 1.163 W/m^2 = 27.78 × 10^{-6}cal/s cm^2

1 cal/cm^2 min = 221.4 Btu/ft^2 h

1 W/ft^2 = 10.76 W/m^2

1 W/m^2 = 0.86 kcal/h m^2 = 0.23901 × 10^{-4}cal/s cm^2 = 0.137 Btu/h ft^2

1 Btu/h ft = 0.96128 W/m

(xv) Heat transfer coefficient, W/m^2 °C

1 Btu/ft^{2}h °F = 4.882 kcal/m^{2}h °C = 1.3571 × 10^{-4}cal/cm^2 s °C

1 Btu/ft^2 h °F = 5.678 W/m^2 °C

1 kcal/m^{2}h °C = 0.2048 Btu/ft^{2}h°F = 1.163 W/m^{2}0C

1 W/m^2 K = 2.3901 × 10^{-5}cal/cm^2 s K = 1.7611× 10^{-1} Btu/ft^2 °F = 0.86 kcal/m^{2}h °C

(xvi) Thermal conductivity, W/m °C

1 Btu/ft °F = 1.488 kcal/m h °C = 1.73073 W/m °C

1 kcal/m h °C = 0.6720 Btu/ft h °F = 1.1631 W/m °C

1 Btu in/ft^{2}h °F = 0.124 kcal/mh °C =0.144228 W/m °C

1Btu/in h °F = 17.88 kcal/mh °C

1 cal/cm s °F = 4.187 × 10^2 W/m °C = 242 Btu/h ft°F

1 W/cm °C = 57.79 Btu/h ft °F

(xvii) Angle. rad

2 π rad = 360 degrees

1 degree = 0.0174533 rad = 60' (min)

1'= 0.290888 × 10^{-3} rad = 60''(s)

1''= 4.84814× 10^{-6} rad

1^0 (h angle) = 4 min (time)

(xviii) Illumination

1 lx (lux) = 1.0 lm (lumen)/m^2

1 lm/ft^2 = 1.0 foot candle

1 foot candle = 10.7639 lx

100 lux = 1 W/m^2

(xix) Time, h

1 week = 7 days = 168 h = 10080 minutes = 604800 s

1 mean solar day = 1440 minute = 86400 s

1 calender year = 365 days = 8760h = 5.256 × 10^5 minutes

1 tropical mean solar year = 365.2422 days

1 sidereal year = 365.2564 days (mean solar)

1 s (s) = 9.192631770 × 10^9 hertz (Hz)

1 day = 24 h = 360^0 (hour angle)

(xx) Concentration, kg/m^3 and g/m^3

1 g/l = 1 kg/m^3

1 lb/ft^3 = 6.236 kg/m^3

(xxi) Diffusivity, m^2/s

1 ft^2/h = 25.81 × 10^{-6} m^2/s

APPENDIX-3

Monthly Averaged Daily Solar Radiation

Monthly averaged daily global radiation on horizontal surface (G), monthly average daily diffuse radiation on horizontal surface (D) and monthly average daily global radiation on the surface tilted to latitude of the location (GL). G, D and GL are in kWh/m^2- day (kilowatt-hour/ m^2-day).

January to April

City	Jan.			Feb.			Mar.			Apr.		
	G	D	GL	G	D	GL	G	D	GL	G	D	GL
Ahmedabad	5	1.2	6.8	5.9	1.3	7.3	6.6	1.6	7.2	7.3	1.8	7.2
Bangalore	5.6	1.6	6.4	6.4	1.6	7.0	6.8	1.9	7.0	6.8	2.2	6.6
Bhubaneswar	5.2	1.4	6.7	5.9	1.4	7.0	6.3	2	6.8	6.5	2.4	6.4
Bhopal	4.8	1	6.6	5.9	1	7.4	6.3	1.6	6.9	7	1.8	6.9
Chandigarh	3.6	1.6	5.4	4.7	2	6.2	5.6	2.4	6.4	6.6	2.6	6.6
Chennai	5.4	1.8	7.5	6.3	1.7	6.9	6.6	2	6.8	6.8	2.2	6.6
Delhi	4.3	1.3	6.3	5	1.4	6.6	6	2	6.8	6.8	2.4	6.8
Gwalior	4.5	1	6.5	5.5	1	7.1	6.2	1.6	7.0	7.5	1.8	7.4
Goa	5.6	1.4	6.7	6.3	1.4	7.1	6.6	1.8	6.9	6.8	2.4	6.6
Guwahati	3.8	1.6	5.2	4.8	2	6.0	5.4	2.4	5.9	5.8	2.8	5.7
Hyderabad	5.5	1.4	6.7	6.2	1.5	7.1	6.5	2	6.8	6.9	2.4	6.7
Indore	5.1	1.1	6.9	5.9	1.3	7.3	6.4	1.6	6.9	7.4	2	7.3
Jabalpur	4.9	1.1	6.6	5.7	1	7.1	6.2	1.6	6.8	6.9	1.8	6.8
Jamnagar	4.9	1	6.5	5.8	1.2	7.1	6.2	1.8	6.8	7	2.2	6.9
Jodhpur	4.6	1	6.7	5.6	1.1	7.3	6.6	1.7	7.4	7.3	1.8	7.2
Kolkata	4.7	1.4	6.1	5.6	1	6.7	6.2	1.6	6.7	6.6	1.7	6.5
Lucknow	4.3	1.4	6.2	5.2	1.2	6.7	5.9	2	6.6	6.8	2.4	6.7
Mumbai	5.2	1.4	6.6	5.9	1.4	6.9	6.5	1.8	6.9	6.9	2.2	6.7
Nagpur	5	1.2	6.6	5.9	1.2	7.1	6.3	1.8	6.8	6.8	2.2	6.7
Patna	4.4	1.6	6.0	5.3	1.7	6.6	6.1	2.2	6.7	6.7	2.5	6.6
Pune	5.3	1.4	6.6	6	1.4	7.0	6.6	1.8	7.0	6.8	2.2	6.6
Ranchi	4.7	1.2	6.4	5.6	1	7.0	6.4	1.8	7.0	7	2.2	6.9
Solapur	5.6	1.4	6.9	5.3	1.4	7.3	6.6	2	7.0	6.8	2.4	6.6
Trivandrum	6	1.8	6.5	6.6	1.8	7.0	6.8	2	6.9	6.5	2.4	6.4
Visakhapatnam	5.3	1.5	6.5	6	1.6	6.9	6.5	2	6.9	6.5	2.5	6.3

May to August

City	May			June			Jul			Aug		
	G	D	GL	G	D	GL	G	D	GL	G	D	GL
Ahmedabad	7.6	2	6.9	6.6	3	5.9	5	3.4	4.6	4.6	3.3	4.4
Bangalore	6.4	2.6	6.0	6	3	5.6	4.6	2.9	4.4	4.8	3	4.7
Bhubaneswar	6.4	2.6	5.8	5.3	3	4.8	4.6	3	4.3	4.8	3	4.6
Bhopal	7.2	2	6.5	6.2	2.9	5.5	4.7	3.2	4.4	4.4	3.1	4.2
Chandigarh	7.3	2.8	6.7	7	3.2	6.2	6.2	3.2	5.6	5.8	3	5.5
Chennai	6.3	2.4	5.9	5.5	2.8	5.1	5.2	2.9	4.9	5.6	2.8	5.4
Delhi	7.2	2.8	6.6	6	3.1	5.3	5.7	3	5.1	5.6	2.9	5.3

[Table Contd.

Contd. Table]

City	May			June			Jul			Aug		
	G	D	GL	G	D	GL	G	D	GL	G	D	GL
Gwalior	7.1	2	6.4	6.5	2.8	5.7	5.2	3	4.7	5	3	4.8
Goa	6.6	2.6	6.1	4.9	3	4.6	3.8	3.1	3.6	4.4	3	4.2
Guwahati	5.6	3	5.1	4.6	3.2	4.2	4.8	3.2	4.4	5	3	4.8
Hyderabad	6.9	2.5	6.4	5.8	3.2	5.3	4.9	3.3	4.6	5.2	3.2	5.0
Indore	7.4	2.1	6.7	6.4	3	5.8	4.9	3.4	4.6	4.4	3.2	4.2
Jabalpur	6.6	2.2	6.0	5.8	2.8	5.2	4.8	3	4.5	4.4	3	4.2
Jamnagar	7.4	2.1	6.7	6.2	3.1	5.6	5	3.4	4.7	4.8	3.2	4.6
Jodhpur	7.8	2.2	7.1	7.4	3	6.5	6.2	3	5.6	6	3	5.7
Kolkata	6.4	2	5.9	5.2	2.8	4.7	4.8	3.1	4.4	4.8	3.1	4.6
Lucknow	7.2	2.6	6.5	6.4	3.2	5.7	5.6	3.1	5.1	5.6	3	5.3
Mumbai	7.2	2.4	6.6	5.4	3	4.9	4.4	3.3	4.1	4	3.2	3.8
Nagpur	6.8	2.5	6.2	5.6	3.1	5.1	4.4	3.2	4.1	4.4	3.2	4.2
Patna	6.8	2.7	6.2	6.2	3.2	5.5	5.2	3.2	4.7	5.6	3	5.3
Pune	7.2	2.4	6.6	5.6	3.4	5.1	4.4	3.4	4.2	4.4	3.4	4.2
Ranchi	6.9	2.3	6.3	6	2.9	5.4	4.8	3	4.4	5	3	4.8
Solapur	6.6	2.6	6.1	5.7	3.3	5.2	4.6	3.4	4.4	4.8	3.3	4.6
Trivandrum	6.8	2.6	6.5	5.2	2.6	5.0	5.4	2.8	5.2	5.8	2.8	5.6
Visakhapatnam	6.4	2.6	5.9	5	3	4.6	4.8	3	4.5	5	3	4.8

September to December

City	Sep			Oct			Nov			Dec		
	G	D	GL	G	D	GL	G	D	GL	G	D	GL
Ahmedabad	5.6	2.7	5.8	6	1.7	7.0	5.3	1.1	7.0	4.8	1	6.7
Bangalore	5.2	2.9	5.2	5.4	2.6	5.7	5	2	5.1	4.9	2	5.6
Bhubaneswar	5.2	2.8	5.3	5.2	2.2	5.8	5.3	1.5	6.6	5.1	1.4	6.7
Bhopal	5.1	2.6	5.2	5.8	1.6	6.8	5.3	1	7.1	4.7	1	6.6
Chandigarh	5.8	2.5	6.2	5.2	1.9	6.6	4.3	1.4	6.3	3.5	2	5.4
Chennai	5.8	2.8	5.8	5	2.4	5.3	4.9	2.2	5.4	4.7	2	5.3
Delhi	5.7	2.2	6.0	5.4	1.5	6.7	4.7	1.1	6.8	4	1.2	6.1
Gwalior	5.6	2.3	5.9	5.8	1.5	7.1	5	1	7.0	4.3	1	6.4
Goa	5	2.8	5.0	5.4	2.4	5.8	5.7	1.6	6.7	5.4	1.4	6.5
Guwahati	4.6	2.8	4.7	4.6	2.2	5.4	4.5	1.6	6.1	4	1.4	5.7
Hyderabad	5.2	3	5.2	5.8	2.3	6.3	5.4	1.8	6.4	5.1	1.4	6.3
Indore	5.2	2.8	5.3	5.8	1.8	6.7	5.4	1.2	7.1	4.8	1.6	6.6
Jabalpur	5	2.5	5.1	5.6	1.6	6.5	4.9	1	6.4	4.5	1	6.3
Jamnagar	5.6	2.6	5.8	6	1.6	7.0	5.3	1.1	6.9	4.8	1.1	6.6
Jodhpur	6.2	2	6.5	6	1.2	7.4	5.1	1.1	7.2	4.5	1.1	6.8
Kolkata	5	2.4	5.1	5.2	1.4	5.9	4.5	1	5.7	4.2	1.5	5.6
Lucknow	5.5	2.4	5.8	5.2	1.6	6.3	4.8	1.3	6.7	4.2	1.2	6.2
Mumbai	5.2	3	5.3	5.6	2	6.3	5.3	1.4	6.5	4.9	1.1	6.3
Nagpur	5	2.8	5.1	5.6	1.8	6.4	5.3	1.2	6.5	4.8	1	6.5
Patna	5.4	2.8	5.6	5.2	2	6.1	4.9	1.6	6.6	4.3	1.5	6.1
Pune	5.4	3	5.4	5.6	2.3	6.2	5.4	1.5	6.6	5	1.3	6.4
Ranchi	5.2	2.6	5.3	5.2	1.8	6.0	5.4	1	7.2	4.8	0.9	7.0
Solapur	5.3	3	5.3	5.6	2.2	6.1	5.4	1.7	6.4	5.5	1.4	6.5
Trivandrum	6	2.6	6.0	5.4	2.4	5.6	5.3	2.4	5.6	5	2.1	5.4
Visakhapatnam	5.2	2.8	5.3	5.2	2.4	5.7	5.2	1.8	6.2	5	1.4	6.3

APPENDIX-4

Heating values of various combustibles and their conversion efficiencies

Fuel	Heating value(kJ/kg)	Conversion efficiencyof device, h_F
Coal coke	29000	70
Wood	15000	60
Straw	14000-16000	60
Gasoline	43000	80
Kerosene	42000	80
Methane (natural gas)	50000	80
Biogas (60% methane)	20000	80
Electricity	-	95

Typical power ratings of energy appliances

Item name	Rating in watts
Air-conditioner (room)	1000
Air-conditioner (central)	2000-5000
CD player	15-30
Electrical dryer (clothes)	4000
Dryer (gas heated)	300-400
Ceiling fan	10-50
Laptop computer	20-75
Desktop PC	80-200
Printer	100
Frying pan	1200
Portable heater	1500
Stock tank heater	100
Water bed heater	400
Microwave oven	600-1500
Refrigerator/freezer	200-600
Table fan	10-25
Vacuum cleaner	200-800
Washing machine	500
Lights	
Incandescent	100, 75, 60, 40
Fluorescent (or compact fluorescent lightbulb)	30, 20, 16, 11
Coffee maker	800
Dishwasher	1200-1500

Specifications of solar-cell Material (at solar Intensity 1000 W/m^2 and cell temperature 25^0 C) and cost

Cell technology	Efficiency (%)	Fill factor (FF)	Aperture are (10^{-4}×m^2)	Life time* (years)	Manufacturing cost ($/kWp in 2007)	Selling price ($/kWp in 2007)
Monocrystalline silicon	24.7±0.5	0.828	4.0	30	2.5	3.7
Multicrystalline silicon	19.8 ± 0.5	0.795	1.09	30	2.4	3.5
Copper indium Diselenide(CIS/CIGS)	18.4 ± 0.5	0.77	1.04	5	1.5	2.5
Thin silicon cell	16.6 ± 0.4	0.782	4.02	25	2.0	3.3
Cadmium telluride (CdTe)	16.5 ± 0.5	0.755	1.03	15	1.5	2.5
Amorphous silicon (A-si)	10.1 ± 0.2	0.766	1.2	20	1.5	2.5

Physical properties of some materials.

S.N.	Material	Density (ρ)(kg/m^3)	Thermal conductivity (W/m K)	Specific heat (J/kg K)
1	Air	11.177	40.026	1006
2	Alumina	3800	29.0	800
3	Aluminium	41-45	211	0.946
4	Asphalt	1700	0.50	1000
5	Brick	1700	0.84	800
6	Carbon dioxide	1.979	0.145	871
7	Cement	1700	0.80	670
8	Concrete	1458	11.28	879
9	Clay	2400	1.279	1130
10	Copper	8795	385	-
11	Cork	240	0.04	2050
12	Cotton Wool	1522	-	1335
13	Fibre board	300	0.057	1000
14	Glass-crown	2600	1.0	670
	Window	2350	0.816	712
	Wool	50	0.042	670

[Table Contd.

Contd. Table]

S.N.	Material	Density (r)(kg/m^3)	Thermal conductivity (W/m K)	Specific heat (J/kg K)
15	Ice	920	2.21	1930
16	Iron	7870	80	106
17	Lime stone	2180	1.5	-
18	Mudphuska	-	-	-
19	Oxygen	1.301	0.027	920
20	Plaster-board	950	0.16	840
21	Polystyrene-expanded	25	0.033	1380
22	P.V.C.-rigid foam-Rigid sheet	25-80	0.035-0.041	-
		1350	0.16	-
23	Saw dust	188	0.57	-
24	Thermocole	22	0.03	-
25	Timber	600	0.14	1210
26	Turpentine	870	0.136	1760
27	Water (H$_2$O)	998	0.591	4190
	(Sea)	1025	-	3900
	(Vapour)	0.586	0.025	2060
28	Wood wool	500	0.10	1000

Absorptivity of various surface for sun's rays.

Surface	Absorptivity
White paint	0.12-0.26
Whitewash/glossy white	0.21
Bright aluminium	0.30
Flat white	0.25
Yellow	0.48
Bronze	0.50
Silver	0.52
Dark aluminium	0.63
Bright red	0.65
Brown	0.70
Light green	0.73
Medium red	0.74
Medium green	0.85
Dark green	0.95
Blue/black	0.97

[Table Contd.

Contd. Table]

Surface	Absorptivity
Roofs	
Asphalt	0.89
White asbestos cement	0.59
Copper sheeting	0.64
Uncoloured roofing tile	0.67
Red roofing tiles	0.72
Galvanised iron, clean	0.77
Brown roofing tile	0.87
Galvanised iron, dirty	0.89
Black roofing tile	0.92
Walls	
White/yellow brick tiles	0.30
White stone	0.40
Cream brick tile	0.50
Burl brick tile	0.60
Concrete/red brick tile	0.70
Red sand line brick	0.72
White and tone	0.76
Stone rubble	0.80
Blue brick tile	0.88
Surroundings	
Sea/lake water	0.29
Snow	0.30
Grass	0.80
Light-coloured grass	0.55
Light green shiny leaves	0.75
Sand gray	0.82
Rock	0.84
Green leaf	0.85
Earth (black ploughed field)	0.92
White leaves	0.20
Yellow leaves	0.58
Aluminium foil	0.39
Unpainted wood	0.60

[Table Contd.

Contd. Table]

Surface	Absorptivity
Metals	
Polished aluminium/copper	0.26
New galvanised iron	0.66
Old galvanised iron	0.89
Polished iron	0.45
Oxidised rusty iron	0.38

SUBJECT INDEX

For Product Safety Concerns and Information please contact our EU
representative GPSR@taylorandfrancis.com
Taylor & Francis Verlag GmbH, Kaufingerstraße 24, 80331 München, Germany